13th National Congress on Theoretical and Applied Mechanics (NCTAM 2017)

MATEC Web of Conferences Volume 145 (2018)

Sofia, Bulgaria
6 - 10 September 2017

Editors:

Vassil M. Vassilev
Svetoslav G. Nikolov
Maria D. Datcheva
Yonka P. Ivanova

ISBN: 978-1-5108-5582-3

Printed from e-media with permission by:

Curran Associates, Inc.
57 Morehouse Lane
Red Hook, NY 12571

Some format issues inherent in the e-media version may also appear in this print version.

Printed by Curran Associates, Inc. (2018)

For additional information, please contact EDP Sciences – Web of Conferences
at the address below.

EDP Sciences – Web of Conferences
17, Avenue du Hoggar
Parc d'Activité de Courtabœuf
BP 112
F-91944 Les Ulis Cedex A
France

Phone: +33 (0) 1 69 18 75 75
Fax: +33 (0) 1 69 28 84 91

contact-edps@webofconferences.org

Additional copies of this publication are available from:

Curran Associates, Inc.
57 Morehouse Lane
Red Hook, NY 12571 USA
Phone: 845-758-0400
Fax: 845-758-2633
Email: curran@proceedings.com
Web: www.proceedings.com

TABLE OF CONTENTS

13th National Congress on Theoretical and Applied Mechanics
Institute of Mechanics – Bulgarian Academy of Sciences
6th – 10th September 2017, Sofia, Bulgaria

Since its inception in 1969, the quadrennial National Congress of Theoretical and Applied Mechanics has become major scientific event, gathering Bulgarian and foreign scientists and engineers from academia and industry.

The Thirteenth Congress in this series fulfilled its main goal – continuing the traditions of the previous twelve congresses – to present the latest results in all areas of mechanics.

The Congress provided opportunities to around 100 Bulgarian and over 40 foreign scientists from Europe, Asia and North and South America to popularize their scientific results, share new ideas, discuss their current research, and lay the foundations for future cooperation.

More than 150 reports were presented in five main thematic areas: General Mechanics, Solid Mechanics, Fluid Mechanics, Biomechanics, Non-Destructive Testing and seven mini-symposia: Multiphase Flows in Science and Engineering; Mechatronics; Scientific Session of the Bulgarian Section of GAMM; Biofluids Modeling: Numerical Simulations and Application in Medicine; Computational Modeling of Arterial Aneurysms in Humans; Biodynamics and Biorheology; Constitutive and Numerical Modeling in Soil Mechanics.

SPONSORS

ФОНД НАУЧНИ ИЗСЛЕДВАНИЯ
МИНИСТЕРСТВО НА ОБРАЗОВАНИЕТО И НАУКАТА
ФНТС 1885
"АЕЦ КОЗЛОДУЙ" ЕАД
KOZLODUY NPP PLC
PROLAB
PROLAB INSTRUMENTS EOOD
chemical products
insulation solutions
78, Sokol Str., Yambol 8600, P.O. Box 382, Bulgaria
Tel/fax: 00359/ 46 663 335; 663 835 fax: 00359/ 46 664 068
E-mail: office@chemprod.eu www.chemprod.eu
Bulgarisch - Deutsche Gesellschaft
SPESIMA

CO-ORGANIZERS

UNIVERSITY OF ARCHITECTURE CIVIL ENGINEERING AND GEODESY
Технически Университет Варна

Prof. D.Sc. Vasil I. Kavardzhikov

Chairman of the Organizing Committee

edp sciences — Web of Conferences

Statement of Peer review

In submitting conference proceedings to *Web of Conferences*, the editors of the proceedings certify to the Publisher that

1. They adhere to its **Policy on Publishing Integrity** in order to safeguard good scientific practice in publishing.
2. All articles have been subjected to peer review administered by the proceedings editors.
3. Reviews have been conducted by expert referees, who have been requested to provide unbiased and constructive comments aimed, whenever possible, at improving the work.
4. Proceedings editors have taken all reasonable steps to ensure the quality of the materials they publish and their decision to accept or reject a paper for publication has been based only on the merits of the work and the relevance to the journal.

Title, date and place of the conference

NCTAM 2017 - 13th National Congress on Theoretical and Applied Mechanics,
September 6 - 10, 2017, Sofia, Bulgaria

Proceedings editor(s):

Vassil M.Vassilev

Svetoslav G. Nikolov

Maria D. Datcheva

Yonka P. Ivanova

Date and editor's signature

December 14,
2017

17, avenue du Hoggar - PA de Courtabœuf – BP 112 - 91944 Les Ulis Cedex A (France)
Tél. +33 (0)1 69 18 75 75 - Fax +33(0)1 69 07 45 17 – www.edpsciences.org

Analysis of the susceptibility in a fluid system with Neumann – plus boundary conditions

Peter Djondjorov[1,⋆], *Vassil* Vassilev[1,⋆⋆], and *Daniel* Dantchev[1,⋆⋆⋆]

[1] *Institute of Mechanics, Bulgarian Academy of Sciences, Acad. Georgi Bonchev Street, Block 4, 1113 Sofia, Bulgaria.*

Abstract. The behaviour of the local and total susceptibilities of a fluid system bounded by different surfaces is studied in the framework of the Ginsburg-Landau Ising type model. The case of a plain geometry, Neumann-infinity boundary conditions under variations of the temperature and an external ordering field is considered. Exact analytic expressions for the order parameter, local and total susceptibilities in such a system are presented. They are used to analyse the phase behaviour of fluids confined in regions close to the bulk critical point of the respective infinite system.

1 Introduction

It is well known that the phase behaviour of a fluid system (simple, or mixture) limited in its spatial extends depends on its interaction with material bodies bounding it, immersed in and/or surrounding this fluid system. In general, this phase behaviour essentially differs from that of the infinite, unbounded, system. On the other hand, from a practical point of view, the phase behaviour of a fluid system is important for the stability and consistency of various devices immersed in it, especially for micro- and nano-technologies.

This contribution concerns the behaviour of fluid systems which has a limited spatial extend in one dimension and forming a film geometry $\infty^2 \times L$ (L being the characteristic length of the system), including the fluid mediated interactions between the boundaries of the system.

In the vicinity of the bulk critical temperature T_c of the bulk system, one observes a diversity of surface phase transitions [1, 2] of different kind in which the surface orders before, together, or after ordering in the bulk of the system, which are known as normal (or extraordinary), surface-bulk and ordinary surface phase transitions. For a simple fluid or for binary liquid mixtures the wall generically prefers one of the fluid phases or one of the components. In the vicinity of the bulk critical point the last leads to the phenomenon of critical adsorption [3–13]. This can be modelled by considering local surface field h_1 acting solely on the surfaces of the system. When the system undergoes a phase transition in its bulk in the presence of such surface ordering fields one speaks about the "*normal*" transition [14]. It has been shown that it is equivalent, as far as the leading critical behaviour is concerned, to the "*extraordinary*" transition [12, 14] which is achieved by enhancing the

⋆e-mail: padjon@imbm.bas.bg
⋆⋆e-mail: vasivas@imbm.bas.bg
⋆⋆⋆e-mail: danieldantchev@gmail.com

surface couplings to become stronger enough than the bulk couplings. In the remainder of this article we will use the surface field picture. There is also a special case of surface enrichment, when the surface orders simultaneously with the bulk. This is normally termed surface-bulk, or special phase transition. One way to achieve it is to enhance the coupling on the surface and very near to it, so that one achieves the delicate balance needed to get the both orderings to become simultaneously. On a mean field level this can be described by imposing Neumann boundary conditions on the boundary.

In the current article a generic mean-field model system in which on one surface the system is subjected to strong surface adsorption, while the other is such that the surface orders exactly at the same temperature as the bulk, is considered. Mathematically, within the continuum mean-field approach such a system is considered to be characterized with the (Neumann,+) boundary conditions, where the + signs means that the corresponding order parameter tends to $+\infty$ at the boundary with strong adsorption. This type of boundary conditions can be achieved on a magnetic system. In principle, such boundary conditions can be implemented for a fluid system as well, but an extremely precise arrangement for one of the surfaces has to be performed.

In the current work the order parameter profile and its response functions to an externally applied ordering field h, as well as the free energy will be studied as functions of both the temperature T and the field h within the three-dimensional continuum mean-field Ising model. Analytical results for the h – dependence of the model in the form of the full $(T;h)$ dependences of the order parameter, local and total susceptibilities in the $(T;h)$ plane are presented for the case of above mentioned Neumann – plus infinity boundary conditions.

2 Ginzburg-Landau mean-field model

2.1 Basic definition of the problem considered

A critical system of Ising type in a film geometry $\infty^2 \times L$ is considered, where L is supposed to be along the z axis. Its state is described by the minimizers of the standard ϕ^4 Ginzburg–Landau functional

$$\mathcal{F}[\phi;\tau,h,L] = \int_0^L \mathcal{L}(\phi,\phi')\mathrm{d}z, \tag{1}$$

with

$$\mathcal{L}(\phi,\phi') = \frac{1}{2}\phi'^2 + \frac{1}{2}\tau\phi^2 + \frac{1}{4}g\phi^4 - h\phi, \tag{2}$$

where $\phi(z;\tau,h,L)$ is the order parameter, $\tau = (T - T_c)/[T_c\,(\xi_0^+)^{-2}]$ is the bare reduced temperature with ξ_0^+ being the nonuniversal amplitude of the correlation length along the temperature axis, h is the external ordering field, g is the bare coupling constant and the primes indicate differentiation with respect to the variable z.

2.2 Euler-Lagrange equation for the order parameter

The extremals of the functional $\mathcal{F}$ are determined by the solutions of the corresponding Euler-Lagrange equation

$$\frac{d}{dz}\frac{\partial\mathcal{L}}{\partial\phi'} - \frac{\partial\mathcal{L}}{\partial\phi} = 0, \tag{3}$$

which, on account of Eq. (2), reads

$$\phi'' - \phi\left[\tau + g\phi^2\right] + h = 0. \tag{4}$$

Multiplying the above equation by ϕ' and integrating once over z, one obtains the following first integral of the system

$$\frac{1}{2}\phi'^2 = \frac{1}{2}\tau\phi^2 + \frac{1}{4}g\phi^4 - h\phi + c = P[\phi], \tag{5}$$

where c is the constant of integration. The value of the integration constant c determines three types of roots of the polynomial $P[\phi]$ – four real roots, two real and two complex conjugated roots (because the coefficients of $P[\phi]$ are real numbers) and two couples of complex conjugated roots as well as various combinations of multiple roots.

2.3 Local and total susceptibilities

By definition, the local layered susceptibility in the case under consideration is

$$\chi_l(z|\tau, h) \equiv \frac{\mathrm{d}\phi(z|\tau, h)}{\mathrm{d}h} = \frac{\partial\phi(z|\tau, h)}{\partial h} + \frac{\partial\phi(z|\tau, h)}{\partial x_0}\frac{\partial x_0}{\partial h}, \tag{6}$$

where $\mathrm{d}a/\mathrm{d}h$ stands for the total derivative of the function a with respect to the variable h. Differentiating all terms in Eq. (4) with respect to h, one obtains an equation for χ_l of form

$$-\chi_l'' + (\tau + 3g\phi^2)\chi_l = 1. \tag{7}$$

Based on the knowledge of the local susceptibility, one can determine the total susceptibility $\chi_T(x_t, x_h)$ simply by integrating

$$\chi_T(x_t, x_h) = \frac{1}{L}\int_0^L \chi_l(z|t, h)\, dz. \tag{8}$$

3 Analytical expressions

3.1 Explicit expressions for the order parameter

Hereafter we specialise to boundary conditions of form

$$\phi'(0) = 0, \qquad \phi(L) = +\infty, \tag{9}$$

that is of Neumann type on the left boundary and $+\infty$ on the right one. We are looking for real-valued solutions of Eq. (5), corresponding to given values of the parameters τ and h, which are smooth in the open interval $(0, 1)$, and satisfy the Neumann–infinity boundary conditions (9). In the case of $(+\infty, +\infty)$ boundary conditions, analytical results for the order parameter and the local and total susceptibilities are presented in our previous paper [15]. Using the results in that paper, we are going to obtain solutions for the case of Neumann–infinity boundary conditions by an appropriate scaling of the results in [15]. Thus, in terms of the scaling variables

$$\zeta = z/L, \qquad x_t = \tau L^{1/\nu}, \qquad x_h = \sqrt{2g}hL^{\Delta/\nu}, \qquad \phi(z) = \sqrt{\frac{2}{g}}L^{-\beta/\nu}X_m(\zeta|x_t, x_h), \tag{10}$$

with $\beta = \nu = 1/2$ and $\Delta = 3/2$, Eq. (4) for the order parameter takes the form

$$X_m''(\zeta) = X_m(\zeta)\left[x_t + 2X_m^2(\zeta)\right] - \frac{x_h}{2}. \tag{11}$$

In the cases when there exist multiple real-valued solutions that meet the differential equation (11) and boundary conditions (9), we seek the one that provides minimal value of the energy functional

$$\mathcal{E} = \int_0^1 f(X_m(\zeta), X'_m(\zeta))d\zeta, \tag{12}$$

where

$$f(X_m(\zeta), X'_m(\zeta)) = [X'_m(\zeta)]^2 + X_m^4(\zeta) + x_t X_m^2(\zeta) - x_h X_m(\zeta) \tag{13}$$

is the energy density. In Eq. (11) and below, the primes indicate differentiation with respect to the variable ζ which, as follows from Eq. (10), varies in the closed interval $[0, 1]$. According to Eq. (5) the first integral of Eq. (11) reads

$$[X'_m(\zeta)]^2 = P[X_m], \qquad P[X_m] = X_m^4(\zeta) + x_t X_m^2(\zeta) - x_h X_m(\zeta) + \varepsilon, \tag{14}$$

where ε denotes the respective constant of integration.

First, let us recall that we are interested in real-valued solutions $X_m(\zeta)$ of Eq. (11), corresponding to given values of the parameters x_t and x_h, which are smooth in the open interval $(0, 1)$ and satisfy the Neumann-infinity boundary conditions. The first boundary condition (9) implies that $P(X_m(0)) = 0$. Denoting $X_m(0) = X_{m0}$, we can express the integration constant in the form

$$\varepsilon = -X_{m0}\left(X_{m0}^3 + x_t X_{m0} - x_h\right). \tag{15}$$

Now, given values of the parameters x_t, x_h and X_{m0}, each real-valued solution $X_m(\zeta)$ of Eq. (14) can be expressed, following [16, §20.22, §21.73], in the form

$$X_m\left(\zeta | x_t, x_h, X_{m0}\right) = X_{m0} + \frac{6X_{m0}\left(x_t + 2X_{m0}^2\right) - 3x_h}{12\wp\left(\zeta; g_2, g_3\right) - \left(x_t + 6X_{m0}^2\right)}. \tag{16}$$

Here, $\wp\left(\zeta; g_2, g_3\right)$ is the Weierstrass elliptic function, g_2 and g_3 are the invariants of the polynomial $P[X_m]$, which according to [16, §20.22, §21.73] and Eq. (15) read

$$g_2 = \frac{x_t}{12} + X_{m0}\left(x_h - x_t X_{m0} - X_{m0}^3\right), \tag{17}$$

$$g_3 = -\frac{1}{432}\left(2x_t^3 + 9\left(3x_h^2 - 8x_t x_h X_{m0} + 8x_t X_{m0}^2\left(x_t + X_{m0}^2\right)\right)\right). \tag{18}$$

In the case of parameters (x_t, x_h) such that more than one solution of form (16) exist, than the order profile of minimal energy (12) is the physical one.

3.2 Explicit expression for the local and total susceptibilities

In scaling variables (10) the local susceptibility (6) reads

$$\chi_l(z|x_t, x_h) = L^{\gamma/\nu} X_\chi(\zeta|x_t, x_h), \tag{19}$$

where

$$X_\chi(\zeta|x_t, x_h) = 2\frac{\mathrm{d}}{\mathrm{d}x_h} X_m(\zeta|x_t, x_h). \tag{20}$$

Then, denoting by $\dot{X}_{m0}(x_h)$ the partial derivative of X_{m0} with respect to x_h, Eq. (6) can be written in the form

$$X_\chi(\zeta|x_t, x_h, X_{m0}) = 2\frac{\partial}{\partial x_h} X_m(\zeta|x_t, x_h, X_{m0}) + 2\dot{X}_{m0}(x_h)\frac{\partial}{\partial X_{m0}} X_m(\zeta|x_t, x_h, X_{m0}), \tag{21}$$

and since $X_\chi(1|x_t, x_h, X_{m0}) = 0$, this equation implies

$$\dot{X}_{m0}(x_h) = -\frac{\frac{\partial}{\partial x_h} X_m(1|x_t, x_h, X_{m0})}{\frac{\partial}{\partial X_{m0}} X_m(1|x_t, x_h, X_{m0})}. \tag{22}$$

Thus, equations (21), (22) provide a tool for analysis of the dependence of the local susceptibility $X_\chi(\zeta|x_t, x_h)$ with respect to the parameters x_t and x_h.

Definition (8) in scaling variables reads

$$\overline{X}_\chi(x_t, x_h) = \int_0^1 X_\chi(\zeta|x_t, x_h)\, d\zeta. \tag{23}$$

4 Results

4.1 Plots of the order parameter profile

Typical examples of the order parameter profiles in the case of Neumann – infinity boundary conditions are shown in Fig. 1. Here, Fig. 1(a) is for $x_t = -8.45$, $x_h = -15.63$, and in this case there exists only one order parameter that satisfies the boundary conditions, and it takes positive values only. Fig. 1(b) represents the case when a new order parameter that satisfies equation (14) and boundary conditions (9) appears. In this case, the blue order parameter is of lower energy meaning that the system is in a liquid state. For a bit lower values of the ordering field, three distinct profiles exist, that are illustrated in Fig. 1(c). Here, the profile of lowest energy is also the blue one, and it indicates that the system is in gas state near the left boundary and in the liquid state in the right one. Further lowering the ordering field causes the system to goes back to a state with a single order parameter, see the case Fig. 1(d), taking both – negative and positive values. Consequently, lake the case in Fig. 1(c), here the system is in gas state near the left boundary and in liquid state near the right one.

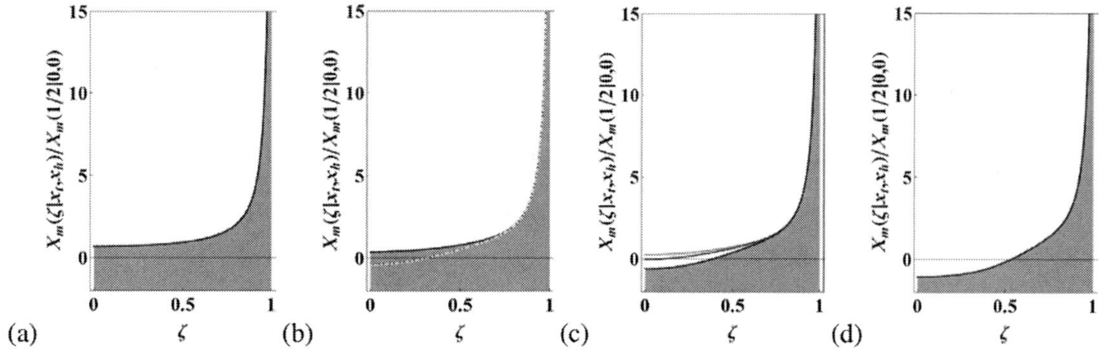

Figure 1. Order parameter profiles in the cases $x_t = -8.45$, and: $x_h = -15.63$ (a), $x_h = -30.12$ (b), $x_h = -32$ (c) and $x_h = -62.5$ (d).

4.2 Plot of the phase diagram

The phase diagram is shown on Fig. 2. It is a straight line starting from $x_t = -6.42458$ and $x_h = -29.5044$. The four dots on Fig. 2 indicate the cases in Fig. 1, namely: the most right black dot is Fig. 1(a), the green one is Fig. 1(b), the blue dot is Fig. 1(c), and the most left black dot is Fig. 1(d).

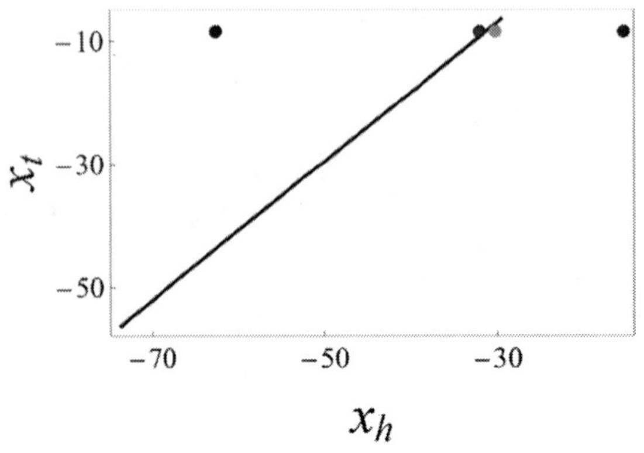

Figure 2. Phase diagram for the Neumann – infinity case. Dots indicate the four cases in Fig. 1.

Our observation is that there exists a short segment in the phase diagram between its top point $T_{cL}(x_t = -6.42458, x_h = -29.5044)$ and the point $T_{cap}(x_t = -6.61822, x_h = -29.67)$, crossing which the system jumps between a less dense gas and a more dense gas. Thus, the very capillary condensation point is T_{cap}, bellow which system jumps between a gas and a liquid states. The same findings concerning the case of (+,+) boundary conditions are reported in [15].

4.3 Plots of the local susceptibility

Fig. 3 represents several examples of profile curves of the local susceptibility. It is seen, that the profile curve of the local susceptibility corresponding to $x_t = -8.45$, $x_h = -62.50$ is not a monotonic one, whereas the profile curves of the local susceptibility corresponding to $x_t = -8.45$ and $x_h = -15.63$, $x_h = -32.00$ are monotonic.

4.4 Plots of the total susceptibility

Finally, two plots of the scaling function of the total susceptibility (23) are shown in Fig. 4. The left figure shows the appearance of the total susceptibility near the point $T_{cL}(x_t = -6.42458, x_h = -29.5044)$, where several solutions of Eq. (14) exists. The right figure represents the classic point of capillary condensation $T_{cap}(x_t = -6.61822, x_h = -29.67)$. It should be noted, that when the system traverses the segment in the phase diagram between T_{cL} and the capillary condensation point T_{cap}, the system jumps between a states of a less dense gas and a more dense gas. Below the capillary condensation point, the system jumps between gas and liquid states.

5 Conclusions

The behaviour of the local and total susceptibilities of a fluid system in a film geometry are studied. The Ginsburg-Landau Ising type model and Neumann-infinity boundary conditions are employed.

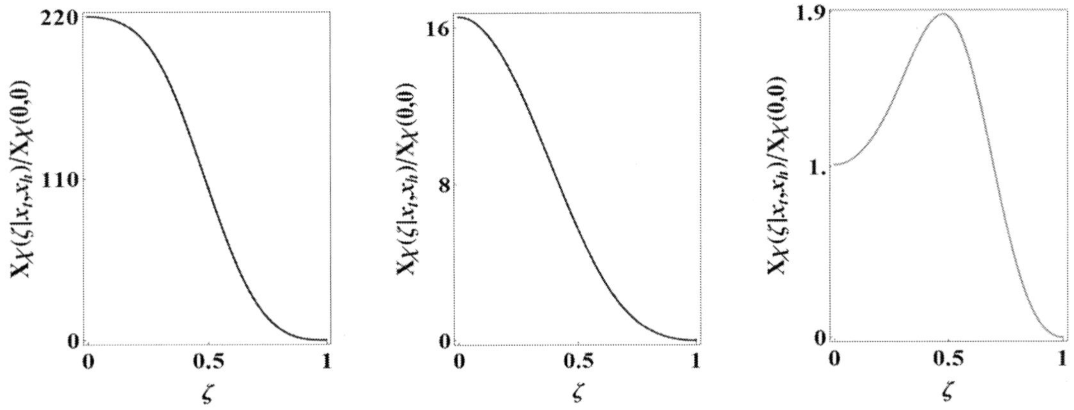

Figure 3. Local susceptibility profiles in the cases $x_t = -8.45$, and: $x_h = -30.12$ (left), $x_h = -32$ (centre), and $x_h = -62.5$ (right).

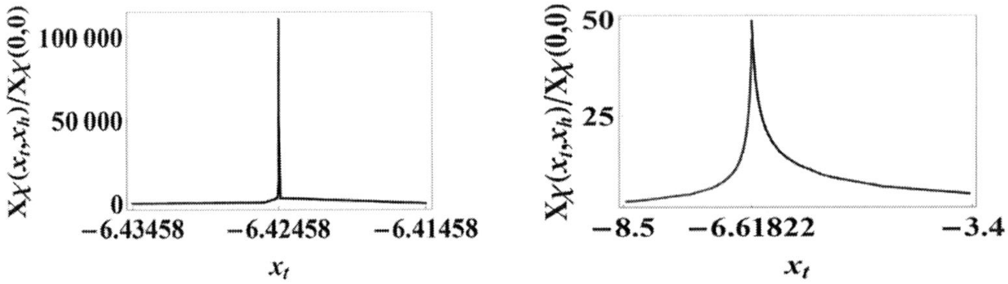

Figure 4. The temperature behavior of the total susceptibility near the point T_{cL}, where the system experiences jumps between states of a less dense gas and a more dense gas (left), and the classic point of a beginning of the capillary condensation T_{cap} (right).

Exact analytic results for the order parameter, as well as local and total susceptibilities are presented. Plots of the order parameter profiles, phase diagram, and local susceptibilities profiles are presented. In this problem, the fluid system is always in a liquid state at the right boundary, and achieves gas of liquid states near the left boundary, depending on the values of the parameters x_t and x_h (see, e.g. [17, 18]).

It is found that the phase diagram is an approximately straight line, starting from the point $T_{cL}(x_t = -6.42458, x_h = -29.5044)$, see Fig. 2. As it is seen in Fig. 4 (left), the total susceptibility there achieves extremely high values and theoretically diverges.

Our observation is that below the point $T_{cap}(x_t = -6.61822, x_h = -29.67)$, the system jumps between a gas and a liquid states. This means, that T_{cap} is the true capillary condensation point, but the total susceptibility achieves finite values there, see Fig. 4 (right). The total susceptibility diverges at the point T_{cL}, and along the segment starting from T_{cL} and ending to T_{cap}, the system jumps between a less dense gas and a denser one.

References

[1] K. Binder, *Critical Behaviour of Surfaces; vol 8 in Phase Transitions and Critical Phenomena, eds C. Domb and J. L. Lebowitz* (Academic, London, 1983)
[2] H. W. Diehl, *Field-theoretical approach to critical behavior of surfaces; vol 10 in Phase Transitions and Critical Phenomena, eds C. Domb and J. L. Lebowitz* (New York: Academic, 1986)
[3] R. Evans, J. Phys.: Condens. Matter **2**, 8989 (1990)
[4] M. Tröndle, L. Harnau and S. Dietrich, J. Chem. Phys. **129**, 124716 (2008)
[5] Z. Borjan and P. Upton, Phys. Rev. E **63**, 065102 (2001)
[6] R. Evans and U. M. B. Marconi, J. Chem. Phys. **84**, 2376–99 (1986)
[7] R. Okamoto and A. Onuki, J. Chem. Phys. **136**, 114704 (2012)
[8] D. Dantchev, F. Schlesener and S. Dietrich, Phys. Rev. E **76**, 011121 (2007)
[9] A. Drzewiński, A. Maciołek, A. Barasiński and S. Dietrich, Phys. Rev. E **79**, 041145 (2009)
[10] J. Cahn, J. Chem. Phys. **66**, 3667–72 (1977)
[11] P.-G. de Gennes, Rev. Mod. Phys. **57**, 827–63 (1985)
[12] F. P. Toldin and S. Dietrich, J. Stat. Mech. **11**, P11003 (2010)
[13] D. Dantchev, J. Rudnick and M. Barmatz, Phys. Rev. E **75**, 011121 (2007)
[14] T. W. Burkhardt and H. W. Diehl, Phys. Rev. B **50** 3894–3898 (1994)
[15] D. Dantchev, V. Vassilev and P. Djondjorov, Journal of Statistical Mechanics: Theory and Experiment **2015**, P08025 (2015)
[16] E. T. Whittaker and G. N. Watson, *A Course of Modern Analysis* (Cambridge University Press, London, 1963)
[17] R. Evans, *Liquids at interfaces* (Elsevier, Amsterdam, 1990)
[18] H.-J. Butt, K. Graf and M. Kappl, *Contact Angle Phenomena and Wetting; Chapter 7 in Physics and Chemistry of Interfaces* (Wiley-VCH Verlag & Co. KGaA, Weinheim (FRG), 2003)

Study of soccer ball flight trajectory

Juliana Javorova[1,*], *Anastas* Ivanov[2]

[1] University of Chemical Technology and Metallurgy, Sofia, Bulgaria
[2] Todor Kableshkov University of Transport, Sofia, Bulgaria

Abstract. In this paper the trajectories of a soccer ball for the most important kicks in the football game - a corner kick and a direct free kick are studied. The soccer ball is modelled as an ideal rigid hollow spherical body with six degrees of freedom, which performs a general motion in an immovable air environment with constant parameters. The ball 3D orientation is determined by the three Cardan angles. The aerodynamic forces and moments with which the air environment acts to the ball are taken into account. Two of the most dangerous areas of the football goal are defined. Differential equations which describe the motion of the soccer ball are solved numerically by MatLab-Simulink.

1 Introduction

Football is one of the most popular ball games. The footballer skills to give the soccer ball the necessary direction, speed and rotation is an extremely important quality that can be improved and perfected if the physical nature of the soccer ball motion in the air is known, as well as the factors that affect this flight.

In the last twenty years, there has been a strong rise in the researches and publications of various types of sports balls, in particular to balls for tennis, golf, baseball, football and volleyball [1-6, etc.]. And although there are many common laws among them, each ball has specific features.

The soccer ball, compared to the ones listed above, has got the largest dimensions and weight. It also differs from them in that its free motion is achieved by the players' feet and sport shoes. In the papers [7, 8, etc.] the footstep of the footballer together with his shoe and the soccer ball are modelled by Finite Element Method.

Most of the studies can be grouped mainly in the following directions.

For one big part of papers it is assumed that the fluid environment is under steady state conditions and the soccer ball moves with a certain speed in it. Here the theoretical knowledge of aerodynamics is complemented and refined by modern experiments [9-14].

In the second group of researches [15-18, etc.] the ball is considered as an immovable and it is flowed by the moving fluid environment. These studies are classical and are related to the presence of a fluid tube (tunnel). Different aerodynamic coefficients are determined, in particular the drag coefficient and the lift force coefficient in presence and absence of ball rotation.

* Corresponding author: july@uctm.edu

A mathematical study of the soccer ball aerodynamic interaction with the fluid environment based on the boundary layer theory is treated in details in [19, 20].

One important clarification of the effect of Magnus is done in the paper [21, 22].

In the researches [23-26] the most important kicks in the football game /direct free kick and corner kick/ are studied.

There are also many publications, sporting and methodical developments, in which the soccer ball strokes are studied from biomechanical point of view [27, 28].

The present study is directed namely to the above mentioned two soccer kicks. The main objectives of this paper can be formulated, as follows:

- To define the most dangerous zone in the upper left or right corner of the goal, which is farthest from the goalkeeper and therefore the ability to score is greatest;
- To find the initial linear and angular velocities also initial angles which are enough for soccer ball to go in the above-mentioned dangerous zone using two kind of soccer strokes;
- To illustrate the trajectories of potential successful kicks graphically.

2 Characteristics of soccer balls and football fields

The soccer balls used in professional leagues and in the FIFA and World Cup have five different sizes. Dimensions 1 and 2 are for promotional soccer balls, size 3 is used for children less than 8 years of age, while size 4 is used for young men and young women aged between 8 and 12 years. Size 5 is used for official football matches for men and women over 12 years of age.

Soccer ball size 5 is described by the following technical characteristics: circumference from 27 to 28 *in* (from 69 to 71 *cm*); diameter from 8.6 to 9 *in* (from 22 to 23 *cm*); weight from 14 to 16 *oz* (from 400 to 450 *gr*); internal pressure between 8.5 and 15.6 *PSI* (from 0.6×10^5 to $1.1\times10^5\,Pa$). The modern soccer balls have a different type of external relief and colours (Fig.1).

Fig. 1. Some professional soccer balls.

However, the most popular soccer ball consists of 12 pentagons and 20 hexagons from special artificial leather. They are joined together so that the soccer ball is an ideal sphere. The skin is extremely smooth and varnished. The connecting points of the individual leather pieces represent indentations which (analogical of the golf balls) further reduce the air environment resistance.

The football fields (called also football pitches or football grounds) are rectangular and usually vary in size. They have a length from 90 to 120 m and a width from 45 to 90 m.

For international matches after March 2008 the football pitches have official standard sizes. They are: 105 m long and 68 m wide. For example, the Allianz-Arena Stadium in Munich has just the same sizes. However, the grass covers a larger area. It is about 111 m long and 72 m wide (Fig. 2).

Fig. 2. Allianz-Arena Stadium in Munich (http://sportslens.com/allianz-arena/92352/).

In the past football fields in South America was constructed with maximum sizes. Today, these playgrounds are rebuilt and now they have standard sizes. For example, these are the stadiums of Maracana (Rio de Janeiro), Antonio Vespuco Liberty and Jose Amalitani (Buenos Aires), Azteca (Mexico) and others.

3 Mechanics of ball flight

One of the basic assumptions related to the soccer ball mathematical model is that it is considered as a perfectly rigid sphere. Therefore in all of the time of football play, its dimensions remain constant.

3.1 Kinematics

The soccer ball has six degrees of freedom during the flight. It performs a general motion in the three-dimensional space. This motion can be counted with respect to a conditionally accepted motionless reference system $Oxyz$.

Each general motion of a rigid body is complexly composed. It consists of a translation component that depends on the selected pole and a spherical component that does not depend on that choice.

The ball centre C is selected for the pole. Therefore, the translation component of body motion is defined by the three functions of time (Fig.3):

$$x = x(t)\ ,\quad y = y(t)\ ,\quad z = z(t)\ . \tag{1}$$

In order to define the spherical component of the motion, two main reference systems are introduced. The first one $CXYZ$ has the mass centre C of the ball. It performs translational motion to the $Oxyz$ system. The second one $C\xi\eta\zeta$ is invariably fixed to the ball and it is called movable reference system. Moreover, for describing of the spherical motion, two additional reference systems $Cx_1y_1z_1$ and $Cx_2y_2z_2$ are used. The orientation of the axes $C\xi$, $C\eta$ and $C\zeta$ in the space is defined by the three Cardan angles: rotation to an angle $\psi(t)$ around the axis $X \equiv x_1$, rotation to an angle $\theta(t)$

around the axis $y_1 \equiv y_2$, and finally, rotation to an angle $\varphi(t)$ around the axis $z_2 \equiv \zeta$. Therefore, the spherical component of the body motion is defined by the three functions:

$$\psi = \psi(t)\ ,\quad \theta = \theta(t)\ ,\quad \varphi = \varphi(t)\ . \tag{2}$$

By functions (1) and (2) the law of general motion of the soccer ball is defined.

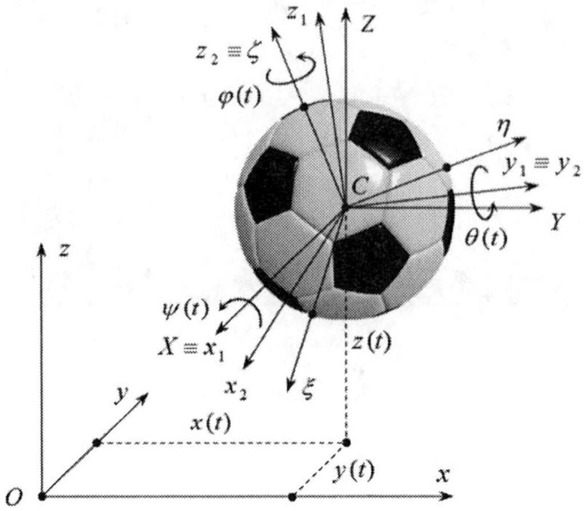

Fig. 3. Generalized coordinates of the soccer ball describing its position in the space.

The soccer ball angular velocity $\boldsymbol{\omega}(t)$ can be determined by its algebraic projections with respect to the axes of $CXYZ$ system. These formulas are the so-called kinematic Cardan equations and they have the following form:

$$\omega_X = \dot{\psi} + \dot{\varphi}.\sin\theta, \tag{3}$$

$$\omega_Y = \dot{\theta}.\cos\psi - \dot{\varphi}.\cos\theta.\sin\psi\ , \tag{4}$$

$$\omega_Z = \dot{\varphi}.\cos\theta.\cos\psi + \dot{\theta}.\sin\psi\ . \tag{5}$$

3.2 Aerodynamics

The following assumptions are introduced in advance. The air environment has steady-state parameters and the influence of the wind actions is not taken into account.

It is known that the complex aerodynamic interaction between the air and the flying sport ball can be analysed analytically by the boundary layer theory. However, in the practice only the main conclusions of this theory are used, as primarily all is based on the experimental researches. The latter are imposed because of very complicated mathematical description of this aerodynamic interaction as well as the great variety of the sports (in particular soccer) balls that does not allow unification of equations.

The main force acting on the ball is the force of gravity (Fig. 4):

$$\mathbf{G} = m.\mathbf{g}\ , \tag{6}$$

where m is the soccer ball mass, $\mathbf{g}$ is a gravitational acceleration.

The second force acting on the ball is the force of Archimedes. It is directly opposite to the force of gravity and is determined by the formula:

$$\mathbf{W}_A = -(0.40).\pi.R^3.\rho_a.\mathbf{g} \,, \qquad (7)$$

where $R = 0.11\ m$ is the ball mean radius, $\rho_a = 1.205\ kg/m^3$ is the air density at temperature about $t = 20\,^0C$.

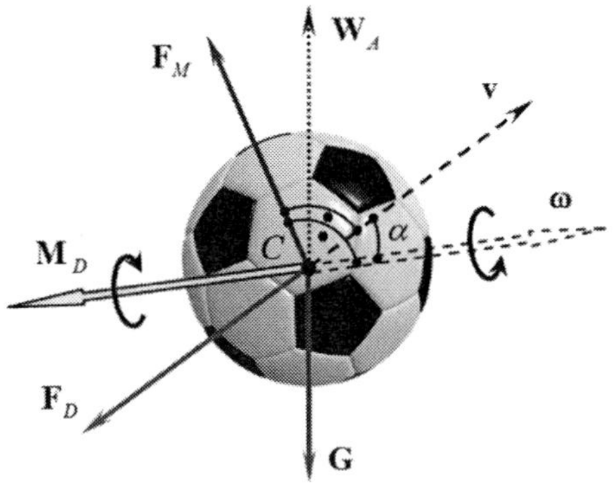

Fig. 4. Forces acting on the soccer ball during its flight in the air

The next force that acts on the soccer ball is a drag force. Its action is in the opposite direction of the velocity of the ball centre of mass (Fig. 4). This force depends on the square of that velocity and it is determined by the formula:

$$\mathbf{F}_D = -\frac{1}{2}.C_D.\rho_a.A.v^2.\frac{\mathbf{v}}{|\mathbf{v}|} \,, \qquad (8)$$

where $A = \pi.R^2$ is the middle cross-section area of the ball, C_D is a drag coefficient.

The values of the non-dimensional drag coefficient depend on the kind of motion of the flying soccer ball. For the case of a translation without spin, for example straight central stroke during the football play, this coefficient C_D depends mainly on the Reynolds number $Re = v.D/\vartheta$. The coefficient of kinematic viscosity is $\vartheta = 1.50\times10^{-5}\ m^2/s$ at environment temperature $t = 20\,^0C$. Then, at the soccer ball velocity in the range $v = 20 \div 30\ m/s$, the Reynolds number ranges from $Re = 2.93\times10^{5}$ to $Re = 4.40\times10^{5}$. The Re values show that the motion performs under turbulent conditions, (Fig. 5).

From Fig. 5 it can be seen that for the above-mentioned values of the velocity v and Reynolds number Re, the drag coefficient in the absence of rotation is approximately equal to $C_D = 0.20$.

If the soccer ball motion has a rotational component also together with translational one, the drag coefficient depends both on the Reynolds number Re and on the spin factor $S_p = \omega.R/\vartheta$. In order to obtain the appropriate rotation of the ball, it is necessary the angular velocity to be within the following limits $\omega = 60 \div 120\ s^{-1}$. Therefore, the spin factor values must be ranged as $S_p = 0.33 \div 0.44$.

As it can be seen on Fig. 6, for the given above limits of values of the Reynolds number Re and spin factor S_p the drag coefficient must vary in the range $C_D = 0.30 \div 0.35$.

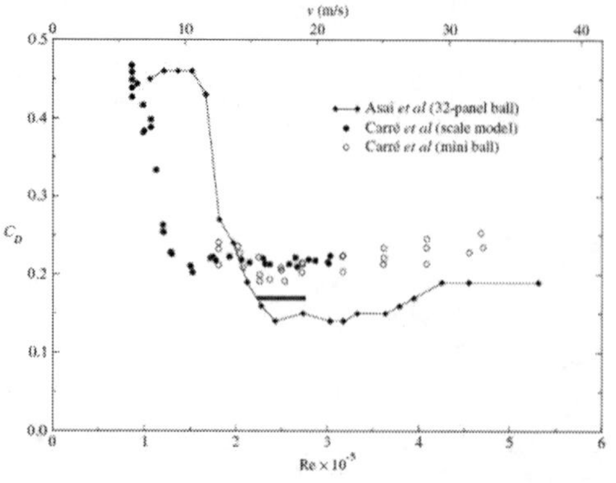

Fig. 5. Wind-tunnel data for C_D as a function of Re and v, according to [13].

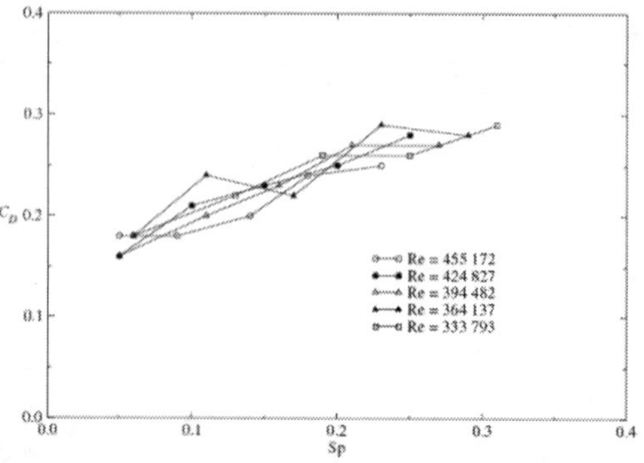

Fig. 6. Wind-tunnel data for C_D as a function of S_p, according to [13].

The fourth force that acts on the soccer ball is the lift force which is very often called Magnus force (Fig.4). It also depends on the square of velocity but it acts in the direction of the vector cross-product $\boldsymbol{\omega} \times \mathbf{v}$:

$$\mathbf{F}_M = \frac{1}{2}.C_L.\rho_a.A.v^2.\frac{\boldsymbol{\omega} \times \mathbf{v}}{|\boldsymbol{\omega} \times \mathbf{v}|}. \tag{9}$$

Here the dimensionless lift coefficient C_L also depends on the Reynolds number Re and on the spin factor S_p. From the results presented in [14] it can be concluded that for the above mentioned limits of Re and S_p the coefficient C_L values must be ranged as follows $C_L = 0.25 \div 0.35$.

Furthermore, in order to be the aerodynamic study more accurate, it is also necessary to introduce the so-called drag moment. It has to act in the opposite direction of the angular velocity $\boldsymbol{\omega}$ at the presumption that the investigated soccer ball is an ideal sphere:

$$\mathbf{M}_D = -\frac{1}{2}.C_D^{(m)}.\rho_a.A.v^2.\frac{\boldsymbol{\omega}}{|\boldsymbol{\omega}|}. \tag{10}$$

It was not found references with experimental data for determination of the coefficient $C_D^{(m)}$. By reason of this and because the soccer ball is quite smooth and polished (at least at the beginning of the game), it can be concluded that this coefficient has comparatively low values. Therefore, in the present study it was accepted for $C_D^{(m)} = 0.05$.

4 Differential equations

The differential equations, which describes the three dimensional general motion of a soccer ball, have the following form:

$$m.\ddot{x} = -k_D.\dot{x}.\sqrt{\dot{x}^2+\dot{y}^2+\dot{z}^2} + k_L.\left(\omega_Y.\dot{z} - \omega_Z.\dot{y}\right), \tag{11}$$

$$m.\ddot{y} = -k_D.\dot{y}.\sqrt{\dot{x}^2+\dot{y}^2+\dot{z}^2} + k_L.\left(\omega_Z.\dot{x} - \omega_X.\dot{z}\right), \tag{12}$$

$$m.\ddot{z} = -k_D.\dot{z}.\sqrt{\dot{x}^2+\dot{y}^2+\dot{z}^2} + k_L.\left(\omega_X.\dot{y} - \omega_Y.\dot{x}\right), \tag{13}$$

$$J.\dot{\omega}_X = -k_D^{(m)}.\omega_X, \tag{14}$$

$$J.\dot{\omega}_Y = -k_D^{(m)}.\omega_Y, \tag{15}$$

$$J.\dot{\omega}_Z = -k_D^{(m)}.\omega_Z. \tag{16}$$

These equations must be solved together with the kinematic equations (3), (4), (5) at the respective initial conditions. Introduced above auxiliary multipliers can be written as:

$$k_D = \frac{1}{2}.C_D.\rho_a.A, \tag{17}$$

$$k_L = \frac{1}{2}.C_L.\rho_a.A.\left(\dot{x}^2+\dot{y}^2+\dot{z}^2\right).\frac{1}{|\boldsymbol{\omega}\times\mathbf{v}|}, \tag{18}$$

$$k_D^{(m)} = \frac{1}{2}.C_D^{(m)}.\rho_a.A.\frac{1}{|\boldsymbol{\omega}|}. \tag{19}$$

5 Goal target and soccer kicks modelling

5.1 Goal target

To all footballers including the goal keepers are well known the two most dangerous zones of the goal. If the soccer ball goes to these areas, the probability of a score is extremely high. These are small areas around the upper two corners (left and right) of each goal, as in the current paper they are defined as a quarter of an ellipse whose centre is located in the inner corner of the goal (Fig. 7). The main dimensions of this ellipse are, as follows:

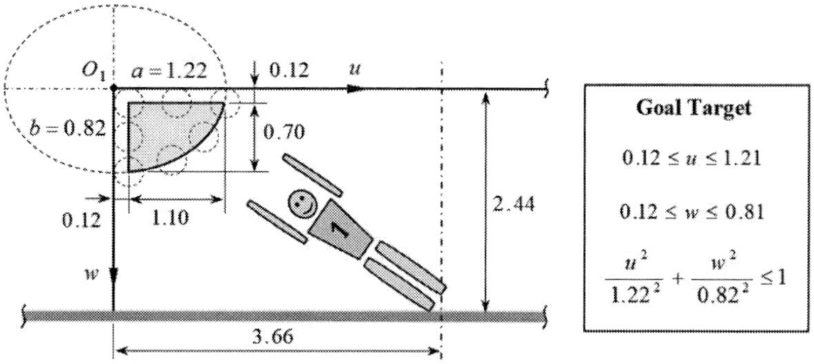

Fig. 7. Goal target – most dangerous zone for goal keeper.

long half-axis $a = 1.22\ m$, short half-axis $b = 0.82\ m$. Two strips (horizontal and vertical) of width $c = 0,12\ m$ are removed from the considered part of the ellipse as the size c is chosen to be larger than the maximum radius R of all models of soccer balls. In the paper namely above-mentioned parts of ellipse are regarded as a goal target (left or right).

5.1 Corner kick

Usually, when a corner kick is performed, the football striker kicks the ball strongly and high so that it reaches the penalty area where his teammates are located. There they attempt to control the falling ball and by head or by foot shot (volley, back scissors or a simple kick) try to direct the ball to the opponent's goal. So, the classic implementation of the corner kick is composed by two separated but connected each other elements.

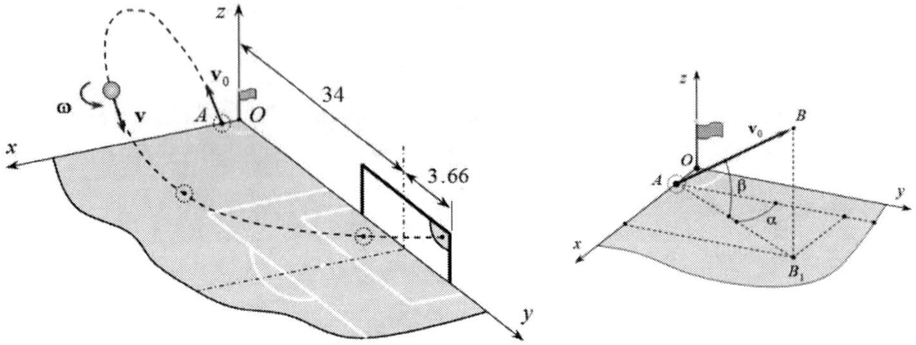

Fig. 8. Corner kick.

Current paper deals with another popular case about a corner kick. The corner striker gives the soccer ball such an initial linear and angular velocity that, thanks to the Magnus effect, the ball flies over half the width of the football field as deviating from the straight line trajectory and finally goes exactly to the remote upper corner of the opponent's goal (Fig. 8). In this way, it is achieved a direct score by the corner kick.

5.2 Free kick

In the paper is considered a direct free kick that aims to the score. The starting point from which the footballer performs the free kick (point O on Fig. 9) is located at a distance $20\,m$ from the goal line on the direction of the central longitudinal line of the football pitch. A wall of defenders is built at a distance $9.14\,m$ on the same direction. It is assumed

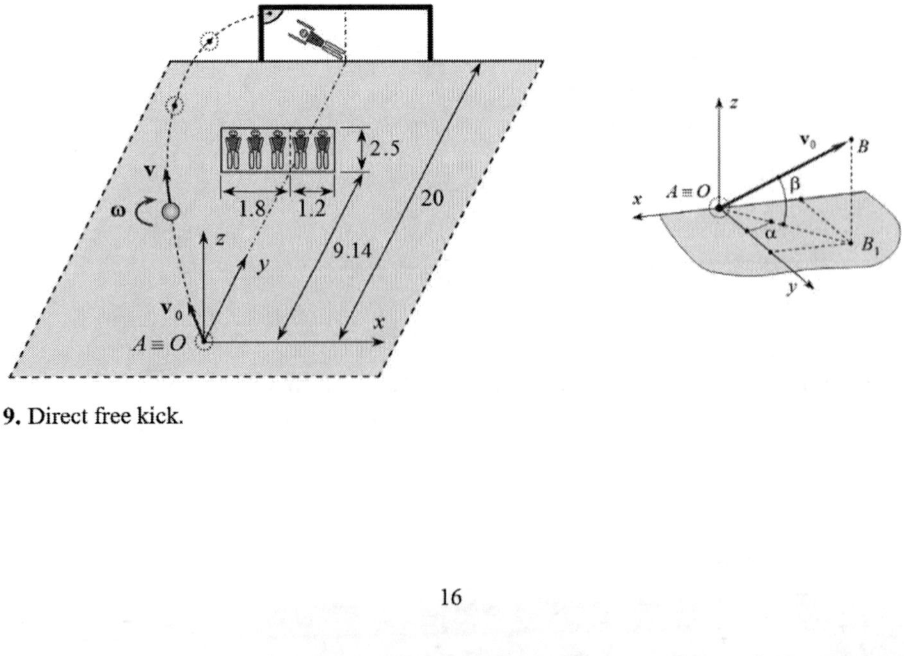

Fig. 9. Direct free kick.

that the defenders wall is wide $3m$ and the guarded height is $2.50m$. The wall is positioned asymmetrically with respect to the central longitudinal line of the football field – 1.2 m in one direction and 1.8 m in the other (Fig.9).

The striker, who performs the direct free kick, attempt to strike the ball with an appropriate in direction and magnitude linear velocity in order to transfer it over the wall. Simultaneously with the required linear velocity, the striker must give to the soccer ball the required angular velocity, i.e. side rotation. Namely on the base of this rotation the Magnus effect starts to act as the ball goes "turning" in its trajectory as it approaches the goal. At this deviation of the trajectory the ball is directed into the upper corner of the goal where the dangerous zone is located (Fig.9).

6 Numerical solution

For the numerical solution of the differential equations $(11)\div(16)$ a program in the MatLab-Simulink environment is created. As initial conditions in the program are introduced the following geometric and kinematic parameters of the ball – initial position, initial angle, initial linear and angular velocities.

By the program can be obtained all kinematic characteristics of the soccer ball motion: the lows of motion about the six coordinates $x(t)$, $y(t)$, $z(t)$, $\psi(t)$, $\theta(t)$, $\varphi(t)$, the trajectory projections $z(x)$, $z(y)$, $y(x)$ on the coordinate planes Oxz, Oyz, Oxy, the linear velocities $\dot{x}(t)$, $\dot{y}(t)$, $\dot{z}(t)$, $v(t)$, the angular velocities $\dot{\psi}(t)$, $\dot{\theta}(t)$, $\dot{\varphi}(t)$, $\omega_X(t)$, $\omega_Y(t)$, $\omega_Z(t)$, $\omega(t)$, the linear accelerations $\ddot{x}(t)$, $\ddot{y}(t)$, $\ddot{z}(t)$, $a(t)$, the angular accelerations $\ddot{\psi}(t)$, $\ddot{\theta}(t)$, $\ddot{\varphi}(t)$, $\varepsilon_X(t)$, $\varepsilon_Y(t)$, $\varepsilon_Z(t)$, $\varepsilon(t)$.

7 Results and discussions

Fig. 10 shows the trajectory of a successful corner kick (CK) with outside spin to the remote target (in the right upper corner of the goal) at the initial parameters: linear velocity $v = 29.54\ m/s$, angular velocity $\omega = 97.88\ rad/s$ ($935\ rpm$), $\alpha = 5.71^o$, $\beta = 17.74^o$.

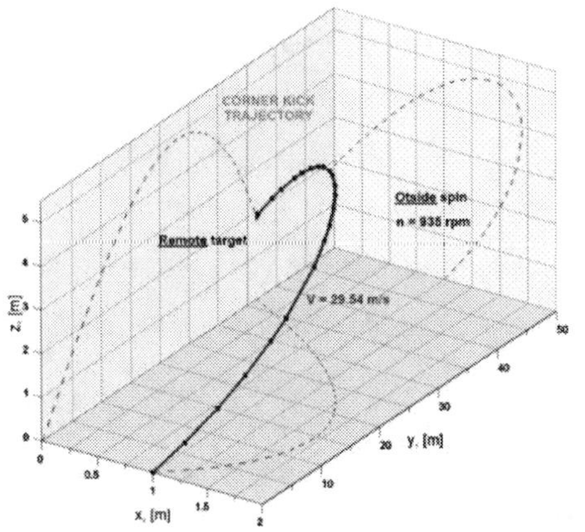

Fig. 10. Successful corner kick to remote target.

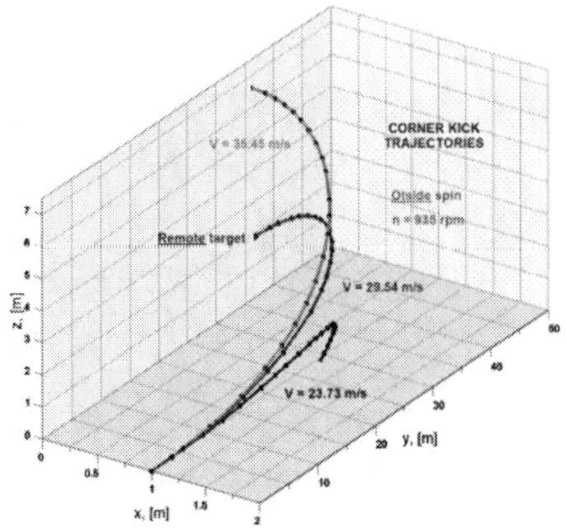

Fig. 11. Three kicks with different velocities.

On the next Fig. 11 are presented three different corner kick trajectories, which are related to: the above mentioned (see Fig. 10) successful kick (red line) and two unsuccessful (failed) ones. Both failed kicks have the same initial angular velocity and angles as the successful one. The linear velocity of the first of them (green line) is $v = 35.45\ m/s$ while for the second (blue line) $v = 23.73\ m/s$. The trajectory projections of the presented on the Fig.11 kicks are given also on Fig. 12 and Fig.13, where they are drawn in the planes yz and xz, respectively. It can be concluded that at the lower linear velocity (blue dotted line) the ball can't reach at all to the goal while with higher velocity (green dotted line) the soccer ball goes very high and away from the goal target.

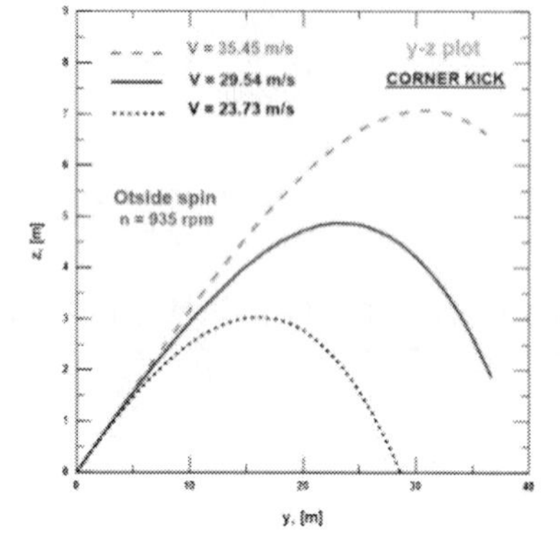

Fig. 12. Trajectory projections of three different corner kicks in yz plane.

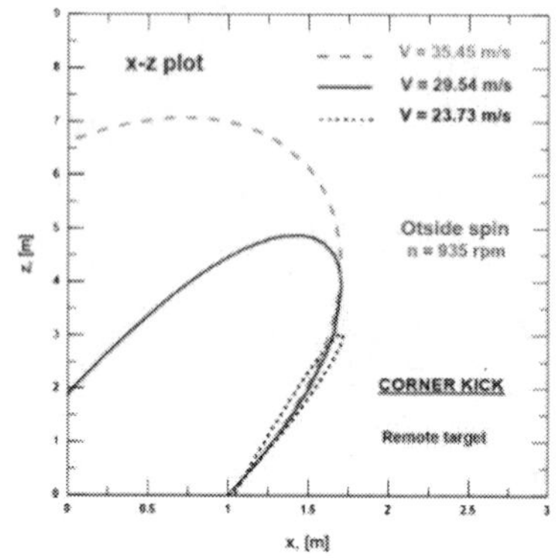

Fig. 13. Trajectory projections of three different corner kicks in xz plane.

The trajectories of two successful free kicks (FK) to the left (green line) and right (red line) target (left and right upper corners of the goal) are shown on Fig. 14. Both of them are realized with outside spin. The initial parameters are: linear velocity $v = 24.15\ m/s$, angular velocity $\omega = 61.86\ rad/s$ ($591\ rpm$), $\alpha = 15.74^o$, $\beta = 18.84^o$.

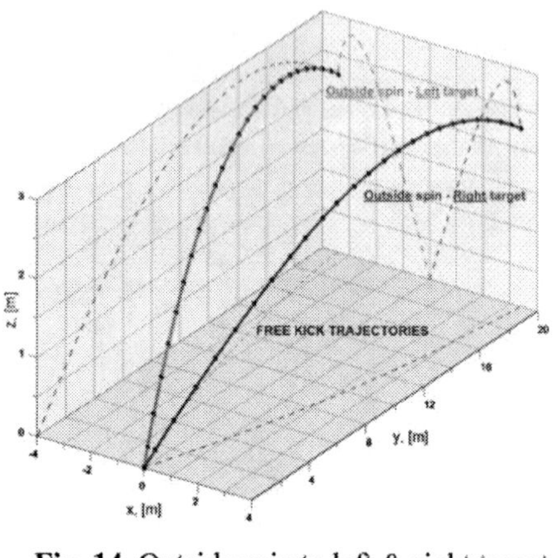

Fig. 14. Outside spin to left & right target.

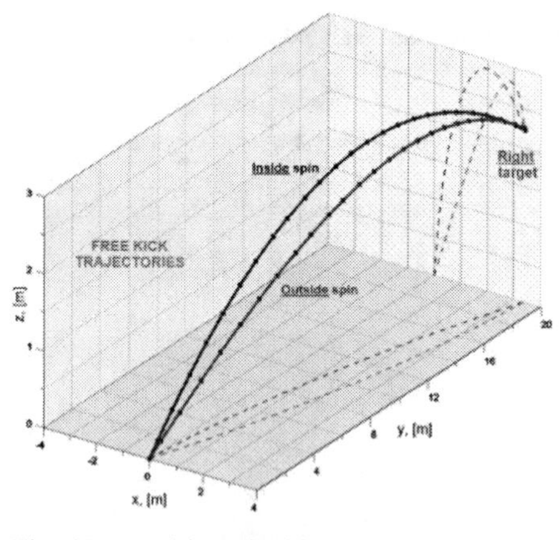

Fig. 15. Outside & inside spin to right target.

On the next Fig. 15 are presented the trajectories of two successful free kicks to the right target of the goal. The first one is realized with outside spin (red line) at the same as above-mentioned initial angular velocity and angles. The second of them is related to inside spin (blue line) at the following initial parameters: linear velocity $v = 23.42\ m/s$, angular velocity $\omega = 58.95\ rad/s$ ($563\ rpm$), $\alpha = 3.64^o$, $\beta = 19.72^o$.

The complex solution of finding the trajectories of four possible successful free kicks, respectively to the left and right targets with outside and inside spins, are illustrated on Fig. 16. The situated on the right side two free kicks are already described and they are the same as shown on Fig. 15. The other two free kicks are directed to the left target, as they are also realized with inside and outside spin. For them it would be specially mentioned that they are symmetrical related to the first pair of free kicks on the right side.

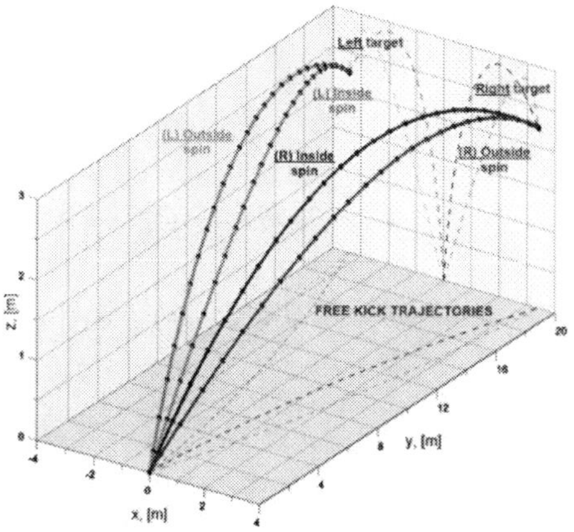

Fig. 16. Outside and inside spins to both targets.

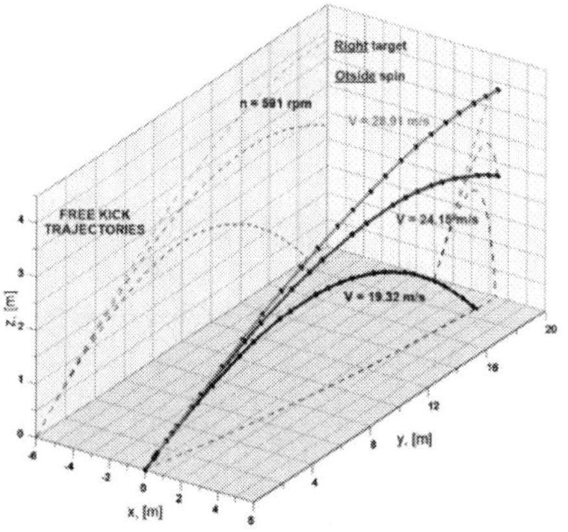

Fig. 17. Three different kicks with outside spin to right target.

On the next Fig. 17 are presented three different free kick trajectories to the right target. The first one is successful (red line) and it is done at the same initial conditions as the above-mentioned ones (see Fig. 14). Both other free kicks are unsuccessful, as their initial angular velocity and angles are the same as for the successful one. The linear velocity of the first failed kick (green line) is $v = 28.91\ m/s$ as for the second failed kick (blue line) this velocity is $v = 19.32\ m/s$.

From the numerical results it is observed as general that a soccer ball can reach the target only under proper values of the initial conditions. These are correctly defined direction and magnitude of the linear velocity also the corresponding angular velocity. So, at proper initial conditions and on the base of the Magnus effect, the trajectory of the soccer ball deviates in the appropriated direction, as the ball goes to the target of the goal. The initial conditions of the successful kicks in fact are limited to a very small range, so only the best football players, thanks to their skills and experience, can realize them. By the obtained results it can be seen that corner or free kicks, which are done with very low or very high values of initial linear velocities are unsuccessful.

8 Conclusions

In this paper a study from the mechanical point of view of the most important kicks in the football game is presented. These are the corner and direct free kick.

The characteristics of soccer ball size five are assumed and the latter is considered as an ideal rigid sphere. The orientation of the ball in the space is defined by Cardan angles. The aerodynamics forces acting to the ball during its flight are taken into account. The system of six nonlinear equations is composed.

Furthermore, in relation with a modelling of the goal target, two of the most dangerous zones of the goal are defined as well as a short description of the geometry and implementations of both investigated soccer kicks is presented.

The numerical solution of the problem is carried out by a program in the MatLab-Simulink environment. All possible kinematic characteristics of the soccer ball flight, as lows of motion, linear and angular velocities, linear and angular accelerations and trajectories are obtained.

The effects of several initial parameters, as velocities and angles, on the ball flight in the football game are studied. Based on the analysis of the results, the recommended values of initial parameters for successful corner and free kicks are found. All of the trajectories are illustrated by 3D or 2D graphics.

The study is suitable and useful for scientists in the field of sport engineering as well as for soccer trainers.

References

1. R. Cross, Am. J. Phys. **70**(5), 482-489 (2002)
2. R. D. Mehta, J. M. Pallis, Materials and Science in Sports, 185-197 (2001)
3. A. M. Nathan, Am. J. Phys. **76**(2), 119-124 (2008)
4. P. Jalilian, P. K. Kreun, M. M. Makhmalbaf, W. W. Liou, Am. J. Sport Sc. **2**(5), 115-121 (2014)
5. J. Leela, D. M. G. Comissiong, K. Rahaman, Lat. Am. J. Phys. Educ. **8**(4), 4505-1-4505-9 (2014)
6. A. Ivanov, J. Javorova, Tehnomus J. **24**, 54-61 (2017)
7. T. Asai, M. J. Carre, T. Akatsuka, S. Haake, Sports Eng. **5**, 183-192 (2002)
8. T. Asai, T. Akatsuka, S. Haake, Phys. World **11** (6), 25-27 (1998)
9. M. J. Carre, T. Asai, T. Akatsuka, S. Haake, Sports Eng. **5**, 193-200 (2002)
10. T. Asai, K. Seo, O. Kobayashy, R. Sakashita, Sports Eng. **10**, 101-110 (2007)
11. J. Kim, H. Park, H. Choi, J.Y. Yoo, J. Fluid Mech. **754**, R2, doi:10.1017/jfm.2014.428 (2014)
12. T. Yamagata, T. Nagasawa, N. Fujisawa, T. Asai, J. Flow Control, Measur. & Visual. **1**, 86-93 (2013)
13. J. E. Goff, M. J. Carre, Am. J. Phys. **77**(11), 1020-1027 (2009)
14. J. E. Goff, M. J. Carre, Eur. J. Phys. **31**, 775-784 (2010)
15. L. Oggiano, L. Saetran, Procedia Eng. **2**, 2473-2479 (2010)
16. J. E. Goff, J. Kelley, C. M. Hobson, K. Seo, T. Asai, S. B. Choppin, Eur. J. Phys. **38**, 044003 (2017)
17. T. Asai, S. Ito, K. Seo, S. Koike, Procedia Eng. **34**, 122-127 (2012)
18. D. Rogers, M.A. Passmore, A. Harland, R. Jones, C. Holmes, T. Lucas, Procedia Eng., Elsevier **2**(2), 2481-2486 (2010)
19. J. C. Cooke, The mathematical Gazette, **39**, 196-199 (1955)
20. S. Verma, Int. J. Scientific Research **4**(1), 420-424 (2015)
21. F. Reid, J. Parabola, **33**(1), 1-4 (1997)
22. K. E. Kenyon, Natural Science **8**, 49-52 (2016).
23. K. Bray, D. G. Kerwin. J. Sports Science **21**, 75-85 (2003)
24. M. J. Carre, S. R. Goodwill, S.J. Haake, Proc. IMechE Part C: J Mech Eng Sci, **219**(7), 657-666 (2005)
25. B. G. Cook, J. E. Goff, European J. Physics **27**, 865-874 (2006)
26. J. E. Goff, Physics Today, **63**(7), 62-63 (2010)
27. A.M. Alcock, W. Gilleard, A.B. Hunter, J. Baker, N. Brown, J Sports Sci **30**(4), 387-394 (2012)
28. A. Katis, E. Kellis, A. Lees, Sports Biomech. **14** (3), 287-299 (2015)

Dynamical behaviour of a compound elastic pendulum

Svetoslav Nikolov[1,2,*] and *Daniela* Zaharieva[2]

[1]Institute of Mechanics-BAS, Biomechanics Department, Acad. G. Bonchev Str., Bl. 4, 1113 Sofia, Bulgaria
[2]University of Transport, Mechanics Department, G. Milev Str. 158, 1574 Sofia, Bulgaria

Abstract. The aim of the paper is a comprehensive study of the compound elastic pendulum (CEP) with two degrees of freedom to point out the main complex (chaotic) dynamics that it can exhibit. The simplest way to find complex behavior in a nonintegrable Hamiltonian system is firstly to look for homoclinic (heteroclinic) orbit(s). Here, under suitable assumptions, we detect the existence of a homoclinic orbit of CEP and present the equation for it. Moreover, we show that for any value of the small parameter the system has a hyperbolic periodic orbit, whose invariant manifolds intersect themselves transversally.

1 Introduction

Nonlinear dynamics studies some conditions that are not observed in linear systems. Nonlinear behavior (movement and evolution) is typical for different dynamical systems. The investigation of nonlinear mechanical systems is an important and very active area [1]. For example, the pendulum is interesting as a paradigm of contemporary dynamics and, more importantly, since the differential equation of the pendulum is frequently encountered in various branches of modern mechanics [2, 3].

The study of complex phenomena in deterministic nonlinear dynamical systems has attracted the attention of applied scientists. This study poses two fundamental questions: 1) by what mechanisms does chaos occur and 2) how can one predict when chaos will occur in a specific dynamical system?

Dynamical systems can be separated into two main kinds: i) systems with conservation of phase volume and ii) systems with decrease of phase volume – called dissipative systems. Systems from kind (i) can be Hamiltonian and non-Hamiltonian systems. Systems are called Hamiltonian if their equations can be written in canonical form by means of a Hamiltonian $H(q,p,t)$, where q and p are generalized coordinates and momenta, and t is the time.

Generally, bifurcations can be local or global. Global bifurcations are qualitative change in the orbit structure of an extended region of phase space. Typical examples are homoclinic and heteroclinic bifurcations.

[*] Corresponding author: S.Nikolov@imbm.bas.bg

In recent years, it has become apparent that homoclinic and heteroclinic orbits are often the mechanism for the chaos (or transient to chaos) in different kinds of dynamical systems – mechanical, biological and et al. [4-8].

Definition 1. (Wiggins [9]) *Let Φ be an invariant set of a dynamical – map or flow. Let p be a point in the phase (space) of the dynamical system and the orbit of p approaches Φ asymptotically as $t \to -\infty$ and Φ asymptotically as $t \to +\infty$. Then the orbit of p is said to be homoclinic to Φ.*

Definition 2.(Wiggins [9]) *Let Φ_1 and Φ_2 be disjoint invariant sets of a dynamical system, and the orbit of p approaches Φ_1 asymptotically as $t \to -\infty$ and Φ_2 asymptotically as $t \to +\infty$. Then the orbit of p is said to be heteroclinic to Φ_1 and Φ_2.*

Definitions 1 and 2 (D.1 and D2) are too general, as in case that Φ, Φ_1 and Φ_2 passes stable and unstable manifolds (e.g. they are hyperbolic or normally hyperbolic), they can be alternately stated as follows.

Definition 3. *Let Φ, Φ_1, Φ_2 and p be denoted as D.1 and D.2. If p lies in both the stable and unstable manifolds of Φ, then the orbit p is said to be homoclinic to Φ.*

If p lies in the unstable manifold of Φ_1 and the stable manifold of Φ_2, then the orbit of p is heteroclinic to Φ_1 and Φ_2.

Note here that the point p is a homoclinic (resp. heteroclinic) point.

For mechanical systems which are modeled by Hamiltonian equations of motion, integrability is a central problem. Most Hamiltonian systems are not integrable but if a Hamiltonian dynamical system is integrable, then its first integrals provide a description of all features of the system [10-12].

It is well-known that all Hamiltonian systems with one degree of freedom and analytical Hamiltonian functions are integrable, and there is no splitting of separatrices. If separatrices of two saddles coincide to a form of heteroclinic orbit then we have a connection bifurcation. This is a global bifurcation as a necessary condition is that the two saddles have the same energy [13]. On the other hand, integrability is not a typical property of Hamiltonian systems with more than one degree of freedom. The phase spaces of such systems are difficult to visualize and a lower dimensional representation of the qualitative behavior of the system must be sought.

The importance of the study of homoclinic (heteroclinic) orbits in nonintegrable Hamiltonian systems has been appreciated since the time of Poincare [14, 15]. Now, it is fairly well-known that Hamiltonian systems with two (or more) degrees of freedom are generally nonintegrable, which usually involves chaotic dynamics. All bounded trajectories of an integrable Hamiltonian system with $n = 2$ degrees of freedom evolve in the $2n$ phase space on manifolds which are diffeomorphic (equivalent) to 2-tori [16]. If integrability is broken, KAM theory guarantees that tori (from resonances) will survive and the motion bears a nonregular (chaotic) character [9]. Furthermore, in Hamiltonian systems with two degrees of freedom there are five distinct cases to consider according to the different possibilities for the eigenvalues of the linearized vector field at a hyperbolic fixed point [9]. They are:

Real eigenvalues $\begin{cases} 1)\ \chi_1, \chi_2 > 0;\ \chi_3, \chi_4 < 0. \\ 2)\ \chi_1 > 0;\ \chi_2, \chi_3, \chi_4 < 0. \end{cases}$ *Complex eigenvalues*

$3)\ n_1 \pm i m_1, n_2 \pm i m_2; n_1 > 0,$

$n_2 < 0, m_1, m_2 \neq 0.$

$$\text{Real and complex eigenvalues} \begin{cases} 4)\, n_1 \pm i m_1, \chi_1, \chi_2 > 0;\, n_1 < 0, m_1 \neq 0. \\ 5)\, n_1 \pm i m_1, \chi_1 < 0, \chi_2 > 0; n_1 < 0, m_1 \neq 0. \end{cases}$$

Other cases may be obtained from these via time reversal. If nonhyperbolic fixed points are considered, then even more cases must be considered.

In generic systems with two degrees of freedom the following types of bifurcations of periodic and homoclinic orbits are possible: i) the periodic saddle-centre bifurcation [5, 6, 13] and ii) the Hamiltonian period-doubling bifurcation. In case (i), under variation of the energy a hyperbolic and elliptic periodic orbit meet at a parabolic periodic orbit with all Floquet multipliers equal to one. In case (ii), under variation of the energy an elliptic periodic orbit turns hyperbolic when passing through a parabolic periodic orbit with Floquet multipliers minus one. Simultaneously, from the parabolic periodic orbit, a family of periodic orbits with the double period emerges, inheriting the normal linear behavior from the initial periodic orbit. The behavior near the phase-space curves separating regions with different type of motion (so called separatrices) is generic for nonlinear Hamiltonian systems.

In this paper we consider a compound elastic pendulum (CEP) with two degrees of freedom- see Figure 1. For this pendulum the kinetic energy T and the potential energy U have the form

$$\begin{aligned} T &= \frac{m_1}{2} L^2 \dot{\theta}^2 + \frac{m_2}{2}\left[(l-\xi)^2 \dot{\theta}^2 + \dot{\xi}^2\right], \\ U &= -m_1 g L(1-\cos\theta) - m_2 g(l-\xi)(1-\cos\theta) + \frac{c}{2}\xi^2, \end{aligned} \tag{1}$$

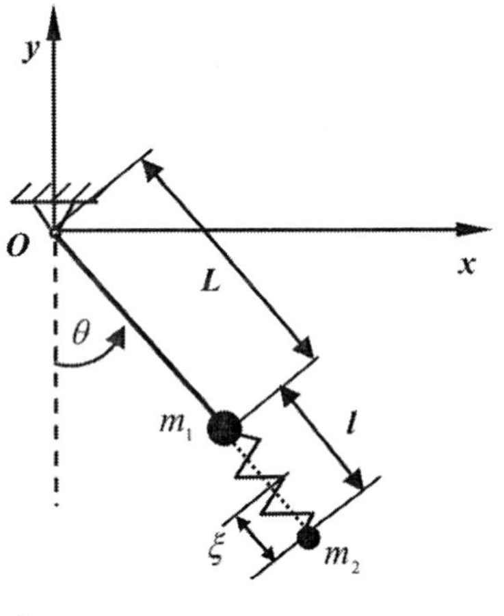

Fig. 1. The mass-spring pendulum system.

where m_1 and m_2 are the masses, θ is the corresponding angle of the pendulum with the gravity vertical (defined $\mathrm{mod}\, 2\pi$), ξ is the deviation of mass m_2 from equilibrium state, l is the distance from mass m_1 to equilibrium of mass m_2 (while both rod and spring are massless), L is the length of the pendulum (Om_1), c is the spring constant and g is the acceleration of the gravity.

Here, we are interested in the dynamical behavior of the pendulum shown in Figure 1, and specially in the homoclinic orbit equation for it.

The paper is organized as follows: In Section two, we obtain the analytical results for the Hamiltonian of the compound elastic pendulum. In Section 3, we present the analytical results for homoclinic loop of CEP. Finally, Section 4 summarizes our results.

2 Hamiltonian form of the compound elastic pendulum

In the Hamiltonian formulation, the state of a system is characterized not only by its positions q_i $(i=1,...,n)$, but also by its momenta p_i. Thus, to obtain the Hamiltonian $H(\theta, p_1, \xi, p_2)$ of CEP, we firstly pass to the canonical momentum representation

$$p_1 = \frac{\partial T}{\partial \dot{\theta}} \quad , \quad p_2 = \frac{\partial T}{\partial \dot{\xi}} . \tag{2}$$

Hence, the autonomous Hamiltonian has the form

$$\begin{gathered} H(\theta, p_1, \xi, p_2) = T - U = \frac{1}{2m_2\psi}\left[m_2 p_1^2 + \psi p_2^2\right] + \\ + m_1 g L(1-\cos\theta) + m_2 g(l-\xi)(1-\cos\theta) - \frac{c}{2}\xi^2 , \end{gathered} \tag{3}$$

where $\psi = m_1 L^2 + m_2 (l-\xi)^2$. The time-evolution of the system is then governed by equations of motion

$$\begin{gathered} \dot{\theta} = \frac{\partial H}{\partial p_1}, \dot{\xi} = \frac{\partial H}{\partial p_2}, \\ \dot{p}_1 = -\frac{\partial H}{\partial \theta}, \dot{p}_2 = -\frac{\partial H}{\partial \xi}. \end{gathered} \tag{4}$$

An immediate consequence is

$$\dot{H} = \frac{\partial H}{\partial \theta}\dot{\theta} + \frac{\partial H}{\partial p_1}\dot{p}_1 + \frac{\partial H}{\partial \xi}\dot{\xi} + \frac{\partial H}{\partial p_2}\dot{p}_2 = 0 , \tag{5}$$

the conservation of energy. Here we note that the system (4) is with two degrees of freedom and belongs to the class of nonintegrable Hamiltonian systems. The four-dimensional phase space $Z = \{(\theta, \xi, p_1, p_2)\}$ is characterized by an extremely complex behavior of the phase trajectories and by existence of instability zones in which the motion has chaotic (nonregular) character [9, 16].

The fixed points of the system (4) in the (θ, ξ, p_1, p_2) plane defined as

$$\frac{\partial H}{\partial p_1} = \frac{\partial H}{\partial p_2} = \frac{\partial H}{\partial \theta} = \frac{\partial H}{\partial \xi} = 0 \tag{6}$$

are

$$\bar{p}_1 = \bar{p}_2 = 0 \, , \bar{\theta} = k\pi \; (k = 0, \pm 1, \pm 2, ...) , \bar{\xi}^{(1)} = 0, \bar{\xi}^{(2)} = -\frac{2m_2 g}{c} . \tag{7}$$

According to [9], the law of energy conservation $E = T - U$ defines an invariant three-dimensional hypersurface $H = E$ about the system phase flow. Here we are interested in a special phase curve- homoclinic orbit, lying in the three-dimensional manifold with level of energy E_h. This level corresponds to the 'upper' position (the masses standing upright vertically) of the compound elastic pendulum, i.e. $\bar{\theta} = \pm\pi$, $\bar{\xi} = -\frac{2m_2 g}{c}$, $\bar{p}_1 = \bar{p}_2 = 0$. Hence, the manifold E_h has the form

$$H(\bar{\theta}, \bar{p}_1, \bar{\xi}, \bar{p}_2) = E_h = 2g\left[m_1 L + m_2\left(l + \frac{m_2 g}{c}\right)\right]. \tag{8}$$

3 Homoclinic solution

In the previous section, we obtained some analytical results that we shall use in our calculations so as to find a homoclinic solution (orbit) $(\theta_h(t), p_{1h}(t), \xi_h(t), p_{2h}(t))$ in the three-dimensional manifold E_h. Note that this homoclinic solution is transversal intersection of the two-dimensional invariant surfaces W^s (stable manifold) and W^u (unstable manifold) of the hyperbolic fixed point $\bar{\theta} = \pm\pi$, $\bar{\xi} = -\frac{2m_2 g}{c}$, $\bar{p}_1 = \bar{p}_2 = 0$. In our considerations, we introduce the small parameter $\varepsilon = {m_2}/{m_1} << 1$, i.e. $0 < m_2 < m_1$.

To find the equation of the homoclinic orbit we use the Lindstedt-Poincare perturbation method (procedure). For a small perturbation parameter, ε, the system (4) can be written in the form

$$\begin{aligned} &\dot{\theta} = \frac{1}{m_1\left[L^2 + \varepsilon(l-\xi)^2\right]} p_1 \quad , \quad \dot{\xi} = \frac{1}{m_1 \varepsilon} p_2 , \\ &\dot{p}_1 = -m_1 g \sin\theta\left[L + \varepsilon(l-\xi)\right], \\ &\dot{p}_2 = -\frac{\varepsilon(l-\xi)}{m_1\left[L^2 + \varepsilon(l-\xi)^2\right]^2} p_1^2 + m_1 g \varepsilon(1-\cos\theta) + c\xi . \end{aligned} \tag{9}$$

The homoclinic solution (trajectory) of system (9) satisfies the boundary conditions

$$\begin{aligned} &\theta \to \pi , \xi \to -\frac{2m_2 g}{c}, p_1 \to 0 , p_2 \to 0 , \ \ for \quad t \to +\infty, \\ &\theta \to -\pi , \xi \to -\frac{2m_2 g}{c}, p_1 \to 0 , p_2 \to 0 , \ \ for \quad t \to -\infty. \end{aligned} \tag{10}$$

The functions θ, ξ, p_1 and p_2 as series in powers of ε can be presented in the form

$$\begin{aligned} &\theta = \theta_0 + \varepsilon\theta_1 + \varepsilon^2\theta_2 + \ldots \quad , \quad \xi = \xi_0 + \varepsilon\xi_1 + \varepsilon^2\xi_2 + \ldots, \\ &p_1 = p_{10} + \varepsilon p_{11} + \varepsilon^2 p_{12} + \ldots \quad , \quad p_2 = p_{20} + \varepsilon p_{21} + \varepsilon^2 p_{22} + \ldots. \end{aligned} \tag{11}$$

After substituting of (11) into (9) and accomplishing some transformations and analytical calculations we obtain the following equations for the functions θ_0, ξ_1, p_{10}, p_{22}

$$\begin{aligned} &\dot{\theta}_0 = \frac{1}{m_1 L^2} p_{10} \quad , \quad \dot{\xi}_1 = \frac{1}{m_1} p_{22}, \quad \dot{p}_{10} = -m_1 g L \sin\theta_0 , \\ &\dot{p}_{22} = \frac{1}{m_1 L^4}\xi_1 p_{10}^2 + 2\frac{m_1 g l^2}{L^2}(1-\cos\theta_0) + 2\frac{cl^2}{L^2}\xi_1 + c\xi_2 , \end{aligned} \tag{12}$$

where $\xi_2 = const$. Here we note that ξ_0, p_{20} and p_{21} are equal to zero.

The function θ_0 is defined by the equation

$$\ddot{\theta}_0 + \omega^2 \sin\theta_0 = 0, \tag{13}$$

yielding the homoclinic solution

$$\theta_0(t) = \pm 2arctg\left(sh(\omega t)\right), \tag{14}$$

where $\omega^2 = {g}/{L}$ (for details see Appendix 1). Hence, for function p_{10} we have

$$p_{10}(t) = \pm m_1 L^2 \frac{2\omega}{ch(\omega t)} . \tag{15}$$

The solution (15) is called also a soliton (self-dependent wave) – see Fig. 2. It is well-known that solitons are nonlinear waves which after a fully nonlinear interaction do not change their speed and shape. For the function ξ_1 respectively we have

$$\ddot{\xi}_1 = a(t)\xi_1 + b(1-\cos\theta_0) + \alpha, \tag{16}$$

where $a(t) = \left(\frac{2\omega}{ch(\omega t)}\right)^2 + \frac{2cl^2}{m_1 L^2}$, $b = \frac{2gl^2}{L^2}$ and $\alpha = \frac{c}{m_1}\xi_2$.

The equation (16) is investigated in the Appendix 2 of this work. Thus, after acceptance that $a(t) = A(t) = const. > 0$, $\theta_0 \in (0, \pi]$ and $\xi_2 = 0$, we obtain (without loss of generality) that (16) has a homoclinic solution in the form

$$\xi_1(t) = C_1 sh(t\sqrt{A}) + C_2 ch(t\sqrt{A}) - C_3, \tag{17}$$

where $C_1 = \frac{1}{\sqrt{A}}\dot{\xi}_1(t_0)$, $C_2 = \xi_1(t_0)$ and $C_3 = \frac{B}{A} = \frac{m_1 g l^2 ch^2(\omega t_0)(1-\cos\theta_0)}{2m_1\omega^2 L^2 + cl^2 ch^2(\omega t_0)}$ are constants.

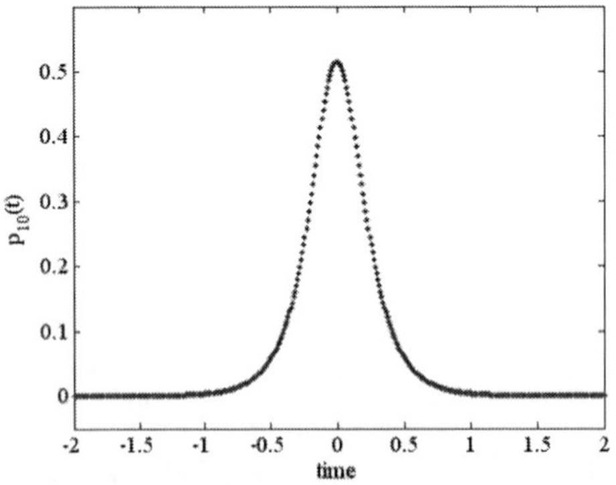

Fig. 2. A soliton solution for the velocity of the separatrice Eq. (15) (in (θ_0, p_{10}) plane) when: $L = 0.3[m]$, $m_1 = 0.5[kg]$ and time $t \in [-2, 2]$ is in seconds. Note that when we have sign '+' into (15), then the soliton motion is to the right (upper part of the separatrice).

For the function p_{22}, using (15) and (17) (when $B = 0$ and $p_{10}(t) = p_{10}^{+}(t)$), we obtain the following first order differential equation

$$\dot{p}_{22} = m_1\tilde{A}\left(C_1 sh\left(t\sqrt{\tilde{A}}\right) + C_2 ch\left(t\sqrt{\tilde{A}}\right)\right). \tag{18}$$

The homoclinic solution of the equation (18) has the form

$$p_{22} = A_1 ch\left(t\sqrt{\tilde{A}}\right) - A_2 sh\left(t\sqrt{\tilde{A}}\right), \tag{19}$$

where

$$A_1 = m_1 C_1\sqrt{\tilde{A}}\ ,\ A_2 = m_1 C_2\sqrt{\tilde{A}}\ ,\ \tilde{A} = \left(\frac{2\omega}{ch(\omega t)}\right). \tag{20}$$

4 Numerical calculations

In this section we investigate numerically the system (9) in terms of the Poincare section. The first step is finding the hyperbolic periodic points(s) for the Poincare map [17, 18]. The approximate position of some hyperbolic point, for $\varepsilon = 0.2$, was obtained visually from the picture of the phase portrait. The second step is the construction of separatrices. In this connection, the pictures of corresponding stable W^s and unstable manifolds W^u are shown in Figure 3. The last step is to prove the existence of homoclinic transversal intersection(s). If the curves W^s and W^u are not tangent at a homoclinic point q ($q \in W^s \cap W^u$), then the intersection is transversal, and q is said to be a transversal homoclinic orbit. Here we note that this step is the central and most complicated part of the procedure for the proof of the existence of chaotic motion, and will be considered in our forthcoming work.

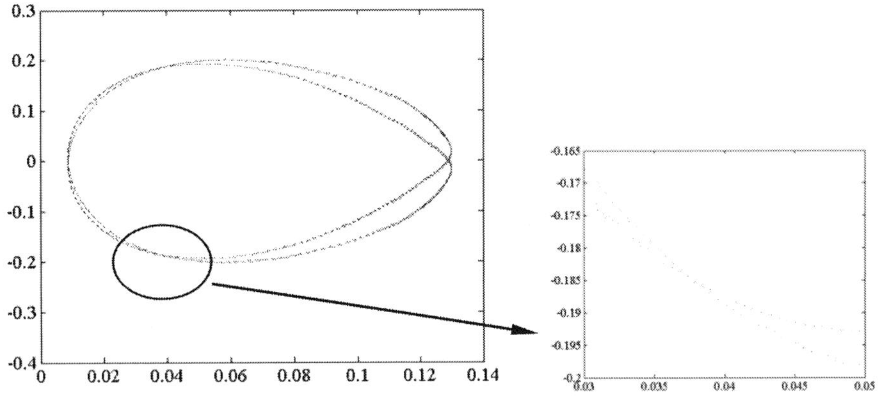

Fig.3. Poincare section and transverse intersection of stable and unstable manifolds of the system (9) for the perturbation $\varepsilon = 0.2$ and $c = 32.15\left[N/m\right]$.

5 Conclusions

In this paper were presented some analytical and numerical results for existence of a homoclinic loop in Hamiltonian systems with two degrees of freedom - compound elastic pendulum (CEP). It is to be noted that the main conjecture from the results obtained in analytical calculations is that CEP with two degrees of freedom is nonintegrable for all nontrivial (non equal to zero or infinity) values of the parameters.

For Hamiltonian systems, there are two types of hyperbolic fixed points: 1) saddle-focus and 2) saddle. If these systems are from fourth order, the homoclinic orbit lies in the transversal interaction of the stable W^s and unstable W^u manifolds of the origin. For both types of fixed points, horseshoes exist on the level sets $H^{-1}(\varepsilon)$ for ε sufficiently small due to the structural stability of the horseshoes. Detecting of homoclinic (heteroclinic) orbits in a nonintegrable Hamiltonian system is not an easy task, but the existence of horseshoes forms conditions for the developing of chaotic behavior [19]. In this connection, in section 4, we show that for any value of the small parameter ε the system CEP has a hyperbolic periodic trajectory, whose invariant manifolds intersect transversally.

This work was supported by grant No 110/18.04.2017 of the University of Transport "T. Kableshkov" Sofia, Bulgaria.

References

1. M. Brin, G. Stuck, *Introduction to dynamical systems* (CUP, Cambridge, 2003)
2. S. Nikolov, Ju. Genov, N. Nachev, Mechanics, Transport, Commun. **8**, 0472 (2010)
3. L. Shilnikov, A. Shilnikov, D. Turaev, L. Chua, *Methods of qualitative theory in nonlinear dynamics* (Part II, World Scientific, Singapore, 2001)

4. S. Nikolov, Mechanics, Transport, Commun. **10**, 0507 (2012)
5. V. Afraimovich, S. Gonchenko, L. Lerman, A. Shilnikov, D. Turaev., Regular and Chaotic Dynamics **19**, 435-460 (2014)
6. R. Driesse, A. Homburg, J. of Differential Equations **246**, 2681-2705 (2009)
7. S. Nikolov, D. Zaharieva, Mechanics, Transport, Commun. **14**, 1352 (2016)
8. S. Nikolov, O. Wolkenhauer, J. Vera, Molecular BioSystems **10**, 172-179 (2014)
9. S. Wiggins, *Global bifurcations and chaos. Analytical methods* (Springer-Verlag, NY, 1988)
10. M. Brack, K. Tanaka, Physical Review E **77**, 046205 (2008)
11. G. Villasi, *Hamiltonian dynamics* (World Scientific, Singapore, 2001)
12. J. Lowenstein, *Essentials of Hamiltonian dynamics* (CUP, NY, 2012)
13. H. Hansmann, *Local and semi-local bifurcations in Hamiltonian dynamical systems* (Springer, Berlin, 2007)
14. H. Poincare, *Les methodes nouvelles de la mecanique celeste*, 1-3 (Gauthier-Villars, 1892; 1893; 1899)
15. R. MacKay, J. Meiss, *Hamiltonian dynamical systems* (Adam Hulger, Boston, 1987).
16. V. Arnold, *Mathematical methods of classical mechanics* (Springer-Verlag, Heidelberg, 1978)
17. V. Gelfreich, D. Sharomov, Physics Letters A, **197**, 139-146 (1995)
18. A. Ivanov, J. of Physics A **34**, 11011-11031 (2001)
19. G. Van Der Heijden, K. Yagasaki, Z. Angew. Math. Phys. **65**, 221-240 (2014)
20. A. King, A., J. Billingham, S. Otto, *Differential equations. Linear, nonlinear, ordinary, partial* (Cambridge University Press, NY, 2003)
21. H. Josephs, R. Huston, *Dynamic of mechanical systems* (CRC Press, NY, 2002)
22. G. Murphy, *Ordinary differential equations and their solutions* (Van Nostrand, NY, 1960)
23. D. Zwillinger, *Handbook of differential equations* (Academic Press, NY, 1997)
24. G. Baker, J. Blackburn, *The pendulum. A case study in physics* (Oxford University Press, NY, 2005)

Appendix

A.1 Homoclinic solution of (13)

Let us consider

$$\frac{d^2\theta}{dt^2} + \omega^2 \sin\theta = 0, \tag{A.1}$$

where $\omega^2 = g/L$ and t is time. For oscillations of small amplitude (i.e. $\theta < 6^0 \div 10^0$, so that $\sin\theta \approx \theta$) we obtain simple harmonic motion in the form

$$\frac{d^2\theta}{dt^2} + \omega^2\theta = 0\,. \tag{A.2}$$

Here, the general solution of (A.2) is

$$\theta = A_1 \sin \omega t + B_1 \cos \omega t\,, \tag{A.3}$$

where A_1 and B_1 are constants that can be fixed from the initial values of θ and $\dot{\theta} = d\theta/dt$. However, if θ is not small ($\theta > 10^0$) the linearization approximation is no longer valid. Thus, we must study the full equation of motion (A.1), which is nonlinear. It is well-known that, in general, nonlinear differential equations cannot be solved analytically, but for equation (A.1) (where the first derivative $d\theta/dt$ does not appear explicitly), an analytical solution is available [20]. Treating $\dot{\theta}$ as a function of θ instead of t, we have

$$\frac{d^2\theta}{dt^2}=\frac{d}{dt}\left(\frac{d\theta}{dt}\right)=\frac{d\dot{\theta}}{dt}=\frac{d\theta}{dt}\frac{d\dot{\theta}}{d\theta}=\dot{\theta}\frac{d\dot{\theta}}{d\theta}=\frac{d}{d\theta}\left(\frac{1}{2}\dot{\theta}^2\right). \tag{A.4}$$

Hence, for (A.1) we can write

$$\frac{d}{d\theta}\left(\frac{1}{2}\dot{\theta}^2\right)=-\omega^2\sin\theta, \tag{A.5}$$

which we can integrate once to give

$$\frac{1}{2}\dot{\theta}^2=\omega^2\cos\theta+c_1, \tag{A.6}$$

where c_1 is a constant that can be evaluated by initial conditions, i.e. when $t=0, \theta=\theta_0=const.$ and $\dot{\theta}=0$. Thus, using also that $\omega^2={}^{g}\!/_{L}$, the first integral of (A.6) becomes

$$\frac{1}{2}m_1L^2\dot{\theta}^2-m_1gL\cos\theta=E=-\frac{g}{L}\cos\theta_0, \tag{A.7}$$

where E is a statement of conservation of energy. From (A.7) we can see that

$$\frac{d\theta}{dt}=\pm\sqrt{\frac{2E}{m_1L^2}+\frac{2g}{L}\cos\theta}=\pm\sqrt{\frac{2g}{L}\left(\cos\theta-\cos\theta_0\right)}. \tag{A.8}$$

Thus, we have a nonlinear first-order differential equation to solve. By separating the variables, we have

$$dt=\pm\sqrt{\frac{{}^{L}\!/_{2g}}{\cos\theta-\cos\theta_0}}d\theta. \tag{A.9}$$

By integration of (A.9) we obtain

$$t=\sqrt{\frac{L}{2g}}\int_{\hat{\theta}}^{\theta_0}\sqrt{\frac{1}{\cos\theta-\cos\theta_0}}d\theta, \tag{A.10}$$

where $\hat{\theta}$ is some small angle and we have selected the negative sign and eliminated it by interchanging the limits on the integral. It is well-known that the integral into (A.10) cannot be evaluated in terms of elementary functions. However, we could expand the integrand as a binomial series and integrate term by term [21, 22]. Alternatively, we can change the variables and convert the integral into the form of an elliptic integral [23]. Thus, according to [21] we have for (A.10)

$$d\theta=\frac{2k\cos z}{\sqrt{1-k^2\sin^2 z}}dz, \tag{A.11}$$

where the constant $k\overset{def}{=}\sin\left({}^{\theta_0}\!/_{2}\right), |k|<1$, and $k\sin z\overset{def}{=}\sin\left({}^{\theta}\!/_{2}\right)$. Note that regarding the limits of the integral into (A.10), we have that for: i) $\theta=\theta_0$, then $z={}^{\pi}\!/_{2}$, and ii) $\theta=\hat{\theta}$, then $z=\hat{z}$.

It is well-known that three cases are distinguished by the regimes of total energy E [24]: 1) Here $E<2m_1gL$ is the energy required for the pendulum bob to reach the upper vertical position – finite motion; 2) $E=2m_1qL$. Now, the energy level is critical and pendulum behavior is on special phase orbit ($\theta=\pm\pi$) called separatrix (homoclinic one). It marks the boundary between oscillatory and hindered rotary motion of the pendulum; 3) Finally, $E > 2m_1 gL$. The pendulum executes hindered rotary motion – infinite motion because the phase θ grows infinitely.

A.2 Homoclinic solution of (16)

Let us examine the equation (16) which has the form

$$\ddot{\xi}_1 - A(t)\xi_1 - B(\theta_0) = 0, \qquad \text{(A.12)}$$

where $A(t) = \left(\frac{2\omega}{ch(\omega t)}\right)^2 + \frac{2cl^2}{m_1 L^2}$, $B(\theta_0) = \frac{2gl^2}{L^2}(1-\cos\theta_0) + \frac{c}{m_1}\xi_2$ and $\xi_2 = const$. Without loss of generality we restrict our attention to the case when $\xi_2 = 0$. Moreover, according to Appendix 1 (and (13) also) $\theta_0 \in [0,\pi]$ (see Fig. A1) and for $t \in [0.6, +\infty)$ $A(t) = const. > 0$ (see Fig. A1). Thus, (A.12) is

$$\ddot{\xi}_1 - A\xi_1 - B = 0. \qquad \text{(A.13)}$$

Hence, (A.13) is ODE from second order with constant coefficients.

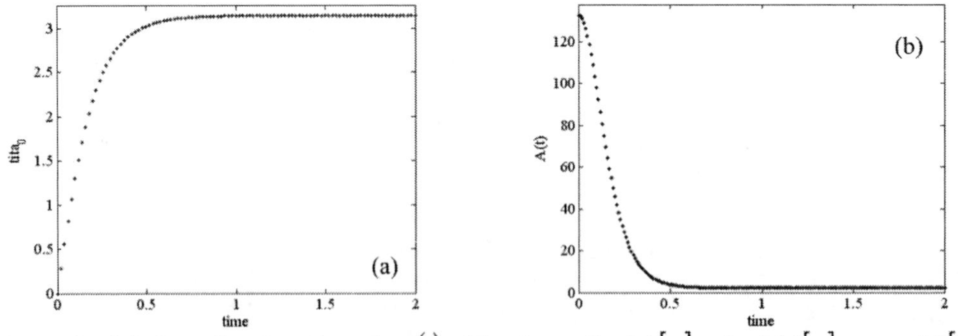

Fig. A1. Solutions for θ_0 (a) and $A(t)$ (b) when: $L = 0.3[m]$, $l = 0.04[m]$, $m_1 = 0.5[kg]$, $c = 30\left[N/m\right]$ and time t is in seconds.

In the case when $l << L$, i.e. $\mu = \left(l/L\right)^2 \approx 0$, we find that $B = 0$ and (A.13) is a homogeneous linear ODE from second order

$$\ddot{\xi}_1 - \tilde{A}\xi_1 = 0, \qquad \text{(A.14)}$$

with the general solution

$$\xi_1(t) = C_1 sh\left(t\sqrt{\tilde{A}}\right) + C_2 ch\left(t\sqrt{\tilde{A}}\right), \qquad \text{(A.15)}$$

where $C_1 = \frac{1}{\sqrt{\tilde{A}}}\dot{\xi}_1(t_0)$, $C_2 = \xi_1(t_0)$, and $\tilde{A} = \left(\frac{2\omega}{ch(\omega t)}\right)$ are constants.

Examination of the possibility of transborder pollution in the days with registered exceedances of pollutant dust in the city of Ruse, Bulgaria

Antoaneta Stefanova[1,*], *Ivanka* Zheleva[1] , *Margarita* Filipova[1,] and *Irina* Tsvetkova[1]

[1] Ruse University, 8 Studentska Str., 7017, Ruse, Bulgaria

Abstract. The presented study of the PM10 concentrations in the atmosphere ground layer emitted from industrial sources in the Industrial zone in the city of Ruse in the days with exceedances for this pollutant is aimed to determining the possibility of cross-border BG-RO PM10 contamination. Climatic conditions measured in these days with PM10 exceedances in January 2017 are also taken into account in Jet Gauss model and in the Langrageo-statistical approach for describing the turbulent diffusion. The results show that the PM10 pollutant from the investigated sources in industrial zone in Ruse is reaching the territory of Republic of Romania with low concentrations and have no potential for impact on air quality in the cross-border area.

1 Introduction

The Municipality of Ruse is one of the municipalities in Bulgaria designated as an area for assessment and management of ambient air quality for dust contaminants. Annually during the winter seasons exceedance of the norms for the protection of human health for pollutant PM10 (particulate matter with a diameter between 2.5 and 10 μm) is registered in the "Vazrazhdane" Automatic Measuring Station (AMS) [7]. The sources of dust from the territory of Ruse are industry, transport and domestic heating. A significant part of the industrial sources in the city of Rousse is concentrated in the territory of the Industrial zone, Ruse (see Fig.1).

The main objective of the present study is to analyse the probability of cross-border pollution of the ambient air as a result of the simultaneous operation of industrial sources concentrated on the territory of Industrial zone in the town of Ruse during January 2017 - period of registered exceedances of the PM10 norms for protection of human health [7].

Subjects of the study are:

- Analysis of weather conditions (temperature, wind rose, prevailing wind direction and speed) in the days of January 2017 with recorded PM10 exceedances.
- Identification of the organized sources of dust on the territory of industrial zone of Ruse;
- Computer simulation to calculate the maximum ground concentrations of PM10 pollutant emitted from the industrial zone of Ruse and the distances to which PM10 pollutant is disseminated;
- Analysis of the dust ground concentration reaching the territory of the Republic of Romania and the possibility of their influence on the air quality in the cross-border area.

* Corresponding author: astefanova@uni-ruse.bg

The territory of the industrial zone - Ruse is located along the Danube riverbank. Numerous industries from the chemical, manufacturing, machine-building and other sectors are concentrated on its territory (see Fig.1).

2 Materials and Methods

The necessary information for the study is provided through:

- Available information from the official website of the competent authority - Regional Inspectorate of Environment and Water (RIEW) - Ruse [1],
- Information provided by the Regional Inspectorate of Environment and Water - Ruse under the Access to Public Information Act [2, 4],
- Verification of data on industrial sources through meetings and contacts with the legal representatives of entities operating the identified sources of dust emissions from the investigated territory [3]

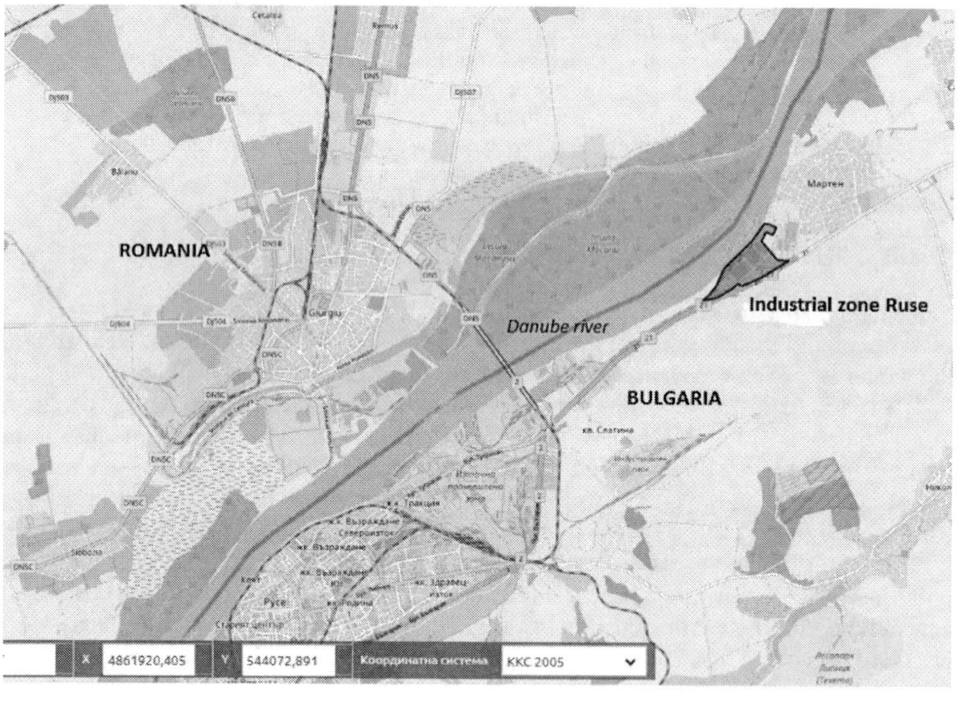

Fig.1. Map of industrial zone of Ruse [5].

For mathematical simulations PLUME software was used. This software is made according the Methodology for calculating the leakage height, dispersion and expected concentrations of pollutants in the ground atmospheric layer from 25 February 1998. It is officially adopted by the Bulgarian Ministry of Environment and Waters, by the Ministry of Regional Development and Public Works and by the Ministry of Health [6].

The software used is based on a Gauss jet gas model and on the Lagrangian-statistical approach for describing the processes of turbulent diffusion. The methodology and the software allow the following problems related to our study to be solved:

- Calculation of the expected concentrations of the harmful substances (PM10) in the ground layer of the atmosphere;

- Calculation of the average concentrations per specified wind rose for the relevant averaging period. In our case we use daily average.

3 Results and Discussions

3.1 Analysis of meteorological conditions in days with registered PM10 exceedances

3.1.1 Days with registered PM10 exceedances in January 2017

Table 1. Days with registered exceedances by pollutant PM10 by automatic measuring station "Vazrazhdane" for January 2017.

Date	Measured concentration of PM10	Exceedance of the daily limit for PM10
	[μg/m³]	(50 μg/m³)
1	2	3
01.1.2017	52.7	1.05
04.1.2017	129.8	2.60
09.1.2017	56.6	1.13
10.1.2017	97.5	1.95
12.1.2017	71.8	1.44
13.1.2017	83.2	1.66
14.1.2017	64.1	1.28
15.1.2017	103.4	2.07
16.1.2017	108.5	2.17
20.1.2017	61.4	1.23
21.1.2017	88.6	1.77
22.1.2017	128.8	2.58
23.1.2017	139.3	2.79
24.1.2017	135.2	2.70
25.1.2017	118.3	2.37
26.1.2017	87.4	1.75
27.1.2017	84.1	1.68
31.1.2017	125.8	2.52

The days with registered exceedances of pollutant PM10 measured by the Vazrazhdane automatic measuring station in Ruse for January 2017 are according to data from the official bulletin for January 2017 of RIEW - Rousse [1]. They are presented in Table 1.

3.1. 2. Meteorological conditions for days with registered exceedances

Data from Vazrazhdane automatic measuring station in Ruse for the period under consideration [4] are presented in Table 2.

Due to the large number of exceedances registered over the period - 18 days with exceedances - for the purposes of the study, selected days were characterized by the highest ground dust concentrations. In order to capture all possible scenarios, the selected days for analysis are characterized by different daily average temperatures (from negative to positive) and by different direction of the predominant wind. Table 3 presents the selected days for the analysis.

Table 2. Meteorological conditions in the days with registered exceedances by pollutant PM10 by automatic station "Vazrazhdane" for January 2017.

Date	Average temperature	Predominant wind direction	Average speed of the prevailing wind
	[°C]		[m/s]
1	2	3	4
01.1.2017	-0.25	SW	2.17
04.1.2017	1.35	SW	1.60
09.1.2017	-9.52	SE	1.35
10.1.2017	-11.46	W	0.92
12.1.2017	-6.19	SW	1.88
13.1.2017	0.54	SE	0.66
14.1.2017	0.59	SW	2.03
15.1.2017	-1.09	SW	1.36
16.1.2017	-6.80	E	1.48
20.1.2017	-5.74	SW	1.47
21.1.2017	-6.67	SW	1.16
22.1.2017	-4.04	NW	0.55
23.1.2017	0.23	NE	0.61
24.1.2017	-1.91	SW	1.15
25.1.2017	-3.03	SW	0.98
26.1.2017	-2.68	E	1.25
27.1.2017	-6.28	SW	1.38
31.1.2017	-2.06	SW	0.44

To obtain the necessary inputs for the software used PLUME for simulating the average daily concentrations of the PM10 pollutant tested for the selected overdrive days (Table 3), the hourly wind data from the AMS "Vazrazhdane" [4] were used. The resulting wind roses for each of the chosen days are shown on Figures 2 – 6.

Table 3. Days with overheads with typical weather conditions.

Date	Average temperature	Predominant wind direction	Average speed of the prevailing wind
	[°C]		[m/s]
1	2	3	4
01.1.2017 г.	-0.25	SW	2.17

09.1.2017 г.	-9.52	SE	1.35
10.1.2017 г.	-11.46	W	0.92
16.1.2017 г.	-6.80	E	1.48
22.1.2017 г.	-4.04	NW	0.55
23.1.2017 г.	0.23	NE	0.61

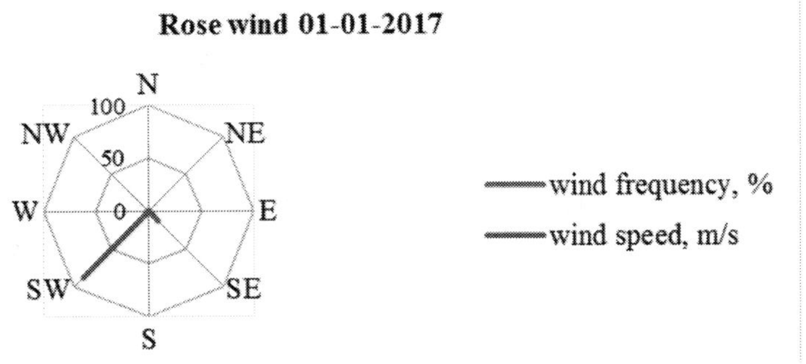

Fig. 2. Rose of the wind on 01.01.2017.

Jan. 1, 2017 is characterized by strong southwest wind (87.5% frequency), changing at the end of the day to southeast. The wind this day is at a low, almost zero average speed.

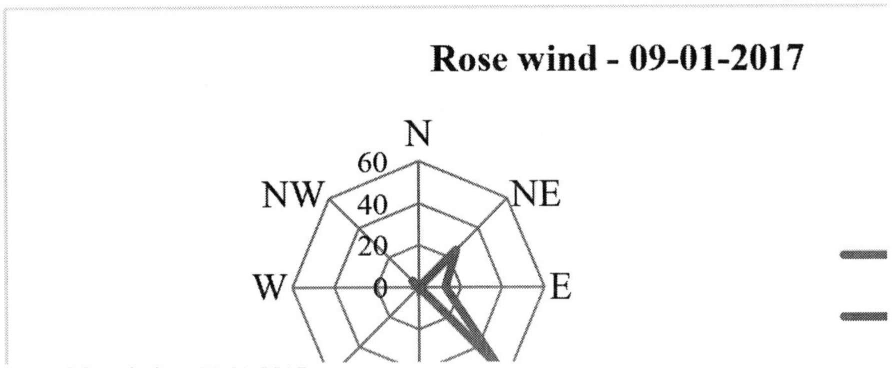

Fig. 3. Rose of the wind on 09.01.2017.

For the 9th of January, 2017 the southeast wind (54.16%) is most pronounced, alternating with the east (12.5%). The wind this day is at a low average speed.

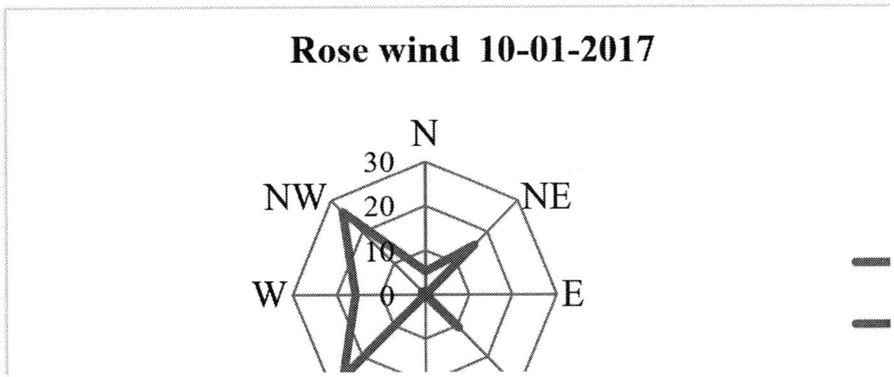

Fig. 4. Wind rose for 10.01.2017.

On January 10, Southwest and Northwest winds (26.32%) were most pronounced, with a sharp shift in direction. Again the winds are at low average speed.

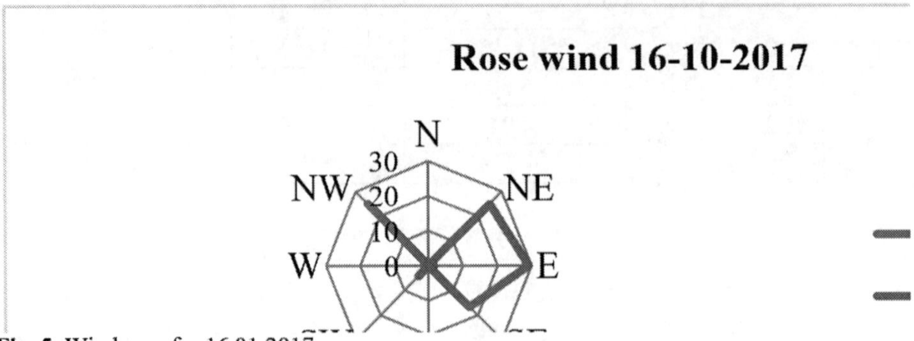

Fig. 5. Wind rose for 16.01.2017.

On January 16 Northwest and Northwest winds (25% frequency) were most pronounced, a sharp shift was observed. Again the winds are at low average speed.

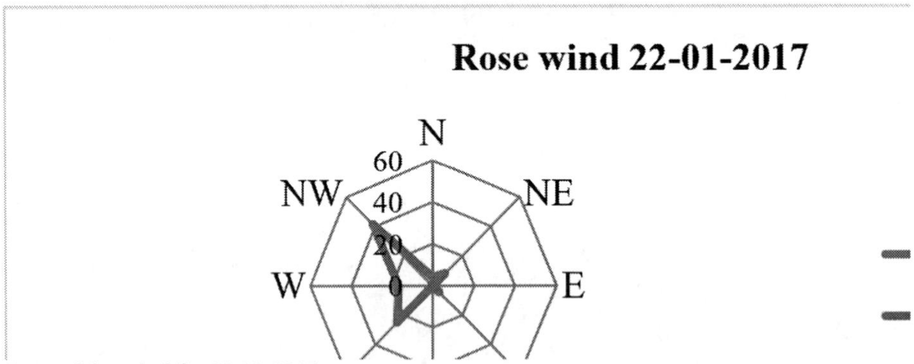

Fig. 6. Rose of the wind for 22.01.2017.

On 22 January, Northwest Wind (41.7%) was most pronounced. Winds throughout the day have very low average speeds (the highest value is 0.67 m / s for the southwest wind).

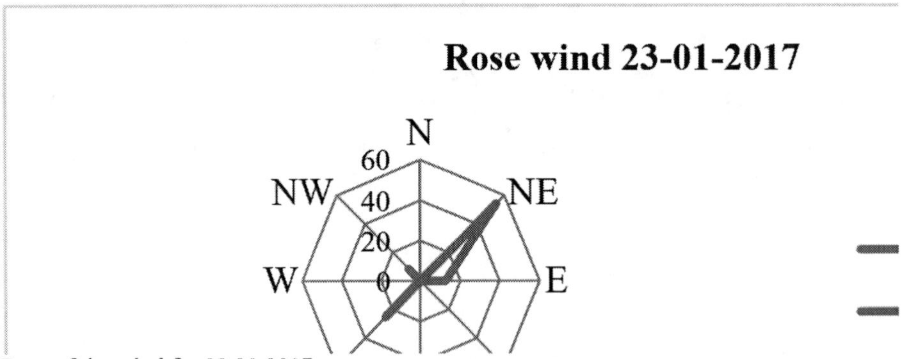

Fig. 7. Rose of the wind for 23.01.2017.

On 23 January the northeast wind was most pronounced (54.2% frequency). Again the winds are at low average speed

3.2 Identification of the dust sources in the industrial zone of Ruse

The firms working on the territory of the industrial zone Ruse, from whose activity dust is emitted are presented in Table 4.

Table 4. Organized sources of dust emissions in the ambient air [2, 3] from Industrial zone, Ruse.

FIRMS	Activity	Source	Number of emitimg devices;
1	2	3	4
TM Technology JSC	Manufacture of metal products	Shot blasting machine	1
Ecogalvanic LTD	Application of surface decorative and corrosion resistant coatings of metals	Local aspiration to production capacities	2
Fibran, Bulgaria JSC	Production of insulating materials	Fuel source, plant for fiber production	2
Oleo protein LTD	Production of raw vegetable oils and groats	Local aspiration to machines	18
Oberosterreich ische Biodiesel Bulgaria LTD	Production of biodiesel	Fuel source	1

3.3 Computer simulation

Computer simulation aimed to calculate the maximum ground PM10 concentrations emitted by the group of industrial sources considered (Table 4) and to calculate the distances to which PM10 air pollutant would be distributed is provided.

A computer simulation was performed to determine the average daily concentration of PM10 of days with exceedances (Table 3) using the established software program PLUME [6].

The program product has limitations on the number of leakages that are entered as incoming data. To avoid the constraint, virtual drives have been created by merging several real chimneys.

The atmosphere area under consideration is presented as coordinate network (x, y) having parameters 16 000 x 16 000 m or 32 steps * 500 m. The model is made with a type of underlying surface "urban area". The industrial zone is located in the regulation boundaries of the town of Rousse.

Input data on weather conditions used are winds for the current day and average daily temperature for that day.

The spread of dust from the territory of the industrial zone of Ruse on January 1, 2017 is presented in Fig. 8.

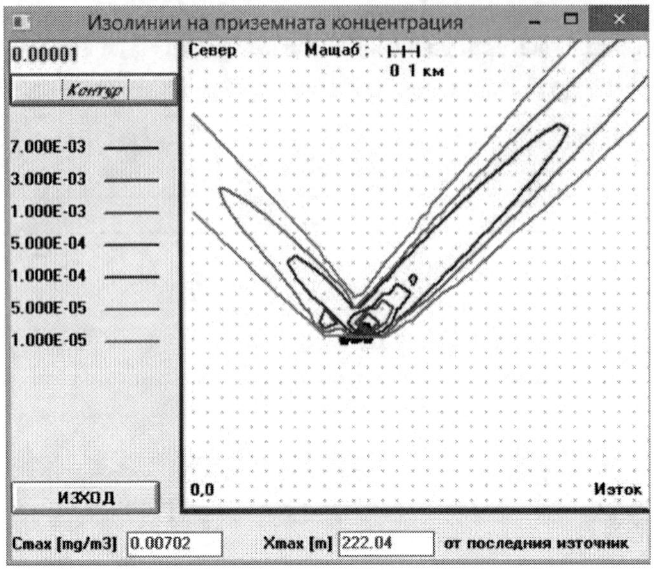

Fig. 8. Isolines of the daily average dust concentrations on January 1, 2017 from the territory of the industrial zone – Ruse.

The spreading of dust from the territory of the industrial zone – Ruse on January 9, 2017 is presented in Fig. 9.

The spreading of dust from the territory of the industrial zone – Ruse on 10 January 2017 is presented in Fig. 10.

The spread of dust from the territory of the industrial zone – Ruse on January 16, 2017 is presented in Fig. 11.

The spread of dust from the territory of the industrial zone – Ruse on 22 January 2017 is presented in Fig. 12.

The spread of dust from the territory of industrial zone – Ruse on January 23, 2017 is presented in Fig. 13.

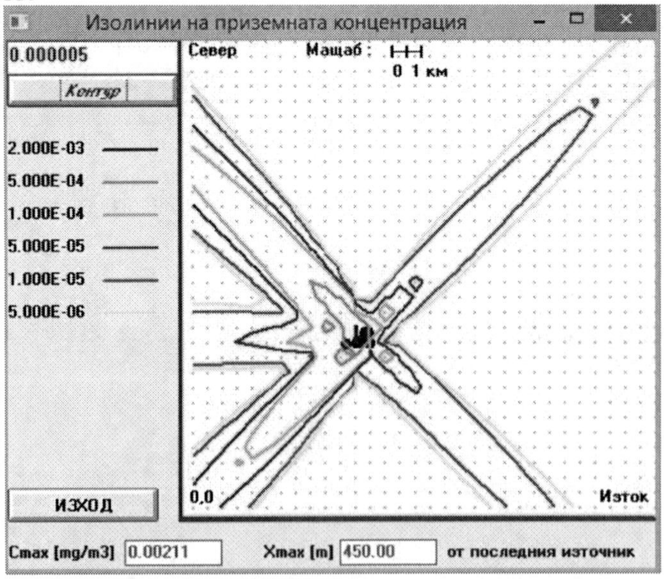

Fig. 9. Isolines of maximum daily average dust concentrations on January 09, 2017 from the territory of the industrial zone – Ruse.

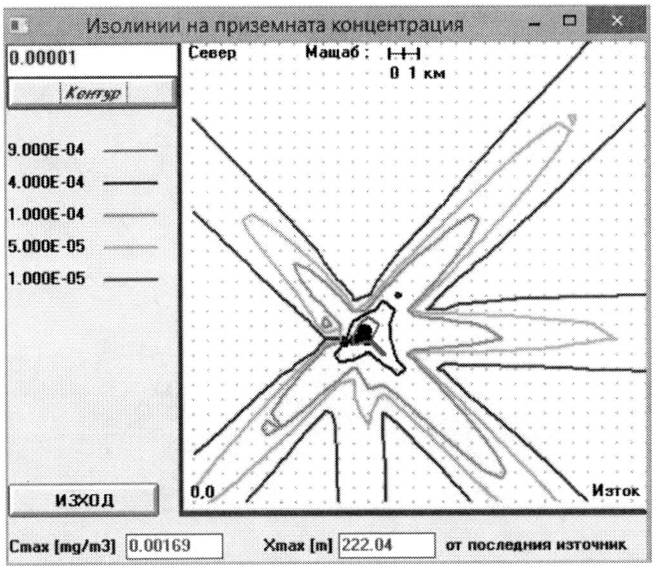

Fig. 10. Isolines of daily average dust concentrations on 10 January 2017 from the territory of the industrial zone – Ruse.

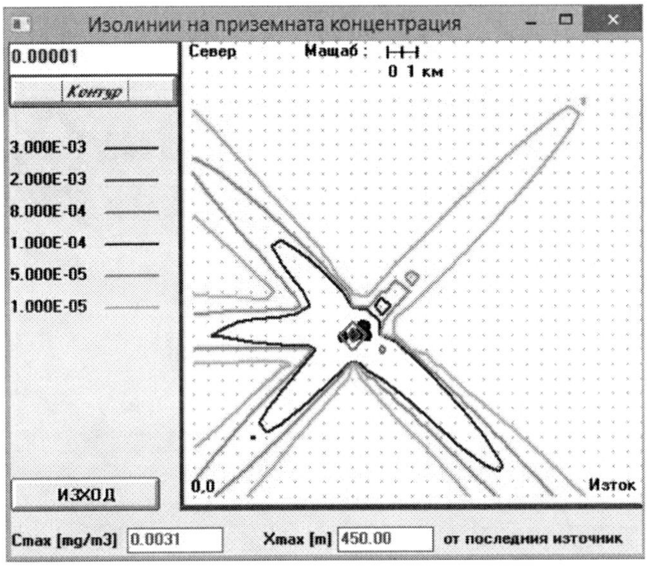

Fig. 11. Isolines of daily average concentrations of dust on January 16, 2017 from the territory of the industrial zone – Ruse.

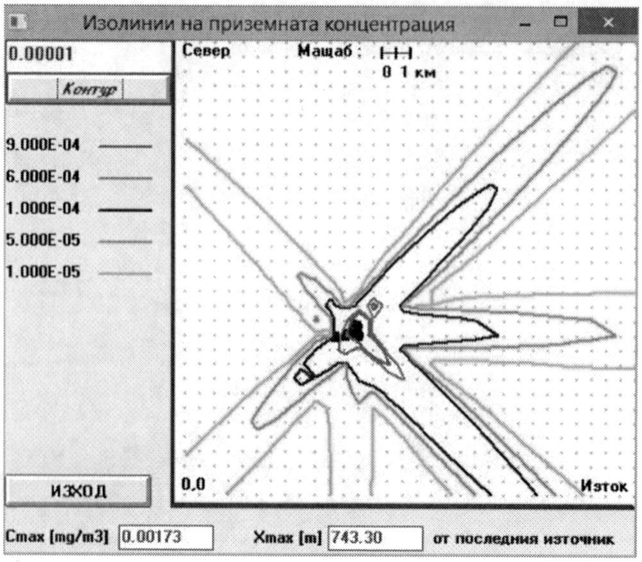

Fig. 12. Isolines of daily average dust concentrations on 22 January 2017 from the territory of the industrial zone – Ruse.

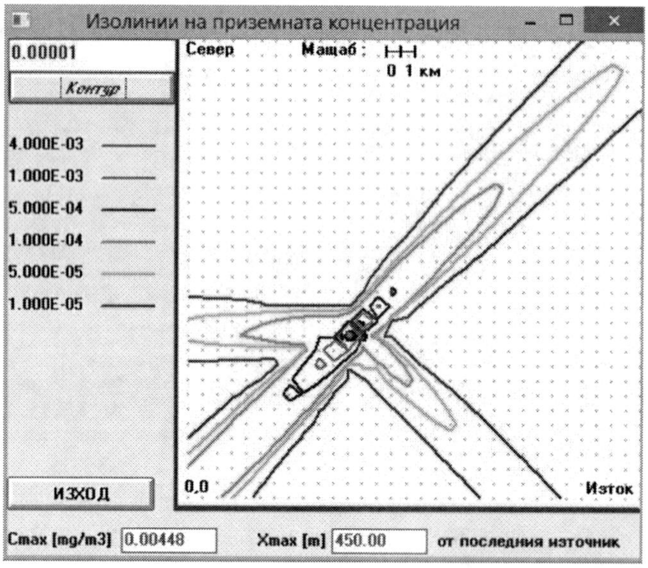

Fig. 13. Isolines of daily average concentrations of dust on 23 January 2017 from the territory of the industrial zone – Ruse.

3.4 Analysis of the values of the ground concentration of dust reaching the territory of the Republic of Romania and the possibility of their impact on the air quality in the cross-border area

The Romanian bank of the River Danube in this area of the river is 910 meters away (see Fig.1). The Romanian town of Giurgiu is situated in the northwest direction of 6,100 km and town Comaska, the county of Comasca, to the north of 4,400 km.

Table 5 Values of PM10 ground concentrations reaching to Romania

Date	Danube river coast, Romania		Gurgue, Romania		Comanska, Romania	
	SDC	MEK	SDC	MEK	SDC	MEK
	mg/m^3	mg/m^3	mg/m^3	mg/m^3	mg/m^3	mg/m^3
01/01/2017	5E-05	5E-05	1E-05	>1E-05	1E-04	1E-05
09/01/2017	5E-05	1E-04	1E-04	1E-05	1E-05	>1E-05
10/01/2017	1E-04	5E-05	1E-05	>1E-05	5E-05	>1E-05
16/01/2017	1E-04	1E-04	1E-05	>1E-05	5E-05	>1E-05
22/01/2017	1E-05	1E-06	1E-05	>1E-06	1E-04	>1E-06
23/01/2017	1E-03	1E-07	1E-05	>1E-07	1E-04	>1E-07

* SDC - average daily PM10 concentration, MEK - maximum PM10 concentration

The results of the computer simulation of the ground concentrations reaching the receptor points on the territory of Rumania are presented in tabular form - Table 5.

To study the possible maximal single concentrations (MEK) reaching the receptors one wind direction simulation (predominant for the day) was conducted. The results are shown in Fig. 14 ÷ 19.

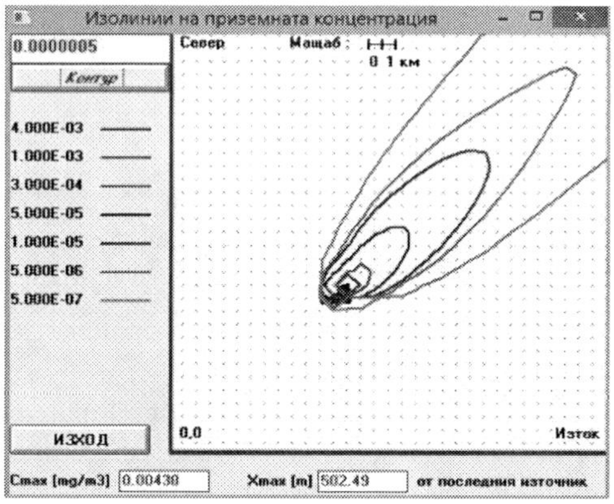

Fig. 14. MEK for 01 Jan. 2017.

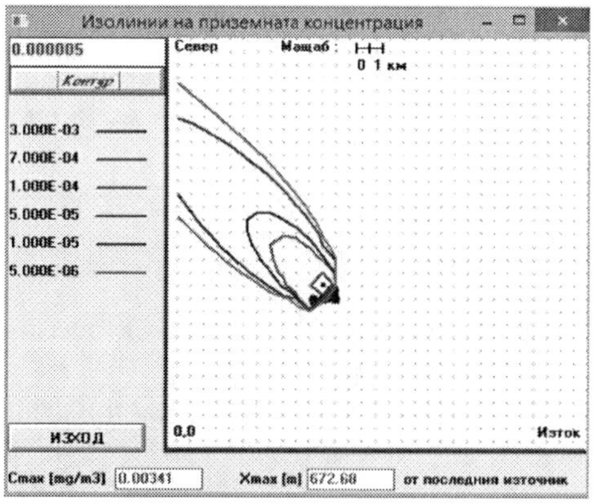

Fig. 15. MEK for 09 Jan. 2017.

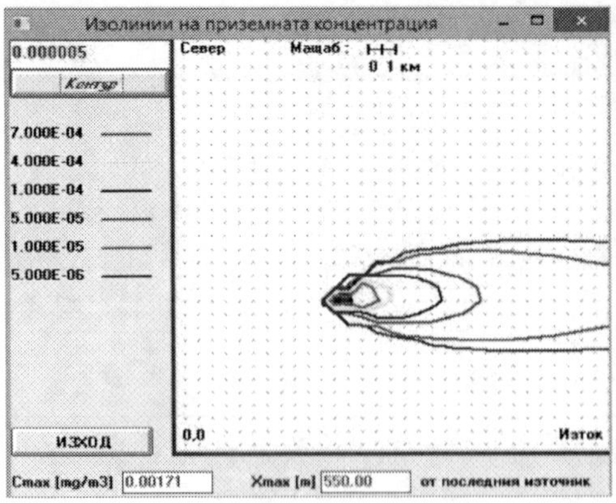

Fig. 16. MEK for 10 Jan. 2017.

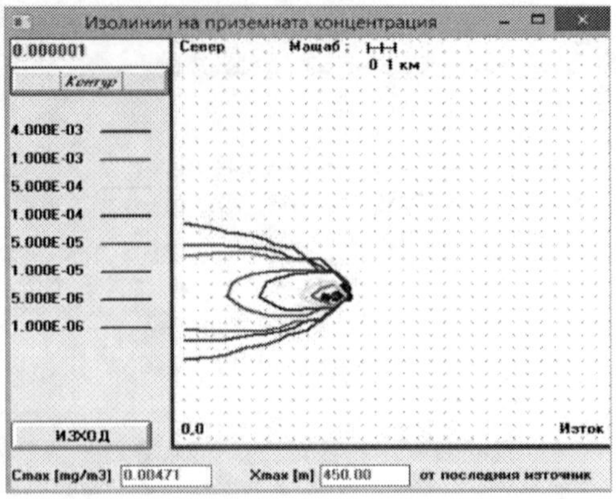

Fig. 17. MEK for 16 Jan. 2017.

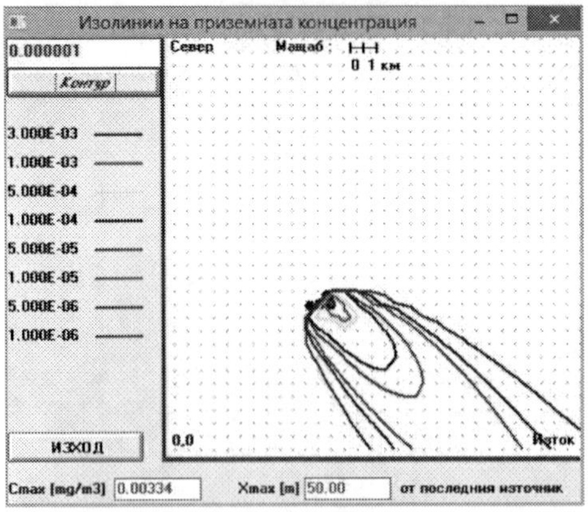

Fig. 18. MEK for 22 Jan. 2017.

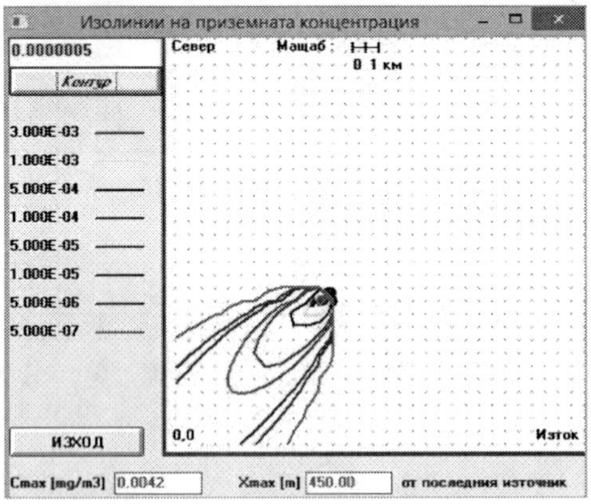

Fig. 19. MEK for 23 Jan. 2017.

The established daily norm for the protection of human health for PM10 is 50 µg / m^3. The ground concentrations of the pollutant reaching the Romanian territory represent only 2 ÷ 0.02% of this norm. They have no potential for disturbing the quality of the ambient air in the area.

4 Conclusions

The survey conducted is limited in time, the representative monthly sample of January 2017 is considered.

Despite the limitation, it is established that the operation of the installations located on the territory Industrial zone - Ruse has no potential for cross-border PM10 contamination of the atmospheric air in the BG – RO cross - border area.

Acknowledgments to the employees of the Regional Inspectorate of Environment and Water - Rousse and the environmentalists of Oleo Protein LTD and Fibran Bulgaria JSC for providing the complete information necessary for the input data of this analysis.

References

1. Data on the measured concentrations of PM10 at the monitoring stations on the territory of the RIEW - Rousse in January 2017, http://www.riosv-ruse.org/mesechna-spravka-za-nivata-na-fpch10.html

2. Reports of own periodic measurements of " Ecogalvanic" LTD, "Oleo Protein" LTD, "Fibran Bulgaria" JSC, "TM Tehnologi" JSC, provided by RIEW-Rousse with Decision for access to public information 329/2017
3. Reports of Own Periodic Measurements of "Oleo Protein" LTD and "Oberosterreichische Biodiesel Bulgaria" LTD, provided by the legal entities.
4. Hourly meteorological report for Rousse, January 2017, provided by the decision for access to public information 330/2017
5. Publicly available map material on the official site of the Agency for Geodesy, Cartography and Cadastre, http://www.cadastre.bg/
6. Methodology for calculation of leakage height, dispersion and expected concentrations of pollutants in the ground layer of 25 February 1998, adopted by the Ministry of Environment and Waters, the Ministry of Regional Development and Public Works and the Ministry of Health, http://www5.moew.government.bg/
7. I. Zheleva, E. Veleva, M. Filipova, AMiTaNS, 2017 (in print)

Potential of multispectral imaging technology for assessment coniferous forests bitten by a bark beetle in Central Bulgaria

Marina Stoyanova[1], *Alexander* Kandilarov[2], *Vesselin* Koutev[3], *Olga* Nitcheva[4,*], *Polya* Dobreva[4]

[1]Agriculture Institute, Industrialna Str. 1, 8400 Karnobat, Bulgaria
[2]Neopterix Ltd, 4000 Plovdiv, Bulgaria
[3]University of Forestry, Kliment Ohridski Str. 10, 1756, Sofia, Bulgaria
[4]Institute of Mechanics - BAS, Acad. G. Bonchev Str., Bl. 4, 1113, Sofia, Bulgaria

Abstract. Much of the coniferous forest in Bulgaria is struck by Ips typographus, Ips acuminatus and Ips sexdentatus beetles. The multispectral camera test near the town of Kalofer, Bulgaria in August 2017, showed the good capabilities of remote-piloted aircraft for observation and mapping of large areas in highly intersected areas. The collected data from the multispectral camera allowed producing maps with different indices of the observed vegetation. Beetles damaged stains were identified and outlining the areas of intervention and their sizing is possible by forestry specialists. The NDVI index in the range of 0.7 to 0.95 characterizes green tree vegetation. Affected forest subjected to stress has lower NDVI index values - 0.65.The NDVI index of damaged areas range from 0.45-0.5. Identifying affected forest areas is the first important step to solve the problem. Taking pictures of drones and multispectral camera can be of great help, and the test conducted clearly demonstrates it. Preventive actions can happen much earlier before trees are completely destroyed and the problem is clearly visible with the naked eye. In view of the rapid spread and the complete destruction of some conifers, immediate action is needed to cut out and remove the diseased and infected trees in order to break the development cycle and prevent subsequent contamination of adjacent plots

1 Introduction

The protection of Bulgarian forests is a priority given in the National Strategy for Development of the Forestry Sector in the Republic of Bulgaria for the period 2013-2020. Forests fulfill many economic, ecological and social functions of particular importance for the sustainable development of the country and are a key factor for the formation and maintenance of a living environment. Much of the coniferous forest is struck by Ips typographus beetles, Ips acuminatus, Ips sexdentatus. This requires urgent measures to

* Corresponding author: olgani@imbm.bas.bg

clean up the damaged terrain and prevent further contamination. In the opinion of Eng. Grigor Gogov (Director of the Executive Agency of Forests), the main reason for the increase of the bark beetles attacks is the good nutritional environment for their development. It is believed that the large amount of wood - over 2 million cubic meters - damaged by abiotic factors - winds, windmills, ice and snow - combined with the impossibility of fast and removal within one year due to the high percentage of inaccessible terrains, are factors that require adequate measures to save the forest.

The only way to deal with the problem is to cut out the damaged trees and immediately remove them from the terrain. Given the percentage of diseased trees and the area of the affected area, the situation can be assessed as disastrous. To deal with this situation serious organizational, human and financial resources are required. In such situations, however, the first step is to carry out quality monitoring to accurately quantify the problem. Till now monitoring of bark beetles was assessed by observation of affected areas or application of pheromone traps [1]. The analysis of damages caused by bark beetle on Norway spruce (Picea abies) forest in Vitosha mountain, Bulgaria was applied using the VIs indicative for forest stress: NDVI, SR, EVI, ARVI, CRI, CSc, and ARI. Main regions for stressed vegetation have been delineated. The CSc has been found to perform better in detecting the pattern of stressed spruce trees compared to ARI. The area affected by Ips typographus was determined by CSc index to 5.97% (0.373 km^2) of the study area [2]. With the aid of the object-oriented image analysis software eCognition classification in the deadwood areas damaged by bark beetle yielded a total classification accuracy of 91.5% [3].

Taking into account the wide spread of bark beetles spots on hard-to-reach forest areas, remote monitoring methods are the most effective way of conducting observations, sizing the affected area, assessing the risk and planning the necessary resources for follow-up. Remote sensing techniques are used to generate maps and assess the different degrees of attack. Spectral and geometric methods of high-resolution remote monitoring serve to characterize the health or vitality of trees. Stress in vegetation leads to changes in the composition and proportion of photosynthetic active pigments such as chlorophyll a and b as well as in plant foliage. This is also the reason why changes in the health status of the forest can be observed at a very early stage [4].

2 Materials and methods

Bark beetles are small beetles, which, depending on the species, range from 2.5 to 8 mm (Figure 1). They develop two generations per year, and are frequent below 800 m above sea level. In the warmer years it is possible to have three generations. The first generation flies in early May, and the second in July and the beginning of August. They feed with coniferous wood, mostly white pine and spruce.

Of the representatives of the bark beetles in Bulgaria, the prevalence and damage of Ips acuminatus is greatest [5 - 6]. It primarily affects white pine (Pinus sylvestris L.). Its color is yellow-brown, the larvae is 6 mm long, white and without legs. Propagation is under the bark of the top of the stems and the thick branches. The male is digging a wedding chamber for the females. Each female makes a motherly move in which they lay eggs. The course of the larvae is short and ends with a cocoon chamber. They develop sister generations. It attacks the top of the stems with a smooth bark. Mothers and lashes move the conductive tissue of the attacked tree and it quickly dries out Figure 2.

Fig. 1. Bark beetle under the removed bark.

Fig. 2. Bark beetle damages.

Although bark beetles tend to colonize dead or weakened trees, there are reports that some species attack healthy trees under epidemic conditions, attacking individual trees or groups of trees, resulting in so-called bark beetle spots - Figure 3.

Fig. 3. Bark beetle spots.

Each type of tree has different reflectivity across the different regions of the electromagnetic spectrum. It can be seen from Figure 4 that the reflectivity of deciduous forests varies in a much larger range at the end of the visible portion of the electromagnetic spectrum (beyond the red end of the spectrum and in the near-infrared wavelength of 700 nm and greater). This can be used to easily distinguish between the two types of vegetation in mixed forests. Other specificity of NDVI indexes from different surfaces are shown on Figures 5 and 6.

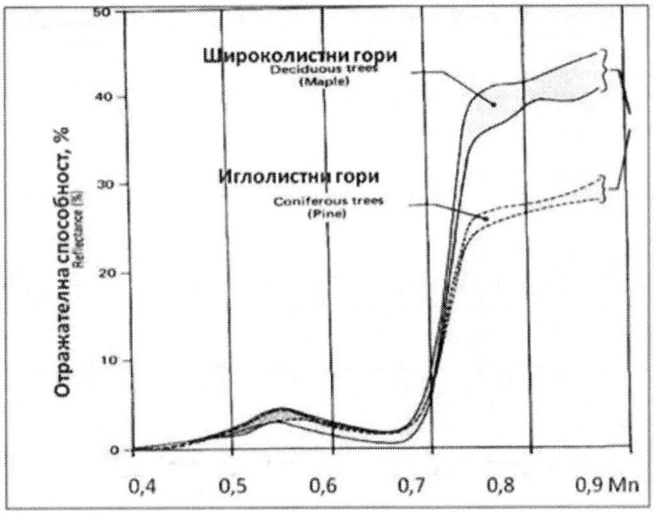

Fig. 4. Difference in the reflectance of deciduous and coniferous trees [7].

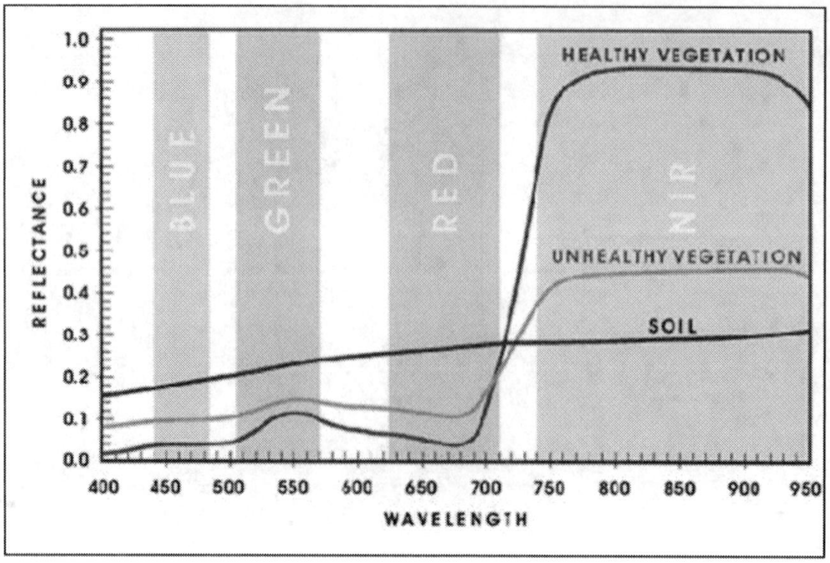

Fig. 5. Combined reflectivity of soil and vegetation - note the difference in the near infrared range between the healthy and diseased vegetation curves [8].

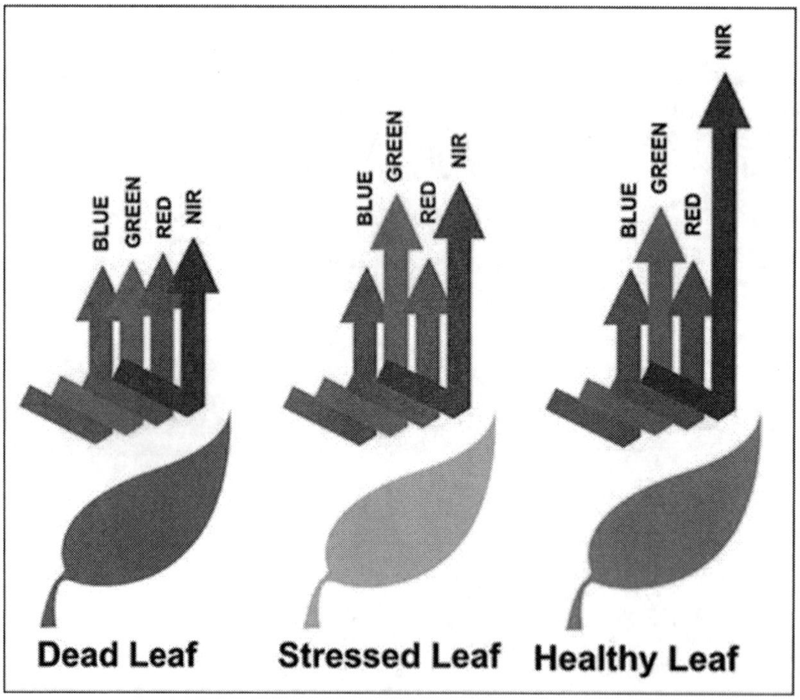

Fig. 6. Schematic presentation of the reflection ability of dead, stressed and healthy vegetation [9].

Healthy plants generally have a higher reflectance in the near infrared area than diseased and stressed plants, which can be used to distinguish one from the other. In remote studies, the so-called **NDVI** (Normalized Difference Vegetation Index) which is a numeric indicator using the visible **(RED)** and near infrared areas **(NIR)** of the electromagnetic spectrum for the analysis of vegetation and vegetation strength. Typically, healthy vegetation takes up much of the visible light that falls on it, and reflects much of the nearby infrared area. Damaged or rare vegetation reflects more than visible light and less than the near infrared.

$$NDVI = (NIR - RED) / (NIR + RED) \quad (1)$$

The study used one aircraft of Neopteryx Ltd. with identification number NTRX002. The NTRX002 is the DJI Matrix 600 Pro Hexcopter. It is equipped with a 5-channel multi-spectral camera, consisting of 16 megapixel standard optical sensor (RGB) and 4 separate sensors capturing light in green, red, extreme red and near infrared of the electromagnetic spectrum. It is additionally equipped with a 16 megapixel optical camera. Its weight in configuration with the two cameras is 12 kg, its radius of action is 4.5 km and the dynamic flight ceiling is 4500 m.

The resulting raw RGB camera results are in JPEG format, and from the multi-spectral camera - in a georeferenced TIFF format.

The survey was accomplished hear the city of Kalofer in Central Bulgaria, Figure 7.

Fig. 7. Satellite image of the surveyed area. (Google Earth Satellite Imagery).

Objectives of the study are to capture damaged forest area with a drone equipped with a five-strip multi-spectral camera and an additional optical camera and to generate maps with different vegetation indices.

3 Results

The results of the field survey are images from the multispectral camera. After processing digital maps are obtained in different bands of the electromagnetic spectrum.

The whole area studied, with the exception of the access roads and the access to it, is a mixed forest, which was clearly visible at the time of the study, and several damaged areas were observed.

To separate the deciduous vegetation, a reflection capacity map is used at the far red end of the visible portion of the electromagnetic spectrum, and the vegetation classification by height and slope of the 3D cloud from points (Figure 8).

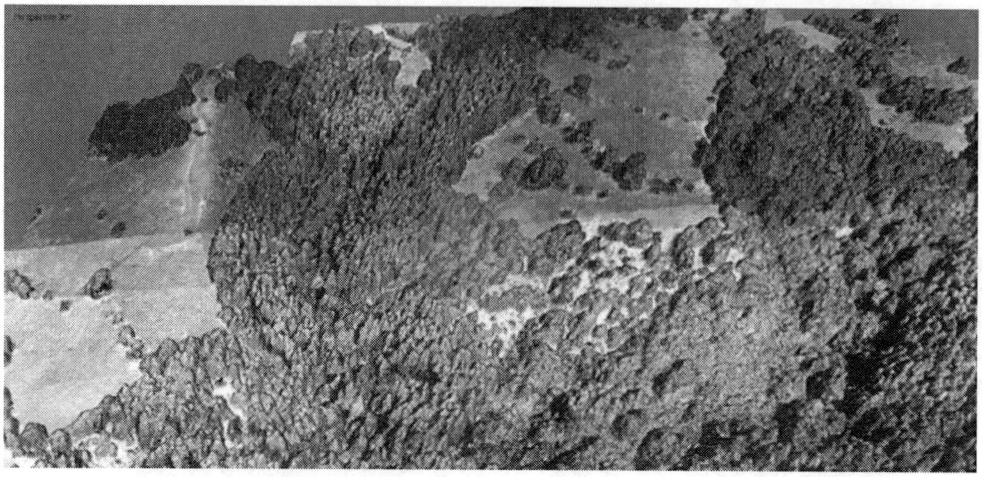

Fig. 8. Part of a 3D model of the terrain, showing a bark beetle spot in the northern part.

A map of the NDVI index (Formula 1) is used to assess the strength of forest vegetation. Photos and maps of the northern and southern parts of the surveyed area are given in Figures 9, 10, 11, 12. For analysis, these two parts of the site are selected because they contain all the elements of the problem that the test to decide - this is the presence of mixed deciduous and coniferous forest vegetation, damaged by bark beetle coniferous trees and damaged spots, the presence of pastures, meadows, arable land and low buildings, which are hindering factors in the analysis of altitude data.

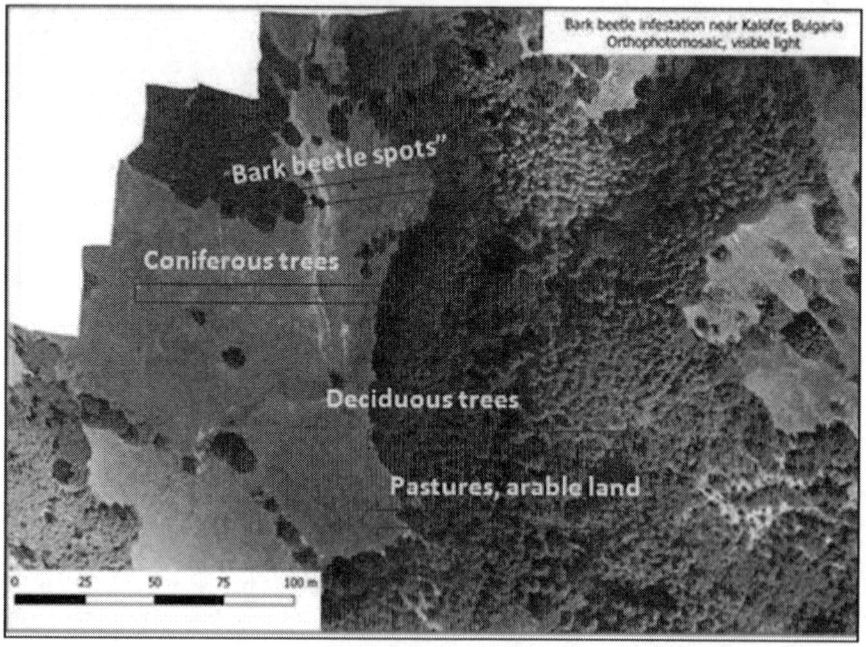

Fig. 9. Part of the orthophotomap at the northern end of the surveyed forest massif.

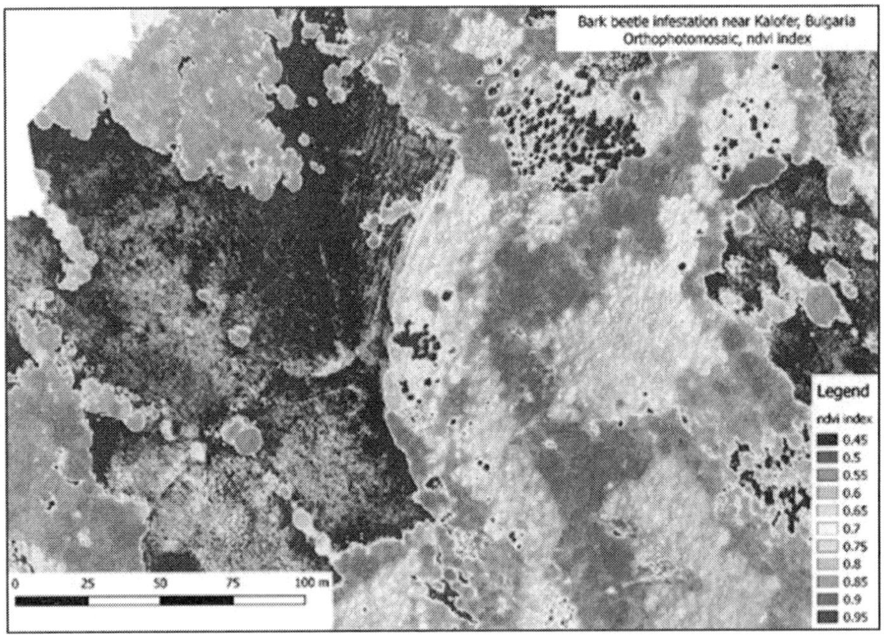

Fig. 10. Map of the NDVI index in the northern part of the surveyed terrain.

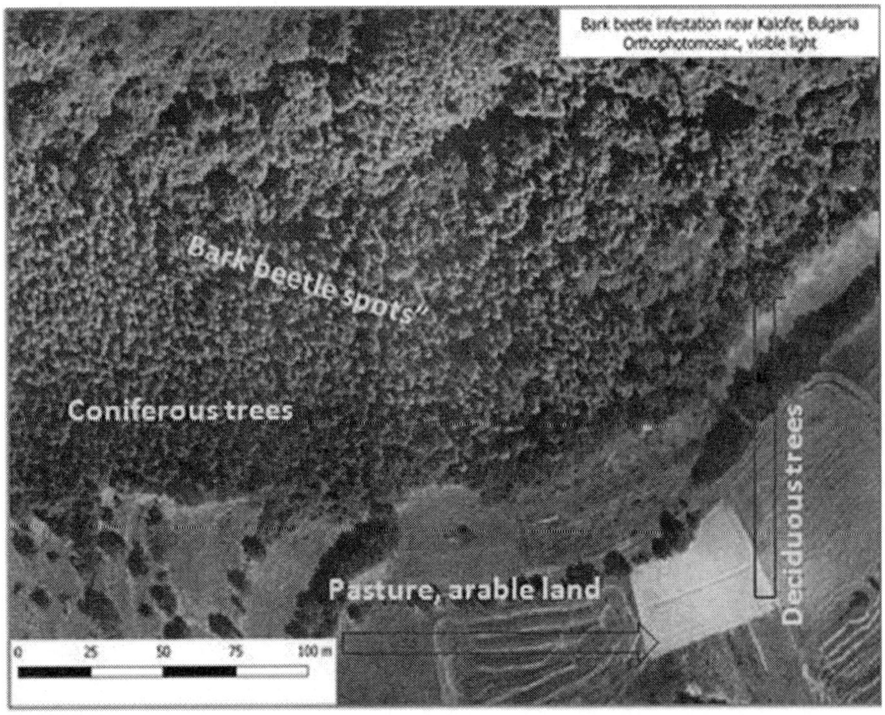

Fig. 11. Part of the orthophotomap at the southern end of the investigated forest.

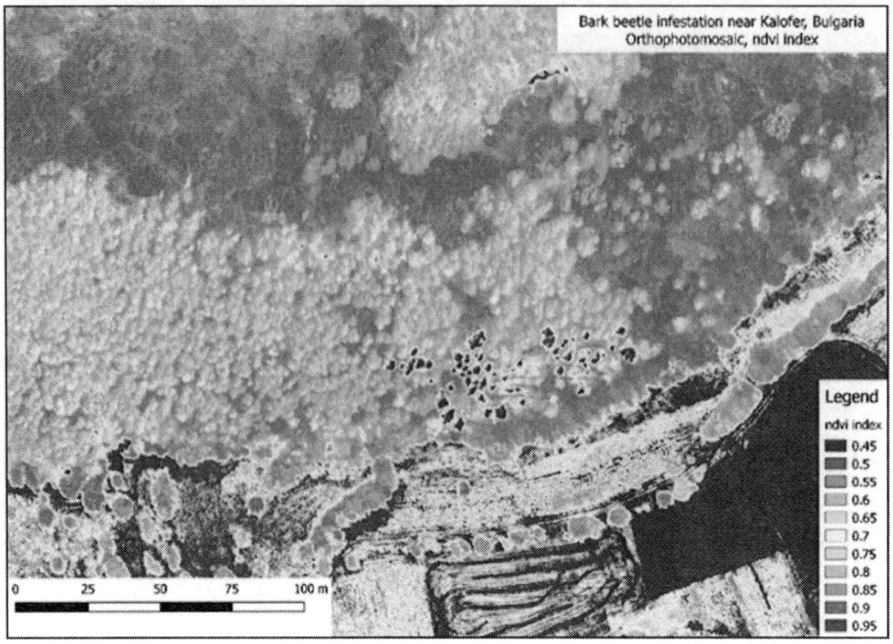

Fig. 12. Map of the NDVI index in the southern part of the surveyed terrain.

The various indexes generated by the multispectral camera data together with the 3D cloud of points help to distinguish deciduous forests from coniferous forests and subsequently identify bark beetle spots and areas of varying degrees of damage from bark beetles and other pests. Quantitative dimensioning of the affected area can be done in the Geographic Information Systems (GIS) and, after consultation with forestry specialists the intervention areas can be outlined.

In the middle of August 2017, when the study was conducted, the vegetation development phase was such that the NDVI field index was below or around 0.45-0.5 (red in the corresponding figures). The NDVI index in the range of 0.7 to 0.95 characterizes green tree vegetation. On both sides of the coniferous forest are fields in clear red color (Figures 9 and 10). In the middle of the terrain are coniferous trees, the spots from red to yellow outline the bark beetle affected trees. The NDVI index score is estimated to range from 0.45-0.5. Spots in red on coniferous stretches outline areas with affected or completely destroyed trees - a deep red color is a sign of it - these are the affected spots.

Similar is the situation in the south. In the lower part of the studied area there are fields and pastures (NDVI 0.45-0.55). Yellow and green is forest vegetation, with a healthy NDVI index of between 0.85 and 0.95. The forest that has been subjected to stress has less reflectivity in the red and near infrared, resulting in lower NDVI index values, and it drops to 0.65. Damaged spots are slightly above 0.5, as a result of which they are easily distinguishable.

4 Achieved goals

The multispectral camera test near the town of Kalofer showed the good capabilities of remote-piloted aircraft for observation and mapping of large areas in highly impassable terrain. Performing such a study on the same area at the same time by foot across the terrain would be impossible.

The collected data from the optical camera allowed the generation of orthophotomaps and a 3D model of the terrain, and data from the multispectral camera allowed to produce maps with different indices of the observed vegetation.

A problem with mixed forests is the separation of areas with deciduous trees from those with coniferous vegetation. In this test, this problem was solved in two steps - by classifying the 3D cloud by points at relative altitudes and gradients, and by classifying the areas in the reflectance map at the red end of the visible portion of the electromagnetic spectrum.

Conclusion

The forest vegetation in the studied terrain is subject to bark beetle stress. Identifying affected forest areas is the first important step to solve the problem. Taking pictures of drones and multispectral camera can be of great help, and the test carried out clearly demonstrates it. The elevation data from it shows that, in addition to bark beetle damaged spots that can be correctly identified this can happen much earlier than when the trees are completely destroyed and the problem is clearly visible with the naked eye. Earlier identification of problem areas will help to take timely measures and prevent the spreading of the bark beetle in adjacent areas. The study found that so far the damaged spots occupy a small part of the area surveyed. In view of the rapid spread of bark beetles and the complete destruction of some conifers, immediate action is needed to cut out and remove the diseased and infected trees in order to break the bark beetle development cycle and prevent subsequent contamination of adjacent plots.

References

1. B. Chandler, *Monitoring Abundance of Ips Bark Beetles and Determining Related Tree Mortality in Arkansas and Texas* (Dissertations, paper 21, 2015)
2. L. Filchev, *Proceedings of 1st European SCGIS Conference with International Participation "Best Practices: Application of GIS Technologies for Conservation of Natural and Cultural Heritage Sites"* (SCGIS-Bulgaria, Sofia, 2012)
3. M. Heurich, T. Ochs, T. Andresen, T. Schneider, Eur J Forest Res **129**, 313 (2010).
4. A. Lausch, M. Heurich, D. Gordalla, H. Dobner, S. Margianto, C. Salback, Forest Ecology and Management **308**, 76 (2013)
5. B. Rosnev, P. Mirchev, G. Georgiev, P. Petkov, Y. Naydenov, G. Tsankov, D. Ovcharov, St. Mirchev, A. Pencheva, D. Doychev, M. Matova, M. Georgieva, *Handbook of forest protection. Part I. Diseases, insects and other pests and damages on forest trees and bushes* (Obrazovanie i nauka, Sofia, 2006)
6. S. Lopes, A. Goldarazena, Journal of Entomology **2012**, Article ID 145930 (2012)
7. T.M. Lillesand , R.W. Keiefer, J. Chipman, *Remote sensing and Image Interpretation 7ed.* (2015)
8. B. Warzak, Vision Systems Design, November (2016)
9. Agribotix.com, Misconceptions about UAV-collected NDVI imagery and the Agribotix experience in ground truthing these images for agriculture (2017)

MATEC Web of Conferences **145**, 01006 (2018) https://doi.org/10.1051/matecconf/201814501006
NCTAM 2017

Simulations and experimental evaluation of an active orthosis for interaction in virtual environments

Mihail Tsveov[1], *Pavel* Venev[1], *Dimitar* Chakarov[1], and *Ivanka* Veneva[1,*]

[1]Bulgarian Academy of Sciences, Institute of Mechanics, 1113 Acad. G. Bonchev Str., bl. 4, Bulgaria

Abstract. In this work, the development of a human arm active orthosis is presented. The orthosis is designed primarily for training and rehabilitation in virtual environments.The orthosis system is intended for embodiment in virtual reality where it is allowing human to perceive forces at different body parts or the weight of lifted objects. In the paper the choice of a mechanical structure is shown equivalent to the structure of the human arm. A mechanical model of the orthosis arm as haptic device is built, where kinematic and dynamic parameters are evaluated. Impedance control scheme is selected as the most suitable for force refection at the hand or arm. An open-loop impedance controller is presented in the paper. Computer experiments are carried out using the dimensions of a real arm orthosis. Computer experiments have been carried out to provide force reflection by VR, according to virtual scenario. The conducted simulations show the range of the forces on the operator hand, orthosis can provide. The results of additional measurements and experimental evaluations of physical quantities in the interaction in a virtual environment are revealed in the paper.

1 Introduction

Exoskeleton is outside wearable robot with joints and limbs corresponding to those in the human body. It transmits torques to human joints by actuators allocated in its mechanical structure. Exoskeletons are used for four basic functions by means different control algorithms [1]: a) rehabilitation – fit closely to his or her body, fulfils tasks of physical therapy in an active or passive working mode; b) Haptic device – the subject physically interacts with virtual objects, as the interaction forces are applied by the exoskeleton opposite the operator; c) Master device – for tele-communication robot control „master-slave" as the interaction forces are applied by the exoskeleton opposite the operator: d) Assistive device - human body amplifier, operator feels diminished the loadings, accepted by the exoskeleton.

The first modern exoskeleton arm was designed by PERCRO (Perceptual Robotics Laboratory) for replication of sensations of contacts and collisions [2]. This is a seven degree of freedom (7-DoF) ungrounded device, attached to operator's shoulder and torso.

* Corresponding author: veneva@imbm.bas.bg

The operator holds onto the device with his/her palm. Hence, the device can only exert forces at the palm of the user. It uses DC motors with a cable transmission system for actuation. The authors of PERCRO have developed arm exoskeletons for haptic interaction with virtual environments L-Exos [3, 4]. This is a five-DoF exoskeleton with a wearable structure and anthropomorphic workspace that can cover the full range of motion of a human arm. A 9-DoF under-actuated exoskeleton arm using pneumatic actuators is developed by [5]. Their device allows full reproduction of the human arm's workspace when operating the exoskeleton. An alternate arm exoskeleton developed at the Korea Institute of Science and Technology addresses the limited wearability issues of previous designs by using parallel mechanisms and pneumatic actuators [6]. The wearable Salford arm addresses some of the issues and limitations of earlier designs [7, 8]. For example, nearly ninety percent of the human arm's workspace can be replicated with their device. Pneumatic muscle actuators (PMA) were selected to power the robot due to their high power-to-weight ratio. Soft arm–exoskeleton was used in physiotherapy and training. The design of a seven-DoF powered cable-driven arm exoskeleton for neuro-rehabilitation is presented in [1]. Proximal placement of motors, distal placement of pulley reductions, and open mechanical human–machine interface are a few features that add to the performance and ease-of-use of the device. Additional characteristics include low inertias, high-stiffness links, and backdrivable transmissions without backlash. The design achieves full-workspace ranges of motion.

Recently, the use of exoskeletons as Haptic device significantly increased due to the development of applications related to interaction in virtual environments [2, 3, 4]. Haptic or force-reflecting interfaces are devices used to display touch or force-related sensory information from a virtual or remote environment to the user [2]. Haptic interfaces allow to perceive and transmit motion and forces to real or virtual entities stimulating the somatosensory system, and indeed the participation of the user in the immersive virtual reality [9]. In immersive virtual reality (IVR) it is possible to give people the illusion that they have a different body. Through real time motion capture when the person moves their real body they see the virtual body move correspondingly [10]. This feature of exoskeletons is combined with their other applications - as a device for rehabilitation using virtual scenarios [4].

The objective of the work presented in this paper is to introduce the design of an active orthosis of upper-limb. Device can be used for application where both motion tracking and force feedback are required, such as human interaction with virtual environments or rehabilitation tasks. Computer simulations and real experiments to be carried out to evaluate the possibility of developed exoskeleton to provide force reflection on the hand of the operator.

2 Mechanical structure and actuation system

The exoskeleton mechanical structure as a wearable device must fulfil the following design requirements [7]: light with low mass/inertia; safety; comfort of wearing; anthropomorphic workspace arm, extensive range of motion, etc.

The device, which can be seen to function as a powered exoskeleton, usually is build up of a serial kinematics structure consisting of rotational joints. A kinematics structure is selected to be similar to the structure of the human arm [11] (Fig. 1).

Each arm structure has 4DoF corresponding to the natural motion of the human arm from the shoulder to the elbow, but excluding the wrist and the hand. In the chosen solution, the first two rotation joints are incident and mutually orthogonal in order to emulate the kinematics of a universal joint with the same centre of rotation of the human shoulder (Fig. 1). The third and the fourth joints are selected to be two rotational joints, incident and mutually orthogonal in order to emulate the kinematics of universal joint with the same centre of rotation of the human elbow. The third and the fourth joints emulate the effective motion of the elbow flexion and human shoulder rotation without decoupling the motion in a way typical for human arm structure. The mechanical structure of exoskeleton system includes two arms, that are grounded. Unlike other solutions, circular guide and three axes joints are avoided. Each universal joint placed in the shoulder and in the elbow consists of a cross shaft and pair of rotational joint oriented at 90 ° to each other. The cross shaft is formed as an angle, in which the human hand is situated, as the axes of rotation intersect at a point, coinciding with the point of rotation of the human natural joint.

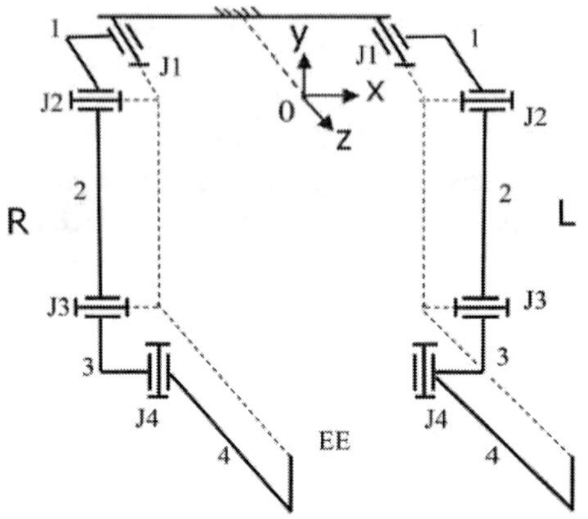

Fig. 1. Exoskeleton structure, similar to the structure of the human arm.

The basic parts of the exoskeleton arm are made of aluminium. The arm is constructed for use by a typical adult. All arm units are designed that allow quick and easy adjustment of length, so making it easy to accommodate a range of users. Plastic shells with bands are placed over the units for attachment to human limb. This design suggests a comfortable, light and low cost mechanical structure. In Figure 2 a picture of two arm prototype is shown. High precision rotation sensors are mounted in the joints.

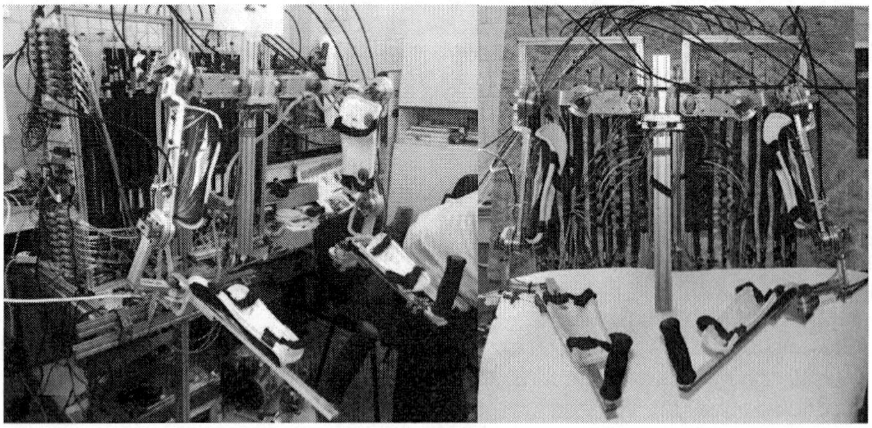

Fig. 2. Two arm exoskeleton prototype.

The masses of four basic units (Fig.1) are m_1=0.437 kg, m_2=0.594 kg, m_3=0.364kg, m_4=0.434kg. Lengths of exoskeleton arm and forearm in the initial setup are 0.286 m and 0.370 m respectively.

Actuation system of exoskeleton arm should have the following advantages: excellent power/weight ratio with inherent safety, natural compliance, low cost. Self-made braided pneumatic muscle actuators (PMA) are used to achieve these advantages. PMA consists of two layers: an inner one, representing rubber liner and outer one, representing braided nylon. Endcaps are allocated at both ends, where the two layer are clamped. A pipeline is located in one endcap for supplying or getting out the pressurized air. Muscles possess a maximum diameter D_0 = 0.016 m and nominal length L_n =0.390 m. They can be used not only singly, but also in a parallel group, as shown in Figure 3. The muscles from the bundle are fed in parallel with air at a maximum pressure 500 kPa (5 bar). All the muscles of the bundle are connected together mechanically in their both ends. High precision pressure sensors are mounted for each bundle.

Fig.3. Sheaf pneumatic muscle actuators.

Joint motion/torque on the exoskeleton arm is achieved by antagonistic actions through cables and pulleys, driven by the pneumatic actuators. The pulleys are attached on the arm segments following joint shafts. All actuators are mounted on the exo-shell on the operator's back. A selection of bundles with different number of muscles for every joint is performed to ensure nominal torques in the joints [11]. Selected bundles antagonists with muscles number m of the four arm joints are as follows: m_{1a}=7, m_{1b}=3; m_{2a}=6,m_{2b}=3; m_{3a}=4, m_{3b}=2; m_{4a}=4, m_{4b}=3.

3 Mechanical model and simulations

Mechanical model of the right arm exoskeleton is build up, according to the kinematics scheme shown in Fig. 1. The kinematics structure includes 4 rotation joints with h = 4 DoF. Independent parameters for evaluation are the rotations in the four joints q_1, q_2, q_3 and q_4, corresponding to the four basic motions of the upper limb: shoulder abduction/adduction, shoulder flexion/extension, elbow flexion and shoulder med./lat. rotation. Reference frames are selected in the manner shown in Fig.4.

Coordinate systems (c.s.) $X_0Y_0Z_0$,..., $X_4Y_4Z_4$, are attached to all links, and $X_5Y_5Z_5$ is to the end effector (EE). In the initial position, all local c.s. are selected to be in the same direction for convenience, and one of the axes have a rotating joint. The rotation is set with the vector $\varphi_i = | \varphi_{ix}, \varphi_{iy}, \varphi_{iz} |^T$, one of whose components is equal to the angle q_i of the selected axis X, Y or Z and the other two in this case are zeros, but the model can be used also for joints with 3 rotations at the same time if needed. The position of each $X_iY_iZ_i$ c.s. is set with the displacement vector $l_i = | l_{ix}, l_{iy}, l_{iz} |^T$ respectively to the former. Two of the units are L-shaped and are therefore given in two lengths. Thus the selected kinematic description parameters are given in Table 1.

Table 1. Parameters used to describe the exoskeleton arm model.

c.s.№	l_{ix}	l_{iy}	l_{iz}	φ_{ix}	φ_{iy}	φ_{iz}
1	**0**	**0**	**-L_2**	**0**	**0**	**q_1**
2	**-L_1**	**0**	**+L_2**	**q_2**	**0**	**0**
3	**0**	**-L_3**	**0**	**q_3**	**0**	**0**
4	**L_5**	**0**	**-L_4**	**0**	**0**	**q_4**
5	**0**	**-L_6**	**L_7**	**0**	**0**	**0**

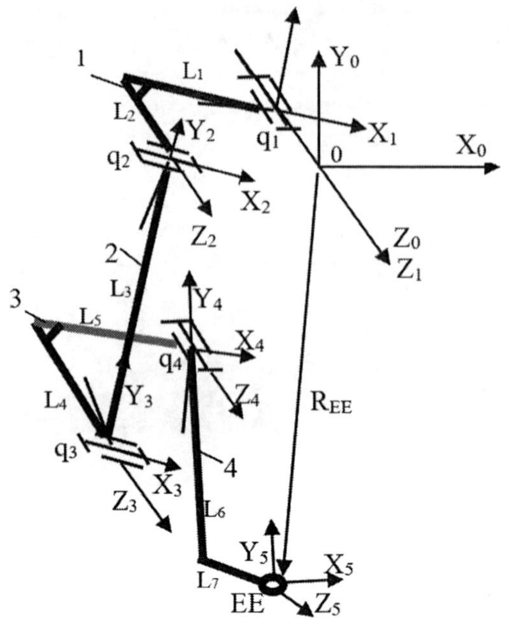

Fig.4. Kinematics scheme with reference frames of the upper limb exoskeleton.

The lengths are chosen: L_1= 0.1; L_2= 0.150; L_3= 0.286; L_4= 0.130; L_5= 0.095; L_6= 0.37; L_7= 0.130 m. Table 1 shows coordinate systems numbers as well as the parameters of displacement and rotation. With this data the transformation (4x4) matrices $T_{i-1,i}$, i= 1,…,5 are compiled, which derive the position of frame 5 with respect to frame 0.

$$\mathbf{T}_{0,5} = \mathbf{T}_{0,1} \ldots \mathbf{T}_{4,5} \, , \tag{1}$$

The joints parameters are accepted to be the generalized parameters:

$$\mathbf{q} = [q_1, \ldots, q_h]^T \tag{2}$$

The coordinates of the end-effector are presented by the vector

$$\mathbf{X} = [X_1, \ldots, X_v]^T, v \leq 6 \tag{3}$$

The relations between the generalized parameters (1) and the coordinates of the end-effector (2) are set by the direct kinematics. Regarding positions and velocities, we formulate that by the following equalities:

$$\mathbf{X} = \Psi(\mathbf{q}) \tag{4}$$

$$\dot{\mathbf{X}} = \mathbf{J}\dot{\mathbf{q}} \tag{5}$$

Here

$$\mathbf{J} = \left[\frac{\partial \mathbf{X}}{\partial \mathbf{q}} \right] \tag{6}$$

is the (v x h) Jacobian matrix.

Assuming that the operator hand is connected to a point EE, we will further consider exoskeleton positions only, and we will accept the operation space as 3 dimensional (v=3). The vector of end-effector coordinates (3) will be (3 x 1) dimensional, and Jacobian (6) will

be (3x4) dimensional. A MATLAB application was created to calculate the components of the matrices $\mathbf{T}_{i-1,\,i}$, and uses MATLAB function **Jacobian** ([τ_{14}, τ_{24}, τ_{34}], [q_1, q_2, q_3, q_4]) to calculate the private derivatives and finds the components of **J**.
Denote by

$$\mathbf{Q} = [Q_1,\ldots,Q_h]^T \tag{7}$$

the (h x 1) vector of the generalized torques in the joints in the basic chain corresponding to the generalized parameters (2) and by

$$\mathbf{F} = [F_x, F_y, F_z]^T \tag{8}$$

the (3 x 1) vector of the external forces applied at the end effector and corresponding to the coordinates (3) of the end effector. The link between the external forces (8) and the effective generalized torques (7) can be defined according to the principle of virtual work as follows

$$\mathbf{Q} = \mathbf{J}^{\mathbf{T}}\mathbf{F} \tag{9}$$

In virtual reality tasks, the vector of the external forces (8) represents the desired force on the end effector, when user's movements correspond to contact with a virtual surface.

The gravity of exoskeleton links and the gravity of human arm influences on the behavior of the end-effector. The (4 x 1) vector

$$\mathbf{G} = [G_1,\ldots,G_4]^T \tag{10}$$

is the vector of gravity torques, generated at the exoskeleton joints. Its components:

$$G_j = \sum_{i=1}^{n} m_i g_o^T \frac{\partial \rho_i}{\partial q_j}, \; j = 1,\ldots,h \tag{11}$$

are determined by the mass of links m_i , i=1…n, the gravity acceleration vector in the base frame $g_0=[g_x,g_y,g_z]^T$ and the position vector of the i-th link centre of mass in the base frame $\rho^i= [\rho^x, \rho^y, \rho^z]^T$.

To make the operator feel force reflection from the virtual environment, an open-loop force controller with feedforward gravity compensation is selected [2]. This type of force control is called open-loop because there is no force feedback from the device to the controller to regulate the force output of the exoskeleton end effector as it is illustrated by the control block diagram in Fig. 5.

Virtual Engine in the controller on Figure 5 generates a desired force command $\mathbf{F_d}$ used to calculate force command in the exoskeleton joints. A software block **G**(q) implementing gravity compensation has been introduced in order to take into consideration the effects of the exoskeleton weight in the calculation of the joint force commands. Generalized torques in the joints as joint force commands are given by the following equation

$$\mathbf{Q} = \mathbf{J}^{\mathbf{T}}\mathbf{F_d} + \mathbf{G(q)} - \mathbf{Q_H} \tag{12}$$

Forces that human exerts on the exoskeleton end effector $\mathbf{Q_H}$ represent a physical, not control input to the device. According to control block diagram on Fig. 6, the vector representing force command in joint space $\mathbf{Q} = [Q_1,\ldots,Q_4]^{\mathrm{T}}$ includes the joints torques

$$Q_j = (P_{jb} - Pj_a)r, \; j=1, \ldots ,4 \tag{13}$$

where P_{jb}, P_{ja} are the driving forces of the pneumatic actuators and r is the pulley radius [11].

The driving forces can be presented by the equation:

$$P = (k_o + tp)(L_n - C)(C_{max} - C) \tag{14}$$

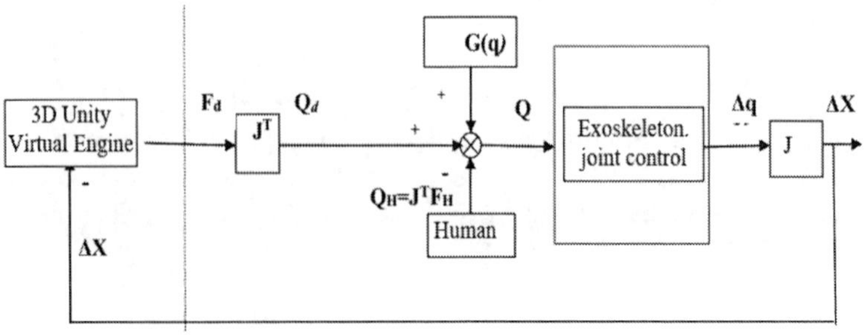

Fig.5. Block scheme of force control with feedforward gravity compensation.

They are a function of the muscle contraction C, as well as a function of the pressure p, the nominal muscle length L_n and the number of muscles in the bundles m, determined by the coefficients k_0 and t, where k_0 = -39 + 321 m, t = -0.21+ 4.41 m. Equations (13), (14) allow the calculation of the pressure for the pneumatic actuators in one joint. To obtain maximum torque in the joint, it is appropriate one of the actuators (antagonist) to be passive (doing zero force) and the pressure of the second actuator to be controlled. When changing the direction of the force, the actuators change their roles.

4 Experiments for ensuring force reflection from VR

An application of the exoskeleton as a VR haptics device is further discussed. 3D Unity Virtual Engine is used for data transfer between the exoskeleton and the avatar in a virtual scenario. Optical tracker, head-mounted display Oculus Rift is used for visualization of the virtual system and to track the head position and orientation. The exoskeleton is designed to track the position and orientation of the human body and to provide force feedback to the human arms. Fig. 6 shows a scenario of virtual gymnastics when: a) - an exercise "Lifting dumbbells" is visualized; b) a scheme of the right-hand motions is presented corresponding to positions in computer graphics. The controller in Figure 5 creates power commands in the joints of exoskeleton, that form the force action on the user's hands, equivalent to the power impact simulated in the virtual scene. In the scene of virtual gymnastics, power influence is the weight of the dumbbells which is applied to the hand. The magnitude of this force is a constant, generated in virtual machine.

Simulations were conducted for assessing the developed exoskeleton ability to provide forceful impact on the operator, in the described scene of virtual gymnastics. Experiments were carried out in the following manner. A desired force $\mathbf{F_d}$ on axis Y is set, simulating the force which exoskeleton exerts on the human hand. Simulations are carried out as exoskeleton arm performs lifting movements in OYZ plane as it is shown on Fig.6 b). For this purpose an experiment was performed with the real exoskeleton, moving up and down along the Y axis, while the joint angles and the position of the end-effector were recorded. Torques in the exoskeleton joints (12), (13) were calculated by taking into account the desired force and the gravitational component (10). For each joint the pressure of the active muscle actuator is calculated according to (14), when the pressure of the passive one is set zero.

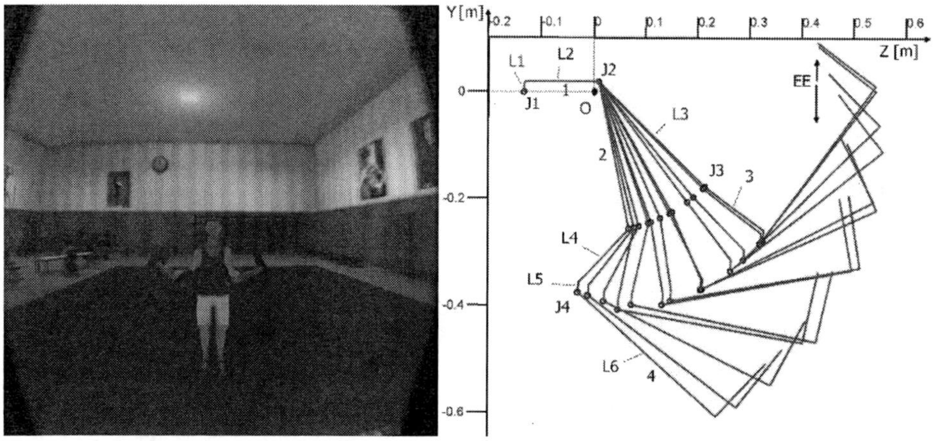

Fig. 6. Scene of virtual gymnastics: a) an exercise "Lifting dumbbells"; b) scheme of the arm motion.

A Matlab simulation model was developed for calculating the joint pressures corresponding to different forces applied to the endefactor along the axis Y for the given positions. For each joint it is calculated for three cases: for demanding lowest possible pressure; for average pressure; or for maximal pressure. For the first case, this is achieved by calculating the minimum pressure, so calculate the minimum forces in the muscles P_{min} and then the joint muscle force command $Pj = Qj / r$ is added to the minimum forces P_{min} of the muscle agonist, and only P_{min} is applied to the other muscle antagonist , for example

$$P_b = P_{a\,min} + P_j, \qquad P_a = P_{a\,min} \tag{15}$$

or
$$P_b = P_{b\,max}, \qquad P_a = P_{b\,max} - P_j. \tag{16}$$

In this paper, the minimum required pressure is considered. The minimum force is applied to the passive muscle and the force of the active muscle is calculated according to (15) and its pressure according to (14).

Fig. 7 shows the variation of the calculated pressure in the active muscles of the four joints for the arm movement from Fig. 6 (b) "lift up-down" by Y axis. On Fig. 7 a) the set force F_y is 24 N and in Fig. 7b) it is 16N. As can be seen from Fig.7, at F_y = 24 N, the pressure in the second joint exceeds the maximal possible supply pressure of 500 kPa and the desired output can not be performed.

At F_y = 16 N, pressures in all joints are below the 500 kPa limit. The simulation shows that setting a force from the virtual system towards the exoskeleton have to be done only after evaluating the possibility of executing it. The performance also depends on the position of the hand. If this is not evaluated, the force effect on the operator will be different than the desired in the direction and value.

In the next experiment, the effect of the virtual desired end-efactor force on the muscles action and on the resulting real end-effector force were researched. The experiment is carried out in a position of the end-effector (X = -0.158; Y = -0.073; Z = 0.550 m), with a force sensor on the Y axis and a computer interface. The desired force 25 N is set in one step and the resultant force is recorded over time (Fig. 8).

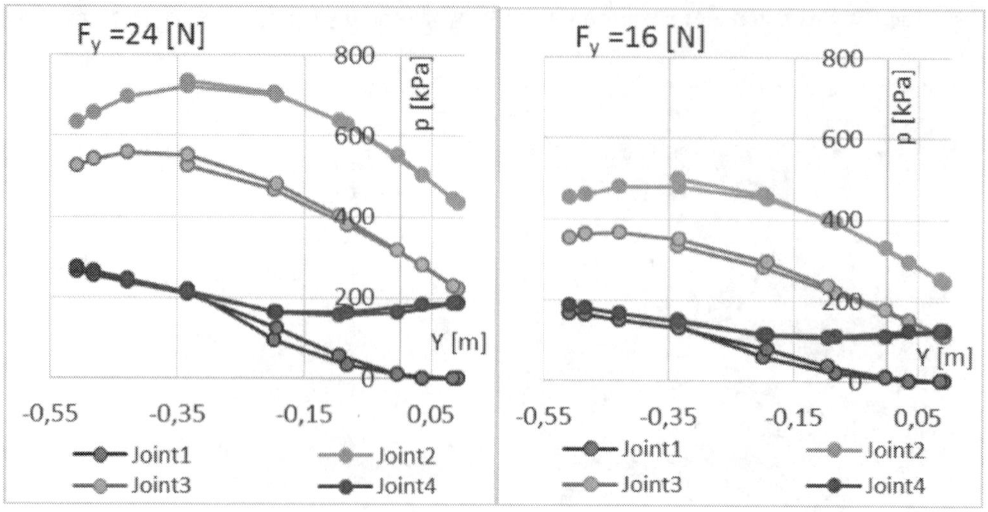

Fig. 7. Calculated pressure variation in the four joints actuators while the hand performs lifting movements, for: a) the set end-efector force is 24 N, b) the set force is 16 N.

The measured real force on the Y-axis of the end-efactor changes with a delay, which is determined by the characteristics of the valves and the pneumatic feed system. The joint pressure variation is measured, as shown in Fig.10. As can be seen from the graph, in the selected position the desired force of 25N is well achieved and the calculated values of the active muscle pressures are in the allowable limits of the power supply.

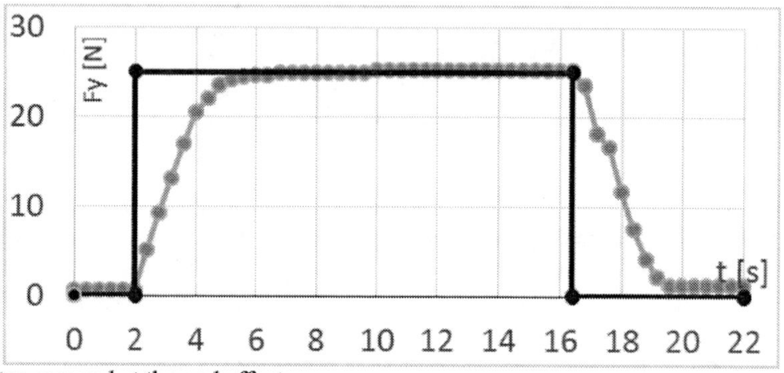

Fig. 8. Force step respond at the end effector.

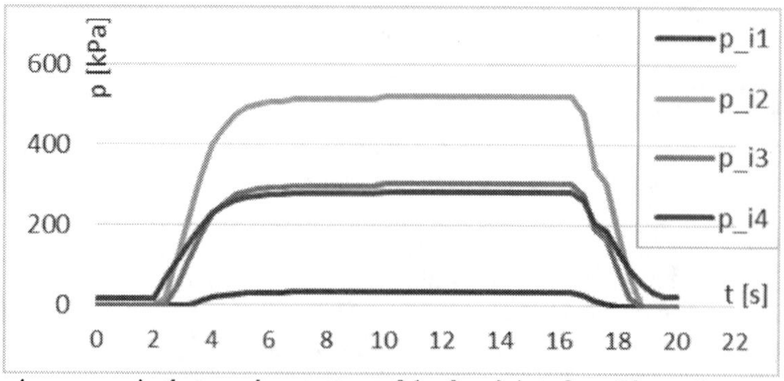

Fig. 9. Variation in pressure in the muscle actuators of the four joints from Figure 8.

6 Conclusions

In this work, the development of a new exoskeleton of the upper limb is presented. The choice of a mechanical structure is kinematics equivalent to the human arm. A drive system is selected based on braided pneumatic muscle actuators and antagonistic drive scheme for each joint. In the work, a mechanical model of the exoskeleton arm as haptic device is built up and a force control scheme is selected for force reflection at the arm. Computer experiments have been carried out to provide force reflection by VR, according to scenario of virtual gymnastics. The conducted simulations show the possibility of developed exoskeleton to provide force reflection on the hand of the operator. The experiment shows that the force assignment must be made after an assessment of the possibility of the exoskeleton to execute it at a given feed pressure and position. Experiments with real exoskeleton prototype were conducted, where force step respond was recorded, which confirms the conclusions of the simulations.

This work was funded by the Bulgarian Science Found, Call: 2016, through Project AWERON – DN 07/9, to which the authors would like to express their deepest gratitude.

References

1. J. Perry, J. Rosen, S. Burns, IEEE/ASME Transactions on Mechatronics **12**, 408–417 (2007)
2. M. Bergamasco, B. Allotta, L. Bosio, L. Ferretti, G. Perrini, G. M. Prisco, F. Salsedo, G. Sartini, IEEE Int'l Conf. Robot. Automat. **2**, 1449–1454 (1994)
3. A. Frisoli, F. Rocchi, S. Marcheschi, A. Dettori, F. Salsedo, M. Bergamasco. A new force-feedback arm exoskeleton for haptic interaction in virtual environments, *WHC 2005. First Joint Eurohaptics Conference and Symposium on Haptic Inter-faces for Virtual Environment and TeleoperatorSystems,* 195–201 (2005)
4. A. Frisoli, F. Salsedo, M. Bergamasco, B. Rossi, M.C. Carboncini, Applied Bionics and Biomechanics **6**, 115–126 (2009)
5. S. Lee, S. Park, M. Kim, C.-W. Lee, *IEEE Conference on Robotics and Automation,* 2574–2579 (1998)
6. Y. Jeong, Y. Lee, K. Kim, Y. Hong, J. Park, A 7 DOF Wearable Robotic Arm using Pneumatic Actuators," *in Proceedings of the 32nd ISR International Symposium on Robotics* April 2001, 388-393 (2001)
7. N. G. Tsagarakis, Dg. Caldwell, Autonomous Robots **15**, 21–33 (2003)
8. N. G. Tsagarakis, Dg. Caldwell, Development and control of a physiotherapy and training exercise facility for the upper limb using soft actuators', *Proc. of the IEEE International Conference on Advanced Robotics,* 1092–1097 (2003)
9. M. Slater, M. Sanchez-Vives, IEEE Computer **47**, 24-30 (2014)
10. M. Slater, B. Spanlang, M. Sanchez-Vives, O. Blanke, PLoS One **5**, e10564 (2010)
11. D. Chakarov, I. Veneva, M. Tsveov, D. Trifonov, Design and Control of a Force Reflecting Arm Exoskeleton for Virtual Reality Applications. *Proc.of the 14th International Conference on Informatics in Control, Automation and Robotics"ICINCO 2017*", Madrid, Spain, **2**, 335-342 (2017)

MATEC Web of Conferences **145**, 01007 (2018) https://doi.org/10.1051/matecconf/201814501007
NCTAM 2017

Statistical analysis of ambient air PM10 contamination during winter periods for Ruse region, Bulgaria

Irina Tsvetanova[1], *Ivanka* Zheleva[1,*], *Margarita* Filipova[1], and *Antoaneta* Stefanova[1]

[1] Department Thermo engineering, Hydraulics, Ecology, Ruse University, 7017, Studentska St. 8, Ruse, Bulgaria

Abstract. The sources of dust on the territory of Ruse region, Bulgaria are industry, transport and domestic heating by solid fuel. PM10 (particulate matter with a diameter between 2.5 and 10 μm) levels for Ruse mark a significant increase during the autumn-winter period compared to the levels during the spring-summer period [1]. Obviously there is a relationship between PM10 contamination levels and ambient air temperature. The lower the temperatures are, the higher the PM10 levels are. The biggest peak of PM10 levels for the autumn-winter period is usually observed in January months. It is in January that the number of days in which there is exceedance of the limit values of the PM10 levels is maximum observed. Also in January months the day and night temperatures are the lowest and usually they do not pass 0°C for many days of the month. To understand better this relationship we provide a statistical analysis of ambient air PM10 contamination during winter periods. Correlations between the measured PM10 values and the respective temperatures measured for January months for different years are presented. Descriptive statistics of PM10 and some atmospheric characteristics as well as linear regression analysis are calculated and commented in the paper.

1 Introduction

According to European Commission surveys, every year poor air quality causes the premature death of more people than the road accidents. Ambient air contamination by fine particulate matter (PM10) mainly is result of human activities - transport, industry and domestic heating. Nowadays this pollutant cause more and more respiratory problems, asthma, lung cancer and premature death.

The problem of PM10 air pollution is very serious. For Bulgaria it is clear that sources of PM10 pollution come mainly from old cars usage and from solid fuel stoves for heating. The combination of these factors leads to a worsening of the ecological situation which consequently leads to serious damages to human health.

Besides the above mentioned human factor also some climate and meteorological conditions probably contribute to the increase of PM10 levels. It is known that the factors which affect PM10 levels are air temperature, humidity, atmospheric pressure, presence or absence of wind, wind direction and so on.

The relationship between the PM10 levels measured by the automated measuring station in Ruse, Bulgaria and the ambient air temperatures for several January months is analyzed in this paper. The January months of the period 2011 – 2017 are chosen, as most of the days

* Corresponding author: izheleva@uni-ruse.bg

with the exceedance of the PM10 limit values are usually observed during this period of the year [1].

2 Material and methods

The period selected for the purpose of the survey covers 7 recent years (since 2011 to 2017).

We use official data for measured PM10 concentrations for Ruse, Bulgaria, for January months (2011 – 2017) published by Regional Inspectorate of Environment and Water [3]. We use SPSS software to analyze the data measured in January months in Ruse, Bulgaria for the period 2011 – 2017.

3 Results and comments

On the Figures 1 - 7 the measured levels of PM10 for Ruse are presented [3]. It is obvious that the characteristics of these levels are quite different and there are many days each year when the PM10 level is much more than the daily limit value (50 μg /m^3)

All graphs (Fig. 1 – Fig.7) are summarized on Fig 8.

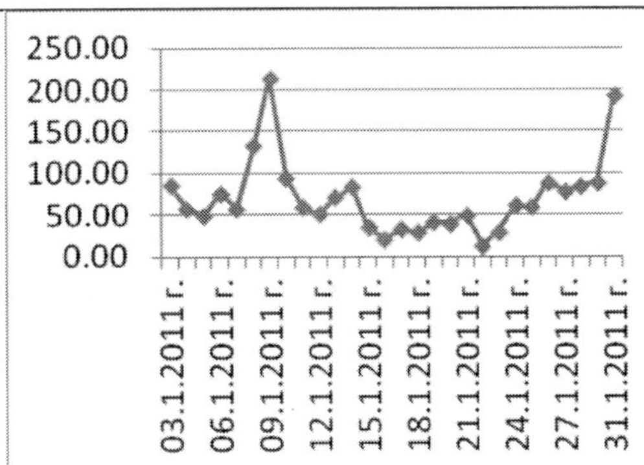

Fig. 1 PM10 concentration (μg /m^3) for January 2011

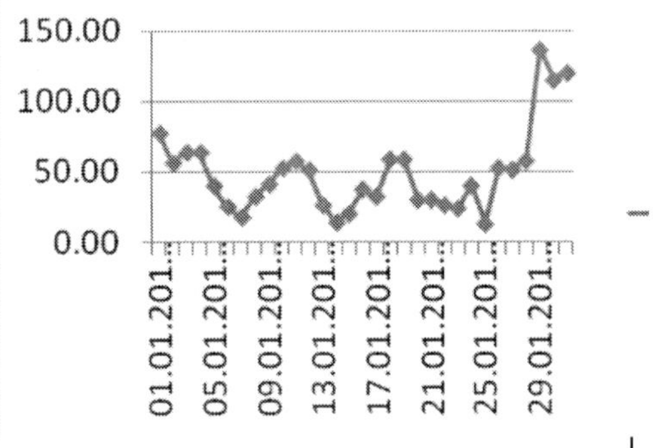

Fig.2 PM10 concentration (μg /m^3) for January 2012

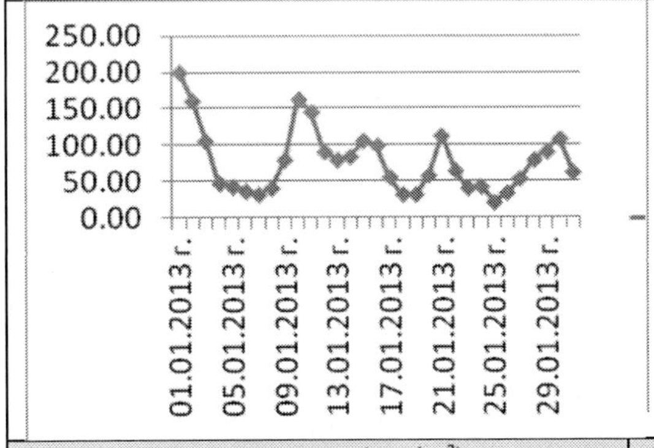

Fig. 3 PM10 concentration (μg /m^3) for January 2013

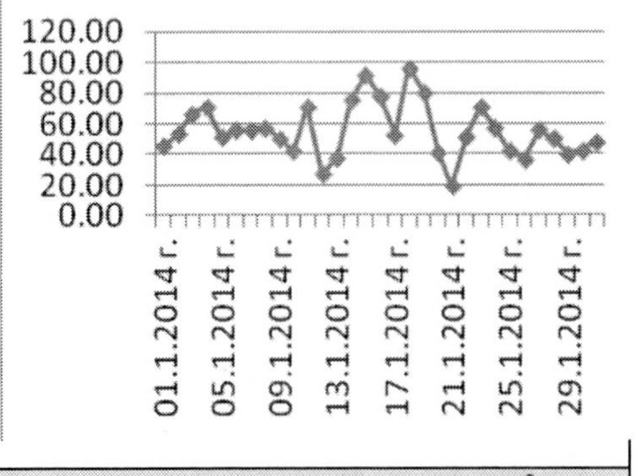

Fig. 4 PM10 concentration (μg /m^3) for January 2014

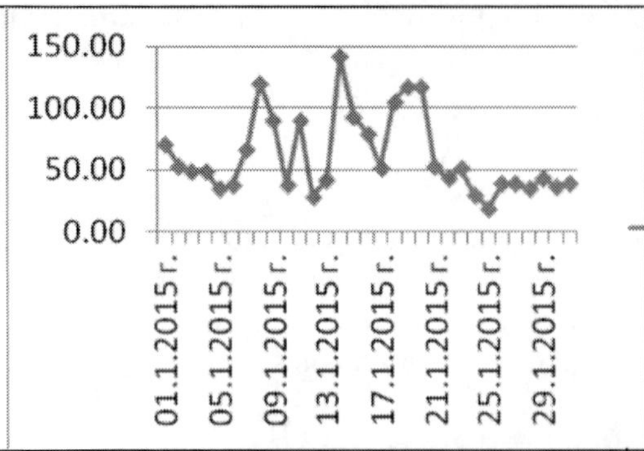

Fig. 5 PM10 concentration (μg /m³) for January 2015

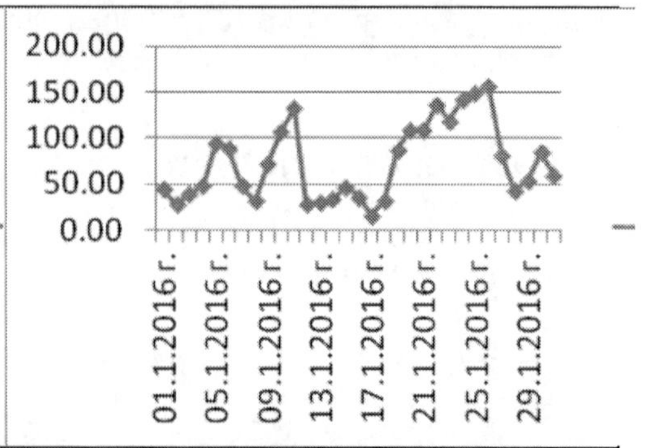

Fig. 6 PM10 concentration (μg /m³) for January 2016

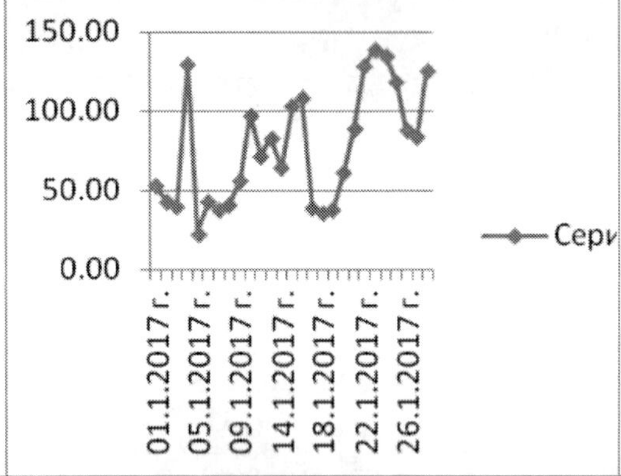

Fig. 7 PM10 concentration (μg /m³) for January 2017

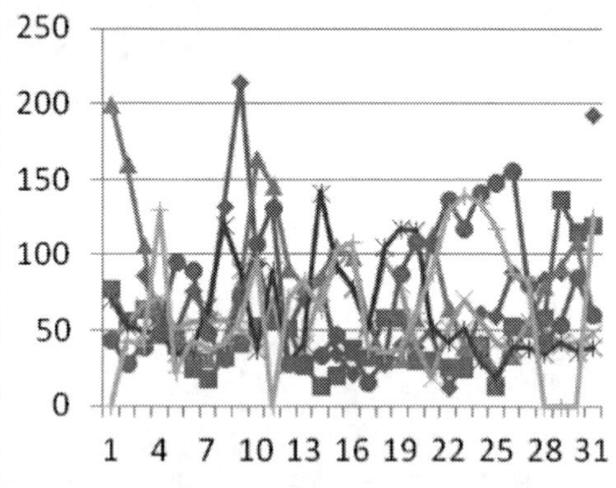

Fig. 8 Daily average PM10 levels (μg /m³) for January months for the period 2011 – 2017

On the next Table 1 is shown the descriptive statistics of PM10. It is very important that the mean value of the PM10 for the period 2011 – 2017 for Ruse is greater than the limit value of PM10 (50 μg /m³). The maximum PM10 value is more than 4 times bigger than the limit value.

Table 1. Descriptive Statistics for PM10

	N	Minimum	Maximum	Mean	Std. Deviation
PM10	210	12,50	214,10	66,1521	37,54768
Valid N (listwise)	210				

The average number of days in January that PM10 levels exceeded the limit value for PM10 is 18 days per a month for the period 2011 - 2017. During 2011, 2013 and 2014 20 days have been reported with PM10 concentration measured bigger than the limit value. During 2012 and 2015 only 15 days were registered with PM10 concentration greater than the limit.

The average excess of the PM10 (50 μg / m³) is 0.86 times for the period.

The largest exceedance of the limit was recorded on 9.01.2011 - 214,1 μg /m³ which is 4,282 times greater than the norm.

In the years with the highest number of days when PM10 levels are above the limits the average minimum air temperature was -3,13°C and the mean maximum air temperature was 3,11°C.

In the years with the lowest number of days when PM10 levels are above the norm the average minimum temperature was -2.45⁰C and the average maximum temperature was 4.14⁰C.

We will examine next the daily average temperatures, measured in Ruse for January months for the period 2011 – 2016, because we suppose that there is a relationship between the PM10 levels and ambient air temperature.

On the Fig.9 the daily average temperatures for January months for Ruse for the period 2011 – 2016 are presented. On Table 2 the descriptive Statistics of these temperatures are presented, too.

The temperatures for the period have small negative mean value, while the interval between the minimum and maximum values is almost 25⁰C.

Next we calculate Pearson Correlation coefficients between PM10 levels and the measured temperatures.

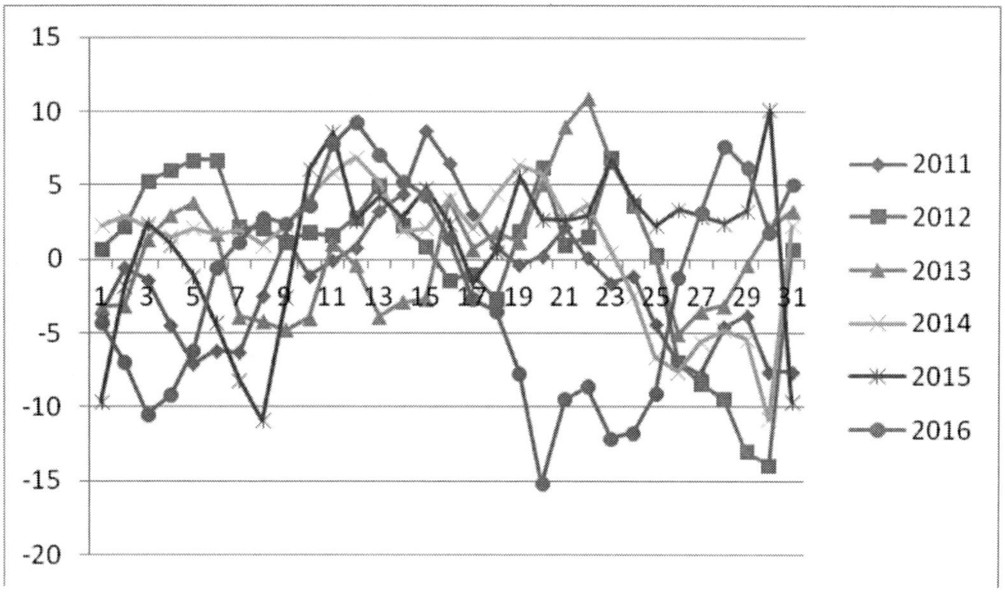

Fig. 9 Daily average temperatures for Ruse for January months for the period 2011 – 2016 [2].

Table 2. Descriptive Statistics for temperature (⁰C) for Ruse for January months for the period 2011 – 2016

	N	Minimum	Maximum	Mean	Std. Deviation
Temperature	186	-15,20	10,90	-0,0177	5,23151
Valid N (listwise)	186				

On the Table 3 the correlation coefficients between PM10 levels and temperature measured for January months for Ruse for the different years can be seen.

Only for 2012 and 2016 correlations are significant. Pearson correlation coefficient for 2012 is -0,574** and it is significant at the 0.01 level (2-tailed) and Pearson correlation coefficient for 2012 is -0,398* and it is significant at the 0.05 level (2-tailed).

If we calculate the Pearson correlation coefficient for the whole period we receive the result, shown on the next Table 4. The Pearson correlation coefficient for the whole period 2011 – 2016 is -0,283** and it is significant at the 0.01 level (2-tailed). As the Pearson correlation coefficient is negative number then PM10 levels grow up when the temperature go down.

Finally we will develop a linear regression model for PM10 levels (Y) and the ambient air temperature (X) also using SPSS software. The result of the modeling is presented on the next Tables 5 – 7 and on Fig.10.

Table 3. The correlation coefficients between PM10 levels and temperature measured for January months for Ruse for the different years

Correlations for 2011

		X1PM	X1T
X1PM	Pearson Correlation	1,000	-0,308
	Sig. (2-tailed)		0,111
	N	28,000	28
X1T	Pearson Correlation	-0,308	1,000
	Sig. (2-tailed)	0,111	
	N	28	31,000

Correlations for 2012

		X2PM	X2T
X2PM	Pearson Correlation	1,000	-0,574*
	Sig. (2-tailed)		0,001
	N	31,000	31
X2T	Pearson Correlation	-0,574**	1,000
	Sig. (2-tailed)	0,001	
	N	31	31,000

Correlations for 2013

		X3PM	X3T
X3PM	Pearson Correlation	1,000	-0,186
	Sig. (2-tailed)		0,318
	N	31,000	31
X3T	Pearson Correlation	-0,186	1,000
	Sig. (2-tailed)	0,318	
	N	31	31,000

Correlations for 2014

		X4PM	X4T
X4PM	Pearson Correlation	1,000	0,237
	Sig. (2-tailed)		0,200
	N	31,000	31
X4T	Pearson Correlation	0,237	1,000
	Sig. (2-tailed)	0,200	
	N	31	31,000

Correlations for 2015

		X5PM	X5T
X5PM	Pearson Correlation	1,000	-0,111
	Sig. (2-tailed)		0,552
	N	31,000	31
X5T	Pearson Correlation	-0,111	1,000
	Sig. (2-tailed)	0,552	
	N	31	31,000

Correlations for 2016

		X6PM	X6T
X6PM	Pearson Correlation	1,000	-0,398**
	Sig. (2-tailed)		0,027
	N	31,000	31
X6T	Pearson Correlation	-0,398*	1,000
	Sig. (2-tailed)	0,027	
	N	31	31,000

* Correlation is significant at the 0.01 level (2-tailed).
** Correlation is significant at the 0.05 level (2-tailed).

Table 4. The correlation coefficients between PM10 levels and temperature measured for January months for Ruse for the period 2011 - 2016

		PM10	Temperature
PM10	Pearson Correlation	1,000	-0,283**
	Sig. (2-tailed)		0,000
	N	210,000	183
Temperature	Pearson Correlation	-0,283**	1,000
	Sig. (2-tailed)	0,000	
	N	183	186,000

** Correlation is significant at the 0.01 level (2-tailed).

Table 5. Model Summary

Model	R	R Square	Adjusted R Square	Std. Error of the Estimate
1	0,283[a]	0,080	0,075	36,08507

a. Predictors: (Constant), Temperature

Table 6. ANOVA[b]

Model		Sum of Squares	df	Mean Square	F	Sig.
1	Regression	20483,376	1	20483,376	15,731	0,000[a]
	Residual	235685,948	181	1302,132		
	Total	256169,323	182			

a. Predictors: (Constant), Temperature

b. Dependent Variable: PM10

Table 7. Coefficients[a]

Model	Unstandardized Coefficients		Standardized Coefficients	t	Sig.
	B	Std. Error	Beta		
1 (Constant)	64,664	2,668		24,240	0,000
Temperature	-2,026	0,511	-0,283	-3,966	0,000

Dependent Variable: PM10

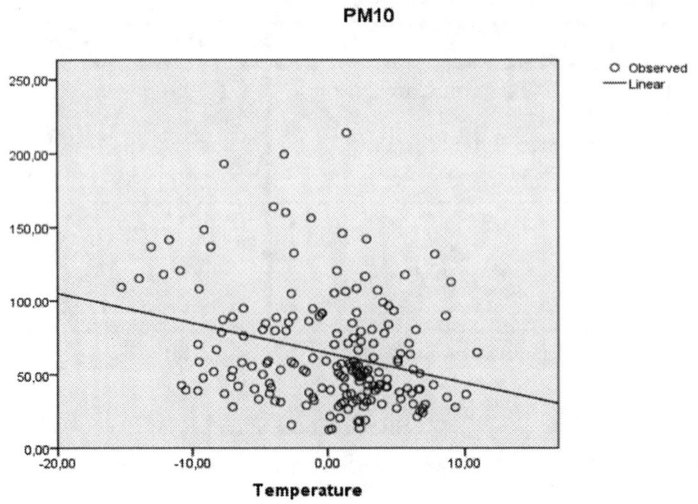

Fig.10. The linear regression model plot (the line) and measure PM10 levels as a function of the temperature (the points)

Therefor the linear regression model is Y = -2,026 + 64,664.X. In the model the both coefficients are statistically significant.

4 Conclusions

There are many factors that influence PM10 levels in the ambient air. Some of them do not depend on human activities but others, such as the use of solid fuel for domestic heating, the industry and the transport affect the PM10 content and can be controlled by man.

For better understudying of the processes which cause contamination of ambient air with PM10 and to realize which are the factors mostly influence the PM10 levels more scientific researches have to be provided.

In order to solve at least partly the air pollution problem, different solutions can be applied. For example by investments in newer urban transport so that the urban transport can be improved to meet better the needs of the population and become a preferred form of transport in the cities and be chosen in front of private cars. Also it is clear that the heat supply in urban zones has to be improved, expand the district heating and gasification network so that people can use their services rather than using solid fuel stoves. Also under certain non-favorable weather conditions, temporary restrictions can be introduced to reduce pollution levels. For example, in such days the use of private cars could be limited.

References

1. I. Zheleva, E..Veleva, M. Filipova Analysis and modeling of daily air pollutants in the city of Ruse, Bulgaria, (2017), AMITANS, 2017 (in print).
2. http://www.stringmeteo.com/synop/bgold_stday.php
3. http://www.riosv-ruse.org/mesechna-spravka-za-nivata-na-fpch10.html
Data on the measured PM10 concentrations by automated measuring station "Vazrazhdane" Ruse (January, 2011; January, 2012; January, 2013; January, 2014; January, 2015; January, 2016; January, 2017).

On the effects from the simultaneous occurrence of the critical Casimir and dispersion forces between conical colloid particle and a thick plate immersed in nonpolar critical fluid

Galin Valchev[1,*]

[1] *Institute of Mechanics, Bulgarian Academy of Sciences, Acad. G. Bonchev St., Block 4, 1113 Sofia, Bulgaria.*

Abstract. Here we study the interplay between the van der Waals (vdWF) and critical Casimir forces (CCF), as well as the *total* force (TF) between a conical colloid particle and a thick planar slab. We do that using general scaling arguments and mean-field type calculations utilizing the so-called "surface integration approach", a generalization of the well known Derjaguin approximation. Its usage in the present research, requires knowledge on the forces between two parallel slabs, confining in between some fluctuating fluid medium characterized by its temperature T and chemical potential μ. The surfaces of the colloid particle and the slab are assumed coated by thin layers exerting strong preference to the liquid phase of a simple fluid, or one of the components of a binary mixture, modeled by strong adsorbing local surface potentials, ensuring the so-called $(+,+)$ boundary conditions. On the other hand, the core region of the slab and the particle, influence the fluid by long-ranged competing dispersion potentials. We demonstrate that for a suitable set of colloid-fluid, slab-fluid, and fluid-fluid coupling parameters the competition between the effects due to the coatings and the core regions of the objects, result, when one changes T or μ, in *sign change* of the Casimir force (CF) *and* the TF acting between the colloid and the slab. Such an effect can provide a strategy for solving problems with handling, feeding, trapping and fixing of microparts in nanotechnology.

1 Introduction

In a recently published work [1], it was shown that the *total* force (TF) between a spherical colloidal particle and a thick planar substrate, as well as between two spherical particles immersed in some critical nonpolar medium, exhibits not only quantitative, but also qualitative control by simple change of some external to the system parameters like the temperature T and the chemical potential μ of the fluid or the minimal separation L between the interacting objects. This study was based on the findings in an earlier research [2], related to the emergence and interplay between the CCF and dispersion vdWF within the gap formed by couple of parallel planar plates, confining a nonpolar fluid which is thought to be at the vicinity of its bulk critical point. There it was shown that the supposed interplay leads to sign change of the net interaction, due to variation of any of the aforementioned external parameters, when the film thickness is less than some threshold value L_{crit}.

*e-mail: gvalchev@imbm.bas.bg

In the present article we are going to be interested in the simultaneous manifestation of both the CCF and the vdWF, which gives rise to the *total* force F_{tot} between a tip like object, say the ending of a cantilever of an atomic force microscope, and a thick smooth plate. In particular, we will resemble such an object with a conical in shape colloidal particle with mesoscopic radius R and height H, immersed in a medium that is either simple fluid or a binary liquid mixture. The mutual orientations which shall be considered include: base and/or apex of the cone facing the plate, together with the case when the height of the cone is parallel to the plane bounding the plate. In all three geometries the minimal surface-to-surface approach between the objects will be designated by L. Furthermore we shall envisage that the fluid is in the vicinity of its bulk critical point, with the critical temperature T_c interpreted either as that at the liquid-vapor critical point of a simple fluid or as that at the critical demixing point of a binary liquid mixture. When such conditions are present, as first suggested by Fisher and de Gennes [3], an additional component adds up to the already acting forces if any, resulting from the critical fluctuation of the medium. This new contribution is of a long-ranged character and strongly depends on the boundary conditions which the bodies immersed in the medium impose on it at their surfaces. Since this fluctuation-induced force (FIF) depends additionally only on some gross features of the fluid medium [4, 5] it can be treated as *universal* in nature. One can make a parallel with the force between neutral bodies due to the quantum and temperature fluctuations of the charge distributions in them, i.e., of the electromagnetic field, which force is known today under the general name of quantum electrodynamic (QED) Casimir force [6–8]. Because of this analogy it became customary to term the fluctuation part F_{Cas} of the total force near T_c *critical Casimir force* [4, 5].

In a medium where both quantum and thermodynamical fluctuations are present, F_{tot} can be thought as a sum of a regular background $F_{\text{tot}}^{(\text{reg})}$ and a singular $F_{\text{tot}}^{(\text{sing})}$ contributions. The first depends in an analytic way on the parameters characterizing the medium, whereas the latter emerges due to the critical fluctuations of the medium, respectively. Therefore

$$F_{\text{tot}} = F_{\text{tot}}^{(\text{reg})} + F_{\text{tot}}^{(\text{sing})}. \tag{1}$$

Undoubtedly, if a system is composed out of objects, where at least one has nonplanar geometry, its excess grand canonical potential $\Omega_{\text{ex}}(T,\mu|L,\cdots)$, or excess free energy, will show dependance on some specific geometrical characteristics like L, R and H in the case considered here. Thus we have

$$F_{\text{tot}} = -\frac{\partial}{\partial L}\Omega_{\text{ex}}(L|T,\mu,\cdots). \tag{2}$$

It is normal to equate

$$F_{\text{Cas}} \equiv F_{\text{tot}}^{(\text{sing})}(L|T,\mu) \quad \text{and} \quad F_{\text{vdW}} \equiv F_{\text{tot}}^{(\text{reg})}(L|T,\mu), \tag{3}$$

since the dispersion van der Waals interactions, ubiquitous for any system, are not influenced by the thermodynamic fluctuations.

In the past few decades the rapid development of the experimental techniques led to the emergence of the field of nanotechnology, where fluctuation-induced phenomena, bearing in mind the vdWF and QED CF, play a dominant role between neutral nonmagnetic objects when placed at distances less than a micrometer. The general theory of the FIFs with quantum character together with the subsequent experiments for its validation, have shown undeniably that these forces lead to irreversible, usually undesirable phenomena, such as stiction (i.e., irreversible adhesion) or pull-in due to mechanical instabilities [9–11] when the interactions take place in vacuum or gas medium. Moreover the weight of a particle becomes negligible when its size is scaled below a micrometer. As a result, when one tries to release such a neutral particle from, say, the surface of whatever handling device in air or vacuum, the particle will not drop down under the gravity but, instead, will stick to the surface due to the effect of the omnipresent vdWF. In that respect it seems adequate to introduce some fluid medium, which properties can lead to modification of the net interaction between the manipulated particle and the operating device, sometimes called gripper, in such a way that it is repulsive at short distances between the handling surfaces and the particle and attractive at larger ones. In the light of a potential experimental realization the choice of a chemically inert fluid medium absent of free changes dissolved in it seems suitable, since from one side any complications with isolation of all possible circuitry at small distances will be minimized, and on the other no chemical reactions with the substances of the particles will take place. That leads us to choose as a fluid a nonpolar liquefied noble gas that has critical parameters as close as possible to the normal ones.

In the current article we will demonstrate that by proper choice of the materials (cores) of the colloid particles and the handling surface of a gripper it is indeed possible to achieve control over the net interaction (TF) as well as the CCF between the surface and the particle by simply changing T and μ. We must note that so far the behaviour of the CCF has never been studied in the geometry under consideration here, namely the interaction between a cone and a plate. Let us also stress that due to its unique temperature dependence, the CCF allows *in situ* control of reversible assembly in soft matter and nanoscience. A further advantage of the force is that both its magnitude and range of action depends on the separation between the objects and the thermodynamic parameters of the fluctuating medium. The last facts can potentially be used in controlling the properties of colloidal suspensions and for governing the behavior of objects at small, below micrometer, distances.

The content of the article is arranged as follows. In Subsec. 2.1 we present some basic results of the finite-size scaling theory. In Subsec. 2.2 we recall and comment on the finite-size behaviour of systems with dispersion forces extending the known facts to the expected behaviour of the CCF, vdWF and TF when they act between pair of parallel plates. By doing so, we introduce the general technique, namely the "surface integration approach" (SIA) (Subsec. 2.3), within which we study the commented spectrum of forces. Section 3 briefly presents the corresponding lattice-gas model suitable for the investigation of fluid media with account of the long-ranged van der Waals interactions. Here we present the equation for the equilibrium profile of the finite-size order parameter, identify the main coupling parameters characterizing the interactions in the systems, which enter in it, and give the general expressions used to calculate the CCF and TF. The article ends with Sec. 4, where we present the exact equations (Subsec. 4.1) used to obtain the numerical results for the behaviour of the investigated forces, and comment them in details in Subsec. 4.2.

2 Theoretical background

2.1 General predictions of the finite-size scaling theory

If an object of certain shape is introduced in a medium, it alters its thermodynamic behaviour in such a way that any quantitative occurrence shows dependance on the penetration depth of the symmetry breaking effect into the volume. The range to which this effect is felt within the system depends both on the scope of the interactions and on that of the correlations between the fluctuations in the fluid, which mediate the interactions between the bodies. The range of the correlations is set by the correlation length ξ of the order parameter of the medium. This quantity becomes large, and theoretically diverges, at the vicinity of the bulk critical point (T_c, μ_c): $\xi(T \to T_c^+, \mu = \mu_c) \simeq \xi_0^+ t^{-\nu}$, where $t = (T - T_c)/T_c$, and $\xi(T = T_c, \mu \to \mu_c) \simeq \xi_{0,\mu} |\Delta\mu/(k_B T_c)|^{-\nu/\Delta}$, where $\Delta\mu = \mu - \mu_c$. Here ν and Δ are the usual critical exponents which, for classical fluids, are those of the three-dimensional Ising model, and ξ_0^+ and $\xi_{0,\mu}$ are the corresponding nonuniversal amplitudes of the correlation length along the t and μ axes. When ξ becomes comparable to the characteristic dimension of the system, say the separation L between the objects, the size dependence of the thermodynamic functions enters into the thermodynamic potentials through the ratio L/ξ, and takes a scaling form given by the finite-size scaling theory [5].

In the remainder we are going to study a system governed by long-ranged dispersion interactions utilizing the recently introduced "surface integration approach" [12]. For this purpose we will need the corresponding results for a system with a film geometry that is governed by the same interactions in question. That is why we are next going to concisely recall these topics. We start with the behavior of the thermodynamic Casimir force in nonpolar fluid systems with plane parallel geometry, in which the dispersion forces are accounted for.

2.2 Influence of the dispersion forces on the thermodynamic Casimir one in nonpolar fluid systems with film geometry

We consider a fluid system consisting of a nonpolar medium $\mathbb{M}$ comprising two thick parallel plates of materials $\mathbb{S}_1$ and $\mathbb{S}_2$ which are coated by thin solid films of substances $\mathbb{L}_1$ and $\mathbb{L}_2$, respectively – see Fig. 1(**a**) in [1]. If the fluid medium is in contact with a particle reservoir with a chemical potential μ, the grand canonical potential $\Omega_{\rm ex}(L|T,\mu)$ per unite area $\mathscr{A}$ of this medium in excess to its bulk value $\mathscr{A}L\omega_{\rm bulk}(T,\mu)$ depends on the film thickness L and, thus, one can define the total effective force $F_{\rm tot}(L|T,\mu)$ [in a full accord with Eq. (2)], which is due to fluctuations of the medium and dispersion interactions in it. Here $\omega_{\rm bulk}(T,\mu)$ is the density of the bulk grand canonical potential, and $\mathscr{A}$ is the surface area of the plates.

The contribution of the dispersion forces to the total effective force Eq. (2) can be distinguished from that of the CCF by their temperature dependence, because the leading such of the former does not exhibit a singularity. Thus, one can perform the aforementioned decompositions and identifications given in Eqs. (1) and (3).

If the system is away from its bulk critical point, for the occurring force $f^{\parallel}_{\rm tot}(L|T,\mu)$ per cross section area $\mathscr{A}$ and $k_BT \equiv \beta^{-1}$ is customary to write the following expression

$$f^{\parallel}_{\rm tot}(L|T,\mu) \simeq (\sigma-1)\beta H_A(T,\mu)L^{-\sigma}\vartheta^{\sigma-d}, \tag{4}$$

where, for dimensional reasons, the microscopic length scale ϑ is introduced. Let us note that one normally considers the case $d=\sigma$ and, thus, omits the apparent dependence of this length, that can be taken to be, e.g., the so-called *retardation length* [12, 13] $\xi_{\rm ret}$. In Eq. (4) H_A is the Hamaker term, whose dependence from the temperature and chemical potential is given by the so-called Hamaker constant [14, 15] (for details see the Appendix in [2]).

As already explained above, upon approaching the bulk critical point of the system the fluctuations of the order parameter of the confined fluid medium exhibit strong correlations which gives rise to new contribution to the TF, the CCF (see below). In the vicinity of this point (critical region) Eq. (4) is no longer valid. Following [2], it becomes

$$f^{\parallel}_{\rm tot}(L|T,\mu) \simeq L^{-d}X^{\parallel}_{\rm crit}\left[x_t,x_\mu,x_l,\{x_{s_i},i=1,2\},x_g\right]+(\sigma-1)\beta H_A(T,\mu)L^{-\sigma}\vartheta^{\sigma-d}. \tag{5}$$

Here $X^{\parallel}_{\rm crit}$ is dimensionless, universal scaling function, $x_t = t\left(L/\xi_0^+\right)^{1/\nu}$ and $x_\mu = \beta\Delta\mu\left(L/\xi_{0,\mu}\right)^{\Delta/\nu}$ are the temperature and field relevant in renormalization group sense scaling variables, whereas x_l, x_{s_i}, $i=1,2$ and x_g are irrelevant scaling variables associated with the interactions in the system. Explicitly these variables are defined in the text below Eq. (2.6) in [2].

According to the scaling hypothesis of the CCF one expects that near the bulk critical point

$$f^{\parallel}_{\rm Cas}(L|T,\mu) = L^{-d}X^{\parallel}_{\rm Cas}(x_t,x_\mu,\cdots), \tag{6}$$

where $X_{\text{Cas}}^{\|}$ is a scaling function, that for large enough L with fixed $x_t = \mathscr{O}(1)$ and $x_\mu = \mathscr{O}(1)$ approaches the scaling function of the short-ranged system $X_{\text{Cas,sr}}^{\|}(x_t, x_\mu)$ (for details see Eqs. (2.12) and (4.10) in [2]). From Eqs. (1) and (3), together with Eq. (5) follows that the scaling function of the CCF $X_{\text{Cas}}^{\|}$ is proportional to the sum of $X_{\text{crit}}^{\|}$ and the singular part of the Hamaker term $H_A^{(\text{sing})}(T,\mu)$. The last implies that in order to determine the CF in systems with dispersion interaction one has to decompose the contribution captured through the Hamaker term in a singular and a regular parts, i.e.

$$H_A(T,\mu) = H_A^{(\text{reg})}(T,\mu) + H_A^{(\text{sing})}(T,\mu). \tag{7}$$

Thus, with $d = \sigma$ one has

$$f_{\text{Cas}}^{\|} = L^{-d}\left[X_{\text{crit}}^{\|} + (d-1)\beta H_A^{(\text{sing})}\right], \tag{8}$$

while F_{vdW} from Eq. (3), normalized per unit area, in the general case coincides with Eq. (4), with $H_A \equiv H_A^{(\text{reg})}$.

In what follows, we are going to present results for the CCF, vdWF and TF in the cases of cone-plate system, utilizing the knowledge gained from studies of the corresponding interactions between parallel plates.

2.3 The van der Waals and Casimir forces for a cone-plate geometry within the SIA

The most common approach used to calculate geometry dependent interaction energy or force in systems where at least one of the objects has a nonplanar geometry, is that first proposed by B. Derjaguin in 1934 [16]. The Derjaguin approximation (DA) suggests that one can relate the interaction force/potential between two gently curved colloidal particles with the knowledge for that between a pair of parallel plates, provided that the separation between the colloids is *much smaller* than their geometrical characteristics. This restriction imposes certain limitations on the usage of the DA. The necessity of more general technique as an improvement of the DA, led to the proposal of the co-called "surface integration approach" reported in [12]. It has been used there to study van der Waals interactions between objects of arbitrary shape and a plate of arbitrary thickness. It delivers *exact* results if the interactions involved can be described by pair potentials. The main advantage of this approach over the DA is that one is no longer bound by the restriction that the interacting objects must be much closer to each other than their characteristic sizes. The main result is that for the force acting between a 3D object (say a colloid particle) $B \equiv \{(x,y,z),(x,y,z) \in B\}$ of general shape $S(x,y) = z$ and a flat surface bounded by the $(x,y)-$plane of a Cartesian coordinate system, one has

$$F_{\text{SIA}}^{B,|}(L) = \int_{A_S^{\text{to}}}\int f_{\mathscr{A}}^{\|}[S(x,y)]\mathrm{d}x\mathrm{d}y - \int_{A_S^{\text{away}}}\int f_{\mathscr{A}}^{\|}[S(x,y)]\mathrm{d}x\mathrm{d}y, \tag{9}$$

where A_S is the projection of the surface S of the particle over the $(x,y)-$ plane, $A_S = A_S^{\text{to}} \bigcup A_S^{\text{away}}$. Equation (9) has a very simple intuitive meaning: in order to determine the force acting on the particle one has to subtract from the contributions stemming from surface regions A_S^{to} that "face towards" the projection plane those from regions A_S^{away} that "face away" from it, where A_S^{to} and A_S^{away} are the projections of the corresponding parts of the surface of the body on the $(x,y)-$plane. It is clear that if one takes into account only the contributions over A_S^{to} one obtains expression very similar to the DA. Both expressions in that case will differ only by the fact that while Eq. (9) takes into account that the force on a given point of S is *along the normal* to the surface at that point, the standard DA does not take this into account. Let us recall that Eq. (9) provides exact results for the interaction under the assumption that the constituents of the body interact via pair potentials. This is, of course, *not* the case of CCF. It is, however, clear that under mechanical equilibrium of the colloid in the fluid, the CCF is again along the normal to the surface at the point of the surface where it acts. Thus, one can get a reasonably good approximation to the effect of that force by keeping just the integration over the part of the surface of the body that faces the plane. This leads to

$$\beta F^{B,\|}_{\mathrm{Cas,SIA}}(L) = \int_{A_S^{\mathrm{to}}}\int f^{\|}_{\mathrm{Cas}}[\mathrm{S}(x,y)]\mathrm{d}x\mathrm{d}y. \tag{10}$$

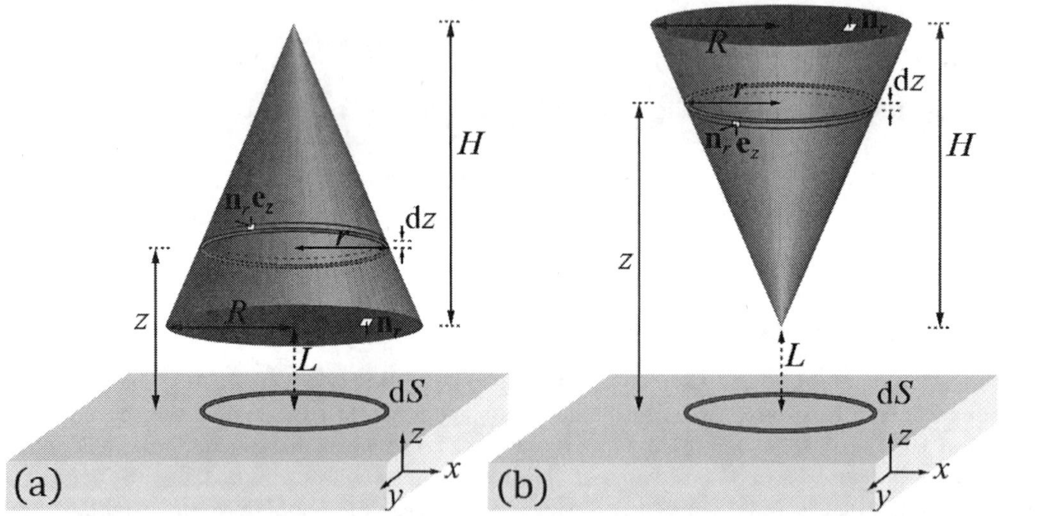

Figure 1. Geometry of the SIA for the interaction between a right circular cone and a plate, in several orientations of the conic particle. The conical shape is characterized by its base radius R and height H. In (a) the cone's base is toward the plate, and the separation between them is L. The z-component $\mathbf{e}_z$ of the normal vector $\mathbf{n}_r$ is with a plus sign on the lateral surface, and with a minus sign on the base where $\mathbf{n}_r \equiv -\mathbf{e}_z$. If one moves along a slant height, from the cone's apex towards its base, he/she'll notice that $\mathbf{e}_z$ changes sign with a jump, when the edge is reached. Thus the general expression for the force has the form Eq. (11). The projection of the infinitesimal area dS of the lateral surface is between two concentric circumference with radiuses r and $r+dr$ and center lying on H. The considerations made is case (a) are altogether applicable for the case where the apex of the cone is toward the plate (b).

Now if one applies Eq. (9) for the case of a conical colloidal particle which *base is toward the plate* ($\lhd|$) [see Fig. 1(a)], with the assumption that the interaction is induced by point-like sources, the result reads

$$F^{\lhd|}_{\mathrm{SIA}}(L) = -2\pi\frac{R^2}{H^2}\int_L^{L+H}(L+H-z)f^{\|}_{\mathscr{A}}(z)\mathrm{d}z + \pi R^2 f^{\|}_{\mathscr{A}}(L). \tag{11}$$

The term $(\cdots)$ in the integrand of Eq. (11) reflects how the projection to the normal to the surface of the cone in this specific orientation changes as a function of z. Substituting Eq. (4) [see also the text below Eq. (8)] in the above expression for the van der Waals cone-plate interaction we can write

$$F^{\lhd|}_{\mathrm{vdW,SIA}}(L) = 2\pi R^2(\sigma-1)H_A^{(\mathrm{reg})}\vartheta^{\sigma-3}\left\{\frac{1}{2L^\sigma} - \frac{(L+H)^{2-\sigma}+L^{1-\sigma}[H(\sigma-2)-L]}{H^2(\sigma-1)(\sigma-2)}\right\}. \tag{12}$$

The corresponding expression for the CCF arising between a cone and a plate for this specific geometry, following Eq. (10), can then be written as

$$\beta F^{\lhd|}_{\text{Cas,SIA}}(L) = \pi R^2 f^{\|}_{\text{Cas}}(L) = \pi R^2 L^{-3} X^{\|}_{\text{Cas}}(x_t, x_\mu, \cdots). \tag{13}$$

The general expression within the SIA, for any geometry conditioned force in a system composed out of a cone which *apex faces the plate* ($\rhd|$) the cone interacts with [see Fig. 1(b)], has the form

$$F^{\rhd|}_{\text{SIA}}(L) = 2\pi \frac{R^2}{H^2} \int_L^{L+H} (z-L) f^{\|}_{\mathscr{A}}(z)\text{d}z - \pi R^2 f^{\|}_{\mathscr{A}}(L+H). \tag{14}$$

Hence for the vdWF with a free varying interaction decay parameter σ one obtains the following expression

$$F^{\rhd|}_{\text{vdW,SIA}}(L) = \pi H_A^{(\text{reg})} \vartheta^{\sigma-3} \frac{R^2}{H^2(\sigma-2)} \left[\frac{2}{L^{\sigma-2}} - \frac{2L^2 + 2LH\sigma + H^2(\sigma-1)\sigma}{(L+H)^\sigma} \right], \tag{15}$$

whereas restricting the integration only over the surface of the cone which faces the plate, i.e., confines the fluctuations of the fluid medium, one writes for the CCF

$$\beta F^{\rhd|}_{\text{Cas,SIA}}(L) = 2\pi \frac{R^2}{H^2} \int_L^{L+H} \frac{z-L}{z^3} X^{\|}_{\text{Cas}}[x_t(z), x_\mu(z), \cdots]\text{d}z \equiv 2\pi \frac{R^2}{H^2} X^{\rhd|}_{\text{Cas,SIA}}(H|x_t, x_\mu, \cdots). \tag{16}$$

Last we deliver the equations describing the interactions in a cone-plate system, with the consideration that *the height of the cone is parallel to the plate* ($\triangle|$) (see Fig. 2). Therefore for the general expression and those for the vdWF and CCF one has

$$F^{\triangle|}_{\text{SIA}}(L) = 2H \left[\int_L^{L+R} \left(1 - \frac{z-L}{R}\right) \mathfrak{F}_+(z) f^{\|}_{\mathscr{A}}(z)\text{d}z - \int_{L+R}^{L+2R} \left(\frac{z-L}{R} - 1\right) \mathfrak{F}_-(z) f^{\|}_{\mathscr{A}}(z)\text{d}z \right], \tag{17a}$$

where

$$\mathfrak{F}_\pm(z) = \ln \left[\frac{R + \sqrt{(z-L)(L+2R-z)}}{\pm(L+R-z)} \right]. \tag{17b}$$

When one considers the standard [non-retarded (London)] van der Waals interaction $\sigma = 3$

$$F^{\triangle|}_{\text{vdW,SIA}}(L|3) = H_A^{(\text{reg})} \frac{2\pi H R^2}{(L+R)[L(L+2R)]^{3/2}}, \tag{17c}$$

while for the retarded (Casimir-Polder) one $\sigma = 4$ and the force becomes

$$F^{\triangle|}_{\text{vdW,SIA}}(L|4) = H_A^{(\text{reg})} \vartheta \frac{\pi H R^2 (2L+R)(2L+3R)}{(L+R)^2[L(L+2R)]^{5/2}}. \tag{17d}$$

In view of Eqs. (10) and (17a) for the CCF in this specific geometry one finds

$$\begin{aligned} F^{\triangle|}_{\text{Cas,SIA}}(L) &= 2H \int_L^{L+R} \left(1 - \frac{z-L}{R}\right) \frac{\mathfrak{F}_+(z)}{z^3} X^{\|}_{\text{Cas}}[x_t(z), x_\mu(z), \cdots]\text{d}z \\ &\equiv 2H X^{\triangle|}_{\text{Cas,SIA}}(R|x_t, x_\mu, \cdots). \end{aligned} \tag{17e}$$

If interested, the reader can refer to Appendix A.2 of [18] for the detail derivation of the infinitesimal area elements appearing in Eqs. (11), (14) and (17a).

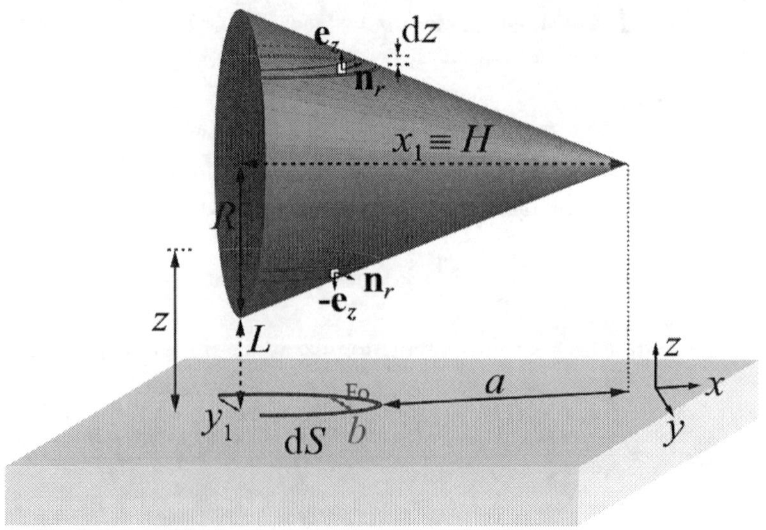

Figure 2. Geometry of the surface integration approach for the interaction between a right circular cone and a plate, when the orientation of the cone is such that H is parallel to the plate. In this scenario the infinitesimal projection dS is the area enclosed between two hyperbolae. In order to obtain an expression for dS, we make use of the expression of the area enclosed between a hyperbola and a line parallel to the y-axis (see p. 323 in [17]) $S = x_1 y_1 - \ln[(x_1/a) + (y_1/b)]$. Here x_1 and y_1 are the coordinates of the point of intersection between the hyperbola and the line, a is the distance between the hyperbola's apex and the beginning of the coordinate frame, and $b = \sqrt{e^2 - a^2}$ (e – the separation between the focus Fo of the hyperbola and the beginning of the coordinate frame).

Since SIA utilize the knowledge of the behavior of the force per unit area arising between a pair of parallel plates, in the following section we present the corresponding model within which we describe such a system. Because the included expressions have already been presented in details in [2], here only key results, as well some notations, are going to be given which will be needed in the remainder of the article.

3 The model

As explained in [2], we consider a lattice-gas model of a fluid confined between two planar plates, separated at a distance L from each other, with grand canonical potential $\Omega[\rho(\mathbf{r})]$ given by

$$\begin{aligned}\Omega[\rho(\mathbf{r})] = k_B T \sum_{\mathbf{r}\in\mathbb{M}} \{\rho(\mathbf{r})\ln[\rho(\mathbf{r})] + [1-\rho(\mathbf{r})]\ln[1-\rho(\mathbf{r})]\} + \frac{1}{2}\sum_{\mathbf{r},\mathbf{r}'\in\mathbb{M}} \rho(\mathbf{r}) w^l(\mathbf{r}-\mathbf{r}')\rho(\mathbf{r}') \\ + \sum_{\mathbf{r}\in\mathbb{M}} \left[V^{(s_1|l|s_2)}(z) - \mu\right]\rho(\mathbf{r}),\end{aligned} \tag{18}$$

where $\mathbb{M}$ is a simple cubic lattice in the region occupied by the fluid medium – $\infty^{d-1} \times [0, L]$ and $V^{(s_1|l|s_2)}(z)$ is an external potential that reflects the interactions between the confining plates and the constituents of the fluid. In Eq. (18) $w^l(\mathbf{r}-\mathbf{r}') = -4J^l(\mathbf{r}-\mathbf{r}')$ is the nonlocal coupling (interaction potential) between the constituents of the confined medium and μ is the chemical potential. All length scales here and in the remainder are taken in units of the lattice constant a_0, so that the particle number density $\rho(\mathbf{r})$ becomes a number which varies in the range $[0,1]$. We recall that in the framework of a mean-field treatment with respect to the critical behavior the effective spatial dimension is $d=4$, irrespective of the actual dimension of the model under consideration.

In Eq. (18)

$$V^{(s_1|l|s_2)}(z) = -\rho_{s_1} J_{\rm sr}^{s_1,l}\delta(z) - \rho_{s_2} J_{\rm sr}^{s_2,l}\delta(L-z) + v_{s_1}(z+1)^{-\sigma} + v_{s_2}(L+1-z)^{-\sigma}, \tag{19}$$

where $v_{s_i} = -G(d,\sigma)\rho_{s_i}J^{s_i,l}$, $i=1,2$, with

$$G(d,\sigma) = 4\pi^{(d-1)/2}\frac{\Gamma\left(\frac{1+\sigma}{2}\right)}{\sigma\Gamma\left(\frac{d+\sigma}{2}\right)}, \tag{20}$$

and $\delta(x)$ is the discrete delta function;

$$J^l(\mathbf{r}) = J_{\rm sr}^l\{\delta(|\mathbf{r}|) + \delta(|\mathbf{r}|-1)\} + \frac{J^l\theta(|\mathbf{r}|-1)}{1+|\mathbf{r}|^{d+\sigma}}, \tag{21}$$

is a proper lattice version of $-w^l(\mathbf{r})/4$ as the interaction energy between the fluid particles, and

$$J^{s_i,l}(\mathbf{r}) = J_{\rm sr}^{s_i,l}\delta(|\mathbf{r}|-1) + \frac{J^{s_i,l}\theta(|\mathbf{r}|-1)}{|\mathbf{r}|^{d+\sigma}},\ i=1,2 \tag{22}$$

is the one between a fluid particle and a substrate particle, $\theta(x)$ is the Heaviside step function with the convention $\theta(0)=0$.

Taking into account the translational symmetry of the system along the bounding surfaces, the variation with respect to $\rho(\mathbf{r})$ leads to an equation of state for the equilibrium density $\rho^*(\mathbf{r}) = [1+\phi^*(\mathbf{r})]/2$, where $\phi(\mathbf{r}) \equiv \phi\left(\mathbf{r}_\parallel, z\right) = \phi(z)$, with $\mathbf{r} = \left\{\mathbf{r}_\parallel, z\right\}$ is the local order parameter profile $\{\phi(z),\ 0\le z\le L\}$. In terms of $\phi(z)$ the equation of state can be written in the following form

$$\begin{aligned}\operatorname{arctanh}[\phi^*(z)] = \frac{\beta}{2}[\Delta\mu - \Delta V(z)] + K\Big\{ &\mathrm{a}_{d,\sigma}(\lambda)\phi^*(z) + \mathrm{a}_{d,\sigma}^{nn}(\lambda)[\phi^*(z+1)+\phi^*(z-1)] \\ &+\lambda\sum_{z'=0}^{L} g_{d,\sigma}(|z-z'|)\theta(|z-z'|-1)\phi^*(z')\Big\},\end{aligned} \tag{23}$$

where $\Delta\mu = \mu - \mu_c$, $K = \beta J_{\rm sr}^l$, $\mathrm{a}_{d,\sigma}(\lambda) = (2d-1) + \lambda(c_{d,\sigma} - d)$ and $\mathrm{a}_{d,\sigma}^{nn}(\lambda) = 1.0 + \lambda(c_{d,\sigma}^{nn} - 0.5)$ with $c_{d,\sigma}^{nn} = g_{d,\sigma}(1) + g_{d,\sigma}^{nn}(\pm 1)$. The functions $c_{d,\sigma}$, $g_{d,\sigma}(|z-z'|)$ and $g_{d,\sigma}^{nn}(|z-z'|)$ are determined in Eqs. (C10), (C11) and (C12) of [19], respectively.

The excess grand canonical potential per unit area, $\omega_{\rm ex} \equiv \lim_{\mathscr{A}\to\infty}[\Omega/\mathscr{A}] - L\omega_{\rm bulk}$, has the form

$$\begin{aligned}\beta\omega_{\rm ex} = \sum_{z=0}^{L}\Big\{&\frac{1}{2}\ln\left[1-\phi^*(z)^2\right] - \frac{1}{2}\ln\left[1-\phi_b^2\right] + \frac{1}{4}\phi^*(z)\ln\left[\frac{1+\phi^*(z)}{1-\phi^*(z)}\right] - \frac{1}{4}\phi_b\ln\left[\frac{1+\phi_b}{1-\phi_b}\right] \\ &+\frac{\beta}{2}\Delta V(z)\phi^*(z) - \frac{\beta\Delta\mu}{2}[\phi^*(z)-\phi_b]\Big\} + \beta\omega_{\rm reg},\end{aligned} \tag{24}$$

where

$$\beta(\sigma-1)\omega_{\rm reg} = \left[\frac{K}{K_c}(s_{1,c}+s_{2,c}) - \frac{1}{4}G(d,\sigma)K\lambda\right]\frac{\vartheta^{\sigma-d}}{L^{\sigma-1}}. \tag{25}$$

Here ϕ_b is the bulk value of the order parameter (see Eq. (4.2) in [2]), $K_c = \beta_c J^l_{\rm sr}$ and $s_{i,c}$, $i = 1,2$ are the values of the plates-fluid coupling parameters evaluated at the bulk critical point of the system $\{\beta = \beta_c = [\sum_{\bf r} J^l({\bf r})]^{-1}, \mu = \mu_c = -2\sum_{\bf r} J^l({\bf r})$, with the sum running over the whole lattice$\}$.

The *effective* surface potential $\beta\Delta V(z)/2$ in Eqs. (23) and (24) is given by

$$\frac{\beta}{2}\Delta V(z) = \frac{s_1}{(z+1)^\sigma} + \frac{s_2}{(L+1-z)^\sigma}, \tag{26}$$

where $1 \le z \le L-1$ and

$$s_i = \frac{1}{2}\beta G(d,\sigma)\left(\rho_{s_i} J^{s_i,l} - \rho_c J^l\right), \; i = 1,2 \tag{27}$$

are the (T- and μ-independent) dimensionless plates-fluid coupling parameters $\propto x_{s_i}$. In Eq. (26) the restriction $z \ge 1$ holds because we consider the layers closest to the substrate to be completely occupied by the liquid phase of the fluid, i.e., $\phi(0) = \phi(L) = 1$, thus ensuring the $(+,+)$ boundary conditions. Physically this can be accomplished by choosing a proper coating of the surfaces of the plates. The coupling parameter $\lambda \propto x_l$ probes the importance of the long-ranged parts of the interaction potential within the fluid medium

$$\lambda = J^l / J^l_{\rm sr}. \tag{28}$$

In Eq. (27) $s_i > 0$, i.e., $\rho_{s_i} J^{s_i,l} > \rho_c J^l$ corresponds to plates "preferring" the "liquid" phase of the fluid, while $s_i < 0$, or $\rho_{s_i} J^{s_i,l} < \rho_c J^l$ mirrors the one with affinity to its "gas" phase. If the interactions between the constituents of the fluid are of Lennard-Jones type one has $\lambda = 2$, as commented in Refs. [2, 19]. The marginal case $s_i = 0$ together with $\lambda = 0$ describes a pure short-range system (for details refer to Eqs. (4.10) and the text therein in [2]).

The expressions given so far provide the basis of the model that will be used to determine the cone-plate interactions in the remainder.

Using Eqs. (2) and (24) for the *total* force $f^{\parallel}_{\rm tot}(L|T,\mu)$ (per unit area $\mathscr{A}$ and $k_B T$) acting between parallel plates bounding the fluid medium the following expression can be written

$$f^{\parallel}_{\rm tot}(L|T,\mu) = -\frac{\beta}{2}\left[\omega_{\rm ex}(L+1|T,\mu) - \omega_{\rm ex}(L-1|T,\mu)\right] - \frac{4K s_{1,c} s_{2,c}}{G(d,\sigma) K_c^2 \lambda} L^{-\sigma} \vartheta^{\sigma-d}, \tag{29}$$

where the last term represents the direct interaction between the plates (for details see the Appendix in [2]). On the other hand, if one subtracts from the potential $\omega_{\rm ex}$ its regular part $\omega_{\rm reg}$, i.e., if we consider the quantity

$$\Delta\omega \equiv \lim_{\mathscr{A}\to\infty}\left[(\Omega - \Omega_{\rm reg})/\mathscr{A}\right] - L\omega_{\rm bulk}, \tag{30}$$

then, in accord with Eqs. (1) – (3), the L dependence of $\Delta\omega$ via Eq. (2) provides the singular part of the total force, i.e., $f^{\parallel}_{\rm Cas}(L|T,\mu)$. Explicitly, one has

$$f^{\parallel}_{\rm Cas}(L|T,\mu) = -\frac{\beta}{2}\left[\Delta\omega(L+1|T,\mu) - \Delta\omega(L-1|T,\mu)\right]. \tag{31}$$

Thus, near T_c the TF and the CCF are related via the expression

$$f^{\parallel}_{\rm tot}(L|T,\mu) = f^{\parallel}_{\rm Cas}(L|T,\mu) + (\sigma-1)\beta H_A^{\rm (reg)} L^{-\sigma}\vartheta^{\sigma-d}, \tag{32}$$

where the last term is the mathematical equivalent of $F_{\rm vdW}$ in Eq. (3) per unit area $\mathscr{A}$ and $k_B T$ for a film geometry with $H_A^{\rm (reg)}$ being the nonsingular (regular) part of the Hamaker term given by

$$(\sigma-1)\beta H_A^{\rm (reg)} = -\frac{4K s_{1,c} s_{2,c}}{G(d,\sigma) K_c^2 \lambda} + \frac{K}{K_c}(s_{1,c} + s_{2,c}) - \frac{1}{4} G(d,\sigma) K\lambda. \tag{33}$$

For the singular part of H_A within the presented model the following expression was derived (see Eq. (A6) in [2])

$$(\sigma-1)\beta H_A^{(\mathrm{sing})}(T,\mu)=\frac{K}{K_c}(s_{1,c}+s_{2,c})\phi_b-\frac{1}{4}G(d,\sigma)K\lambda\phi_b^2, \tag{34}$$

where the T and μ dependance is carried by the bulk order parameter ϕ_b (see Eq. (4.2) in [2]).

In the next section, based on the results reported in Subsec. 2.2 and Sec. 3, we present numerical results for the behaviour of the above discussed forces in cone-plate fluid system for the cases $d=\sigma=3$.

4 Results

In this section, using the results for $d=\sigma=4$ from the mean-field type numerical study discussed in the text above, together with the SIA approximation, we will present some results for the behavior of the CCF, vdWF and TF between a conical particle and a plate for several orientations of the colloid with respect to the smooth planar substrate in $d=3$. As it became clear from the above shown equations, the key knowledge which is required for the desired calculations to be performed, is the force [scaling function(s)] per unit area between two parallel plates, for many different separations L, and at various values of T and μ of the fluctuating fluid medium.

Within the mean-field theory the T and μ dependance of the corresponding forces near the bulk critical point ($T=T_c, \Delta\mu=0$) is given by the temperature and field scaling variables x_t and x_μ, respectively [see the text beneath Eq. (5)], with $\nu=1/2$ and $\Delta=3/2$. In our numerical treatment we take these variables to range in the intervals: $x_t\in\left[-24^2;24^2\right]$ and $x_\mu\in\left[-24^3;24^3\right]$. For the study of the scaling function of the CCF, the separation L between the set of parallel plates is varied from 20 to 60 with step 2, from 25 to 95 with step 5 and from 70 to 200 with step 10. In order to demonstrate the effect of the sign change of the considered forces and in view of potential experimental realization of the predicted effects we choose the following values of the coupling parameters: $s_1=1.0$, $s_2=-0.01$ and $\lambda=2.0$. The physical argumentation of the so chosen values is explained in Section IV C of [1].

4.1 Calculation of the forces in the cone-plate system within the SIA in $d=3$

From Eq. (12), considering only genuine van der Waals interactions ($d=\sigma$) in $d=3$, one has that within the SIA the force is calculated via the following expression

$$L\beta F_{\mathrm{vdW,SIA}}^{\lhd|}(L)/\pi\Xi_R^2=2\beta H_A^{(\mathrm{reg})}\frac{\Xi_H}{1+\Xi_H}, \tag{35}$$

where we have introduced the notations $\Xi_R\equiv R/L$ and $\Xi_H\equiv H/L$. Accordingly for the CCF from Eq. (13) one can write

$$L\beta F_{\mathrm{Cas,SIA}}^{\lhd|}(L)/\pi\Xi_R^2=X_{\mathrm{Cas}}^{\|}(x_t,x_\mu,\cdots), \tag{36}$$

i.e. its behavior coincides with that of the plate-plate scaling function obtained for a desired separation L.

For the forces of interest in the remaining configurations one has:

(i) from Eq. (15) with $\sigma=3$ the following

$$L\beta F_{\mathrm{vdW,SIA}}^{\rhd|}(L)/\pi\Xi_R^2=2\beta H_A^{(\mathrm{reg})}\frac{\Xi_H}{(1+\Xi_H)^3}. \tag{37a}$$

(ii) With the use of Eq. (16), one can evaluate the scaling function entering there and hence the CCF vie

$$X^{\rhd|}_{\mathrm{Cas,SIA}} = \int_1^{\Lambda_1} I^{\rhd|}(\bar{z}|\Lambda_1) X^{\|}_{\mathrm{Cas}}[\mathfrak{a}_1(\bar{z})]\mathrm{d}\bar{z} + \sum_{i=j+1}^{j_{\max}-1} \int_{\Lambda_{i-1}}^{\Lambda_i} I^{\rhd|}(\bar{z}|\Lambda_i) X^{\|}_{\mathrm{Cas}}[\mathfrak{a}_i(\bar{z})]\mathrm{d}\bar{z}$$
$$+ \int_{\Lambda_{\max}}^{1+\Xi_H} I^{\rhd|}(\bar{z}|\Lambda_{\max}) X^{\|}_{\mathrm{Cas}}[\mathfrak{a}_{\max}(\bar{z})]\mathrm{d}\bar{z}, \quad \text{with} \quad L\beta F^{\rhd|}_{\mathrm{Cas,SIA}}/\pi\Xi_R^2 = \frac{2}{\Xi_H^2} X^{\rhd|}_{\mathrm{Cas,SIA}} \tag{37b}$$

where

$$I^{\rhd|}(\bar{z}|\Lambda_j) = \frac{\bar{z}-1}{\bar{z}^3}\theta(\Xi_H - \Lambda_j), \tag{37c}$$

with $\Lambda_j \equiv L_j/L,\ j = 1, j_{\max}$; $\bar{z} \equiv z/L$ is dimensionless variable; $L_{\max} \equiv L_{j_{\max}}$ is the largest system for which numerical data are available, the arguments of the scaling function $[\mathfrak{a}_j(\bar{z})] \triangleq [x_t(L_{z;j}), x_\mu(L_{z;j}), \ldots]$, $L_{z;j} \equiv \bar{z}L_j$ and θ is the Heaviside step function with the convention $\theta(0) = 0$. In the calculations performed in the current article $L_{\max} = 200$.

(iii) Last we give the expressions used to evaluate the vdWF and the CCF for the case when the height of the cone is parallel to the plate. Indeed, performing the dimensionalization with respect to L in Eq. (17c) we have

$$L\beta F^{\triangle|}_{\mathrm{vdW,SIA}}(L)/\pi\Xi_R^2 = 2\beta H_A^{(\mathrm{reg})} \frac{\Xi_H}{(1+2\Xi_R)^{3/2}(1+\Xi_R)}. \tag{38a}$$

(iv) For the scaling function of the CCF from Eq. (17e), one has

$$X^{\triangle|}_{\mathrm{Cas,SIA}} = \int_1^{\Lambda_1} I^{\triangle|}(\bar{z}|\Lambda_1) X^{\|}_{\mathrm{Cas}}[\mathfrak{a}_1(\bar{z})]\mathrm{d}\bar{z} + \sum_{i=j+1}^{j_{\max}-1} \int_{\Lambda_{i-1}}^{\Lambda_i} I^{\triangle|}(\bar{z}|\Lambda_i) X^{\|}_{\mathrm{Cas}}[\mathfrak{a}_i(\bar{z})]\mathrm{d}\bar{z}$$
$$+ \int_{\Lambda_{\max}}^{1+\Xi_R} I^{\triangle|}(\bar{z}|\Lambda_{\max}) X^{\|}_{\mathrm{Cas}}[\mathfrak{a}_{\max}(\bar{z})]\mathrm{d}\bar{z}, \quad \text{with} \quad L\beta F^{\triangle|}_{\mathrm{Cas,SIA}}/\pi\Xi_R^2 = \frac{2\Xi_H}{\pi\Xi_R^2} X^{\triangle|}_{\mathrm{Cas,SIA}} \tag{38b}$$

where

$$I^{\triangle|}(\bar{z}|\Lambda_j) = \left(1 - \frac{\bar{z}-1}{\Xi_R}\right)\frac{\mathfrak{F}_+(\bar{z})}{\bar{z}^3}\theta(\Xi_R - \Lambda_j). \tag{38c}$$

In the above expressions the critical component $X^{\|}_{\mathrm{crit}}$ of the scaling function $X^{\|}_{\mathrm{Cas}}$ calculated within the mean-field theory is normalized by the procedure described in-and-out in Section V of [2], so that it contributes properly to the CCF and hence to the total force of interaction when $d < 4$. The need of such normalization is explained in details in [19] (see there Sections IV.A.1 and IV.A.3).

After presenting the mathematical means to calculate the CCF and vdWF, we now pass to the detailed discussion of the results and argumentation of the experimental feasibility of the parameters used in the model calculations utilizing the SIA.

4.2 Discussion and concluding remarks

In the present section for physical reasons related to future experimental studies we will assume that the conical particle has either ruthenium (Ru) or platinum (Pt) core, while the planar substrate is made either of carbon (C) or silica (SiO_2) aerogels. The contact surface of any of the interacting objects is considered coated by monolayer of lead (Pb) or thallium nitride (TlN) to ensure the (+,+) boundary conditions. Finally, we complete the so assembled system by adopting xenon (Xe) as the

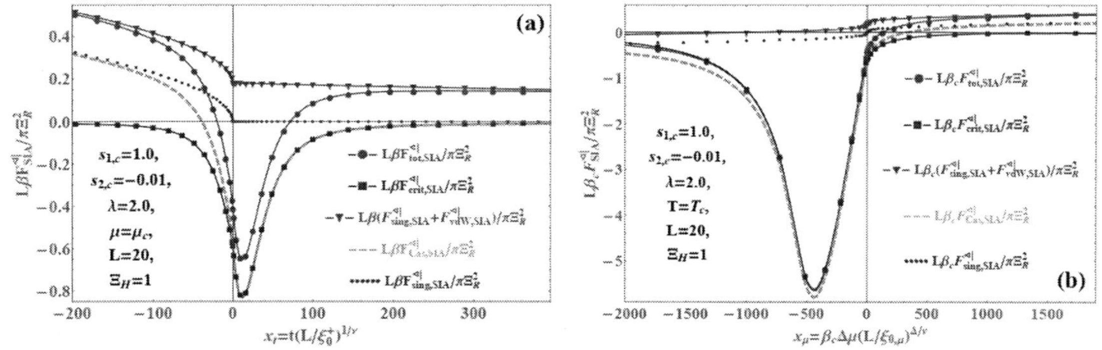

Figure 3. Behaviour of the discussed forces (multiplied by the separation L and normalized by $k_B T$ and the area of the cones's base $\pi\Xi_R^2$) in $d = 3$ for the case where the base of the cone is facing the plate the cone interacts with. On both figures the interplay between the critical $F^{\triangleleft|}_{\rm crit,SIA}$ (-■-) and singular $F^{\triangleleft|}_{\rm sing,SIA}$ (• • •) components of the net Casimir force $F^{\triangleleft|}_{\rm Cas,SIA}$ (– – –) is shown, as well as that between $F^{\triangleleft|}_{\rm Cas,SIA}$ and the van der Waals interaction $F^{\triangleleft|}_{\rm vdW,SIA}$ (-▼-) resulting in the *total* force $F^{\triangleleft|}_{\rm tot,SIA}$ (-●-). On (**a**) the temperature dependance at $\Delta\mu = 0.0$, $(x_\mu = 0.0)$ is depicted, while on (**b**) that of the field one at $T = T_c$, $(x_t = 0.0)$. The minimal separation between the objects confining the fluid medium is taken $L = 20$, where the cones's normalized height is taken $\Xi_H = 1$ and the parameters characterizing the interactions in the systems have the following values: $\lambda = 2.0$, $s_{2,c} = -0.01$ and $s_{1,c} = 1.0$.

critical fluctuating medium in which the interacting objects are immersed. For detail comment and argumentation on the choice of the presented substances the reader can refer to [1].

We open the discussion with the results presented on Fig. 3. Here the interplay between all of the aforementioned forces is clearly depicted. For the chosen separation $L = 20a_0 \simeq 12$ nm and substances, characterized by the set of coupling parameters $\{s_{1,c}; s_{2,c}; \lambda\}$, $F^{\triangleleft|}_{\rm crit,SIA}$ is negative for any value of x_t and x_μ, i.e., it corresponds to attraction between the cone and the plate in the entire temperature and filed intervals. Here a_0 is the distance between the xenon atoms at the critical point. From the data used here one can estimate that at $T = 1.0058T_c$ this component of the Casimir force reaches its minimum, with a value of $-0.9\Xi_R^2$ pN. With regard only to Fig. 3(**a**), at $\Delta\mu = 0$ for $T > T_c$ one has $\phi_b = 0$ and hence, substituting Eq. (34) instead of $H_A^{(\rm reg)}$ into Eq. (35) we see that $F^{\triangleleft|}_{\rm sing,SIA}$ is zero. On the other hand for $T < T_c$ the bulk order parameter $\phi_b \neq 0$ as the singular part of the Hamaker term (and consequently $F^{\triangleleft|}_{\rm sing,SIA}$) corresponds to repulsion. Therefore, the behavior of net Casimir interaction $F^{\triangleleft|}_{\rm Cas,SIA} = F^{\triangleleft|}_{\rm sing,SIA} + F^{\triangleleft|}_{\rm crit,SIA}$ coincides with that of its critical component above T_c, but one observes a single sign change at $T = 0.9805T_c$ in the "liquid" phase of the fluid. The superposition between the vdWF $F^{\triangleleft|}_{\rm vdW,SIA}$ and $F^{\triangleleft|}_{\rm sing,SIA}$, leads to interaction which is positive for any temperature and shows explicit decrease with the increase of the temperature like T^{-1} [see Eqs. (33) and (34)]. As a result the total force $F^{\triangleleft|}_{\rm tot,SIA}$ becomes repulsive outside the critical region, changes sign twice at $T_1 = 0.9901T_c$ and $T_2 = 1.036T_c$ after which turns attractive only near the critical point. It reaches minimum for $T_{\rm min} = 1.0061T_c$, which magnitude is $-0.71\Xi_R^2$ pN. The nature of this observation is dictated by the increased correlations between the fluctuations in the fluid close to T_c.

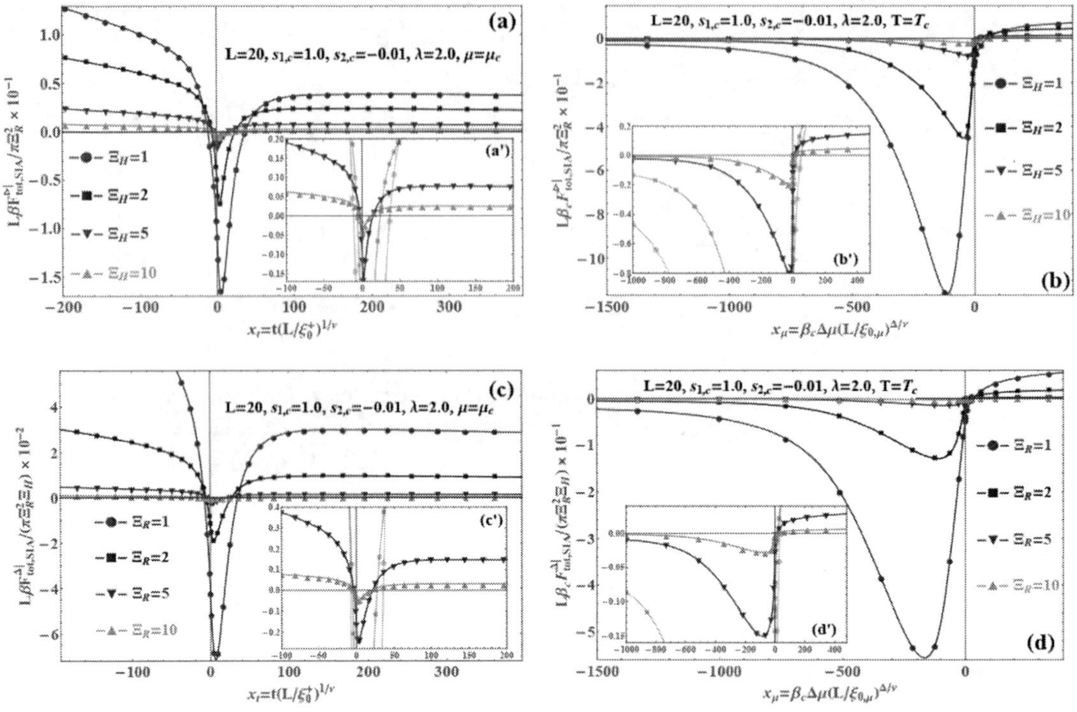

Figure 4. Behavior of the total force $F_{\text{tot,SIA}}$ (multiplied by the separation L and normalized by $k_B T$ and the area of the cones's base $\pi\Xi_R^2$) in $d=3$ for the cases where: [(**a**) and (**b**)] the cone's apex is toward the plate and [(**c**) and (**d**)] the height of the cone is parallel to the plate the cone interacts with. On (**a**) and (**c**) the temperature dependance at $\Delta\mu = 0.0$, $(x_\mu = 0.0)$ is depicted, while on (**b**) and (**d**) that of the field one at $T = T_c$, $(x_t = 0.0)$. The upper two subfigures illustrate how $F^{\rhd|}_{\text{tot,SIA}}$ changes with the increase of the cone's height Ξ_H, and on the lower ones how $F^{\triangle|}_{\text{tot,SIA}}$ is influenced by the variation of the base radius Ξ_R. There the force is additionally normalized by Ξ_H. The minimal separation between the objects confining the fluid medium is taken $L = 20$ and the parameters characterizing the interactions in the systems have the following values: $\lambda = 2.0$, $s_{2,c} = -0.01$ and $s_{1,c} = 1.0$.

The typical behavior of $F^{\lhd|}_{\text{sing,SIA}}$ as a function of the chemical potential at $T = T_c$ is presented on Fig. 3(**b**). The force is attractive in the "gas" phase of the fluid, i.e., for $x_\mu < 0$ and repulsive otherwise with an infinite slope at the critical point, where a sign change occurs. Hence, the CCF changes sign once in the "liquid" phase $(x_\mu > 0)$ at $\beta_c\Delta\mu \simeq 1.2495 \times 10^{-3}$ and its minimum increases in magnitude slightly in comparison to $F^{\lhd|}_{\text{crit,SIA}}$, as $(F^{\lhd|}_{\text{Cas,SIA}}/F^{\lhd|}_{\text{crit,SIA}})|_{\min} = 1.0253$, where $F^{\lhd|}_{\text{crit,SIA}}(x_{\mu,\min}) = -6.1077\Xi_R^2$ pN. From Eq. (33) follows that the vdWF does not depend on μ, as a result of which the singular component of the CCF only "shifts up" and changes sign for $\mu < \mu_c$ at $\beta_c\Delta\mu \simeq -3.293 \times 10^{-3}$. When add up to the Casimir force component stemming from the finite-size contributions, the total force exhibits behavior similar to that of $F^{\lhd|}_{\text{crit,SIA}}$, but now the sign change occurs closer to the critical point at $\beta_c\Delta\mu \simeq 0.4794 \times 10^{-3}$.

Some results about the behavior of the total force in a cone-plate system with geometry $(\rhd|)$ are presented on Fig. 4(**a**) and (**b**). The first obvious result we notice is that for fixed radius of the base, the force decreases with the increase of the height like Ξ_H^{-2} [see Eqs. (37a) and (37b)]. This is easily understood given that the CCF strongly depends on the confining area of the object's surface. When Ξ_H is increased at fixed Ξ_R the area of the surface elements, which are close to the plate and hence have higher contribution weight, decreases which in turn results in critical Casimir force with lower strength. The same is in a way true for the vdWF, but in this case the decrease is caused from the decline of the number of constituents in the volume near the cone's apex which are near the surface of the plate. As a function of x_t the force changes sign twice, as the temperature for which it occurs in the "gas" phase decreases towards T_c with the increase of Ξ_H, and viceversa for $T > T_c$. A simple quantitative assessment indicates that $t_1 = -4.633 \times 10^{-3}$ and $t_2 = 12.776 \times 10^{-3}$, where $t_i \equiv [T_i(\Xi_H = 1) - T_i(\Xi_H = 10)]/T_i(\Xi_H = 10)$, $i = 1, 2$ with $T_1(\Xi_H = 1) = 0.9944T_c$ and $T_2(\Xi_H = 1) = 1.0183T_c$. The temperature at which the minimum of the force is expected to occur also decreases towards T_c when Ξ_H is increased. In addition, for the difference in the magnitude of the force at the point where it reaches its minimum for the lowest and highest values of Ξ_H considered here we found that $[F^{\triangleright|}_{\text{tot,SIA}}(\Xi_H = 1)/F^{\triangleright|}_{\text{tot,SIA}}(\Xi_H = 10)] = 36.1428$, where $F^{\triangleright|}_{\text{tot,SIA}}(\Xi_H = 10) = -5.0155\Xi_R^2$ fN.

For the same geometry, but as a function of x_μ, $F^{\triangleright|}_{\text{tot,SIA}}$ changes sign only once for $\Delta\mu > 0$, as the value of the chemical potential for which this is observed decreases towards μ_c with the increase of Ξ_H. The strength of the force at its minimum also decreases as does the value of x_μ at which it emerges. This case is quantitatively describes as follows: $x_\mu(\Xi_H = 1)|_{F_{\text{tot}}=0} = 52.763$, $x_\mu(\Xi_H = 10)|_{F_{\text{tot}}=0} = 4.096$, $x_\mu(\Xi_H = 1)|_{F_{\text{tot,min}}} = -117.649$, $x_\mu(\Xi_H = 10)|_{F_{\text{tot,min}}} = -19.683$, $[F^{\triangleright|}_{\text{tot,SIA}}(\Xi_H = 1)/F^{\triangleright|}_{\text{tot,SIA}}(\Xi_H = 10)]_{\text{min}} = 58.905$, with $F^{\triangleright|}_{\text{tot,SIA}}(\Xi_H = 10) = -21.346\Xi_R^2$ fN.

As far as the geometry $(\triangle|)$ is concerned, here we see that its differences with the one designated by $(\triangleright|)$ are only quantitative, and hence no special comments are needed. The results are depicted on Fig. 4(**c**) and (**d**).

For the sake of completeness, we give the critical temperature of xenon $T_c = 289.765$ K and the value of the critical chemical potential per $k_B T_c$: $\beta_c \mu_c = -16.213$, calculated using Eqs. (2.2b), (5.37b) and (8.8) from [20].

Although the magnitude of the TF together with the CCF may seems rather negligible, a comparison with say the weight of a single particle proofs otherwise. For instance, the weight of a platinum conical in shape particle of height $\Xi_H = 100$ and base radius $\Xi_R = 10$ is approximately $F_{\text{w}}^{\text{cone}} \approx 1.2 \times 10^{-2}$ pN. In the liquid phase of xenon at $T = 206.4$ K and $\Delta\mu = 0$ one finds that $F^{\triangleleft|}_{\text{tot}} \approx 120$ pN, $F^{\triangleright|}_{\text{tot}} \approx 1.2 \times 10^{-2}$ pN and $F^{\triangle|}_{\text{tot}} \approx 17$ pN. Therefore the repulsive part of the total force in one such concrete system is indeed capable of levitating a single particle or a system of such. Thus, one can indeed make use of the interplay between the quantum and thermodynamical fluctuations for governing the behavior of objects, say colloidal particles, at small distances. It can also provide a strategy for solving problems with handling, feeding, trapping and fixing of micro parts in nanotechnology resolving the issues related to sticking of the particles on the surface of the mechanical manipulator utilizing, e.g., the *reversible dependence* on the forces under minute changes of the temperature of the critical medium. One can perform grabbing of particles for small values of x_t, where the force is attractive and release them at a given spacial position after slightly increasing or decreasing of temperature achieving in that way a value of x_t with a repulsive TF.

Acknowledgements

The author gratefully acknowledges the financial support via Contract No. DN 02/8 of Bulgarian NSF.

https://doi.org/10.1051/matecconf/201814501008

References

[1] G. Valchev, D. Dantchev, Phys. Rev. E **96**, 022107 (2017)
[2] G. Valchev, D. Dantchev, Phys. Rev. E **92**, 012119 (2015)
[3] M.E. Fisher, P.G. de Gennes, C. R. Seances Acad. Sci. Paris Ser. B **287**, 207 (1978)
[4] M. Krech, *Casimir Effect in Critical Systems* (World Scientific, Singapore, 1994)
[5] J.G. Brankov, D.M. Dantchev, N.S. Tonchev, *The Theory of Critical Phenomena in Finite-Size Systems - Scaling and Quantum Effects* (World Scientific, Singapore, 2000)
[6] H.B. Casimir, Proc. K. Ned. Akad. Wet. **51**, 793 (1948)
[7] H.B.G. Casimir, D. Polder, Phys. Rev. **73**, 360 (1948)
[8] H.B.G. Casimir, Physica **19**, 846 (1953)
[9] E. Buks, M.L. Roukes, Europhys. Lett. **54**, 220 (2001)
[10] E. Buks, M.L. Roukes, Phys. Rev. B **63**, 033402 (2001)
[11] H.B. Chan, V.A. Aksyuk, R.N. Kleiman, D.J. Bishop, F. Capasso, Phys. Rev. Lett. **87**, 211801 (2001)
[12] D. Dantchev, G. Valchev, J. Coll. Interf. Sci. **372**, 148 (2012)
[13] R. Garcia, M.H.W. Chan, Phys. Rev. Lett. **83**, 1187 (1999)
[14] V.A. Parsegian, *Van der Waals Forces* (Cambridge University Press, 2006)
[15] J.N. Israelachvili, *Intermolecular and surface forces* (Academic, London, 2011)
[16] B. Derjaguin, Kolloid Z. **69**, 155 (1934)
[17] H.J. Bartsch, *Mathematische Formeln* (Fachbuchverlag Leipzig, 1991)
[18] G. Valchev, Ph.D. thesis (2015)
[19] D. Dantchev, F. Schlesener, S. Dietrich, Phys. Rev. E **76**, 011121 (2007)
[20] D.C. Johnston, *Advances in Thermodynamics of the van der Waals Fluid* (Morgan & Claypool Publishers, 2014)

MATEC Web of Conferences **145**, 01009 (2018) https://doi.org/10.1051/matecconf/201814501009
NCTAM 2017

Order parameter profiles in a system with Neumann – Neumann boundary conditions

Vassil M. Vassilev[1,*], *Daniel M.* Dantchev[1,**], and *Peter A.* Djondjorov[1,***]

[1] *Institute of Mechanics, Bulgarian Academy of Sciences, Acad. G. Bonchev St., Bl. 4, 1113 Sofia, Bulgaria.*

Abstract. In this article we consider a critical thermodynamic system with the shape of a thin film confined between two parallel planes. It is assumed that the state of the system at a given temperature and external ordering field is described by order-parameter profiles, which minimize the one-dimensional counterpart of the standard ϕ^4 Ginzburg–Landau Hamiltonian and meet the so-called Neumann – Neumann boundary conditions. We give analytic representation of the extremals of this variational problem in terms of Weierstrass elliptic functions. Then, depending on the temperature and ordering field we determine the minimizers and obtain the phase diagram in the temperature-field plane.

1 Introduction

In the current article we consider one generic mean-field model describing a phase transition in a film in which the surfaces order exactly at the same temperature as the bulk. Mathematically this is described by imposing (Neumann, Neumann) boundary conditions on the order parameter. This type of boundary conditions can be achieved in a magnetic system. They can, in principle, be also accomplished for a fluid, but a very precise arrangement for the surfaces shall be performed so that the wall-fluid interaction exactly matches that one of the interactions inside the fluid system. Let is recall that in the vicinity of the critical temperature T_c of the bulk system, one observes a diversity of surface phase transitions [1, 2] of different kind in which the surface orders before, together, or after ordering in the bulk of the system, which are known as normal (or extraordinary), surface-bulk and ordinary surface phase transitions. Mathematically these surface transitions correspond to different boundary conditions imposed on the order parameter characterizing the system. For a simple fluid or for binary liquid mixtures the wall generically prefers one of the fluid phases or one of the components. In the vicinity of the bulk critical point the last leads to the phenomenon of critical adsorption [3–15]. This can be modelled by considering local surface field h_1 acting solely on the surfaces of the system. When the system undergoes a phase transition in its bulk in the presence of such surface ordering fields one speaks about the "*normal*" transition [16]. It has been shown that it is equivalent, as far as the leading critical behavior is concerned, to the "*extraordinary*" transition [2, 16] which is achieved by enhancing the surface couplings stronger than the bulk couplings. The usual way one mathematically

*The authors gratefully acknowledge the financial support via contract DN 02/8 with Bulgarian National Science Fund.
e-mail: vasivas@imbm.bas.bg
**e-mail: danieldantchev@gmail.com
***e-mail: padjon@imbm.bas.bg

describes these boundary conditions is to require that the order parameter diverges and, thus, one has the so called $(+,+)$ or $(+,-)$ boundary conditions which differ by the equal, or opposite behaviour of the order parameter in vicinity of the borders of the system. There is also a special case of surface enrichment, when the surface orders simultaneously with the bulk. This is normally termed surface-bulk, or special phase transition. One way to achieve it is to enhance the coupling on the surface and very near to it, so that one achieves the delicate balance needed to get the both orderings to become simultaneously. On a mean field level this can be described by imposing Neumann boundary conditions on the boundary. These are the boundary conditions studied in the current article.

2 Statement of the problem

Let us consider an Ising type critical thermochemical system having the shape of a thin film confined between two parallel planes placed at a distance L from one another. Then, within the framework of the Ginzburg–Landau theory of phase transitions, see e.g. [17, 18], the state of the system is described by the minimizers of the one-dimensional counterpart

$$\mathcal{F}[\phi;\tau,\eta,L] = \int_0^L \mathcal{L}(\phi,\phi')dz, \qquad \mathcal{L}(\phi,\phi') = \phi'^2 + \phi^4 + \tau\phi^2 - \eta\,\phi, \tag{1}$$

of the standard ϕ^4 Ginzburg–Landau Hamiltonian in terms of the order parameter $\phi(z)$, $z \in (0, L)$ being the independent variable associated with the position of a layer perpendicular to the bounding planes. Here, the parameters τ and η represent the temperature of the system and the applied ordering field, respectively, and the prime indicates differentiation with respect to the variable z.

Given three real numbers $L > 0$, τ and η, we are interested in the minimizers $\phi = \phi(z)$ of the functional $\mathcal{F}[\phi;\tau,\eta,L]$, which are:

(*i*) continuous together with their derivatives up to the second order in the interval $[0, L]$;

(*ii*) satisfy the so-called Neumann – Neumann boundary conditions

$$\phi'(0) = \phi'(L) = 0. \tag{2}$$

Actually, (2) are the natural boundary conditions associated with the regarded variational problem, which means that any variation of the order parameter ϕ is allowed at both end points $z = 0$ and $z = L$.

It is clear that each such minimizer should satisfy the corresponding Euler-Lagrange equation, which, on account of expressions (1), reads

$$\phi'' - \phi\left(\tau + 2\phi^2\right) + \frac{\eta}{2} = 0. \tag{3}$$

Obviously, the order parameter $\phi_b(\tau,\eta,L)$ of the bulk system, which is a constant, with respect to z, solution of equation (3) determined as the real root of the cubic polynomial

$$Q(\phi) = \phi\left(\tau + 2\phi^2\right) - \frac{\eta}{2} \tag{4}$$

for which

$$\mathcal{F}[\phi_b;\tau,\eta,L] = L\left(\phi_b^4 + \tau\phi_b^2 - \eta\,\phi_b\right) \tag{5}$$

attains its minimum, meets the requirements (*i*) and (*ii*). Therefore, a minimizer of the regarded functional with the required properties always exists and the question is whether it is ϕ_b or some other (but non-constant) solution $\phi(z)$ of equation (3) such that the conditions (*i*) and (*ii*) hold and

$$\mathcal{F}[\phi;\tau,\eta,L] \le \mathcal{F}[\phi_b;\tau,\eta,L]. \tag{6}$$

Equation (3) possesses a first integral $\varepsilon(\phi)$ of the form

$$\varepsilon(\phi) = \phi'^2 - \phi^4 - \tau\phi^2 + \eta\,\phi, \tag{7}$$

i.e. $\varepsilon = \varepsilon(\phi)$ is a certain real number on any non-constant sufficiently smooth real-valued solution of equation (3). Thus, the problem we are interested in can be formulated as follows. Given three real numbers $L > 0$, τ and η, find the non-constant real-valued solutions $\phi(z)$ of the equation

$$\left(\frac{d\phi}{dz}\right)^2 = P(\phi), \qquad P(\phi) = \phi^4 + \tau\,\phi^2 - \eta\,\phi + \varepsilon \tag{8}$$

corresponding to certain $\varepsilon \in \mathbb{R}$, which meet the conditions (*i*), (*ii*) and (6).

3 Solution of the problem

Let us first express the general solution of equation (8) taking advantage of the fact that the polynomial $P(\phi)$ should have at least one simple real root ρ because of the considered boundary conditions (2). Then

$$\varepsilon = \rho\left(\eta - \rho^3 - \rho\tau\right) \tag{9}$$

as $P(\rho) = 0$, and since the right-hand side of equation (8), that is $P(\phi)$, is a polynomial of fourth degree with respect to the variable ϕ, we can, following [19, pp. 452–455], express each solution of equation (8) in the form

$$\phi(z) = \rho + 3\frac{\eta - 2\rho\left(2\rho^2 + \tau\right)}{6\rho^2 + \tau - 12\wp(z; g_2, g_3)} \tag{10}$$

where $\wp(z; g_2, g_3)$ is the Weierstrass elliptic function with elliptic invariants

$$g_2 = \eta\rho - \rho^4 - \rho^2\tau + \frac{\tau^2}{12}, \qquad g_3 = -\frac{1}{432}\left[27\eta^2 + 72\rho\tau\left(\rho^3 + \rho\tau - \eta\right) + 2\tau^3\right]. \tag{11}$$

Now, taking into account the well known properties of the Weierstrass elliptic functions (see, e.g., [20]), it is evident that each function of form (10) is real-valued provided that $z, \rho, g_2, g_3 \in \mathbb{R}$, just as in the considered case. Consequently

$$\phi'(z) = 36\wp'(z; g_2, g_3)\frac{\eta - 2\rho\left(2\rho^2 + \tau\right)}{\left[6\rho^2 + \tau - 12\wp(z; g_2, g_3)\right]^2} \tag{12}$$

and hence, bearing in mind that $\eta - 2\rho\left(2\rho^2 + \tau\right) \neq 0$, since ρ is a simple root of the polynomial $P(\phi)$, $\lim \wp(z; g_2, g_3)|_{z\to 0} = \infty$ and the Weierstrass equation

$$\wp'^2 = 4\wp^3 - g_2\wp - g_3, \tag{13}$$

we see that each function of form (10), which is such that

$$\wp'(L; g_2, g_3) = 0 \tag{14}$$

meets the conditions (*i*) and (*ii*).

4 Determination of minimizers and phase diagram

Hereafter we assume $L = 1$ and solve numerically for ρ the transcendental equation (14) for given values of the parameters τ and η. In this way we find the functions of form (10) that satisfy the conditions (*i*) and (*ii*). Then, we compare the energy of each such state of the system with the energy corresponding to the constant solution $\phi_b(\tau, \eta, 1)$ and determine the minimizers.

Two minimizers of the functional $\mathcal{F}[\phi; \tau, \eta, 1]$ corresponding to $\tau = -400$ and $\eta = -97.52$ are depicted in Figure 1. They both have energy equal to the energy $\mathcal{F}[\phi_b; \tau, \eta, 1]$ of the "finite" bulk system. As a result of a number of lengthy computations, we have obtained the phase diagram shown

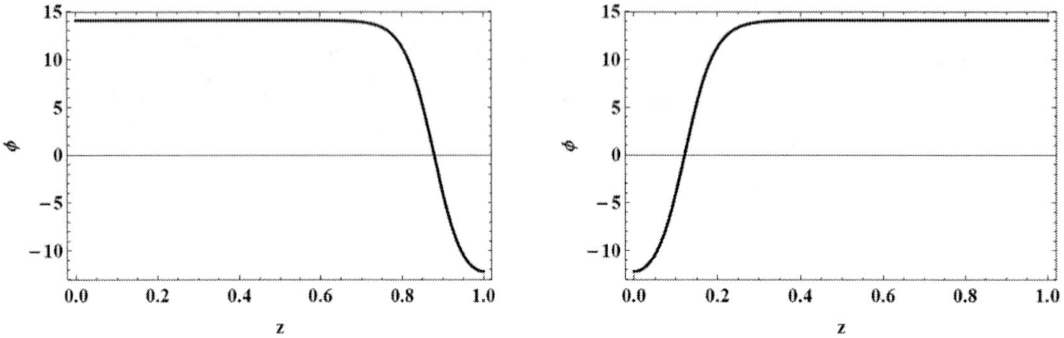

Figure 1. Two minimizers of the functional $\mathcal{F}[\phi; \tau, \eta, 1]$ corresponding to $\tau = -400$, $\eta = -97.52$.

in Figure 2.

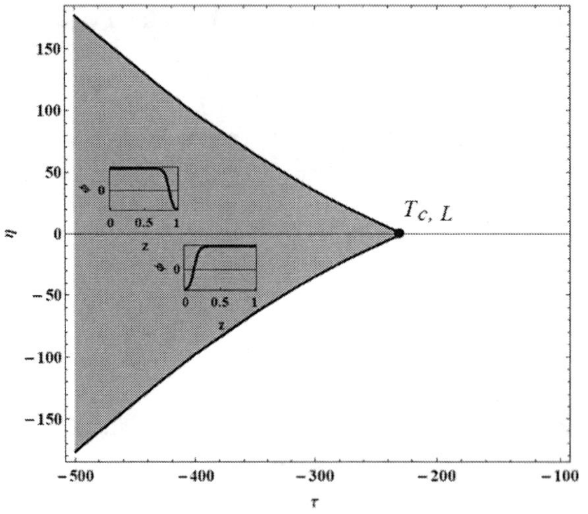

Figure 2. Phase diagram. The blue region in the (τ, η)-plane comprises the values of the parameters τ and η for which there co-exists two non-constant minimizers, while in the white region we find only the constant solution $\phi_b(\tau, \eta, 1)$. All these three types of minimizers co-exists at the border of the foregoing regions indicated by the black thick curves. The point $T_{c,L}$ of coordinates $\tau = -227.556$, $\eta = 0$ marks the critical point of the finite system.

5 Concluding remarks

This is a work in progress, which is focused mainly on the mathematical aspects of the problem. The physics behind this study will be discussed elsewhere. In the current article we have found the general form of the solution, see Eq. (10), of the equation (3) for the order parameter profile of a system with a film geometry subjected to the (Neumann-Neumann) boundary conditions – see Eq. (2). The usual form of the order parameter profile as a function of the orthogonal variable z is shown in Fig. 1. One observes that they are mirror-symmetric with respect to the middle of the system. We have determined the phase diagram of the film system – it is shown in Fig. 2. The two branches of the diagram below the point $T_{c,L}$ are symmetric with respect to the $\eta = 0$ line. For such a system it would be interesting to determine the response functions – the local and the total ones, as well as the force acting between the plates of the system which near $T_{c,L}$ is known as the Casimir force [21–24].

References

[1] K. Binder, *Phase Transitions and Critical Phenomena* vol 8 (Academic, London, 1983), pp. 1–144.

[2] H. W. Diehl, *Phase Transitions and Critical Phenomena* vol 10 (Academic, New York, 1986), p. 76.

[3] R. Evans, J. Phys.: Condens. Matter **2**, 8989 (1990).

[4] G. Flöter, S. Dietrich, Z. Phys. B **97**, 213–232 (1995).

[5] M. Tröndle, L. Harnau, S. Dietrich, J. Chem. Phys. **129**, 124716 (2008).

[6] Z. Borjan, P. J. Upton, Phys. Rev. E **63**, 065102 (2001).

[7] R. Evans, U. M. B. Marconi, J. Chem. Phys. **84** 2376–2399, (1986).

[8] R. Okamoto, A. Onuki, J. Chem. Phys. **136**, 114704 (2012).

[9] A. Maciòłek, A. Ciach, J. Stecki, J. Chem. Phys. **108**, 5913–5921 (1998).

[10] D. Dantchev, F. Schlesener, S. Dietrich, Phys. Rev. E **76**, 011121 (2007).

[11] A. Drzewiński, A. Maciołek, A. Barasiński, S. Dietrich, Phys. Rev. E **79**, 041145 (2009).

[12] J. W. Cahn, J. of Chem. Phys. **66**, 3667–3672 (1977).

[13] P. G. de Gennes, Rev. Mod. Phys. **57** (3), 827–863 (1985).

[14] F. P. Toldin, S. Dietrich, J. Stat. Mech **11**, P11003 (2010).

[15] D. Dantchev, J. Rudnick, M. Barmatz, Phys. Rev. E **75**, 011121 (2007).

[16] T. W. Burkhardt, H. W. Diehl, Phys. Rev. B **50**, 3894–3898 (1994).

[17] A. O. Parry, R. Evans, Physica A **181** 250–296 (1992)

[18] P. C. Hohenberg, A. P. Krekhov, Physics Reports **572**, 1–42 (2015)

[19] E. Whittaker, G. Watson, *A Course of Modern Analysis* (Cambridge University Press, Cambridge, 1927)

[20] D.F. Lawden, *Elliptic functions and applications* (Springer, New York, 1989)

[21] M. E. Fisher, P. G. de Gennes, C. R. Seances Acad. Sci. Paris B, 287–207 (1978)

[22] M. Krech, *Casimir Effect in Critical Systems* (World Scientific, Singapore, 1994)

[23] J. G. Brankov, D. M. Dantchev, N. S. Tonchev *The Theory of Critical Phenomena in Finite-Size Systems – Scaling and Quantum Effects* (World Scientific, Singapore, 2000)

[24] N. B. G. Casimir, On the Attraction Between Two Perfectly Conducting Plates. A Proceedings of the Koninklijke Nederlandse Akademie van Wetenschappen, Series A, **51**, p. 793 (1948).

MATEC Web of Conferences **145**, 01010 (2018) https://doi.org/10.1051/matecconf/201814501010
NCTAM 2017

Statistical modeling of particle mater air pollutants in the city of Ruse, Bulgaria

Evelina Veleva[2] and *Ivanka* Zheleva[1,*]

[1] Department Applied Mathematics and Statistics, Ruse University, 8 Studentska Str., 7017, Ruse, BULGARIA, www.uni-ruse.bg
[2] Department Thermo engineering, Hydraulics, Ecology, Ruse University, 8 Studentska Str., 7017, Ruse, BULGARIA, www.uni-ruse.bg

Abstract This paper is devoted to examine the PM10 and PM2.5 pollution in Ruse region, Bulgaria. It is a continuation of our previous work [1] where we presented a statistical analysis and modelling of the level of PM10 air pollution in Ruse using data from one of two monitoring stations (station 2) in the city. Now in this paper we present statistical analysis of the level of PM10 pollutant on the basis of data from the another monitoring station (station 1). The measurements cover the period since 2015 up to now. The results from analysis and modelling of PM2.5 air pollutant are also presented and commented in the paper.

1 Introduction

Ruse is a city in the north-eastern part of Bulgaria on the Danube river coast, which is the north border between Bulgaria and Romania. The climate of the region of Ruse is temperate continental, characterized by cold winters and dry, warm summers. Spring and autumn are short. Recently it becomes clear that the mean values of the temperature in Ruse region are slightly goes up for the last 40 years and they are bigger than the mean temperature for Bulgaria [6]. Air pollution especially by particle matter (PM10 and PM2.5), which in Ruse region is going up recently, maybe affects temperature, atmospheric pressure and humidity in the region. All this in our opinion may be a reason for the increase in average temperatures nowadays for Ruse region. That is why studies on the air pollution by particle matter is needed for better understanding of these processes and for taking adequate measures to protect environment.

2 Material and methods

In the town of Ruse, Bulgaria the regular measurement of air pollution starts on 24.10.2015 [2, 3]. It is carried out by an automatic measuring station Vazrazhdane (station 1) and by a stationary station in the Zdravets - east district of the town (station 2). Since 30.09.2016 measurements are performed also by a mobile automatic station. The two stationary stations

* Corresponding author: izheleva@uni-ruse.bg

are located about 2.20 kilometers apart (the distance between them is measured by the Google Earth program) (Fig.1).

A map of the location of the stations is published on the site [3] of the Regional Inspectorate of Environment and Water in Ruse, which shows that each one covers an area with a radius of about 1,7 km.

For this study we use official data for particle matter air pollutant PM10 from stations 1 and 2 and for PM2.5 – from station 1 [3].

We use SPSS software to analyze and model data from the measurement stations in Ruse.

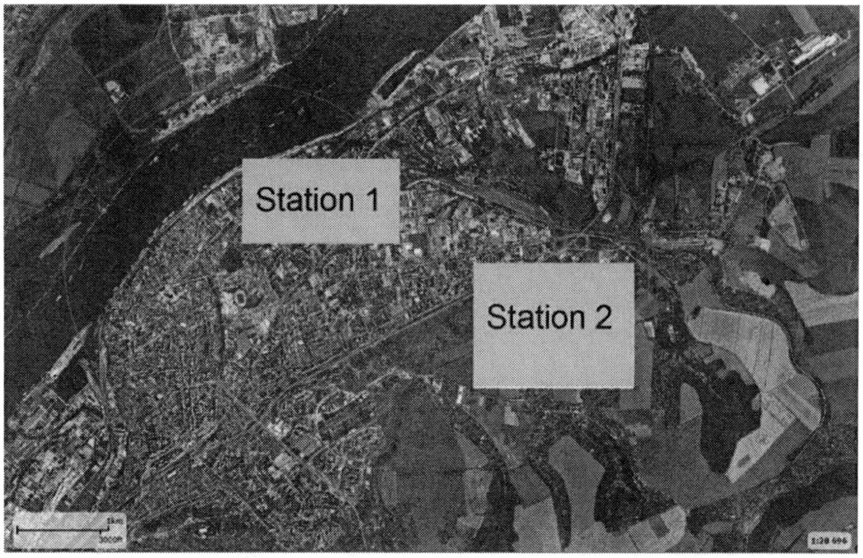

Fig 2 . Locations of the atmospheric air monitoring stations in the town of Ruse, Bulgaria [3]

3 Results and discussions

Station 1 monitors 8 pollutants: sulfur dioxide (SO_2), nitrogen dioxide (NO_2), nitrogen oxide (NO), carbon monoxide (CO), ground level ozone (O_3), particulate matter less than 10 micrometers in diameter (PM10), benzene (C_6H_6) and particulate matter less than 2.5 micrometers in diameter (PM2.5). Their main descriptive statistics for the period 24.10.2015 - 31.07.2017 are given in Table 1 below.

Table 1. Descriptive statistics for pollutants monitored at Station 1

Polutant	N	Minimum	Maximum	Mean	Std. Deviation	Skewness	Kurtosis
SO_2	277	5.60	22.20	8.83	2.46	2.33	6.54
NO_2	609	0.00	65.40	21.78	9.95	1.42	2.66
NO	609	0.00	211.90	9.79	15.35	6.10	58.32
CO	609	0.00	3.50	0.48	0.35	2.75	12.31
O_3	274	7.70	94.20	42.92	18.44	0.22	-0.46
PM10	584	12.00	176.00	43.10	25.75	2.03	5.02
C_6H_6	605	0.00	12,90	0.92	1.01	4.90	42.41
PM2.5	604	3.90	161,10	26.20	22.04	2.55	7.74

Above the norm are only particulate matter PM10 and PM2.5. The norm for PM10 is: average annual rate below 40 (μg/m^3) and average daily rate below 50 (μg/m^3). For year 2016 the average annual rate for PM10 is 41.17 (μg/m^3) and for the one-year period 01.08.2016 – 31.07.2017 it is 41.82 (μg/m^3) – both above the average annual norm. 146 of 584 valid observations on PM10 are above the average daily norm 50 (μg/m^3), i.e. 25% of all monitored days. The maximum value 176.00 (μg/m^3) exceeds 3.52 times the average daily norm 50 (μg/m^3).

The norm for PM2.5 is average annual rate under 25(μg/m^3). The average annual rate of PM2.5 for 2016 is 23.52, i.e. within the norm. For the one-year period 01.08.2016 – 31.07.2017 that includes the last 365 days of the sample it is 26.05, i.e. above the norm.

Four of the values of PM10 and two of the measurements of PM2.5 in august 2016 are suspiciously recorded as zero. We treated them as missing as they probably are.

As it can be seen from Figure 2, the values of PM10, measured in Station 1 (blue line), Station 2 (red line) and PM2.5 (green line) in Ruse, are highly and positively correlated. The Pearson product-moment correlation coefficients for the three variables: PM10 at Station 1 (PM10_s1), PM2.5 (measured only at Station 1, PM2_5) and PM10 at Station 2 (PM10_s2) are given in Table 2. All are statistically significant ($p < 0.0005$).

Table 2. Correlations between the three factors out of norm in Ruse

	PM10_s1	PM2_5	PM10_s2
PM10_s1	1	0.815	0.773
PM2_5	0.815	1	0.813
PM10_s2	0.773	0.813	1

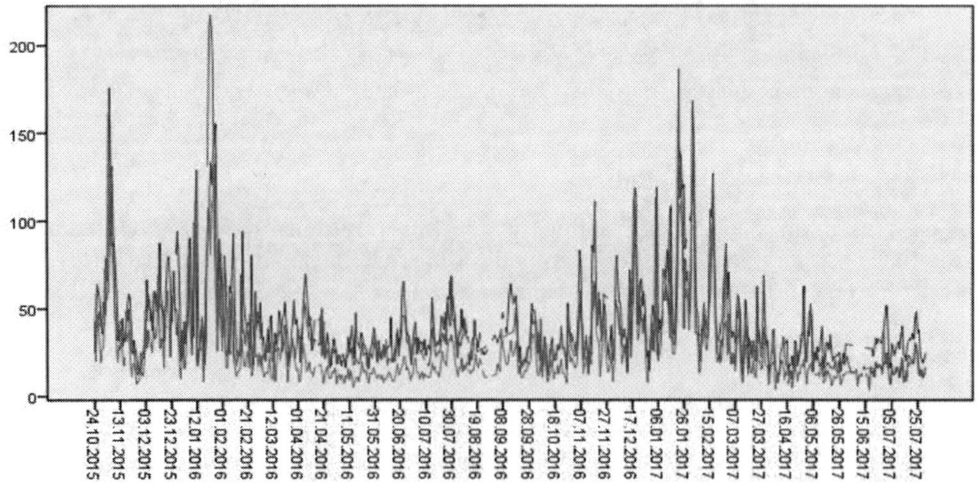

Fig. 2 Values of PM10 at Station 1 (blue line), Station 2 (red line) and PM2.5 (green line)

The distributions of PM10 at Stations 1 and PM2.5 are probably unimodal as can be seen in Figure 3.

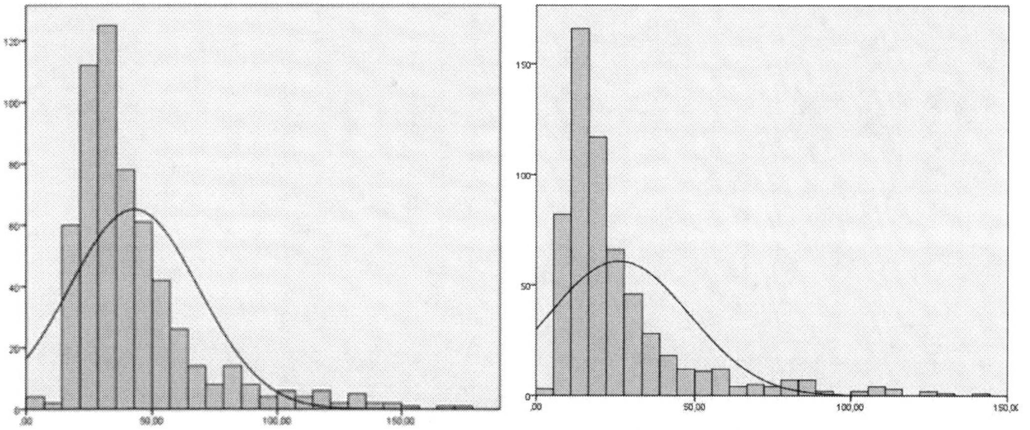

Fig. 3 Histograms with normal curve of PM10 at Station 1 (left) and PM2.5 (right)

The Kolmogorov – Smirnov and Shapiro – Wilk tests for normality are both significant ($p < 0.0005$) so we have to reject the hypotheses of normality of the distributions of PM10 and PM2.5.

3.1 Analysis and modelling of PM10 values measured at Station 1

In this section we will look in more detail at the values of the dangerous air pollutant PM10 ($\mu g/m^3$) measured in Station 1 in Ruse for the period 24.10.2015 - 31.07.2017. The study covers 647 days. Values for 584 days were measured, i.e. for 90.26% of the days, and the values for 63, i.e. for 9.74% of them are missing.

The sample quartiles are Q1=27.10; Q2=35.70 and Q3=50.75. All 41 outliers lie above 86.225, i.e. 1.5*(Q3-Q1) from the third quartile Q3; 14 of them can be classified as extreme – they exceed 121.70, i.e. 3*(Q3-Q1) from the third quartile Q3. The outliers are distributed only in winter months: 21 in January (8 extreme), 9 in February (3 extreme), 6 in November (3 extreme) and 5 in December (0 extreme). Almost all values above the norm, in particular 118 out of 146, are measured in the winter months: 32 in January, 24 in February, 10 in March, 21 in November and 31 in December. In fact, 57.14% of the valid measurements in January are above the norm. For February, November and December these percentages are 42.11%, 35% and 53.45% respectively. The monthly averages of PM10 are shown on Figure 4. Although seasonality is clearly expressed, the observation period is not long enough to examine seasonality with respect to the months of the year.

Figure 5 presents the mean values of PM10 for weekdays. The analysis of variance performed did not show statistically significant ($p = 0.94$) differences in the means over the days of the week.

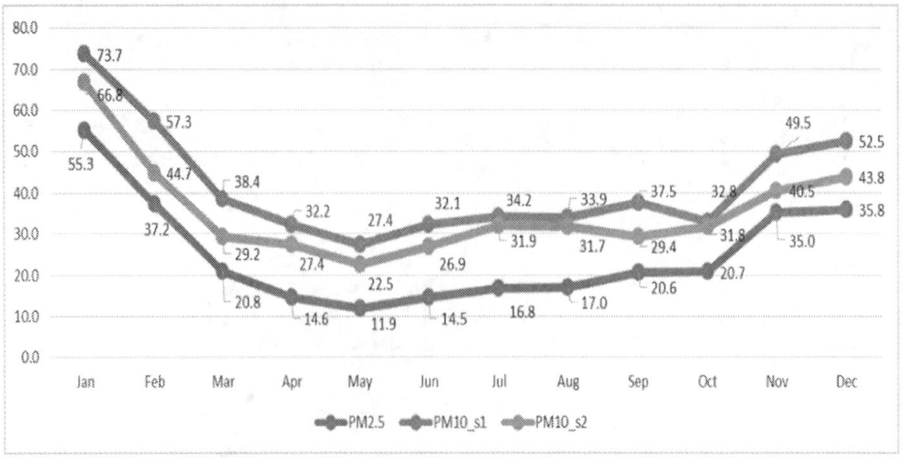

Fig. 4 Monthly averages of PM2.5 and PM10 at Stations 1 and 2

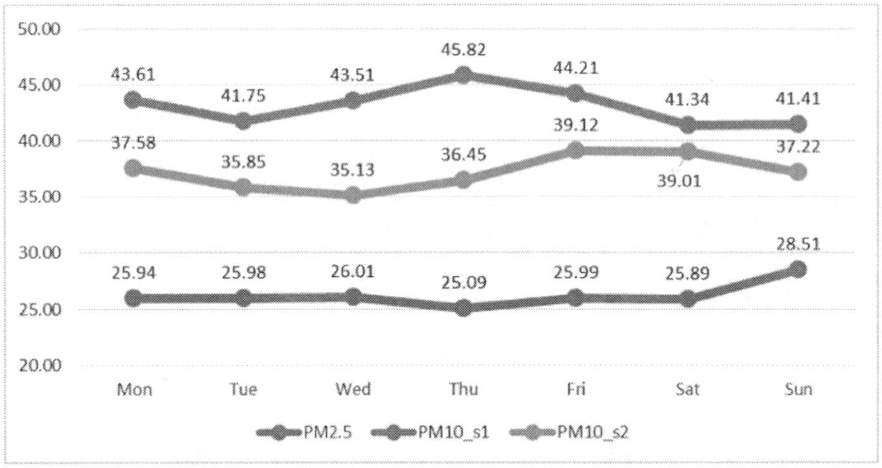

Fig. 5 Mean values of PM2.5 and PM10 at Stations 1 and 2 for weekdays

We will model the time series of values of PM10 at Station 1 using the method of ARIMA (Auto-Regressive Integrated Moving Average) as one of the better known short-term forecasting methods (see [4, 5]). We replace the 63 missing values using linear interpolation. Because in the future it is expected new extreme values to be observed, it is not appropriate to exclude the 41 outliers from the study.

The best reported model using Expert modeler mode in Forecasting procedure (Create Traditional Models) of SPSS gives the following autoregressive ARIMA(1,0,2)(1,0,1)$_7$ model:

Model	Model Fit statistics						Ljung-Box Q(18)			Number of Outliers
	Stationary R-squared	R-squared	RMSE	MAPE	MAE	Normalized BIC	Statistics	DF	Sig.	
1	0.599	0.632	15.305	25.535	10.229	5.516	30.231	14	0.007	1

Model	Transformation	Model Parameters		Estimate	SE	t	Sig.
1	Natural Logarithm	Constant		3.612	0.055	65.597	0.000
		AR	Lag 1	0.821	0.026	31.021	0.000
		MA	Lag 2	0.152	0.046	3.338	0.001
		AR, Seasonal	Lag 1	-0.792	0.188	-4.222	0.000
		MA, Seasonal	Lag 1	-0.738	0.208	-3.553	0.000

Outliers		Estimate	SE	t	Sig.
05.01.2017	Additive	1.186	0.234	5.061	0.000

Using the Ljung – Box test for randomness of the residuals we have evidence to conclude that the fitted Model 1 is not adequate. The model determines the observation on 05.01.2017 as an additive outlier, i.e. it corresponds to a surprisingly large (in our case) or small value, but the subsequent observations are unaffected by it. We introduce in the model one intervention variable I,

$$I = \begin{cases} 0, & t \neq 05.01.2017 \\ 1, & t = 05.01.2017 \end{cases} \tag{1}$$

and run again the Expert modeler, which gives a model without outliers, but having identical characteristics with Model 1, i.e. still not adequate. Undoubtedly, the natural logarithm transformation is necessary, because the variance of the time series is obviously non-stationary (see Fig.3). Let us denote by Z_t, $Z_t = \ln(Y_t)$, the transformed series of the original data Y_t. The new series Z_t are shown on Figure 6, it has already stationary variance.

The autocorrelation (ACF) and partial autocorrelation (PACF) functions of Z_t are given on Figure 7.

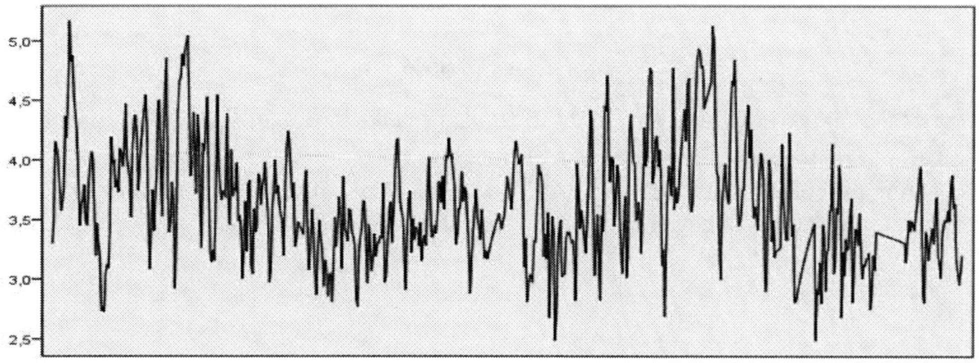

Fig. 6 Time series plot of Zt

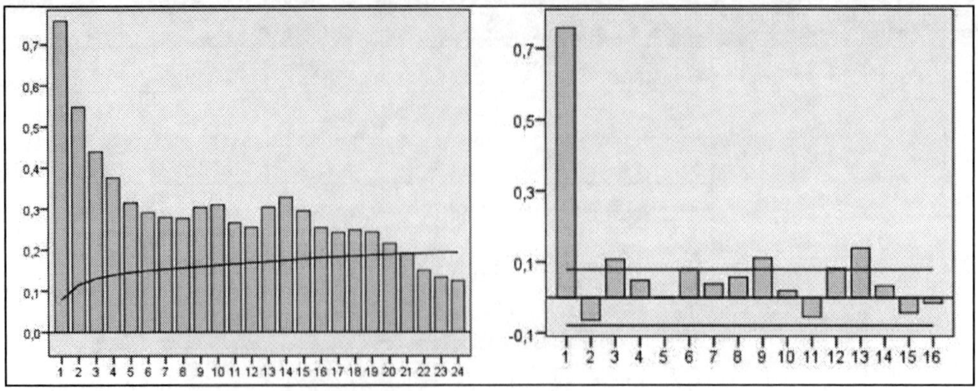

Fig.7 The ACF (left) and PACF (right) of transformed series Zt

The autoregressive part of the model, i.e. the AR component is probably 1, since the PACF "cuts off" at lag 1. In [1] we show that the best model for PM10, measured at Station 2 in the period 24.10.2015 - 25.05.2017, is the AR(1) model for the natural logarithm transformed series Z'_t of the original data,

$$Z'_t = 3.544 + 0.749 Z'_{t-1} + \varepsilon'_t,$$

where ε'_t is a shock, innovation, or random error at time t. This model has 3 outliers: Innovational at 12.01.2016, Additive at 15.12.2016 and Level Shift at 07.04.2017. We used three intervention variables

$$I_1 = \begin{cases} 0, & t < 12.01.2016 \\ 1, & t \geq 12.01.2016 \end{cases}, \quad I_2 = \begin{cases} 0, & t \neq 15.12.2016 \\ 1, & t = 15.12.2016 \end{cases}, \quad I_3 = \begin{cases} 0, & t < 7.04.2017 \\ 1, & t \geq 7.04.2017 \end{cases}$$

to obtain a model free of outliers:

$$Z'_t = 3.867 + 0.739 Z'_{t-1} + \varepsilon'_t - 0.393 I_1 + 0.928 I_2 - 0.565 I_3$$

which explains 60% of the variation in the series Z'_t (Stationary R-squared equals to 0.596).

Let us use the same model - AR(1), for the values of PM10 measured in Station 1. We obtain the model

$$Z_t = 3.613 + 0.765 Z_{t-1} + \varepsilon_t,$$

where ε_t is the error term at time t for the transformed series Z_t of PM10 at Station 1. The model is not adequate. It identifies the observation on 05.01.2017 as an Additive outlier. Analyzing the charts of the ACF and PACF of the residuals of this model (Figure 8), we see that at lags 2, 10 and 14 they are outside of the defined limits.

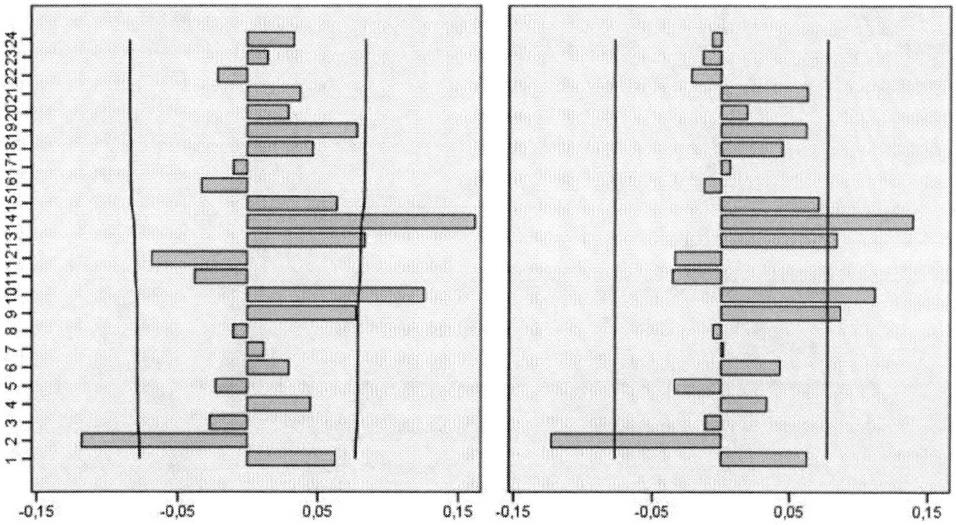

Fig.8 Residual ACF (left) and PACF (right) for the model AR(1) for Zt

The graphics of ACF on Figure 7 shows that the autocorrelations begin to decay toward zero only after the lag 13 (13=14-1). Based on these considerations, it can be assumed that a model ARIMA(1,0,14) with MA terms only at lags 2, 10 and 14 could be appropriate for describing the time series Z_t. Really, when we start Expert modeler again, but now with the option to consider only no seasonal models, it reports this model as the best one.

The obtained Model 4 has the same number of estimated parameters as Model 1, but is already adequate. It has better characteristics than Model 1 at all points. To get rid of the additive outlier on 05.01.2017 we use again the intervention variable I in (1) and get the following Model 5:

$$Z_t = 3.611 + 0.780 Z_{t-1} + \varepsilon_t - 0.111\varepsilon_{t-2} + 0.107\varepsilon_{t-10} + 0.130\varepsilon_{t-14} + 1.162 I .$$

The Model 5 has identical characteristics with Model 4, but it does not contain outliers. Stationary R-squared equals to 0.607 indicating that the model explains 61% of the variation in the transformed series Z_t of PM10 at Station 1. The forecasting for the next 10 days is given in table 3 below. The root-mean-square error for the next 7 days is 2.95 and for the next 10 days is 6.35.

Figure 9 shows the plot of the observed and fitted values of PM10 by Model 5. It can be well seen that the model gives good results and can be used for prediction of PM10 pollution in Ruse.

Model	Model Fit statistics						Ljung-Box Q(18)			Number of Outliers
	Stationary R-squared	R-squared	RMSE	MAPE	MAE	Normalized BIC	Statistics	DF	Sig.	
4	0.607	0.640	15.144	25.114	10.022	5.495	18.556	14	0.183	1

Model	Transformation	Model Parameters		Estimate	SE	t	Sig.
4	Natural Logarithm	Constant		3.611	0.061	59.455	0.000
		AR	Lag 1	0.780	0.029	26.871	0.000
		MA	Lag 2	0.111	0.045	2.455	0.014
			Lag 10	-0.107	0.039	-2.740	0.006
			Lag 14	-0.130	0.039	-3.314	0.001

Outliers		Estimate	SE	t	Sig.
05.01.2017	Additive	1.162	0.235	4.950	0.000

Table 3. Forecast values for model 5

	01.08.	02.08.	03.08..	04.08.	05.08.	06.08.	07.08.	08.08.	09.08.	10.08.
Real values	31,9	37,4	36,4	38,9	35	37,2	36,7	22,8	27,1	41
Forecast	28,00	32,05	34,49	35,91	36,73	36,83	37,95	36,86	37,90	35,69
UCL	48,56	63,44	71,41	76,25	79,16	80,08	82,97	80,85	83,29	78,53
LCL	14,71	13,95	14,03	14,05	14,03	13,88	14,18	13,70	14,05	13,20

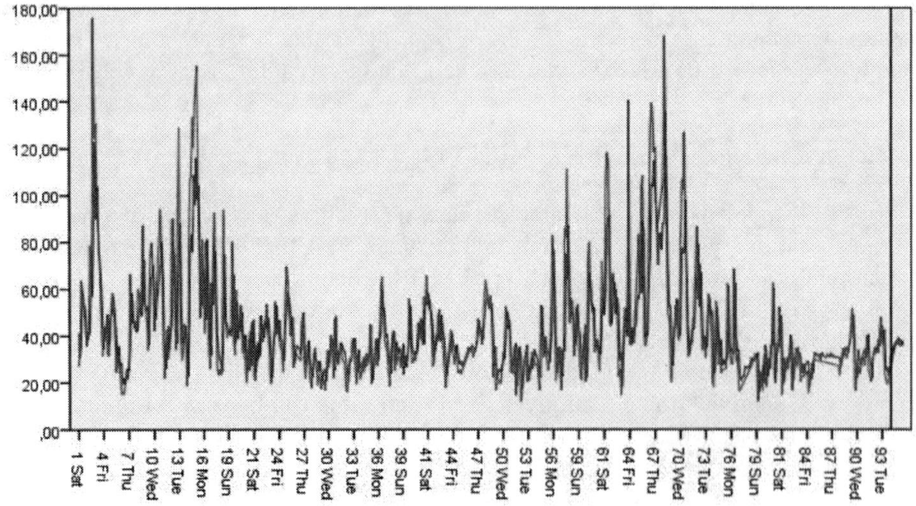

Fig. 9 Observed (red line) and fitted values (blue line) of PM10 at Station 1

3.2 Analysis and modelling of PM2.5 values

The best reported model using Expert modeler is the ARIMA(1,0,2) model:

Model	Model Fit statistics						Ljung-Box Q(18)			Number of Outliers
	Stationary R-squared	R-squared	RMSE	MAPE	MAE	Normalized BIC	Statistics	DF	Sig.	
6	0.654	0.651	12.859	32.646	7.771	5.208	22.766	15	0.089	4

Model	Transformation	Model Parameters		Estimate	SE	t	Sig.
6	Natural Logarithm	Constant		3.027	0.137	22.123	0.000
		AR	Lag 1	0.958	0.014	63.914	0.000
		MA	Lag 1	0.241	0.042	5.752	0.000
			Lag 2	0.357	0.041	8.690	0.000

The model is adequate since the Ljung-Box statistics is not significant at alpha = 0.05. Four of the observations on PM2.5 are identified as outliers: Transient at t_1=12.01.2016 and t_2=16.11.2016, Additive at t_3=04.01.2017 and t_4=07.04.2017. The effect of a Transient (temporary) change outlier diminishes exponentially over the subsequent observations and the series returns to its normal level after several steps. We introduce four intervention variables

$$J_i = \begin{cases} 0, & t \neq t_i \\ 1, & t = t_i \end{cases}, \quad i = 1,2,3,4$$

and obtain an adequate model (Model 7) with similar characteristics, but without outliers:

$$W_t = 3.037 + 0.960W_{t-1} + a_t - 0.241a_{t-1} - 0.361a_{t-2} - 1.511\delta_{1,t} + 1.450\delta_{2,t} + 1.467J_3 - 1.471J_4,$$

where $W_t = \ln U_t$, U_t is the time series of values of PM2.5 at Station 1; a_t is a shock, innovation, or random error at time t and $\delta_{1,t}$, $\delta_{2,t}$ are defined by

$$\delta_{1,t} = \begin{cases} 0, & t < t_1 \\ 0.802^{(t-t_1)}, & t \geq t_1 \end{cases}, \quad \delta_{2,t} = \begin{cases} 0, & t < t_2 \\ 0.753^{(t-t_2)}, & t \geq t_2 \end{cases}. \tag{2}$$

The forecast of Model 7 for the next 10 days is given in Table 4 below. Inspecting the residual ACF and PACF of Model 7, only these of Lag 10 are beyond the defined limits. When we introduce in Model 7 an MA term of Lag 10, we obtain the following model (Model 8):

$$W_t = 3.028 + 0.925W_{t-1} + a_t - 0.188a_{t-1} - 0.323a_{t-2} + 0.122a_{t-10}$$
$$-1.498\delta'_{1,t} + 1.389\delta'_{2,t} + 1.498J_3 - 1.374J_4,$$

where $\delta'_{1,t}$ and $\delta'_{2,t}$ are the same as $\delta_{1,t}$ and $\delta_{2,t}$ in (2), but 0.802 is replaced by 0.798 and 0.753 by 0.778 respectively. The Model 8 has better characteristics than Model 7 at all points:

Model	Model Fit statistics						Ljung-Box Q(18)			Number of Outliers
	Stationary R-squared	R-squared	RMSE	MAPE	MAE	Normalized BIC	Statistics	DF	Sig.	
8	0.657	0.666	12.604	32.479	7.698	5.178	16.465	14	0.286	0

It also has a better performance in the next 10 days.

Table 4. Forecast values for models 7 and 8

	01.08.	02.08.	03.08.	04.08.	05.08.	06.08.	07.08.	08.08.	09.08.	10.08.
Real values	19,2	20,3	22,5	20,4	20,7	23,9	-	-	-	20,5
Model 7	16,20	17,15	17,48	17,81	18,12	18,42	18,71	18,98	19,25	19,51
Model 8	16,59	18,79	19,50	20,07	18,95	19,57	19,21	19,05	19,04	19,32

The root-mean-square error of the prediction for the available real values of PM2.5 (the values for the period 07 - 09 august are missing) for Model 7 is 3,555 and for Model 8 - 2,430. Figure 10 shows the plot of the observed and fitted values of PM2.5 by Model 8.

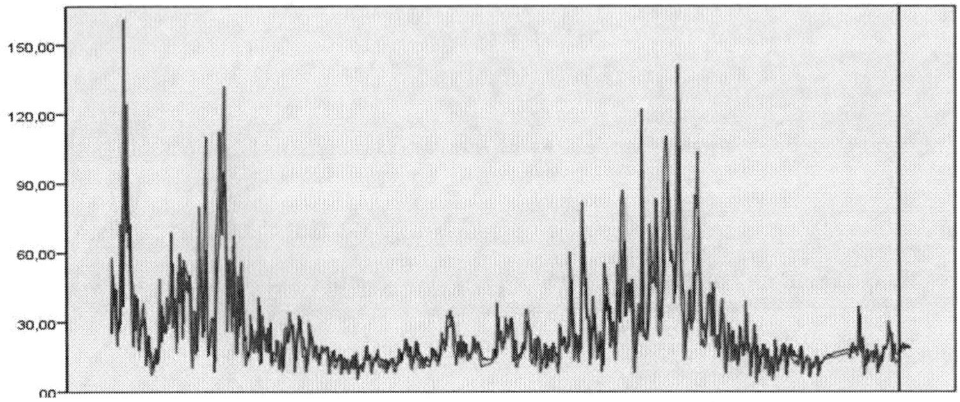

Fig. 10 Observed (red line) and fitted by Model 8 values (blue line) of PM2.5

4 Conclusions

The most dangerous air pollutants (PM10 and PM2.5) for Ruse region in Bulgaria measured in the period 24.10.2015 - 25.05.2017 by two different measuring stations were statistically analyzed in the paper. The measurements of PM10 show that during winter time the concentration of PM10 is above the norm, reaching a maximum value of 217.30 ($\mu g/m^3$) at norm 50 ($\mu g/m^3$) for the average daily rate. Next to examine better these air pollutants we create ARIMA statistical models of PM10 and PM2.5 measured by station1. The values of PM10 and PM2.5 given by the models are in a good agreement with the measurements.

The main atmospheric characteristics for Ruse region (temperature, atmospheric pressure and humidity) measured for more than 40 year period were examined in the paper [6]. For the period 1988 – 2012 the average temperature for Ruse region is higher than the mean temperature for Bulgaria. At the same time in Ruse region it is observed that water of river Danube and the air pollution increased [7, 8, 9]. Obviously the atmospheric characteristics temperature, atmospheric pressure and humidity are interrelated and maybe they are correlated with the air pollution. Especially strong is the air pollution during the autumn-winter season with particulate matter in the last decade [1]. The reasons for this are not only transportation and industry but the use of solid fuel heating. Air pollution maybe affects temperature, atmospheric pressure and humidity and this influence may cause the climate change in the Ruse region. This influence has to be studied in details further.

This paper contains results of the work on project No 2017 - FNSE – 05 and No 2017 – AIF - 3 financed by „Scientific Research" Fund of Ruse University.

References

1. I. Zheleva, E. Veleva, M. Filipova Analysis and modeling of daily air pollutants in the city of Ruse, Bulgaria Application of Mathematics in Technical and Natural Sciences AIP Conf. Proc. (2017) (in print);
2. National report on the status and protection of the environment in Bulgaria in 2012, Executive Environment Agency, Ministry of Environment and Water, available at www.eea.government.bg
3. http://www.riosv-ruse.org/kachestvo-na-atmosferniya-vazduh.html
4. A. Pankratz, Forecasting with dynamic regression models, John Wiley and Sons, New York , 1991.
5. A. Pankratz, Forecasting with Univariate Box - Jenkins Models: Concepts and Cases, John Wiley and Sons, New York , 1983.
6. I.Zheleva, M.Filipova Atmospheric Characteristics Statistic Study of Ruse Region, Bulgaria Application of Mathematics in Technical and Natural Sciences AIP Conf. Proc. 1773, 110019 (2016); doi: 10.1063/1.4965023
7. Filipova M., I. Zheleva, P. Rusev, Characteristics of PM air pollution along Bulgaria - Romania Danube region ECOLOGICA, 2013, 71, p.215-217P.
8. M. Filipova, I. Zheleva, A. Lecheva, P. Rusev, Analysis of Key Pollutant Surface Waters of the Tributaries of the Danube River in Bulgarian Section, Pliska Stud. Math. Bulgar. 24 (2015), 151-162
9. Filipova M., I. Zheleva, P. Rusev, Investigation of monitoring systems for water quality of the Danube River in the border region Romania – Bulgaria, ECOLOGICA 2015, Vol. 22 №77, p. 13-18

Theoretical-experimental method of multiscale modelling of two-component materials

Anguel Baltov[1,*], *Ana* Yanakieva[1]

[1]Institute of Mechanics, Bulgarian Academy of Sciences, 1113 Sofia, 4 G.Bonchev str., Bulgaria

Abstract. Modern mechano-mathematical models of materials are multilevel ones combining macro-, meso- and micro-level models. A major problem however is how to present a clear link between the levels. Moreover, structural models should be well verified, reflecting material behavior at meso-level and its microstructure. Hence, we propose a theoretical-experimental approach to multilevel modeling, treating discs which operate under plane stress. The approach comprises the following stages: I stage: Identify a system of characteristic points on the disc surface. Employing an appropriate optical method, find point displacements using time steps in the course of deformation. II stage: Based on displacement development in time, rationally find the variation of the strain tensor components using strains directed from the main point to the closest points of its vicinity. III stage: Define a meso-representative element in the disc plane as an ellipse with diameters related to the principal strains. IV stage: Having prepared metallographic micro-sections, define systems of characteristic structural elements in the meso-representative ellipses. V stage: Averaging the structural systems at meso-level, define macro-parameters involving material deformation parameters. VI stage: Find material macro-deformation parameters on the basis of an appropriate design of micro-representative elements and calculate the stressed-strained state of the latter applying representative loading. A test example is give in order to illustrate the approach.

1 Introduction

A modern tendency in the mechanics of materials is the design of mechano-mathematical models accounting for the material structure at three scales: micro-, meso- and macro-ones [1, 2]. The arising major problems of multiscale modeling are: (1) Design of a particular model with specific parameters and development of a method of parameter identification at each scale; (2) Identification of the link between respective models and parameters, as well as verification of scale transition.

Multi-scale models with their plausibility and reliability are especially useful in modeling and calculation of safe structures and as a result- in saving energy needed for structure erection and exploitation. We focus in what follows on the analysis of materials with grain structure consisting of two components: a matrix and grain inclusions, which

* Corresponding author: anguelbaltov@gmail.com

generally reinforce the material. Models at all the three scales are designed within the frames of continuum mechanics. Such an approach seems reasonable enough, since we do not analyse sub micro-scale of a material with a discrete atomic or molecular structure. In view of a possible application of some experimental methods we consider thin disks under plane stress undergoing static external loading. Disc thickness is much smaller than that of the grains. We treat disks as structural elements operating in exploitation conditions, i.e. they deform purely elastically without the development of damage and inelastic strains. The elastic properties are assumed to be isotropic and non-homogeneous, and they may change as a result of eventual structural changes occurring at meso- level.

The paper offers a theoretical-experimental method of multi-scale modelling of the materials concerned. It is based on generalization and further elaboration of a number of results published in refs. [3-6], adopting stages of an eventual practical realization.

2 First stage: Macro-experiments on disks

Consider thin elastically deformed disks undergoing static loading which generally varies in time $t \in [t_0, t_f]$. Select n characteristic points on the disk surface with projections on the disk middle surface - $M(p,t)\begin{Bmatrix} X_M(p,t) \\ Y_M(p,t) \end{Bmatrix}$, $p = 1,2\ldots n$ where a coordinate system $OXYZ$ is introduced, plane OXY coincides with the disk middle surface and axis OZ is directed along the disk thickness, Fig. 1.

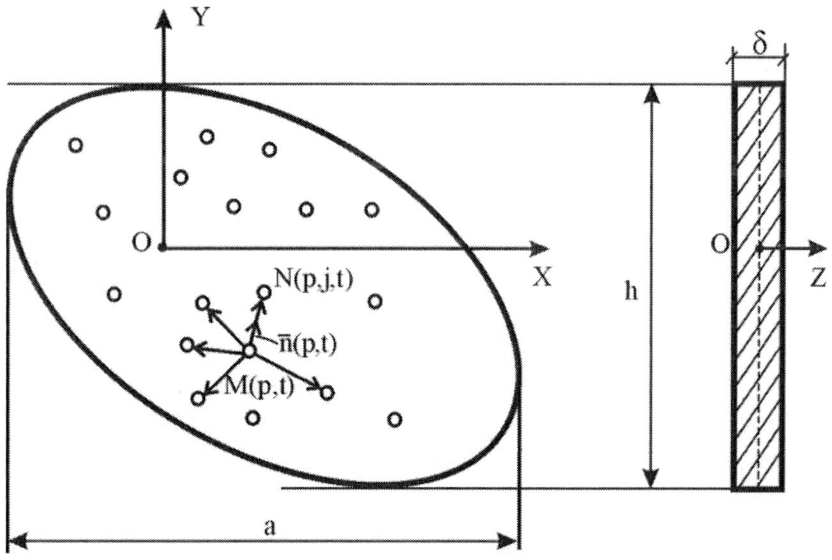

Fig. 1. A thin disk with characteristic points on its middle surface

Consider each point $M(p,t)$ together with the close characteristic points $N(p,j,t)$, $j = 1,2\ldots s$ that bound it. Discretize the process in time t_k, $k = 0,1,2\ldots m$, where t_0 is process start and $t_f = t_m$ is process finish. Adopt identical intervals $\Delta t = t_k - t_{k-1}$ for each k. Then, one can find the location of points $M(p,t_k)$ (respectively $N(p,j,t_k)$) at each time t_k (this can be done using different optical methods to locate the point position [11]) and specify vector $\overrightarrow{MN}$ with magnitude

$$|MN| = C(p,j,t_k) = \sqrt{[X_N(p,j,t_k) - X_M(p,t_k)]^2 + [Y_N(p,j,t_k) - Y_M(p,t_K)]} \tag{1}$$

The respective unit vector is

$$\vec{n}(p,j,t_k) = \begin{cases} n_X(p,j,t_k) = \dfrac{X_N(p,j,t_k) - X_M(p,t_k)}{C(p,j,t_k)} \\ n_Y(p,j,t_k) = \dfrac{Y_N(p,j,t_k) - Y_M(p,t_k)}{C(p,j,t_k)} \end{cases} \tag{2}$$

$$p = 1,2\ldots n;\; j = 1,2\ldots s;\; k = 0,1,2\ldots m$$

We determine approximately strains directed along $\vec{n}(p,j,t_k)$using the selected characteristic points and the calculated distances $|MN| = C(p,j,t_k)$ for each p,j and k

$$E_n^{\exp}(p,j,t_k) = \frac{C(p,j,t_k) - \overline{C}(p,j,t_{k-1})}{\overline{C}(p,j,t_{k-1})} \tag{3}$$

where $\overline{C}(p,j,t_k)$ is the magnitude of $\overrightarrow{MN}(p,j,t_{k-1})$, the vector rotating as an ideal solid.

The experimentally found strain $E_n^{\exp}(p,j,t_k)$ is compared to the theoretically calculated one $E_n^{th}(p,j,t_k)$, using a known formula [9]

$$\begin{aligned} E_n^{th}(p,j,t_k) = E_{XX}(p,t_k)[n_X(p,j,t_k)]^2 + E_{YY}(p,t_k)[n_Y(p,j,t_k)]^2 + \\ + E_{XY}(p,t_k)[n_X(p,j,t_k) n_Y(p,j,t_k)] \end{aligned} \tag{4}$$

Design functional $L(p,t_k)$ of the least square deviation between the experimental (eq. 3) and the theoretical strains (eq. 4).

$$L(p,t_k) = \sqrt{\frac{\sum_{j=1}^{S}\left[E_n^{\exp}(p,j,t_k) - E_n^{th}(p,j,t_k)\right]^2}{S-1}} \tag{5}$$

Using condition for $L_{\min}$ we find the components of the strain tensor $E_{XX}(p,t_k)$, $E_{YY}(p,t_k)$ and $E_{XY}(p,t_k)$ at point $M(p,t_k)$ and time t_k

$$\frac{\partial L}{\partial[E_{XX}(p,t_k)]}=0,\quad \frac{\partial L}{\partial[E_{YY}(p,t_k)]}=0,\quad \frac{\partial L}{\partial[E_{XY}(p,t_k)]}=0 \tag{6}$$

To find transversal strain $E_{ZZ}(p,t_k)$ we should measure the change of the thickness $\delta(p,t_k)$ at point $M(p,t_k)$ and time t_k, i.e.

$$E_{ZZ}(p,t_k)=\frac{\delta(p,t_k)-\delta(p,t_{k-1})}{\delta(p,t_{k-1})} \tag{7}$$

We can calculate the increment $\Delta E_{ZZ}(p,t_k)$ in the interval Δt through (7). On the other hand, the increments of $E_{XX}(p,t_k), E_{YY}(p,t_k)$ and $E_{XY}(p,t_k)$ are found through a system of equations involving two times - t_{k-1} and t_k, $k=0,1,2\ldots m$.

Following a method described in [5] we can find the stress tensors with components $\Sigma_{XX}(p,t_k)$, $\Sigma_{YY}(p,t_k)$ and $\Sigma_{XY}(p,t_k)$, and the respective increments $\Delta\Sigma_{XX}(p,t_k)$, $\Delta\Sigma_{YY}(p,t_k)$ and $\Delta\Sigma_{XY}(p,t_k)$. We apply our methodology starting from points close to the boundary of the disk, whereas some points $N(p,j,t_k)$ lie on that boundary. Then, using Hooke's law, we can also find the elasticity characteristics $E_{mac}(p,t_k)$, $\nu_{mac}(p,t_k)$ and $\Delta E_{mac}(p,t_k)$, $\Delta\nu_{mac}(p,t_k)$, respectively.

If the material is homogeneous within the frames of the statistical error and there are no structural changes, the elasticity characteristics will be constant- $E_{mac}=const$ and $\nu_{mac}=const$. Then, it is easy to obtain the macro elastic characteristics using a different experimental technique [10].

3 Second stage: Specification of a meso-representative element and meso-structural parameters

We found the components of the macro tensor at selected characteristic points $\widetilde{M}(p,t_k)$ and at each discrete time t_k. Hence, we can also find the principal strains in the disk middle plane, $E_N(p,t_k)$, $(N=I,II)$. We design meso-representative elements at point $\widetilde{M}(p,t_k)$. They are ellipses with diameter $D_N=1+E_N(p,t_k)$, with axes directed along the principal strains and concluding angle $\alpha_M(p,t_k)$ with OX. The height of an elliptic cylinder thus formed is $\delta(p,t_k)$, whose section cut by the middle plane is an ellipse, Fig. 2.

Considering each representative element at point $\widetilde{M}(p,t_k)$, we grind the surface around the representative ellipse, which is a projection of the elliptical cylinder. Using an electron microscope, we identify the grain inclusions. Each grain is approximated as an elliptic cylinder, whose section cut by the middles plane is an ellipse with point $Q(p,t_k;q)$ as a center. Moreover, its diameters are $d_1(p,t_k;q)$ and $d_2(p,t_k;q)$, its height is $\delta(p,t_k)$, q is the consecutive grain number, $q=1,2\ldots q_n$, q_n is the general number of grains within the representative meso-element at $\tilde{M}(p,t_k)$.

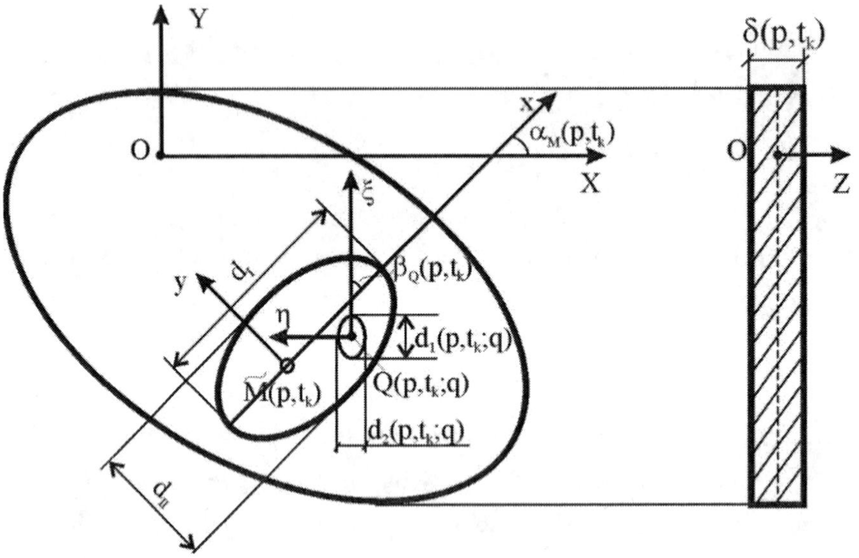

Fig. 2. A meso-representative element. Approximated element grains

Consider the middle plane of the cross section of the meso-representative element. Introduce a coordinate system $\tilde{M}xy$ with meso-strains $\varepsilon_{xx}(p,t_k)$, $\varepsilon_{yy}(p,t_k)$ and $\varepsilon_{xy}(p,t_k)$ and meso-stresses $\sigma_{xx}(p,t_k)$, $\sigma_{yy}(p,t_k)$ and $\sigma_{xy}(p,t_k)$. Introduce into each grain a local coordinate system $Q\xi\eta$ with micro-strains $e_{\xi\xi}(p,t_k;q)$, $e_{\eta\eta}(p,t_k;q)$ and $e_{\xi\eta}(p,t_k;q)$ and micro-stresses $s_{\xi\xi}(p,t_k;q)$, $s_{\eta\eta}(p,t_k;q)$ and $s_{\xi\eta}(p,t_k;q)$. Axis OX and line $\tilde{M}x$ conclude angle $\alpha_M(p,t_k)$, while $\tilde{M}x$ and $Q\xi$ conclude angle $\beta_Q(p,t_k;q)$.

Divide the grains of the meso-representative element at point $\tilde{M}(p,t_k)$ into grain systems (a) with identical slope $\beta_Q(a)$, $a=1,2\ldots n_a$, where n_a is the number of grain systems in that element. Each system (a) has r_a number of grains where $\sum_{a=1}^{n_a} r_a = n_q$ is the total number of grains in (a) system. Each grain has different diameters $d_1(g_a)$, $d_2(g_a)$ and the distance $l(g_a)$ between it and the neighboring grains of every system (a) is minimal. We calculated for system (a) the mean values $\bar{d}_1(a)$, $\bar{d}_2(a)$ and $\bar{l}(a)$, Fig. 3.

4 Third stage: Identification of the structural parameters at the three scales

4.1 Micro-scale structural parameters

We select a micro-structural element for system (a): a prism with rectangular cross section $\bar{l}(a)\text{x}\,\bar{l}(a)$ in the disk middle plane. Its center is $Q_a(p,t_k)\subset Q(p,t_k;q)$ located in the middle of the representative inclusion- an ellipse with diameters $\bar{d}_1(a)$, $\bar{d}_2(a)$. Its height is $\delta(p,t_k)$, being one and the same for the entire meso-representative element at point $\widetilde{M}(p,t_k)$, Fig. 3.

The micro-representative element at point $Q_a(p,t_k)$ is loaded conditionally by a uniformly distributed load applied along the element border- see the representative loading in [5]. This loading yields uniformly distributed meso-stresses at $Q_a(p,t_k)$, according to the scheme in [5].

Using nanoindentation [9] we find material micro-elasticity characteristics: $E_{mic}(\text{mat})$, $\nu_{mic}(\text{mat})$ of the matrix and $E_{mic}(\text{incl})$, $\nu_{mic}(\text{incl})$ of the inclusions. Micro-stresses and micro-strains at point $Q_a(p,t_k)$ are determined via FEM [12]. Micro-strains are presented as $e_{\xi\xi}\left(\pm\frac{\bar{l}(a)}{2},\eta,t_k\right)$, $e_{\eta\eta}\left(\xi,\pm\frac{\bar{l}(a)}{2},t_k\right)$, $e_{\xi\eta}\left(\pm\frac{\bar{l}(a)}{2},\eta,t_k\right)$ at the element border, where $\xi=\eta=\pm\frac{\bar{l}(a)}{2}$.

Thus, meso-strains at point $Q_a(p,t_k)$ are found, performing averaging according to the following expressions:

$$\begin{aligned}
\varepsilon_{\xi\xi}(p,t_k;a)&=\int_{-\bar{l}(a)/2}^{\bar{l}(a)/2} e_{\xi\xi}\left(\frac{\bar{l}(a)}{2},\eta,t_k\right)d\eta\\
\varepsilon_{\eta\eta}(p,t_k;a)&=\int_{-\bar{l}(a)/2}^{\bar{l}(a)/2} e_{\eta\eta}\left(\frac{\bar{l}(a)}{2},\eta,t_k\right)d\eta\\
\varepsilon_{\xi\eta}(p,t_k;a)&=\int_{-\bar{l}(a)/2}^{\bar{l}(a)/2} e_{\xi\eta}\left(\frac{\bar{l}(a)}{2},\eta,t_k\right)d\eta
\end{aligned}\tag{8}$$

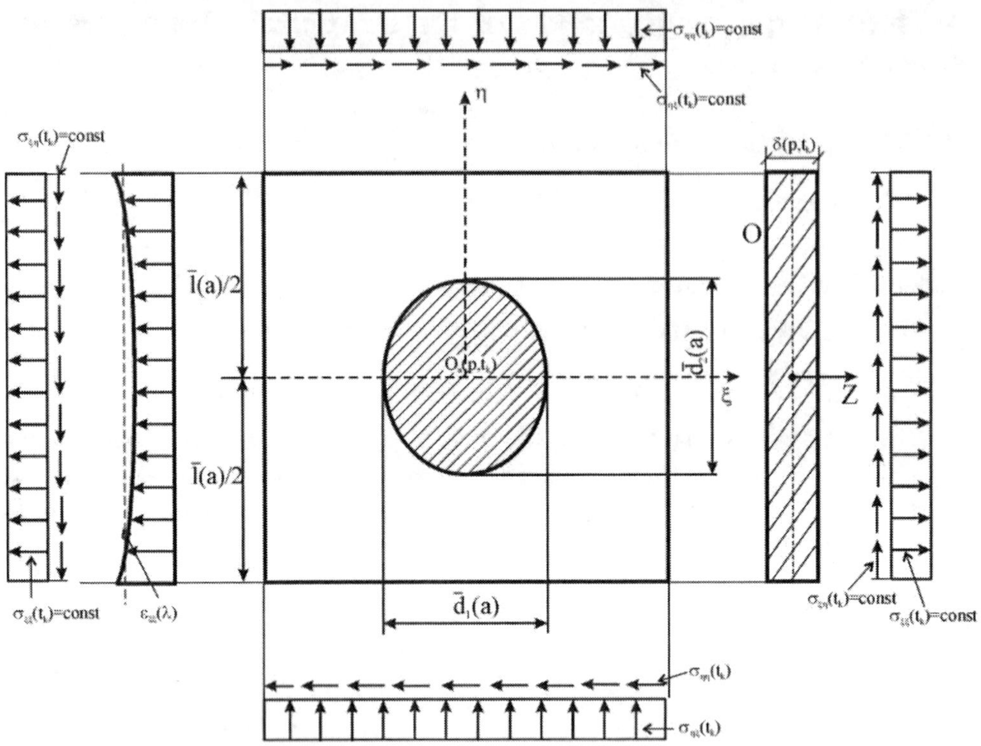

Fig. 3. Micro-representative element and essential constant loading by meso-stresses

Meso-scale stresses and strains at point $Q_a\left(p,t_k\right)$ are linked to each other via Hooke's law, using the meso-elastic characteristics $E_{mes}\left(a\right)$ and $\nu_{mes}\left(a\right)$. We have for system (a) that

$$\begin{aligned}\varepsilon_{\xi\xi}\left(p,t_k;a\right)&=\frac{1}{E_{mes}\left(a\right)}\left[\sigma_{\xi\xi}\left(p,t_k;a\right)-\nu_{mes}\left(a\right)\sigma_{\eta\eta}\left(p,t_k;a\right)\right]\\ \varepsilon_{\eta\eta}\left(p,t_k;a\right)&=\frac{1}{E_{mes}\left(a\right)}\left[\sigma_{\eta\eta}\left(p,t_k;a\right)-\nu_{mes}\left(a\right)\sigma_{\xi\xi}\left(p,t_k;a\right)\right]\end{aligned}\tag{9}$$

Using system (9) we find those meso-elastic characteristics, assuming that they are valid for all systems (a). If we want to obtain the real meso-stresses at the point $M\left(p,t_k\right)$, the meso-ellipse can be enveloped by a rectangle, Fig 2. Then, the new rectangular element can be uniformly loaded by $\Sigma_{XX}\left(p,t_k\right)$, $\Sigma_{YY}\left(p,t_k\right)$ and $\Sigma_{XY}\left(p,t_k\right)$, according to the scheme in [5], Fig. 3, and the calculations can be performed using FEM [12].

The following parameter characterizes the meso-structure

$$K_a=\frac{E'_{mes}\left(a\right)}{E'_{mac}};\quad E'_{mes}\left(a\right)=\frac{E_{mes}\left(a\right)}{1-\nu_{mes}\left(a\right)},\qquad E'_{mac}=\frac{E_{mac}}{1-\nu_{mac}}\tag{10}$$

It presents the effect of the microstructure (divided into grain systems) on the model parameters at the meso-scale.

4.2 Meso-scale structural parameters

We introduce a meso-scale structural parameter of the grain system (a) as a vector having the form

$$J_\zeta(a) = m_\zeta(a)\varphi_a(p,t_k)r_a K_a(p,t_k), \quad \zeta = \xi,\eta \tag{11}$$

where $m_\zeta(a)$ gives the grain orientation of the system (a) e.g.

$$m_\zeta(a), \begin{cases} m_\xi(a) = \cos\beta_Q(a) \\ m_\eta(a) = \sin\beta_Q(a) \end{cases} \tag{12}$$

We have $\varphi_a(p,t_k) = \dfrac{\overline{F}_{mic}(a)}{\overline{F}_{el}(p,t_k)}$, $\overline{F}_{mic}(a) = \pi\,\overline{d}_M(a)$, $\overline{d}_M(a) = 0{,}5(\overline{d}_1(a) + \overline{d}_2(a))$ and $\overline{F}_{el}(p,t_k) = \pi\,d_M^2(p,t_k)$, $d_M = \dfrac{1}{2}(d_I(p,t_k) + d_{II}(p,t_k))$, Fig. 2, which gives the effect of the grain size on different systems (a). The mechanical coefficient K_a is found using expression (10).

Based on $J_\zeta(a)$ we introduce a meso-structural parameter $I_\zeta(p,t_k)$ which expresses the prevailing grain orientation:

$$I_\zeta(p,t_k) = \sum_{a=1}^{n_a} J_\zeta(a), \ \zeta = \xi,\eta \tag{13}$$

It unifies all systems (a) within the meso-representative element at point $\widetilde{M}(p,t_k)$.

4.3 Macro-scale structural parameters

Based on the vector parameter $I_\xi(p,t_k)$ we introduce its magnitude as a macro scale structural parameter

$$I_0(p,t_k) = \sqrt{I_\xi^2(p,t_k) + I_\eta^2(p,t_k)} \tag{14}$$

The macro elasticity characteristics E_{mac} and ν_{mac} depend on $I_0(p,t_k)$. If real structures presented at macro-scale by different parameters $I_0(p,t_k)$ are considered, the dependence can be found theoretically [32] determining the macro elasticity characteristics $E_{mac}(p,t_k) = \Phi(I_0(p,t_k))$ and $\nu_{mac}(p,t_k) = \Psi(I_0(p,t_k))$ where Φ and Ψ are approximation functions. For elastically deforming bodies that keep their structure, we have $I_0 = I_0(p)$. If the material is homogeneous I_0 does not depend on p. So far we identified the scale structural parameters. Note that energetic links exist between the three scales (at point $\widetilde{M}(p,t_k)$) which have been discussed in [4]:

$$\Sigma_{IJ}(p,t_k)E_{IJ}(p,t_k)=\int_{F_{el}}\sigma_{ij}(p,t_k)\varepsilon_{ij}(p,t_k)dF\,,\ (I,J=X,Y;i,j=x,y) \quad (15)$$

Table 1. Grain systems.

a	$\beta_a(a)$	r_a	$\bar{d}_1(a)[\mu m]$	$\bar{d}_2(a)[\mu m]$	$F_{incl}(a)[\mu m^2]$	φ_a	K_a	$J_\xi(a)$ $J_\eta(a)$
1	0^0	15	0,224	0,101	0,0834	0,0033	0,86	0,0043 0,0000
2	15^0	3	0,205	0,128	0,0876	0,0035	0,87	0,0088 0,0024
3	30^0	2	0,285	0,205	0,1900	0,0076	0,90	0,0012 0,0007
4	45^0	3	0,137	0,082	0,0380	0,0151	0,95	0,0030 0,0030
5	60^0	2	0,225	0,090	0,0780	0,0031	0,85	0,0029 0,0051
6	120^0	6	0,167	0,081	0,0490	0,0120	0,92	-0,0055 0,0095
7	150^0	12	0,206	0,101	0,0740	0,0030	0,84	-0,0261 0,0151
8	165^0	4	0,363	0,148	0,2042	0,0032	0,91	-0,0287 0,0077

5 Test example

We use the method proposed to illustrate the efficiency of Stage 3 and Stage 4, mainly, treating a two-component nanocomposite with a matrix consisting of polypropylene and polymer glue and reinforced by means of grains composed of carbon tube bundles. The material was analyzed in detail in [7]. We use a representative microstructure pattern found by preparing a metallographic micro-section and magnifying it via an electron microscope type SEM. No damage and change of the structure are observed in the range of the elastic strains.

For example we consider a meso-structural element with 47 grains ($\eta_q = 47$) distributed within eight grain system (a) ($n_a = 8$). The accuracy of $\beta_Q(a)$ is 2^0, Table 1. The table presents: systems angles $\beta_Q(a)$; number of grains r_a in each system (a); mean diameters $\bar{d}_1(a)$ and $\bar{d}_2(a)$; coefficients φ_a; structural parameters K_a and the components of the meso-structural parameters $I_\xi(a)$ for all systems (a).

We calculate the following vector measure

$$I_\zeta,\begin{cases} I_\xi = -0{,}0167 \\ I_\eta = +0{,}0435 \end{cases} \quad (16)$$

The macro structural measure is $I_0 = +0{,}0466$.

6 Conclusions

The method presented in the study is based on available experimental data and theoretical studies and results. It provides means of the design of physically clear structural parameters at the three scales: macro-, meso- and micro-ones, specifying a link between them. This makes the mechano-mathematical model a physically well-defined tool.

The employment of a multi-scale model in the mechanics of materials is an important issue enabling one to describe adequately material mechanical behavior and explicitly account for the effect of material structure. The model is exceptionally useful in structural mechanics since it allows a plausible record of the behavior of structural elements. It is also useful in modern energy saving structures compounded by surface-reinforced elements [8] providing an appropriate modeling of the reinforcement.

The study has been supported by the National Fund "Scientific Research", Project DFNI E02/10 121214.

References

1. A. Roos, et al, Int. J of Plastisity, **20**, 811-830, (2004)
2. F. D. Ficher,, E.Werner, ZAMM, **77**, 341-348, (1977)
3. A. Baltov, J. Th. Appl. Mech., **40** (4), 41-54, (2010)
4. A. Baltov, *12th Nat. Congress on Th. and Appl. Mechanics*, Varna, (On CD), (2013)
5. A. Baltov, A. Yanakieva, J. of Th. and Appl. Mech., **43** (3), 43-50, (2011)
6. A. Baltov, A. Yanakieva, N. Nikolov, *14th Int. Conf. VSU'2014*, **2**, 19-25, (2014)
7. Y. Dong, et al, Compos. Sci. Technol., **68**, (14), 2864-2875, (2008)
8. A. Baltov, A. Yanakieva, G. Nikolova, *Springer Proc. Materials, 3rd International Congress on Energy Efficiency and Energy Related Materials*, Oludeniz, Turkey, 167-174, (2015)
9. X. Li, B. Bhushan, Materials Characterization, **48**, (1), 11-36, (2002)
10. W. Szczepinski, *Experimental Methods in Mechanics of Solids*, Amsterdam, Elsevier, (1990)
11. V. Kavardzhikov, I. Nikolov, D. Pashkouleva, *Series in Applied Mathematics and Mechanics*, **5**, Institute of Mechanics - BAS, Sofia, (2015)
12. K. Kazakov, Theory of Elasticity, *Stability and Dynamics of Structures*, 2nd ed., Academic Publishing House of BAS, Sofia, (2011)

Study of the mechanical properties of single-layer and multi-layer metallic coatings with protective-decorative applications

Sabina Cherneva[1,*], *Desislava* Guergova[2], *Roumen* Iankov[1], and *Dimitar* Stoychev[2]

[1]Institute of Mechanics, Bulgarian Academy of Sciences, Acad. G. Bonchev St., Bl.4, 1113 Sofia, Bulgaria
[2]Institute of Physical Chemistry, Bulgarian Academy of Sciences, Acad. G. Bonchev St., Bl.11, 1113 Sofia, Bulgaria

Abstract. Single thin coating of matt nickel (N_{imat}), a mirror bright copper (Cu_{bright}), a mirror bright nickel (Ni_{bright}) and their combinations were electrochemically deposited on brass substrate with thickness 500 μm. The basic aim was electrodeposition of two-layer Cu_{bright}/Ni_{mat} and Ni_{bright}/Cu_{bright} systems, and three-layer Ni_{bright} Cu_{brigh}/Ni_{mat} system, which are among the most widely applied protective and decorative systems in light and medium operating conditions of corrosion. The thicknesses of the obtained films varied from 1 μm to 3.25 μm. They were investigated via nanoindentation experiments, in order to characterize their basic physical and mechanical characteristics, related with their good adhesion and corrosion protective ability, as well as ensuring the integrity of the system "protective coating/substrate" to possible mechanical, dynamic and/or thermal stresses. As a result, load-displacement curves were obtained and indentation hardness and indentation modulus were calculated using the Oliver & Pharr approximation method. The dependence of the indentation modulus and the indentation hardness on the depth of the indentation, surface morphology and structure of the obtained coatings, their texture and surface roughness were investigated too. The obtained results showed that the three-layer Ni_{bright}/Cu_{bright} $/Ni_{imat}/CuZn37$ system has highest indentation modulus and indentation hardness, following by two-layer Ni_{bright}/Cu_{bright} system and single layer coatings.

1 Introduction

It is considered that among the most reliable protective and decorative coatings of metal details (made of different types of steel or zinc, copper and aluminium alloys, which are exploited under different corrosion conditions) are electrodeposited coatings of copper, nickel and chromium [1], which can be combined into different double or triple systems. By properly combining the physico-chemical and physico-mechanical properties of these layers/systems, as well as the thickness and sequence of the arrangement/deposition, it can be substantially predetermined lifetime of the devices, as well as their functional operation.

* Corresponding author: sabina_cherneva@yahoo.com

Along with the physico-chemical properties that determine the corrosion resistance of the protective coatings, the physico-mechanical properties of the deposited coatings/coating systems are of particular importance, as they have influence on their adhesion to the substrate and between them, their hardness and wear resistance, and etc.

In connection with this, it is important to have data about their properties, because in the case of electrochemically deposited protective metal layers, depending on the composition of working electrolytes and regimes in which are obtained, they differ too strongly from the physico-mechanical properties of metallurgically obtained metals (for which reference data are usually published).

In this connection the aim of present work was determination of indentation hardness (H_{IT}) and indentation modulus (E_{IT}) of single-layer thin films of mat nickel (Ni_{mat}), mirror bright copper (Cu_{bright}), mirror bright nickel (Ni_{bright}), as well as their combinations, designed to obtain the Cu_{bright}/ Ni_{mat} and Ni_{bright}/Cu_{bright} double layer systems and the Ni_{bright}/Cu_{bright}/ Ni_{mat} three-layer systems, electrochemically deposited on brass substrates (CuZn37), which are among the most widely used protective/decorative metal coatings/systems in light and medium operating/corrosion conditions.

2 Experimental part

The layers of electrochemically obtained fine crystalline copper, characterized by a mirror surface reflectance (96% vs. Silver mirror), were deposited on a 4 x 2 cm substrates, cut from brass (CuZn37) sheet with a thickness of 0.5 mm, which were carefully chemically degreased in organic solvent followed by etching and surface activation in aqueous solution of HNO_3 (50 wt. %) for 20 s at room temperature. The composition of the working electrolyte was: $CuSO_4.5H_2O$ – 220 g/l; H_2SO_4 – 50g/l; NaCl – 0, 090g/l; Brightening additive THB-6-II [2] - 5 ml/l. Elliptic phosphorus-containing (0.07 wt.%) copper anodes were used. The electrolyte temperature was 25°C. The working standard two-electrode electrochemical cell in which the electrolysis was carried out was equipped with an air-agitating device for the electrolyte, which was performed with specially purified compressed air, supplied at a rate of 13 $m^3/m^2/h$. The cathode current density was 4.5 A/dm^2 and the thickness of the obtained bright copper layer was ~2 μm.

The investigated mat nickel film was obtained electrochemically by means of cathode deposition on analogous substrates (with sizes 4 x 2 cm, cut from brass sheets with thickness 0.5 mm), which were pre-treated chemically as described above. We used a working electrolyte with a following composition: $NiSO_4.7H_2O$–270 g/l; $NiCl_2.6H_2O$–70 g/l; H_3BO_3 – 40 g/l; and as soluble anodes - rolled de-passivated sheet nickel with a purity of 99.7%. Electrochemical deposition was performed in a standard two-electrode electrochemical cell at a cathode current density of 2 A/dm^2 and air agitation of the electrolyte at a pH ~ 4.5, temperature of 55°C and a deposition time of 2.5 minutes. The thickness of the obtained nickel coating was ~1 μm.

The deposition of the bright nickel coating (necessary for the obtaining of the three-layer Nibrigth/Cubright/Nimat/CuZn37 system) was performed electrochemically by cathodic deposition on 4 x 2 cm substrates, cut from brass (CuZn37) sheet with a thickness of 0.5 mm, which were chemically pretreated as described above. We used a working electrolyte with a following composition: $NiSO_4.7H_2O$–270 g/l; $NiCl_2.6H_2O$–70 g/l; H_3BO_3 – 40 g/l; saccharin - 1.1 g/l; butanediol - 0.3 g/l, and as soluble anodes - rolled de-passivated sheet nickel with a purity of 99.7%. Electrochemical deposition was performed in a standard two-electrode electrochemical cell at a pH ~ 4.5, temperature of 55°C, cathode current density of 5 A/dm^2, air agitation of the electrolyte and a deposition time of 1.7 min.

The deposition of two-layer system - mat nickel (with thickness ~1 μm), followed by bright copper (with thickness ~2 μm) and three-layer system of: mat nickel (with thickness

~1.4 μm), followed by bright copper film (with thickness ~2 μm) and bright nickel film (with thickness ~1,4 μm) on chemically degreased in organic solvent and etched in a suitable solution of HNO_3 brass substrates, was performed sequentially according to the procedures described above for depositing of single-layer coatings (rinsing with deionised water was performed after each step).

Surface morphology and structure of the obtained coatings were investigated by JEM 200-CX scanning electron microscopy in secondary electron imaging regime (SEI), their texture by Philips X-ray diffractometer, the thickness of the layers was determined using X-ray fluorescence analysis (FISHERSCOPE X-RAY XDVM), and their micro profile - with Perthen profilometer.

Table 1. Input parameters for nanoindentation experiments

Parameter	Unit	Value
Percent To Unload	%	90
Surface Approach Velocity	nm/s	10
Delta X For Finding Surface	μm	-50
Delta Y For Finding Surface	μm	-50
Maximum Load	gf	50
Load Rate Multiple For Unload Rate	[-]	1
Number Of Times To Load	integer	10
Allowable Drift Rate	nm/s	0.05
Approach Distance To Store	nm	1000
Peak Hold Time	s	10
Time To Load	s	15
Surface Approach Distance	nm	5000
Surface Approach Sensitivity	%	40
Frame Stiffness Correction	N/m	0
Percent Unload In Stiffness Calculation	%	50
Perform Drift Correction	1/0	1
Poisson's Ratio	[-]	0.3

Mechanical properties of the obtained films and brass substrate were investigated by nanoindentation experiments, using Nano Indenter G200 (Keysight Technologies, USA). The nanoindenter is equipped with a Berkovich three-sided diamond pyramid with centerline-to-face angle 65.3° and a 20 nm radius at the tip of the indenter. The minimum allowed load is 10 mN, and the maximum load is 500 mN. Displacement recording resolution is 0.01 nm and the load recording resolution is 50 nN. The device is equipped with an optical microscope with 2 objectives with magnifications of 250x and 1000x. We made series of 25 or 12 indentations on each sample probe in order to have better statistics. We used an indentation method which prescribes series of 10 loading/unloading cycles in a single indentation experiment [3]. The maximum load was 50 gf and we had 10 s peak hold time at maximum load for each loading-unloading cycle. Indentation hardness and modulus are determined using stiffness calculated from the slope of the load–displacement curve during each unloading cycle. As a result of nanoindentation experiments, load–displacement curves were obtained and two mechanical characteristics of substrate and investigated films – indentation hardness (H_{IT}) and indentation modulus (E_{IT}) were calculated using Oliver & Pharr approximation method [4]. Dependence of indentation modulus and indentation hardness on depth of indentation was investigated as well. Basic input parameters, used in this indentation method are given in Table 1.

3 Results and discussions

In Figures 1 and 2 are presented SEM images (obtained in BEI regime) and TEM images of the structure of investigated copper layer. It can be seen from Figure 1 that it is extremely smooth (measured values for R_a and R_z are respectively 0.1 µm and 0.14 µm) and it is made of isotropically arranged crystallites with cubic symmetry (Fig. 2), the dimensions of which are in the range 20 – 80 nm. On the basis of the conducted X-ray investigations it was found that the coatings had a mixed texture, with a low degree of preferable orientation, in the directions [311] and [110]. Figure 3 shows a SEM micrograph (JEM 200-CX in SEI mode) of the structure of the electrochemically deposited matt nickel, at a magnification 5,000x.

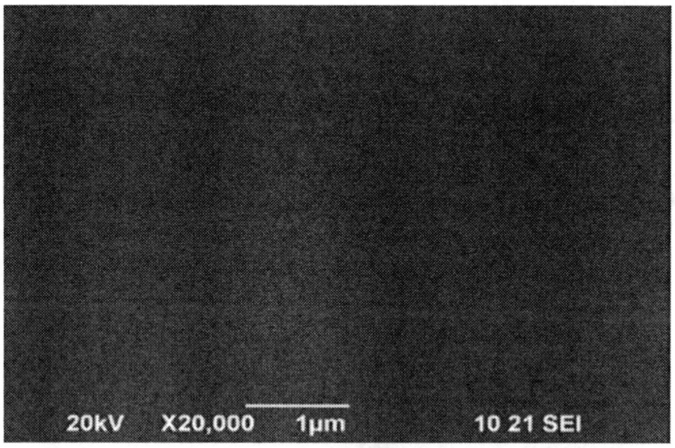

Fig.1 SEM micrograph of electrochemically deposited copper film *Cu(1,8 µm)/CuZn37 (X20 000)*

It can be seen from obtained results, that the nickel layers are characterized by a relatively large crystalline homogeneous pyramidal structure made up of building aggregates with dimensions ~ 50-500 nm. The measured values for Ra and Rz, which characterize the surface roughness of these layers, are respectively 0.1 µm and 0.9 µm. X-ray investigations showed that the coatings have a mixed texture in the directions [110] and [211].

Figure 4 shows a SEM image (JEM 200-CX in SEI mode) of the structure of the electrochemically deposited bright nickel coating at a magnification of 20,000×. The nickel layers are extremely uniform, dense and finely dispersed, and are characterized by building blocks with sizes of 40-100 nm. Coatings have a texture with a major axis of predominant orientation in the direction [211].

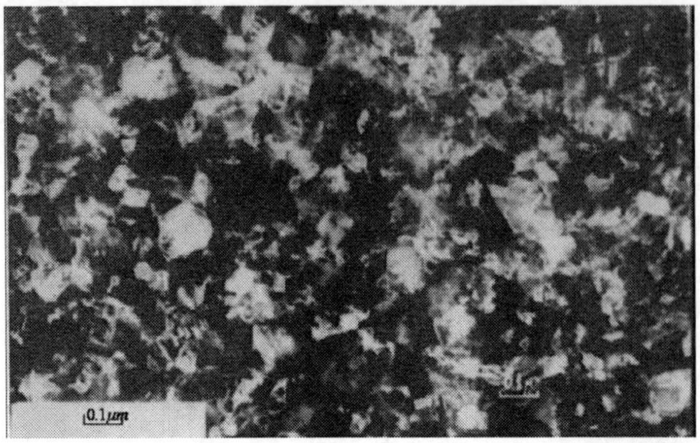

Fig.2 TEM micrograph of electrochemically deposited copper film *Cu(1,8 μm)/CuZn37 (X100 000)*

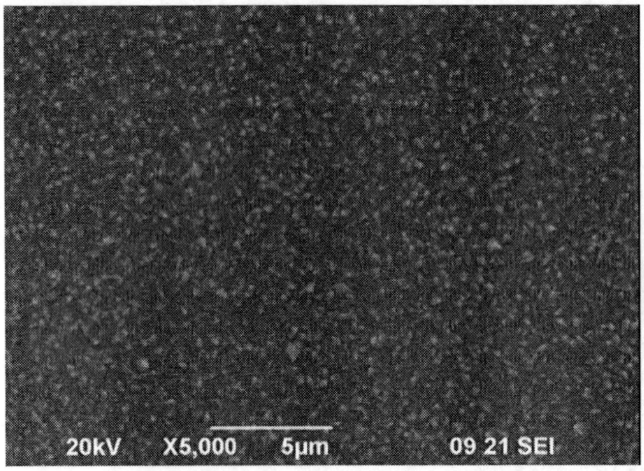

Fig.3 SEM micrograph of electrochemically deposited mat nickel film Ni_{mat} (1 μm)/CuZn37 (X5 000)

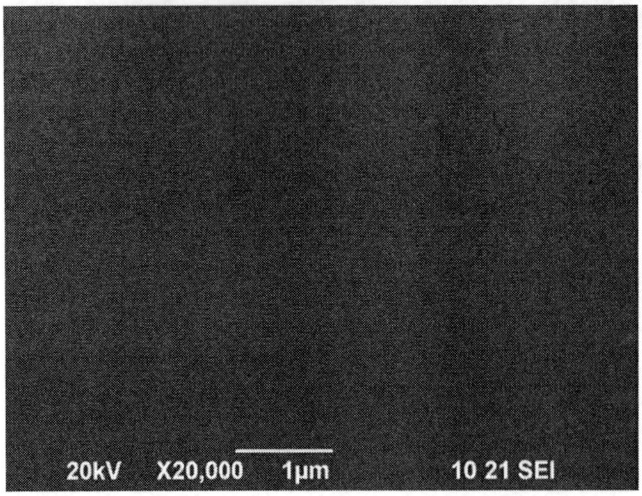

Fig.4 SEM micrograph of electrochemically deposited bright nickel film *Ni_{bright} (1,8 μm)/CuZn37 (X20 000)*

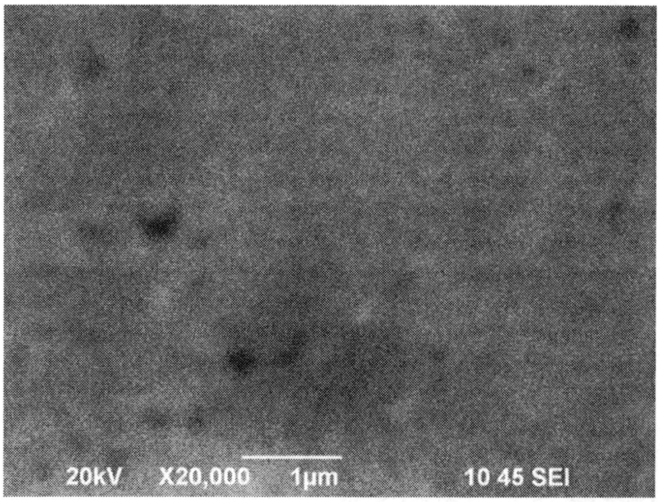

Fig. 5 SEM-photograph of the surface of the "upper" electrochemically deposited bright copper layer on the matt nickel layer, electrodeposited on a brass substrate $Cu_{bright}(2,3\ \mu m)/Ni_{mat}(1,8\mu m)/CuZn37$ *(X20 000)*

Figure 5 shows an SEM image (JEM 200-CX in SEI mode) of the microstructure of the upper bright copper layer, deposited electrochemically on the matte nickel coating (20 0000×). Since the thickness of the copper layer is very small (in this case 2.3 μm), its leveling is not sufficient to flatten the microwells (dark spots) of the matt nickel coating (in this case the thickness is 1.1 μm). The surface morphology of the specimen, however, is dominated by the actual microstructure/habitus of the bright copper coating. The measured values for R_a and R_z, that characterize the surface roughness due to the two layers are 0.1 μm and 0.8 μm respectively.

The deposition on a brass substrate of a three-layer coating of matt nickel (with thickness 1.4 μm), followed by a bright copper coating (with thickness 2 μm) and a bright nickel coating (with thickness 2 μm) was carried out consecutivelly (including intermediate distillated water washing operations and acid activation).

Figure 6 shows an electron microscopic photo (JEM 200-CX in SEI mode) of the microstructure of the upper bright nickel`s layer surface, deposited electrochemically on the Cu_{bright} coating of the system $Cu_{bright}/Ni_{mat}/CuZn37$ (20,000 ×). The picture shows that the nickel coating is extremely smooth, dense and finely dispersed, which results both from the conditions of its production and the epitaxial effect of the extremely fine-grained and flattened intermediate mirror-bright copper coating. The measured values for R_a and R_z, which characterize the surface roughness of the top bright nickel coating, are respectively 0.1 μm and 1.1 μm. Undoubtedly, the Cu_{bright} and Ni_{mat} interlayers, as well as the brass substrate, have contributed to these values. For comparison, Fig. 7 shows the SEM-microphotography on the surface of the brass (CuZn37) substrate, for which the measured R_a and R_z are respectively 0.2 μm and 1 μm.

Data on thicknesses and micro-irregularities (R_a and R_z) of investigated individual layers, as well as for systems are given in Table 2.

As a result of nanoindentation experiments, load–displacement curves were obtained and two mechanical characteristics of substrate and investigated films – indentation hardness (H_{IT}) and indentation modulus (E_{IT}) – were calculated using Oliver & Pharr approximation method [5,6]. The results we receive are objectively limited by the load / penetration potential of the used indenter(maximum load=500mN). That's why, the maximum load on the indenter did not allow a complete drilling of the three-layer coating to the substrate.

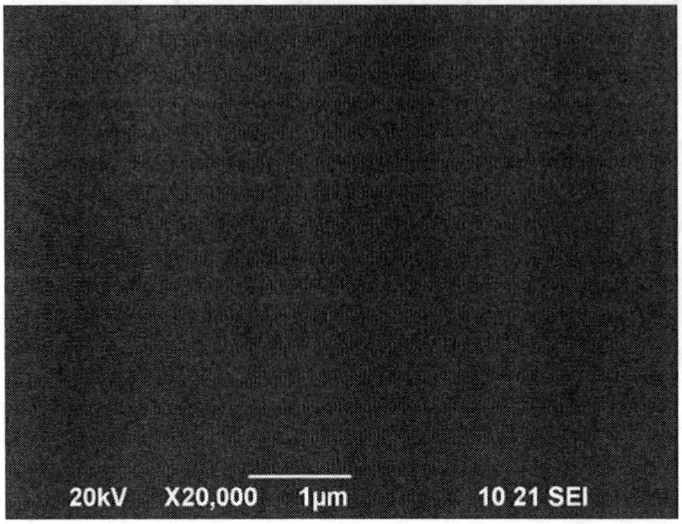

Fig.6 SEM image of the surface of the "upper" electrochemically deposited bright nickel layer of the "sandwich" system Ni_{bright} (1,4 μm)/Cu_{bright} (2 μm)/Ni_{mat} (1,4 μm)/CuZn37 (X20 000)

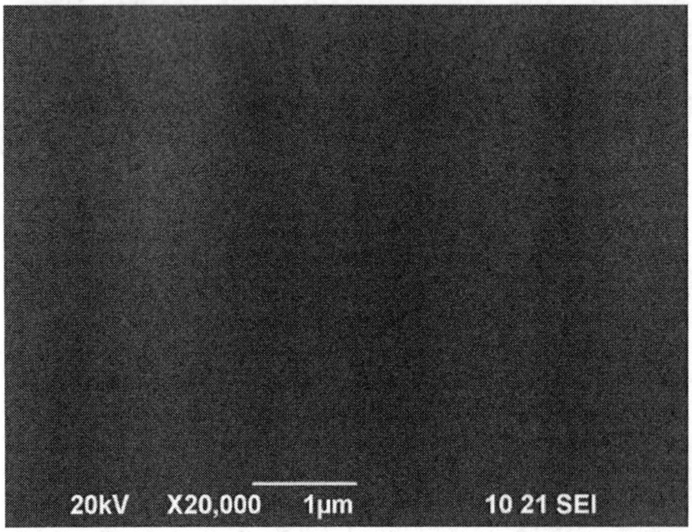

Fig.7 SEM photograph of the surface of the used CuZn37 substrates (X20 000)

Table 2. Thickness and surface roughness of the investigated samples.

No	Type of coatings	Thikness, μm	R_a, μm	R_z, μm
1	Cu_{bright}/ CuZn37	1.78/500	0.2	1.4
2	Ni_{mat}/ CuZn37	1.00/500	0.1	0.9
3	Cu_{bright}/Ni_{mat}/CuZn37	3.24/1.16/ 500	0.1	0.8
4	Ni_{bright}/Cu_{bright}/Ni_{mat}/CuZn37	1.41/1.98/1.40/500	0.1	1.1
5	Degreasing and etching treated substrate CuZn37	500	0.2	1.0

The obtained results for indentation hardness and indentation modulus of investigated films are shown in Figs.8 and 9.

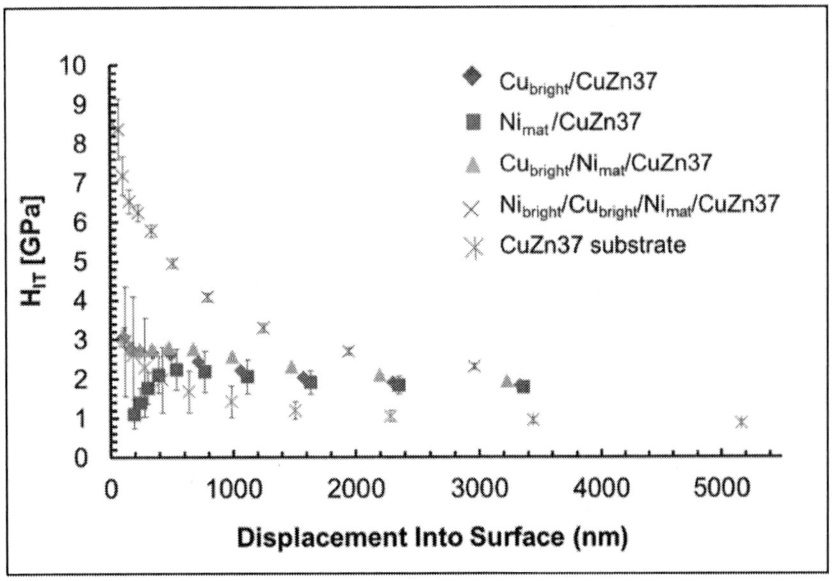

Fig.8. Indentation hardness of investigated multilayer films

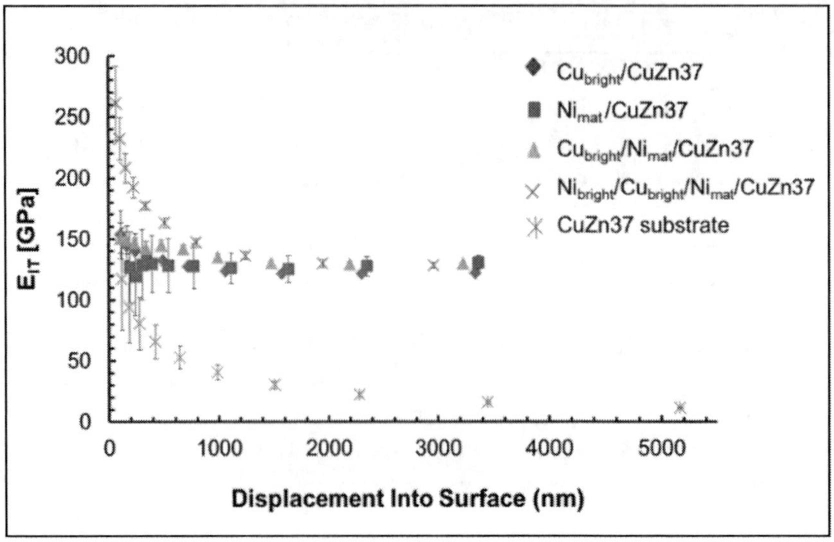

Fig.9 Indentation modulus of investigated multilayer films

It can be seen that the $Ni_{bright}/Cu_{bright}/Ni_{mat}$/CuZn37 three-layer coating has the highest indentation hardness and modulus, followed by the two-layer Cu_{bright}/Ni_{bright} system and single layer coatings. The single Cu_{bright}/CuZn37 layer has higher indentation hardness and indentation modulus than single Ni_{mat}/CuZn37 layer, which is in good agreement with Hall-Petch relationship [7], because Cu_{bright}/ CuZn37 layer is more fine grained (20 – 80 nm) than Ni_{mat}/ CuZn37 layer (50-500 nm). With increasing depth of indentation, the indentation hardness and indentation modulus of both Cu_{bright}/CuZn37 and Ni_{mat}/CuZn37 layers becomes constant, but still higher than these of the substrate. This could be explained with pile-up of material around the indenter and indentation-induced densification of material under the indenter tip [8].

In triple layer Ni_{bright}/Cu_{bright} / Ni_{mat}/CuZn37 coating with the inclusion of the third (Ni_{bright}) layer, indentation hardness and indentation modulus is drastically increased, which among the higher own hardness of brilliant nickel coating [1] may be associated with the presence of an additive effect, driven by the higher microhardness of the bilayer system (green curve) and its screening (reducing) of the influence of the soft brass substrate. In the same time, a decrease in indentation hardness and indentation modulus is recorded, with an increase in the penetration depth, which may be related to the influence of the softer Cu_{bright} layer. The obtained results are in good agreement with the obtained from other authors [9,10].

4 Conclusions

In the present work we investigated by means of nanoindentation experiments mechanical properties of electrochemically deposited on brass substrate single-, two- and three-layer metal coatings, including nickel and copper layers. Surface morphology and structure of the obtained coatings, their texture and surface roughness were investigated too. The results show that the Ni_{bright} / Cu_{bright} / Ni_{mat} / CuZn37 three-layer coating has the highest indentation hardness and modulus, followed by the two-layer Cu_{bright}/Ni_{bright} system and single layer coatings. The conducted study indicates that in the assembly of electrochemically deposited multi-layer metal coatings, formation of protective and decorative systems with high corrosion-protective ability [1] is achieved, which are characterized by significantly better mechanical properties, compared to those of the individual layers. This fact is a guarantee of their better adhesion to the protected substrate in case of mechanical, thermal and corrosion attack.

Authors gratefully acknowledge the financial support of Bulgarian National Science Fund under Grant No. T02-22/12.12.2014.

References

1. Y. Matulis, *Bright electrolytic coatings*, ("Mintis" Publ. House, Vilnius, 1969).
2. D. Stoychev et al., Bulg. Pat. No No 43112.
3. M. Datcheva, S. Cherneva, D. Stoychev, R. Iankov, M. Stoycheva, Mater. Sci. Appl. **2**, 1452 (2011).
4. W. Oliver, G. Pharr, J. Mater. Res. **19,** 3 (2004).
5. J.C. Hay, A. Bolshakov, G.M. Pharr, J. Mater. Res. **14**, 2296 (1999).
6. G.M. Pharr, A. Bolshakov, J. Mater. Res. **17,** 2660 (2002).
7. W. Smith, J. Hashemi, *Foundations of Materials Science and Engineering*, 4th ed., McGraw-Hill, New York, 2006.
8. S. Yoshida, H. Sawasato, T. Sugawara, Y. Miura, J. Matsuoka, J. Mater. Res. **25**, 2203 (2010).
9. H.C. Barshilia, K.S. Rajam, Surf. Coat. Technol. **155,** 195 (2002).
10. D. P. Cole, H. A. Bruck, Proceedings of the SEM Annual Conference, June 1-4, 2009 Albuquerque, New Mexico, USA.

Study of the mechanical properties of Ti-and Cr-based multicomponent hard coatings

Tetiana Cholakova[1,*], *Vasilliy* Chitanov[1], *Lilyana* Kolaklieva[1], *Roumen* Kakanakov[1], *Konstantin* Balashev[2], *Bogdan* Ranguelov[3], *Velko* Rupetchov[4] and *Georgi* Mishev[4]

[1]Central Laboratory of Applied Physics, Bulgarian Academy of Sciences, Plovdiv, Bulgaria
[2]Laboratory of Biophysical Chemistry, Department of Physical Chemistry, Sofia University, Bulgaria
[3]Institute of Physical Chemistry, Bulgarian Academy of Sciences, Sofia, Bulgaria
[4]Paisii Hilendarski University, Plovdiv, Technical College Smolyan, Bulgaria

Abstract. TiCrAlN and CrTiAlN multicomponent coatings have been developed using closed-field unbalanced magnetron sputtering technique (CFUBMS) in a gas mixture of Ar + N2. The nitrogen level was varied by using the feedback control of plasma optical emission monitor (OEM). An investigation into the effect of the CFUBMS process parameters on the properties of the coatings was undertaken. The main coatings parameters such as thickness, surface morphology, nanohardness, strength of adhesion and wear resistance were studied by means of ball-cratering method, atomic force microscopy, scanning electron microscopy, scratch tests and nanoindentation measurements. The study revealed strong dependency of the mechanical properties on the nitrogen flow rate. Analysis of the experimental results showed that Cr-based multicomponent coatings possess better mechanical properties than Ti-based coatings at a nitrogen flow rate of 21 sccm: higher value of hardness ($\leq$ 31GPa) and higher scratch resistance (> 30 N).

1 Introduction

Unbalance magnetron sputtering technique was developed to improve the homogeneity of deposition films by increasing the region of homogeneous plasma and the substrate – target distance. This technique [1] has been used successfully to deposit various coating materials on different kind of substrates and tools. Hard coatings based on Ti and Cr nitrides are widely used as protective coatings. Cubic(c)-TiAlN and c-CrAlNcoatings have exhibited superior mechanical properties and oxidation resistance in comparison to traditional binary TiN and CrN coatings [2], and have been successfully used in the industry. Films based on the Ti-Cr-N system were also deposited using a variety of deposition methods in order to improve the hardness and wear resistance [3, 4]. Many studies were reported on the deposition and properties of TiAlN coating produced by various techniques. TiAlN coatings were developed by Physical vapour deposition (PVD) as an alternative to TiN, because of their higher oxidation resistance (>700°C), enhanced hardness (30–35 GPa) and higher corrosion resistance [5-10]. The later concept has been found to have promising

* Tetiana Cholakova: ipfban-dve@mbox.digsys.bg

performance compared to that of ternary films. Ti–Al–Cr–N system offers a high variability ranging from Cr rich to Al rich [11, 12] coatings. These films demonstrated improved hardness and thermal stability, oxidation resistance and exhibit lower friction and wear coefficient. However, as these coatings become more complex and the number of possible compositions increases dramatically, it is important to understand the role of each element in order to optimize the final composition, which determines their mechanical and tribological properties.

The present study focuses on the investigation of the Ti/TiN/TiAlCrN and Cr/CrN/CrTiAlN coatings obtained by a CFUBMS technique. The goal of this work is to establish optimal technological regimes which provide stable mechanical properties of the system "substrate - multicomponent coating" and to explore its potential for industrial application.

2 Experimental

2.1 Material and preparation

Ti-Cr-Al-N and Cr-Ti-Al-N coatings were deposited onto substrates of high-speed steel (HSS) and stainless steel (SS) using a closed-field unbalanced magnetron sputtering system UDP850 (Teer Coatings Ltd., UK) as shown in Figure 1. CFUBMS is equipped by four rectangular cathodes with high purity (99,99%) Cr, Ti and two Al targets operated in an unbalanced magnetron mode. The closed magnetic field coupling results in a highly degree of ionization and bias current density. Prior to the deposition, the substrates were cleaned in an ultrasonic bath for 5 min. at temperature of 65°C using commercial cleaning solution to remove oils, residues and protection against corrosion. After that the substrates were rinsed in distilled water for 2 min. followed by ultrasonic cleaning in hot isopropanol for 5 min. and dried up. Generally, ultrasonically cleaned metal substrates contain thin top oxide layer which affects the adhesion. After wet chemical cleaning the substrates were loaded in the deposition chamber and the system was pumped down to the base pressure less than $2{,}5\times10^{-3}$ Pa. In order to remove the top oxide layer the substrates were plasma cleaned in argon atmosphere ($1.7\text{x}10^{-1}$ Pa) for 30 min. at negative pulse substrate bias voltage of 500 V and frequency of 250 kHz. In all the deposition processes the substrates were biaxially rotated at a speed of 5 rpm in order to obtain homogenous film composition. The distance between the substrates and the targets was kept 150 mm.

All the experiments were realized without additional substrate heating at a bias voltage of -70 V and frequency of 150 kHz, in DC work regime of the Ti and Cr cathodes and pulsed regime of the Al cathodes. A mixture of argon and nitrogen gases was introduced into the chamber during the deposition process with a fixed Ar flow rate of 25 sccm. Nitrogen content was controlled by a plasma optical emission monitor with a feedback control of 95-50% (N_2 flow rate being in the range between 5 and 24 sccm).

Two series of experiments were performed under the same deposition conditions for deposition of Ti-based and Cr-based multicomponent coatings. In the first series, TiCrAlN coatings were deposited using Ti as the adhesive layer and TiN as an interlayer. In the second series, Cr and CrN were the adhesive layer and interlayer, respectively. Deposition started with a Ti or Cr adhesion layer. Over it a Ti-N or Cr-N transition film with gradually increasing nitrogen was layered. Following these two sub-layers, a compositionally graded Ti-Cr-Al-N or Cr-Ti-Al-N top layer was deposited. The relative concentration of Cr, Ti and Al in the coatings was adjusted through the sputtering power (P) applied to the targets during deposition: P_{Cr} = 1.5 kW; P_{Ti} = 2.4 kW; P_{Al} = 0.6 kW.

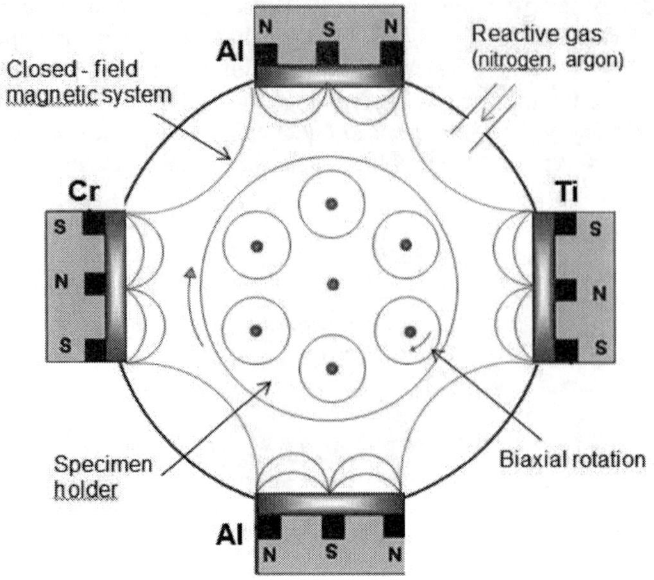

Fig.1. Cross-section schematic drawing of the CFUBMS sputtering system

2.2 Characterization

The coating thickness measurements were performed by a ball-cratering method (Calotest) which is suitable supplemental method providing information on configuration of individual layers and resistance to wear. Rotating ball of known diameter (20 mm or 30 mm) produces crater on the specimen surface, the so-called calotte which is used for determination the thickness of the layers.

Surface observations and morphology are performed on JEOL JSM 6390 electron microscope, equipped with INCA Oxford EDS energy dispersive detector. Surface images are obtained in secondary electrons (morphology contrast) and back-scattered electrons (density contrast) as well.

The Atomic force microscopy (AFM) studies were carried out by NanoScope VAFM (Bruker Inc.) in air in tapping mode. Silicon cantilevers with reflective aluminium coating with thickness of 30 nm, Tap 300Al-G (Budget Sensors, Innovative solutions Ltd, Bulgaria) were used. The characteristics of the cantilevers are as follows: spring constant, from 1.5 to 15 N/m, resonance frequency, 150 ± 75 kHz and radius of the tip curvature less than 10 nm. The scanning rate was set at 1 Hz.

The mechanical properties of the deposited coatings were investigated using Compact Platform CPX (MHT/NHT) CSM Instruments equipment which includes a Nanoindentation module (NHT), a Micro Scratch module (MST) and an Optical video microscope with CCD camera, installed together on the same platform. Nanoindentation was performed by a triangular diamond Berkovich pyramid in the loading interval of 10 - 200 mN.

The adhesion strength of the coatings to the substrate material and a friction coefficient were evaluated with MST equipped with a spherical Rockwell indenter with radius of 200 µm under the following test conditions: progressively increasing pressing force in a range of 1-30 N; loading rate 0.5 N/min; indenter movement speed 0.02 mm/min; acoustic emission detector sensitivity – 9. The character of the damages formed was assessed by observations with an Optical video microscope equipped with a CCD camera.

The wear resistance examinations were performed using a Ball on Flat Sliding Wear Test in a horizontal orientation of the test surface. A mineral-ceramic ball of Al_2O_3 with a diameter d = 3.0 mm, fixed in a holder, was used as a counter -body. The ball was rubbed on a Reciprocal Drive in air at room temperature, without lubricants.

3 Results and discussion

The study revealed that the same technological conditions have different effect on Ti –and Cr- based multicomponent coatings, which is, probably, due to different nucleation properties and growth rates during deposition. Below the results for the coatings deposited at a nitrogen flow rate of 21 sccm, which showed best mechanical properties, are presented.

3.1 Thickness and morphology

The thickness of the deposited coating was calculated from measurements of the diameters of the craters obtained in the coating and substrate using a steel ball with a diameter of 20 mm and diamond slurry with particles of 0.25 μm in diameter. The coating was abraded until the substrate was reached by the ball. This method provides an opportunity to observe continuous transition from layer to layer down to the substrate. The calculated total thickness (d) of deposited coatings is in the range of 1.8–2.5 μm. Figure 2a depicts architecture of the CrTiAlN coating, deposited at a nitrogen flow rate of 21 sccm.

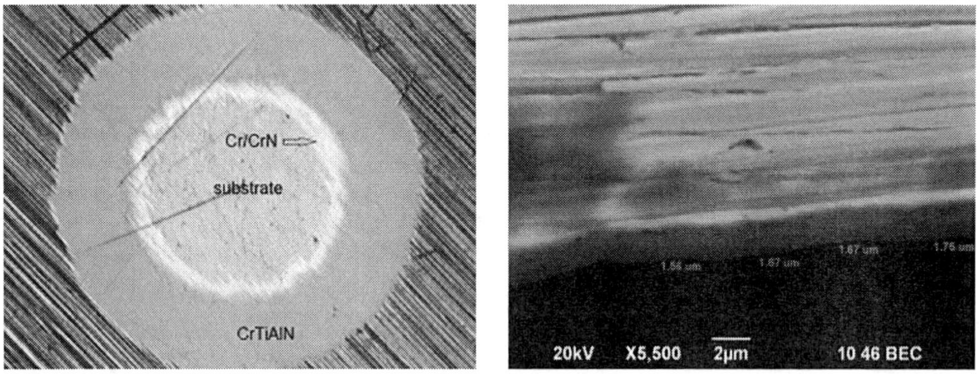

(a) (b)

Fig.2. An optical microscopy image of the CrTiAlN coating architecture, obtained by ball-cratering test (a) and the cross-sectional SEM micrograph (b) of same the coating sample

Determination of the coating thickness was performed also on a cross-section of the CrTiAlN coating using a scanning electron microscope (Fig.2b). The obtained result (d=1.75 μm.) is in good agreement with the Calotest method calculation (d=1.8 μm).

Several scans were performed by the AFM in tapping mode over different areas. The roughness analysis gives the value Ra, which is an arithmetic average of the absolute values Zj of the surface height deviations measured from the mean plane, while the image Rq is the root mean square average of height deviations taken from the mean image data plane. The AFM data revealed that the coating, obtained at a nitrogen flow rate of 21 sccm exhibited a densely packed structure [13], consisting of well separated grains with apparently spherical form and predominantly equal sizes (Fig.3a-b). The average roughness (Ra) and root mean square (Rq) surface roughness were obtained on a scanned area of 10 μm x 10 μm as follows: Ra = 16.4 nm and Rq = 22.0 nm. On the coating surface were observed sporadic small defects, which lead to an increase of the roughness (Fig.3a, box): Ra=19.9 nm, Rq= 26.4 nm.

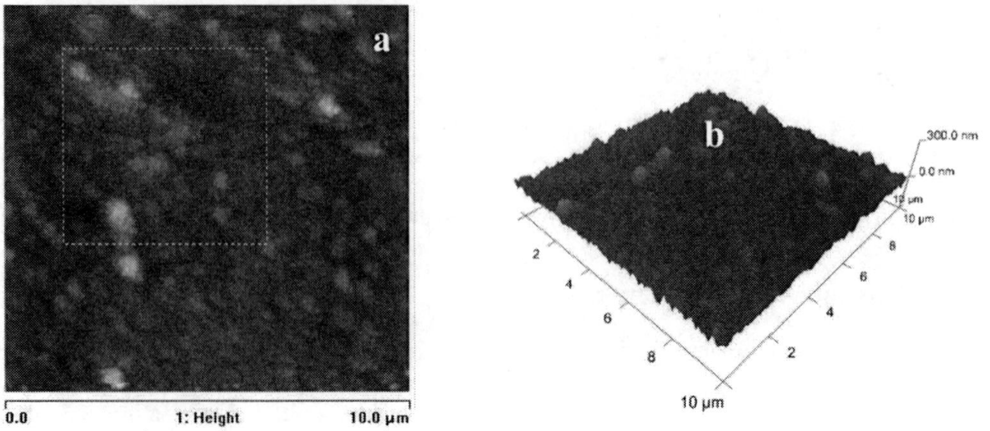

Fig.3. Surface topography of the CrTiAlN coating (SS substrate) obtained by AFM: a) 2D image - box dimension 5x5 µm; b) 3D image.

3.2 Hardness

The nanohardness (H) and elastic modulus (E) were calculated from nanoindentation load–displacement data at a periodic loading and unloading of the samples at a successively increasing load using an Oliver & Pharr method [14], executed by the software. To eliminate influence of the substrate on the measurement a load of 15 mN was applied, which corresponded to a maximum indentation depth of about 170 nm (less than 10% of the coating thickness). The maximum nanohardness of 31 GPa and elastic modulus of 418 GPa were measured for CrTiAlN coatings at the applied load of 15 mN. This value is higher as compared to the best value of the TiCrAlN coating (28 GPa and 358 GPa). The obtained nanohardness is significantly higher than those reported in [15, 16] and coincides with that obtained in [17].

3.3 Scratch strength

During the tests the load increased progressively in linear mode from 1N to 30 N at scratch lengths of 1 mm and 3 mm and constant scratch speeds of 0.02 mm/min and 0.05 mm/min. The measuring device registers the penetration depth and the acoustic emission along the length of the track. Three scratches on the coated HSS substrates for each sample were carried out in order to determine the adhesion strength. The critical load (*Lc*) values show relationship between the normal load, coefficient of friction and scratch length were determined after the test by optical microscopy observation of the damages formed in the scratch tracks and from the recorded acoustic emission (AE) and friction force (Ft) signals. The critical loads *Lc1* and *Lc2* indicate the appearance of the first cohesive and adhesive cracks, respectively [16].

 https://doi.org/10.1051/matecconf/201814502003

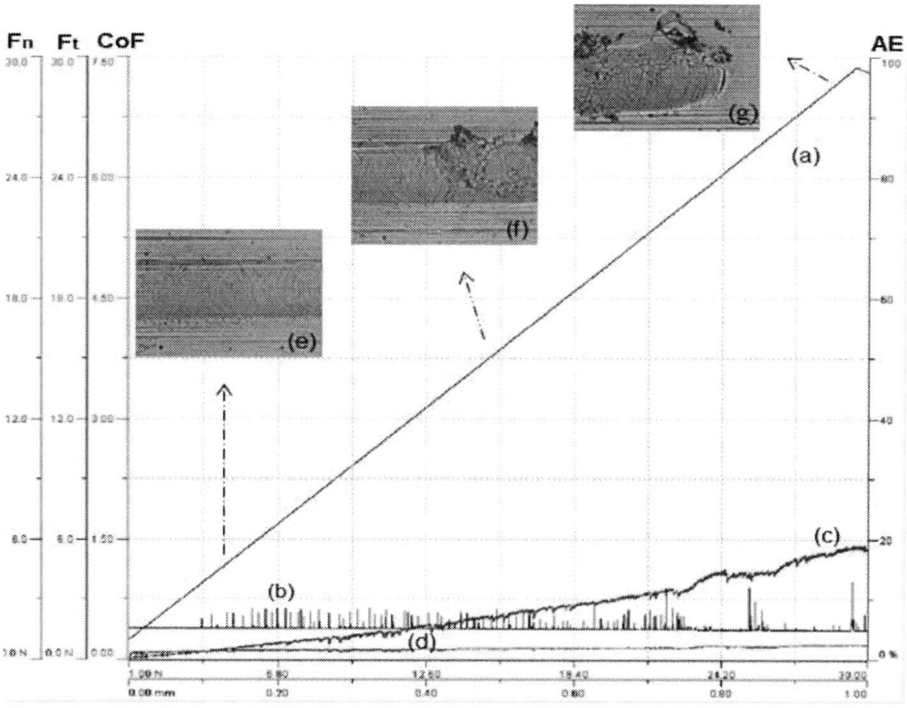

Fig.4. Typical scratch graphs on the TiCrAlN/HSS coating and optical micrographs (e, f and g) of the main parts of scratch track with cracks at different locations: a) Normal load; b) Acoustic emission; c) Friction force; d) Coefficient of friction.

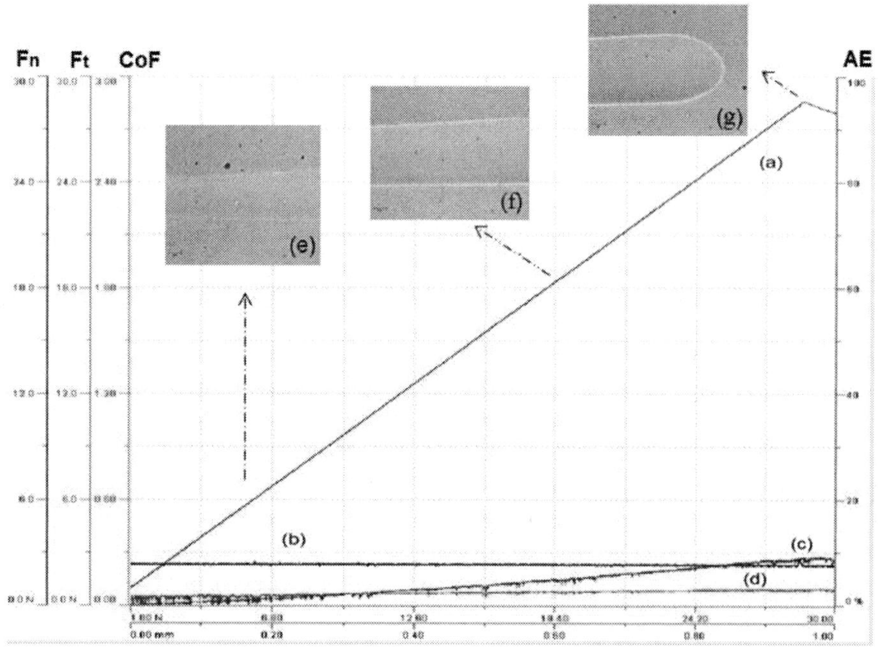

Fig.5. Typical scratch graphs on the CrTiAlN/HSS coating and optical micrographs (e, f and g) of the main parts of scratch track at different locations: a) Normal load; b) Acoustic emission; c) Friction force; d) Coefficient of friction.

The test consists of applying a continuously increasing load on the system coating-substrate at a constant speed. The scratching point causes increase the elastic and plastic deformation until damage occurs in the surface region. In most of the experiments the normal load at which the coating failure was initiated was detected by an increase in acoustic emission and confirmed by microscopic examination. Figure 4 shows scratch graphs of the TiCrAlN/HSS coating and the optical micrographs of the main parts of the scratch track with cracks at different locations. It is observed that with increase of the scratch length and the normal load respectively, the values of the friction force and acoustic emission are change. Very slight semi-circular cracks were observed at the normal load of about 5 N (Fig.4e). The first critical load is registered at Fn = 12,5 N (Fig.4f) and the second one at Fn = 23 N (Fig.4g). A distortion of the line of the friction force and the acoustic emission is observed at a higher normal load Fn = 23 N. However, it should be noted that the total removal of the coating did not occur within the loading interval.

The best adhesion result without any visible damages within loading range of 1-30 N possessed the CrTiAlN coating. An optical micrograph of the scratch track parts in the CrTiAlN coating and scratch test results of the acoustic emission (AE), friction force (Ft) and coefficient of friction (Cof), are shown in Figure 5.

As a result of the tests carried out, it was established that the critical loads Lc1 and Lc2 are observed for the Ti-based multicomponent coatings, while Cr-based coatings exhibit very good adhesion properties in the loading range 1 -30 N. The coefficient of friction of the Cr-based coatings against diamond indenter was measure to be $\mu = 0.1$, while the Ti-based coatings had a higher coefficient of friction ($\mu \approx 0.2$).

3.4 Wear and wear rate

Investigations of the effect of normal force on the wear resistance of the CrTiAlN coating (d=2μm) at a load of 1, 1.5, 2, 2.5 and 3N and an average speed of sliding $V_{av} = 20$ mm/s = const. were carried out. The width of the groove was measured with a microscope TESA VISO-300 non-contact measuring system with magnification x100 (resolution 0.001 mm). The average value of the width of the groove (b_{av}) is determined by the equation:

$$b_{cp} = \frac{1}{n}\sum_{i}^{n} b_i \text{ , mm.} \quad (1)$$

The wear rate (coefficient) I_w was determined by the formula:

$$I_w = \frac{V}{F.L} \text{ , mm}^3\text{/N.m} \quad (2)$$

where: V is the volume of the removed coating material (track) in mm^3; F is the normal load in N; L is the sliding distance of the sample in m.

The track volume is determined according to the method described in [18, 19]. The wear coefficient is based on the assumption that the volume wear varies directly with the contact load and the sliding distance. The data of the wear volume and the wear rate as a function of the loading at $V_{av} = 20$ mm/s and $L = 50$ m are presented in Table 1.

The experimental results show that the normal load significantly influences on the wear of the Cr/CrN/CrTiAlN multilayer coating. The worn volume increases with the increase of the normal load while the wear intensity changes slightly. The experiments with loads from 1N to 3N passed completely in the coating without reaching the substrate. The maximum penetration depth (h) is 1.56 μm. It should be noted that the coating has a complex vertical construction (from the surface to the substrate) and, under greater load, it passes through layers having different mechanical properties. Consequently, the volume variation at different loads is a direct criterion for the evaluation of the coating wear, while the wear rate I_w is an universally accepted measure for the wear of different coatings at the same working conditions.

Table 1: The calculated value of wear and wear rates of the CrTiAlN coating, deposited on an unhardened stainless steel substrate (1.2343).

F [N]	V x 10^{-6} [mm^3]	I_w x 10^{-6} [mm^3/N.m]	h x 10^{-3} [mm]
1	497.343	9.947	0.724
1.5	675.627	9.008	0.888
2	892.216	8.922	1.068
2.5	1128.545	9.028	1.249
3	1583.803	10.559	1.564

4 Conclusion

Two series of multicomponent coatings without additional substrate heating were successfully obtained under the same deposition conditions using the closed-field unbalanced magnetron sputtering technique. The study revealed strong dependency of the mechanical properties on the nitrogen flow rate. The coatings based on Ti/TiN/TiAlCrN and Cr/CrN/CrTiAlN exhibited hardness and an elastic modulus of 26 ± 5 GPa and 350 ± 70 GPa, respectively, at a nitrogen flow rate of 21 sccm. Further increase of the N_2 flow rate caused decreases of the coating hardness. The results of the scratch tests showed different states of adhesion to the HSS substrate of these coatings. The Cr-based coatings demonstrated excellent adhesion in the loading interval from 1N to 30N and a low coefficient of friction (μ=0.1), while the Ti-based coatings had worse adhesion expressed by presence of higher coefficient of friction ($\mu \approx$ 0.2) and a critical load Lc1 at loading of 12 N.

The research showed that further optimization of the deposition process of Ti-based multi-component coatings is needed in order to obtain better mechanical properties.

References

1. B.Window, N.Savvides, J.Vac.Sci.Technol. A, Vol. **4**, Iss.2, 196 (1986)
2. O.Knotek, M. Bohmer, T. Leyendecker, J. Vac. Sci. Technol. A, Vol. **4**, Iss. 6, 2695 (1986)
3. J.H. Hsieh, W.H. Zhang, C. Li, C.Q. Sun, Surf. & Coat. Technol., **146 –147**, 331–337 (2001)
4. W. Aperador, A. Delgado, J. C. Caicedo, Int. J. Electrochem. Sci., **12**, 4502 – 4514 (2017)
5. S. PalDey, S.C. Deevi, Mater. Sci. Eng. A, **342,** 58–79 (2003)
6. P.W. Shum, K.Y. Li, Z.F. Zhou, Y.G. Shen, Surf. & Coat. Technol., **185**, 245– 253 (2004)
7. S.K. Wu, H.C. Lin, P.L. Liu, Surf. & Coat. Technol., **124**, 97–103 (2000)
8. L. García-González, M. G. Garnica-Romo, J. Hernández-Torres and F. J. Espinoza-Beltrán, Braz. J. of Chemical Eng., **24,** 2, 249–257 (2007)
9. H. C. Barshilia, K. Yogesh, K.S. Rajam, Vacuum, **83**,427–434 (2009)

10. T. M. Cholakova, V. A. Chitanov, L. P. Kolaklieva, R. D. Kakanakov, D.G. Kovacheva, P. K.Stefanov, S. N. Rabadzhiyska, E. A. Korina, V. I. Kopanov, Bulg. Chem. Commun, **48**, Spec. Iss. 384 – 390 (2016)
11. P.L. Tam, Z.F. Zhou, W. Shum, K.Y. Li, Thin Solid Films, **516,** 5725-5731 (2008)
12. G.S. Fox-Rabinovich ,A.I. Kovalev, M.H. Aguirre, B.D. Beake, K. Yamamoto, S.C. Veldhuis, J.L. Endrino, D.L. Wainstein, A.Y. Rashkovskiy, Surf. & Coat. Technol., **204**, 489–496 (2009)
13. H. Zhou, J. Zheng, B. Gui, D. Geng, Q. Wang, Vacuum **136,** 129-136 (2017)
14. Oliver, W.C.; Pharr, G.M., J. Mater. Res. **7,** 1564-1583 (1992)
15. M. S. Kabir, P. Munroe, Z. Zhou, Z. Xie, Surf. & Coat. Technol. **309**, 779–789 (2017)
16. Y. Shi, S. Long, S. Yang, F. Pan, Appl. Surf. Sci. **254,** 7342–7350 (2008)
17. L. Lu, Q. Wang, B. Chen, Y. Ao, D. Yu, C. Wang, S. Wu, K. Ho Kim, Trans. Nonferrous Met. Soc. China, **24,** 1800−1806 (2014)
18. G. Mishev, V. Rupetsov, S. Dishliev, Ch Pashinski., *11th International Conference in Manufacturing Engineering Proceedings*, 1-3 October 2014, Thessaloniki, Greece, 139-147 (2014)
19. V. S. Rupetsov, J. of Food and Pack. Sci., Technique and Technologies, **1,** 4, 60-65 (2014)

Tensile behavior of humid aged advanced composites for helicopter external fuel tank development

Mihaela Condruz[1*], *Alexandru* Paraschiv[2], *Cristian* Pușcașu[2], *and Ionuț Sebastian* Vintilă[1]

[1]COMOTI - Romanian Research & Development Institute for Gas Turbines, Composite Materials Laboratory for Aeronautical Field and Space, 220 D Iuliu Maniu Av., Bucharest, Romania
[2]COMOTI - Romanian Research & Development Institute for Gas Turbines, Gas Turbine Special Components. Physics and Mechanic Testing Laboratory, 220 D Iuliu Maniu Av., Bucharest, Romania

Abstract. Influence of humid aging on tensile properties of two polymeric composites was studied. The purpose of the study was to evaluate the suitability of the materials for a naval helicopter external fuel tank. Due to the application, the humid environment was kerosene and saline solution to evaluate the sea water effect on the composite tensile strength. The composite samples were immersed in kerosene for 168 hours, respective 1752 hours and in saline solution for 168 hours. Tensile tests were performed after the immersion. The composite sample tensile tests showed that kerosene and saline solution had no influence on the elastic modulus of the materials, but it was observed a slight improvement of the tensile strength of the two polymeric composites.

1 Introduction

Nowadays, fibre reinforced polymers (FRP) are a common material used in the aerospace field (for commercial/military aircrafts, helicopters, satellites etc.) due to their remarkable properties, such as strength, stiffness, low density and low coefficient of thermal expansion [1, 2]. An important factor that stays behind the greater use of composites is the cost reduction experienced by carbon fibres [1]. Also, in case of aerospace applications, the reduction of weight, costs and fuel along with the increase of resistance and durability of the components are key aspects. Usually, in helicopter applications, composite materials are used to manufacture rotor blades, which ensure a significant weight reduction over aluminium and provide a better control over the vibration characteristics of the blades [3]. Excepting rotor blades, polymeric composites were integrated in tank assemblies for launch vehicles and helicopters. Researches were conducted to manufacture and validate external composite fuel tanks for helicopters to replace the metallic tanks [4]. External fuel tanks have been introduced to extend the range of a helicopter or to provide refuelling capacity at a remote location (it can provide at least an additional hour of flight depending on its capacity [5]). The first attempt to replace metallic external tanks was in the middle of 1970 and they were replaced by a hybrid model, a Composite Overwrapped Pressure Vessel -

* Corresponding author : raluca.condruz@comoti.ro

COPV [4] (structure with a thin metal liner that acts as a leakage barrier, overwrapped by composite winding layers [6]). Afterwards, the COPV tanks were replaced by all composite external tanks. For researchers, an opportunity is represented by designing and manufacturing of an external composite fuel tank for F-35 Lightning II fighters [4,7]. An important aspect for designing and manufacturing external and auxiliary fuel tanks is to study structure's requirements and specification, which are stated in MIL-T-7378A [8].

Plastic fuel tanks should be fuel and environmental resistant (corrosion due to salt, moisture, ultraviolet radiation, thermal cycling in atmospheric conditions during normal service use and storage between -54°C and +71°C) [8-11]. Durability of FRP can be affected by the contact with water or other fluids [12]. Most polymeric materials (in the form of a composite matrix or a polymeric fibre) are capable of absorbing relatively small but potentially significant amounts of moisture from the surrounding environment. The physical mechanism for moisture gain, assuming there are no cracks or other wicking paths, is generally assumed to be the mass diffusion following Fick's Law (the moisture analogue to thermal diffusion) [13]. The humid aging is one of the main causes of long-term failure of organic composites, which can affect the material in several modes: matrix microcracking, plasticization, hydrolysis, delamination, blistering, void growth, swelling, localized damage at the fibre-matrix interface [12-18]. Epoxy resin is known to have affinity for water [12], moisture has significant effects on physical and chemical properties of the resin as well as on the final composite structures performance, especially in long-term utilization [19]. The moisture absorption effect on polymeric composites mechanical properties had been studied for decades [20,21], but until now it's still a challenge to predict how a fibre reinforced composite will behave under a certain humid environment.

In early technical studies regarding environmental influence on composites, it was observed that in case of carbon/epoxy composites, the filament properties show very little environmental sensitivity, while epoxy resin is affected by the absorbed moisture and temperature effects which can lead to a change in failure mode [20], also the mechanical properties of composites are reduced after humid exposure, but all moisture effects were completely reversed upon re-drying the specimens [21]. Most of the authors studied water humid aging influence on polymeric composites [22-28] and as a general conclusion, the water causes mechanical properties degradation of thermosetting composites in case of extended period of immersion which can be noticed by matrix osmotic, cracking, swelling, mass loss, fibre exposure, chemical alterations etc. Many information about the effect of other fluids than water on composite properties are not available in the literature.

Souza et.al. [29] evaluated the creep behaviour of an E-glass/polyester composite in water and lubricant oil and it was observed that after 6 and 14 months of exposure, Young's modulus of the composite was reduced by 20% regardless the aging fluid when the test was performed at room temperature and in case of tests performed at 60°C, the effects of immersion in lubricant oil were more severe than those after immersion in water. Areef [30] performed an experimental study on different solution absorption behaviour of a sandwich glass fibres/epoxy composite. The results showed a maximum value of mass gain of the samples immersed in organic solutions (kerosene, benzene) compared with samples immersed in other solutions, the samples immersed in distilled water had a minimum mass gain. Curliss [31] studied the effect of jet fuel JP-4 on thermoplastic and thermosetting composites. The thermoplastic matrix composites absorbed more fuel compared with the thermoset matrix composites, but the absorption appeared to be reversible without notable degradation of the matrix polymer. There was no evidence that any component of the matrix polymers was leaching into the jet fuel or that any fraction of the polymers was dissolving into the jet fuel. Taghavi [32] studied the diffusion of moisture into glass/epoxy, carbon/epoxy and carbon/polyimide composites immersed in water and kerosene. The glass epoxy composites showed a loss of about 21% of their transverse tensile strength after 100 hours of immersion in 80°C kerosene, but after this point no further changes were observed up to 3500 hours. The transverse modulus was not significantly changed even after 3500 hours of immersion. For carbon polyimide composites, even after 3500 hours of immersion in 80°C kerosene neither the transverse modulus nor the strength changed.

The present research was focused on studying the kerosene and saline solution influence on the tensile properties of polymeric composite materials suitable for the manufacturing of an external fuel tank for a naval helicopter (IAR 330 PUMA).

2 Experimental procedure

Two types of polymeric composite laminates were manufactured for this experiment. The laminates were manufactured by using metallic moulds and an autoclave (curing process under temperature, vacuum, pressure). The composite consisted in two prepreg precursors, a HexPly M49/42%/200T2X2/CHS-3K (referred as HexPly) and an EP 127-C20-45 T2 (referred as EP). The HexPly consisted in a twill 2x2 carbon fabric 3K HS preimpregnated with an epoxy resin system (42% resin content), and the EP was a twill 2x2 carbon fabric 3K HTA preimpregnated with a blend of epoxy resin and cyanate ester (45±3% resin content). After curing, samples were cut from laminates using a water jet cutting machine.

The samples dimensions (250x25x3 mm with 1 mm thick tabs, type 3) were established based on SR EN ISO:527-4/2000 [33], and ASTM D3039/D 3039M/2000 [34]. To ensure the water removal from the cutting process and the curing of the tabs, all the samples were post-cured in a vacuum assisted oven (curing cycle: heating for 35 minutes until 120°C were reached, heating rate of 2°C/min., 90 minutes at 120°C and oven assisted cooling).

All the samples were weighed before the immersion in saline solution and in kerosene at room temperature. To analyse the sea water effect on the composite tensile strength a saline solution was made by dissolving 35 g sea salt in a litre of distilled water to obtain a saline solution with 35 ‰ salinity (typical seawater salinity). The specimens were immersed in saline solution in an open glass recipient and in kerosene in a sealed container. The samples were kept at room temperature (20±2°C) in kerosene for 168 hours (7 days), respective 1752 hours (73 days) and in saline solution for 168 hours. In Fig. 1 can be observed the immersed samples in kerosene and saline solution.

After immersion, the samples were wiped out with a cloth, weighed out to calculate the moisture content (or percent weight gain) of the material to establish if the material has kept its dimensional measurements (or has swollen).

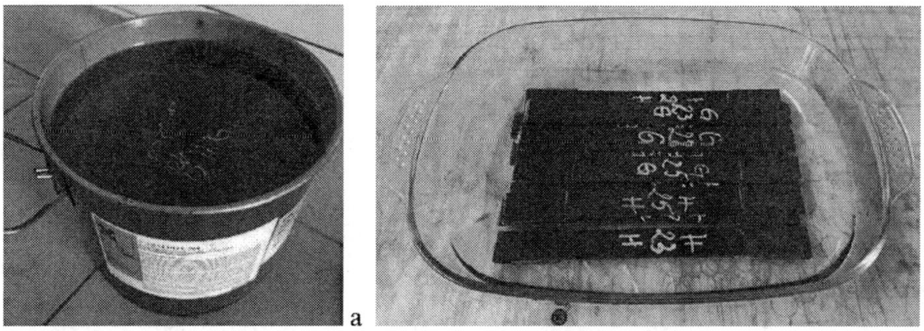

a b

Fig. 1. Samples immersed in kerosene (a) and saline solution (b)

Tensile tests were carried out by using a Universal mechanical testing machine Instron 8802 with a displacement rate of 2 mm/min, according to SR EN ISO: 527-4 [33]. For evaluating the influence of the immersion in kerosene and salted water as a function of time it were used five specimens for each type of sample and for comparison were used reference samples from each lot of samples (not immersed). The fracture's modes were coded using typical modes from ASTM D3039/D 3039M/2000 [34], in Table 1 are presented the tensile test failure codes. In Table 2 can be observed the fracture's modes of composite specimens after tensile tests.

Table 1. Tensile test failure codes/typical modes [34]

First character	Code	Second character	Code	Third character	Code
Angled	A	Inside grip/tab	I	Bottom	B
edge Delamination	D	At grip/tab	A	Top	T
Grip/tab	G	<1W from grip/tab	W	Left	L
Lateral	L	Gage	G	Right	R
Multi – mode	M	Multiple areas	M	Middle	M
long. Splitting	S	Various	V	Various	V
eXplosive	X	Unknown	U	Unknown	U
Other	O	-		-	-

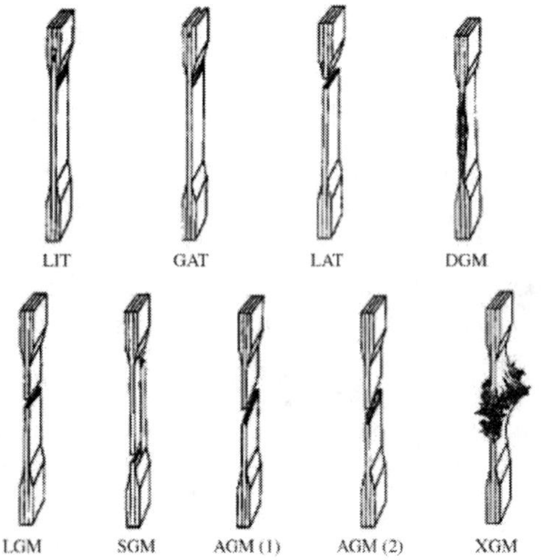

Fig. 2. Example representation of typical failure modes in tensile tests of composites [34]

3 Results and discussions

Both materials (HexPly and EP) showed a moist weight gain of 0,10% after the first 168 hours of immersion in kerosene and 0,16% after the 1752 hours of immersion in kerosene. Regarding the immersion in saline solution, the moist weight gain of both composites was 0,13% after 168 hours of immersion. The composite materials gained more weight after the immersion in saline solution compared with kerosene immersion, those were different results compared with the findings of Areef [30] who observed a higher weight gained by the composite immersed in organic solution (kerosene) then in distilled water (this could be influenced by the fact that the composite material used by Areef [30] consisted in glass fibre reinforced polymer and the samples were immersed only in distilled water not in distilled water with salt).

The moisture absorption of the HexPly and EP materials could increase after more hours of immersion, but the overall capacity of absorption of the composite wasn't the goal of this study. The composite samples maintained their dimensional measurements after the immersion in kerosene and saline solution immersion, they didn't modify the physical aspect of the samples (no peels or cracks were observed) and the polymeric materials weren't dissolved after immersion in saline solution or kerosene. Similar physical observations regarding the effect of jet fuel on polymeric composite materials were made by Curtis et.al. [31].

Table 2. Fracture's modes of composite specimens after tensile tests

Immersion conditions	HexPly	EP
Reference - unimmersed	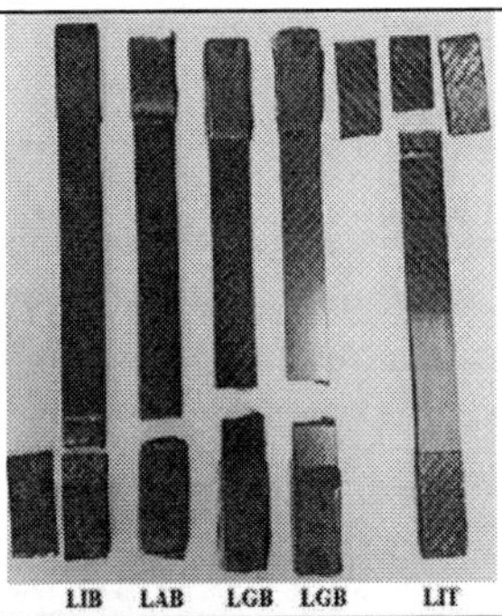	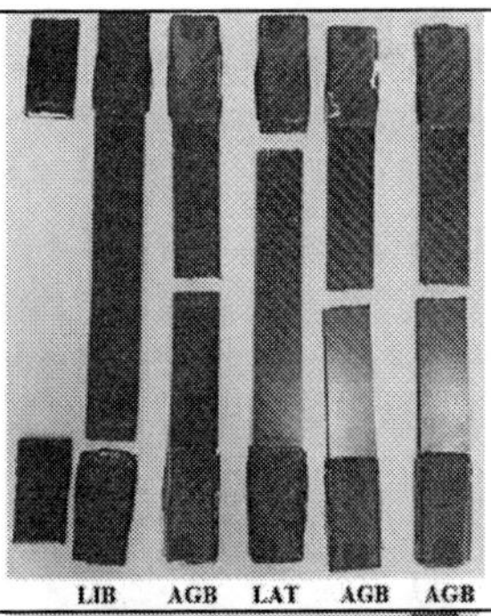
168 h saline	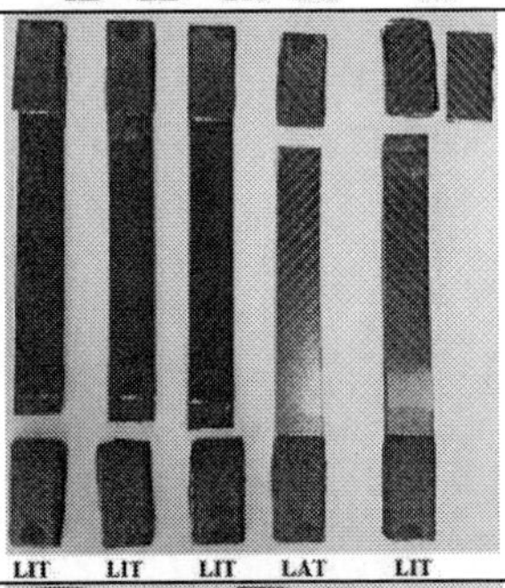	
168 h kerosene	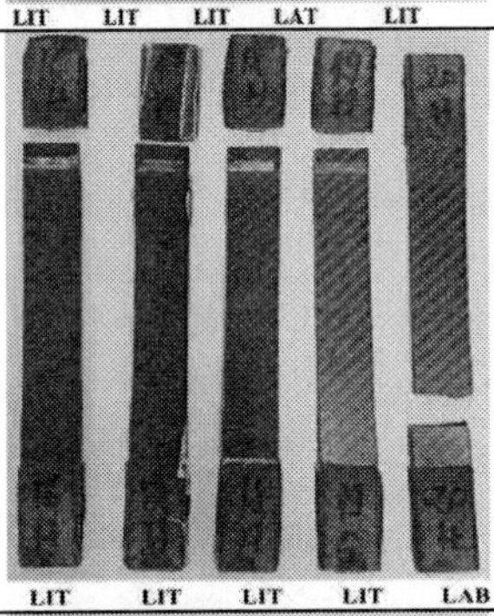	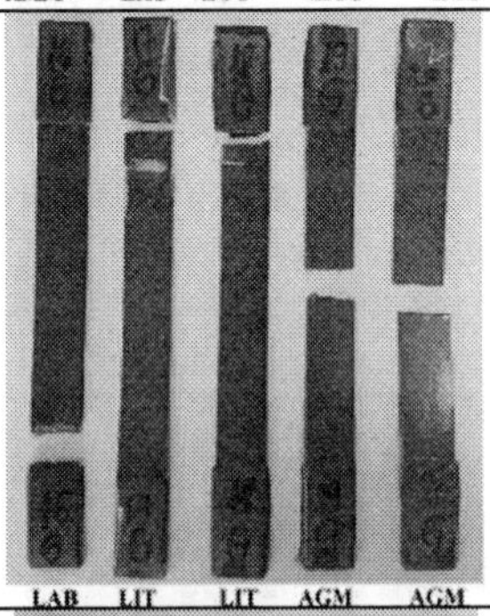
1752 h kerosene		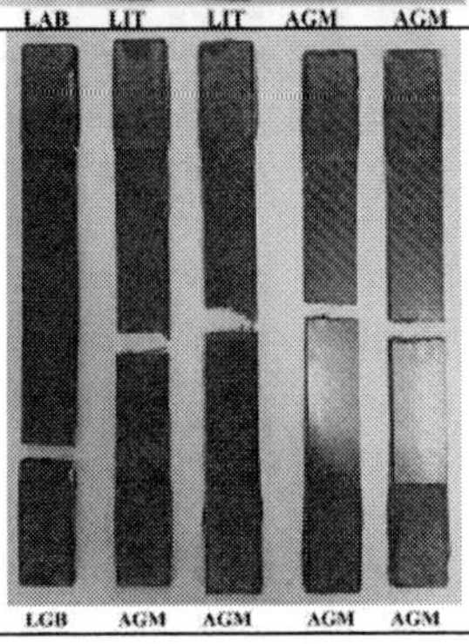

The tensile test results showed a slight increase of the tensile strength of laminates after each immersion phase in kerosene.

In comparison with their reference samples, the HexPly specimens showed an increase of the tensile strength with 3% after 1752 hours of kerosene immersion at room temperature. EP specimens showed an increase of the tensile strength with 5% after 1752 hours of kerosene immersion at room temperature.

A critical factor that can affect the tensile strength of the polymeric composite materials immersed in kerosene is the temperature of kerosene, fact proved by Taghavi [32] in his thesis, in which the specimens immersed for 100 hours in kerosene at 80°C showed a 21% reduction of transverse tensile strength, but after this point no further changes were observed up to 3500 hours. Thereby, in case of room temperature immersion for this period of time (max. 1752 h), the kerosene had an opposite effect, it leads to a slight increase of tensile strength.

It can be concluded that kerosene absorption could slightly improve the tensile strength on short term due to modifications of the internal structure of the polymer and on long term it could weaken the polymer due to breaking of hydrogen bonds because a decrease number of hydrogen bonds is a factor which influences the mechanical properties of the polymeric matrix.

Regarding the influence of the immersion on saline solution the tensile properties of composites weren't affected by the saline solution immersion on short term (168 h), results confirmed also by the literature.

The studies regarding saline solution/seawater immersion of composite materials are limited, Roy Xu et. al. [28] observed that compression-after-impact strengths of the wet specimens were reduced by around 10% compared to the baseline dry specimens.

Visco et. al. [22] studied seawater absorption ability and mechanical performance of two glass–fibre reinforced composites with different thermoset matrices (isophthalic polyester, vinyl ester) and concluded that vinyl ester resin composite had a lower water diffusion coefficient but higher chemical stability, stiffness and strength after immersion in seawater than the isophthalic one.

Even if the tensile strength of the composite materials was increased it was observed that they maintain their Young's modulus value, 68 GPa for HexPly samples, respective 62 GPa for EP samples. The elastic modulus could be maintained over a period of time in moist environment and afterwards it could be altered, the polymer becoming more brittle and easy to break.

Regarding the analysis of the tensile failure modes, it was observed that in case of HexPly samples, the majority presented a lateral inside grip top (LIT) typical failure mode followed by lateral gage top (LGT) failure mode, while EP samples presented an angle gage middle (AGM) failure mode followed by angle gage bottom (AGB) failure mode.

In Table 3 are specified the results of tensile test failure mode analysis of the composite specimens. It can be observed that in case of HexPly, the tested specimens present less scattered failure modes compared with the EP specimens. The predominant failure modes are typical for the matrix shear failure and brittle failure of polymeric matrix composites reinforced with fibres along with fibre pull-out.

In Fig. 3 a, b it can be summarized the influence of kerosene immersion on the tensile strength of the two composite materials and in Fig. 3 c can be observed the influence of saline solution on their tensile strength after 168 hours of immersion.

Table 3. Results of the tensile test failure mode analysis

Material	Typical failure mode	Number of specimens
HexPly	LIT	10 specimens
HexPly	LGB	2 specimens
HexPly	LGT	3 specimens
HexPly	LAB	2 specimens
HexPly	LAT	1 specimen
HexPly	LIB	1 specimen
HexPly	MIT	1 specimen
EP	AGM	6 specimens
EP	AGB	3 specimens
EP	LIB	1 specimen
EP	LAT	2 specimens
EP	AMM	1 specimen
EP	LGT	1 specimen
EP	AGT	1 specimen
EP	LGB	2 specimens
EP	LIT	2 specimens
EP	LAB	1 specimen

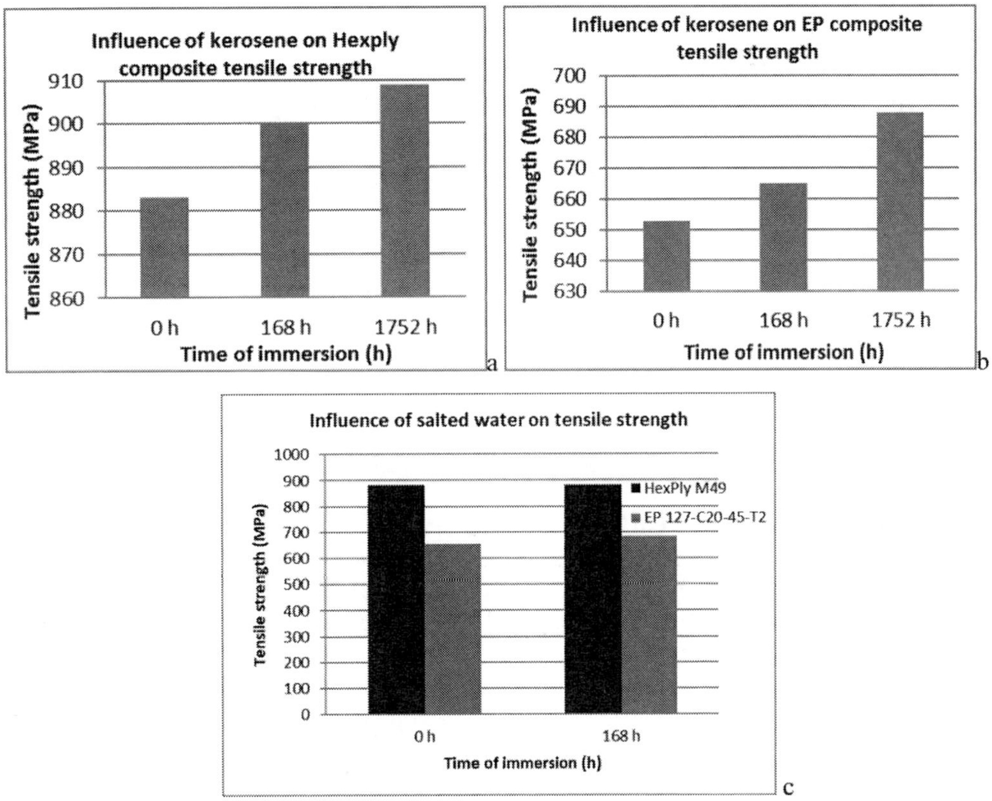

Fig. 3. Influence of kerosene and saline solution immersion on composite tensile strength: a) in case of HexPly; b) in case of EP; c) results in case of saline immersion of the composites

4 Conclusions

The present study was focused on the tensile behaviour of two composite materials before and after aging in humid environment (saline solution and kerosene). The humid environments selected for the tests were kerosene due to the fact that jet fuels consist in kerosene and different types of additives and saline solution due to the fact that the tank is designed for a navel helicopter. Two types of composite laminates were tested and it was observed that kerosene slightly improves the tensile properties on short term exposure.

HexPly M49/42%/ 200T2X2/CHS-3K showed an increase in tensile strength of 3% after 1752 hours of kerosene immersion and EP 127-C20-45 T2 showed an increase in tensile strength of 5% after 1752 hours of kerosene immersion. The saline solution exposure on short term (168 h) didn't influence the tensile properties of the HexPly M49/42%/ 200T2X2/CHS-3K, but in case of EP 127-C20-45 T2 it was observed that the saline absorption increased the tensile properties with 5%. However, it was observed that the composite laminates maintain their Young modulus 68 GPa for HexPly samples, respective 62 GPa for EP samples. Peeling, cracking or matrix dissolution of polymeric composite immersed in saline solution or kerosene weren't observed.

Due to the fact that it can't be predicted how a fibre reinforced composite will behave under a certain humid environment further studies will be made regarding the fuel compatibility with polymeric composites.

Acknowledgment

This study was carried out within Grant no. 293/2014, supported by the Romanian Minister of Research and Innovation.

References

[1] A.P. Mouritz, *Introduction to aerospace materials*, Woodhead Publishing, 303-337 (2012)
[2] R.J.H. Wanhill, *Aerospace Materials and Material Technologies : Aerospace Materials*, **1**, Springer (2017)
[3] P.K. Mallick, *Fiber-reinforced composites: Materials, manufacturing and design*, CRC Press, 19 - 48 (2007)
[4] M.R. Legault, http://www.compositesworld.com/articles/carrier-capable-all-composite-external-fuel-tank(2) (2011)
[5] UH-60 A/L/M CRASHWORTHY EXTERNAL FUEL SYSTEM (CEFS) Brochure, http://www.robertsonfuelsystems.com/defense/crashworthy-external-fuel-system-cefs/
[6] M. Ren, X. Chang, H.Y. Xu, T. Li, Compos Struct, **160,** 1339-1347 (2017)
[7] E. Elkhershi, *The Telegraph,* http://www.telegraph.co.uk/news/2016/03/18/inside-the-f-35-lightning-ii---the-invisible-fighter-jet/ (2015)
[8] MIL-T-7378A. Military specification Tanks, Fuel, Aircraft, external, Auxiliary, Removable (1958)
[9] MIL-DLT-5578D. Detail Specification Tanks, Fuel, Aircraft, Self-sealing (2008)
[10] B. Wei, H. Cao, S. Song, Corros Sci, **53,** 426-431 (2011)
[11] S. Lampman, *Characterization and Failure Analysis of Plastics*, ASM International, 146-152 (2003)
[12] J.A. Hinkley, J.W. Connell, *Long-Term Durability of Polymeric Matrix Composites*, Springer, 1-38 (2012)

[13] MIL-HDBK-17-1F. *Composite Materials Handbook*, Polimeric Matrix Composites Guidelines for Characterization of Structural Materials, **1**, (2002)
[14] B.C. Ray, J. Reinf. Plast. Compos., **25**, (2016), pp. 1227-1240
[15] P. Vaddadi, T. Nakamura, R.P. Singh, Acta Mater, **51**, (2013), pp. 177-193
[16] B.G. Kumar, R.P. Singh, T. Nakamura, J. Compos. Mater., **36**, (2002), pp. 2713-2733
[17] S. Roy, *Long-Term Durability of Polymeric Matrix Composites*, Springer, (2012), pp. 181 - 236
[18] I. Merdas, F. Thominette, A. TCharkhtchi, J. Verdu, Compos. Sci. Technol., **62**, (2002), pp. 487-492
[19] E. Perez-Pacheco, J.I. Cauich – Cupul, A. Valadez-Gonzalez, P.J. Herrera-Franco, J. Mater. Sci, **48,** (2013), pp. 1873-1882,
[20] C.E. Browning, C.E. Husman, J.M. Whitney, *Moisture effects in epoxy matrix composites*, Interim Report AFML-TR-77-171976, (1976)
[21] J.M. Augl, *The effect of moisture on carbon fibre reinforced epoxy composite - Mechanical property changes*, Technical Report, Naval Surface Weapons Center, White Oak Laboratory, Maryland, USA, (1977)
[22] A.M. Visco, N. Campo, P. Cianciafara, Compos. Part A-Appl S, **42**, (2011), pp. 123-130
[23] P. Krishnan, M.S. Abdul Majid, M. Afendi, S. Yaacob, A.G. Gibson, Compos. Struct., **148**, (2016), pp. 1-11
[24] A. Chilali, W. Zouari, M. Assarar, H. Kebir, R. Ayad, Compos. Struct., Article in press, (2017)
[25] V. Fiore, L. Calabrese, G. DI Bella, T. Scalici, G. Galtieri, A. Valenza, E. Proverbio, Compos. Part B-Eng., **93**, (2016), pp. 35-42
[26] A.P.C. Barbosa, A.P.P. Fulco, E.S.S. Guerra, F.K. Arakaki, M. Tosatto, M.C.B. Costa, J.D.D Melo, Compos Part B-Eng., **110**, (2017), pp. 298-306
[27] J. Naveen, T.P. Sathish Kumar, S. Satheesh Kumar, **3**, (2014), pp. 1091-1095
[28] L. Roy Xu, A. Krishnan, H. Ning, U. Vaidya, Compos Part B-Eng, **43**, (2012), pp. 2480-2486
[29] L.R. de Souza, A.T. marques, J.R.M. D'Almeida, Compos Struct, **168**, (2017), pp. 285-291
[30] S.R. Areef, Eng. & Technology, **26**, (2008), pp. 1235-1239
[31] D.B. Curliss, *The effect of Jet fuel exposure on advanced aerospace composites I: Thermal and Chemical Analysis*, Report, Materials Directorate, Wright Laboratory, Air Force System Command, Wright-Patterson Air Force Base, Ohio, USA, (1991)
[32] S.G. Taghavi, *Moisture effects on high performance Polymer Composites*, Master Thesis, Department of Chemical Engineering and Applied Chemistry, University of Toronto, 2000
[33] SR EN ISO:527-4/2000 Plastics – Determination of tensile properties-Part 4: Test conditions for isotropic and orthotropic fibre-reinforced plastic composites
[34] ASTM D3039/D 3039M/2000 Standard Test Method for Tensile Properties of Polymer Matrix Composite Materials

Indentation hardness and modulus of electrochemically deposited triple protective-decorative Cu-Ni-Cr system

Maria Datcheva[1,*], *Gergana* Chalakova[1], *Antonio* Nikolov[2], *Rayna* Dimitrova[2], *Roumen* Iankov[1], *Valentin* Kamburov[2], *Dessislava* Guergova[3], *Emilia* Stoyanova[3], and *Dimitar* Stoychev[3]

[1]Institute of Mechanics, Bulgarian Academy of Sciences, 1113 Sofia, Bulgaria
[2]Technical University of Sofia, 1000 Sofia, Bulgaria
[3]Institute of Physical Chemistry, Bulgarian Academy of Sciences, 1113 Sofia, Bulgaria

Abstract. With respect to the corrosion-protective and decorative characteristics of the Cu-Ni-Cr systems the advance of the technology resulted in products covering the required technical and economic norms and indicators. However, the knowledge about the mechanical characteristics such as elastic and plastic properties of the individual layers and the whole multilayered system is still sparse. The objective of this study is to obtain by means of nanoindentation experiments the mechanical characteristics of standard triple Cu-Ni-Cr coating system whose morphology, structure, texture, composition, and the copper-, nickel- and chrome-layer thickness are predefined. A series of nanoindentation tests in vertical direction to the layered coating and in the lateral direction (along a vertical cut) is analyzed to assess the changes of the mechanical properties of the investigated system in depth and along the layer as well as at layer-to-layer and layer-to-substrate interfaces.

1 Introduction

Multi-layered protective-decorative Cu-Ni-Cr systems became well accepted after an operating period of over 90 years [1-3]. Nowadays this type of systems is widely recommended by national and international standards being an inherent component in the automated production lines for plating with application to the motor-car construction and other branches of the modern engineering industry.

With respect to the corrosion-protective and decorative characteristics of the Cu-Ni-Cr systems the advance of the technology resulted in products covering the required technical and economic norms and indicators. However, the knowledge about the mechanical characteristics such as elastic and plastic properties of the individual layers and the whole multi-layered system is still sparse.

* Corresponding author: datcheva@imbm.bas.bg

The plating process involves specific organic and non-organic agents to the electrolytes resulting in co-deposition of these additives (S, C, N, P, H, etc.) into the metallic layers and their embedding in the metal lattice. This leads to lattice deformation and severe deviation of the mechanical and physical properties of the metal layers as compared to their metallurgically obtained counterparts. This may lead to high internal stresses in and between the layers, degradation of the adhesion of the coating to the substrate as well as between the coating layers themselves, unexpected and unwanted changes in the hardness and the wear resistance. In this context, the objective of this study is to obtain by means of nanoindentation experiments the mechanical characteristics of standard triple Cu-Ni-Cr coating system whose morphology, structure, texture, composition, and the copper-, nickel- and chrome-layer thickness are predefined. A series of nanoindentation tests in vertical direction to the layered coating and in the lateral direction (along a vertical cut) is analysed to assess the changes of the mechanical properties of the investigated system in depth and along the layer and layer-to-substrate interfaces.

2 Experimental

2.1 Materials and coating process

The multilayer Cu-Ni-Cr system was obtained by consecutive electrodeposition on a steel substrate of each of the individual copper, nickel and chromium layers. The working specimens were of rectangular form with sizes 3 x 6 cm, and the substrate samples were cut from 08KP steel sheet with a thickness of 1 mm. All samples were subject of standard chemical degreasing in organic solvent and inorganic solution followed by etching and surface activation in aqueous solution of HCl [1-3].

These pre-treatment operations of the steel substrate were completed with a deposition of a very thin copper layer from cyanide electrolyte in order to ensure the necessary adhesion for the multilayer system to the steel substrate.

The ultrafine crystalline copper layer was deposited in an electrolyte whose composition was: $CuSO_4.5H_2O$ – 220 g/l; H_2SO_4 – 50g/l; NaCl – 0.090g/l; Brightening additive THB-6-II [4] – 5 ml/l. Elliptic phosphorus-containing (0.07 wt.%) copper anodes were used. The electrolyte temperature was 25°C. The standard two-electrode electrochemical working cell in which the electrolysis was carried out was equipped with an air-agitating device for the electrolyte. The air-agitating was performed with specially purified compressed air, supplied at a rate of 13 $m^3/m^2/h$. The cathode current density was 4.5 A/dm^2 and the thickness of the obtained bright copper layer was ~8.92 μm. The electrochemically obtained ultrafine crystalline copper layer is characterized by a mirror reflective surface (96% vs. Silver mirror).

The deposition of the bright nickel layer was performed electrochemically by cathodic deposition using working electrolyte with the following composition: $NiSO_4.7H_2O$–270 g/l; $NiCl_2.6H_2O$–70 g/l; H_3BO_3 – 40 g/l; saccharin - 1.1 g/l; butanediol - 0.3 g/l, at a pH ~ 4.5. The working temperature was 55°C. Rolled de-passivated sheet nickel with a purity of 99.7% was used for the soluble anodes. The electrochemical deposition was performed in an analogous standard two-electrode electrochemical cell at cathode current density of 5 A/dm^2 and air agitation of the electrolyte. The thickness of the obtained bright nickel layer was ~19.6 μm.

The deposition of the chromium layer was performed electrochemically by cathodic deposition using a working electrolyte with the following composition: CrO_3–250 g/l; H_2SO_4 – 2.5 g/l. The working temperature was 55°C. In this case unsoluble Pb anodes were employed. The electrochemical deposition was performed in an analogous standard two-electrode electrochemical cell at cathode current density of 12 A/dm^2. The thickness of the obtained chromium layer was ~0.234 μm.

2.2 Characterization

The surface morphology and the structure of the obtained coatings were investigated by JEM 200-CX scanning electron microscopy in secondary electron imaging regime (SEI). The texture of the coating system was studied by Philips X-ray diffractometer. The micro profile of the samples was analysed with Perthen profilometer. The interpretation of the nanoindentation data rely on the information about the thickness of the layers obtained using X-ray fluorescence analysis (FISHERSCOPE X-RAY XDVM).

2.3 Sample IPC-3-6

Figure 1 depicts the sample IPC-3-6 and the area where the indentations were done can be seen there marked with black dots.

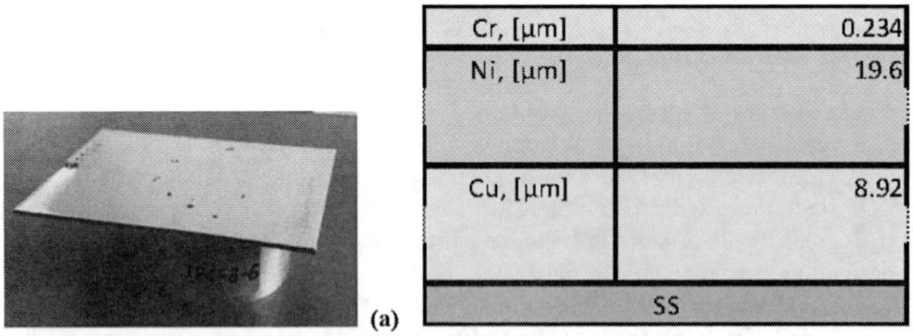

Fig. 1. (a) Sample IPC-3-6 mounted on the punch; (b) sketch of the cross section.

2.4 Sample IPC-4-4 (cross section of IPC-3-6)

Figure 2a presents schematically the way the sample IPC-3-6 had been cut in order to prepare probes for depth-sensing indentation tests along the sample cross section. The red cut side in Fig. 2a indicates the cut used to prepare sample IPC-4-4, while the solid black line shows the side used to prepare sample IPC-1-1. The mechanical properties of the Cu layer, the interface between the Ni and Cu layers and the steel substrate were investigate using this sample IPC-4-4 and IPC-1-1. Here we will discuss only the tests and the results obtained using sample IPC-4-4.

3 Nanoindentation

The mechanical properties of the obtained triple-layer protective-decorative coating and of the 08KP steel substrate were investigated by depth-sensing nanoindentation at room temperature, using Nano Indenter G200 (Keysight Technologies, USA) with a sharp diamond Berkovich tip. The geometrical characteristics of the tip are: centerline-to-face angle 65.3°◦ and a rounding radius of 20 nm. The maximum load that can be applied is 500 mN. The recording resolution for displacement is 0.01 nm and for the load is 50 nN.

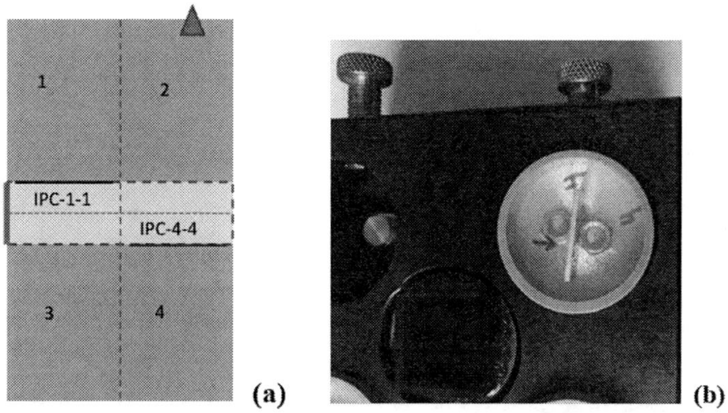

Fig. 2. Cutting scheme for obtaining the cross section samples;(b) sample IPC-4-4 in the tray.

3.1 Indentation methods and experimental program

The basic indentation methods employed in this study are Hardness, Modulus via Cyclic Load Control and Basic Hardness, Modulus at a Depth. For all test the unloading was set to be up to 90% of the reached maximum load. The thermal drift effects were corrected for each test using a holding segment before the indentation and after unloading up to 90% of the maximum applied load. The Poisson ratio was predefined to be equal to 0.25. Basic input parameters, used in this indentation method are given in Table 1. The loading programs were selected based on observations and employing the following experimental design strategy. First the loading programs A (load control) and E (displacement control) were applied to the top of the sample (sample IPC-3-4). This way the maximum penetration depth and the force to have maximum displacements not exceeding 25% of the thickness of the upper Cr-layer were determined. It has been taken into account that because of the surface roughness indentation tests below 50 nm penetration depth are not reliable. The average force applied to have a penetration of the indenter ~50 nm (program D) when performing nanoindentations on the top of the examined multi-layer system under the program E was afterwards used to test the response to the indenter penetration of each of the layers. This ensure for having reliable comparison of the mechanical properties. The cyclic loading test programs B and C allow of obtaining the indentation hardness and modulus at various depths from a single indent, from where the variation of these two parameters can be analysed.

Table 1. Input parameters for the experimental programs

Input parameter	unit	Programs				
		A	B	C	D	E
Max Load	mN	500	200	500	1.5	n/a
Number of Times to Load	-	1	10	3	1	n/a
Time to Load	s	10	10	10	10	n/a
Strain Rate Target	1/s	n/a	n/a	n/a	n/a	0.05
Depth Limit	nm	n/a	n/a	n/a	n/a	50

The indentation tests on sample IPC-3-4 consist of 10 indentations under program E, 9 indentations under each of the programs A and B, while the tests performed on sample IPC-4-4 followed the program D (16 indents for the Ni layer and 16 indents for the Cu layer) and the program C (total 17 indents including indents along the Ni and Cu layers, at the interface between them and 9 indents in the steel substrate). The distance between the

 https://doi.org/10.1051/matecconf/201814502005

indents varies but was not less than 20 µm insuring the independent measurements of the deformation at each particular indentation test.

3.2 Nanoindentation results

Figure 3 shows the load-displacement curves (P-h curves) and the imprint images for testing programs A and B (sample IPC-3-6, indentations from the top of the multi-layer system) and C (indentations along the cross section of the multi-layer system - sample IPC-4-4). The indentation data from the tests at the top and in the cross section of the Ni layer show very small scatter indicating the high homogeneity of the Ni layer. The average maximum displacements that can be achieved applying the maximum load of 500 mN to the top of sample IPC-3-6 is 1800 nm. Therefore, considering the thickness of the Ni-layer (~19.6 µm) it is not possible to investigate the properties of the whole multi-layer system using indentation tests from the top of sample IPC-3-6. This is the reason to cut the sample and to prepare samples (Fig. 2b) allowing to perform indentations along the cross section of the triple layer system (IPC-4-4).

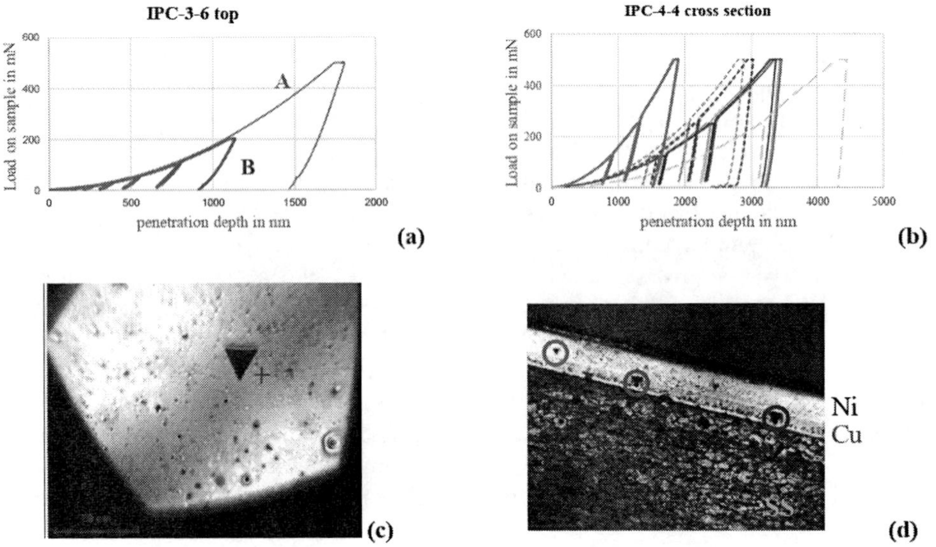

Fig. 3. P-h curves after programs (a) A (sample IPC-3-6) and (b) C (sample IPC-4-4). Imprint images: (c) sample IPC-3-6 (1000x) and (d) sample IPC-4-4 (250x).

The resulting load-displacement curves after indentation tests along the cross section of the layered structure are shown in Fig. 3b, where the black curve represents test data from indentations in the Ni-layer, red curves are from tests in the Cu-layer, blue dashed curves give the response of the interface between the Ni- and Cu- layers and the orange curve depicts the results from the tests in the steel substrate.

The series of nanoindentation tests on the top of the triple protective system (program E) aims to evaluate the mechanical properties of the upper very thin Cr-layer. Figure 4 depicts representative load-displacement curves showing the scatter observed in these test data. The reason for this scatter can be the roughness and the thermal drift correction.

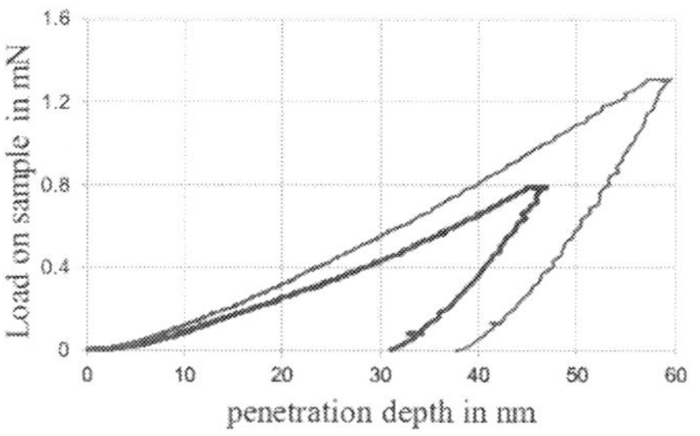

Fig. 4. Load-displacement curves for depths ~25% of the Cr layer's thickness (sample IPC-3-6).

The comparison between the indentation modulus and hardness of the three layers is illustrated in Fig.5. The indentation modulus and hardness are calculated based on Oliver&Pharr approximation method [5] and using for the Ni and Cu layers the data from indentation tests at different locations along the cross section of the layered coating employing the experimental program D, while for the Cr-layer the data from on-top indentions under program E were used. Therefore, the mechanical characteristics of the different layers are determined and compared for an average indentation force of 1.5 mN.

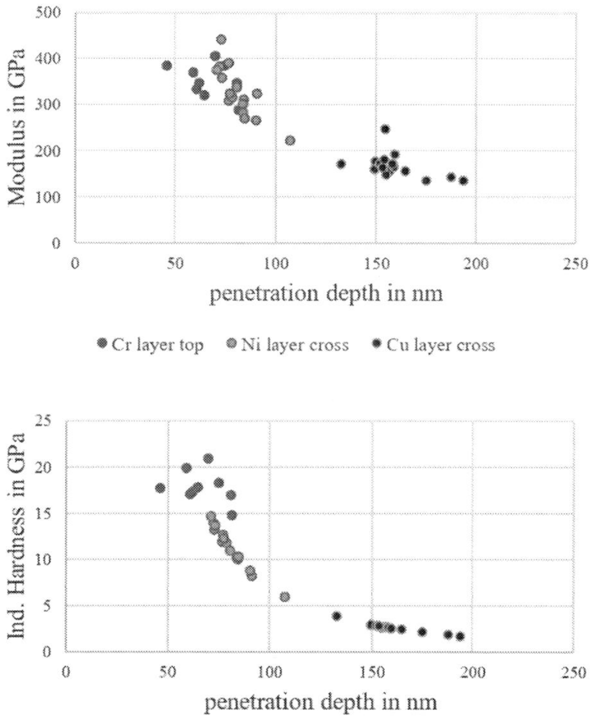

Fig. 5. Indentation modulus and hardness for Cr-, Ni- and Cu-layers at maximum load of 1.5 mN.

Conclusion

In conclusion, nanoindentation measurements were performed on Cr-Ni-Cu triple layered protective coating. In order to assess the mechanical characteristics of all of the layers, the specimen of the coated steel plate was cut and a sample was prepared to allow nanoindentation tests along the cross section of the multi-layered system. A series of nanoindentation tests employing four different experimental programs was performed and the indentation data was used to calculate the indentation modulus and hardness of each of the layers composing the multi-layered coating. The average value of the Cr-layer's nanohardness is superior compared to the corresponding hardness measure for the other two layers. The indentation modules of the Cr- and Ni layers have similar values, while the indentation modulus of the Cu-layer is almost twice lower.

The financial support of the Bulgarian National Science Fund under the grant T02/22/12.12.2014 is gratefully acknowledged.

References

1. Yu. Matulis, Ed., *Bright electrolytic coatings* (Mintis Publ. House, Vilnius, 1969)
2. H.Silman, Galvanotechnik, **76** (1985)
3. L.A.Lowenheim, H. Brown, B. Knapp, *Modern Electroplating* (Third Edition A Willey-Interscience Publication, NY, 1974)
4. D. Stoychev et al., Bulg. Pat. No 43112
5. W. Oliver, G. Pharr, J. Mater. Res. **19**, 3 (2004)

Evaluation by nanoindentation of technological products manufactured by pulse injection molding process

Margarita Natova[1,*], *Ivan* Ivanov[2] , *Sabina* Cherneva[1], *Maria* Datcheva[1], and *Roumen* Iankov[1]

[1]Institute of Mechanics of BAS, Acad. G. Bochev St., Bl. 4, 1113 Sofia, Bulgaria
[2]Himicheski Produkti Ltd., 7-b, Sokol St., 8600 Yambol, Bulgaria

Abstract. During conventional polymer injection molding, flow- and weld lines can arise at the molded parts caused by disturbed polymer melt flow when it crosses different parts of the equipment. Such processed plastic goods have discrete zones of inhomogeneities of very small dimensions. In order to stabilize the melt flow and to equalize dimensions of such defective products, an approach for pulse injection molding is applied during production of polymer packagings. Testing methods used for evaluation of macromechanical performance of processed polymer products are not readily applicable to estimate the changes in visual surface obtained during pulse injecting. To overcome this testing inconvenience the performance of processed packagings is evaluated by nanoindentation. Using this method, a quantitative assessment of the polymer properties is obtained from different parts of technological products.

1 Introduction

When using special effect pigments for producing polymer packagings by injection molding flow and weld lines are frequent aesthetic confronts. Over the past years extensive research has been completed to reduce visibility of such lines repeatedly displayed during the process of injection molding. Further investigation reveals that in injection molded parts various orientations of the melt layers can be seen due to the well-known "fountain flow" in the mold cavities. The reason is established to be the fact that near the mold surface part of the polymer chains is oriented almost parallel to the mold surface and below this layer the chains in the body encompass a "fountain" distribution related to the flowing melt. This orientation in the mold equipment makes areas to have trim med gates and to lack brightness that are representative for microcrystalline polymer materials (**Fig. 1**). Such a different polymer performance results in a greatly reduced concentration of the pigment, which makes further tool design more challenging.

* Corresponding author: mnatova@yahoo.com

Fig 1. Disposition of visual defects on the injection moulded products

The use of proper molding conditions can help in making produced parts of improved quality. Any changes made to the molding equipment design intended for reducing flow- and weld lines are desired innovation. In producing of injection molded products the uses of sequential, cascade or pulse gating are between the most often used inventions that may reduce visual lines seen and identified with an unaided eye. They are predominantly approved as companies' recommendations and have been included in their prospects and patents. Up to now, no available scientific information is known to explore any new testing methods for numerical estimation of the technological design performance. Transformed technologies for manufacturing of polymeric products involve new melt processing steps that can alter the polymer structure and, through it, the final material performance. In addition, changes in the micro-molecular architecture have a direct impact on the material processability, possibly inducing flow instability [1].

Considering the mechanical behavior of melt boundaries in a micro- and nanocrystalline polymer material there is a large variety of possible relative orientations of the crystals. It is likely that they would be representative micro- and nanometric examples of these visually inhomogeneous boundaries. Thus by examining the properties of those boundaries in a crystalline material, an understanding of their nanomechanical behavior can be obtained [2-4].

Processed polymer products are inhomogeneous on microscale and appropriately lower load must be used for studying their individual parts. The nanoindentation at low load under specified loading program is used to examine micro-mechanical behavior of polymer products. Reducing the influence of viscoelastic after-effects is achieved by a suitable dwell held for a proper time in between the loading and unloading [5]. The apparent contact stiffness, *S*, is determined from the slope of the upper part of unloading curve.

In order to compare properties of different parts located near to and far from the weld lines of products manufactured with and without the use of a pulsed type of injection the load-displacement curves have been obtained using nanoindentation experiments on Agilent G200 Nanoindenter fitted with a diamond Berkovich type tip. Thus, the nanomechanical response is studied for gaining information for products' indentation hardness and indentation modules. The mechanical response of the specimens can be evaluated comparing their indentation modules and hardness from tests performed on the samples surface.

2 Experimental

2.1 Material and preparation

Two different brands of polypropylene, Borealis RJ 901 MO and Ineos 450-HP 90, intended for injection moulding were used without further treatment. Manufactured products are transparent and opaque commercial round-shaped containers in three different sizes - 400 ml, 180 ml and 100 ml, respectively. Containers for pharmaceutical and cosmetic products were made with two types of colorants, giving a pearl shade of the products (light and dark purple).

Standardized specimens for macro-mechanical testing in a dogbone-shaped form were cut directly from the processed polymer packagings. Thus, the necessity to use method-oriented test standards was fulfilled (according to EN ISO 3167:2014) that defines only specimen geometry and dimensions in addition to test condition requirements [6].

Testing probes for the nanoindentation are cut from the samples without special form and size aiming only to fit the size of the holder in the nanoindenter's tray (a cylindrical punch with a diameter $\phi = 31.8\ mm$).

2.2 Characterization

Agilent G200 Nanoindenter fitted with a Berkovich three-sided diamond pyramid tip with a rounding of 20 *nm* and centreline-to-face angle 65.3° was used for nanoindentation examinations. The nanoindentation tests were performed under conditions and using a testing program described in details in [7].

Based on the measured load-displacement curves and the methodology described in [8], the indentation modulus and indentation hardness are calculated for each test and the mean values are extracted for comparison.

The hardness is determined from the maximum load and the projection of the contact area $A(h_c)$:

$$H_{IT} = P_{\max} / A(h_c) \tag{1}$$

It is to note that for micro-indentation, the contact area in equation 1 is the surface area of the tip-faces that are in contact with the sample. This explains the difference between the calculated microhardness and indentation hardness. However, the conversion is straightforward and will not be discussed here. It is obvious, therefore, that the trends observed for the indentation hardness hold for the microhardness.

The indentation modulus, E_{IT} that may be related to Young's modulus can be calculated by:

$$E_{IT} = \left(1 - \nu^2\right) \Big/ \left[\frac{2\beta\sqrt{A(h_c)}}{S\sqrt{\pi}} - \left(1 - \nu_i^2\right) / E_i \right], \tag{2}$$

where β is a correction factor, whose value for Berkovich indenter is $\beta = 1.03$,
ν is the Poisson's ratio of the tested material ($\nu = 0.42$ in our case),
E_i and ν_i are the indenter's elastic parameters and
S is the elastic contact stiffness defined as the slope of the upper (50%) portion of the unloading part of the load–displacement curve [9].

3 Results and discussion

In order to improve polymer processing a new pulse melt injecting has been applied. A device engagement was designed and plugged in the production equipment for polymer packagings fabrication (**Fig. 2**).

Conventional mechanical tests used for evaluating the mechanical properties are not readily applicable to produced polymer goods. To overcome this testing deficiency the mechanical characteristics at the visual surface were assessed from nanoindenter load-displacement curves. Indentation hardness, H_{IT}, and indentation modulus, E_{IT}, have been compared with macromechanically obtained characteristics of the material itself by applying tensile load to samples from the same processed packagings as those subjected to nanoindentation testing.

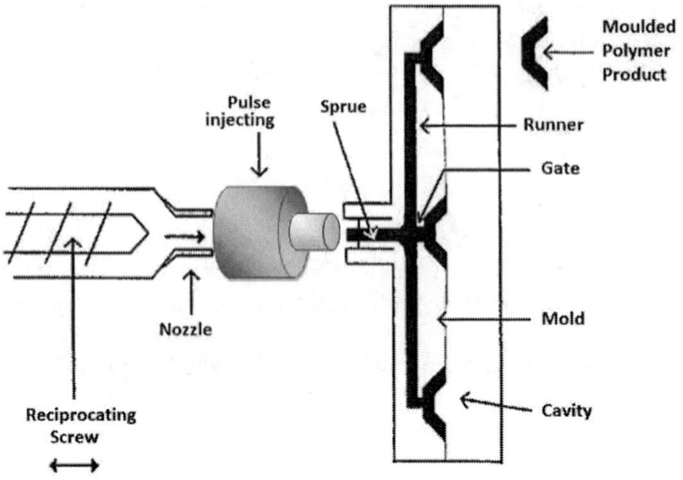

Fig. 2. New device plugged in for pulse injection molding

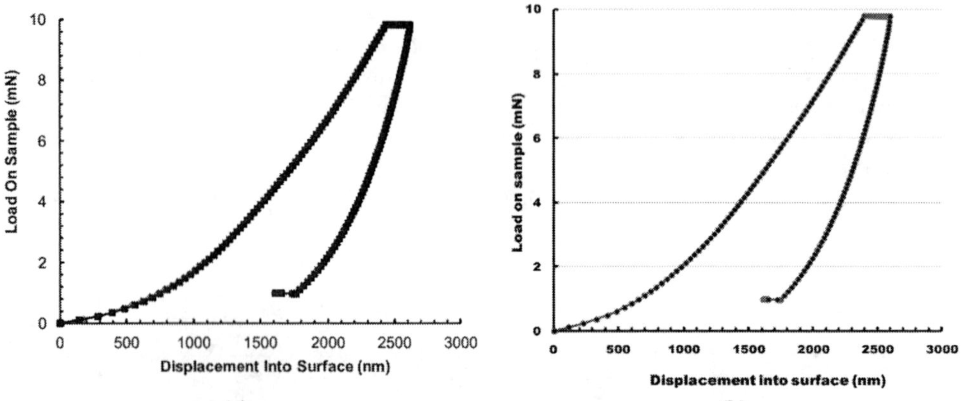

Fig. 3. Load-displacement curves for Borealis samples obtained far from the weldline without **(a)** and with **(b)** pulse injection

To obtain the onset of plasticity with the present loading procedure the study is started by performing nanoindentation at a reasonably small load. Before the onset of plasticity, the load-displacement response is expected to be essentially linear with the slope determined by the elastic properties of the polymer. The unloading part of the measured load–displacement curve is supposed to represent pure elastic material response. The further goal of this research is by microscopical examining of various parts of wall structures – near to and far from the gating system of the equipment and consequently from the weld lines, to provide evidence that yield strength is determined by the structure obtained during processing [9].

Figures 3 ÷ **6** show load versus displacement data and images of four representative indentation imprints on sample surfaces manufactured without and with pulse injecting (maximum indention depth is up to 3 μm) (load, P, in mN $vs.$ penetration depth, h, in nm). The magnification for the imprint images is 250 times.

What one in fact measures are the local material properties on the surface area of the polymer samples, and then from the obtained results the polymer bulk mechanical behavior can be inferred.

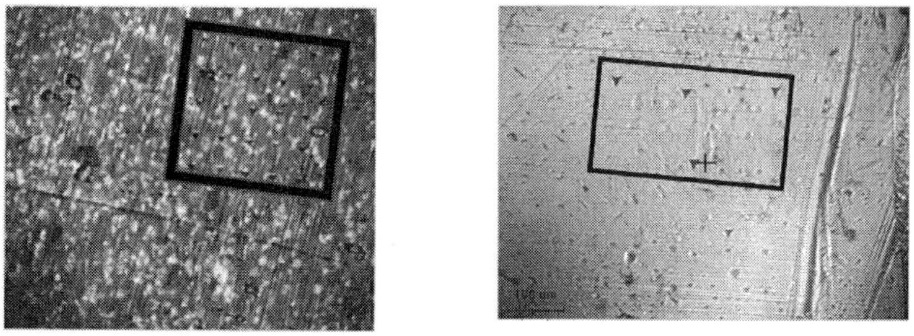

(a) **(b)**

Fig. 4. Images for Borealis samples obtained far from the weldline without (**a**) and with (**b**) pulse injection

From the data in **Table 1** it can be concluded that there is a slight reduction of the mechanical properties for the sample located near the weldline due to the manufacturing instability. However, the results show minor inhomogeneity along the 1200×1200 area that is more pronounced for the sample near to the weldline. **Fig. 7** shows as well the variation of mechanical properties for samples with and without pulse injection located near to and far from the weldline along an indentation testing line of 3 *mm* length.

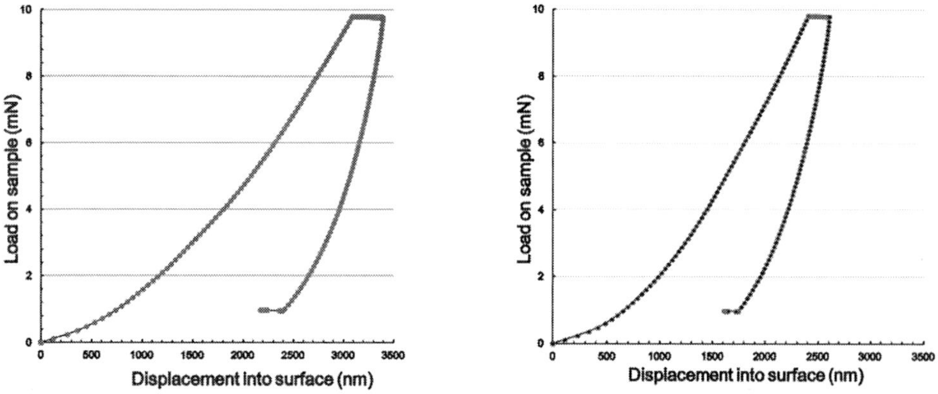

(a) **(b)**

Fig. 5. Load-displacement curves for Ineos samples obtained near to the weldline without (**a**) and with (**b**) pulse injection

An example of the results obtained from macroscopic tensile tests performed without pulse injecting is shown in **Fig. 7a**. Normal stress–strain behavior is observed for both the samples – near to and far from the weldlines. After passing a maximum tensile stress the stress–strain curves show a commonly occurring decrease, reaching a final failure. **Fig. 7b** is an illustration of similar results obtained using pulse injection molding of products.

The difference in absolute values is assumed to be caused by a processing effect governing strengthening mechanism and inducing an amorphization: the higher the inhomogeneity (near to the weldline), the weaker the level of the yield stress. The effective stiffness of products with pulse injecting (Young's modulus × sample thickness), S_{eff}, of the polymer located far from the weldline exceeds the values for samples near to the weldline on average by 8-10 % (the right column of Table 1).

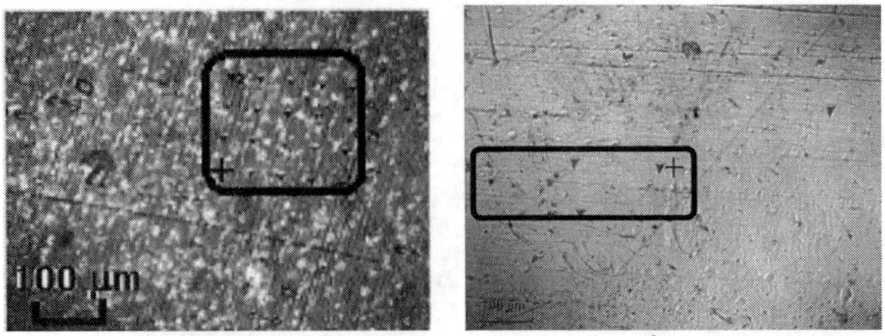

a b

Fig. 6. Images for Ineos samples obtained near to the weldline without (**a**) and with (**b**) pulse injection

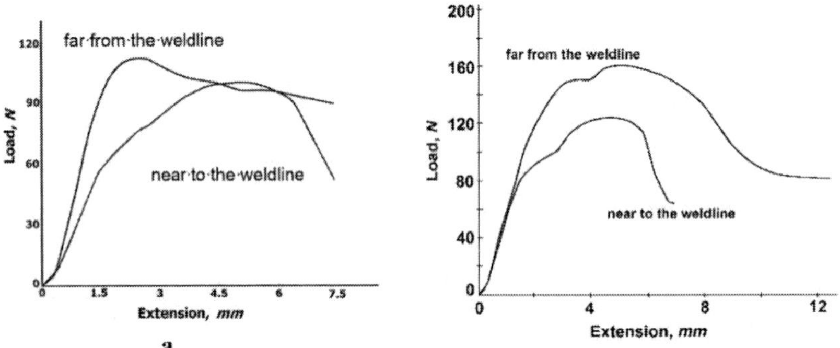

a b

Fig. 7. The tensile fracture of samples located near to and far from the weldlines prepared without (**a**) and with (**b**) pulse injecting

It is well-established fact that PP is a semicrystalline polymer, and as such it could be meant as a two-phase system. In the simplest case the intrinsic microhardness of the amorphous phase is considerably low than that of the crystalline one, and is approaching zero. Consequently, the microhardness is due to plastic deformation taking place only in the crystalline regions [10]. The increasing transition to an amorphous nature in examined by us PP products is proved by X-ray diffraction experiments (not presented here [11]), and is caused by the pulse type injection molding. It leads to a decrease in microhardness. Since microhardness is related to the packing of the chains in the crystals, polymorphic polymer changes may be expected to influence the strength of the intermolecular forces and, hence, the microhardness.

Specific microhardness measurements were carried out on samples prepared by the temperature-gradient method by the authors Fujiwara, Y. et al. [12, 13]. Their investigations allow separating of the α (monoclinic) and β (hexagonal) phases within the single PP sample. The authors of [14] showed that microhardness decreases when passing from α to the β phase, and is partially connected with a crystallinity decrease. Obviously, this is as well the main reason for the microhardness reducing in our experiments carried out using pulse injection molding.

The calculated indentation hardness and modules at the local area of some specific samples are given in **Table 1**.

Table 1. Calculated indentation hardness and modules compared to the macromechanical properties for Ineos and Borealis samples with and without pulse injection

Product \ Measure		E_{IT} (MPa)	St. dev. (MPa)	H_{IT} (MPa)	St. dev. (MPa)	Tensile modules, E_{tens} (MPa)	Effective stiffness, $S_{TensEff}$ (MPa × mm)
Ineos 450-HP90	**Wall**	1129	58	55	4	404	426
	Bottom	1290	34	72	1	413	392
Ineos 450-HP90 with pulse injection	**Wall**	933	11	47	1	347	356
	Bottom	1392	40	77	3	431	426
Borealis RJ 901 MO	**Wall**	1003	43	54	3	328	295
	Bottom	1147	13	55	1	385	366
Borealis RJ 901 MO with pulse injection	**Wall**	1103	23	60	1	316	278
	Bottom	1133	31	56	3	380	327

The authors of [15] have found an experimental relationship between indentation hardness and modulus and tensile modulus and/or yield stress for injection-molded semicrystalline plastics. According to the classical theory of polymer plasticity the expected indentation hardness should be approximately equal to three times the yield stress. Therefore, we might expect such a relationship with a certain variation of the microhardness/tensile yield ~ 3 ratio caused by the proportion of the participating amorphous and crystalline phases. On the **Table 1** it can be noted that in our case the elastic modulus measured by nanoindentation technique is 2.7-3.3 times higher than that obtained by tensile tests what is in excellent agreement with the well-known empirical relationship. The difference between these results again reflects the fact that under tension loads full breaking of polymer bonds occurs, whereas in nanoindentation, which applies a sensibly low load, the bonds are only disrupted and not destructed. Thus, the nanometric technique underestimates the mechanical properties.

4 Conclusions

The Agilent G200 NanoIndenter was successfully used to determine the indentation modules and indentation hardness of two types of technological polypropylene samples containing or free of manufacturing defects. It has been clearly demonstrated that irregularities in the sample topography can produce little but notable changes in the perceived properties.

The indentation hardness of the products varies depending on the position of the sample examined. The local inhomogeneities that occur as a changing proportion between amorphous and crystalline phases obviously affects the indentation hardness. The incipient plasticity is observed to be not quite different but noticeable among the different parts of the samples.

The inhomogeneous sample surface did not introduce substantial differences into the near-surface region of the sample and did not considerably affect the measurement by the

Berkovich indenter tip. A reason may be that the used here indenter tip is a very sharp pyramid that may cut the polypropylene rather than to deform it.

Based on the obtained nano- and macro-mechanical results an assumption can be made for the assessment of polymer processing effect occurring after application of the pulse injection moulding. It is proved that enhanced product performance is due to the fact that pulse injecting intensifies the molecular diffusion across the weld lines so as to reduce the polymer crystallinity thus decreasing the weldline weakness without creating any visible marks.

The experimental data can be successfully used for evaluation of dissimilarity in different areas of the polymer product and to estimate their mechanical characteristics, for post-produced sorting and classification before exploitation of polymer packagings.

Authors gratefully acknowledge the financial support of the Bulgarian National Science Fund under Grant No. T02-22/12.12.2014.

References

1. G. Filippone, S. C. Carroccio, R. Mendichi, L. Gioiella, N. Tz. Dintcheva, C. Gambarotti, Polymer, 72, 134-141 (2015).
2. K. T. Ramesh, *Nanomaterials: Mechanics and Mechanisms*, Springer Science + Business Media LLC (2009).
3. *Materials Characterization: Modern Methods and Applications*, Edited by Narayanaswami Ranganathan, Taylor & Francis Group LLC (2016).
4. W. D. Nix, *Metallurgical transactions a – physical metallurgy and materials science*, **20**, 11, 2217-2247(1989).
5. *Nanoindentation in Materials Science*, Ed. J. Němeček, InTech, Rijeka, Croatia (2012).
6. W. Oliver, G. Pharr, J. of Mat. Research, **19**, 1, 3-20 (2004).
7. M. Natova, Iv. Ivanov, R. Iankov, M. Datcheva, S. Cherneva, 16th Int. Multidisciplinary Scientific Geoconf. SGEM 2016, 28 June – 7 July 2016, Volume Nano, Bio and Green - Technologies for a sustainable future, Albena Co. – Varna, Bulgaria, 111-118 (2016).
8. *Polymer Testing*, 2nd Edition. Eds. W. Grellmann, S. Seidler, Carl Hanser Verlag, Munich (2013).
9. N. M. Jennett, R. Ghisleni, J. Michler, Appl. Phys. Lett., **95**, 12, 123102 (2009).
10. F. J. Baltá Calleja, S. Fakirov, *Microhardness of polymers*, Cambridge University Press, New York (2000).
11. M. Natova, Iv. Ivanov, V. Georgiev, F. Ublekov, Hr. Penchev, 46th IUPAC World Polymer Congress MACRO 2016, Istanbul, Turkey (2016).
12. Y. Fujiwara, Koll. Zeitsch. und Zeitsch. für Polymere, **226,** 2, 135-138 (1968).
13. T. Asano, Y. Fujiwara, Polymer, **19**, 1, 99-108 (1978).
14. F. J. Baltá Calleja, J. Martinez-Salazar, T. Asano, Mater. Sci. Lett., **7**, 2, 165-166 (1988).
15. J. Bowman, M. Bevis, Colloid Polym. Sci., **255**, 10, 954-966 (1977).

A correlation between fracture toughness and cohesion strength of molybdenum thermal sprayed coatings

Alexandru Paraschiv[1,*], *Gheorghe* Matache[1], *Cristian* Puscasu[1] and *Mihaela* Grigorescu[1]

[1]Romanian Research and Development Institute for Gas Turbines COMOTI, 220D Iuliu Maniu Ave., Bucharest 6, Romania

Abstract. The Molydenum thermal sprayed coatings on steel substrate using electric arc, atmospheric plasma spray and high velocity oxy-fuel spray were investigated. The tensile bond strength tests were made in order to examine the degree of adhesion/cohesion strength of coatings. The Vickers indentation fracture toughness tests were made on parallel and perpendicular direction to the coating-substrate interface in order to examine the fracture toughness of the material in relationship with the microstructure, indentation direction, hardness and the length of microcracks developed under different loads. It was found that the fracture toughness is influenced by the direction and load of indentation. It was established a correlation between the fracture toughness and cohesion strength of Mo coatings and a general trend was described

1 Introduction

The Molybdenum thermal spray coatings are used for many applications in sectors of the transport industry, such as automotive and aerospace industries. These are used to increase the corrosion resistance, friction control and thermal insulation to substrate and in particular, for wear resistance [1]. The molybdenum coatings can be also used in railway industry for dimension reconstruction and protection against abrasive, adhesive, surface fatigue, fretting or corrosion of axles of train [2].

Two critical aspects which affect the quality and lifetime of a coating-substrate system are the propagation behavior of the microcracks and the bonding strength (cohesion/adhesion strength) of the coatings. Firstly, the microcracks in these inhomogeneous thermal coatings are propagating in an unpredictable way and finally will lead to a delamination of the coating. The morphology and propagation behavior of microcraks in thermal sprayed coatings, which can occur when mechanical stresses are applied or as a result of residual stresses generated by the cooling and solidifications of hot Mo splats, define a mechanical property of coatings, called critical stress intensity factor or fracture toughness that is generally used for predicting of the mechanical behavior of brittle materials [3]. The fracture toughness is the ability of the coating to withstand at the highest level of stress intensity without fracturing and can be evaluated by using several methods

* Corresponding author: alexandru.paraschiv@comoti.ro

such as scratch testing, double cantilever beams loaded in tension, Chevron notch bar and single-edge notched beams tested in three or four point bending [3-4]. One of the most used methods in recent years, providing good results for the evaluating of fracture toughness of thermal sprayed coatings, is the indentation method by Vickers indenter [3-12]. Beside the accessibility of this method, the major advantage is that the fracture toughness of coatings can be tested locally, on several levels such as, at the coating-substrate interface, also called interfacial fracture toughness, on the perpendicular and parallel direction to the coating-substrate interface, between layers or individual splats. The general approach is by correlating the length of microcracks initiated in the corners of the Vickers impressions with different constants and mechanical characteristics of the material, such as hardness and modulus of elasticity.

On the other hand, the functionality of a thermal sprayed coating/substrate system depends on the cohesion strength of the layers and lamellar splats and also the adhesive strength of the coating on its substrate. These two types of bonding strength of thermal sprayed coatings can be identified after a visual or microscopic examination of the failure of coatings tested by tensile bond strength tests, according to ASTM Standard C633. Beside the disadvantages of this method in terms of costs and time for preparation of specimens (machining, gluing, curing, etc.), the penetration of the bonding agent into the coating due to the porosity of the coating, modifying its physical properties, and the impossibility to measure the adhesion strength of the coating when the rupture is cohesive and vice versa, represent the main limitations of this method [5-6]. Apart of this, more than eighty methods are reported to measure the coating adhesion, but many of them are also energy consuming in terms of time, cost and equipment [5]. Therefore, an additional method that is increasingly used for evaluating the adhesion strength due to its simplicity, accuracy and low costs is the Vickers indentation testing at the coating-substrate interface, with results that can be expressed either as a stress intensity factor, with units of $MN \cdot m^{3/2}$ or as fracture surface energy, with unit of $J \cdot m^{-2}$ [5-6]. In the same vein, in the case when the rupture of thermal sprayed coatings is produced only within the layers and splats, the Vickers indentation testing can be used as an additional method for evaluating the cohesion strength by locally testing the coating into splats and between layers.

Based on the propagation behavior of the microcracks developed under the Vickers indentation tests and the fracture modes of the coatings tested by tensile bond strength tests, it can be made an association between two of the most important mechanical properties of thermal sprayed coatings such as, the fracture toughness and cohesion bond strength.

So, the basic aim of this paper is to evaluate and correlate the fracture toughness by indentation tests on parallel and perpendicular to the coating-substrate interface with cohesive strength by tensile bond strength tests of molybdenum coatings obtained by three thermal spray methods, electric arc (EA), atmospheric plasma spray (APS) and high velocity oxygen fuel (HVOF).

2 Materials and procedures

The support material of the coating-substrate system is an A5-T steel that is currently used for the railway axle manufacturing. This steel support was coated with high purity (99.9%) Mo by electric arc (EA), atmospheric plasma spray (APS) and high velocity oxy-fuel spray (HVOF). For EA method it was used an AWS 400 Arc Spray System using Mo wire of 1.6 mm diameter (Sulzer Metco 8228), and for APS and HVOF method it were used a plasma spray torch (Metco 7M with F4 gun) and a HVOF powder spray system (MachJet 2700) respectively, using Mo powder with nominal particle size distribution of -75+25 μm (Metco 63NS). For the evaluating of the bonding strength of these coatings it were used two cylindrical specimens of Ø 25,4 x 25,4 mm for each type of coating with a coated frontal face that was machined until a thicknesses of around 350 μm for coatings deposited by APS and EA method and 150 μm for coatings deposited by HVOF method. The finished surfaces had a surface roughness Ra of around 2 μm for each type of coating and were measured by using a surface roughness tester MarSurf PS10.

Generally, for the evaluating of the adhesion strength of thermal sprayed coatings, the surface of coatings is better to be as-received, with a rougher surface for a good mechanical anchoring between the coating and glue, but in our case the coatings surface was machined because in the real work conditions of the component parts that will be protected by these coatings the thickness and surface roughness are critical technical specifications.

The cylindrical parts with coated surfaces (bond cap) and reference samples with uncoated surface (for testing just the bonding strength of bonding agent) were joined with the cylindrical counter parts (mating cap) using an adhesive agent HTK Ultra Bond 100®, were fixed by using loading weights on a self-aligning device (Fig.1) and were cured at elevated temperature of 150°C for 80 minutes, according to ASTM C633 (Standard Test Method for Adhesion or Cohesion Strength of Thermal Spray Coatings, 2008). The testing specimens presented in Fig. 1a and schematically shown in Fig. 1b were loaded along their longitudinal axis with a displacement rate of 0.75 mm/min by using the Universal mechanical testing machine INSTRON 8802. After the tensile bond strength tests, the rupture surfaces of coated specimens were visual and microscopic inspected to establish the fracture mode. Based on the specifications of ASTM C633, the adhesion strength of the coating is given if failure is entirely at the coating-substrate interface and also the cohesion strength if rupture is only within the coating layers. For the Mo coatings investigated in this paper, the rupture was cohesive (Fig. 1b) and the cohesion strength values were calculated by dividing the ultimate force by the cross-sectional area of the cylindrical specimens.

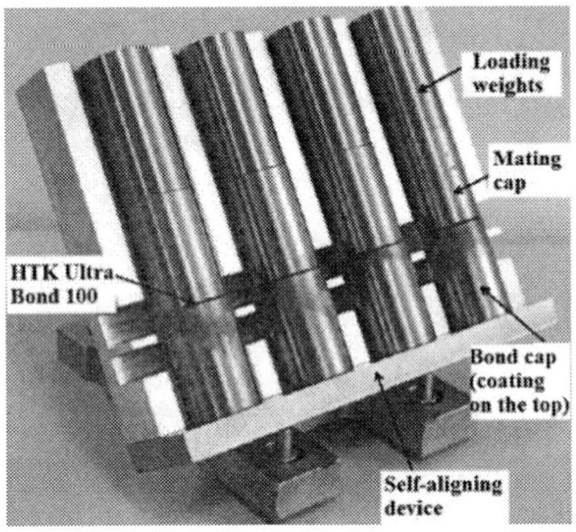

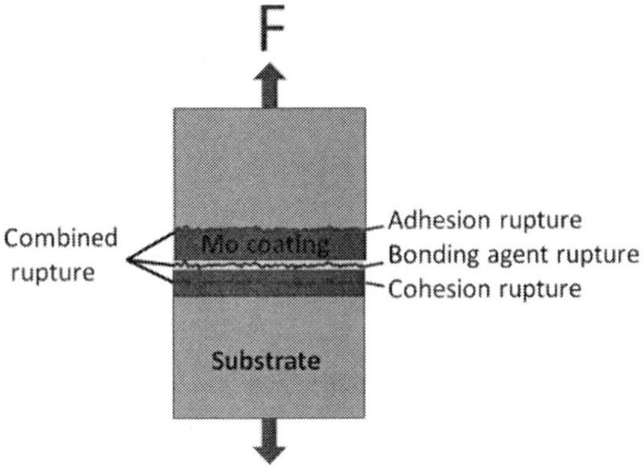

a) b)

Fig. 1. The preparation of the cylindrical tensile specimens by glueing the coating to the face of a loading fixtures (a) and schematic presentation of different fracture modes generally found in tensile bond strength tests of thermal spray coatings (b)

For the evaluating of the microcracks fracture behavior by calculating the fracture toughness of these anisotropic thermal coatings, we have proposed to test the Mo coatings, on the cross section (perpendicular to the coating-substrate interface or between splats) and on the surface (parallel to the coating-substrate interface or into splats) (Fig. 2a). An average of the local fracture toughness values obtained on both parallel and perpendicular direction to the coating-substrate interface were made, in order to correlate the fracture toughness results with the results obtained by the tensile bond strength tests. For the Vickers indentation tests three steel support samples with 30 mm in diameter and 10 mm height were coated on a face by each thermal method with a thickness of around 400 μm for EA and APS methods and 150 μm for HVOF, were cut in half parts and metallographically prepared, according to the indentation direction. By Vickers indentation method were made five measurements of micro and macroindentation for each of the twelve selected loads ranging from 0.098 to 98.1 N. Vickers microindentation hardness tests were performed with a Wilson Hardness™ 402MVD tester for loads ranging from 0.098 to 9.8 N (10 to 1000 gf), while Vickers macroindentation hardness test were performed with an EMCO TEST M4C/R for loads ranging between 19.6 to 98.1 N (2 to 10 kgf), according to ASTM E384 [7]. The lengths of microcracks, which were developed near the impression of indentation, were measured by using SEM images (FEI Inspect F50) and SEM image analysis software (Scandium) (Fig.4). It was observed that the microcraks generated by Vickers indentation occur after a critical load has been reached and depending on the type of coating and indentation direction they usually propagate on the directions of the indent diagonals. This mechanism of propagation of microcracks appears usually in homogeneous materials and in particular, in brittle materials [3], [8], [9]. Generally, after the Vickers indentation two types of microcracks can be generated, described as Half-penny cracks (median/radial cracks) or Palmqvist cracks (radial cracks) (Fig.2b), where l is the average length of the cracks, a is the half average length of the diagonal and c is the sum of l and a.

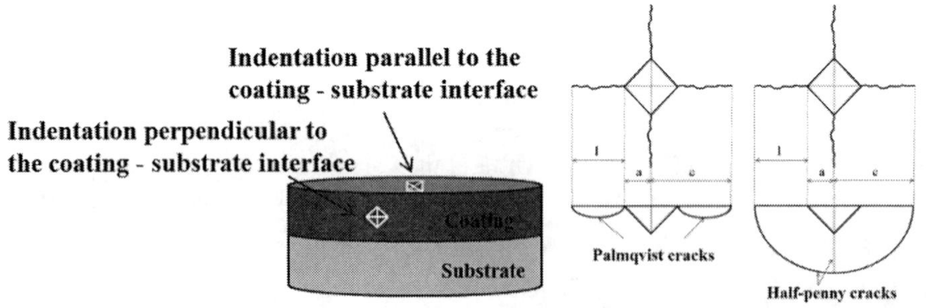

a) b)

Fig. 2. Schematic illustration of the indentation directions (a) and Palmqvist and interfacial cracks system (b)

To distinguish between these two types of cracks, it can be applied a criterion based on the ratio of cracks length/diagonal indentation [9]. A general rule admits that if $c/a \geq 2$, the model is the type Half-penny and if $c/a < 2$, the model is Palmqvist [3]. Other authors assert that the cracks are Palmqvist type only if $0.25 \leq l/a \geq 2.5$ [10] and they are Half-penny only if $c/a \geq 2.5$ [11]. In the case of Mo thermal coatings the average measurements of diagonals and microcracks initiated on the parallel and perpendicular coating-substrate interface indicate that all the microcracks are the Palmqvist type by complying with criteria listed above. The fracture toughness Kic both in parallel and perpendicular to the coating-substrate interface was evaluated by using an equation (1) established by Brantner et al. [12] for flame sprayed molybdenum coatings, where k is a coefficient which depends on the geometry of the used indenter and the tested material, and P is the load (N).

$$K_{IC} = k \cdot \left(\frac{P}{a \cdot l^{1/2}}\right) \tag{1}$$

3 Results and Discussions

The experimental results of the tensile bonding tests indicate that the Mo coatings deposited by HVOF (HV) method had the highest value of bonding strength (43.3 MPa), followed by APS method (26.7 MPa) and by EA method (19.2 MPa). These differences can be explained by the higher temperature and velocity of sprayed particles which impact the substrate and generate a higher level of compressive residual stresses in coatings creating a stronger bond between the layers. Another factor that influences the bonding strength is the dimensions of the Mo splats, which in the case of EA-sprayed coatings are approximately 55 µm and 37 µm for APS [13] that generates a lower porosity with decreasing of the splats dimension and a higher total contact surface.

The microfractographic investigations of the rupture surfaces for all tested specimens have shown that the failure is produced within the coating layers (Fig. 3b) and individual splats (Fig. 3c), with small adhesive parts removed (no more 2% of total area) (Fig. 3a) and without a rupture within the glue, which means that the bonding strength calculated is actually the cohesion strength of Mo coatings and it is lower than the adhesion and the glue strength.

a) b) c)

Fig. 3. The general aspect of the coated surfaces after tensile bond strength tests (a) with details of cohesion rupture of Mo layers (b) and fracture of Mo splats (c)

The Vickers indentation method was used for evaluating the propagation behavior of the microcracks and also used as an alternative method for evaluating the cohesion strength following the known rule of interfacial fracture toughness and adhesion strength of coating-substrate systems. The Vickers indentations within the specified loads range were performed on parallel (into splats) and perpendicular (between splats) direction to the coating-substrate interface to initiate and propagate microcracks in splats and between layers of coating. In most investigated indentation impressions, two main microcracks grew in two of the four corners of the indentation on both parallel and perpendicular directions (Fig. 4a-b).

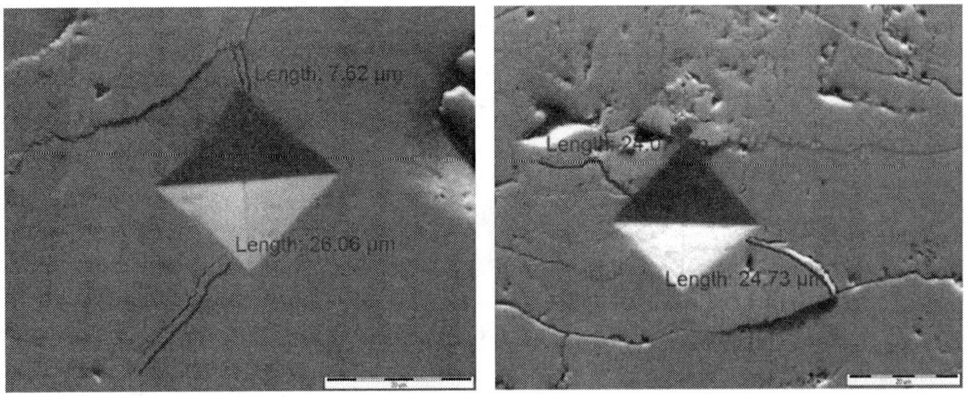

a) b)

Fig. 4. Vickers indentations on parallel (a) and perpendicular (b) to the coating-substrate interface

It was observed a minimum load (or lower critical load) at which the first microcrack occured and a maximum critical load at which the delamination apperead (e.g. HV2 coating after indentation load of 4.9 N, Tabel 1) which depending on the type of coating, indentation direction and thickness of coating. The main limitation of this method was the thickness of coating which has to be higher than 150 μm to avoid the delamination. Beside the indentation loads and mechanical properties of coatings, the pre-existing cracks, diameter and thickness of the splats and interface splats play a critical role in propagation of microcracks in a such way that in some cases the cracks propagate randomly (not liniar) and for loads higher than 9.8 N small secondary microcraks develop near the impression of indentation. All the microcracks that were developed under different indentation loads were of the Palmqvist type and based on their lengths, material hardness and the *k* parameter according to the indentation direction, it was determined the fracture toughness of the Mo coatings in both parallel an perpendicular direction to the coating-substrate interface.

Table 1 summarizes the fracture toughness values on the parallel (EA1, AP1 and HV1 samples) and perpendicular direction (EA2, AP2, HV2 samples) to the coating-substrate interface for Mo coatings samples deposited by EA (EA1, EA2), APS (AP1, AP2) and HVOF methods (HV1, HV2).

Table 1. Fracture toughness of Mo coatings deposited by APS, EA and HVOF

F[N]	0.1	0.24	0.49	0.98	1.96	4.9	9.8	19	29.4	49	98
EA1	*	0.28	0.33	0.33	0.34	0.41	0.49	0.52	0.53	0.66	0.76
EA2	0.2	0.27	0.28	0.30	0.32	0.39	0.43	0.46	0.48	0.51	0.58
APS1	*	*	0.35	0.38	0.43	0.51	0.57	0.81	0.90	1.02	1.17
APS2	*	0.36	0.43	0.51	0.55	0.57	0.60	0.69	0.73	0.74	0.76
HV1	*	0.70	0.85	0.88	1.00	1.15	1.40	1.59	1.81	2.05	2.27
HV2	*	0.45	0.55	0.59	0.63	0.69	~	~	~	~	~

* no initiated microcraks
~ delamination appeared

As it can be seen in Table 1, there were significantly differences between the fracture toughness values obtained on these three types of Mo coating, but in all cases the values obtained on the parallel direction were higher than those on perpendicular direction to the coating-substrate interface. This is due to the fact that on parallel direction the most indentations are in the homogeneous splats while on the cross section (or perpendicular to the coating-substrate interface) the indentations are mostly on the interface between the splats which represents the weakest link in a thermal spray coating. Unlike the homogeneous materials for which the fracture toughness is relatively constant irrespective of the indentation load [8], for inhomogeneous coatings, containing pores, oxides, boundaries of splats or containing pre-existing microcracks, the tendency of fracture toughness is to increase with the increase of the indentation loads. This tendency can be explained by the deformation field which is not self similar for different indentation and thus leads to microcrack driving stress field that varies with indentation load [9] or by the R-curve effect of the toughness [12]. Two other assumptions can explain this local fracture behavior of thermal sprayed Mo coatings. The first assumption is that by increasing indentation loads the microcracks propagate towards the boundaries of splats, pores or the pre-existing microcracks. In such case, it is encountered an increase of indentation diagonals length by increasing the indentation load inversely proportional to the length of the developed microcracks. As a result, a higher value of fracture toughness is calculated. The second assumption can be taken into consideration, in particular for indentation loads higher than 9.8 N (1 kgf), when the main microcracks which are initiated at the corners of Vickers indentation branch into smaller secondary microcracks. Therefore, the inhomogeneity of Mo coating can prevent the increase in length of microcracks and favors the increasing of fracture toughness values on both indentation directions. The propagation behavior of the microcracks in thermal sprayed coatings in terms of fracture toughness indicates that the interface between splats is the weakest link in a thermal spray coating. This aspect is also confirmed by the tensile bond strength tests due the major fracture mode within the Mo layers. Between these two mechanical properties it can be made an association in terms of resistance to crack propagation because before the failure of a thermal sprayed coating, the microcracks appear, grow and propagate through the layers of coatings and finally, will contribute to a cohesive fracture, in the case of tensile bond strength testing, or to a delamination, in the case of indentation testing.

However, similar associations between results from the tensile adhesive tests and interfacial indentation tests were made for thermal sprayed coatings and it was concluded that the interfacial fracture toughness is a suitable parameter to characterize the adhesion of thermal sprayed coatings [5] [14]. To correlate the local fracture toughness with the cohesion strength of Mo coatings investigated, it was considered an average fracture toughness of the values obtained for each thermal spray method (0.419 Mpa·√m for EA, 0.623 Mpa·√m for APS and 1.081 Mpa·√m for HVOF). Therefore, in Fig. 5 it is presented a correlation between the fracture toughness Kic and cohesion strength of Mo coatings deposited by EA, APS and HVOF methods. The linearity of the cohesion strength as a function of fracture toughness can be expressed as the slope of the line that describes the increasing of cohesive strength with increasing of fracture toughness.

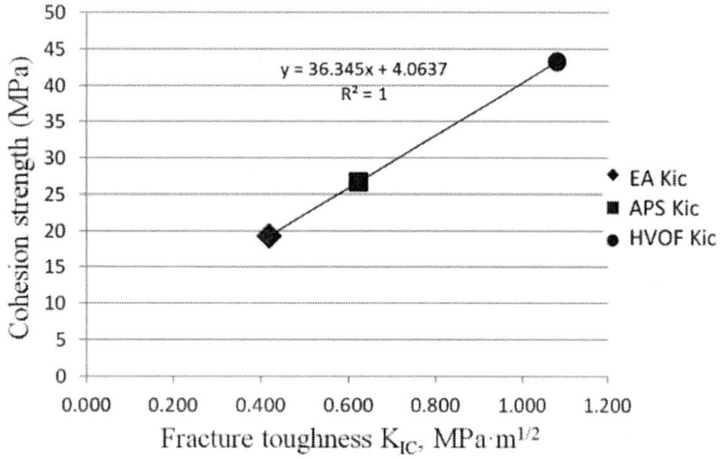

Fig. 5. Correlation between fracture toughness Kic and cohesion strength

This trend shows a superior resistance to crack propagation of Mo coatings deposited by HVOF and consequently, a higher cohesion strength value than those of the coatings deposited by APS and EA methods. The results obtained by indentation tests for evaluating the fracture toughness of Mo coatings are in the range 0.419-1.081 MPa·$m^{1/2}$, which is in good agreement with the values obtained by Brantner et al. for flame sprayed molybdenum coatings [12], but the results obtained by tensile bond strength tests for similar coatings and testing conditions are very scattered. For example, for Mo coatings deposited by plasma spraying with a thickness between 300-350 μm, it was calculated an adhesion strength values of 33-35 MPa [15], a value of around 52 MPa for Mo coatings with a thickness of around 400 μm [16] or values between 18-52 MPa in another case [17]. Also, for Mo coatings deposited by high velocity combustion wire spray process (similar to the HVOF process) it was reported a cohesive failure of Mo coatings and tensile bond strength values between 19.7-27.9 MPa [18]. Beside the influence of thickness, which is well know that influence the bonding strength of thermal sprayed coatings, these differences can originate due to many causes. Firstly, the thermal spray equipment, spray distance parameters, raw material for spraying and substrate preparation can differ from case to case and generate these different performances of bonding strength. Other factors which can influence the bonding strength values are the bonding agent that can cause an overestimate of the adhesion strength due penetration into the coating towards the interface [5], and also the state and the roughness of the surface. In our case, the coated surfaces were machined and a surface roughness value Ra of around 2 µm was obtained for all tested coatings, because in the real work conditions of the component parts coated by Mo coatings, such as shafts and axles, this is the critical technical specification. During the tensile bond strength tests such a smooth coating surface had a critical role for the mode of rupture and bonding strength performance because it did not allow a good mechanical anchoring between surface coated and bonding agent and just a few superior layers of coating experienced rupture by delamination. Also, internal tension stresses were induced in coating during machining of surface that weakened the link between Mo layers, in particular on the superior part of the coating-substrate system where a partial rupture of the coating (cohesion rupture) was generated and not a total rupture (adhesion rupture).

Therefore, as the relationship between the interfacial fracture toughness and adhesion strength can be employed to evaluate and anticipate the mechanical properties of coatings in terms of fracture behavior and bonding strength, the general trend which has been found in this paper for the relation between the fracture toughness and cohesion strength can be used in the cases when the rupture of coatings is produced within the layers, in particular for thermal sprayed coatings where the interface splats is the weakest link in the coating.

4 Conclusions

Based on the experimental results obtained in this study, general conclusion can be drawn.

Mo coatings deposited by three thermal spray techniques such as EA, APS and HVOF methods are investigated by tensile bond strength and Vickers indentation tests in order to evaluate and correlate the two of most important mechanical properties of the thermal sprayed coatings such as, cohesion strength and fracture toughness.

The micro fractographic investigations of the rupture surfaces of specimens tested at tensile bond strength tests, indicated that the failure of all coatings is mostly produced within the layers and individual splats and therefore, the cohesion strength was calculated according to ASTM Standard C633. It was obtained a mean value of cohesion strength of 43.3 MPa for HVOF-sprayed coatings, 26.7 MPa for APS-sprayed coatings and 19.2 MPa for EA-sprayed coatings.

The Vickers indentations of Mo coatings on both parallel and perpendicular direction to the coating-substrate interface generate Palmqvist cracks at the corners of impressions under different indentation loads and depending on the Mo coating type and the indentation direction. The fracture toughness as a function of microcracks size and indentation load of Mo coating obtained by HVOF method is higher than the fracture toughness of Mo coatings obtained by APS and EA methods, under each indentation load, and indentation directions due the splats dimensions, porosity, oxides, boundaries splats or pre-existing microcracks.

It was shown that between these two mechanical properties, cohesion strength and fracture toughness, it can be made a similar correlation that it usually is made between adhesive strength and interfacial fracture toughness of thermal sprayed coatings. The general trend is expressed as the slope of the line that describes the tendency of fracture toughness to increase with the increase of cohesive strength and it was concluded that the fracture toughness is a suitable parameter to characterize the cohesion strength of thermal sprayed coatings.

This work has been done within Partnerships in Priority Areas Program – PN II, implemented with the support of MEN – UEFISCDI, project no. 81/2014.

References

1. J. Shields, *Applications of Molybdenum Metal and its Alloys*, IMOA (2013)
2. F. S. M. Kasparova, *18.-20.5 Brno Metals 2011*, Czech Republic (2011)
3. A. Moradkhani, H. Baharvandi, M. Tajdari, H. Latifi, J. Martikainen, J. Adv. Ceram. **2**, 1 (2013)
4. Z. Xia, W.A. Curtin, B.W. Sheldon, Acta Materi. **52,** 3507–3517 (2004)
5. G. Marot, J. Lesage, Ph. Démarécaux, M. Hadad, St. Siegmann, M.H. Staia, Surf. Coat. Technol. **201,** 2080–2085 (2006)
6. Y. Yamazaki, M. Arai, Y. Miyashita, H. Waki, and M. Suzuki, JTTEE5 **22,** 1358–1365 (2013)
7. ASTM Standard ASTM E384-10e2. Standard Test Method for Knoop and Vickers Hardness of Materials, ASTM International (2010)
8. R. Spiegler, S. Schmauder, L.S.Sig., J. Hard. Mater. **1,** 147-158 (1990)
9. T. Lube, J. Eur. Ceram. Soc. **21**, 211-218 (2001)
10. E. E. Medeiros, A.M.S. Dias., IJRRAS **17,** 9-18 (2013)
11. K. K. Bamzai, P. N. Kotru, B. M Wanklyn, J. Mater. Sci. Technol. **16**, 405-410 (2000)
12. H. P. Brantner, R. Pippan and W. Prantl., J. Therm. Spray. Techn. **12,** 560-571 (2003)
13. G. Matache, C.Puscasu, A. Paraschiv and O.Trusca, AMM **811,** 19-23 (2015)
14. A. Kishi, S. Kuroda, T. Inoue, T. Fukushima, and H.Yumoto, JTTEE5 **17,** 228-233 (2008)
15. J. Ružbarský, A. Panda, SpringerBriefs Appl. Sci. Technol. (1st ed. 2017) (2016)
16. A. Vencl, S. Arostegui, G. Favaro , F. Zivic, M. Mrdak, S. Mitrovic, V. Popovic. Tribol. Int. **4,** 1281-1288 (2011)
17. M. Mrdak, A. Vencl, M. Ćosić, FME Trans. **37,** 27-32 (2009)
18. S.C. Modi & A. Calla, J. Therm. Spray. Techn. **10,** 480-486 (2001)

Some considerations on modelling and homogenization of multiphase light alloys

Ludmila Parashkevova[1,*]

[1] *Institute of Mechanics, BAS, Acad. G. Bonchev Str. bl.4, 1113 Sofia, Bulgaria*

Abstract. The modelling strategies aimed to access the thermo-mechanical behavior of the up-to date light and superlight alloys should take into account characteristics of each particular manufacturing technology. In the present study the light alloys and materials of high porosity are regarded as multiphase composites for which the content of embedded phases is predominant. A new modification of differential effective medium theory is presented accounting for inclusion's size effects. The applied approach is based on the size-sensitive dilute homogenization procedure with Cosserat material matrix. Numerical examples for elastic properties of a close cell foam materials with different Poisson's ratio are presented and discussed.

1 Introduction

In recent years, the development of enhanced technologies for preparing of composites with higher volume fractions of secondary phases shows a significant increase. Special attention has been attracted to light alloys and other materials of metal or ceramic matrix [1] , [2] , [3]. Metal foams are materials of high porosity which possess specific combination of thermo-mechanical and physical properties, [4]. For their very low specific weights and thus high specific stiffnesses, they are able to absorb significant amount of deformation energy while guaranteeing other properties such as high fire and heat resistance, noise attenuation and shielding of electromagnetic devices. The unique combination of useful characteristics makes the application of metal foams an rising industry in many engineering fields from interior design and equipment to civil engineering and vehicle construction, see [5].

There exist several approaches to obtain porosity – properties relations for such composites. Among them one could differentiate the methods of micro-mechanics as differential [6], [7], [8] and self-consistent methods [9], FEM [10], minimum solid area models [11] , as well as commonly used semi empirical models [12], [13], [14], [15], and etc. Recently a comparative study has been published in [5] covering most elastic modulus–porosity relationships suggested in the literature.

In [10] Garbozci and Berryman presented a variant of Differential Effective Medium (DEM) theory where the size effect is taken into account introducing a composite particle (an inner particle surrounded by a thin shell with different properties). This new slightly larger inclusion is firstly homogenized following Christensen's method, [9]. In the mentioned approach the size sensitivity of the model has been expressed via the ratio between the diameters of the inner particle and outer spherical shell. In such way the size effect vanishes if the thickness of the shell tends to zero. From our point

*e-mail: lusy@imbm.bas.bg

of view this variant can't be recommended for foams and other composites of high porosity. The volume fracture of the shell is comparable with matrix volume fraction and it is very hard to avoid the overlapping among homogenized effective inclusions. Besides, the asumption that the shell of dense material being obviously a part of the matrix around the inclusion should be incorporated to a void, is open to criticism too.

This paper is aimed to develop a variant of DEM theory accounting for size effects in multi-phase composites with minor matrix fraction, including closed cells foams. Approximate analytical formulae are derived in order to assess and analyze the influence of pore size on the elastic properties of porous composite with higher porosity. The homogenization technique based on DEM theory and modified Mori - Tanaka two-phase model [16],[17] for the case of spherical inclusions is further developed incorporating an additional size dependent term to one of the classical Eshelby's tensor components. The consequences of this new assumption are demonstrated on a few limiting cases. For them the system of differential equations is explicitly solved and numerical computations elucidate the influence of pores-matrix microstructure interaction on overall elastic moduli.

2 Modelling

For obtaining the overall properties of a multiphase composite at higher concentration of embedded phases two–steps homogenization procedure is suggested. On the first step the Representative Volume Element (RVE) of the material is distributed into pseudo grains, following the ideas from [18]. According to this approach the multiphase RVE, consisting of matrix and n_f phases, is equivalent to a RVE, consisting of n pseudo grains, $n \geq n_f$. Each pseudo grain is a two-phase composite, containing a part of the matrix and all inclusions with a particular size and elastic properties. Each pseudo grain should be homogenized as a two phase composite following a proper scheme which depends on the total volume fraction of the inclusions C_{sum}. For composites with low volume fraction of matrix phase as closed cell foams or porous metals a new variant of DEM is presented and discussed in this paper. The proper choice of a homogenization method at the fist step is a key point for the overall properties predictions of entire model. On the second step the RVE's agglomerate of the already homogeneous Cauchy -type pseudo-grains has to be subjected to the final homogenization performing Hill-Budiansky scheme. For a composite with known size distribution function of inclusions the variant developed in [19] is appropriate at the second step.

2.1 Limits of size effect

In the case of low volume fraction, not exceeding 20-30 %, the hypothesis of dilute inclusions is valid and updated size sensitive Mori-Tanaka homogenization could be applied [16],[19]. For the RVE of the composite with inclusions entirely surrounded by a matrix the moduli of $i-th$ pseudo-grain are obtained by means of equations:

$$K_{ci} = K_m + \frac{C_{sum}K_m(K_i - K_m)}{C_m a_m(K_i - K_m) + K_m}, \quad G_{ci} = G_m + \frac{C_{sum}G_m(G_i - G_m)}{C_m(b_m - b_{0i})(G_i - G_m) + G_m} \tag{1}$$

where: $a_m = \frac{3K_m}{3K_m+4G_m}$, $b_m = \frac{6(K_m+2G_m)}{5(3K_m+4G_m)}$. According to the variant of Cosserat theory adopted, [17], the following relations holds among the parameters of initially micropolar matrix:

$$(l_0)^2 = \frac{\beta}{G_0} = \frac{\gamma}{\kappa} \quad , \quad \frac{\kappa}{G_0} = \frac{\gamma}{\beta} = p \tag{2}$$

In (2) the moduli G_0 and β connect symmetric parts of stresses - strains, and couple stress- curvature tensors, respectively, and κ and γ connect antisymmetric parts of corresponding tensors at elastic state of the initial micropolar matrix, $p \geq 0$. The size sensitivity of the model depends on the dimensionless parameter D_i/l_0 through the average Eshelby tensor component $(b_m - b_{0i})$, [16],[17], where l_0 is intrinsic length of the initial matrix and D_i is inclusion's diameter.

$$b_{0i} = \frac{6p}{5(p+1)} R_i(\eta_i) ,$$

$$R_i(\eta_i) = e^{-\eta_i}\left(\eta_i^{-2} + \eta_i^{-3}\right)(\eta_i ch\eta_i - sh\eta_i) , \tag{3}$$

$$\eta_i = \frac{D_i}{2h} , \quad h = \frac{p+1}{\sqrt{p}} \frac{l_0}{2} , \quad \eta_i = \frac{\sqrt{p}}{p+1} \frac{D_i}{l_0}$$

It is important to emphasize that after the procedure, described by (1) the homogenized material is isotropic Cauchy - type medium and the influence of the initial matrix as centrosymmetric micropolar material is explicitly included in the term b_{0i}. Now for the case of spherical pores (closed cells) limiting values, which serve as bonds for assessment of the size effects will be presented. At the beginning appling some algebra it is proven that the function $R_i(\eta_i)$ behaves as demonstrated on Fig.1.

$$R_i\big|_{\eta_i \to 0} = 1/3 \quad , \quad R_i\big|_{\eta_i \to \infty} = 0 \tag{4}$$

As far as $\eta_i \geq 0$ the maximum size effect is expected when η_i tends to zero.

Case I $p = 0$, $(\kappa = 0)$

$$\begin{aligned} \eta_i &= 0 \quad , \quad R_i(0) = 1/3 \\ b_{0i} &= 0 \end{aligned} \tag{5}$$

In this case there is no size effect and results coincide with the classical Mori-Tanaka ones.

Case II $p = 1$, $(\kappa = G_0)$

$$\begin{aligned} \eta_i &= \frac{D_i}{2l_0} \quad , \quad 0 < R_i = R_i\left(\frac{D_i}{2l_0}\right) \leqq 1/3 \\ b_{0i} &= \frac{3}{5} R_i\left(\frac{D_i}{2l_0}\right) \quad , \quad 0 < b_{0i} \leqq 1/5 \end{aligned} \tag{6}$$

Herein it is assumed (as in many investigations [16], [17],[20] concerning the application of centrosymmetric micropolar theory) that the Cosserat parameter κ is in order of the shear modulus of initial matrix. If the inclusions (pores) are much smaller than the characteristic length l_0, the term b_{0i} converges to $1/5$.

Case III $p = \infty$, $(\kappa = \infty)$;

$$\begin{aligned} \lim \eta_i\,|_{\kappa \to \infty} &= 0 \quad , \quad R_i(0) = 1/3 \\ \lim b_{0i}\,|_{\kappa \to \infty} &= 2/5 \end{aligned} \tag{7}$$

Case I and Case III trace out the bounds from minimum to maximum susceptibility of the matrix to the inclusions fitting in it.

2.2 Differential approach

The differential effective medium approach requires an explicit homogenization scheme for a dilute inclusion problem to be postulated at the beginning. Let the equations (8) represent such a model for the bulk and shear modulus of composite, respectively:

$$K = F_K(K_m, K_i, c, G_m, G_i) \quad , \quad G = F_G(K_m, K_i, c, G_m, G_i) \tag{8}$$

It is supposed that functions F_K and F_G satisfy the requirements of Taylor's theorem and are expanded in Maclaurin series with respect of concentration c. Then:

$$\begin{aligned} K &= F_K(0) + \frac{dF_K}{dc}(0)\,c + O_K\left(c^2\right), \\ G &= F_G(0) + \frac{dF_G}{dc}(0)\,c + O_G\left(c^2\right) \end{aligned} \tag{9}$$

where $F_K(0) = K_m$ and $F_G(0) = G_m$.

The linear parts of (9) as dilute relations are used now to generate approximate differential equations suitable to estimate the elastic moduli of the composite when arbitrary amount of the included phase is embedded into the matrix, [10]. Suppose that RVE of the composite material V contains c volume fraction of inclusions and $\phi = (1 - c)$ volume fraction of matrix. Let one remove from the RVE a differential volume element dV. In this volume element the matrix part is $dV_m = (1 - c))\,dV$ and after that a new portion of the inclusions is added to the composite replacing the differential volume element with the same small volume consisting of inclusions only. Than, the volume fraction of the matrix will change as follows:

$$\phi + d\phi = \frac{V_m - dV_m}{V - dV + dV} = \phi - (1 - c)\frac{dV}{V} \Longrightarrow \frac{dc}{1 - c} = \frac{dV}{V}. \tag{10}$$

The main idea of the DEM is to consider the new inclusion enriched material as a composite consisting of matrix of equivalent material and second phase of inclusions with a very low volume fraction dV/V. For such composite the dilute hypothesis holds and the overall elastic properties can be obtained by simplified linear formulae (9). The linear system (9) is rewritten in the following form: the moduli K, G are expressed as $\left(K_{eq} + dK_{eq}\right), \left(G_{eq} + dG_{eq}\right)$ and the matrix index "m" is replaced by matrix index "eq" (equivalent):

$$\begin{aligned} K_{eq} + dK_{eq} &= K_{eq} + F_K'(eqv, 0)\frac{dV}{V}, \\ G_{eq} + dG_{eq} &= G_{eq} + F_G'(eqv, 0)\frac{dV}{V}; \end{aligned} \tag{11}$$

$$\begin{aligned} dK_{eq} &= \frac{dF_K\left(K_{eq}, K_i, G_{eq}, G_i\right)}{dc}\Big|_{c=o}\frac{dc}{1 - c}, \\ dG_{eq} &= \frac{dF_G\left(K_{eq}, K_i, G_{eq}, G_i\right)}{dc}\Big|_{c=o}\frac{dc}{1 - c}. \end{aligned} \tag{12}$$

2.3 Size - sensitive differential approach

It is worth noting that the system of differential equations (12) can be obtained for every dilute homogenization (8) which allows Taylor series expansion to be performed. Further the homogenization scheme (1) is used as a representative of (8), see [17], [16], leading to the system:

$$\begin{aligned} dK_{eq} &= \frac{\left(3K_{eq} + 4G_{eq}\right)\left(K_i - K_{eq}\right)}{\left(3K_i + 4G_{eq}\right)}\frac{dc}{1 - c}, \\ dG_{eq} &= \frac{G_{eq}\left(G_i - G_{eq}\right)}{\left(G_i - G_{eq}\right)\left[\frac{5}{6}\frac{\left(K_{eq}+2G_{eq}\right)}{\left(3K_{eq}+4G_{eq}\right)} - b_{0i}\right]}\frac{dc}{1 - c}. \end{aligned} \tag{13}$$

The term b_{0i} appearing in (3) and (13) is the only trace of Cosserat properties of the initial matrix in the average Eshelby tensor. The new assumption addopted herein is that the term b_{0i} could be regarded as an internal model parameter which expresses in average maner the sensitivity of the genuine matrix to the presence of spherical inclusions of diameter D_i. It depends on dimentionless model parameters p and D_i/l_0. For metal based composites internal lenght l_0 is often related to the grain size, i.e. to the matrix microstructure self-organization. From the DEM's point of view such assuption seems resonable since DEM supposes that accounting for addition of another small portion of inclusions the simple linear homogenization procedure (11) should be executed over and over again. Every time the eqivalent material, obtained at the previous step has to play the roll of matrix at the next step. As a complementary agrument one may point out that after such kind of homogenization (1) the overall modulae of the composite define an ordinary isotropic Cauchy - tipe material without any micropolar features.

2.3.1 Size - sensitive DEM theory in case of porous inclusions

In the case of spherical pores inclusions or closed cells foams the system of differential equations (13) expressed in terms of Poisson's ratio and Young's modulus can be uncoupled. To this end we consider the moduli ν_{eq} and E_{eq} obey the common relations $\nu_{eq} = \frac{(3K_{eq}-2G_{eq})}{2(G_{eq}+3K_{eq})}$, $E_{eq} = \frac{9K_{eq}G_{eq}}{(G_{eq}+3K_{eq})}$, $\frac{G_{eq}}{K_{eq}} = \frac{3(1-2\nu_{eq})}{2(1+\nu_{eq})}$ as $\nu_m = \frac{(3K_m-2G_m)}{2(G_m+3K_m)}$, $E_m = \frac{9K_mG_m}{(G_m+3K_m)}$, $\frac{G_m}{K_m} = \frac{3(1-2\nu_m)}{2(1+\nu_m)}$. Then the system (13) is transformed into the equations:

$$\frac{2}{3}\frac{\left(7-5\nu_{eq}\right)+15\left(1-\nu_{eq}\right)b_{0i}}{\left(1-\nu_{eq}{}^2\right)\left[\left(1-5\nu_{eq}\right)-5\left(1-\nu_{eq}\right)b_{0i}\right]}d\nu_{eq} = \frac{dc}{1-c}, \tag{14}$$

$$\frac{dE_{eq}}{E_{eq}} = \frac{1}{2\left(2\nu_{eq}-1\right)}\left[3\left(1-\nu_{eq}\right)+4F_\nu{}'\left(eq\upsilon,0\right)\right]\frac{dc}{1-c}. \tag{15}$$

where:

$$F_\nu{}'\left(eq\upsilon,0\right) = \frac{3}{2}\frac{\left(1-\nu_{eq}{}^2\right)\left[\left(1-5\nu_{eq}\right)-5\left(1-\nu_{eq}\right)b_{0i}\right]}{\left(7-5\nu_{eq}\right)+15\left(1-\nu_{eq}\right)b_{0i}}. \tag{16}$$

Taking into account that $F_\nu{}'(eq\upsilon,0)\,\frac{dc}{1-c} = d\nu_{eq}$ and if $F_\nu{}'(eq\upsilon,0) \neq 0$ the equation (15) can be now represented by the varables E_{eq} and ν_{eq} :

$$\frac{dE_{eq}}{E_{eq}} = \frac{1}{2\left(2\nu_{eq}-1\right)}\left[\frac{(3\left(1-\nu_{eq}\right)}{F_\nu{}'\left(eq\upsilon,0\right)}+4\right]d\nu_{eq}. \tag{17}$$

$$F_\nu{}'\left(eq\upsilon,0\right) = 0 \quad \Leftrightarrow 1-5\nu_{eq}-5\left(1-\nu_{eq}\right)b_{0i} = 0 \tag{18}$$

The condition (18) gives particular solutions of the equations (14) and (15), respectively. For any admissible parameter $0 \leqq b_{0i} \leqq 2/5$ there exists a single value $\nu_m{}^*$ satisfying the condition (18) so that:

$$\nu_m{}^* = \left(1-5b_{0i}\right)/5\left(1-b_{0i}\right) \tag{19}$$

and eqns. (21) are solutions of the system (20):

$$d\nu_{eq} = 0 \quad ; \quad \frac{dE_{eq}}{E_{eq}} = \frac{3\left(1-\nu_m{}^*\right)}{2\left(2\nu_m{}^*-1\right)}\frac{dc}{1-c}. \tag{20}$$

$$\nu_{eq}(c) = \nu_m{}^* = const \quad ; \quad E_{eq}(c, \nu_m{}^*) = E_m (1-c)^{\frac{3(1-\nu_m{}^*)}{2(1-2\nu_m{}^*)}} \ . \tag{21}$$

The bulk and shear moduli of the composite corresponding to (21) are:

$$K_{eq}(c, \nu_m{}^*) = \frac{E_m (1-c)^{\frac{3(1-\nu_m{}^*)}{2(1-2\nu_m{}^*)}}}{3(1-2\nu_m{}^*)} \quad ; \quad G_{eq}(c, \nu_m{}^*) = \frac{E_m (1-c)^{\frac{3(1-\nu_m{}^*)}{2(1-2\nu_m{}^*)}}}{2(1+\nu_m{}^*)} \ . \tag{22}$$

3 Results and discussions

Differential equations (14) and (17) should be solved consecutively one after another. For some special values of b_{0i} the analytical solutions are presented in the next sections.

3.1 Size - sensitive DEM theory in case I

Case I $\quad b_{0i} = 0, \quad (p = 0)$. Differential equations (14) and (17) take now the form (23), (24):

$$\frac{2}{3} \frac{\left(7 - 5\nu_{eq}\right)}{\left(1 - \nu_{eq}{}^2\right)\left(1 - 5\nu_{eq}\right)} d\nu_{eq} = \frac{dc}{1-c} \ , \tag{23}$$

$$\frac{dE_{eq}}{E_{eq}} = \frac{1}{\left(1 - 2\nu_{eq}\right)} \left[\frac{\left(5\nu_{eq} - 7\right)}{\left(1 + \nu_{eq}\right)\left(1 - 5\nu_{eq}\right)} - 2 \right] d\nu_{eq} \ ; \tag{24}$$

with solutions:

$$\left(\frac{1 - \nu_m}{1 - \nu_{eq}}\right)^{1/6} \left(\frac{1 + \nu_m}{1 + \nu_{eq}}\right)^{2/3} \left(\frac{1 - 5\nu_{eq}}{1 - 5\nu_m}\right)^{5/6} = 1 - c \tag{25}$$

$$E_{eq} = E_m \left(\frac{1 + \nu_m}{1 + \nu_{eq}}\right)^{2/3} \left(\frac{1 - 5\nu_{eq}}{1 - 5\nu_m}\right)^{5/3} \tag{26}$$

$$\nu_m{}^* = 1/5 \quad ; \quad E_{eq}(c, \nu_m{}^*) = E_m (1-c)^2 \tag{27}$$

$$K_{eq}(c, \nu_m{}^*) = \frac{5}{9} E_m (1-c)^2 \quad ; \quad G_{eq}(c, \nu_m{}^*) = \frac{5}{12} E_m (1-c)^2 \ . \tag{28}$$

Case I describes those type of matrix –inclusions interactions when embedded phase has much bigger size than the internal parameter l_0 dealing with matrix microstructure. In other words the microstructure of the matrix is so fine, that the matrix does not "feel" the presence of the coarse added phase. Of course, case I also represents the classical solution of DEM when size effects are not considered at all, see [8].

3.2 Size - sensitive DEM theory in case II

Case II $\quad b_{0i} = 1/5, \quad (p = 1)$. Differential equations (14) and (17) take now the form (29), (30):

$$-\frac{3}{2} \frac{\left(13 - 11\nu_{eq}\right)}{\left(1 - \nu_{eq}{}^2\right)\nu_{eq}} d\nu_{eq} = \frac{dc}{1-c} \ , \tag{29}$$

$$\frac{dE_{eq}}{E_{eq}} = \frac{1}{\left(1 - 2\nu_{eq}\right)} \left[\frac{\left(5 - 4\nu_{eq}\right)}{2\left(1 + \nu_{eq}\right)\nu_{eq}} - 2 \right] d\nu_{eq} \ ; \tag{30}$$

and the solutions are:

$$\left(\frac{1-\nu_m}{1-\nu_{eq}}\right)^{1/6}\left(\frac{1+\nu_m}{1+\nu_{eq}}\right)^{3/2}\left(\frac{\nu_{eq}}{\nu_m}\right)^{5/3} = 1-c \tag{31}$$

$$E_{eq} = E_m\left(\frac{1+\nu_m}{1+\nu_{eq}}\right)^{3/2}\left(\frac{\nu_{eq}}{\nu_m}\right)^{5/2} \tag{32}$$

$$\nu_m{}^* = 0 \quad ; \quad E_{eq}\left(c,\nu_m{}^*\right) = E_m\left(1-c\right)^{3/2} \tag{33}$$

$$K_{eq}\left(c,\nu_m{}^*\right) = \frac{1}{3}E_m\left(1-c\right)^{3/2} \quad ; \quad G_{eq}\left(c,\nu_m{}^*\right) = \frac{1}{2}E_m\left(1-c\right)^{3/2} \ . \tag{34}$$

Case II in this subsection is a special variant of the case II described in the section 2.1. It represents the limiting case when the diameter of inclusions tends to zero and Cauchy initial shear modulus G_0 is of the same order than Cosserat modulus κ. Expressed by b_{0i} this case is situated just in the middle between Case I and Case III.

3.3 Size - sensitive DEM theory in case III

Case III $b_{0i} = 2/5$, $(p = \infty)$. Differential equations (14) and (17) take now the form (35), (36):

$$-\frac{2}{3}\frac{\left(13-11\nu_{eq}\right)}{\left(1-\nu_{eq}^2\right)\left(1+3\nu_{eq}\right)}d\nu_{eq} = \frac{dc}{1-c}\ , \tag{35}$$

$$\frac{dE_{eq}}{E_{eq}} = \frac{1}{\left(1-2\nu_{eq}\right)}\left[\frac{\left(13-11\nu_{eq}\right)}{\left(1+\nu_{eq}\right)\left(1+3\nu_{eq}\right)}+2\right]d\nu_{eq}\ ; \tag{36}$$

with solutions given by:

$$\left(\frac{1-\nu_m}{1-\nu_{eq}}\right)^{1/6}\left(\frac{1+\nu_m}{1+\nu_{eq}}\right)^{4}\left(\frac{1+3\nu_{eq}}{1+3\nu_m}\right)^{25/6} = 1-c \tag{37}$$

$$E_{eq} = E_m\left(\frac{1+\nu_m}{1+\nu_{eq}}\right)^{4}\left(\frac{1+3\nu_{eq}}{1+3\nu_m}\right)^{5} \tag{38}$$

$$\nu_m{}^* = -1/3 \quad ; \quad E_{eq}\left(c,\nu_m{}^*\right) = E_m\left(1-c\right)^{6/5} \tag{39}$$

$$K_{eq}\left(c,\nu_m{}^*\right) = \frac{1}{5}E_m\left(1-c\right)^{6/5} \quad ; \quad G_{eq}\left(c,\nu_m{}^*\right) = \frac{3}{4}E_m\left(1-c\right)^{6/5} \ . \tag{40}$$

If one remember that for any anisotropic material Poisson's ratio should lie in the interval $-1 < \nu < 1/2$, providing positive values of the other Cauchy elastic moduli, the current value $\nu_m{}^* = -1/3$ is admissible and corresponds to a material of auxetic type, see [21], [4]. Theoretically at $p \to \infty$ i.e. at such ratio between mentioned Cauchy and Cosserat moduli the material of the initial matrix is the most sensitive to the presence of inclusions. This case also corresponds to the variant when $G_0 \ll \kappa$. Case III manifests the limited "resistance" of the matrix when very small extraneous objects give rise to obstacles blocking the matrix to develop its own structures with size $l_0 >> D_i$.

3.4 Size - sensitive DEM theory in case IV

Case IV $b_{0i} = 1/10$, $(p = 1)$. Differential equations (14) and (17) take now the form (41), (42):

$$\frac{2}{3}\frac{\left(17 - 13\nu_{eq}\right)}{\left(1 - \nu_{eq}^2\right)\left(1 - 9\nu_{eq}\right)} d\nu_{eq} = \frac{dc}{1 - c}, \tag{41}$$

$$\frac{dE_{eq}}{E_{eq}} = \frac{1}{\left(1 - 2\nu_{eq}\right)}\left[\frac{\left(13\nu_{eq} - 17\right)}{\left(1 + \nu_{eq}\right)\left(1 - 9\nu_{eq}\right)} - 2\right] d\nu_{eq}\,. \tag{42}$$

Derived solutions are as follows:

$$\left(\frac{1 - \nu_m}{1 - \nu_{eq}}\right)^{1/6}\left(\frac{1 + \nu_m}{1 + \nu_{eq}}\right)\left(\frac{1 - 9\nu_{eq}}{1 - 9\nu_m}\right)^{7/6} = 1 - c \tag{43}$$

$$E_{eq} = E_m\left(\frac{1 + \nu_m}{1 + \nu_{eq}}\right)\left(\frac{1 - 9\nu_{eq}}{1 - 9\nu_m}\right)^2 \tag{44}$$

$$\nu_m^* = 1/9 \quad ; \quad E_{eq}(c, \nu_m^*) = E_m (1 - c)^{12/7} \tag{45}$$

$$K_{eq}(c, \nu_m^*) = \frac{3}{7} E_m (1 - c)^{12/7} \quad ; \quad G_{eq}(c, \nu_m^*) = \frac{9}{20} E_m (1 - c)^{12/7}\,. \tag{46}$$

Case IV (another variant of the case II described in section 2.1 does not represent a limiting case. It is an eligible example which illustrates in complementary way the features of the solutions of the system of equations (14), (15) for composite elastic moduli ν_{eq}, E_{eq} and their connection with bulk and shear moduli as well.

The lower and the upper bounds of elastic moduli for metal matrix composite saturated with spherical voids, corresponding to Case I and Case III are shown on Fig.2.

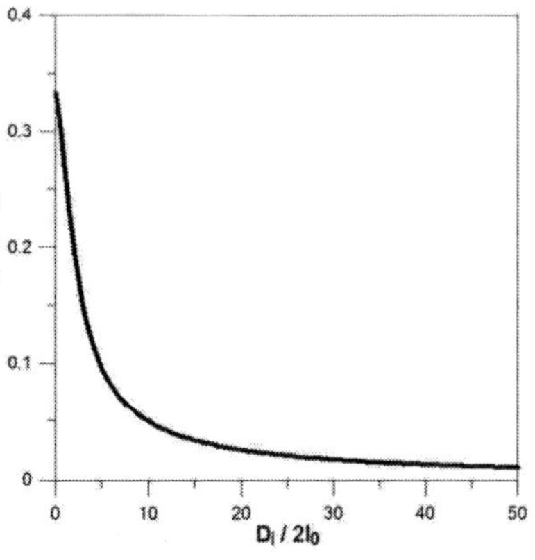

Figure 1. Size sensitivity of function $R_i(\eta_i)$.

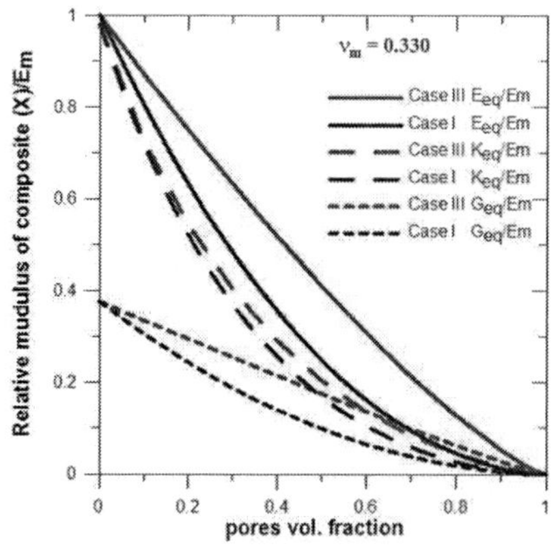

Figure 2. Lower and upper bound of composite moduli accounting for size sensitivity of a metal matrix with $\nu_m = 0.33$. Case I - black curves; Case III - red curves.

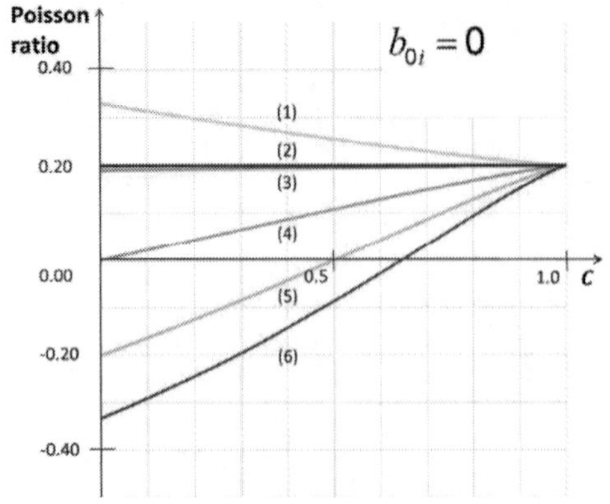

Figure 3. Dependence of Poisson's ratio on inclusions vol. fraction c at different initial matrix Poisson's ratio, Case I; (1) : $\nu_m = 0.33$; (2) : $\nu_m = \nu_m^* = 1/5$; (3) : $\nu_m = 0.19$; (4) : $\nu_m = 0$; (5) : $\nu_m = -0.2$; (6) : $\nu_m = -0.333$.

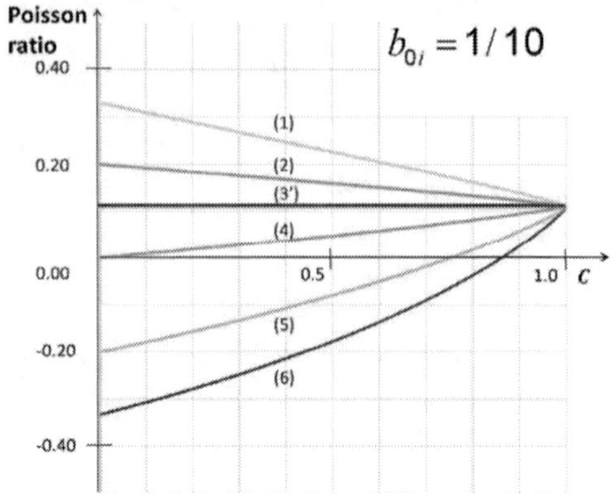

Figure 4. Dependence of Poisson's ratio on inclusions vol. fraction c at different initial matrix Poisson's ratio, Case IV; (1) : $\nu_m = 0.33$; (2) : $\nu_m = 0.2$; (3') : $\nu_m = \nu_m^* = 1/9$; (4) : $\nu_m = 0$; (5) : $\nu_m = -0.2$; (6) : $\nu_m = -0.333$.

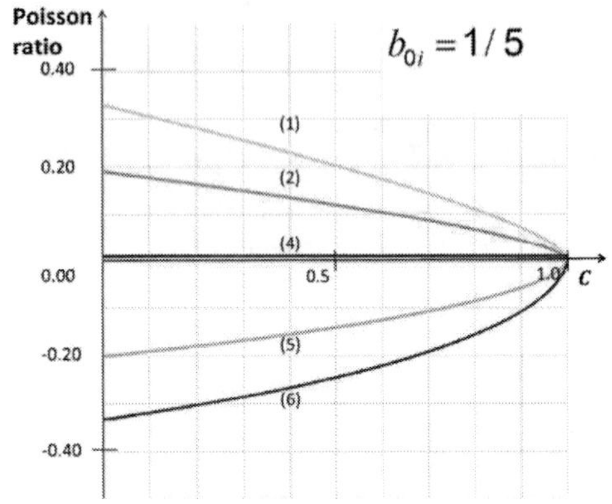

Figure 5. Dependence of Poisson's ratio on inclusions vol. fraction c at different initial matrix Poisson's ratio, Case II; (1) : $\nu_m = 0.33$; (2) : $\nu_m = 0.19$; (4) : $\nu_m = \nu_m^* = 0$; (5) : $\nu_m = -0.2$; (6) : $\nu_m = -0.333$.

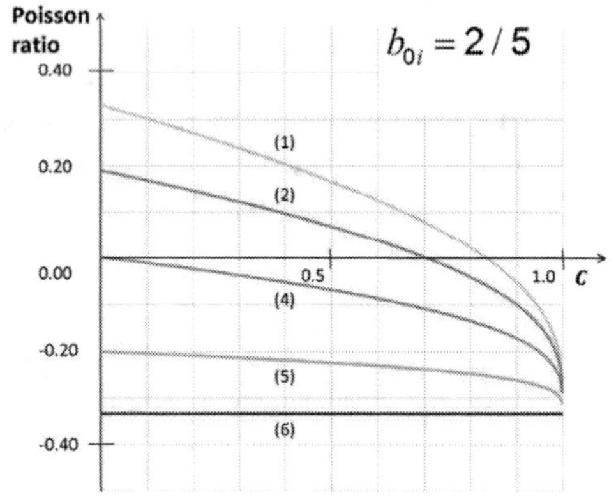

Figure 6. Dependence of Poisson's ratio on inclusions vol. fraction c at different initial matrix Poisson's ratio, Case III; (1) : $\nu_m = 0.33$; (2) : $\nu_m = 0.19$; (4) : $\nu_m = 0$; (5) : $\nu_m = -0.2$; (6) : $\nu_m = \nu_m^* = -1/3$.

As far as the ratios "shear modulus"/"bulk modulus", "Young's modulus"/"bulk modulus" and "Young's modulus"/"shear modulus" are functions of Poisson' ratio exclusively, the solution $\nu_{eq}(c) = \nu_m^* = const$ leads to two remarkable consequences.

1) No matter what amount of pores is added to a matrix with initial elastic moduli (ν_m^*, E_m), where E_m has an arbitrary positive value, the ratio between each pair of moduli of such composite (except for pairs including Poisson's ratio) remains constant for every pore inclusions volume fraction c , $0 \leqslant c \leqslant 1$.

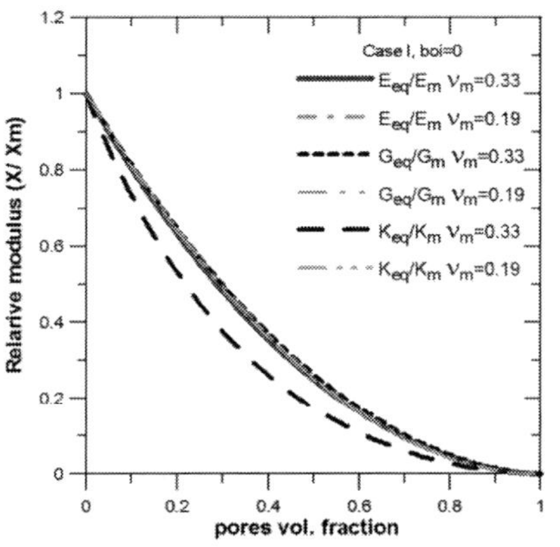

Figure 7. Case I: Dependence of Relative moduli E_{eq}/E_m ; G_{eq}/G_m ; K_{eq}/K_m on inclusions vol. fraction at different initial Poisson's ratio: $\nu_m = 0.33$ - for metal matrix; $\nu_m = 0.19$ - for ceramic matrix.

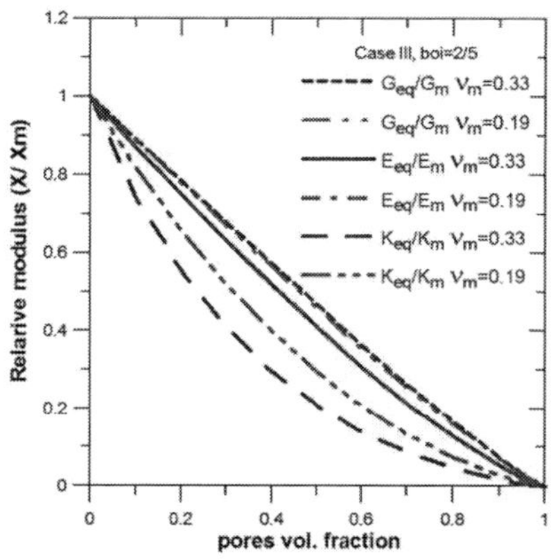

Figure 8. Case III: Dependence of Relative moduli E_{eq}/E_m ; G_{eq}/G_m ; K_{eq}/K_m on inclusions vol. fraction at different initial Poisson's ratio: $\nu_m = 0.33$ - for metal matrix; $\nu_m = 0.19$ - for ceramic matrix.

2) For any admissible b_{0i} there exists a special value of Poisson's ratio $\nu_m{}^* = (1 - 5b_{0i})/5(1 - b_{0i})$ which is connected to the main solution of eqn. (14) for overall Poisson's ratio . For a given b_{0i} and arbitrary chosen matrix moduli (ν_m , E_m) the composite Poisson's ratio tends to $\nu_m{}^*$ at very high pores volume fraction:

$$\lim_{c\to 1}\left[\nu_{eq}(c, b_{0i})\right] = \nu_m{}^*(b_{0i}) \ . \tag{47}$$

This tendency for the case $b_{oi} = 0$ has been noticed in [8] without any relation to the solution $\nu_{eq}(c) = \nu_m{}^*$ of the eqn. $d\nu_{eq} = 0$, see (20), (21). Earlier in [22] also it has been discussed that at sufficiently high porosity Poisson's ratio converges to a fixed non-zero value, irrespective of the matrix Poisson's ratio. Mentioned limiting behavior of equivalent Poisson's ratio is illustrated on Fig.3,4,5,6 and could be seen directly from the analytical results, presented for cases I-IV by equations (23),(29),(35),(41), respectively. If the initial reactivity of a matrix via the parameter b_{oi} is known, than even without information about its classical elastic moduli, Poisson's ratio bound of the corresponding porous material at very high porosity can be predicted apriory by means of eqn. (19).

The cases I, II, III and IV show up the role of Poisson's ratio in this model. The analytic solutions for $\nu_{eq}(\nu_m, b_{0i}, c)$ and corresponding $E_{eq}(E_m, b_{0i}, c)$ enable relationships porosity - elastic moduli of all kinds to be dipper investigated and many new aspects of elastic behavior of light composite to be discovered. For example, it is useful to compare two very popular groups of composites - one based on metal matrix, the other - based on ceramics. Despite of observable differences between $\nu_{eq}(\nu_m = 0.3, c)$ and $\nu_{eq}(\nu_m = 0.19, c)$, (see yellow curves against green curves on Figures 3 - 6) the results obtained indicate that relative Young's moduli and shear moduli have very similar behavior with respect to increasing porosity but relative bulk moduli are highly distinguished for both groups. Such calculations are illustrated on Fig. 7 for the Case I and on Fig. 8 - for the Case III, respectively.

4 Conclusions

A new modification of DEM is presented accounting for inclusion's size effects. The applied approach is based on the size-sensitive dilute homogenization procedure of [16],[17] for two-phase composites.

Analytical solutions ate obtained for elastic properties of a composite containing higher volume fraction of spherical pores, which are suitable for modelling of closed sell foams. Special formulae have been presented which estimate the influence of pore size on elastic behavior of a porous composite by the lower bound (Case I, no size effect) and the upper bound restricting the effective moduli from above, (Case III).

The new variant of DEM described above takes into account inclusions size effects practically in all ranges through the initial internal parameter b_{0i}, see eqn. (13). This model parameter enables materials with equal Cauchy moduli but different microstructure arrangement to be compared as potential matrices of composites with spherical inclusions.

Analytical solutions presented outline some new areas where the desired improvement of mechanical properties of the light composite materials can be achieved and controlled by a proper combination of volume fraction, properties and size parameters of matrix and embedded phases. The model is applicable for metal matrix as well as for non-metal matrix composites and will be developed further for composites with high volume fraction of inclusions of various shapes.

Acknowledgement

The financial support from BG FSI through the grant DH 07/17/ 15 Dec 2016 (project H 07/37/2016) 'New approach for structure and properties design of amorphous and nanostructured metallic foams' is gratefully acknowledged.

References

[1] Peter, I., Rosso, M., Chapter 1, In: Zaki Ahmad (ed.) New Trends in Alloy Development, Characterization and Application, pp. 3-37, InTech (2015)
[2] Lefebvre, L.Ph., Banhart, J., Dunand, D., Adv Eng Mat, **10**, 77–787 (2008)
[3] Stanev, L., Kolev, M., Drenchev, B., Drenchev, L., J Mat Sci Technol, **23**, 303–344 (2015)
[4] Kolken, H.M.A., Zadpoor, A., RSC Adv, **7**, 5111-5129 (2017)
[5] Koudelka, P., Jirousek, O., Doktor, T., Zlamal, P., Fila, T., In: 18th International Conference "Engineering Mechanics", May 14 – 17, 2012, Paper no 218, pp. 691–701, Svratka, Czech Republic (2012)
[6] McLaughlin, R.: A study of the differential scheme for composite materials. Int J Eng Sci **15**, 237–244 (1977)
[7] Berryman,J.G., In: Ahren, T.,J. (ed.) Rock Physics and Phase Relations—A Handbook of Physical Constants, pp. 205-228, American Geophysical Union, Washington DC (1995)
[8] Giordano, S., Eur J Mech A-Solid, **22**, 885–902 (2003)
[9] Christensen, R.M., J Mech Phys Solids, **38**, 379–404 (1990)
[10] Garboczi, E.J., Berryman, J.G., Mech of Mat, **33**, 455–470 (2001)
[11] Rice, R.W., J Mater Sci, **31**, 1509—1528 (1996)
[12] Archie, G.E., Trans AIME, **146**, 54–67 (1942)
[13] Phani, K. K., Niyogi, S. K., J Mater Sci, **22**, 257–263 (1987)
[14] Gibson, L.J., Ashby, M.F.: Cellular Solids - Structure and Properties, second ed. Cambridge University (1997)
[15] Pabst, W., Gregorova, E., J Mater Sci, **39**, 3501–35031 (2004)
[16] Hu, G., Liu, X. , Lu, T.J., J Alloy Compd, **37**, 407–425 (2005)
[17] Parashkevova, L., Bontcheva, N., Babakov, V., Comp Mater Sci, **50**, 527–537 (2010)

[18] Pierard, O., Friebel, C., Doghri, I., Compos Sci Technol, **64**, 1587–1603 (2004)
[19] Parashkevova, L., Bontcheva, N., Comp Mater Sci, **67**, 303–315 (2013)
[20] Chen, Huan, Hu, G., Huang, Zhuping, Int J Solid Struct, **44**, 8106–8118 (2007)
[21] Lakes, R., Adv Mater, **5**, 293–296 (1993)
[22] Roberts, A.P., Garboczi, E.J., J Am Ceram Soc, **83**, 3041–3048 (2000)

Some specifics considering the urban territories river discharge determination

Mila Chilikova-Lubomirova[1,*], and *Daniel Constantin* Diaconu[2]

[1]Institute of Mechanics – Bulgarian Academy of Sciences, Fluid Mechanics, Acad. G. Bonchev str., Block 4, 1113 Sofia, Bulgaria
[2]University of Bucharest, Faculty of Geography, Nicolae Bălcescu Blv., 1 District, Bucharest, 010041, Roumanie

Abstract. Urban territories are specific territories with a significant anthropogenic influence on the natural environment. As a result most of the existing natural conditions have been modified. Parts of them cover the natural forms of river beds and floodplains. Concerning to the humans safety, comfort and needs, while keeping ecosystems healthy function, different artificial structures also have been created. The process is connected to the well understanding and good quality data obtaining about the existing conditions and river flow behaviour, that are interconnected and relevant to the river discharge determination and its variations description – key issue for the entire river structures project, water extremes mitigation and maintaining a healthy state of the ecosystems. For the purpose various contact measurements and monitoring procedures are implemented. To clarify the process this material aims to present some specifics connected to the urban territories river discharge determination and the possibility for related monitoring networks creation. It is focused on the most used methods, their specifics and possible challenges for practical application. Main specifics connected to the related decision support systems creation and implementations are also presented. Main purpose is such state of the art dissemination, in help of decision makers and professionals in the area.

1 Introduction

Urban territories are very specific territories where natural and artificial conditions are joined in a unique complex, suitable for the man needs, interests and perspectives and connected representatives of the urban ecosystems well being. Such territories function with regard to the concrete urban specifics and natural existences forming various scale settlements and conglomerates. In many cases river systems are part of them lending to the settlements additional recreation areas, nice views and tourist attractions. But the most important part is that rivers ensure a certain amount of water resources and related goods connected to sectors as water supply and sewage, energy, agriculture, transportation, sport, etc. The better utilization of such resource requires becoming a reliable and precise data and

* Corresponding author: milasemail@yahoo.com

information. But in many cases this is inaccessible from the national level monitoring networks that cover only restricted number of monitoring points.

With regard to the increasing impacts of water extremes and floods in particular, connected impacts and harmful consequences state the more important question about the possibility for advanced or timely mitigation measures obtaining. For the purpose in many cases implementation of appropriate decision support systems is applied.

To clarify the problems this article will consider the existing experience in the area focusing on the possibilities for the river discharge measurements and determination. Problems connected to the continuous monitoring conducting and monitoring networks creation will be also considered. For the need of related water extremes mitigation with respect to the floods some considerations connected to the possible decision support systems (DSS) creation and operation will be also considered. Presented material will be purposed to the experts and decision makers in the area accounting their huge role in related processes management and mitigation.

2 River systems specifics with regard to the urban territories

Rivers are natural formations serving for the surface water transportation. In the past they were the main transport corridors thus in many cases settlements appear nearby. Nowadays their function is extended taking part in the daily life and leisure, economy and industry. Presented roles closely relate to the local conditions. From one side there are the natural existing water and environment conditions and from the other side - urban specifics and interests. Both vary by scale and spatial specifics.

Urban territories are territories with a high population density. With regard to population socioeconomic structure following main structures are forming: town, city, metropolitan, municipality, and region. They differ by type, inner connections and population size and following classification is widely accepted [1]:

- small urban areas – if their population is between 50 000 and 200 000;
- medium size urban areas – with population between 200 000 and 500 000;
- metropolitan areas – with population between 500 000 and 1.5 millions and
- large population areas – with a population of 1.5 million or more.

Presented information aim to emphasize on some possible differences, connected to the river systems development with regard to the urban communities interests.

Rivers in urban territories are closely connected to the local conditions. Their shape, slope, hydraulic properties and water ability are closely connected to the concrete area specifics and urban perspective. Of importance are connected factors such as climate and relief specifics, time and seasonal variations and climate change impacts that are timely natural variations. Other man initiated effects also exist. They are connected both to the watercourse and riverflow time and spatial distribution. Such influences can be connected to the rivers canalization, urban riverflows management and maintainance, etc. For illustration various scale urban areas and watercourses are presented on Fig. 1.

a) b)

c)

d)

Fig. 1. Various scale urban territories and related watercourses: a) river channel in a village, b) river watercourse in the Bansko resort, c) Osam River flowing through Lovech city, d) Dâmbovița river flowing through the Bucharest city.

Presented on Fig. 1 urban territories and related watercourses aim to illustrate part of the huge cases variety that can be observed in practice. Small or big rivers can be spread in small or huge urban areas. In almost all cases they are influenced by man with regard to the river bed and surrounding floodplain areas. Additional artificial structures in many cases are also created. These interferences are closely connected to the man interests. In such cases two main aspects are in a leading position:

- water resources and urbanization area utilization and
- enhancement and maintaining of safety conditions providing.

For the purpose various improvement facilities and structures are created and operate in practice. Appropriate local related measures can be also provided. The basis in both cases is the reliable and secure data that afterwards serve for the streamflow and related river channel conditions and surrounding floodplain area description, best experienced by direct contact measurements. For the purpose various direct hydraulic properties measurements are conducted aiming such data obtaining that afterwards are used by the hydrological practice or for the need of various type preparedness or decision support systems creation.

3 River discharge specifics with regard to contact measurements and determination

River streamflow is main river specification. Its variations are connected to the concrete topological, hydraulic, hydrological, climatological conditions, etc. It can be described as the water quantity that flow through the concrete river bed section. With regard to its nature it occupies the related river bed conditions, as it does not own a specific form. It can not be directly measured, thus it is computed from other measured variables [2-10]. They are descriptive for the volume rate of water that is transported through the cross-sectional area, described by the equation:

$$W(t) = \int_{\omega} dW \tag{1}$$

where W is the water flowrate $(L^3; m^3)$; dW are the elementary volume flowrates flowing through a given cross-section $(L^3T^{-1}; m^3/s)$, ω is the cross-sectional area occupied by the flow $(L^2; m^2)$.

This quantitative characteristic represents the total water volume that in case of one direction flow summarize all momentum t cross-sectional flowrates W_t. For practical reasons connected to the hydrological, construction, water economy, etc. purposes more often is used its momentum value, presented by (2):

$$Q(t) = W'_t = \frac{dW_t}{dt} = \frac{dW}{dt} \tag{2}$$

where Q is the water discharge $\left(L^3T^{-1}; m^3/s\right)$. It can be also described by the equation:

$$Q = \int_\omega dQ = \int_\omega u d\omega \tag{3}$$

In case of uniform flow flowrate estimations can be provided by:

$$Q = v\omega \tag{4}$$

where Q is water discharge $\left(L^3T^{-1}; m^3/s\right)$; u $\left(LT^{-1}; m/s\right)$ is the flow velocity of the parcel; v $\left(LT^{-1}; m/s\right)$ is the average flow velocity.

In practice the conditions rarely allow availability of a uniform flow. For practical reasons in such conditions the river cross-section is divided on subsections by representative number verticals allowing the streamflow determination by the velocity-area method, graphically illustrated on fig. 2. Thus:

$$Q = k_i \sum_{i=1}^{n} u_i f_i \tag{5}$$

where k_i is the number of verticals; u_i $\left(LT^{-1}; m/s\right)$ is the mean i-th vertical flow velocity and f_i is the subsection area $\left(L^2; m^2\right)$ and

$$f_i = b_i h_i \tag{6}$$

where b_i is the width corresponded to the *i*-th vertical $(L; m)$, h_i is the vertical depth $(L; m)$; u_i is the mean *i*-th vertical flow velocity $\left(LT^{-1}; m/s\right)$.

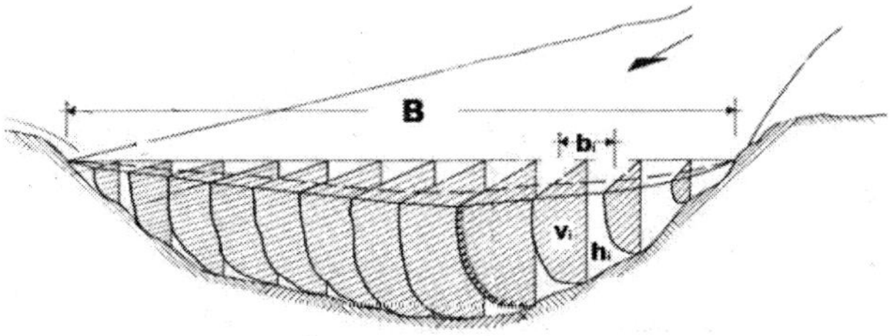

Fig. 2 Scheme of the velocity-area model

With regard to presented principles and the concrete conditions different measurement methods and equipments are implemented in practice. Some of them are used for concrete water discharge measurement and determination but other are implemented continuously that differentiate methods in two main categories [2-6]:

- Discharge measurement methods and
- Discharge monitoring or observation methods.

Discharge measurement methods fundamentally are used for unknown streams quantification. Some of them are applicable in case of discharge monitoring purposes. Discharge monitoring or observation methods are used for continuous determination of the flowrate. Most of them are based on some theoretical or empirical correlation, allowing the streamflow evaluation on the base on some streamflow physical parameters measurement. Most of methods are widely developed and in some cases work as measurement or monitoring products. Their practical implementation is connected to proper monitoring point establishment. This requires being familiar with the concrete river conditions and appropriate monitoring method and related measurement scheme and equipments for implementation.

4 Urban territories discharge measurement and monitoring. Monitoring networks and DSS specification and challenges

River streamflow quantification is a routine work for the hydrometrical practice. It is connected both to the river measurements and observations. For the purpose appropriate monitoring point is established with regard to the selected measurement method and applicable equipments. Practically the procedure is connected to some preliminary works that cover:

- Inspection of the river segment for selection of appropriate for the measurements work site;
- Selection of the most appropriate measurement or monitoring method with regard to the available territorial specifics, energy and road infrastructure, property management and ownership;
- Selection of the most appropriate measurement equipments with regard to the available budget, staff and operation specifics;
- Ordering and purchase of needed gages, arrangement of needed staff, needed licenses obtaining.

Afterwards within the required data needs the particular measurement or monitoring site, method, equipments and staff are selected.

With regard to the practice following wide used methods and procedures are implemented [3-12]:

- Measurement of discharge by current meter approaches

Measurements are conducted at the monitoring point to determine the flow velocity and related water stage and river channel profile. Afterwards related discharge is estimated with regard to the accepted measurement scheme. In case of measurements by verticals presented in the previous section and on fig. 2 procedure is implemented. Other approaches are also applicable. Discharge is estimated by (5). Often conventional current-meters are used, but in the practice are implemented also the acoustic current meters, electromagnetic current meters, side looking Doppler sensors, etc. [11-12] They are illustrated on fig. 3. Connected to the presented method implementation of additional equipments such timers, counting equipments, electronic field notebooks, sounding equipments, width-measuring are also needed for the related water stage and river bed profile determination.

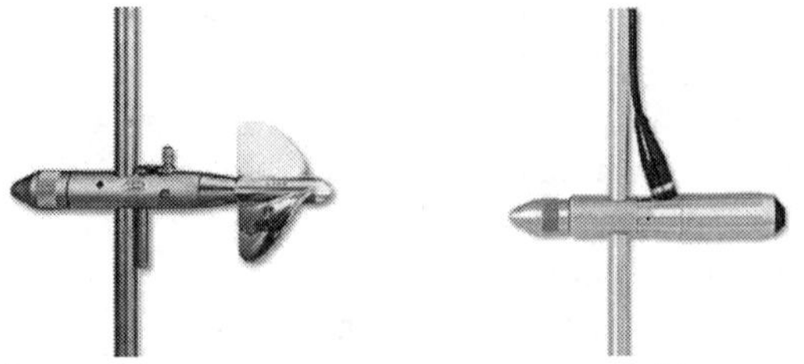

a) b)

c) d)

Fig. 3. Velocity meters: a) Conventional current meter, b) Acoustic Digital Current Meter, c) electromagnetic current meter, d) SLD, produced by OTT Hydromet [11].

With regard to measurement needs in practice the Acoustic Doppler Velocimeters (ADV) and Acoustic Doppler Current Profiler (ADCP) for surface water discharge measurements are also implemented.

a)

b)

Fig. 4. Acoustic Doppler Velocimeter (ADV), SonTek and Acoustic Doppler Current Profiler (ADCP), OTT Hydromet [11-12]

- Stream gauging [3-12].

Stream gauging is useful procedure with regard to the river flow stage determination. It is of especial interest in case of various watercourse structures project and maintenance – bridges, embankments, levees, etc. In case of river discharge determination the stage-discharge relation or some well grounded empirical or semi-empirical relation is used. For the purpose related flow stage is measured and afterwards from the connected rating curve corresponding discharge is determined. Two main measurement approaches exist:

o Non-recording type. It is connected to permanent or portable gauges for measurement of the water stage use and personal observations providing. Equipments can be graduated vertical staff gauge; ramp or inclined gauge; wire-weight gauge, installed on existing structure over the flow; graduated rod, tape, wire or point gauge for measuring the distance to the water surface; etc. It is rarely used for conventional measurements in practice. Usually is used for control measurements.

o Recording gauges. Implemented recording type equipments allow continuous measurement of water stage. Practically complement stage sensor, stage recorder and gauge height retrieval, for measured data extraction are assembled. It is possible the recorder to be mechanical. Following main gauges are used for the stage measurements:

- Wire-weight gauges, chain gauges and float-tape gauges. These gauges consist of a graduated disk or floats that are connected by wire or chain by pulley to a strip-chart recorder. They are used for the water level fluctuations recording with regard to the graduated disk or float position movement detection. They can operate directly in the flow or in a stilling well to reduce the hydraulically inducted errors.
- Float-driven sensors. The operation principle is common to the already presented one. The difference is connected to the sensors detecting system and used strip-chart recorder.
- Pressure-actuated recording gauges (bubble gas system). Operate with regard to the principle that static pressure at a fixed point in the stream is directly proportional to the head of liquid above the point:

$$H = (P_{static} - P_{atm})C \tag{7}$$

where H is the water stage, P_{static} is the pressure in bar at the fixed point in the water column, P_{atm} is the atmospheric pressure in bar at the water column surface and C is a factor on the waters net weight.

o Ultrasonic gauges. Measurement is realized by high-frequency acoustic transducer that propagates a sound wave through the air to water surface. The reflected acoustic wave is received at the transducer, converted to an electrical signal and processed into a water surface stage. The time from the moment of emission of the pulse and the moment of replication of the echo by the sensor is directly proportional to the distance between the sensor and the water surface, and inversely proportional to the speed of the pulse in the air [3, 5]:

$$T = \frac{2d}{v} \tag{8}$$

where T is the time that passes from the moment of emission of the pulse and the moment of reception of the echo by the sensor, d is the distance between the sensor and the water surface and v is the speed of the pulse in the air.

Based on the presented principle for practical implementation are developed various equipments. Main challenge in their implementation is the appropriate monitoring point and monitoring network creation. This is connected to the following circumstances:

- main purpose of the measurements realization;
- available budget;
- local river hydraulic specifics.

Afterwards with regard to the results needs and implementation various type decision support system (DSS) can be created. This is connected to appropriate telemetry connections creation. With regard to the interest additional functions and interconnections can be also established. They can serve for preventive or mitigation measure needs with regard to the extreme events occurrences such floods or drought in particular, water scarcity detection, ecological flow maintenance, etc. and in case of temporary or permanent river and river floodplain structures creation. All mentioned events are closely connected to the river flow state, thus its proper quantification is of great importance. Such quantification is the basis for their proper past or future events characterization. But of especial importance is to become reliable and accurate data. This is achievable in case of measurement method, equipment and staff well balanced selection and following always must be taken into account:

- There is a huge variety of river discharge measurement methods and equipments. Part of most used has been upward briefly presented. Other methods also exist and can be considered in the future for more detail problems consideration;
- Each river discharge measurement method and related equipments have some advantages and disadvantages and a field of practical implementation with respect of the hydraulic properties of the flow;
- Each practice applicable discharge measurement method is characterized by the method accuracy that is connected to the uncertainty of the method, uncertainty of connected equipments and the accuracy and skills of connected staff;
- All measurement equipments and needed for their operation staff are cost and supplies consuming;
- Only the existence of good and appropriate measurement method and related technique does not make the correct measurement result. Reliable measurement results can be obtained only in case of proper river hydraulic conditions ensuring, correct river bed hydraulic parameter properties application into the estimation procedure, permanent monitoring point constant conditions maintenance, well qualified staff implementation.

Presented circumstances are appropriate both for urban and non-urban territories. In case of urban territories discharge measurement procedures are connected to one huge advantage. In most cases urban river bed conditions are modificated and exchanged by artificial one. In most cases they are stabile and with constant properties. But with regard to the real condition operation some variations into the connected hydraulic properties of the river bed is likely to occur. For illustration on fig. 5 it is presented a river channel segment during different seasons: on the Balkans and in many countries with a continental climate during the winter period with regard to the lowest temperatures partial or entire river bed ice covering can be observed and in spring, summer and autumn periods various vegetation occurrence can be also noticed. As a result the cross-sectional area, river bed gradient, river bed roughness, etc. are changed. This can significantly affect the final results accounting the attendance of presented hydraulic parameters in discharge estimation relations with regard to already presented principles. For example when selected measurement scheme is based on the Manning equation related discharge Q is estimated by [5-7, 10]:

$$Q = \frac{k}{n} \omega R_h^{2/3} S^{1/2} \tag{9}$$

where k is a constant dependent upon units, n represent the surface roughness (based on the channel material and conditions), ω is the cross-sectional area of flow, R_h is the hydraulic radius (cross-section area divided by the wetted perimeter and S is the slope of the channel at the point of measurement.

If the Chezy equation is used, discharge is determined by [5-7, 10]:

$$Q = C \omega R_h^{1/2} S^{1/2} \tag{10}$$

where C is the Chezy resistance coefficient.

With regard to the implemented measurement method and connected equipments presented hydraulic properties can significantly affect the final result thus following factors must be carefully appraised: surface roughness, vegetation, silting, obstructions, size and shape of the channel, seasonal changes, suspended material, bed load, flow depth, velocity and discharge.

a) b) c)

Fig. 5 River bed segment in urban conditions: a) winter period flow, with a partial ice covering; b) spring period flow with no vegetation; c) summer period flow with existence of a vegetation in the stream.

5 Urban territories discharge measurements and networks with regard to the national level monitoring systems and River basin management

Cities development connected both to the population growth and related society and infrastructure aspect certainly affects surface waters. This is especially connected to rivers, and small rivers in particular, in two ways: by increasing the surface that becomes impermeable because it is covered by buildings, roads, pavements and parking lots, and by introducing drainage systems in the storm that brings superficial water leakage from paved and covered surfaces, directly to the drainage channels. As a result urban environment appears to be a predilection for the location of risks, which leads to the recognition that urban development and risk development are indispensable. Consequently preserving, protecting and improving the quality of the environment and rivers are closely connected to the knowledge about the existing water state. River systems discharge determination is part of such procedures and play significant role in the water management programs creation and implementation on basin, sub-basin, sector, or particular aspects of water management. On national and river basin level in help of decision makers in the area with respect to the legal requirements surveillance and operative monitoring for selected water quality elements including water quantity is conducted, and in some cases additional investigative monitoring is also provided [13- 16]. Such measurements are implemented to ensure the coherent and comprehensive review of the status of waters within each river basin management district.

As it was already presented urban territories are very specific territories. They differ by size, population density, perspective for future development, available resources, including water resources. Connected to them hydrography strongly depend on local conditions. Some of territories are “reach” on waters thus other can be considered as “water scarce” territories. With respect to the proper identification of all related pressures and impacts with regard to climate and anthropogenic influence following is often considered:

- Urban territories borders rarely coincide with existing river basin districts. In some cases it is possible urban territories to cover parts of more than one watershed or river basin district;
- Urban territories cover huge areas with a high risk of significant water impacts connected to floods, droughts, water scarcity events, technogenic water risk events, impacts on river ecosystems, etc., directly connected to the related discharge state;
- Most of presented events are rapidly developed. For mitigation and preparedness purposes in many cases local measures implementation is the best guarantee for most effective decision obtaining;
- It is possible existing national level monitoring system data to be insufficient for the presented events characterization;
- Connected activities, involving the population in case of resulted events disasters, are carried out on municipal, district and national level [17-18];
- Only the good collaboration between municipal, district, national and water basin district level measures can guarantee best obtained results in case of water harmful impacts, extreme or risk events occurrences.

Accounting the existing knowledge about the water risk vulnerably territories [14-25] we find appropriate local urban water information or risk events preparedness systems creation. Such systems can be built on the base of real operating contact river discharge monitoring systems and expanded with regard to presented events characterization. For the purpose following methodological consistency is appropriate:

- o Analysis and investigation of the probable urban territory significant water impacts;
- o Identification of most vulnerable territories;
- o Analysis and selection of appropriate monitoring procedure;
- o Preliminary investigations about the potential monitoring sites;
- o Selection of appropriate monitoring points;
- o Selection of appropriate monitoring method, equipments and procedure for the event detection;
- o Monitoring network creation;
- o Establishment of a telemetry system with connections to all interested sides, including decision makers, key players, public.

This approach is proposed with regard to the practice applicable principles for discharge monitoring and DSS systems creation [3, 26-27]. It is appropriate for implementation also for discharge measurement network creation on sub-basin, basin and national levels.

6 Conclusion

Urban territories development is a complex and complicated process. It is connected to territorial occupation marked by the specialization of human activities, population diversity, which involves exchanges within urban areas and adjacent spaces. As a result significant impact on the environment and ecosystems is observed. To mitigate such challenges urban risk management becomes an appropriate solution. It tends to entrust the study of preventive solutions to health professionals in terms of danger. Concerning rivers and connected impacts resulting of floods, droughts, water scarcity events, technogenic water risk events, impacts on river ecosystems, etc. river discharge state play a significant role. Its well recognition is a key factor for such events characterization thus its proper determination becomes an important task. In help of professionals in the area and connected decision makers in the material briefly are presented some of most used methods applicable for contact measurements. They are considered with regard to the possibility of appropriate monitoring network creation. Accounting the adopted legislative requirements and measures with regard to river basin management and national monitoring requirements, and urban territories specifics a proposal for local urban water information or risk events preparedness systems is presented. For the purpose connected methodological frame is briefly presented accounting the importance and relevance of such tool both for decision makers, key players and society.

References

1. OECD, Urban population by city size indicator. doi: 10.1787/b4332f92-en (2017)
2. E. Monev, and M. Chilikova, Water Problems **32**, 37-50, (2002)
3. WMO No. 1044. MANUAL ON STREAM GAUGING. (2010). World Meteorological Organization. Geneva, Switzerland. ISBN 978-92-63-11044-2.
4. Herschy R. *Streamflow Measurement* (Elsevier, Cambridge, 1995)
5. M. Chilikova-Lubomirova, Metrological evaluation of the methods for discharge measurements and observation, PhD Thesis, IWP-BAS, Sofia (2009)
6. E. M. Wilson, *Engineering Hydrology* (The Maximilian Press Ltd., London,1980)
7. G. Jolankai, *Hydrometric Field Studies for Catchments Basin Exploration* (VITUKI, Budapest, 2002)
8. B. Marchinkov, *Hydrology* (Technika, Sofia,1973)
9. E. M. Shaw, *Hydrology in Practice* (Chapman &Hall, London, 1994)
10. E. Marinov, B. Kazakov, M. Maradjieva, and I. Minkov. *Hydraulics*, (VIAS, Sofia, 1994)
11. http://www.ott.com/products/
12. http://www.sontek.com/productsdetail.php?10-MHz-ADV-2
13. Directive 2000/60/EC of the European Parliament and of the Council establishing a framework for the Community action in the field of water policy, EC, 22.12.2000
14. Directive 2007/60/EC of the European Parliament and of the Council of 23 October 2007 on the assessment and management of flood risks (Text with EEA relevance). OJ L 2228/27, 6.11.2007
15. Water act. SG67/27.07.1999, last amend. S.G.61/11.08.2015
16. Water Law no. 107/1996 with subsequent amendments and completions
17. Disaster Protection Act. SG. 102/19 Dec 2006, last amend. SG. 80/14 Oct 2011
18. Government Emergency Ordinance no. 21/2004 on the National Emergency Situation Management System, approved by Law 15/2005
19. L. Gaňová, M. Zeleňáková, P. Purcz, D. C. Diaconu, T. Orfánus, Ž. Kuzevičová, Urbanism. Arhitectură. Construcţii **8** (4) 365-376 (2017)
20. G. N. Zaimes, and D. Emmanouloudis, J. of Engineering Science and Technology Review **5**(1) 77-82 (2012)
21. M. Zeleňáková, L. Gaňová, P. Purcz, M. Horský, L. Satrapa; P. Blišťan; and D. C. Diaconu, Nat. Hazards Rev. **18**(4): 1-15 (2017)
22. Flood Risk Management Plan 2016-2021 (2016) Danube River Basin Directorate, http://www.bd-dunav.org/content/upravlenie-na-vodite/upravlenie-na-riska-ot-navodneniia/plan-za-upravlenie-na-riska-ot-navodneniia/
23. Flood Risk Management Plan 2016-2021 (2016) East Aegean River Basin Directorate, http://earbd.org/indexdetails.php?menu_id=611
24. Flood Risk Management Plan 2016-2021 (2016) West Aegean River Basin Directorate, http://www.wabd.bg/index.php/2015-06-25-12-30-02/2016-2021

25. Flood Risk Management Plans 2016-2021. (2016) Black Sea River Basin Directorate, http://www.bsbd.org/bg/page_purn_bsbd.html
26. R. H. Sprague, MIS Q. **4** (4), 1-26 (1980)
27. M. Chilikova-Lubomirova, SGEM Conference proceedings, **17** (51) 321-328 (2017)

River Hydraulics during Flood Events: The Balkan Experiences

Mila Chilikova-Lubomirova[1,*] and *George* Zaimes[2]

[1]Institute of Mechanics – BAS, Fluid Mechanics Department, Acad. G. Bonchev Str., Block 4, 1113 Sofia, Bulgaria
[2] Eastern Macedonia and Thrace Institute of Technology (EMaTTECH), Dept. of Forestry and Natural Environment Management, Drama, 66100, GREECE

Abstract. Nowadays the appearances of water extremes are frequently observed mainly because of anthropogenic and climate change influences. These extremes cause various impacts on the environment. With respect to river systems, floods are such events that are characterized with increasingly untypical high water levels that overflow river beds. In many cases, they cause extremely harmful impacts to human societies, accompanied with serious environmental damages and modifications. This paper focus on the problems connected to flood appearances and some of their possible related impacts on river systems. To better understand these, phenomenon main specifics with regard to its occurrence, behavior and possible consequences are summarized. Most commonly used methods for flood assessments are presented. Some real case floods and their impacts on the river hydraulics from the Balkans are briefly presented. To better explain floods and the reason-outcome-mitigation nexus related recent required and adopted EC, Bulgarian and Greece policies and practices in the area are outlined concisely with regard to prevention measures in the form of early warning systems, river restoration etc. The presented information address the experts in regards to knowledge, practices dissemination and enhancement, and the connection between experts and responsible decision makers that deal with flood mitigation and prevention.

1 Introduction

An increase in flood occurrences signifies the importance mitigating the phenomenon. To mitigate floods effectively all aspects regarding its genesis and behavior, its temporal and spatial distribution, severity and interconnection in regards to negative impacts concerning the environment and ecosystems need to be accounted for. Once the above are determined for a specific case study, the most appropriate measures can be selected. The adopted

* Corresponding author: milasemail@yahoo.com

practices and policies of each country also play a key role in selecting the appropriate measures. When implementing measures both practices and policies need to be taken into account that will allow for the most efficient decisions to be taken and to lead to the most successful outcomes. To increase the awareness of these procedures while taking into account the importance of the local phenomenon features, this study will observe flood interconnection of the Balkans by emphasizing in Bulgaria and Greece. To achieve this some real cases studies from each country will be analyzed to better understand the casual connection between extreme events appearances and the related consequences. In addition the procedures, practices and legislation of flood management of both countries will be summarized. Of course this will be observed in the context of the EC adopted legislation such as the Water Directive 2000/60/EC [1] and the Flood Directive 2007/60/EC [2]. Taking into account the potential harmful results of floods, effective measures will be presented. In regards to the practices in the region the main focus will be on the possible prevention and mitigation measures in the context of early warning systems and river restoration practices implementation. This will be accomplished by clarifying the flood phenomenon, with regards to the developed processes nature, propagation and consequent impacts effects on the environment. Presented aspects will be examined from the river hydraulic side, considering it is the basis science in their physical processes development.

2 Floods genesis and characteristics

Floods are considered as an overflow of a large amount of water beyond its normal limits damaging normally dry land. They are a rapidly developed process evolved in a period of several minutes to several days, affecting different territories with regard to the local conditions, ranging from the mountains and plains to coastal areas, and from natural to urban areas. The events are temporally and spatially distributed irregularly. It is very hard to predict the exact place and severity of their occurrences although with the use of modern technologies this is becoming more efficient in large rivers but not as effective in small streams and flashy torrents.

Floods are generated by abnormal amounts of water. Its origin and state differs and depend on the local topography, ground cover, existing soil type, its physical properties and state, available artificial structures, etc. Typically, extreme precipitations are the main cause of such events. Another factor with a natural origin can be the sudden snowmelt. The two mention factors are also closely related to infiltration rates and when they are slow or negligible, as a result of the saturated or frozen ground, rock, artificial surface or road covering, they can significantly enhance flooding phenomena.

Floods are often associated with river systems (rivers, streams and torrents) where the flow rate exceeds the water transport capacity of the channel overflowing channel banks and flowing in to the adjacent floodplains and estuaries. Additional factors that contribute to flood appearances are the riverbed aggradation as observed in many cases by backwater effects. Aggradation is observed when the channel sediment volume is more than the flow can remove (sediment transport capacity). Sediment is therefore deposited into the channel making the existing river bed shallower, causing changes in the river channel gradient and decreasing flow velocities. Once the riverbanks height is reduced smaller discharge events can overflow them compared to past conditions. If substantial amounts of sediment are deposited in the channel, the river may be heavily modified and the river can create a new channel or divert to an old one to bypass the deposited sediment. Backwater effect occurs when the river flow meets obstructions that restrict its flow. Such effect can be observed in case of in-channel sediment deposits, river confluences, landslips of river banks, levees and dams, culverts, bridges, etc. Measures in regard to the natural or artificial structures e.g. lakes, reservoirs, wet zones, etc. can also lead to floods. Such events are connected to the exploitation and maintenance of river systems and related to artificial structures such as dams, levees, etc. and their control or emergency drainage. Natural disasters such as earthquakes or volcanic eruptions or severe artificial modification over the related existing structures areas can be a reason for flood events too. Floods caused by anthropogenic reasons should also not be ignored.

3 Floods propagation

Rivers are natural or man influenced geomorphologic formations that gather and transport all the watershed surface runoff and groundwater recharge to the river mouth. These processes are connected to the main river and its tributaries and are connected to natural or artificial structures such as lakes, reservoirs, wetlands, etc.

In a simple way, river systems physically can be considered as water and sediment transportation systems. With regard to the water (hydrological) cycle specifics, climate and anthropogenic impacts their flow rate fluctuate. The upper extremes of these fluctuations are considered as floods while the overall yearly normal flow is the baseflow. Observing the behavior of such extremes at a given point can be specified as the water amount that rises to a unique maximum (flood peak) and then recedes slowly. The progress of a flood event is presented on Fig. 1.

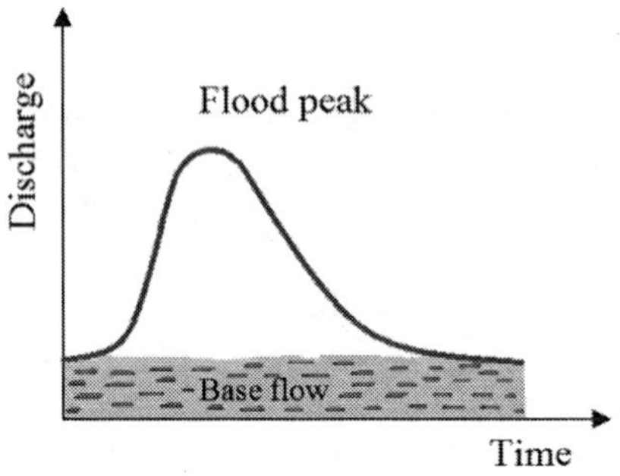

Fig. 1. Flood wave propagation at a certain point of the river

After the generation, flood is propagated as a wave along the river, its velocity and depth are continuously changing with time and distance with regard to the initial flood causes and the specifics of the river channel and/or floodplain area in case of an overflow.

3.1 Physical description of the river flood propagation and hydraulic specifics

Floods are temporally and spatially developed processes. Their physical behavior can be described by the main principle of hydrodynamics, utilizing the proper temporal-spatial scale and travelling time steps of the floods flow.

Flood water travelling downstream natural streams or artificial ones have typically unsteady flows. In case of significant precipitations, high snowmelt or floods that are created during the normal operation of the functioning and emptying of the hydraulic structures or canals, flow rises gradually. In regards to the basic hydraulic principles, river flood flows are considered as gradually varied unsteady flows, characterized by high temporal and spatial variation. Flood flow is three dimensional. It fluctuates with regard to the down-stream distance, depth and lateral position thus its specifics are closely connected to the river bed cross section properties and the channel hydraulic state. The flow surface profile also varies. It changes with regard to the downstream distance and lateral position.

Each fluid flow is a continuous phenomenon. A tree-dimensional description of the Eulerian concept is appropriate to study and understand flows. For this purpose the flow is observed as a control volume considering the behavior of its flow pressure, velocity, acceleration, and all other flow properties fields [3 - 9]:

$$\vec{v} = \vec{v}(x, y, z, t),\quad p = p(x, y, z),\quad T = T(x, y, z), \tag{1}$$

where $\vec{v}$ is the velocity vector, determined by the time t and space coordinates x, y, z; pressure is denoted by p and flow temperature by T.

In case of real fluid studies, connected to river flows and flood waves, the following considerations must be assumed [3-9]: a) water is an isotropic Newton fluid b) river and floods flows are considered as an incompressible fluid and c) the temperature variations during the processes are neglected, because they are very small.

For such studies the following main equations are used: Continuity equation (preservation of mass); Momentum equations (preservation of momentum); Energy equation (preservation of energy).

Based on the description of the viscosity fluid flow behavior of the Navier-Stokes equations are derived as [3 - 9]:

$$\begin{aligned}
\rho\left(\frac{\partial u}{\partial t}+u\frac{\partial u}{\partial x}+v\frac{\partial u}{\partial y}+w\frac{\partial u}{\partial z}\right) &= -\frac{\partial p}{\partial x}+\mu\left(\frac{\partial^2 u}{\partial x^2}+\frac{\partial^2 u}{\partial y^2}+\frac{\partial^2 u}{\partial z^2}\right)+\rho g_x,\\
\rho\left(\frac{\partial v}{\partial t}+u\frac{\partial v}{\partial x}+v\frac{\partial v}{\partial y}+w\frac{\partial v}{\partial z}\right) &= -\frac{\partial p}{\partial x}+\mu\left(\frac{\partial^2 v}{\partial x^2}+\frac{\partial^2 v}{\partial y^2}+\frac{\partial^2 v}{\partial z^2}\right)+\rho g_y,\\
\rho\left(\frac{\partial w}{\partial t}+u\frac{\partial w}{\partial x}+v\frac{\partial w}{\partial y}+w\frac{\partial w}{\partial z}\right) &= -\frac{\partial p}{\partial x}+\mu\left(\frac{\partial^2 w}{\partial x^2}+\frac{\partial^2 w}{\partial y^2}+\frac{\partial^2 w}{\partial z^2}\right)+\rho g_z,
\end{aligned} \tag{2}$$

where u, v, w, are the velocity components in regards to the coordinates x, y, z and time t; p is the pressure, ρ represents the density, μ represents the shear viscosity and $g = g(x, y, z)$ represents the acceleration, acting on the continuum.

The Navier-Stokes equations represent the balance of momentum. To estimate the flow description and practical solutions, additional data and information is needed. This information can be gain by the continuity equation and boundary data introduction.

Final solutions can be obtained within the help of numerical modeling and computer simulations with regard to the adopted assumptions and initial conditions. For this reason the following approaches are used in practice [4-11]: a) the method of characteristics b) the finite difference methods and c) the finite elements methods.

In the case of floods wave studies the length of propagation usually exceeds significantly the depth and width of the flow. In such conditions the more simple two or one dimensional models can be used. If we assume that the horizontal and vertical accelerations (and velocities) compared to the longitudinal ones can be neglected, a one dimensional investigation approach is applicable. Its theoretical basis follows:

The river floods wave, is described as a gradually varied unsteady flow, is characterized with significant variations on the flow depth and velocity thus the floods flow behavior can be described by the following [4 - 9]. Based on the Reynolds transport theorem, assuming water as a incompressible fluid the Continuity equation for a controlled volume can be presented as:

$$\frac{d}{dt}\int_{x_1}^{x_2} A dx + A_2 v_2 - A_1 v_1 , \tag{3}$$

where A is the flow area and v represents the flow velocity at sections 1 and 2, denoted by the subscripts respectively.

Application of the Leibritz`s theorem and the mean value theorem, assuming both A and $\frac{\partial A}{\partial t}$ continuous with respect to both x and t yields that:

$$\frac{\partial A}{\partial t}(x_2 - x_1) + A_2 v_2 - A_1 v_1 = 0 . \tag{4}$$

Assuming Av and $\frac{\partial vA}{\partial t}$ as continuous with respect to x and t, and letting $\Delta x = x_2 - x_1$ tend to zero gives:

$$\frac{\partial A}{\partial t} + \frac{\partial Av}{\partial x} = 0 . \tag{5}$$

Accounting that flow rate is expressed by:

$$Q = Av \tag{6}$$

it changes as following:

$$\frac{\partial A}{\partial t} + \frac{\partial Q}{\partial x} = 0 . \tag{7}$$

This equation represents the continuity equation in its *conservation form*.

For prismatic channels after some transformations equation (7) can be presented as:

$$\frac{\partial y}{\partial t} + \left(\frac{A}{B}\right)\frac{\partial v}{\partial x} + v\frac{\partial y}{\partial x} = 0 , \tag{8}$$

where B is the top surface width of the water.

Based on the Reynolds transport theorem the momentum equation for the control volume can be presented as:

$$F_{\mathrm{Res}} = \frac{d}{dt}\int_{x_1}^{x_2} v\rho A dx + \rho v_2^2 A_2 - \rho v_1^2 A_1 , \tag{9}$$

where F_{Res} is the resultant force acting on the control volume in the direction of flow.

After the application of both the Leibritz`s and mean value theorems (9) we get:

$$\frac{F_{\mathrm{Res}}}{\rho(\Delta x)} = \frac{\partial (Av)}{\partial t} + \frac{\partial}{\partial x}\left(Av^2\right). \tag{10}$$

Taking into account equation (6) we gave the following equation:

$$\frac{F_{\mathrm{Res}}}{\rho(\Delta x)} = \frac{\partial Q}{\partial t} + \frac{\partial}{\partial x}(Qv). \tag{11}$$

The resulting force F_{Res} on the controlled volume that is evaluated considering that: river channel is assumed prismatic therefore forces do not arise due to changes in the cross section; waves that are set up by the wind are not considered, therefore shear stress on the flow surface due to wind is neglected; open channel flow is considered. Flows in large water bodies such as estuaries and oceans are not considered here. Coriolis forces are also neglected. Net forces on the control volume comprise of: a) pressure force at section 1, b) pressure force at section 2, c) the component of weight of water in the flow direction and d) the friction force due to shear between water and the channel sides and the channel bottom. These forces are estimated as follows:

Pressure forces at section 1 and 2 are presented by the following equations:

$$F_1 = \rho g A_1 \overline{y_1} \tag{12}$$

and

$$F_2 = \rho g A_2 \overline{y_2}\,, \tag{13}$$

where $\overline{y_1}$ and $\overline{y_2}$ are depths to the centroid of A_1 and A_2, F_1 act in the positive direction thus F_2 acts in the negative direction.

The component of the water weight in the direction of flow acts in the positive x-direction and can be presented as:

$$F_3 = \rho g \int_{x_1}^{x_2} A S_0 dx\,. \tag{14}$$

Friction force is:

$$F_4 = \rho g \int_{x_1}^{x_2} A S_f dx\,, \tag{15}$$

where S_0 is the channel bottom slope and S_f is the friction slope. Friction slope may be estimated by using any friction loss equation such as the Manning equation. It acts in the negative x-direction.

The forces substitution in eq. (11) leads to:

$$\frac{g\left(A_1 \overline{y_1} - A_2 \overline{y_2}\right)}{\Delta x} + gA\left(S_0 - S_f\right) = \frac{\partial Q}{\partial t} + \frac{\partial}{\partial t}\left(Av^2\right) \tag{16}$$

or

$$\frac{\partial Q}{\partial t} + \frac{\partial}{\partial x}\left(Qv + gA\overline{y}\right) = gA\left(S_0 - S_f\right). \tag{17}$$

The final equation represents the momentum equation in the *conservation form*. Taking into account eq. (6) it can be presented as:

$$\frac{\partial Q}{\partial t} + \frac{\partial}{\partial x}\left(\frac{Q^2}{A}\right) + gA\overline{y} - gA\left(S_0 - S_f\right) = 0\,. \tag{18}$$

Both equations (7) and (18), are known as the as Saint Venant equations and are widely implemented for flood wave propagation assessments. The presented equations are of great interest for surface hydrology and flood phenomena studies. The differential equations are practically applicable to calculate surface flows, considering the boundary and initial conditions in regard to river flows. Numerical modeling in general and the numerical hydraulics in particular are used commonly in practice, and are becoming popular methods in hydrology when taking into account the existing anthropogenic influence and human intervention on the natural state of water. The implementation of these numerical methods is widely applied in various computer programs and assessment tools [12, 13].

3.2 Additional aspects connected to river floods characterization

River floods behavior is a complicated phenomenon. In order to characterize it properly, following aspects must be determined: a) floods wave origin and related initial flood parameters; b) the river bed and floodplain area specifics; c) floods wave discharge, time of appearance and propagation; and d) flood impacts.

Related data and information are used for the floods events and affected zones description and the event severity evaluation. For this purpose various types of initial data and information must be gained. Some of it can be gained by the implementation of direct measurements and monitoring, other can be obtained by additional assessments.

- Specification of floods origin

This is an important task connected to the related meteorological situation in the observed area. For this purpose the available precipitation data must be gathered and analyzed. Such data can be collected from the precipitation, weather and climatic monitoring stations by installing different types of rain gauges such as the weighing type, float type, tipping-bucket gauge, rainfall-intensity recorder, acoustic type gauges, radars (Doppler, multi-parameter, weather radars) or using satellites data [13-17].

Considering the particular river bed state and its hydraulic properties is also important. The conditions of the surrounding environment must also be considered – elevation, ground coverage - state and hydraulic parameters, existing structures and related possible floods conductive circumstances.

It is possible outer non natural reasons to be floods cause. For the purpose all floods connected factors must be considered.

- Floods wave quantification

This procedure deals with the event related water quantity determination and its time and spatial propagation. Floods wave quantity is the amount of water that passes during the occurrence of the event. It can be obtained conventionally for river discharge monitoring data or determined by other methods. Discharge measurements and observations are conducted in specific monitoring points. Taking into account the local specifics and related river bed hydraulic conditions, the most suitable discharge measurement method, measurement procedure and equipment are selected [13, 15-22]. The result of these types of measurements is flow rate for a specific point. In case of a continuous monitoring the river hydrograph can be obtained along with the discharge rate, volume and stage, time of beginning and time of duration of the event.

In practice discharge monitoring networks cover a number monitoring points and it is very likely the obtained measurements data to be insufficient to describe the overall flood event. In such cases the flood wave specification can be obtained by implementation of various modeling programs based on the already presented hydraulic theoretical base.

- Hydrological evaluations

When investigating the natural character of the flood phenomenon the focus is on the overflow events over a specific threshold followed afterwards by determining the recurrence intervals and probabilities of occurrences [13, 15-17, 23-25]. The floods are classified by the recurrence intervals, probabilities of occurrences and the related thresholds (see Table 1).

Recurrence intervals are based on the probability of the peak streamflow occurrence at a given location in any year. Initial information for such studies are data obtained from the river gauging stations. In case of data absence for ungauged sites, regionalized streamflow statistics methods are used. Most commonly used methods are the: a) index-flood procedure, b) ordinary-least-squares regression procedure, c) weighted- and generalized-least-squares regression procedure and d) region-of-influence procedure. Recently some methods using satellite images for floods events characterization are also developed. Of importance is to account that the flow at any particular point depends on local river bed and floodplain conditions with regard to the riverbed slope, floodplain shape, impoundments of streamflow, existing river regulation structures or reservoirs, water intakes and transfers, etc. At the locations where the stream channel widens or contains dense vegetation, the water velocity may decrease, the flood peak moves downstream, the water can overflow related floodplain where it can be stored until the water level begins to recede. Once the water level decreases, the stored water in the floodplain typically slowly re-enters the stream from the banks. As a result downstream flood flow peak may be modified and less in magnitude but longer in duration. These presented specifics must always be considered in case of floods wave propagation studies with regard to the most appropriate investigation scheme that should be selected.

Table 1. Flood recurrence intervals and probabilities of occurrences

Recurrence intervals and probabilities of occurrences			
Recurrence interval, in years	Probability of occurrence in any given year	Percent chance of occurrence in any given year	Annual exceedance percentage
100	1 in 100	1	1
50	1 in 50	2	0,50
25	1 in 25	4	0,25
10	1 in 10	10	0,10
5	1 in 5	20	0,05
2	1 in 2	50	0,02

3.3 Floods impacts. Balkan experiences

Floods are events that could occur every year, especially after large precipitation events or snowmelt. In natural conditions they are part of the river flow regime and the water cycle. Anthropogenic works and the operation or destruction of artificial structures can also cause floods. Severe nature phenomenon, such as earthquakes and volcanoes, or man caused environmental modifications, e.g. detonations or explosions, can also lead to floods. So some of the events are natural but others completely differ and would not appear in "normal" river and floodplain conditions. The result of all presented causes depends on the floods severity and local conditions. Overall, flood impacts always must be studied at a local scale.

Floods impacts can have a positive impact, considering that the floodplain is watered and enrichment by the minerals from rivers sediments. They can also have a negative impact, considering the severe impacts and possible damages they cause. The following results after a flood on the affected river channel and surrounding floodplain areas can be observed:

- rivers aquatic and related terrestrial ecosystems disturbance;
- river banks erosion and landslides;
- sediment introduction by landslides and downstream aggradation;
- river bed hydraulics modification;

- existing artificial structure damage or collapse (bridges, levees, roads, etc.);
- artificial dam failure;
- local areas isolation.

Additionally floods flow can cause damages on the surrounding environment and forests, on the pastures, reduce agricultural production, fences, farm buildings and houses, equipment and animals and urban areas (existing buildings and houses, roads, electricity, water supply and sewage infrastructure, communications, etc.). Some other related impacts are: evacuation of people and animals, people and animals victims, social disruption and financial costs. Their range closely depends on the floods severity, thus the proper event characterization must account for them in the related assessment process.

To illustrate the above mentioned points flood event cases from Bulgaria and Greece are presented. These were selected to illustrate different aspects of the floods phenomenon and related consequences.

3.3.1 Batulijska River: Bulgarian Case Study

The flood event on Batulijska River that occurred on August 1st 2014 was selected. Purposefully a large river was not selected. The idea was to present how flood event caused by extremely heavy rain precipitations, impact very seriously the environment at a medium scale to observed river.

Batulijska River is a tributary to the Iskar River. It runs near Murgash hut, situated in Murgash Mountain, part of Stara Planina Mountain and flows to north-west. After passing the Batulija village it meets Iskar River (Fig. 2).

Fig. 2. The location of Batukijska River.

The river length is 40.2 km and the watershed area is 256 km^2. Its annual average flow near Batulija village is 2.7 m^3/s. Four villages are situated downstream the river: Jablanitza, Bakjovo, Batulija, Rebrovo. In Fig. 3 the “normal” flow conditions are illustrated.

The heavy rainfall resulted in the Batulijska River streamflow significantly increasing in several hours. In some places the water level raised up to 1 m. Large areas were flooded including river banks, streets, houses, whole section of villages. Some infrastructure was destroyed such as roads, bridges, electric infrastructure while water supply was interrupted. On August 1st 2014 a state of natural disaster was declared in the area for the villages: Vlado Trichkov, Lukovo, Rebrovo, Batulija, Bakjovo, Ogoja, Jablanitza, Bukovetz, Tompsan, Redina and Leskov dol.

a) b)

Fig. 3. Batukijska River: a) River streamflow near Jablanitza village during the summer, b) Batulijska River close to the point, where it joins Iskar River during summer flow conditions.

To better understand the process of the flood event and some of related impacts, images are illustrated in Fig. 4-7.

Fig. 4. The Batulijska River flood event in Rebrovo village zone

b)

a)

c)

Fig. 5. The Batulijska River flood wave development and its impacts: a) stream water rising in a river segment, b) and c) the same river segment after the flood event – some impacts are evident.

Fig. 6. The Batulija village bridge after the flood event

Fig. 7. Flood event damages by the Leskovska River in Leskov dol village, on August 1st 2014 that is near the Batulijska River.

These images in Fig. 4-7 illustrate the flood wave time development by showing some of the consequences. The aim is to describe the floods influences on river hydraulics. It is clear that floods can seriously impacts the river bed since after the event it has been totally modified. The significant impacts on the river hydraulics can also be detected. River banks are eroded, huge amounts of the stream load material were deposited downstream, related river bed vegetation is totally destroyed, the river bed profile is changed to a new one with totally different hydraulic properties (cross-sectional area, river bed gradient, river bed roughness, etc.). During the flood some river related and floodplain situated structures and buildings were destroyed. It is possible such structures to lead to a backwater effect and cause additional flood wave rising. The sediment and debris that were transported downstream by the streamflow were either deposited on the specific floodplain area or stopped by some other existing permanent structures. During the process other obstructive occurrences are likely to have been observed causing additional damages. As a result the related flood area was also heavily modified. Consequently the connected natural aquatic and terrestrial ecosystems were also impacted.

The presented images (Fig. 4-7) are to illustrate the changes that occur in the river bed hydraulics and related hydraulic parameters after a flood event. These are of great importance considering that the various parameters are related to the quantitative measurements and assessments with regard to the water level, streamflow rate, etc.

3.3.2 Evros/Marista River: Greek Case Study1

The plains that are located in the downstream section or traversed by large rivers (such as Evros, Axios, Nestos, Aliakmon and Peneios) typically face floods impacts in Greece and the Balkans [26]. The discharge volumes of the large rivers often exceed the capacity of the river channels. In this case study, the large floods of the Evros/Marista River were investigated. The total length of the river is 528 km with 310 km in Bulgaria and 208 km determining the borders between Greece and Turkey. In regards to the basin 66.2 % belongs to Bulgaria, 37.5 % to Turkey and 6.3% to Greece [27]. The transboundary nature of the river indicates that only with collaboration of the riparian countries can effective flood mitigation can be achieved.

The last major flood event that started in December 2014 and lasted till May 2015 (Fig. 8). This flood caused 54 houses to be evacuated and 23,000 ha of agricultural fields where flooded because the river levees were broken at 84 points. The flood had a devastating effect on the local economy that depends primarily on agriculture. Major floods have also occurred along the Evros River in 1996, 1997, 1998, 2005, 2006, 2010 and 2012 that indicates that is a repeated problem. Typically these floods and flood waves are more predictable although the overall flood events can last for long periods of time. Still despite being more predictable these problem continues.

Fig. 8. Breaking of the levees near Sufli led to flooding of the agricultural and rural areas (05/02/2015) that originated from Evros River (source: http://papaioannou-giannis.net).

3.3.3 Attica Flash Floods: Greek Case Study 2

Another major category of floods in the Balkans is within the urbanized areas [26]. More than 40% of flood-related casualties reported in Europe between 1950 and 2006 were from flash floods [28]. In Greece, Athens, Thessaloniki, Patras and Heraklion are located in coastal areas and had above-average flood frequency and an above average number of events [26]. The significant increase in impermeable surfaces in urban catchments as the population exponentially increased in these centers has altered the hydrologic regime (Fig. 9). Infiltration decreases with surface runoff increasing and leading to peak flows (Fig. 1). These types of flood events appear to be on the rise because of the high population density in urban areas, the associated reduction of natural retention areas and the lack of land-use regulations on construction works that are allowed adjacent to the river channels [29].

Fig. 9. Urbanization has altered the hydrologic regime and cause flash floods in these environments (source: http://www.zougla.gr).

Focusing on the prefect of Attica in Greece it had 95 flood events during 1887-2010 that translated in 1 per 40 km^2 that was the highest number of flood events per area in the Greece [26]. In addition, this district had 1 flood event per 13 years that was again the highest frequency in Greece for the same period. The most recently flash floods in Athens were in 27/11/2016 and 04/06/2017. The high recurrence and number of flash floods per area is related to rapid with no planning and widespread housing developments. In addition, the mitigation strategy for the prefect seems to be reactive instead of proactive and preventive with little organization and coordination that is unlikely to stop flash floods [30]. At the same responsible authorities appears to not take under consideration the anticipated increase in intensity and rapidity of rainfalls due to climate change at the Attica basin. Large forest fires that frequently happen in the Prefect increase erosion and surface to further enhance flash floods.

The characteristics of these floods are the highly irregular stream flow that can be from zero (no flow) to overflowing the channel in hours. Typically the channels are torrents a characteristics type of water courses in the Mediterranean [31]. While the flooding lasts for a very short period of time, the flood wave can develop very fats and cause major erosional and depositional and because they are difficult to predict can be very catastrophic and lead to losses of human lives.

4 Prevention and preparedness measures

To decrease flood impacts and societies vulnerability various measures and tools have been developed. These measures and tools can be separated in two main categories: proactive and mitigation.

Proactive measures aim to characterize flood events in advance in order to be able help direct the mitigation measures. They are two main types – flood watch and early warning systems [24, 32-33]. Such systems deal with the detecting, measuring and determining the phenomenon and its parameters and/or forecasting it. In regards to the starting period and time of occurrence of the phenomenon the following forecasting methods temporal scales are applicable: a) real time (few minutes up to 3 hrs before), b) short-term (2 hrs to 5 days before), c) mid-term forecasts (2 to 10 days before) and d) long term (more than 10 days before). In addition, seasonal forecast for several months while ultra-long term forecasts are for up to year or even greater periods. The ultra long forecasts are connected more to the climate change effect studies. They are not directly connected to the early warning systems for flooding.

Other measures that deal directly with the floods physics and its propagation processes are connected to the river systems condition maintenance. This must include the entire river channel and floodplain with respect to: natural rivers, channelization and straightening, longitudinal constructions (dykes and levees), bank reinforcement and embankments, cross-profile structures (dams, weirs, locks/sluices, culverts, and impoundments), etc. The well-maintaining of their physical state and hydraulics conditions is connected to the river bed hydraulics and channel morphometry state with respect to dredging and mineral extractions, land drainage and sealing, etc.

The importance of mitigating flood events is also evident in the large amount of legislation and administrative works that are currently enforced and elaborated. For the EU Countries the flooding issues has been addressed mainly with two Directives. Specifically the first was the Water Framework Directive 2000/60/EC [1] that addressed flooding issues indirectly. The Flood Directive 2007/60/EC [2] focus on the assessment and management of flood risks. Of course there are additional documents that address this problem [34-37]. Respectively flood related problems are also considered in the national legislation too [38-40]. As a result, complex related documents and measures are elaborated in the context of the River Basin Management Plans and their second stage, Flood Risk Management Plans, Flood Risk Maps, etc. [41-48]. In Bulgaria floods related specifics were considered for the whole country waters with regard to the River basin management plans, 2016-2021 and Flood risk management plans [41-44] prepared by the four River basin directorates. In Greece, the identification of the highly vulnerable to flooding areas has been completed in the 14 Water Districts of the country [45]. All presented documents are prepared following the scientific and expert works in the area, considering local problems and national specifics.

In order for the above legislations and regulations to be more effective implementation at the national, regional and local level taking into to consideration the local watershed specifics and vulnerability assessments and developing the appropriate structural and non-structural measures. To achieve this, preliminary field investigations at the specific vulnerable reaches must be done. The measures that are implemented need to be based on the hydro-morphology of the river reaches and the related floodplain. In regards to rivers the main activities, are inspection and maintenance of the existing channel and adjacent infrastructure, and implementing restoration measures when needed. These measures can include river channel and bank maintenance, such as removing the riparian debris, growing trees and shrubs and deposited sediment, preventing erosion from the river banks, trimming the grass and vegetation, clearing banks from obstructions, providing adequate sediment capacity for the downstream flow, estimating ecological flows, constructing protective measures for the existing structures, etc. In case the flood event creates a new channel (alluvial deposit, existing river meanders rupture, etc.) it is possible to restore the original river channel. Some natural water retention measures are also applicable [46]. All presented measures need to be implemented with respect to the local ecosystems well being, thus additional measures such as the ecological river flows should be considered.

5. Conclusion

Floods are episodic events that occur by natural or anthropogenic reasons. Their nature follows the main principles of hydrodynamics. To better understand such events the main triggers connected to the flood occurrence, behavior and characteristics and the applicable methods need to be determined. In addition, the theoretical basis for the phenomenon propagation in time and space and main methods that are used were presented. The focus was on applicable practical systems for flood watch and early warning. Accounting the importance of recognizing the river hydraulic parameters when implementing assessment methods, a special attention on flood impacts on the river channel and the related floodplain hydraulics should be taken. The main problems caused by floods are summarized and to enhance clarification, some real cases from the Balkans, specifically from Bulgarian and Greece are presented. Recognizing the phenomenon's importance and its social impacts are also related to the legislative and administrative aspects. A better interconnection between flood regulation and legislation and applicable practical solutions is needed and should be obtained at the national, regional and local scale. Another important aspect to mitigate floods is river maintenance and restoration. Finally, state of the art methods should be implemented by experts in the field while taking into account the appropriate policy and decision making. Such material can help develop early warning and decision support systems that enhance flood mitigation by improving preventive and preparedness measures.

Acknowledgement

We want to express our gratitude to all colleagues from the Institute of Mechanics – BAS for their support and assistance in the article preparation and good partnership creation.

References

1. Directive 2000/60/EC of the European Parliament and of the Council establishing a framework for the Community action in the field of water policy, EC, 22.12.2000;
2. Directive 2007/60/EC of the European Parliament and of the Council of 23 October 2007 on the assessment and management of flood risks (Text with EEA relevance). OJ L 2228/27, 6.11.2007
3. G. K. Batchelor, *An introduction to fluid dynamics* (Cambridge University Press, (2000)
4. R. W. Jeppeson, *Simulation of Steady and Insteady Flows in Channels and Rivers*. (1974). Reports. Paper 301. http://digitalcommons.usu.edu/water_rep/301
5. I. Popesku, *Computational Hydraulics. Numerical Methods and Modelling* (IWA Publishing, 2014).
6. M. Feistauer, *Theory and Numerics for Problems of Fluid Dynamics* (Charles University, Prague, 2006)
7. H. Lamb, *Hydrodynamics* (Cambridge Univercity Press, 6th ed., 1994)
8. Y. Nakayama, and R. F. Boucher, *Introduction to Fluid Mechanics* (Butterworth-Heinemann Linacre House, member of the Reed Elsevier plc group, 2000)
9. E. Marinov E., B. Kazakov, M. Maradjieva, and I. Minkov, *Hydraulis* (VIAS, Sofia, 1994)
10. O. Gunduz, *Numerical Solution of Flow Equations*. ENV5056 Numerical Modelling of Flow and Contaminant Transport in Rivers.
11. Th. M. Walski, Th. E. Barnard, S. R. Durrans, and M. E. Meadows, *Computer Applications in Hydraulic Engineering* (Heisted Methods Inc., Waterbury-USA, 1997-2002)
12. M. S. Horritt and P. D. Bates, J. of Hydrology **268**, 87-99. Elsevier (2002)
13. WMO No. 1044 *Manual on stream gauging. Vol. I Fieldwork.* (WMO, G., Switzerland, 2010)
14. NASA. Global Precipitation Measurement. https://pmm.nasa.gov/science
15. E. M. Shaw, *Hydrology in Practice* (Chapman &Hall, L. 1994)
16. E. M. Wilson, *Engineering Hydrology* (The Maximilian Press Ltd., L., 1980)
17. B. Marchinkov, *Hydrology* (Technika, Sofia, 1973)
18. R. Herschy, *Streamflow Measurement* (Elsevier Applied Science Publishers Ltd., Cambridge, 1995)
19. M. Chilikova-Lubomirova, 2009. Metrological evaluation of the methods for discharge measurements and observation, PhD Thesis. IWP-BAS. Sofia.
20. V. D. Bykov, *Hydrometrics* (Hydrometrical edt., Leningrad, 1949)
21. G. Jolankai, *Hydrometric Field Studies for Catchments Basin Exploration* (VITUKI, Budapest, 2002)
22. Str. Gerasimov, *Methods for analysis and evaluation of the maximum river discharge*. (NIMH, Sofia, 1988)
23. R. R. Jr. Holmes and K. Dinicola. *100-Year Flood – It`s all about Chance. Haven`t we already had one this* century? (USGS, USA, 2010)
24. J. Szolgay, K. Hlavcova, S. Kohnova and G. Balint. *Hydrological modelling and forecasting. Handouts for lectures* (VITUKI, Budapest, 2001)
25. E. Monev and M. Chilikova-Lubomirova, *Flood risk – hydrotechnical aspect. Research project* (IWP-BAS, Sofia, 2008)
26. K. Karagiorgos, Fuchs, S., Thaler, T., Chiari, M., Maris and F., Hübl, J., A flood hazard database for Greece. Wildbach Lawinenverbau 170, 264–277 (2013)
27. K. Chouvardas K. and Ch. Papapostolou, Geografies **27**, 44-71 (2016)
28. J. I. Barredo, Nat. Hazards **42** (1) 125–148 (2007)

29. J. Ganoulis, Int. J. of River Basin Management **1**, 1-41 (2003)
30. O. Lasda, A. Dikou, and E. Papapanagiotou, Ambio, Springer **39** (8) 608–611 (2010)
31. D. Emmanouloudis, J. L. García Rodríguez, G. N. Zaimes, M. C. Giménez Suárez, and E. Filippidis, K.E. Richards (Ed.) *Euro-Mediterranean torrents: Case studies on tools that can improve their management, In: Mountain Ecosystems: Dynamics, Management and Conservation* (Nova Science Publishers, Inc, Hauppauge New York, USA, 1-44, 2011)
32. R. Sprague, MIS Quarterly. **4** (4), 1-26 (1980)
33. M. Chilikova-Lubomirova, SGEM Conference proceedings, 3 (1), 621-628 (2016)
34. COM (2004) 472 final. COMMUNICATION FROM THE COMMISSION TO THE COUNCIL, THE EUROPEAN PARLIAMENT, THE EUROPEAN ECONOMIC AND SOCIAL COMMITTEE AND THE COMMITTEE ON THE REGIONS. Flood risk management. Flood prevention, protection and mitigation. Brussels, 12.07.2007
35. TR-2009-040. CIS FOR THE WATER FRAMEWORK DIRECTIVE (2000/60/ec). Guidance document No. 24. RIVER BASIN MANAGEMENT IN A CHANGING CLIMATE. EC. 2009
36. TR – 2013-071 Guidance for Reporting under the Floods Directive (2007/60/EC), EC, 2013
37. TR-2014-082, EU policy document on Natural Water Retention Measures, EU, 2014
38. Water act. SG67/27.07.1999, last amend. S.G.61/11.08.2015
39. Environment protection Law (latest amendm. SG No. 98/28.11.2014)
40. Gazetter of the Hellenic Republic Government. 2nd Issue, No. 1108. Published July 21, 2010
41. Flood Risk Management Plan 2016-2021 (2016) Danube River Basin Directorate, http://www.bd-dunav.org/content/upravlenie-na-vodite/upravlenie-na-riska-ot-navodneniia/plan-za-upravlenie-na-riska-ot-navodneniia/
42. Flood Risk Management Plan 2016-2021 (2016) East Aegean River Basin Directorate, http://earbd.org/indexdetails.php?menu_id=611
43. Flood Risk Management Plan 2016-2021 (2016) West Aegean River Basin Directorate, http://www.wabd.bg/index.php/2015-06-25-12-30-02/2016-2021
44. Flood Risk Management Plans 2016-2021. (2016) Black Sea River Basin Directorate, http://www.bsbd.org/bg/page_purn_bsbd.html
45. Greek Ministry Of Environment, Energy And Climate Change. Special Secretariat Of Waters. 2012. Implementation Of Directive 2007/60 /EC. Preliminary Assessment of Flood Risk. Athens, GREECE
46. M. Zelenakova, D. Diaconu and K. Haarstad. ScienceDirect **190**, Elsevier Ltd., 420-426 (2017)

Experimental and numerical investigation of plasma parameters in the magnetosheath

Polya Dobreva[1,*], *Monio* Kartalev[1], *Olga* Nitcheva[1], *Natalia* Borodkova[2], and *Georgy* Zastenker[2]

[1]Institute of Mechanics, Bulgarian Academy of Sciences, Acad. G. Bonchev Str., block 4, Sofia, Bulgaria
[2]Space Research Institute, Russian Academy of Sciences, Profsoyuznaya Str., 84/32, 117997 Moscow, Russia

Abstract. We investigate the behaviour of the plasma parameters in the magnetosheath in a case when Interball-1 satellite stayed in the magnetosheath, crossing the tail magnetopause. In our analysis we apply the numerical magnetosheath-magnetosphere model as a theoretical tool. The bow shock and the magnetopause are self-consistently determined in the process of the solution. The flow in the magnetosheath is governed by the Euler equations of compressible ideal gas. The magnetic field in the magnetosphere is calculated by a variant of the Tsyganenko model, modified to account for an asymmetric magnetopause. Also, the magnetopause currents in Tsyganenko model are replaced by numericaly calulated ones. Measurements from WIND spacecraft are used as a solar wind monitor. The results demonstrate a good agreement between the model-calculated and measured values of the parameters under investigation.

1 Introduction

The Earth magnetosheath is a transitional region between the bow shock and the magnetopause in which the solar plasma is heated, decelerated and compressed. It plays a crucial role in solar-terrestrial relations, because the solar wind energy is transmitted through it and propagates to the magnetosphere and ionosphere. That is why it is important to understand how characteristics of the interplanetary medium are modified in this transitional region. The picture is further complicated by the presence of a variety of structures and waves within a wide frequency, inherent to the magnetosheath itself.

The first gas-dynamic description of the magnetosheath was given in the pioneer model of Spreiter [1]. The model presents two-dimensional picture of flow properties in the magnetosheath for given input data, which are the solar wind Mach number, polytropic index and a predetermined magnetopause position. Later, the model was further developed to calculate the magnetic field in the magnetosheath [2]. Although the model takes into account the solar wind variations in some extend, it was not adapted to practical purposes and has some serious shortcomings. For example, it does not allow for the north-south and

* Corresponding author: polya2006@yahoo.com

dawn-dusk asymmetries in bow shock and magnetopause geometries and does not take into account the indentations around the cusp area at the magnetopause [3].

The numerical magnetosheath-magnetosphere model provides a three-dimensional description of the plasma flow in the magnetosheath, based on the gas-dynamics. The magnetosheath boundaries are self-consistently determined in the process of solution. The model takes into account the cusp indentations at the magnetopause, the dawn-dusk and north-south displacement at the bow sock (BS) and magnetopause (MP) positions in the tail. Moreover, the model is relatively simple, it does not require extensive computing resources and is executed on a personal computer.

The model was already applied in interpretation of Interball-1 measurements [4, 5, 6]. The main parameter under investigation was the ion flux, which values were compared in several events, when the satellite crossed the magnetosheath.

The aim of this paper is to investigate the behaviour of the plasma parameters in a case when the satellite stayed in the magnetosheath, crossing the magnetopause only. Parameters to be considered are the ion flux and velocity along the Interball-1 orbit. On January 16-18, 1997 Interball-1 was in the northern tail at about -15 R_E (Earth radii) moving southward and it stayed in the magnetosheath for about 32 hours, several times crossing the magnetopause.

Short description of the model is presented in Section 2. The necessary solar wind data are described in Section 3.1, the magnetosheath ones – in Section 3.2, and the results of the comparison are given in Section 4.

2 Brief description of the numerical magnetosheath-magnetosphere model

The numerical magnetosheath-magnetosphere model is based on the modular approach, that allows separate model to be applied in the each of the two interconnected regions. The relation between these separate regions is established by the conditions, imposed at the boundaries.

The magnetosheath description is based on the gas-dynamic approximation, as we assume that the flow is compressible ideal gas, governed by the Euler's equations. As a result of the simulation the velocity, density and temperature in the magnetosheath are obtained and as a consequence we calculate the gasdynamic pressure and the Mach number distributions. The numerical scheme [7], which is a variant of the method of characteristics, is applied in solving the system of equations.

The magnetic field in the magnetosphere is sum of several contributions: filed of the Earth's dipole, the field, due to the tail, Birkeland and ring currents and the field of the Chapman-Ferraro currents. Our model allows for the calculation of the magnetopause currents, confining the magnetic field into the magnetosphere with an arbitrary geometry. The field of the external sources – the ring, Birkeland and tail currents, is calculated by a variant of Tsyganenko model – T95 [8] or T01 [9]. The Tsyganenko model is modified as the shape of its magnetopause is not used, instead, it is to be found in the process of simulation. Besides, the magnetopause currents, related to that particular form of the MP are replaced by ones, calculated via the finite element method. The penetration of magnetic field across the magnetopause is not allowed and B_n=0.

The interaction between these two models is achieved through the boundary conditions. The Rankine-Hugoniot relations, expressing the conservation of mass, energy and momentum are imposed at the magnetopause and the bow shock. At the points of the magnetopause these conditions are reduced to the equation of pressure balance. The method applied here is capable of describing the real size and geometry of the bow shock and the magnetopause, including the cusp indentations at the magnetopause, the north-sought and dawn-dusk asymmetries in the tail magnetosheath boundaries. Moreover, there is no need to rescale the trajectory of the satellite passage through magnetosheath in order to coincide with the model boundaries.

Input data, needed for the calculations are: the density, velocity and temperature of the solar wind; B_y and B_z components of the IMF (Interplanetary Magnetic Field); Dst index (based on the Kyoto Web page); the dipole tilt angle; polytropic index γ. Description of the magnetosheath properties is based on the single fluid approach, as the density and velocity are that of the ions, but the temperature is a sum of the ion and electron temperatures. A more detailed description of the numerical model, used here, can be found in some previous papers [4, 10, 11]

3 Satellite measurements, used in the study

3.1 Solar wind data

On January 16-18, 1997 WIND spacecraft [12] was situated in the solar wind at about 130-136 R_E from the Earth. Here, we use WIND data as a solar wind monitor. The propagation time of the solar wind from WIND to Interball-1 is in the range 48~55 min. Fig. 1 and Fig. 2 present the input plasma and magnetic field data. The proton number density Fig. 1a) gradually increases from 8 sm^{-3} at the beginning of the interval, reaches its peak of about 28 sm^{-3} at 20:50 UT January 17 then decreases to 12 sm^{-3} after 04:00 UT January 18. The velocity varies in the range -300~-350 km/s – Fig. 1b). The temperature rises after 00:00 UT on January 18, as electron temperature increases twice and proton temperature - more than twice – Fig. 1c).

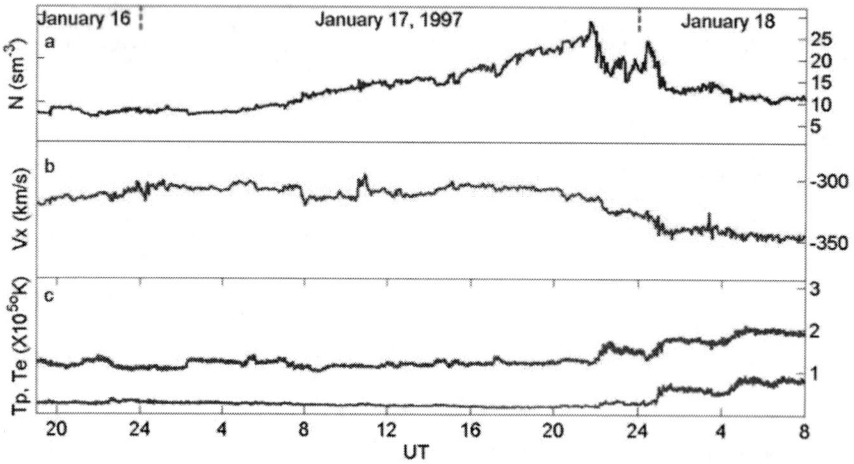

Fig. 1. Input plasma parameters, measured by the WIND spacecraft on January 16-18, 1997: a) proton number density; b) V_x component of the velocity; c) proton temperature T_i (gray) and electron temperature T_e (green). Data are not shifted in time.

The variations of the IMF, based on the WIND spacecraft are presented in Fig. 2. The input magnetic field is northward during the interval 20:00 UT January 16 to 08:00 UT January 17, then it is duskward and mainly southward during the interval [08:00-20:00 UT] on January 17. After that, it changes its direction several times and is strongly dawnward at a four-hour interval [01:00-05:00 UT] January 18.

The data of solar wind density, velocity and electron temperature are based on WIND/SWE (Solar Wind Experiment) instrument [13], the proton temperature - on WIND/3DP (3D Plasma Analyzer) – [14]. The IMF data are measured by WIND/MFI (Magnetic Field Investigator) device [15]. The time resolution is 1.5 min for the density and velocity, 12 sec for the electron temperature, 3s for the proton temperature and 1 min for the magnetic field. However, for our purposes the temperature data are averaged over 1 min intervals.

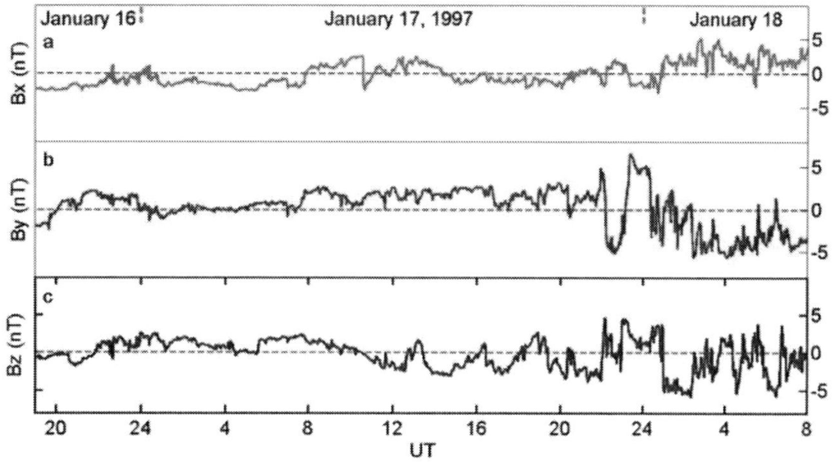

Fig. 2. a), b), c) B_x, B_y and B_z components of the Interplanetary Magnetic Field measured by the WIND spacecraft on January 16-18, 1997. Data are plotted in GSE (Geocentric Solar Ecliptic) system of coordinates and are not shifted in time.

3.2 Magnetosheath data

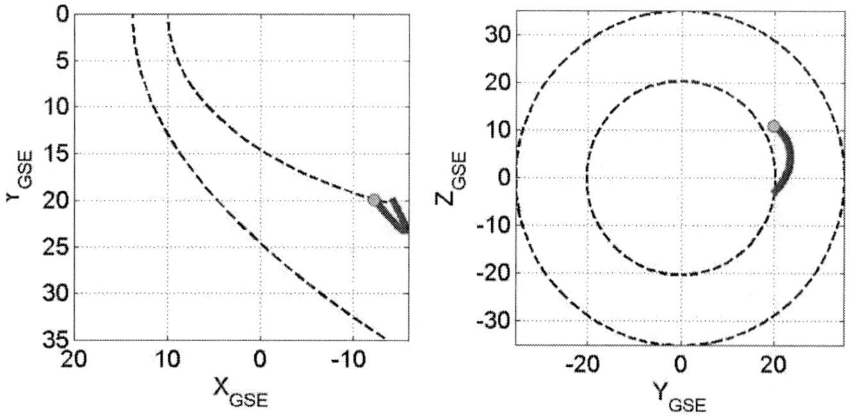

Fig. 3. Interball-1 trajectory from 19:00 UT January 16 to 08:00 UT January 18, 1997. The projections shown are on (X, Y, Z=0) and (X=-14, Y, Z) GSE coordinate planes (the distances are in Earth radii). The inner dashed curve represents the magnetopause of Shue 97 [20] calculated with parameters - dynamic pressure P_d= 2 nPa and IMF B_z=0 nT. The outer dashed curve is the shape of the bow shock, calculated by our model for the given Shue form, mentioned above. The point denotes the beginning of the interval.

Interball-1 is a high apogee satellite, part of the Interball Project [16], which primary objective was to study the solar–terrestrial relations. It was in operation from August 1995 until October 2000 and moved in strictly elliptical orbit with an apogee of about 195000 km and a perigee in the range 500-10000 km – [17]. Measurements of the ion flux are based on the Omnidirectional plasma sensor VDP - [18]. The device consisted of four integral Faraday cups - C1, C2, C3 and C4. The C1 sensor was oriented along the principal axis of the satellite, which deviated up to 10° from the Sun-Earth line, thus collecting the flux coming from the Sun. Directions of the other three sensors –C2, C3 and C4 were perpendicular to this axis (see the device description [18]). The time resolution of the Interball-1 data was 1s, but for our purposes the flux measurements were averaged over 1 min intervals. The velocity data are based on the CORALL device (more details are given in [19]) and have a time resolution of 2 min. During February – September of each year the satellite completed an orbit in four days period, meanwhile crossing the magnetosheath twice.

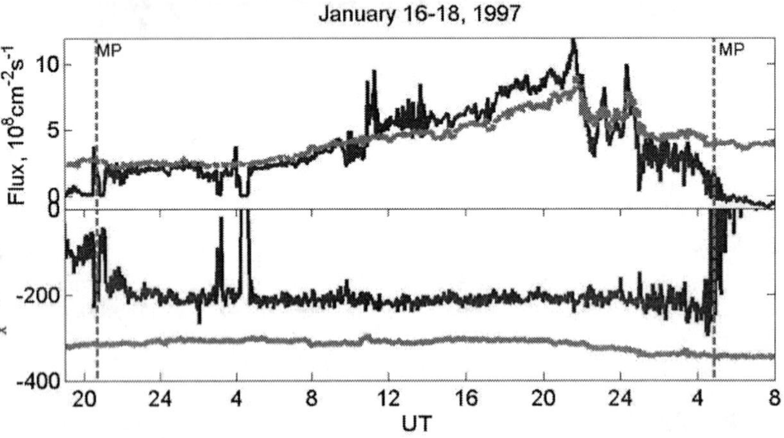

Fig. 4. Ion flux, measured onboard Interball-1 (blue) and onboard WIND (gray) satellites on January 16-18, 1997. Here the MP labels depict the moments of Interball-1 crossings of the magnetopause. WIND data are shifted appropriately.

Interball-1 orbit during the interval 19:00 UT January 16 to 08:00 UT on January 18 is presented in Fig. 3. The satellite is in the dusk sector at about -13~-15 R_E in the tail, moved from north to south, as during the second half of the interval it is close to the equatorial GSE plane. In the interval, we observe multiple magnetopause crossings and the satellite stayed in the magnetosheath for about 32 hours without crossing the bow shock.

The ion flux and ion velocity (both in x direction), conducted onboard Interball-1 satellite, are given in Fig. 4. The corresponding solar wind values, shifted appropriately, are also added in the plot. The satellite first entered the magnetosheath at 20:30 UT January 16 and finally left at 4:53 UT January 18. The flux drop and the corresponding velocity peaks, observed during the intervals [20:47-21:05 UT] on January 16, [03:00-03:13 UT] and [04:13-04:40 UT] on January 17 are connected with the satellite's re-entering in the magnetosphere. During the interval [04:13-04:53 UT] January 18 the device encountered both magnetosheath and magnetospheric plasma. The peak of the flux is associated with the solar wind flux increase, and before 10:00 UT January 17 and after 01:00 UT January 18 the flux in the magnetosheath is lower that the solar wind one.

4 Model results

In the present paper we apply the comparison algorithm, that we have used in earlier papers considering Inteball-1 measurements - [4, 5, 6].

We divide the whole interval into five subintervals [20:30 UT January 16 - 03:00 UT January 17], [03:00-09:00 UT January 17], [09:00-16:00 UT January 17], [16:00 UT January 17 - 01:00 UT January 18], [01:00-04:53 UT January 18]. Then we run the model with sets of input parameters, corresponding to each interval. The parameters at the first and last intervals correspond to the time moments near the magnetopause crossings. The parameters of the solar wind and interplanetary medium applied in the simulations are given in Table 1, the ratio of specific heats is $\gamma=2$.

Table 1. Input data, used in the simulations: dynamic pressure P_d (in nPa), Mach number, B_y and B_z GSE components (in nT), Dst index and dipole tilt angle (in degrees).

UT	P_d	M	B_y	B_z	Dst	tilt
21:05(16/01)	1.26	6.43	1.5	-0.5	-12	-15.6°
06:00(17/01)	1.47	6.33	0.6	2.0	-1	-30.3°
12:00(17/01)	2.08	6.44	2.5	-1.8	6	-17.4°
21:50(17/01)	4.69	6.56	1.3	-3.0	1	-17.2°
04:10(18/01)	1.98	5.06	-4.5	-1.0	-5	-30.0°

The magnetosheath geometry and velocity distribution in the plane that passes through the satellite location at the moment of MP crossing at 04:10 UT (January 18) are presented in Fig. 5a). The tick black curve is the projection of the spacecraft orbit. The satellite registered moment of crossing (denoted by the outer yellow point) agrees well with the model calculated magnetopause.

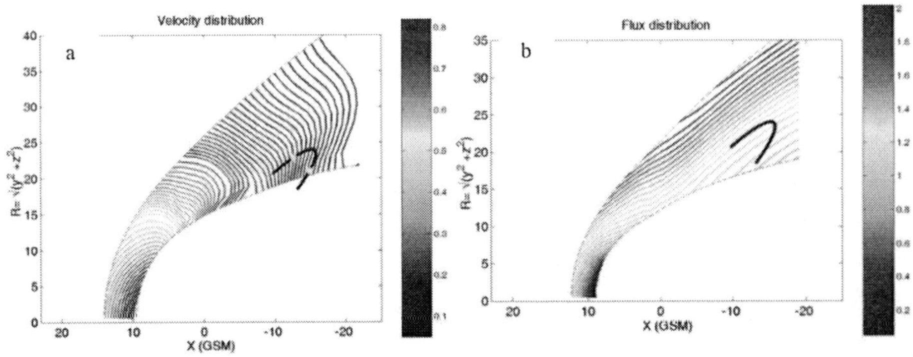

Fig. 5. Left a): Distribution of the model-calculated V_x component of the ion velocity, corresponding to the moment of the MP crossing at 04:10 UT January 18. Right b): Distribution of the model-calculated ion flux (in X direction), based on the moment of 21:50 UT January 17. The black line on both panels depicts the satellite trajectory during [12:00 UT January 16 - 12:00 UT January 18]. The yellow points on the left denote the spacecraft location at the moments of the two MP crossings - 20:30 UT January 16 (inner point) and 4:53 UT January 18 (outer point). The plasma parameters are shown in non dimensional units.

The magnetosheath geometry and the ion flux distribution in the plane that passes through the satellite location at the moment 21:50 UT (January 17) are presented in Fig. 5b). The satellite trajectory during the interval is depicted by a black line. Note that around

21:50 UT January 17 the solar wind density increased to about 28 sm^{-3} that is more than four times higher than the average solar wind density of 6 sm^{-3}. The increase of the solar wind density - Fig. 1a), and respectively the dynamic pressure, resulted in magnetopause compression. Thus, the satellite remained deeper in the magnetosheath - Fig. 5b).

The behaviours of the ion flux and velocity, according to Interball-1 data and to the model calculations are shown in Fig. 6. The model flux most significantly differs from the measured one at the interval close to the MP crossing [02:00- 4:53 UT January 19]. It is lower than the satellite-registered flux at the interval [11:00 -22:00 UT January 17] (17% on average) and is slightly higher at the interval [20:30 UT January 16 – 10:00 UT January 17]. One can see that the flux increases from 2 ($x10^8$ sm^3s^{-1}) at the beginning of the interval to about 10 ($x10^8$ $sm^{-3}s^{-1}$) at 22:00 UT January 17. This increase of the flux is related to the rise of the solar wind flux (added by a thin gray line), which is well represented by the numerical model.

The experimental and model-calculated V_x velocities are plotted in Fig. 6 (bottom). The magnetosheath velocity is about 68-75% from the solar wind one – Fig. 5a), Fig. 6 (bottom). As the trajectory is close to the equatorial GSM (Geocentric Solar Magnetospheric) plane the flow has a strong V_x component - about 200 km/s, as expected. The model velocity differs from the experimental one most in the interval [20:30- 24:00 UT January 16] where the difference is about 20%. In the whole interval the result of the simulations is in good agreement with the measurements, the difference is about 11% on average as the model values been higher (in absolute value) – Fig. 6 (bottom).

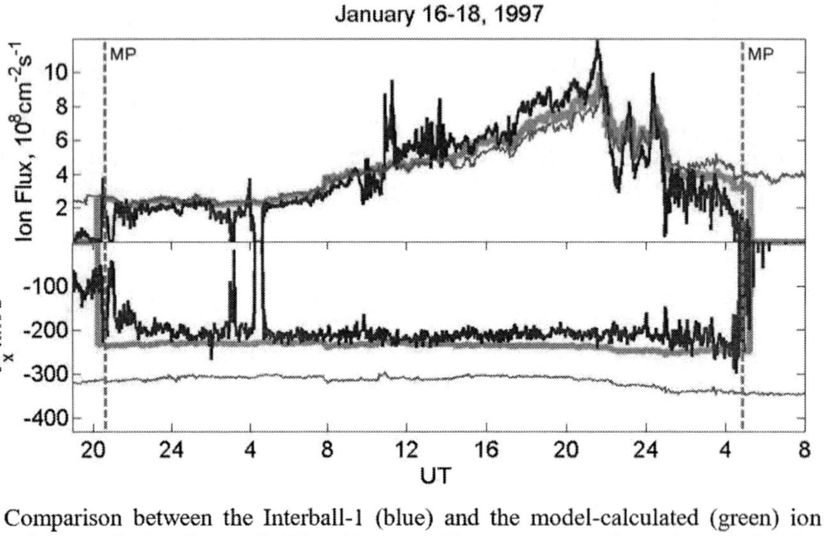

Fig. 6. Comparison between the Interball-1 (blue) and the model-calculated (green) ion flux (top panel) and ion velocity (bottom). The thin grey line represents the solar wind values, shifted accordingly. The MP labels denote the moment of magnetopause crossings, registered by the satellite.

5 Summary

We investigated the behaviour of two parameters in the magnetosheath – the ion flux and velocity. We compared the model estimates with the data, based on satellite's measurements. The satellite stayed for a long time in the magnetosheath, remaining close to the magnetopause. The model adequately represents the velocity values and the strong variation of the flux in the magnetosheath.

The paper is supported by the Bulgarian National Council "Scientific Research" under Project DMU 03/88.

References

1. J.R. Spreiter, A.L. Summers, A.Y. Alksne, Planet. Space. Sci. **14**, 223-253, (1966)
2. J.R. Spreiter, S.S. Stahara, J. Geophys. Res. **85**, 6769-6777, (1980)
3. J.R. Spreiter, S.S. Stahara, Adv. Sp. Res. **7,** 5-19, (1994)
4. P.S. Dobreva, M.D. Kartalev, M.D. Shevyrev, G.N. Zastenker, Planet. Sp. Sci. **53**, 117-125, (2005)
5. G.N. Zastenker, M.D. Kartalev, P.S. Dobreva, N.N. Shevyrev, A. Koval, Cosm. Res. **46**(6), 469-483, (2008)
6. P.S. Dobreva, M.D. Kartalev, N.L. Borodkova, G.N. Zastenker, Adv. Sp. Res. **58**, 188-195, (2016)
7. A.S. Magomedov, K.M. Holodov, Grid Characteristic Numerical Method, Nauka, Moscow, (1988)
8. N.A. Tsyganenko, J. Geophys. Res. **100**, 5599-5612, (1995)
9. N.A. Tsyganenko, J. Geophys. Res. **107**(A8), 10.1029/2001JA000219, (2002)
10. P.S. Dobreva, M.D. Kartalev, D. Koitchev, V.I. Keremidarska, M. Kaschiev, Adv. Sp. Res. **41**, 1279-1285, (2008)
11. M. Kartalev, P. Dobreva, E. Amata, M. Dryer, S. Savin, J. Atm. Sol. Terr. Phys. **70**(2-4), 627-636, (2008)
12. M.H. Acuna, K.W. Ogilvie, D.N. Baker, S.A. Curtis, D.H. Fairfield, W.H. Mish, Space Sci. Rev. **71**, 521, (1995)
13. K.W. Ogilvie, D.J. Chornay, R.J. Fritzenreiter et al., Space Sci. Rev. **71**, 41-54, (1995)
14. R.P. Lin, K.A. Anderson, S. Ashford et al., Space Sci. Rev. **71**(1-4), 125-153, (1995)
15. R.P. Lepping, M.H. Acuna, L.F. Burlaga et al., Space Sci. Rev. **71**, 207-218, (1995)
16. A.A. Galeev, Yu.I. Galperin, L.M. Zelenyi, Cosm. Res. **34**(4), 339-362, (1996)
17. R.S. Kremnev, A.I. Smirnov, S.S. Gorkin, Cosm. Res. **34**(4), 363–370, (1996)
18. J. Safrankova, G. Zastenker, Z. Nemecek, et al., Ann. Geophys. **15**, 562-569 (1997)
19. Yu.I. Yermolaev, A.O. Fedorov, O.L. Vaisberg, Ann. Geophys. **15**(5), 533-541, (1997)
20. J.-H. Shue, J.K. Chao, H.C. Fu et al., J. Geophys. Res. **102**(A5), 9497-9511, (1997)

A case study of the plasma in the magnetosheath using the numerical magnetosheath-magnetosphere model and THEMIS measurements

Polya Dobreva[1,*], *Olga* Nitcheva[1], *Monio* Kartalev[1]

[1] Institute of Mechanics, Bulgarian Academy of Sciences, Acad. G. Bonchev Str., block 4, Sofia, Bulgaria

Abstract. This paper presents a case study of the plasma parameters in the magnetosheath, based on THEMIS measurements. As a theoretical tool we apply the self-consistent magnetosheath-magnetosphere model. A specific aspect of the model is that the positions of the bow shock and the magnetopause are self-consistently determined. In the magnetosheath the distribution of the velocity, density and temperature is calculated, based on the gas-dynamic theory. The magnetosphere module allows for the calculation of the magnetopause currents, confining the magnetic field into an arbitrary non-axisymmetric magnetopause. The variant of the Tsyganenko magnetic field model is applied as an internal magnetic field model. As solar wind monitor we use measurements from the WIND spacecraft. The results show that the model quite well reproduces the values of the ion density and velocity in the magnetosheath. The simlicity of the model allows calulations to be perforemed on a personal computer, which is one of the mean advantages of our model.

1 Introduction

The magnetosheath plays very important role in the solar wind–magnetosphere-ionosphere interaction. In the magnetosheath the solar wind slows down, as its density and temperature increase. The Earth's magnetosheath is a transitional region, through which energy is transferred from the Sun to the Earth's atmosphere. Therefore, understanding the mechanism of energy transfer requires the creation of adequate magnetosheath models.

The pioneering work in description the magnetosheath dynamics was the model of Spreiter [1]. For given input parameters – the Mach number and polytropic index, and a given magnetopause shape, the model calculates the plasma parameters in the magnetosheath - density, velocity and temperature. The bow shock position is also calculated as a part of the solution. Later, the model was further developed including the distribution of the magnetic field in the magnetosheath [2]. As the solution (and in particular the velocity distribution) is known, the magnetic field in the magnetosheath can

* Corresponding author: polya2006@yahoo.com

easily be calculated from the magnetic induction equation. However, the model has some limitations, such as the assumption of axisymmetric bow shock shape and a given prescribed shape of the magnetopause [3].

Execution of large global MHD simulations in the domains of the solar wind to the atmosphere requires significant computing resources. In recent decades, the advances of computer technology made possible the creation of codes, that solve the magnetohydrodynamics equations. Examples of these simulations are the Ogino model [4], GUMICS model (Grand Unified Magnetosphere–Ionosphere Coupling Simulation) [5]. Other well known codes include OpenGGCM (Open General Geospace Circulation Model) [6], LFM model (Lyon–Fedder–Mobarry) [7], SWMF (Space Weather Modelling Framework) [8] which is based on BATS-R-US MHD code [9].

The numerical magnetosheath-magnetosphere model was developed based on a simpler gas-dynamic approach, that is capable of describing the main characteristics of the megnetosheath without the need for extreme computational resources. The model is able to reproduce the real 3D geometry of the magnetosheath boundaries - bow shock (BS) and magnetopause (MP), including indentations around the cusp at the magnetopause. The model also accounts for the down-dusk asymmetries in the tail magnetopause and north-south displacement, induced by the dipole tilt variations. The accuracy and relative simplicity of our model make it an accessible instrument to describe the large-scale magnetosheath properties.

The model was previously applied to study the ion flux distribution in the magnetosheath, using the Interball-1 measurements mainly – [10, 11].

The aim of this paper is to test the model capabilities on recent data from the THEMIS mission. Parameters to be interpreted are the ion velocity and ion density in a case of THEMIS passage through the magnetosheath. On 10-11 June 2009 the satellite crosses the equatorial magnetosheath close to the terminator (x=0) plane.

A brief description of the model will be given in Section 2. In Section 3 we will present the solar wind data, the magnetosheath data and a comparison between the model calculations and the experiment.

2 Short description of the model

Here, we will present only in general outline the numerical scheme that has been applied. The interaction of the solar wind with the Earth's magnetosphere is described by the self-consistent magnetosheath-magnetosphere model. A single-fluid model is implemented in the magnetosheath, while in the magnetosphere we employ a magnetic field model. The flow in the magnetosheath is described by the compressible 3D Euler equations of a perfect gas. The system of equations is solved by a variant of the grid-characteristic methods [12], which are often used in gas-dynamics. The magnetic field is represented as a variant of the magnetosphere model of Tsyganenko -T96 or T01 [13, 14]. The Tsyganenko model is modified to allow for a particular, non-axisymmetric shape of the boundary. The magnetopause is approximated as a surface of a tangential discontinuity.

The imposed boundary conditions are the Rankine-Hugoniot relations. They are reduced to the pressure balance equation at the magnetopause. The equation allows us to obtain specific magnetopause geometry, including indentations at the cusp, north-south and dawn-dusk asymmetries. Thus, there is no need of rescaling the satellite trajectories or the magnetosheath thickness in order to achieve a coincidence with the boundary positions.

Input data are plasma and magnetic field properties of the incoming flow, including the dynamic pressure P_d, sonic Mach number M, B_y and B_z components of the Interplanetary Magnetic Field (IMF), polytropic coefficient γ.

The Dst index, based on the Kyoto Web page, and the dipole tilt are also set at the beginning of a simulation. A more detailed description of the numerical algorithm, used in the calculations, can be found in previous papers [10, 11, 15, 16].

3 Numerical results

3.1 Input data

On 10-11 June 2009 the ACE spacecraft [17] is in orbit around Lagrange L1 point. Since during this interval the ACE density measurements are not available, we will use measurements from the WIND spacecraft [18] as input data. The WIND spacecraft is also

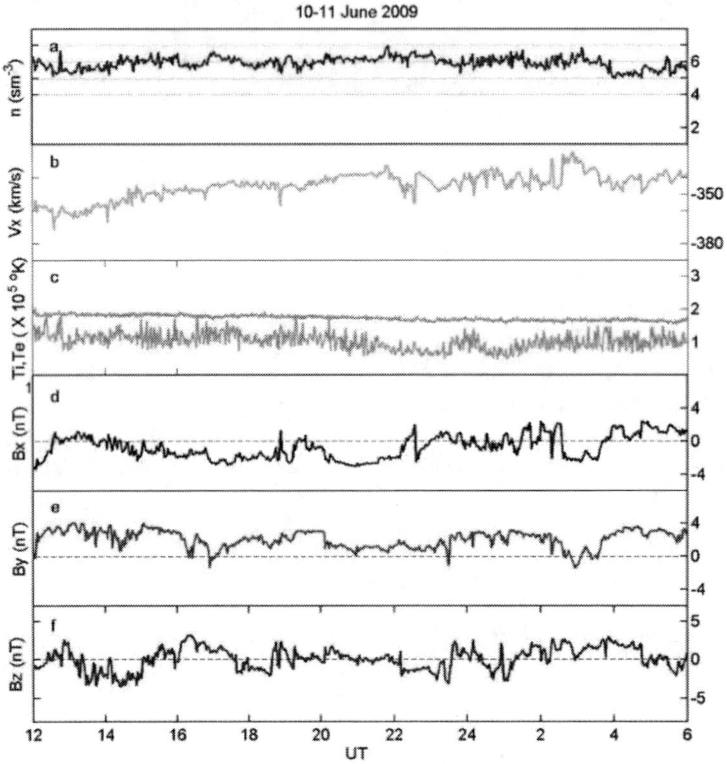

Fig. 1. Input solar wind data on 10-11 June 2009 : a) ion number density; b) V_x component of the ion velocity; c) ion temperature T_i (green) and electron temperature T_e (gray); d) B_x, e) B_y and f) B_z GSM (Geocentric Solar Magnetospheric) components of the interplanetary magnetic field. The data are not shifted in time.

in the pristine solar wind, located at about 261 R_E (Earth radii) from the Earth. The propagation time from WIND to the Earth varies between 79 min (at 17:00 UT) and 82 min (at 2:30 UT on 11 June). The plasma and magnetic field measurements, conducted by the WIND satellite on 10-11 June 2009 are presented in Fig.1. During this interval the solar wind conditions are quiet as the plasma parameters vary in a narrow range. The ion density is almost constant – about 6 sm^{-3}, similarly to the electron temperature, that is about 1.9-1.8 x 10^5 °K. The velocity varies in the range -330~ -350 km/s, the ion temperature is 1.3 x 10^5 °K at the beginning of the interval, drops to about 0.8 x 10^5 °K at the interval [22:00-02:00 UT] on 10-11 June, and rises to 1.0 x 10^5 °K after 2:00 UT on 11 June.

The input density, velocity and electron temperature are based on WIND/SWE (Solar Wind Experiment) device [19], the ion temperature – on WIND 3DP (3D Plasma analyzer) [20]. The Interplanetary magnetic field is determined by WIND/MFI (Magnetic Field Investigation) [21]. The time resolution is 1.5 min for the density, the velocity and the ion temperature, 1 min for the magnetic field and 12s for the electron temperature.

3.2 Magnetosheath measurements

THEMIS (Time History of events and Macroscale Interaction during Substorms) is a constellation of five probes, which primary objective is to study the energy release during the substorms in the magnetosphere [22, 23]. Launched on February 2007, they carried identical instrumentation aboard. For the first months of the mission, they operated in so called coast phase, moving as a tight cluster in elliptical orbit with an apogee of 15.4 R_E. After the preliminary phase, that continued until September 2007, they were separated from each other and their baseline mission of scientific observations began.

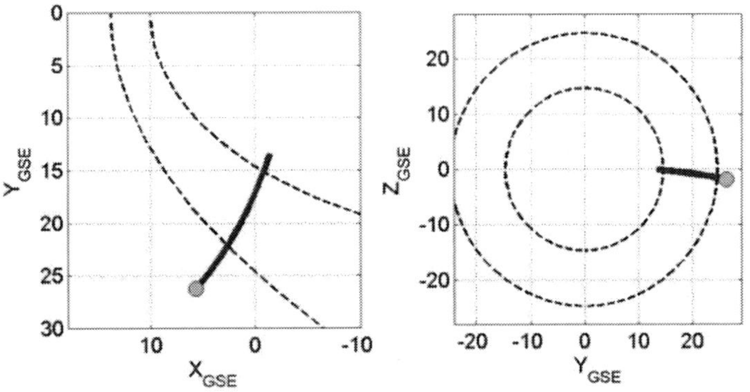

Fig. 2. Projection of the TH B orbit on the (X,Y) and (Y,Z) GSE (Geocentric Solar Ecliptic) coordinate planes, during the interval [12:00-06:00] UT on 10-11 June 2009. The point denotes the beginning of the interval. The inner curve represents the magnetopause from the Shue 97 model [24], calculated with average solar wind parameters of P_d=2 nPa, and B_z=0 nT, while the outer curve is the bow shock, calculated by our model for the above mentioned magnetopause boundary.

From 15 April to 15 June 2009 they were on radiation belt mission and moved in strictly elliptic orbits with apogees on the dusk side. The outer probes P1(TH B) and P2 (TH C) have an apogee of 30 R_E and 20 R_E respectively. The inner three probes P3 (TH D), P4 (TH E) and P5 (TH A) moved close to each other with an apogee of 12 R_E.

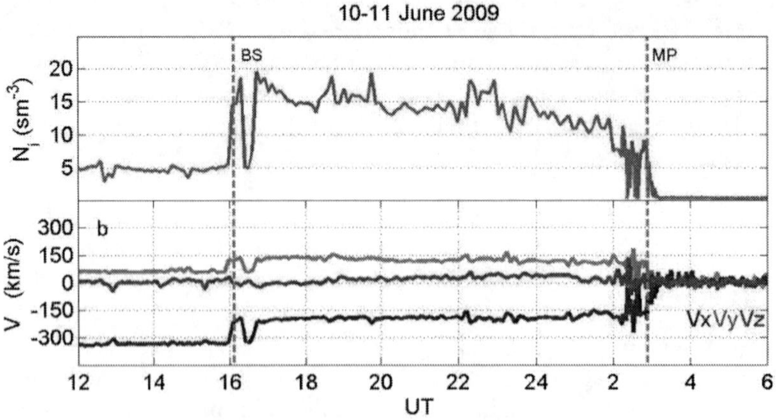

Fig. 3. Plasma measurements, conducted onboard THEMIS B satellite on 10-11 June 2009: a) ion number density; b) V_x (blue) ,V_y (gray) and V_z (red) GSM component of the measured ion velocity. The vertical dashed lines denote the moments of the BS and MP crossings.

On 10-11 June 2009 all five satellites moved on the dusk side, as the inner three probes P3, P4 and P5 stayed in the magnetosphere. The outer P2 probe only partially crossed the magnetosheath. We will use data from the outermost P1 probe, which on 10-11 June crossed the magnetosheath entirely. In this paper we use TH B notation to refer to THEMIS P1 probe, which is most commonly used in the literature.

Fig. 2. shows the orbit of TH B on two perpendicular GSE planes, corresponding to the interval [12:00 – 06:00 UT] on 10-11 June 2009. For clarity, exemplary shapes of the bow shock and magnetopause are also shown. The magnetopause shape is the form of Shue 1997 [24], corresponding to mean values of dynamic pressure P_d=2. nPa and IMF B_z = 0.nT. The bow shock is that calculated by out model for the given magnetopause shape and given input parameters of Mach=8 and polytropic index γ =2. Most of the time TH B stayed close to equatorial GSE plane, meanwhile moving earthward and slowly northward.

The ion density and velocity, based on TH B on 10-11 June 2009 are visualized in Fig. 3. The jump of the density and velocity shows, that the satellite crosses the BS at 16:03 UT and after 2:50 UT (11 June) it stayed entirely in the magnetosphere. The density drop at the short interval [16:20-16:40 UT] is associated with the satellite's re-entering in the solar wind. After 16:40UT the spacecraft remained entirely in the magnetosheath. Since our purpose is not the investigation of multiple crossings we will consider the magnetosheath passage from the moment of the first BS crossing at 16:03 UT. During the interval [2:20-2:50 UT] on 11 June the satellite encountered both magnetosheath and magnetospheric plasma, as after 2:50UT TH B remained in the magnetosphere. The data, shown on the plot are based on TH B/ESA (Electrostatic Analyzer) instrument [25] and have a time resolution of 6.5 min.

3.3 Model results

Our purpose is to calculate the values of the plasma parameters along the satellite trajectory in the magnetosheath and to compare them with the measurements. Here we apply the algorithm, that was already used in our previous papers, considering satellite data interpretation [10, 11, 16].

We divide the whole interval into three subintervals [16:03-18:30 UT], [18:30-22:30 UT], [22:30-2:50 UT] and make the calculations with sets of input parameters, corresponding to the time moments 16:45UT, 22:00UT (on 10 June) and 2:30 UT (on 11

June). Then values in each individual interval are concatenated to give the solution in the entire interval. Parameters, applied in the simulations are given in Table 1, the ratio of specific heats is γ=2.

Table 1. Solar wind data used as an input to the model: dynamic pressure P_d (in nPa), Mach number, B_y and B_z GSM components (in nT), Dst index and dipole tilt angle (in degrees).

UT	P_d	M	B_y	B_z	Dst	tilt
16:45	1.26	4.96	3.0	1.0	4	32.6°
22:00	1.21	4.99	1.2	0.4	1	24.7°
02:30	1.1	5.13	2.5	0.1	4	14.8°

Fig. 4 presents distributions of the density a) and velocity b) in the magnetosheath, calculated at the moment of MP crossing 2:30 UT. The plain, that passes through the satellite location at the moment of the MP crossing is shown. The satellite orbit is also added. The blue point denotes the satellite position when it finally enters into the magnetosphere at 2:50 UT. We see, that the observed moment of MP crossing is in good coincidence with the model-calculated boundary.

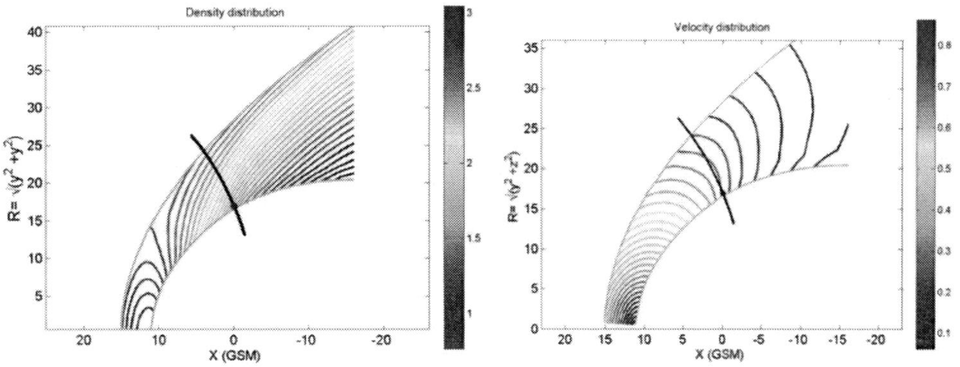

Fig. 4. Distribution of the ion density (left) and the velocity (right) in the magnetosheath. The parameters are presented in non dimensional units. The thick black line depicts the orbit projection of TH B from [12:00-06:00 UT] on 10-11 June 2009.

The comparison of the results of the calculation of the plasma velocities with the ones, measured onboard TH B is shown in Fig. 5. The directional components V_x, V_y, V_z in GSM coordinates and the absolute value of the velocity are shown. As the trajectory is close to the equatorial GSM plane, the flow has strong V_x and V_y components, the V_z component is small and varies in range -50 ÷ 50 km/s. The computed values of the directional components V_x and V_y coincide well with the registered ones – Fig.5 a,b. The model V_z components differs more significantly from the observed values, but this component in itself is less than other components. Before the BS crossing at 16:03UT TH B is in the solar wind and measures the velocity of the solar wind. At the satellite's position, close to the terminator plane, the velocity in the magnetosheath is about 70% from the solar wind velocity – Fig. 4b, 5d.

Fig. 6 presents the observed and numerically calculated ion density at the points of the satellite's trajectory in the magnetosheath on 10-11 June 2009. We observe a good coincidence of the model-calculated and measured values of the density. We see in the magnetosheath an increase of the density by a factor of 2.6 compared to the unshocked solar wind - Fig. 4a. The density reaches its maximum at the bow shock, which is about 1.7 times higher than that at the magnetopause – Fig. 4a, Fig 6.

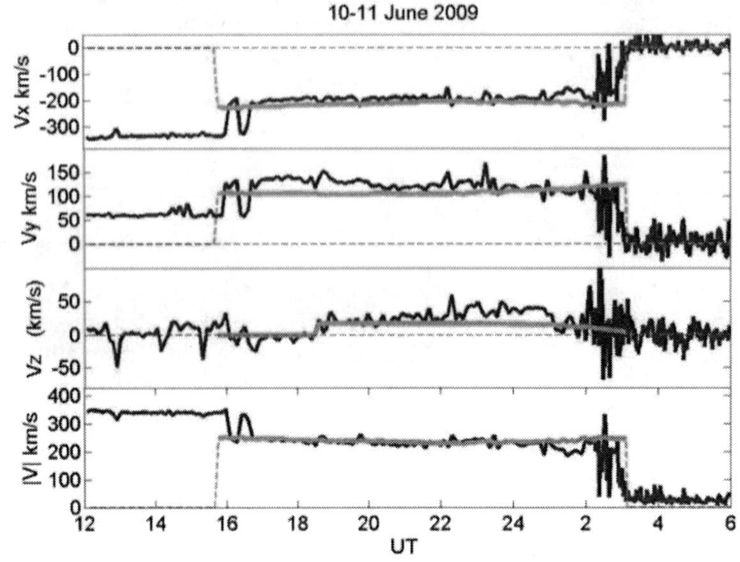

Fig. 5. Comparison between the measured plasma parameters (blue) and those, calculated by the model (green). From top to bottom: V_x, V_y V_z, (GSM) components and |V| of the ion velocity.

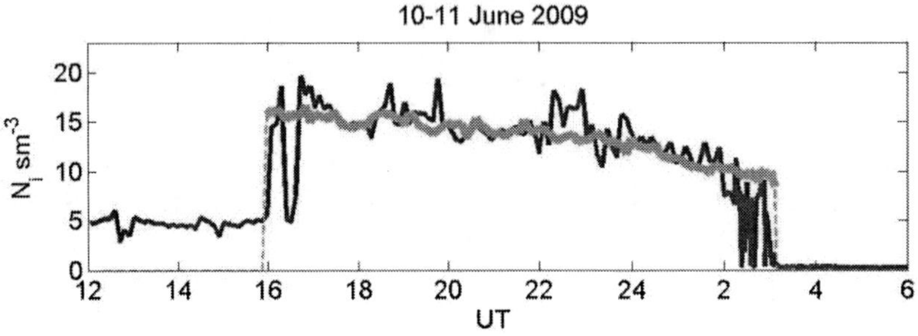

Fig. 6. Comparison between the measured (blue) and the numerically calculated (green) ion density.

4 Summary

This is a test of the model capabilities in the description of the behaviour of the plasma parameters in the magnetosheath. Here, data from the most recent space mission of THEMIS spacecraft were used. The model presented here is capable of describing the measured plasma parameters in the magnetosheath with great accuracy and may be used to interpret the phenomena, observed in the terrestrial magnetosheath.

The paper is supported by the Bulgarian National Council "Scientific Research" under Project DMU 03/88.

References

1. J.R. Spreiter, A.L. Summers, A.Y. Alksne, Planet. Space. Sci. **14**, 223-253, (1966)
2. J.R. Spreiter, S.S. Stahara, J. Geophys. Res. **85**, 6769-6777, (1980)
3. J.R. Spreiter, S.S. Stahara, Adv. Sp. Res. **7**, 5-19, (1994)
4. T. Ogino, R.J. Walker, M. Ashour-Abdalla, J. Geophys. Res. **99**, 11027-11042, (1994)
5. M. Janhunen, T. Palmroth, I. Laitinen et al., J. Atm. Sol. Terr. Phys. **80**, 48-59, (2012)
6. J. Raeder, D. Larson, W. Li, E.L. Kepko, T. Fuller-Rowell, Space Sci. Rev. **141**, 535-555, (2008)
7. J.G. Lyon, J.A. Fedder, C.M. Mobarry, J. Atm. Sol. Terr. Phys. **66**, 1333-1350, (2004)
8. G. Toth, I.V. Sokolov, T.I. Gombosi et al., J. Geophys. Res. **110**, A12226, (2005)
9. K.G. Powell, P.L. Roe, T.J. Linde, T.I. Gombosi, D.L. De Zeeuw, J. Comput. Phys. **154**, 284-309, (1999)
10. P.S. Dobreva, M.D. Kartalev, M.D. Shevyrev, G.N. Zastenker, Planet. Sp. Sci. **53**, 117-125, (2005)
11. P.S. Dobreva, M.D. Kartalev, N.L. Borodkova, G.N. Zastenker, Adv. Sp. Res. **58**, 188-195, (2016)
12. A.S. Magomedov, K.M. Holodov, Grid Characteristic Numerical Method, Nauka, Moscow, (1988)
13. N.A. Tsyganenko, J. Geophys. Res. **100**, 5599-5612 , (1995)
14. N.A. Tsyganenko, J. Geophys. Res., **107**(A8), 10.1029/2001JA000219, (2002)
15. P.S. Dobreva, M.D. Kartalev, D. Koitchev, V.I. Keremidarska, M. Kaschiev, Adv. Sp. Res. **41**, 1279-1285, (2008)
16. M. Kartalev, P. Dobreva, E. Amata, M. Dryer, S. Savin, J. Atm. Sol. Terr. Phys. **70**(2-4), 627-636, (2008)
17. E.C. Stone, A.M Frandsen, R.A. Mewaldt et al., Space Sci. Rev. **86**(1-4), 1-22, (1998)
18. M.H. Acuna, K.W. Ogilvie, D.N. Baker, S.A. Curtis, D.H. Fairfield, W.H. Mish, Space Sci. Rev. **71**, 521, (1995)
19. K.W. Ogilvie, D.J. Chornay, R.J. Fritzenreiter et al., Space Sci. Rev. **71**, 41-54, (1995)
20. R.P. Lin, K.A. Anderson, , S. Ashford et al., Space Sci. Rev. **71**(1-4), 125-153, (1995)
21. R.P. Lepping, M.H. Acuna, L.F. Burlaga et al., Space Sci. Rev. **71**, 207-218, (1995)
22. V. Angelopoulos, Space Sci. Rev. **141**, 5-34, (2008)
23. D.G. Sibeck, V. Angelopoulos, Space Sci. Rev. **141**, 35-59, (2008)
24. J.-H. Shue, J.K. Chao, H.C. Fu et al., J. Geophys. Res. **102**(A5), 9497-9511, (1997)
25. J.P. McFadden, C.W. Carlson, D. Larson, M. Ludlam, R. Abiad et al., Space Sci. Rev. **141**(1-4), 277-302, (2008)

Thermoacoustic waves in pulsed Pirani sensor MEMS. Numerical Investigation

Peter Gospodinov[1], *Dobri* Dankov[1*], *Vladimir* Roussinov[1], *Mirona* Mironova[1]

[1]Institute of Mechanics, Bulgarian Academy of Science, Bulgaria

Abstract. The transient heat transfer process is studied in rarefied gas confined between two stationary concentric cylinders. The inner cylinder (filament) is subjected to a periodically heating-cooling cycle. The energy transfer is modeled with continuous model based on Navier-Sockes Fourier equations of motion and energy transfer and with a statistical DSMC model. Numerical results for the temperature, thermodynamic pressure and pressure difference between thermodynamic pressure and radial stress tensor component are obtained for different circular frequencies of heating cooling cycle of filament and for different filament radii. The pressure variation at the end of any local heating stage of heating-cooling cycle is close to the value of equilibrium thermodynamic pressure. The results are applicable in designing the pulsed Pirani sensors.

1 Introduction

Pirani vacuum gauges usually consist of an electrically heated filament suspended in a cylindrical tube. The filament is kept at a constant temperature. A new gauge is described for measurement of pressures in the range 10^{-4} mbars to 5 bars. The gauge uses a conventional Pirani sensor, but pulses the filament temperature and measures filament heating and cooling rates [4]. The physical foundations of this measurement method are considered in [8]. A novel Pirani transmitter which is operated in pulsed mode reveals significant advantages. Its efficiency can be compared to the classical Pirani system. The filament is no longer operated steadily at a constant temperature, but cyclically heated up to a certain temperature threshold by an increasing voltage ramp [10]. Numerical calculations of heat flux in the Pirani sensor were carried out for a wide range of the gas rarefaction based on the kinetic Boltzmann equation. To take into account a non-diffusible gas-surface interaction, the Cercignani-Lampis scattering kernel was used in the boundary condition [11]. Sound waves, propagating through a rarefied gas confined between two coaxial cylinders and induced by an unsteady heating of the inner boundary, are modeled on the basis of a kinetic model to the linearized and non-stationary Boltzmann equation [16].

We study a rarefied monatomic gas between two coaxial, stationary, unconfined cylinders (one dimensional, axis-symmetrical problem). The temperature of the outer cylinder is constant and equal to the gas equilibrium temperature, while the temperature of

[*] Corresponding author: dobri_dankov@abv.bg

the inner cylinder varies in time. The inner cylinder (filament) is subjected on periodically heating-cooling cycle.

2 Continuous Model (NSF) and Numerical Simulation

The continuous model is based on the Navier-Stokes-Fourier (NSF) equations for compressible fluid, completed with the equations of continuity and energy transport. For details see [15, 17, 19]. The tangential velocity component in the gas flow is $\mathbf{v} \equiv 0$, for $R_1 \le r \le R_2$ The coordinates in radial and tangential direction are (r, φ).

The following notations are used to describe the rarefied gas flow: u is the velocity component in r directions, ρ is the density and T is the temperature, P is the thermodynamic pressure. The nonzero stress tensor components taking part in the equation of motion [7] can be expressed as.

$$\tau_{rr} = -\mu\left(2\frac{\partial u}{\partial r} - \frac{2}{3}\frac{1}{r}\frac{\partial}{\partial r}(ru)\right) \tag{1}$$

$$\tau_{\varphi\varphi} = \mu\left[\frac{2}{3}\left(\frac{1}{r}\frac{\partial}{\partial r}(ru) - 2\left(\frac{u}{r}\right)\right)\right] \tag{2}$$

The dissipation function Φ [7] in the equation of energy transport is

$$\Phi = \frac{4}{3}\mu\left[\left(\frac{\partial u}{\partial r}\right)^2 - \frac{\partial u}{\partial r}\frac{u}{r} + \left(\frac{u}{r}\right)^2\right] \tag{3}$$

For a perfect monatomic gas, the viscosity and the heat transfer coefficients read as [1, 2, 3, 6]:

$$\mu = \mu(T) = C_\mu \rho_0 l_0 V_0 \sqrt{T}\,, C_\mu = \frac{5}{16}\sqrt{\pi} \tag{4}$$

$$\lambda = \lambda(T) = C_\lambda \rho_0 l_0 V_0 \sqrt{T}\,, C_\lambda = \frac{15}{32}\sqrt{\pi} \tag{5}$$

The NSF model is completed with the equation of state for ideal gas:

$$P = \rho R T$$

The governing equations are normalized by using the following scales: for density $\rho_0 = m n_0$ (m is the molecular mass, n_0 – the average number density), for velocity $V_0 = \sqrt{2RT_0}$ – R is the gas constant, for length - the distance between the cylinders $L = R_2 - R_1$, for time $t_0 = L/V_0$, for temperature T_0 – the wall temperature of both cylinders. The Knudsen number is $\mathrm{Kn} = l_0/L$, where the mean free path is l_o and $\gamma = c_P/c_V = 5/3$ (c_P and c_V. are the heat capacities at constant pressure and constant volume respectively). In this way, the characteristic number Kn and the constants C_μ and C_λ take part in the dimensionless model.

For the problem formulated first-order slip boundary conditions are imposed at both walls which can be written directly in dimensionless form as follows [19]:

$$u = 0 \tag{7}$$

$$T \pm \varsigma_T \mathrm{Kn}_{local} \frac{\partial T}{\partial r} = \overline{T}_i \tag{8}$$

at $r=R_i$,: i=1, 2. In Eqs. (7)-(8) $V_i = v_i / V_0$ is the dimensionless wall velocity, (v_i , $i = 1,2$ is the dimensional wall velocity). For diffuse scattering we have used the temperature jump coefficients ζ_T = 2.1904 calculated, respectively, from the kinetic Bhatnagar–Gross–Krook (BGK) equation [2, 14],

The dimensionless wall pulsating temperature on the filament $\overline{T}_1$ depends on the dimensionless time t harmonically:

For $2\pi k \le \varpi t \le 2\pi(k + 1/2)$, and $2\pi(k+1) \le \varpi t \le 2\pi(k+1+1/2)$, $k = 1,2,3,\ldots\ldots$

$$\overline{T}_1 = T_{1,0}\left[1 + A\sin\left|(\varpi t)\right|\right] \tag{9}$$

where $T_{1,0}$ is the inner cylinder (filament) wall temperature value for $\varpi t = \pi/2 + k\pi$, and $A = 0.1$ is the oscillations amplitude. The condition (8)-(9) describes the filament heating stage from the heating-cooling cycle.

For $2\pi(k+1/2) \le t \le 2\pi(k+1)$ instead (8) is applied the adiabatic condition (10), corresponding to the cooling stages of the heating-cooling cycle.

$$\frac{\partial T}{\partial r} = 0, \quad r = R_1 \tag{10}$$

$T_2 = 1$ is the outer cylinder wall constant temperature.

The boundary conditions are modeled by using the local Knudsen number Kn_{local}. [20]

$$\mathrm{Kn}_{local} = \frac{l}{L} = \left(L\sqrt{2}\pi\sigma^2 \rho n_0\right) = \frac{\mathrm{Kn}}{\rho} \tag{11}$$

In (11) with l is denoted the local mean free path, σ - molecular diameter and ρ is the dimensionless density.

Following [9, 12, 15] we introduce a Stokes number for temperature oscillations $\beta = \sqrt{\frac{\omega L^2}{\nu}}$, where ν is kinematic viscosity and ω – temperature circular frequency. The relation between dimensionless temperature frequency ϖ and Stokes number β can be written, using eq. Eq.(4).

$$\varpi = C_\mu \mathrm{Kn} \beta^2 \tag{12}$$

The dimensionless value of thermodynamic pressure P_0 corresponding to initial equilibrium state can be calculated as:

$$P_0 = \frac{1}{2}\rho_0 T_0 = \frac{1}{2} \tag{13}$$

In Eq.(13) ρ_0 and T_0 are the equilibrium dimensionless density and temperature.

The heating-cooling cycle is the reason why the thermoacoustic waves arise in the gas media. In fact, the measured pressure is the pressure difference defined as [16]:

$$\Pi = P - \tau_{rr} \tag{14}$$

3 Direct Simulation Monte Carlo (DSMC) Method

The gas considered is simulated as a stochastic system of N particles [5]. All quantities used are non-dimensional, so that the mean free path at equilibrium is equal to 1. The basic steps of simulation are as follows:

A. The time interval $\left[0;\hat{t}\right]$ over which the solution is found, is subdivided into subintervals with step Δt .

B. The space domain is subdivided into cells with sides $\Delta z, \Delta r$.

C. Gas molecules are simulated in gap G using a stochastic system of N points (particles), having position $z_i(t), r_i(t)$ and velocities $\left(\xi_z^i(t), \xi_r^i(t), \xi_\varphi^i(t)\right)$,

D. N_m particles are located in the m-th cell at any given time. This number varies during the computer simulation by the following two stages:

Stage 1. Binary collisions in each cell are calculated, whereas particles do not move. Collision modeling is realized using Bird's scheme "no time counter".

Stage 2. Particles move with new initial velocities, acquired after collisions and no external forces act on particles. No collisions are accounted for at this stage.

E. Stage 1 and Stage 2 are repeated until $t = \hat{t}$.

F. Flow macro-characteristics (density, velocity, temperature) are calculated as time-averaged.

G. Boundary conditions are: diffusive at the outer cylinder wall; diffusive or adiabatic at the inner cylinder wall depending on the time t; periodic along axis Oz.

We denote the particle velocity before collision with wall with $\left(\xi_z, \xi_r, \xi_\varphi\right)$ and after with $\left(\xi_z^*, \xi_r^*, \xi_\varphi^*\right)$, respectively the velocity magnitude with $c = \sqrt{\xi_z^{\ 2} + \xi_r^{\ 2} + \xi_\varphi^{\ 2}}$ and $c^* = \sqrt{\xi_z^{*2} + \xi_r^{*2} + \xi_\varphi^{*2}}$. This gas-surface interaction model can be described by keeping the particle velocity magnitude invariant and the velocity direction of the reflected particle set based on isotropic scattering boundary conditions in the half unite sphere. We discuss this problem first in the spherical coordinate system (c, α_1, α_2) $\left(\alpha_1 \in \left[0, \frac{\pi}{2}\right], \alpha_2 \in [0, 2\pi]\right)$ [13, 18]. The three velocity components on the inner cylinder for adiabatic case reflected by:

$$\begin{aligned} \xi_r^* &= c.\cos\alpha_1 \\ \xi_z^* &= c.\sin\alpha_1.\cos\alpha_2 \\ \xi_\varphi^* &= c.\sin\alpha_1.\sin\alpha_2 \end{aligned} \tag{15}$$

and for diffusion case reflected by:

$$\begin{aligned} \xi_r^* &= \sqrt{\frac{2k_B\overline{T}_1}{m}}.\cos\alpha_1 \\ \xi_z^* &= \sqrt{\frac{2k_B\overline{T}_1}{m}}.\sin\alpha_1.\cos\alpha_2 \\ \xi_\varphi^* &= \sqrt{\frac{2k_B\overline{T}_1}{m}}.\sin\alpha_1.\sin\alpha_2 \end{aligned} \tag{16}$$

The difference between diffusion and adiabatic reflection is only the coefficient. At the diffusion reflection the coefficient $\sqrt{\frac{2k_B\bar{T}_1}{m}}$ depends on the wall temperature $\bar{T}_1$ and the particle mass m, while at the adiabatic reflection the coefficient c is the particle velocity magnitude before collision on the particle with the wall.

The temperature on the filament $\bar{T}_1$ depends on the dimensionless time t harmonically for $2\pi k \le \varpi t \le 2\pi(k+1/2)$, and $2\pi(k+1) \le \varpi t \le 2\pi(k+1+1/2)$, $k = 1,2,3,......$ The boundary condition is modeled by (16). The adiabatic condition (16) is applies for $2\pi(k+1/2) \le t \le 2\pi(k+1)$ instead (15), corresponding to the cooling stage from the heating-cooling cycle.

4 Numerical results and discussion

The numerical solution of the NSF and DSMC models in detail is described in [15, 17, 19].

We study three main cases with different frequency – $\beta = 2,\ \beta = 4,\ \beta = 8$ In numerical calculation is used the Knudsen number Kn=0.02, inner cylinder radius $R_1 = 0.05, 0.005, 0.001$ and outer cylinder radius respectively $R_2 = 1.05, 1.005, 1.001$. Numerical results for filament temperature (Fig.1, 2 ,3 and 5), pressure P (Fig.4) and pressure difference Π (Fig.1, 2, 3, 4) are presented. The values of the macrocharacteristics for a fixed dimensionless time value are obtained as a mean value in the radial direction.

The filament temperature variation $0 \le t \le 300$ is shown on the Fig.1, 2 and 3. At the lowest frequency, the dimensionless time to reach sustained oscillations for filament temperature and macrocharacteristics is less than the half period – Fig.1, 2, 3, while at higher frequencies this time is in the order of 30 (Fig.5 left) and 50 (Fig.5 right).

The filament temperature pulsation generates thermoacoustic waves in the direction of the radius that are reflected between the two cylinder walls. This causes the deviation of the gas pressure from its equilibrium value. The pressure difference Π is the actual measured value with the gauge [16]. The stress tensor variation limits are negligible and the difference between thermodynamic pressure and pressure difference is negligible (Fig.4). The pressure difference local extreme coincides at the time with the local extreme temperature after reaching the sustained oscillations. No phase difference was observed for the studied cases. It can be concluded that, while maintaining the amplitude of filament temperature, the pressure variation limits are narrowed at increasing the pulses' forced frequency.

The filament radius decreasing decreases the amplitudes of filament temperature and pressure differential oscillations.

The local maxima of sustained oscillations of pressure difference are close to the value of the equilibrium thermodynamic pressure when decreasing the filament radius. (Fig. 6)

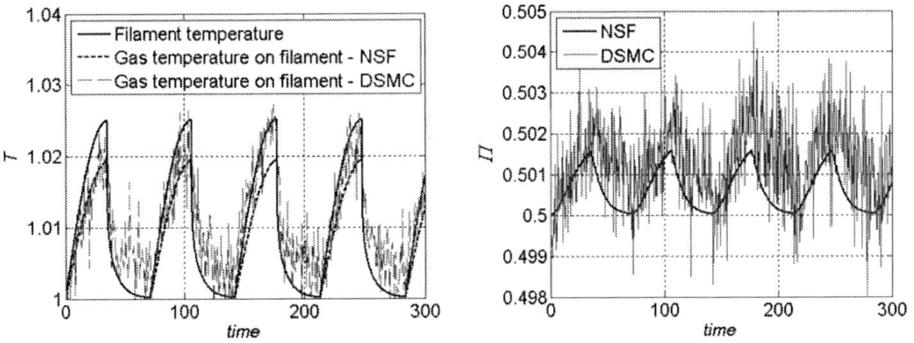

Fig. 1. The temperature and the pressure difference variation at β=2 and $R_1 = 0.05$

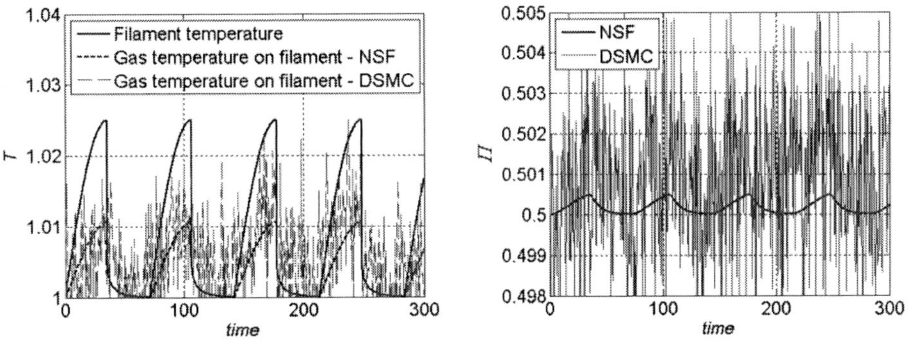

Fig. 2. The temperature and the pressure difference variation at β=2 and $R_1 = 0.005$

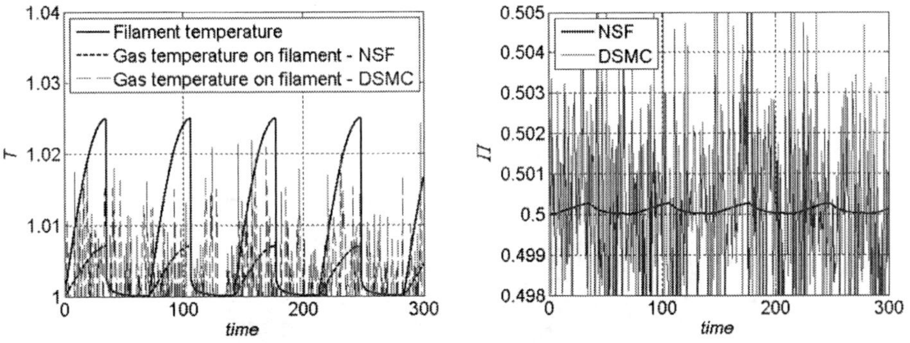

Fig. 3. The temperature and the pressure difference variation at β=2 and $R_1 = 0.05$

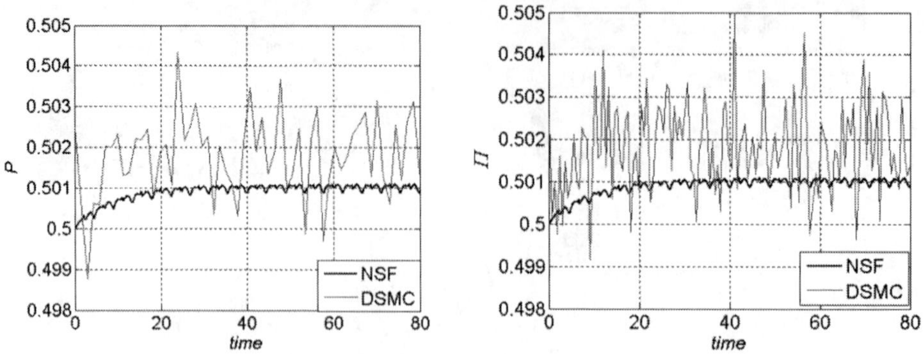

Fig. 4. The temperature and the pressure difference variation at β=2 and $R_1 = 0.05$

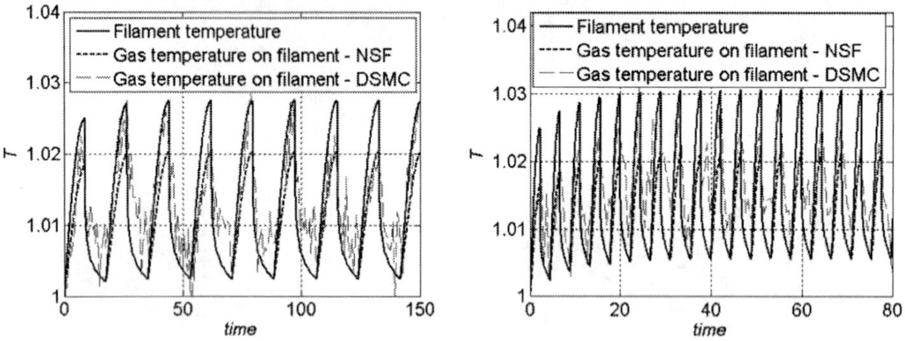

Fig. 5. The fiber temperature variation at β=4 (left figure) and β=8 (right figure).

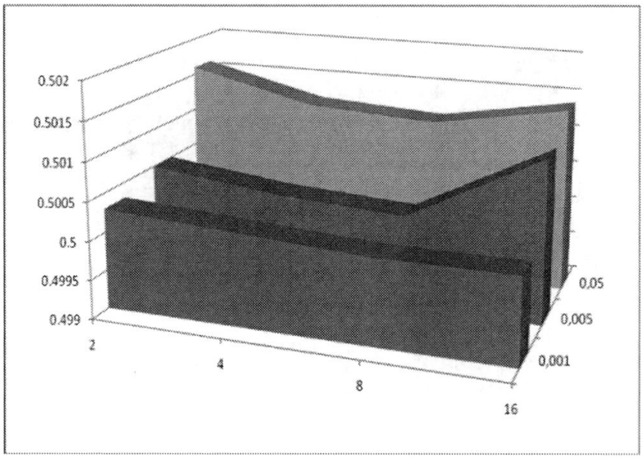

Fig. 5. The local maxima of sustained oscillations of pressure difference.

5 Conclusion

The differences in numerical results between the two models appear visually large but as values are below 0.5%. This means that there were no significant differences in numerical results obtained by two methods – NSF and DSMC.

Numerically is analyzed the operation of the MEMS pulsating Pirani sensor. The actual read pressure (pressure difference) variation is studied. At the heating stage end of heating-cooling cycle the pressure difference is close to the value of equilibrium thermodynamic pressure.

The results can be used to analyze the pulsating Pirani sensor sensitivity, i.e. the time to reach sustained oscillations as well as to adjust the pressure reading to the actual value.

Acknowledgment: V. Roussinov would like to thank for the financial support provided by the NSF of Bulgaria under Grant No DN 02/7–2016.

References

1. S. Chapman, T. Cowling, *The Mathematical Theory of Non-Uniform Gases* (Cambridge University Press, Cambridge, 1952).
2. S. Albertoni, C. Cercignani, L. Gotusso, *Phys. Fluids* **6**, 993–996 (1963)
3. C. Cercignani, F. Sernagiotto, *Physics of Fluids* **10**, 1200–1204 (1967)
4. M. Cole, *Vacuum* **38**(8-10), 897–899 (1988)
5. G. A. Bird, *Molecular Gas Dynamics and the Direct Simulation of Gas Flows* (Oxford University Press, Oxford, 1994).
6. C. Cercignani, *Rarefied Gas Dynamics, From Basic Concepts to Actual Calculations* (Cambridge University Press, Cambridge, 2000).
7. R. Bird, W. Stewart, and E. Lightfoot, *Transport Phenomena*, 2nd edn (John Wiley, New York, 2002).
8. W. Jitschin, S. Ludwig, *Vakuum in Forschung und Praxis* **16**, 23–29 (2004)
9. J. Park, P. Bahukudumbi, A. Beskok, *Phys. Fluids* **16**, 317 (2004)
10. H. Ploechinger, F. Salzberger, Pulsed Pirani vacuum gauge (2005): http://www.researchgate.net/publication/260123890.
11. F. Sharipov, G. Bertoldo, Numerical modelling of Pirani sensor, in *XVIII IMEKO World Congress*, Metrology *for a Sustainable Development, Rio de Janeiro, Brazil, September, 17-22, 2006.*
12. D. Emerson, Gu Xiao-Jun, S. Stefanov, S. Yuhong, W. Barber, *Phys. Fluids* **19**, 107105 (2007)
13. P. Tzeng , W. Chou, C. Liu, W. Li, *Journal of Aeronautics, Astronautics and Aviation, Series A* **43**(3), 147–158 (2011)
14. F. Sharipov, *J. Phys. Chem. Ref. Data* **40**, 023101 (2011)
15. P. Gospodinov, V. Roussinov, S. Stefanov, *European Journal of Mechanics B/Fluids* **33**, 68–77 (2012)
16. D. Kalempa, F. Sharipov, *Vacuum* **109**, 326–332 (2014)
17. P. Gospodinov, V. Rousinov, M. Mironova, Cylindrical nonisothermal oscillatory Couette gas flow in the slip regime: Wall shear stress and energy transfer, numerical investigation, *Intellectual Archive Bulletin*, Feb. 19, pp. 2–19 (2014)
18. A. Mohammadzadeh, A. Rana, H. Struchtrup, *International Journal of Thermal Sciences* **101**, 9–23 (2016)
19. P. Gospodinov, D. Dankov, and V. Roussinov, AIP Conference Proceedings **1773**, paper 080001 (2016)
20. P. Gospodinov, D. Dankov, V. Roussinov, S. Stefanov, AIP Conference Proceedings **1404**, 451–456 (2011)

Wetting of doubly periodic rough surfaces in Wenzel's regime

Stanimir Iliev[1,*], *Nina* Pesheva[1], *Pavel* Iliev[2]

[1]Institute of Mechanics, Bulgarian Academy of Sciences, 1113 Sofia, Bulgaria
[2]ETH Zurich, Computational Physics for Engineering Materials, CH-8093 Zurich, Switzerland

Abstract. In this work we present preliminary results from our numerical study of the shapes of a liquid meniscus in contact with doubly sinusoidal rough surfaces in Wenzel's wetting regime. Using the full capillary model we obtain the advancing and the receding equilibrium meniscus shapes for a broad interval of surface roughness factors. The contact angle hysteresis is obtained when the three-phase contact line is located on one row (block case) or several rows (kink case) of physical defects. We find that depending on the mutual disposition of the contact line and the lattice of periodic defects, different stick-slip behaviors of the contact line depinning mechanism appear, leading to different values of the contact angle hysteresis.

1 Introduction

For various industrial applications one needs to devise solid surfaces with prescribed wettability properties. The surface wettability can be influenced through the formation of regularly spaced surface defects. These defects can be either physical (i.e., rough surfaces), and/or chemical defects (i.e., heterogeneous surfaces). That is why the study of the wetting properties of structured surfaces is very important. The presence of surface defects is known to lead to a set of metastable equilibrium states of the liquid, wetting the solid surface. Due to the presence of these defects, the liquid-vapor interface may move with the solid surface while the three-phase contact line (CL) is pinned at the surface defects and stays immobile relative to the solid surface. This phenomenon can lead to the increase or decrease of the apparent contact angle (CA). For the wetting properties of such solid surfaces is characteristic the so-called stick-slip behavior of the CL and the existence of maximal and minimal apparent contact angles, called the advancing and the receding CAs θ_{adv} and θ_{rec} (ACA and RCA). These CAs are attained just before the CL unpins, thus resulting in a displacement of the whole liquid meniscus. The difference of the advancing and the receding CAs is called CA hysteresis (CAH), which is an important characteristic of the surface wettability. In the case of physical defects, i.e., rough surfaces, the ACA and RCA also depend on whether or not air is trapped between the solid surface and the liquid. The first case, when there is a trapped air, is known as Cassie's regime and the second case,

* Corresponding author: stani@imbm.bas.bg

when there is no trapped air, as Wenzel's wetting regime. There are two possible types of CLs in respect to the rows of defects, these are: the straight line and the stepped contact line (terms, introduced in Ref. [1]), and the mechanisms of transition to the new equilibrium meniscus state are termed the block- and the kink-depining of the CL, respectively [2, 3]. Here, we focus our study on the effect of the CL-depinning mechanism on the CAH in the case of smooth physical defects in Wenzel's wetting regime.

2 Problem formulation

2.1 System

We are concerned here with the study of the following system: a vertical solid plate, immersed in a tank of liquid (see Figure 1).

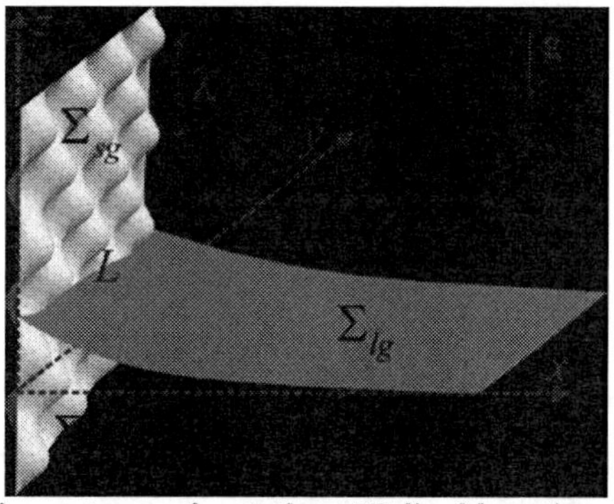

Fig. 1. Schematic image of the system under study: a 3D liquid meniscus, attached to a vertical rough solid plate.

More precisely, our goal is the determination of the set of metastable equilibrium shapes of the liquid meniscus in contact with the vertical plate, whose surface is rough and chemically homogeneous. The liquid in the tank has density ρ, volume V and free surface Σ_{lg}. The rough solid plate, whose surface is denoted by Σ_s, has contact area Σ_{sl} with the liquid ($\Sigma_{sl} \subset \Sigma_s$), and we assume that, the wetting of the plate by the liquid, occurs in the Wenzel's wetting regime. The rough surface of the solid plate in Cartesian coordinates is described by the function $f(y,z)$, i.e., $\Sigma_s \equiv \left(f(y,z), y, z\right)$, where the y-axis is horizontal and the z-axis is directed upward, as shown in Fig. 1. The ($z = 0$)-plane is chosen to coincide with the liquid level far from the moving plate. The function $f(y,z)$, describing the rough surface, is doubly periodic function of its coordinates with the same period λ along the y- and the z-axes. Here we concentrate on two types of such surfaces, for which one has

$$f(y,z) = A\sin(2\pi y/\lambda)\sin(2\pi z/\lambda + z_0). \qquad (1)$$

The values of the ratio A/λ, considered here, are of the order 0 - 0.2 and the parameter z_0 (the vertical displacement of the plate, relative to the pool of liquid) varies in the interval $[0,\lambda)$.

For obtaining the metastable equilibrium states of the meniscus under the action of the gravitational and capillary forces we adopt the integral approach. In this work we consider the case when the meniscus forms with the solid surface a local equilibrium CA $\theta_{eq} = 90^0$. Thus we need to find the local minima of the system's free energy *U*, defined as

$$U = \int_{\Sigma_{lg}} d\Sigma_{lg} + l_c^2 \int_V z\, dV , \qquad (2)$$

where $l_c = \sqrt{\gamma/\rho g}$ is the capillary length, γ is the liquid-gas surface tension, g is the gravitational acceleration.

2.2 Numerical method

We obtain the metastable equilibrium meniscus states, using the minimization algorithm for the free energy *U* (Eq. 2), described in detail in Refs [4-6]. For finding a numerical solution, the liquid free surface Σ_{lg} is approximated by a set of triangles with $N_i \times N_j$ nodes with radius vectors $\mathbf{r}_{ij}$. The nodes $\mathbf{r}_{ij}$ of the triangles, approximating the meniscus shape, are shifted vertically and the nodes $\mathbf{r}_{i1}$, approximating the CL, are shifted along the solid surface, each in direction minimizing the system energy *U*. We check the precision of our numerical method by comparing the results, obtained for different values of the displacements and for different number of the nodes N_i, N_j. On a length λ we use in the present study between 30-90 points. The numerical method is similar to the one, used in the public domain software "Evolver" [7].

2.3 Apparent CA and CAH

For the system, shown schematically in Fig. 1, the apparent CA θ_{ap} can be determined [3] through the observed height $h(L)$ of the CL *L*:

$$\theta_{ap} = \arcsin\left(1 - \langle h(L)\rangle^2 / 2l_c^2\right). \qquad (3)$$

For the system under consideration, θ_{rec} and θ_{adv} are attained, as follows from Eq. 3, for equilibrium meniscus states with the highest and the lowest capillary rise, respectively. Therefore, in order to obtain the whole set of equilibrium meniscus states and the CAH we proceed in the following way. First, we obtain numerically an equilibrium meniscus shape with an initial equilibrium CL, which for the study of the CL block depinning, is situated on one horizontal row of defects and for the CL kink depinning case – on two neighboring horizontal rows of defects. After that we find a sequence of metastable equilibrium meniscus states, simulating plate immersion and withdrawal in the pool of liquid. To this end we move vertically the solid plate in small steps $\pm\Delta z$ ("+" for the study of the receding and "-" of the advancing case). Together with the plate we move also the liquid meniscus, obtained as the equilibrium meniscus state for the previous position of the plate. We use it as an initial state to find through minimization the new equilibrium meniscus for the new plate position. This process is repeated until the CL remains connected with the same row(s) of defects, to which it was connected initially. The last found equilibrium meniscus states (corresponding to withdrawal/ immersion of the plate) before a CL depinning occurs, are the ones determining the hysteresis interval.

We note here, that the CAs θ_{adv} and θ_{rec} in the block-depinning case are obtained analytically [8] when $A/\lambda << 1$:

$$\theta_{adv} = \pi/2 + \pi^2 (A/\lambda)^2 + O(A^3/\lambda^3) \quad (4)$$

$$\theta_{rec} = \pi/2 - \pi^2 (A/\lambda)^2 + O(A^3/\lambda^3) \quad (5)$$

From Eqs. (4) and (5) it follows that

$$\theta_{adv} = \pi - \theta_{rec} \,. \quad (6)$$

This result holds true without the assumption $A/\lambda << 1$ for the system, shown in Fig. 1, with free energy (2) and local equilibrium CA $\theta_{eq} = 90^0$. This key statement is based on the existence of the "Corresponding states" [9].

3 Results and discussion

In Fig. 2 we present θ_{rec} (shown by solid symbols and solid line) as function of the ratio A/λ and $\pi - \theta_{adv} \equiv \theta_{rec}$ (as it follows from Eq. 6).

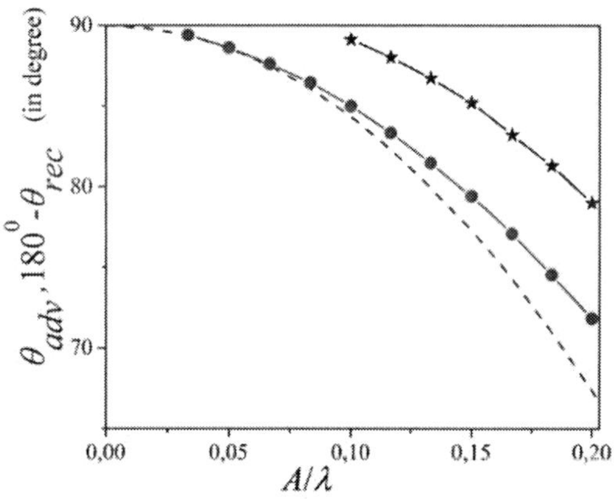

Fig. 2. The ACA and the RCA as functions of the ratio A/λ. In the CL block-depinning regime the results are shown by circles, connected by solid lines and in the CL kink-depinning regime by stars, connected by solid lines. The analytical results (dashed line) are shown too.

The results for the receding CA θ_{rec} are obtained by the procedure, described in Section 2.3. In the CL block depinning regime the results are shown by circles and in the CL kink depinning regime - by stars. The analytical results (Eqs. 4, 5) are shown in Fig. 2 by the dashed line. Our analysis of the presented numerical data shows that all obtained results for

θ_{rec} and θ_{adv} as functions of the ratio A/λ are approximated well by a quadratic functions. In the CL block-depinning case the numerical results for θ_{adv} are smaller and for θ_{rec} are bigger than the respective asymptotic solutions. The asymptotic solutions provide good results for $A/\lambda < 0.1$. The difference between the numerical solutions and the asymptotic solutions increases with the ratio A/λ. The difference in the values of θ_{adv} (and in the values of θ_{rec}) for both regimes is almost constant and is of the order of $5^0 - 7^0$ and slowly increases with the increase of A/λ. The values of the CAH are respectively by $10^0 - 14^0$ smaller for all considered values of the ratio A/λ. When the CAH in the CL block-depinning case is smaller than 10^0, then in the CL kink-depinning case we do not find CAH.

Conclusion

Different stick-slip regimes of the CL appear, depending on the mutual disposition of the CL and the lattice of periodic defects. This leads to different values of the CAH. One needs to investigate both the CL block- and the CL kink-dewetting regimes for a complete study of the CAH in the system, and to take into account also the possible different positions of the defects lattice in respect to the liquid front. When a CL kink-dewetting mechanism is taking place in the system, there is no CAH at small values of the ratio A/λ.

References

1. L. W. Schwartz, S. Garoff, J. Colloid Interface Sci. **106**, 422 (1985)
2. M. Rivetti, J. Teisseire, and E. Barthel, Phys. Rev. Lett. **115**, 016101 (2015)
3. S. Iliev, and N. Pesheva, Phys. Rev. E **93**, 062801 (2016)
4. S. Iliev, Comput. Methods Appl. Mech. Engrg. 126, **251** (1995)
5. S. Iliev, and N. Pesheva, J. Colloid Interface Sci. **301**, 677 (2006)
6. D. Iliev, N. Pesheva, and S. Iliev, Langmuir **29**, 5781 (2013)
7. K. Brakke, Exp. Math. **1**, 141 (1992)
8. R. G. Cox, J. Fluid Mech. **131**, 1 (1983)
9. A. D. Myshkis, V. G. Babskii, N. D. Kopachevskii, L. A. Slobozhanin and A. D. Tyuptsov, *Low Gravity Fluid Mechanics* (Springer-Verlag, Berlin, 1987)

On the modified Reynolds equation for journal bearings in a case of non-Newtonian Rabinowitsch fluid model

Juliana Javorova[1,*] and *Jordanka* Angelova[2]

[1]UCTM, Dept. of Applied Mechanics, 8 Kliment Ohridski Blvd., 1756 Sofia, Bulgaria
[2]UCTM, Dept. of Mathematics, 8 Kliment Ohridski Blvd., 1756 Sofia, Bulgaria

Abstract. In this paper, a theoretical analysis of hydrodynamic plain journal bearings with finite length at taking into account the effect of non-Newtonian lubricants is presented. Based upon the Rabinowitsch fluid model (cubic stress constitutive equation) and by integrating the continuity equation across the film, the nonlinear modified 2D Reynolds type equation is derived in details so that to study the dilatant and pseudoplastic nature of the lubricant in comparison with Newtonian fluid. A dimensionless equation of hydrodynamic pressure distribution in a form appropriate for numerical modeling is also presented. Some particular cases of 1D applications can be recovered from the present derivation.

1 Introduction

Hydrodynamic journal bearings are considered to be a vital component of all rotating machines. Journal bearings are used widely for thousands of years in many areas of mechanical engineering. Currently large numbers of them are applied successfully in machine tools, automotive and aircraft piston engines, turbo-machinery, micro-electro-mechanical systems, etc. Furthermore, it is well known that the performance of bearings is strongly influenced by the composition and rheological characteristics of the lubricant.

In order to meet requirements of the modern machinery systems, it is necessary to look for the enhancement in lubricating performance of all kind of bearings and other types of lubricated contacts. In this relation, currently, the use of Newtonian fluids blended with various additives increases, which is based on the effective improvement in the bearing characteristics as compared to the lubrication with Newtonian lubricants. These additives often are viscosity index improvers which represent high molecular weight polymers such as polyisobutylene, polymethacrylate, ethylene propylene, etc. These kinds of lubricants exhibit pseudoplastic and dilatant behaviour of non-Newtonian lubricants, in which the ratio between the shear stress and shear rate is no longer a constant. According to the experimental work of Wada and Hayashi [1] the non-Newtonian rheological behaviour of such kinds of lubricants with additives can be represented by an empiric cubic stress model, also called Rabinowitsch fluid model.

* Corresponding author: july@uctm.edu

Applying this fluid model, a number of scientific papers concerning the subjects of non-Newtonian fluid film lubrication have been published. For instance, the theoretical study of bearing performance with non-Newtonian lubricants using Rabinowitsch fluid model is done on journal bearings by Wada and Hayashi [1], Rajalingham et al. [2], Bourgin and Gay [3], Sharma et al. [4], Javorova et al. [5, 6], Kushare [7], Lin et al. [8]. Some contributions for squeeze film bearings are presented by Hashimoto and Wada [9], Lin [10], Naduvamani [11], Shigehalli [12]. Furthermore, Lin et al. [13-15] and Singh et al. [16-20] used the model of Rabinowitsch fluid to investigate in details the performance of different types of hydrostatic, hydrodynamic and squeeze film bearing systems. More general lubrication problems for the hybrid bearings can be found in the works of Walicka et al. [21, 22], Ratajczak et al. [23] as the same authors work also on trust bearings [24]. Recently several papers in the field of peristaltic flow in a tube with Rabinowitsch fluid model are also published.

As a whole, according to results in the mentioned studies, the influences of non-Newtonian Rabinowitsch fluids on the lubrication performances of journal bearings, slider bearings, squeeze film bearings, hydrostatic bearings, and other types of lubricated contacts are significantly apparent. It is shown that the steady state and squeeze-film performances are strongly affected by non-Newtonian effects based on the Rabinowitsch fluid model.

The most of above mentioned papers aimed to obtain some modified Reynolds equation for such kind of lubricants as well as to receive an analytical solution of it for more simple cases. For more complicated ones, as behaviour of finite length journal bearings, the authors regularly apply modified Reynolds equation obtained by Wada and Hayashi derived in short in [1]. In the mentioned studies a detailed mathematical output of modified Reynolds equation is not given that motivates authors to present here such one. The current study aims to represent consistently and in details a mathematical procedure to obtain the nonlinear modified two-dimensional Reynolds type equation based on the Rabinowitsch fluid model. By this way it is possible to study the pseudoplastic and dilatant nature of lubricants with long chain polymer additives in comparison with Newtonian fluids.

2 Modeling of the hydrodynamic lubrication in journal bearings

A radial journal bearing with finite length is considered under steady state conditions. The lubricant fluid in bearing clearance has non-Newtonian properties. The rheological fluid law described by the Rabinowitsch model [1, 13, 18, etc.] is presented with the following cubic equation

$$\tau + k\tau^3 = \eta \frac{du}{dy} , \tag{1}$$

where: τ - shear stress in the fluid film ; k - coefficient of pseudoplasticity (parameter responsible for the lubricants non-Newtonian behaviour); η - initial viscosity of the lubricant; u, v, w - velocity components in the directions of Cartesian coordinates x, y, z, respectively.

In dependence of the values of the coefficient of pseudoplasticity k there are three different groups of lubricants: for $k > 0$ - pseudoplastic fluids, if $k = 0$ - Newtonian fluid, as for $k < 0$ - dilatant fluid.

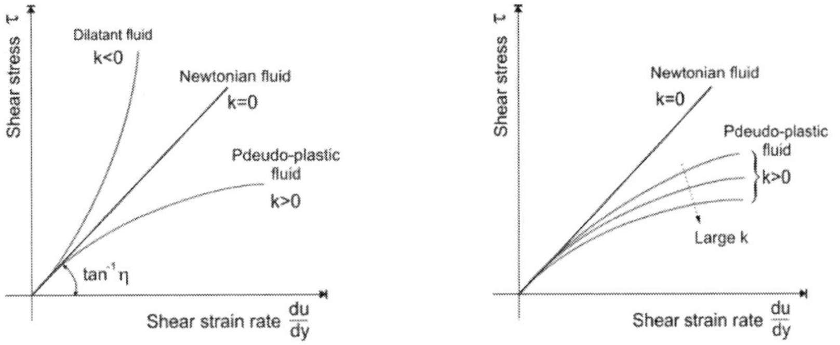

Fig. 1. Flow curves for non-Newtonian Rabinowitsch fluid.

The flow characteristics of these kinds of lubricants are shown in Fig. 1. Here η is a tangent at the original point of flow curves because of which it is called "initial viscosity" by analogy with [1, 13, 17]. If values of η do not vary, the nonlinearity of the flow curve increases with the value of the coefficient of pseudoplasticity k (see Fig. 1 - on the right).

2.1 Bearing geometry

For the geometry of the considered plane journal bearing with 360° range are introduced the following assumptions (see Fig.2): The journal and the bearing have a round shape and parallel axes; their surfaces are perfectly smooth. The bearing gap is filled with a lubricating fluid with a constant pressure equal to the external. The journal rotates with a constant angular velocity ω around its axis. The bearing radius R is approximately equal to the journal radius r $(R \approx r)$. The radial bearing clearance $c = R - r$ is of the order 1.10^{-4}, i.e. $c << 1$, see Fig.1. Therefore the clearance ratio $\beta = c/r \approx 1.10^{-3} << 1$. The bearing eccentricity is also very small, so the eccentricity ratio $\varepsilon = e/r \approx 1.10^{-3} << 1$. Moreover, for all types of bearings the fluid film thickness h is very small compared to the other dimensions of the contact surfaces because of which the ratio $h/l \approx h/r \approx 1.10^{-3} << 1$, where l is the fluid film length in a circumferential direction. All of the above mentioned about the bearing geometry justifies the commonly accepted hypothesis in the theory of hydrodynamic lubrication to neglect the curvature of the lubricated surfaces and, respectively, the curvature of the lubricant film.

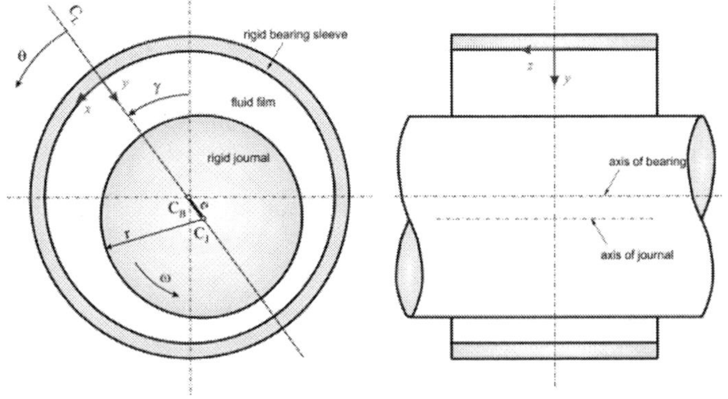

Fig. 2. Hydrodynamic journal bearing.

2.2 Boundary value problem

The constitutive equations between the shear stress and rate of share strain for two dimensional flow of the Rabinowitsch fluid obeys the following nonlinear relationships

$$\tau_{xy} + k\tau_{xy}^3 = \eta \frac{du}{dy}, \tag{2.a}$$

$$\tau_{zy} + k\tau_{zy}^3 = \eta \frac{dw}{dy}, \tag{2.b}$$

where τ_{xy} and τ_{zy} are shear stress components in x and z directions, respectively.

According to the thin film theory of hydrodynamic lubrication [25, 1, 18], the momentum and continuity equations in Cartesian coordinates are represented by the following differential equations:

$$\frac{\partial p}{\partial x} = \frac{\partial \tau_{xy}}{\partial y}; \tag{3.a}$$

$$\frac{\partial p}{\partial y} = 0; \tag{3.b}$$

$$\frac{\partial p}{\partial z} = \frac{\partial \tau_{zy}}{\partial y}; \tag{3.c}$$

$$\frac{\partial u}{\partial x} + \frac{\partial v}{\partial y} + \frac{\partial w}{\partial z} = 0, \tag{3.d}$$

where p is the hydrodynamic pressure.

Related to the bearing configuration, the boundary conditions for the fluid velocity components are

at $y = 0$:

$$u(x,0,z) = u_0 = 0; \quad v(x,0,z) = v_0 = 0; \quad w(x,0,z) = w_0 = 0, \tag{4}$$

at $y = h$:

$$u(x,h,z) = u_h = U = \omega r; \quad v(x,h,z) = v_h = \omega r \frac{\partial h}{\partial x}; \quad w(x,h,z) = w_h = 0, \tag{5}$$

where: ω is the journal angular velocity, $h = h(x,z)$ is the fluid film thickness.

2.3 Mathematical procedure

The integration of (3.a) and (3.c) with respect to y yields:

$$\tau_{xy} = \frac{\partial p}{\partial x} y + \tau_1; \tag{6.a}$$

$$\tau_{zy} = \frac{\partial p}{\partial z} y + \tau_2, \tag{6.b}$$

where τ_1 and τ_2 are integration constants.

By substitution of (6.a) and (6.b) in (2.a) and (2.b), respectively, it is obtained:

$$\frac{df}{dy}=\frac{1}{\eta}\left[qy+\xi+k\left(qy+\xi\right)^3\right], \tag{7}$$

where $f\in\{u,w\}$, $q\in\{\partial p/\partial x,\partial p/\partial z\}$, $\xi\in\{\tau_1,\tau_2\}$. Here the couples $\{..,..\}$ are ordered and mapping between them one-to-one.

Integration of (7) with respect to $y\in[0,h]$ leads to

$$\eta\left[f(h)-f(0)\right]=\int_0^h\left[qy+\xi+k\left(qy+\xi\right)^3\right]dy, \tag{8}$$

where: $f(h)\in\{u_h,w_h\}$, $f(0)\in\{u_0,w_0\}$.

Applying boundary conditions (4) and (5) and performing some elementary transformations the following cubic equation for ξ is obtained:

$$\xi^3+\frac{3}{2}qh\xi^2+\left[(qh)^2+\frac{1}{k}\right]\xi+\frac{(qh)^3}{4}+\frac{qh}{2k}-\frac{\eta f(h)}{hk}=0. \tag{9}$$

On the other hand, reordering (9) in ascending order by power of $\zeta=\xi+\dfrac{qh}{2}$ leads to a semi-cubic equation of ζ:

$$\zeta^3+\left[\frac{1}{k}+\left(\frac{qh}{2}\right)^2\right]\zeta-\frac{\eta f(h)}{hk}=0. \tag{10}$$

If $f(h)=0$ /see (4) for w_h/, then $\zeta=\xi+\dfrac{qh}{2}=0$, and $\xi=-\dfrac{qh}{2}$, i.e. $\tau_2=-\dfrac{qh}{2}$.

According to [1], since the value of $\lambda=\left|\dfrac{\eta u_h}{h\left[1+k(qh/2)^2\right]}\right|$ is enough small within the range of parameters considered, then from (8) follows that $\tau_1=-\dfrac{qh}{2}+\lambda$, $\lambda^2\approx\lambda^3\approx0$.

Furthermore, by integrating (7) with respect to $s\in[0,y]$ it is possible to find an antiderivative of f as

$$f(y)-f(0)=\frac{1}{\eta}\left[\int_0^y(qs+\xi)ds+k\int_0^y(qs+\xi)^3ds\right], \tag{11}$$

where s - variable of integration.

Afterwards, based on (11) the velocity component u is obtained in a form:

$$u=\frac{1}{\eta}\left[q\int_0^y\left(s-\frac{h}{2}\right)ds+\lambda y+k\int_0^y\left(qs-\frac{qh}{2}+\lambda\right)^3 ds\right]=$$

$$=\frac{1}{\eta}\left\{q\Phi_1(y)+\lambda y+k\left[q^3\int_0^y\left(s-\frac{h}{2}\right)^3 ds+3q^2\lambda\int_0^y\left(s-\frac{h}{2}\right)^2 ds+3q\lambda^2\int_0^y\left(s-\frac{h}{2}\right)ds+\lambda^3 y\right]\right\}. \quad (12)$$

Neglecting in (12) terms multiplying by second and third powers of λ yields

$$u\approx\frac{1}{\eta}\left[q\Phi_1(y)+kq^3\Phi_2(y)\right]+\frac{\lambda\left[y+kq^2\Phi_3(y)\right]}{\eta}, \quad (13)$$

where:

$$\Phi_1(y)=\int_0^y\left(s-\frac{h}{2}\right)ds=\frac{1}{2}y(y-h);$$

$$\Phi_2(y)=\int_0^y\left(s-\frac{h}{2}\right)^3 ds=\frac{1}{4}\left(2y^2-2hy+h^2\right)\Phi_1(y); \quad (14)$$

$$\Phi_3(y)=3\int_0^y\left(s-\frac{h}{2}\right)^2 ds=\frac{1}{4}y\left(4y^2-3hy+3h^2\right).$$

The expression for w is similar to (13) without the term multiply by λ . Therefore

$$u\approx\frac{1}{\eta}\left[q\Phi_1(y)+kq^3\Phi_2(y)\right]+\frac{u_h\left[y+kq^2\Phi_3(y)\right]}{h\left[1+k(q/2)^2h^2\right]},\quad q=\frac{\partial p}{\partial x}; \quad (15)$$

$$w=\frac{1}{\eta}\left[q\Phi_1(y)+kq^3\Phi_2(y)\right],\quad q=\frac{\partial p}{\partial z}. \quad (16)$$

Integrating the continuity equation (3.d) with respect to $y\in[0,h]$ leads to

$$\int_0^h\frac{\partial u}{\partial x}dy+\int_0^h\frac{\partial v}{\partial y}dy+\int_0^h\frac{\partial w}{\partial z}dy=0. \quad (17)$$

As the lubricant film thickness $h=h(x,z)$, using the Leibniz integral rule

$$\frac{\partial}{\partial s}\int_{a(s)}^{b(s)}f(x,s)dx=\int_{a(s)}^{b(s)}\frac{\partial f(x,s)}{\partial s}dx+f(b(s),s)b'(s)-f(a(s),s)a'(s)$$

and applying the boundary conditions, for the continuity equation (17) follows

$$\frac{\partial}{\partial x}\int_0^h u\,dy-u_h\frac{\partial h}{\partial x}+v_h+\frac{\partial}{\partial z}\int_o^h w dy=0. \quad (18)$$

Integrating the first term of (18) with respect to $y \in [0,h]$ and rendering into account (15) leads to

$$\int_0^h u\,dy = \frac{1}{\eta}\left[q\int_0^h \Phi_1(y)dy + kq^3\int_0^h \Phi_2(y)dy\right] + \frac{\lambda}{\eta}\left[\int_0^h y dy + kq^2\int_0^h \Phi_3(y)dy\right]. \qquad (19)$$

Analogously, the integral of w in the forth term of (18) is obtained by (16) in a form

$$\int_0^h w\,dy = \frac{1}{\eta}\left[q\int_0^h \Phi_1(y)dy + kq^3\int_0^h \Phi_2(y)dy\right]. \qquad (20)$$

By using (14) the integrals of Φ_1, Φ_2, Φ_3 in (19) and (20) are respectively equal to

$$\int_0^h \Phi_1(y)dy = -\frac{h^3}{12}; \quad \int_0^h \Phi_2(y)dy = -\frac{h^5}{80}; \quad \int_0^h \Phi_3(y)dy = \frac{h^4}{8}. \qquad (21)$$

Then after substitution of the velocity components (15), (16) in (18) and using (21) the following Reynolds type equation is obtained

$$\frac{1}{\eta}\frac{\partial}{\partial x}\left[\frac{h^3}{12}\frac{\partial p}{\partial x} + k\frac{h^5}{80}\left(\frac{\partial p}{\partial x}\right)^3\right] + \frac{1}{\eta}\frac{\partial}{\partial z}\left[\frac{h^3}{12}\frac{\partial p}{\partial z} + k\frac{h^5}{80}\left(\frac{\partial p}{\partial z}\right)^3\right] = \frac{\omega r}{2}\frac{\partial h}{\partial x}. \qquad (22)$$

This is the modified Reynolds equation for a finite length journal bearing lubricated with non-Newtonian Rabinowitsch fluid under steady state conditions. The type of equation coincides with the one obtained in [1] as here the coefficient of pseudoplasticity k is also involved. It is known that for Newtonian fluids $k = 0$ and from (22) can be obtained the classical Reynolds equation for Newtonian fluids. The above equation is more general than classical one, as it includes the case of non-Newtonian lubricants ($k \neq 0$).

This equation has to be represented in a non-dimensional form by introducing the coordinate transformations and corresponding dimensionless variables, according to substitutions:

$$\theta = \frac{x}{R} \approx \frac{x}{r}; \quad z_1 = \frac{z}{L/2}; \quad \alpha = \frac{2r}{L}; \quad \beta = \frac{c}{r}; \quad \Pi = \frac{c^2}{6\eta U r}p; \quad H = \frac{h}{c}, \qquad (23)$$

where: θ - circumferential coordinate, z_1 - dimensionless axial coordinate, L - bearing axial length, α - diameter to length ratio, β - clearance ratio, Π - dimensionless pressure, $U = \omega r$ - journal circumferential velocity, H - dimensionless film thickness.

All variables in (22) are changed with non-dimensional ones (23), as for the partial derivatives the following equalities are in effect

$$\frac{\partial}{\partial x} = \frac{1}{r}\frac{\partial}{\partial \theta}; \qquad \frac{\partial p}{\partial x} = \frac{\partial \Pi}{\partial \theta}\frac{6\eta\omega r}{c^2};$$

$$\frac{\partial}{\partial z} = \frac{1}{L/2}\frac{\partial}{\partial z_1}; \qquad \frac{\partial p}{\partial z} = \frac{\partial \Pi}{\partial z_1}\frac{6\eta\omega r^2}{(L/2)c^2}; \qquad (24)$$

$$\frac{\partial h}{\partial x}=\frac{c}{r}\frac{\partial H}{\partial \theta}.$$

After substitution of (23) and (24) in (22) and applying some mathematical transformations the final dimensionless form of 2D Reynolds equation for pressure distribution is obtained as:

$$\frac{\partial}{\partial \theta}\left[\frac{H^3}{12}\frac{\partial \Pi}{\partial \theta}+\psi\frac{36H^5}{80}\left(\frac{\partial \Pi}{\partial \theta}\right)^3\right]+\alpha^2\frac{\partial}{\partial z_1}\left[\frac{H^3}{12}\frac{\partial \Pi}{\partial z_1}+\alpha^2\psi\frac{36H^5}{80}\left(\frac{\partial \Pi}{\partial z_1}\right)^3\right]=\frac{1}{12}\frac{\partial H}{\partial \theta}. \quad (25)$$

In this equation the non-dimensional nonlinear factor ψ, which takes into account the non-Newtonian behaviour of a fluid, described by the cubic model, is involved. According to [1, 13, 18, etc.] the nonlinear factor is represented by the following relationship

$$\psi=k\left(\frac{\eta U}{c}\right)^2=k\left(\frac{\eta \omega r}{c}\right)^2. \quad (26)$$

When the nonlinear factor ψ is set to zero, the Reynolds classical equation for a journal bearing lubricated with a Newtonian fluid is obtained. At pseudoplastic non-Newtonian fluids $\psi>0$, while at the dilatant ones $\psi<0$ [13, 26, 18]. As it can be seen from (26), the nonlinear factor values depend on the coefficient of pseudoplasticity, the lubricant initial viscosity, the velocity of journal, and the radial clearance. Given the relation (26) between the nonlinear factor and the coefficient of pseudoplasticity, the increase of k increases the values of ψ. For Newtonian fluids $\psi=0$ because $k=0$. Namely by introducing this dimensionless variable into equation (25), the rheological characteristics of the pseudoplastic and dilatant fluids, which can be described by the Rabinowitsch fluid model, are taken into consideration in the mathematical model.

3 Particular cases

As it was written in [1], the analysis of Reynolds type equation as (23) can be performed for small and respectively for large values of the nonlinear factor by using different approaches.

Furthermore, for small nonlinear factor two particular cases of 1D journal bearings lubricated with a Rabinowitsch fluid can be considered at using the long and short bearing approximations, which are by the other side extensively applied in the studies of lubrication theory.

It is well known that for an infinitely long bearing the derivative $\partial p/\partial z=0$, because of which the modified Reynolds equation for such kind of bearings is obtained from (22) and written as

$$\frac{\partial}{\partial x}\left\{\frac{1}{\eta}\left[\frac{h^3}{12}\frac{\partial p}{\partial x}+k\frac{h^5}{80}\left(\frac{\partial p}{\partial x}\right)^3\right]\right\}=\frac{\omega r}{2}\frac{\partial h}{\partial x}. \quad (27)$$

Using (23) and (24) a non-dimensional form of (27) satisfies the equation:

$$\frac{\partial}{\partial\theta}\left[\frac{H^3}{12}\frac{\partial\Pi}{\partial\theta}+\psi\frac{36H^5}{80}\left(\frac{\partial\Pi}{\partial\theta}\right)^3\right]=\frac{1}{12}\frac{\partial H}{\partial\theta}\ . \tag{28}$$

The pressure distribution for the case of an infinitely short bearing is governed by the following equation, which is also a reduced form of (22):

$$\frac{\partial}{\partial z}\left\{\frac{1}{\eta}\left[\frac{h^3}{12}\frac{\partial p}{\partial z}+k\frac{h^5}{80}\left(\frac{\partial p}{\partial z}\right)^3\right]\right\}=\frac{\omega r}{2}\frac{\partial h}{\partial x} \tag{29}$$

Here the second term on the left hand side of (22) is vanished since $\partial p/\partial x$.
Dimensionless form of (29) is obtained by using of (23) and (24) and is given as:

$$\alpha^2\frac{\partial}{\partial z_1}\left[\frac{H^3}{12}\frac{\partial\Pi}{\partial z_1}+\alpha^2\psi\frac{36H^5}{80}\left(\frac{\partial\Pi}{\partial z_1}\right)^3\right]=\frac{1}{12}\frac{\partial H}{\partial\theta} \tag{30}$$

Both of the equations for infinitely long and short bearings, respectively, can be solved by analytical or numerical methods as several solutions of similar type of equations are already published.

4 Conclusions

This paper presents precisely, step by step, with necessary explanation obtaining of a modified Reynolds equation for lubrication of finite length journal bearing lubricated with non-Newtonian Rabinowitsch fluid.

Starting from the momentum and continuity equations, and based on the cubic stress constitutive equation, the velocity components for two dimensional flow of a Rabinowitsch fluid are obtained. For this purpose differentiation and integration techniques are applied.

The constitutive equations for the shear stress and shear strain rate are represented by a cubic function of a special type, which finally, after mathematical treatment, are reduced to semi-cubic equations for their components. An approximation to the real root of semi-cubic equation is used such as in [1]. It is possible to find by Cardano's method all roots of the semi-cubic equation, and in further mathematical processing to operate with some appropriate their approximations (due to the complexity of Cardano's formula).

Afterwards, a standard approach for the lubrication theory is applied. Integrating the continuity equation across the fluid film and substituting the obtained velocity components in it, the nonlinear modified 2D Reynolds type equation is derived. The obtained equation is slightly differing in coefficients from those in [1]. This equation is in the appropriate form for numerical modelling and solving by some mesh methods or adapted computer codes ones.

Based on the presented above, a study of the dilatant and pseudoplastic nature of the lubricant in comparison with Newtonian fluid can be carried out. The performances of journal bearings lubricated with a non-Newtonian Rabinowitsch fluid can be compared with the case of Newtonian lubricant through the variation of the non-Newtonian parameter, i.e. the nonlinear factor ψ . The future analyses can be performed with combinations of other effects in lubrication performances of journal bearings as influence of fluid inertia, elastic deformations effects, surface roughness, etc.

References

1. S. Wada, H. Hayashi, Theoretical and experimental studies, Bulletin of JSME **14**, 279-286 (1971)
2. C. Rajalingham, B.V. Rao, S. Prabu, Wear **50**, 47-57 (1978)
3. P. Bourging, B. Gay, Journal of Tribology **106**, 285–290 (1984)
4. S. Sharma, S. Jain, P. Sah, Tribology International **33**, 507–517 (2000)
5. J. Javorova, A. Radulescu, R. Lovchalieva, N. Nikolov, P. Kostova, Trib. J Bultrib **III**, 248-252 (2012)
6. J. Javorova, A. Mazdrakova, I. Andonov, A. Radulescu, Tribology in Industry, **38**, 186-196 (2016)
7. P.B. Kushare, C.S. Sharma, Tribology International **69**, 84-101 (2014)
8. J. Lin, T. Hung, C. Lin, J of Mechanics, doi.org/10.1017/jmech.2017.48, (2017)
9. H. Hashimoto, S. Wada, J of Tribology **108**, 282–288 (1986)
10. J. R. Lin, Tribology International **52**, 190–194 (2012)
11. N.B. Naduvinamani, M. Rajashekar , A. Kadadi, Tribol. Int. **73**, 78–82 (2014)
12. V. Shigehalli, B. Hanumagowda, C. Rajani, Int J Mathem Arch **8**, 36-46 (2017)
13. J.R. Lin, Tribology Letters **10**, 237-243 (2001)
14. J. Lin, L. Chu, C. Hung, R. Lu, M. Lin, Z. Naturforsch. **68a**, 291-299 (2013)
15. J.R. Lin, Appl Mathem Modelling **38**, 392-404 (2014)
16. U.P. Singh, R.S. Gupta, V.K. Kapur, Tribol. Trans. **54**, 723-29 (2011)
17. U.P. Singh, R.S. Gupta, V.K. Kapur, IMechE Journal of Engineering Tribology **227**, 34-42 (2012)
18. U.P. Singh, R.S. Gupta, V.K. Kapur, Tribology in Industry **34**, 128-137 (2012)
19. U.P. Singh, Archive of Mechanical Engineering **LX**, 247 – 267 (2013)
20. U.P. Singh, Int J Eng Sci Reviews **14**(4), 1411-1420 (2014).
21. A. Walicka, E., Walicki, M. Ratajczak, Proc. 4th Minsk Int. Heat Mass Transfer Forum, Minsk Belarus, 19-29 (2000)
22. M. Ratajczak, A. Walicka, E. Walicki, Problems of Machines Exploitation **44**, 159-170 (2006)
23. A. Walicka, E., Walicki, Int. J. Appl. Mech. Enging **15**, 895-907 (2010)
24. A. Walicka, E. Walicki, P. Jurczak, J. Falicki, Int. J. Applied Mechanics and Engineering **22**, 717-737 (2017)
25. A. Cameron, *Basic lubrication theory* (Chichester, Ellis Horwood, 1976)
26. C. Hung, Education Specialization **97**, 87-97 (2009)

MATEC Web of Conferences **145**, 03008 (2018) https://doi.org/10.1051/matecconf/201814503008
NCTAM 2017

Viscous flow around spherical particles in different arrangements

Tsvetan Kotsev*
Sofia University, Department of Mathematics and Informatics, James Boucher blvd. No. 5, 1014 Sofia, Bulgaria

Abstract. Viscous flow past two or more spherical particles in direction parallel or perpendicular to the line connecting the body centers is studied numerically at Reynolds number up to 200. The FEM is used to simulate the flow structure for different arrangements and diameters of the spheres. The drag and lift coefficients are calculated as functions of the flow velocity and the distance between the spheres. The results show the downstream spheres face a reduced drag compared to a single sphere and that the upstream one changes its drag coefficient according to the distance between the bodies. In the case of side by side arranged spheres, an effect of repulsion exists up to a certain separation distance and after that a weak pulling effect appears. A configuration of three (or more) spherical particles arranged in „V" formation was examined where one body is set to be a leading and the others are located symmetrically up and down behind it. The flow characteristics are studied when the angle behind the leading sphere vary from 40^0 to 180^0. This configuration supposes stronger interaction between the bodies due to the influence of the wakes behind the first sphere to the downstream ones and to the lift forces appeared.

1 Introduction

Investigation of the flow and hydrodynamic interaction between droplets, solid spherical particles is important in various engineering applications and chemical processes, pollution diffusion, energy systems, ship hydrodynamics, etc. Classical problem of Stokes [1] for a single sphere in uniform incompressible flow is the first one among many others that was solved in analytical way when the inertial terms are neglected but relatively less are the studies investigating the flow past two or more particles. Stimson and Jeffrey [2] have studied the flow past two spheres that are close together. Later Kim [3] has suggested a technique for multi-sphere Stokes flow problems that involves the analytical transformation of basic functions between coordinates based on different spheres. An analytical study of hydrodynamic interaction of three spheres in a quiescent viscous fluid was considered by Wilson [4] for spheres with equal and different diameters using a method based on Lamb's solution to Stokes flow and the method of reflections. Fluid dynamic interaction between two spheres in tandem configuration was experimentally studied by Tsuji, Morikawa &Terashima [5] for Reynolds numbers between 100 and 1000 and Zhu, Liang & Fan [6] for Re varying from 20 to 130. They have investigated the evolution of the flow field and drag experienced by the bodies when the distance between them vary. Chen and Wu [7] studied the flow characteristics between the

* Corresponding author: tkotsev@fmi.uni-sofia.bg

two spheres experimentally at Reynolds numbers in the range of $54 \leq Re \leq 118$. Applications of computational fluid dynamics (CFD) have widely increased in the field of heat and fluid flow. The benefits of CFD compared to the experimental methods are that simulation is much cheaper and not time consuming. For moderate Reynolds numbers where both the viscous and convective terms must be taken into consideration the only way to receive an approximately good solution is the numerical calculations and in the literature exist such investigations but as a whole they are not so many especially for complex body configurations. Three-dimensional flow over two spheres placed side-by-side was investigated by Kim et al. [8] at Re=50, 100, and 150 for various separation distances. They considered the lift, moment and drag coefficients on the spheres and investigated the dependence on the distance between the spheres. Schouveiler et al. [9] studied the interactions of the wakes of two spheres placed side-by-side both experimentally and numerically. They applied various separation distances, for example, in the range of $1.0 \leq d \leq 4{,}6$ at Re=300 and revealed that for a given Reynolds number, the wake frequency varies smoothly with separation distance.

In the present study the steady flow past two, three and more spherical particles arranged in line or side by side is simulated at Reynolds numbers up to 200. Comparison both with experimental and numerical results was done to test the model. A complex configuration of three and seven spherical particles arranged in „V" formation was examined. The flow characteristics are studied when the angle behind the leading sphere varies from 40^0 to 180^0. This configuration supposes stronger interaction between the bodies due to influence of the wakes behind first sphere to the downstream ones and to the lift forces appeared.

2 Statement of the problem and method of solution

Two or more rigid spherical particles are immersed in a uniform stream of viscous fluid. The velocity of the ongoing fluid is V_0 and is parallel or perpendicular to the line connecting the body centers. The fluid is accepted to be water with density of 998 [kg/m^3] and dynamic viscosity 0,00101 [Pa.s]. The following arrangements of the spheres according to the flow direction are considered:

a) The flow direction is parallel to the line connecting the body centers. Their diameters may be equal or different. In this case the flow is axisymmetric in respect to the central line that allows the three-dimensional problem to be reduced to a two-dimensional one.

b) The flow direction is perpendicular to the line connecting the centers of the spheres (side by side arrangement) that makes the problem three-dimensional.

c) The flow around three or more spheres arranged in „V" formation where one sphere is set to be a leading one and the others are located symmetrically up and down behind it.

It is obvious both from the experiments and numerical simulations that the flow past a single sphere becomes unstable for Re about 200 when the wake starts to oscillate. Here Re is based on the diameter of the upstream sphere – $Re=2R_1.V_0/\nu$, where ν is the kinematic viscosity and takes values from 0,01 to 200. Another parameter having strong influence on the flow picture is the distance between the centers varying here from 0,25 up to 30 times the radius of the upstream sphere. For the purpose of the present calculations a cartesian coordinate system is used with x axis parallel to the line connecting the sphere centers in the axisymmetric case and perpendicular to this line in the case for side by side arranged bodies. Equations of the flow motion are the continuity equation (1) and the momentum equation for steady Newtonian flows (2).

$$\nabla.\mathbf{v} = 0 \tag{1}$$

$$\rho(\mathbf{v}.\nabla)\mathbf{v} = -\nabla p + \mu\nabla^2\mathbf{v}, \tag{2}$$

where $\mathbf{v}$ is velocity vector, μ – dynamic viscosity, p – pressure, ρ – fluid density.

Boundary conditions are:

- no slip on the sphere surfaces – $\mathbf{v} = 0$ (3)
- uniform velocity at the inlet – $v = V_0$ (4)
- zero pressure at the outlet – $p = 0$ (5)
- symmetry at the other boundaries – $\mathbf{v}.\mathbf{n} = 0$. (6)

The drag and lift forces are calculated integrating the stress tensor on the body surfaces:

$$F_D = \int_S -p\mathbf{n}.\mathbf{i}dS + \int_S \mathbf{n}.\boldsymbol{\tau}.\mathbf{i}dS\,, \quad F_L = \int_S -p\mathbf{n}.\mathbf{j}dS + \int_S \mathbf{n}.\boldsymbol{\tau}.\mathbf{k}dS, \tag{7}$$

where S denotes the surface of the spheres, **i** and **j** are unit vectors in x and y direction respectively, **n** is the outward normal vector at the surface and $\boldsymbol{\tau}$ is the viscous stress tensor.
Non-dimensional drag and lift coefficients C_D and C_L are defined as:

$$C_D = \frac{F_D}{\rho V_0^2 A_{pr}/2} \qquad C_L = \frac{F_L}{\rho V_0^2 A_{pr}/2}\,, \tag{8}$$

where A_{pr} is the area of the sphere projected on a plane perpendicular to the flow direction.
The equations (1) and (2) were solved numerically using COMSOL Multiphysics package based on the finite element method. It supposes covering the computational domain with a mesh of different type of elements. Most used ones are tetrahedral in 3D and triangle in 2D case because these elements approximate in the best way any complicated geometrical domain. The software gives a possibility to cover with a fine mesh the regions near to the bodies where the gradients of the velocity are significant and with a coarse one the area far from them (Fig.1).

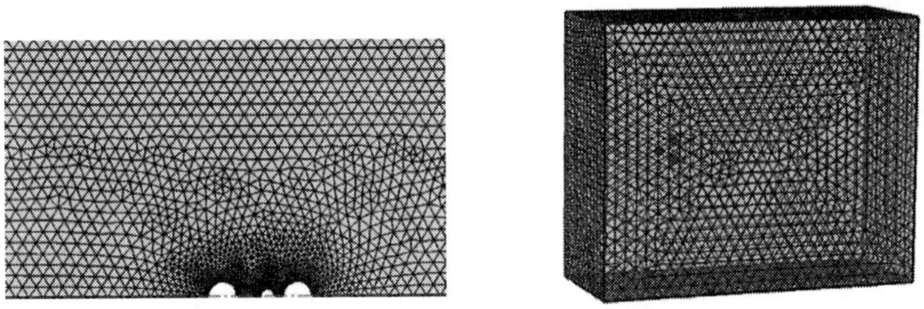

a) b)

Fig. 1 a) Mesh with triangle elements for 2D. b) Mesh with tetrahedral elements for 3D case.

Taking into account the information about the physics of the process, material of the bodies, properties, boundary conditions, elements type and shape, the software assembles a system of equations which need to be solved to get an answer of the finite element problem. The size of the matrix of this system is the number of degrees of freedom (DOF) of the model, where the number of DOF is a function of the number of elements, the discretization order used in each physics, and the number of variables solved for. For example, in the axisymmetric case of flow past two spheres in line the complete mesh consists of 23 352 domain elements and 626 boundary elements and total number of DOF 161370. To solve the system of equations both direct and iterative solvers could be used, this one used here is GMRES – an iterative solver that supposes good convergence, solution time and use of memory. In the case of side by side located spheres an extra fine mesh of elements was used that results in 795 537 domain elements, 18346 boundary elements, and 616 edge elements and 551076 DOF.

3 Results

In order to test the model a comparison with the flow around a single sphere was done. For that purpose the distance between two equal spheres in tandem arrangement was set to be 30 times the radius of the upstream sphere, so the interaction between the bodies to be negligible small. At Fig. 2 a) is shown the experimental result [10] for the laminar flow around a single sphere for Re=118. Numerical results for the same Re and spheres far enough one form another show good accordance with the experiment carried out using water and aluminum dust to visualize the flow.

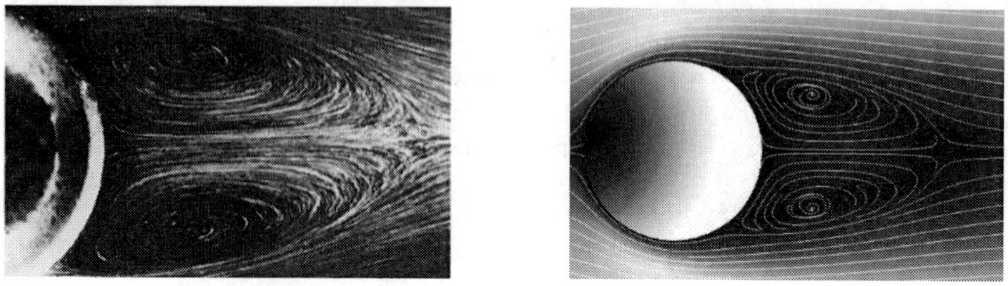

a) b)

Fig. 2 a) Experimental study of the flow around a single sphere, Re=118 b) present calculations

It could be seen that the structure of the recirculating zones behind the spheres is identical and the length of the eddies in both cases is about one sphere diameter. The length of the vorticity zone behind the sphere increases as a logarithm of Reynolds number unlike of the case for a cylinder where the wake grows almost linear with Re. Line representing the drag coefficient of a single sphere measured experimentally for wide range of Re together with Stokes and Oseen's approximations for small Re are shown at Fig.3. The values from the numerical calculations are also noted with black dots. It is seen that our data are in a good accordance with Stokes and Oseen's solutions for small Reynolds numbers as well as with the experimental data for moderate Re.

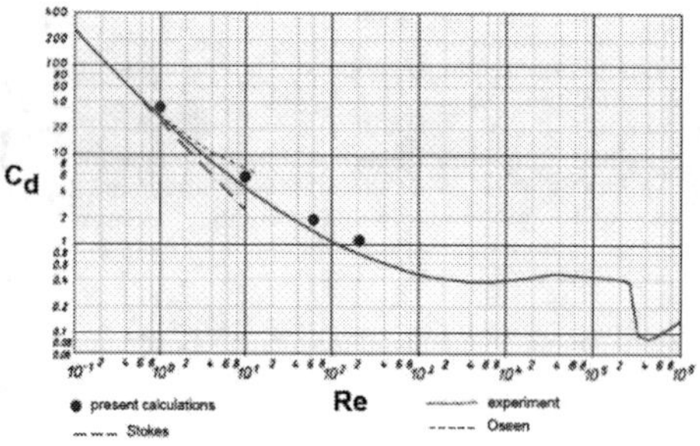

Fig. 3. Comparison with the drag coefficient for a single sphere.

3.1 Flow past spheres in line arrangement

3.1.1 Two equal spheres in tandem

The flow structure past two spheres in tandem for small and moderate Reynolds numbers was a subject of study both analytically and numerically respectively. Existence of a suitable coordinate system allowing to describe the body surfaces as coordinate lines and developing of the numerical techniques has resulted in many studies – Zapryanov et al. [11], Tal et.al. [12], etc. Here an attempt was done to confirm and expand these results and to demonstrate the efficiency of the FEM integrated in COMSOL package to solve complicated flow problems.

Numerical simulations were done for two spheres with radius 0,02 [m] and Reynolds numbers from 0,01 to 200 for different distances between the spheres. In all cases considered, the distance between bodies is denoted with *d* and is normalized by the diameter of the upstream sphere. The flow picture past two equal spheres for distance between them equal to 1 diameter is shown at Fig. 4. Up to Re=10 the flow is similar to Stokes one – there is no separation of the fluid from the surfaces and no forming of vortices behind the bodies. Experiments show that for a single sphere separation of the flow from the surface begins at Re about 25. Here due to a presence of a second body in the wake behind the upstream one the separation from the first sphere appears at lower values of Re – about 20.

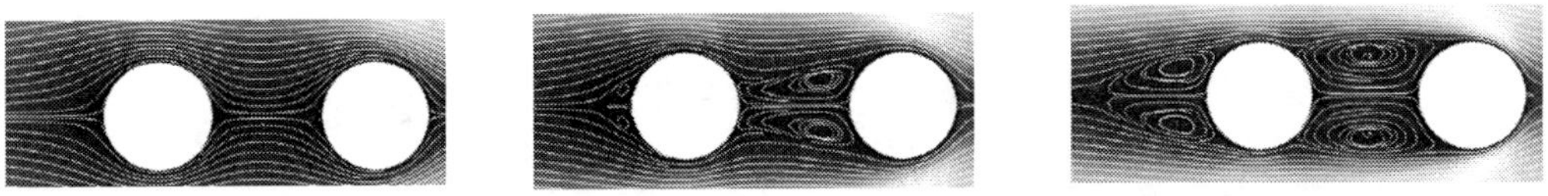

Re=10 Re=47 Re=200

Fig. 4. Flow around two equal spheres at d=1.

The point of separation is at first close to the rear of the sphere and separation results in the formation of a ring eddy attached to the rear surface of the sphere. As Re increases the point of separation moves up to the side of the sphere and the wake in the gap grow and enlarges toward the second body. At Re=30 a small vortex is seen in front of the downstream sphere that merge with this one behind the upstream body when the velocity of the fluid grows. There is no vortex zone behind the downstream sphere, it appears at Re about 40. If the distance between spheres decreases to 0,5 the mention above events begin earlier – separation of the fluid from upstream sphere is visible at Re=10 and the vortex zone in the gap is well formed for lower Reynolds number – Re=25. When the two spheres touch each other, the total drag of this assemblage is less than the drag of a single sphere but this depends on Re and is connected with the flow structure. For example, at Re=10 there is no eddy behind the downstream body and the total drag of two touching spheres is bigger than the drag of a single sphere. At Re=62 when vorticity behind the second body is well formed, then the total drag becomes less than the drag of a single body. This effect disappears for distance more than 0,21 and the sum of the drag of two bodies becomes higher than this one of a single one. The distance between the bodies at which the total drag is less that this one of a single body grows with Re and at Re=200 it is appr. 0,5 but the most important factor that influences the hydrodynamic interaction of the particles is the distance between them.
The drag coefficient of the spheres with respect to Reynolds number is shown at Fig.5.

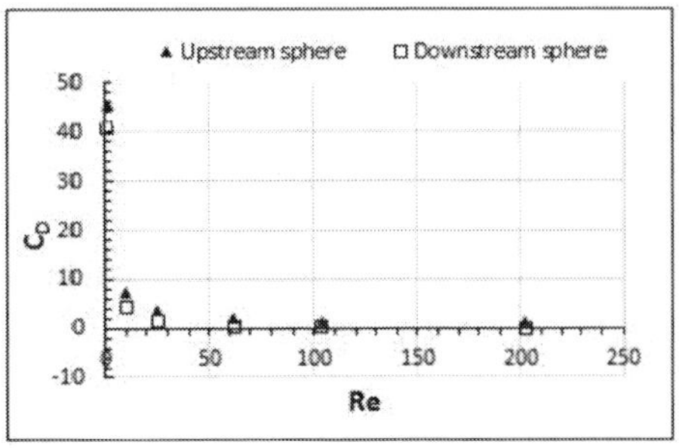

Fig. 5 Drag coefficient of two equal spheres at d=1 vs Reynolds number.

The downstream sphere faces a reduced drag compared to an upstream one. It decreases rapidly with Re, becomes close to zero and for Re=200 is negative even with a small value that means there is a weak attractive effect between bodies. The upstream sphere also changes its drag coefficient – it decreases or increases by approximately 12% according to the drag of a single sphere and depends strongly on the distance between the spheres.

3.1.2. Two different spheres in tandem

A case when the upstream sphere is bigger than the downstream one is considered. The radius of the first sphere is 0,02 [m], this one of the second is twice less – 0,01 [m]. A flow picture for Re=62 and d=0 and d=0,5 is shown at Fig.6. Like in the case of equal spheres at distance d=0,5 the vortex zone behind the bigger upstream sphere appears for Re about 10, but separation from the downstream one could be seen for quite higher values of Reynolds number – Re≈ 90.

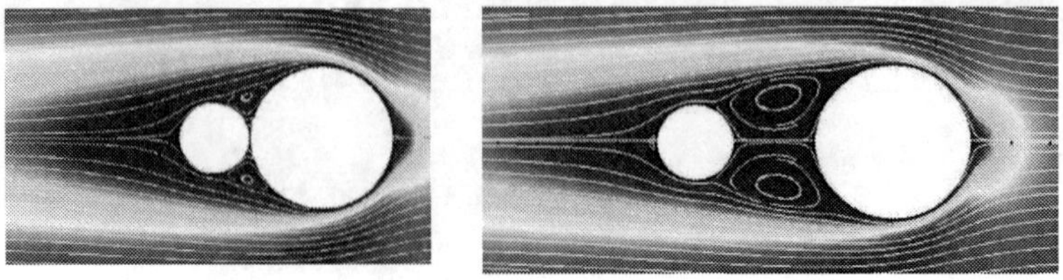

a) b)

Fig. 6. Flow past two different spheres a) d=0, b) d=0,5

To follow the dependency of the drag coefficient with Reynolds number calculations were done for Re=10, 25, 62, 100 and 200 for distances between the spheres equal to 0,5 and 1 (Fig. 7a). The drag of the upstream body falls down with Re but stays almost the one and the same both for d=0,5 and 1 at all values of Re considered. The drag coefficient of the downstream sphere also decreases with Re and for values of Re>50 and d=0,5 becomes negative that indicates attractive force between bodies. The interaction between different spheres with distance d for fixed Reynolds number (Re=62) is shown at Fig. 7 b) where d takes values 0, 0,5, 1, 1,5 and 2,5. For comparison is drawn also the curve representing the drag coefficients of two equal spheres for the same parameters. When the two spheres touch each other the total drag analogous to the case of equal spheres is also less than the drag of a single sphere.

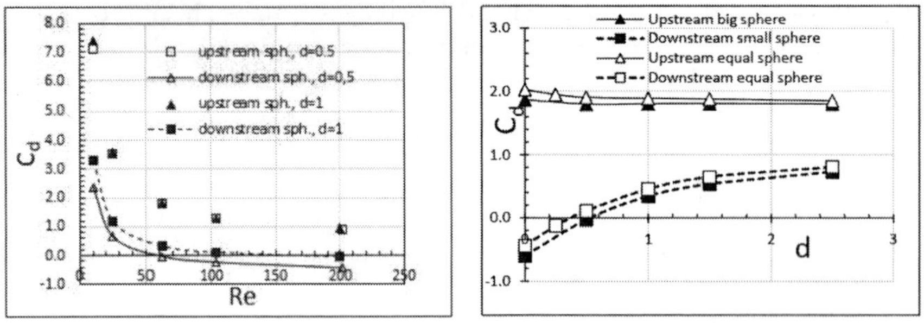

a) b)

Fig. 7 Drag coefficient vs Re and d of two different spheres:
a) d= 0,5; 1 b) Re=62, d= 0; 0,5; 1; 1,5; 2,5

The distance between spheres up to which this trend is valid for Re=62 is appr. d=0,5 while for the case of equal spheres this distance is appr. twice less. The drag coefficient of the first sphere is positive and decreases slightly with d while this one for the second sphere stays negative up to d= 0,5. With further taking away the second body, C_{D1} is almost constant while C_{D2} increases and takes positive values.

3.1.3. Three equal spheres in line

If we denote with d_{12} the distance between upstream and middle sphere and with d_{23} – the distance between the middle and downstream sphere, the following cases are investigated:

d_{12}	0	0	0	1	0.5	1
d_{23}	0	0.5	1	1	0	0

Fig.8 shows the flow at Re=62 and two distances between bodies – d_{12}=d_{23}= 0 and d_{12}= d_{23}= 1. In the gaps between spheres recirculating zones exists even for small Re, but this one behind the downstream body appears at Re about 42 that is almost twice bigger compared to the flow around a single sphere.

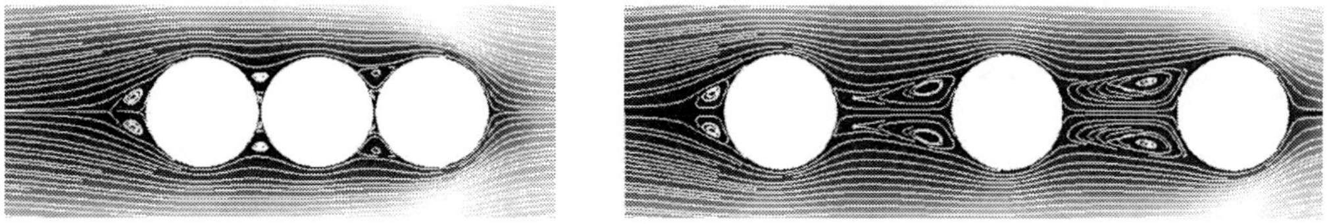

a) b)

Fig. 8. Flow past 3 equal spheres for Re=62: a) touching spheres b) d_1=d_2= 1 diameter

The variation of the drag coefficients with ratio k=d_{12}/d_{23} and for Re=62 is shown at Fig. 9a). When k=1 all spheres are at one and the same distances each from another. Moving the middle sphere towards the first one its drag coefficient rapidly decrease while the drag coefficient of the upstream sphere increase with appr. 9%. At a distance between upstream and middle sphere about 0,25 (k=0,14) the drag coefficient of the middle body takes negative values. If the middle body is moving toward the downstream one (k >1) C_{D1} keeps almost one and the same value while the drag coefficient of the downstream body quickly fall down but is positive even when second sphere touch the third one.

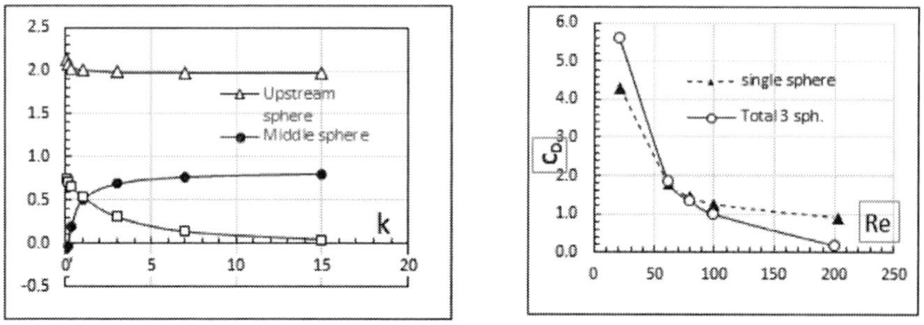

a) b)

Fig. 9. Drag coefficient of three equal spheres: a) Re=62 b) vs k=d_{12}/d_{23}.

Three touching spheres feel less drag than a single sphere with the same diameter for Reynolds numbers higher than 60 (Fig.9b).

3.1.4. Three different spheres in line

The case when upstream and downstream spheres have equal diameters and are placed at a distance d=3 is considered. The middle body has twice less diameter and changes its position. Initially it touches the upstream sphere, then starts to move towards the downstream one. The flow picture is shown at Fig.10 for Re=62.

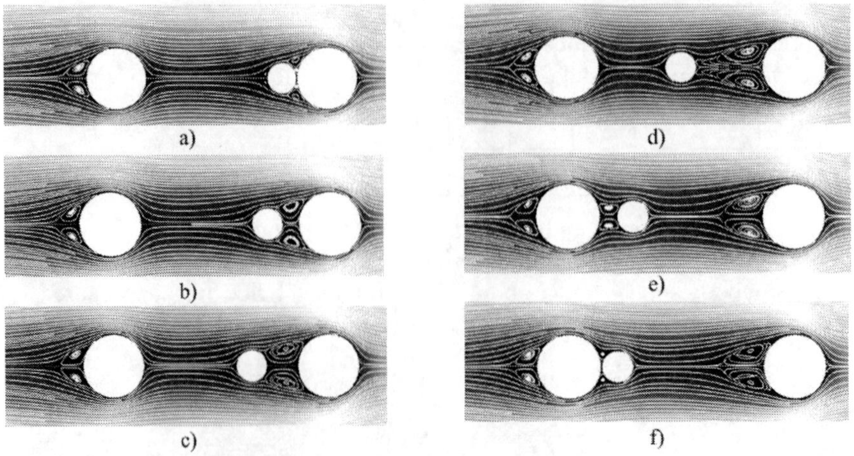

Fig. 10. Flow past three different spheres at Re=62 and various $k=d_{12}/d_{23}$.
a) k=0, b) k=0,14, c) k=0,33, d) k= 1, e) k=7, f) sphere 2 touch sphere 3

Fig.11 shows the variation of the drag coefficients with Reynolds number and with ratio k at fixed Re=62. They decrease with increasing of Re, C_{D1} and C_{D3} stay positive while C_{D2} accept negative values for Reynold numbers higher than 55.

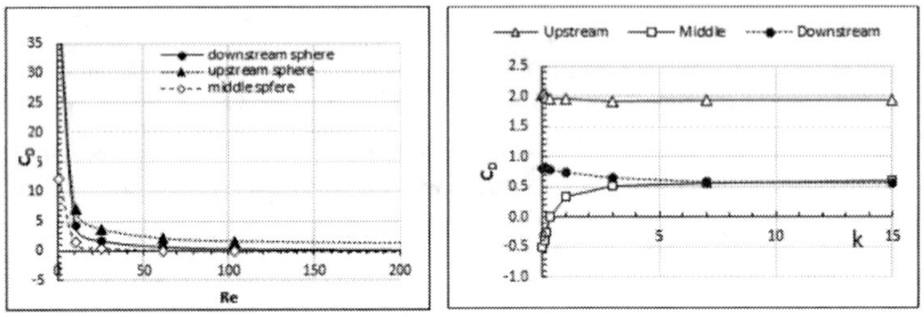

Fig. 11. Drag coefficient of three different spheres vs Re for fixed distance $d_{12}= d_{23}= 0,5$.

When k=0 (upstream and middle sphere are touching) the upstream sphere feels almost the same drag as the corresponding one in the case of equal spheres – C_{D1} is 2,038 in the case of different spheres and 2,12 in the case of equal bodies. Increasing k, the drag falls very slow and keeps almost a constant value. The drag coefficient of the downstream sphere C_{D3} decreases with k and for k>7 takes almost a constant value of 0,5 as opposed when bodies have equal dismeters where C_{D3} takes positive values close to zero. Quite different is the behavior of the drag of the middle sphere. Being twice smaller than the other two ones it has negative C_{D2} for higher valuse of k compared with the previous case. C_{D2} is negative up to k=0,33 (1 radius between spheres 1 and 2) while for the case of equal spheres the drag coefficient takes negative values up to k=0,14. In other words, the smallest it is the middle sphere, the longer is the distance from the upstream one at which $C_{D2}<0$.

3.2 Flow past spheres in side by side arrangement

3.2.1. Two side by side spheres

A steady 3D incompressible laminar flow past two identical spheres held fixed with the line connecting the sphere centers normal to a uniform stream was considered. The simulation of the flow was done for Re=10, 50, 62, 200 and separation distances normalized by the sphere diameter from 0 (touching spheres) up to 12. The hydrodynamic interaction in this case strongly depends on the separation distance and different flow regimes could be seen past the bodies. In the case when the spheres touch each other, only one single wake is visible behind them (Fig. 12a). When the separation distance is small the wakes behind each sphere interact strongly and the combined wake is offset with respect to the center plane between the spheres.

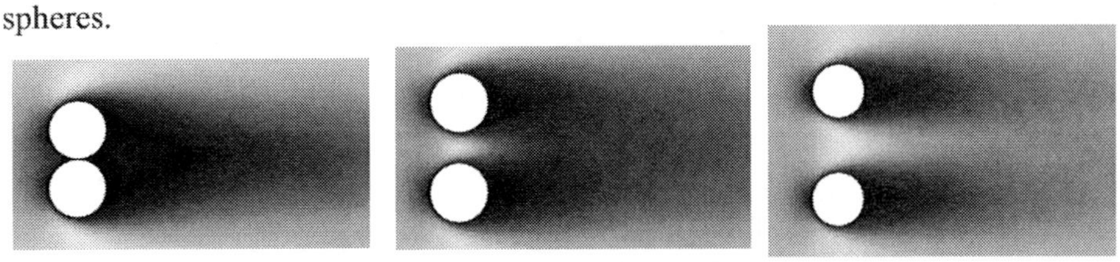

a) b) c)

Fig. 12 Flow picture at Re=50: a) d=0, b) d=0,5, c) d=1

This trend continues up to separation distance about 3 when the wakes are almost independent each other. Comparison for the drag coefficients with results of Folkersma et all [12] was done at Re=10 and 50 and for a separation distance d=0,5. Results in [12] are 4,87 and 1,74 respectively while our values are 4,618 and 1,73. The difference is about 5% at Re=10 and less than 1% at Re=50. For separation distances less than 10-10,5 and Re=50 the lift coefficient is negative that means there is a repulsion of the two spheres. For bigger separation distances the lift coefficient is positive but relatively small that means there is a weak attractive force between the spheres (Fig.13). Further increasing the separation distance leads to a negligible small interaction between spheres.

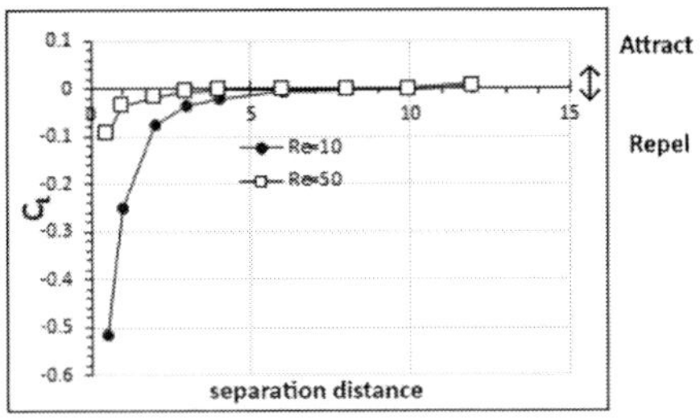

Fig.13. Lift coefficient of two side by side spheres for Re=10 and 50 vs separation distance

3.2.2. Spheres in „V" formation

This configuration of three or more bodies could be seen in different devices in chemical industry, particle suspension flows, fluidized beds, heat exchangers, etc. Of interest in this study is the flow patterns around three and seven equal spheres arranged in a triangular formation. Let us agree the sphere that is in the leading position according to the incoming flow to call leading sphere and to denote it with number 1. The other spheres let be marked with numbers up to 3 or 7. We consider the cases when the distance d between the leading sphere and other ones are equal and could vary from 0 to 7 diameters. Reynolds numbers considered are Re=10 and 50. Let the angle between the lines connecting the centers of the upper and lower spheres denote with α and to consider values of α from 30^0 to 180^0. When all spheres are in contact there is only an one wake behind them (Fig.14). The total drag of this assemblage is 2,562 and is bigger than the drag of a single sphere – 2,116. For α=45^0 the wake is still one and the interaction of the bodies is significant. Increasing the angle wakes behind each sphere is formed and the mutual influence decrease.

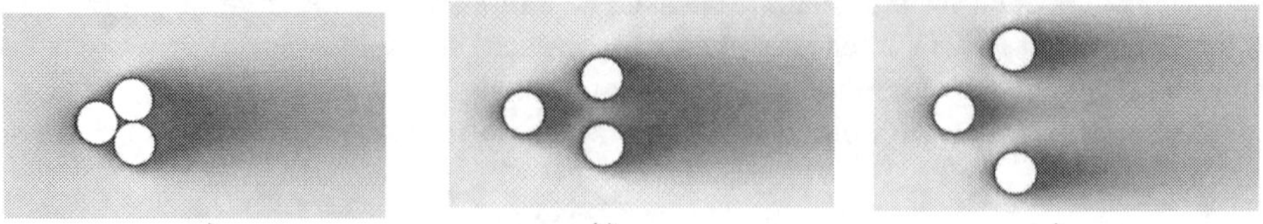

a) b) c)

Fig. 14 Flow past three spheres at Re=50: a) $d_{12}=d_{13}=0$; b) $d_{12}=d_{13}=1$, $\alpha=45^0$; c) $d_{12}=d_{13}=1$, $\alpha=90^0$

Lift and total drag coefficients are shown at Fig. 15.

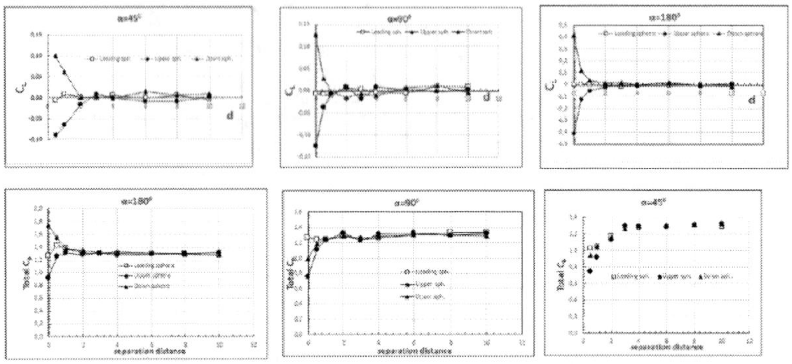

Fig. 15 Lift and total drag coefficient for Re=50 and $\alpha=45^0$, 90^0 and 180^0

The configuration of seven spheres with equal diameters was chosen to analyze and verify the influence between the wakes formed by the particles. Such a configuration could be related with the well known „V" formation of the birds, which in accordance with the scientists, reduces the drag force that each bird experiences compared to if it were flying alone. The spheres were arranged in different configurations by changing the angle α. It takes values 40^0, 60^0, 90^0, 110^0, 130^0, 150^0 and 180^0. The distances between the spheres are equal to d=1 and Re=100 for all simulations. Due to a flow symmetry the spheres with numbers 1, 3, 5 and 7 where chosen to be analyzed. The pictures at Fig.16 shows the flow in the center plane of the configuration for different angles α.

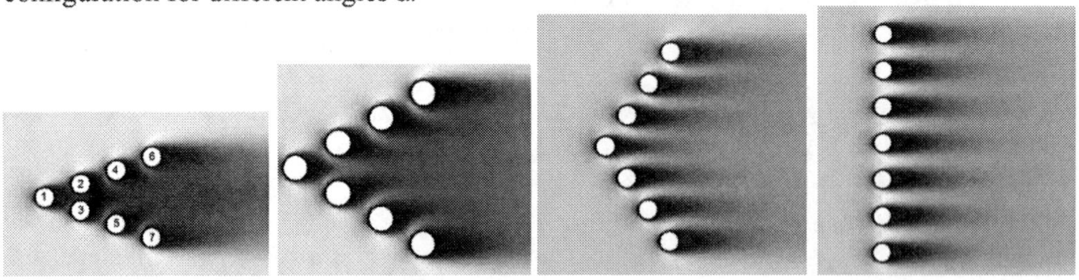

$\alpha = 40^0$ $\alpha = 60^0$ $\alpha = 110^0$ $\alpha = 180^0$

Fig. 16 Flow past 7 spheres at Re=100 and different values of α.

When the angle α is small the wakes behind spheres are not formed well except these ones at the end of the formation. As a whole the wake behind this assemblage looks like a wake behind a single delta body immersed in the fluid. With increase of α more and more clearly visible are the individual wakes behind each sphere that interact strongly each other and as a result the drag force coefficients change. The drag coefficients for the spheres 1, 3, 5 and 7 divided by the value of a single sphere are presented in Fig. 17 for Re=100. For $40^0<\alpha<180^0$ all spheres have drag coefficients lower than for a single sphere ($C_{D\,\mathrm{single}} = 1,083$).

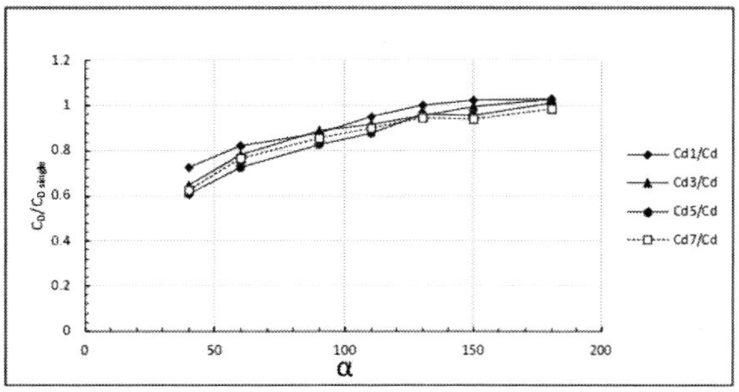

Fig. 17 drag coefficients for the spheres 1, 3, 5 and 7, Re=100.

It is seen that for all spheres the drag increases with α and reach values close to this one of an isolated sphere. These parameters are influenced by the number of spheres and the angle of the configuration.

4 Conclusions

Hydrodynamic interaction of two or more rigid spherical particles in different arrangements have been studied numerically for moderate Reynolds numbers. It is confirmed that one of the most important factors acting on the flow parameters is the distance between the spheres. The strongest effect on the drag force when one or more spheres are included in the flow compared with a single sphere case arise for line arrangement where the difference could reach 12% and more. When particles are side by side due to zones with high and low pressure between them a weak repulsion and attraction forces appear. In the case of „V" arrangement the angle α play significant role on the wake structure behind the particles and drag forces. The software used gives the possibility to study numerically with a high accuracy the flow past any formation of equal or different particles at different flow regimes.

References

1. G.G. Stokes, *Trans. Camb. Phil. Soc.* 9, 8, (1851)
2. M. Stimson, G.B. Jefferey, *Proc. R. Soc. Lond* A **111**, 110-116, (1926)
3. S. Kim, *Phys. Fluids*, **30**, 2309-2314, (1987)
4. H. Wilson, *Journal of Computational Physics*, **245**, 302–316, (2013)
5. Y. Tsuji, Y. Morikawa, K. Terashima, *Intl J. Multiphase Flow*, **8** (1), 71–82, (1982)
6. C. Zhu, S. C Liang, L. S. Fan, *Intl J. Multiphase Flow*, **20** (1), 117–129, (1994)
7. R. C. Chen and J. L. Wu, *Journal of Chem. Eng. Science,* **55**, 1143-1158, (2000)
8. I. Kim, S. Elghobashi, W. Sirignano, *J. Fluid Mech.*, **246**, 465-488, (1993).
9. L. Schouveiler, A. Brydon , T. Leweke, M.C. Thompson, *European Journal of Mechanics B/Fluids* **23**, 137–145, (2004)
10. Van-Dyke, *An Album of Fluid Motion,* The parabolic Press, Stanford California, (1982)
11. Z. D. Zaprynov, E. T. Toshev, *Proceedings of the Eighth International Conference on Heat Transfer,* ASME, New York, **5**, 2549-2553, (1986)
12. R. Tal, D. N Lee, W. A Sirignano, *International Journal of Heat and Mass Transfer,* **27**, No. 11, 1953-1962, (1984)
13. R. Folkersma, H. Stein, F. Van de Vosse, *International Journal of Multiphase Flow* **26**, 877-887, (2000)

Magnetohydrodynamic calculation of the temperature and wind velocity profile of the solar transition region. Preliminary results.

Todor M. Mishonov[1,⋆], *Albert M.* Varonov[1], *Nedeltcho I.* Zahariev[1], *Rositsa V.* Topchiyska[1], *Boian V.* Lazov[1], and *Stefan B.* Mladenov[1]

[1] *St. Clement of Ohrid University at Sofia, 5 J. Bourchier blvd., BG-1164 Sofia, Bulgaria*

Abstract. The sharp almost step like increase the temperature in the transition region (TR) between chromosphere and solar corona is well-known from decades; for first time we are giving a detailed magnetohydrodynamic (MHD) calculation of the height dependence of the temperature. The width of the transition region is evaluated by maximal value of the logarithmic derivative of the temperature. At fixed heating, only MHD can give such a narrow width and in such sense, even the qualitative agreement with the observational data, gives the final verdict what the heating mechanism of the solar corona is. Static profiles of the temperature and wind velocity are calculated for static frequency dependent spectral density of the incoming MHD waves; no time dependent computer simulations. At fixed spectral density of MHD waves, the MHD calculation predicts height dependence of the non-thermal broadening of spectral lines and its angular dependence. For illustration is used one dimensional approximation of completely ionized hydrogen plasma in weak magnetic field, but it is considered that the width of the TR is weakly dependent with respect of further elaboration. The analyzed MHD calculation is a numerical confirmation of the qualitative concept of self-induced opacity of the plasma with respect to MHD waves. The plasma viscosity strongly increases with the temperature. Heated by MHD waves, plasma increases the wave absorption and this positive feedback leads to further heating. The static temperature profile is a result of a self-consistent calculation of propagation of MHD wave through the static background of wind and temperature profile. The numerical method allows consideration of incoming MHD waves with an arbitrary spectral density. Further elaboration of the method are briefly discussed: influence of second viscosity in the chromospheric part of the TR, influence of the magnetic field on the coronal side of the TR and investigation of such type effects on the width of the TR.

1 Alfvén model for corona heating

The discovery of the 5303 Å green[1] of Fe^{13+} and 6374 Å red[2] of Fe^{9+} magneto-dipole transition lines in the solar corona spectrum [1] posed an important problem for the fundamental physics – what is the mechanism of the heating of the solar corona and why the temperature of the corona is 100 times

⋆e-mail: mishonov@gmail.com

[1] Fe XIV, ${}^2P_{3/2} \rightarrow {}^2P_{1/2}$, $3s^23p^1$

[2] Fe X, ${}^2P_{1/2} \rightarrow {}^2P_{3/2}$, $3s^23p^5$

larger than the temperature of the photosphere. Those spectroscopic experiments put perhaps the most living unresolved problem in physics. None of the heating mechanisms has been confirmed and none has received an indisputable refutation. The 75 years old problem is still fresh and green.

The first idea by Alfvén [2] was that now called Alfvén waves (AW) [3] are the mechanism for heating the corona. Perhaps AW are generated by the turbulence in the convection zone and propagate along the magnetic field lines as through wave guides. Absorption is proportional to ω^2 and the heating comes from high-frequency AW which are absorbed and not observed in the corona. Alfvén's idea for the viscous heating of plasma by absorption of AW was analyzed in the theoretical work by Heyvaerts [4]. In support of this idea is the work by Chitta [5] (Figures 8, 9 therein). The authors came to the conclusion that the spectral density of AW satisfies a power law with an index of 1.59. This gives a strong hint that this scaling can be extrapolated in the nearest spectral range for times less than 1 s and frequencies in the Hz range. Furthermore in the work by Tomczyk [6] it is stated that there exist very few direct measurements of the strength and orientation of coronal magnetic fields, meaning that the mechanisms responsible for heating the corona, driving the solar wind, and initiating coronal mass ejections remain poorly understood. After the launch of Hinode, Alfvènic type MHD modes were observed [7] and the well-forgotten spatially and temporally ubiquitous waves in the solar corona [8] came again into the limelight and gave strong support for the idea of Alfvén and the frequency of the notion AW as a key worth significantly increased.

A clear presence of outward and inward propagating waves in the corona was noted. $k - \omega$ diagnostics revealed coronal wave power spectrum with an exponent of $\approx -\frac{3}{2}$ (cf. Fig. 2 of [6]). The low frequency AW, on the other hand, reach the Earth orbit and thanks to the magnetometers on the various satellites we "hear" the basses of the great symphony of solar turbulence.

The observational data for the temperature profile of the solar corona show that the TR is extremely thin compared to the radius of the Sun [9–13]. All those observational data for the quiet-Sun chromosphere and transition region reveal one dimensional and time independent profiles which theoretically have to be derived by time averaged dynamic consideration. The width of the TR λ might be evaluated by the maximum of logarithmic derivative of the temperature, $\lambda = \max(\frac{\mathrm{d}T}{T\mathrm{d}x})$. In order to qualitatively explain this small width, in [14] the idea of self-induced opacity of the plasma for Alfvén waves was introduced. A similar idea was analysed also by Suzuki [15] (see also [16–21]), however none of these articles is helpful for comparison with our manuscript because the width of the TR is not calculated in details in any of them. In the current paper we give a numerical realization of this idea by calculating for the first time the width of the TR using the framework of MHD.

First detailed observations by Skylab half a century ago revealed surprisingly that TR is very thin λ–equal in width to metropolitan Los Angeles [10]. As λ is much smaller that the solar radius we have actually a plane one dimensional (1D) problem. In 1D the constant fluxes of mass, energy and momentum give 3 integrals of motion which significantly alleviate the calculations. We can eliminate the density $\rho(x)$ as dynamic variable and for the static profiles of temperature $T(x)$ and solar wind $U(x)$ we have to solve numerically first order differential equations.

Without a doubt the kinetic approach is indispensable for the treatment of low density solar corona but this problem is beyond the purpose of the present work. In our work we take into account the influence of the magnetic field on the viscosity within the hydrodynamic approach.

The purpose of the present work is to examine whether the initial Alfvén idea is correct and to solve the MHD equations which give the dependence of the temperature on the height $T(x)$ and the related velocity of the solar wind $U(x)$ supposing static spectral density of the incoming AW. We illustrate Alfvén's idea by an MHD calculation for completely ionized hydrogen plasma in the beginning of TR where the density of neutral Hydrogen atoms is already negligible. Due to the high density of the TR, MHD is an adequate tool to analyze the beginning of the process.

Our starting point are the MHD equations [22] for the velocity field $\mathbf{v}$ and magnetic field $\mathbf{B}$

$$\partial_t \rho + \mathrm{div}\mathbf{j} = 0, \qquad \mathbf{j} = \rho\mathbf{v}, \tag{1}$$

$$\partial_t \left(\frac{1}{2}\rho\mathbf{v}^2 + \varepsilon + \frac{\mathbf{B}^2}{2\mu_0} \right) + \mathrm{div}\,\mathbf{q} = 0, \tag{2}$$

$$\partial_t(\rho\mathbf{v}) + \nabla \cdot \mathbf{\Pi} = 0, \tag{3}$$

where

$$\mathbf{q} = \rho\left(\frac{1}{2}\mathbf{v}^2 + h\right)\mathbf{v} + \mathbf{v}\cdot\mathbf{\Pi}^{(\mathrm{visc})} - \varkappa\nabla T + \mathbf{S} \tag{4}$$

is the energy density flux, ρ is the mass density, ε is the internal energy density, $\varkappa$ is the thermal conductivity, h is the enthalpy per unit mass;

$$\mathbf{S} = \frac{1}{\mu_0}\left[\mathbf{B}\times(\mathbf{v}\times\mathbf{B}) - \nu_{\mathrm{m}}\mathbf{B}\times\mathrm{rot}\,\mathbf{B}\right], \tag{5}$$

is the Poynting vector and $\nu_{\mathrm{m}} \equiv c^2\varepsilon_0\varrho$ is the magnetic diffusivity determined by Ohmic resistance ϱ and vacuum susceptibility ε_0; vacuum permeability is μ_0. For hot enough plasma ϱ is negligible. The total momentum flux

$$\mathbf{\Pi} = \rho\mathbf{v}\mathbf{v} + P\mathbb{1} + \mathbf{\Pi}^{(\mathrm{visc})} + \mathbf{\Pi}^{(\mathrm{Maxw})} \tag{6}$$

is a sum of the inviscid part $\rho\mathbf{v}\mathbf{v} + P$ of the fluid, with pressure P and

$$\Pi_{ik}^{(\mathrm{visc})} = -\eta\left(\partial_i v_k + \partial_k v_i - \frac{2}{3}\delta_{ik}\nabla\cdot\mathbf{v}\right) - \zeta\delta_{ik}\nabla\cdot\mathbf{v}, \tag{7}$$

the viscous part of the stress tensor, with viscosity η and second viscosity ζ, and lastly, the Maxwell stress tensor

$$-\Pi_{ik}^{(\mathrm{Maxw})} = \frac{1}{\mu_0}\left(B_iB_k - \frac{1}{2}\mathbf{B}^2\delta_{ik}\right), \tag{8}$$

with δ_{ik} the Kronecker delta. For temperatures above 10 kK we treat coronal plasma as completely ionized hydrogen plasma with the following temperature dependence of kinetic coefficients [23]:

$$\varkappa = 0.9\frac{T^{5/2}}{e^4m^{1/2}\Lambda}, \qquad \eta = 0.4\frac{M^{1/2}T^{5/2}}{e^4\Lambda}, \qquad \zeta \approx 0, \tag{9}$$

$$\Lambda = \ln\left(\frac{r_{\mathrm{D}}T}{e^2}\right), \qquad \frac{1}{r_{\mathrm{D}}^2} = \frac{4\pi e^2 n_{\mathrm{tot}}}{T}, \qquad e^2 \equiv \frac{q_e^2}{4\pi\varepsilon_0}$$

where q_e is the electron charge, m is the mass of electron, M is the proton mass, T is the temperature and $n_{\mathrm{tot}} = n_e + n_p$ is the total density of electrons and protons; $\rho = Mn_p$. We suppose that $\mu_0 = 4\pi$ and $\varepsilon_0 = 1/4\pi$, but in the practical system all formulae are the same; as well as in Heaviside-Lorentz units where $\mu_0 = 1$ and $\varepsilon_0 = 1$. As we mentioned above

$$\nu_{\mathrm{m}} = \frac{c^2}{4\pi}\frac{e^2m^{1/2}\Lambda}{0.6\,T^{3/2}} \ll \nu_{\mathrm{k}} \equiv \frac{\eta}{\rho} = \frac{0.4\,T^{5/2}}{e^4M^{1/2}n_p\Lambda}; \tag{10}$$

i.e. the hot hydrogen plasma is sticky, dilute, and "superconducting". Here ν_{k} is the kinematic viscosity. We can introduce the magnetic Prandtl number as the ratio between the kinematic and the magnetic viscosity,

$$\mathrm{Pr}_{\mathrm{m}} = \frac{\nu_{\mathrm{k}}}{\nu_{\mathrm{m}}} = \frac{0.96\pi}{e^6c^2\sqrt{Mm}\,n_p\Lambda^2(T)}T^4. \tag{11}$$

The coronal heating mechanism can be revealed without additional accessories. Let us mention also the relations $\varkappa\varrho = 1.5\,T/q_e^2$ leading to

$$\varrho = \frac{1}{4\pi\varepsilon_0}\frac{e^2 m^{1/2}\Lambda}{0.6\,T^{3/2}}, \tag{12}$$

and $\eta/\varkappa \approx \frac{4}{9}\sqrt{mM}$.

The applicability of our equations is governed by an additional condition:

$$\nu_{pp} = \frac{1}{\tau_{pp}} = \frac{e^2 n\Lambda}{T^{3/2}\sqrt{M}} \gg \omega_{Be} = \frac{q_e B}{M}. \tag{13}$$

If in the dense plasma the frequency of the proton-proton collisions ν_{pp} is much smaller than the proton cyclotron frequency ω_c, the magnetic field can be considered a perturbation and for the kinetic coefficients we can use the zero magnetic field formulas. An elementary gas kinetic consideration of main notions of plasma as proton-proton collision frequency ν_{pp} mean free time τ_{pp}, thermal velocity $v_{T_p} = \sqrt{T/M}$ mean free path $l_p = v_{T_p}\tau_{pp}$ and other is given in the the appendices.

Lastly, we need to mention the existence of radiative losses. However, the thin TR has width much smaller not only than Earth's orbit radius but than Earth's radius and thus these losses are neglectable when considering the mechanism for coronal heating; continuum radiation may be neglected for all temperatures fewer than a few million degrees [24]. The hot corona exists in broad ranges where the heating mechanism cannot be effective. For this reason, in the narrow range of the transition region the radiation losses are negligible compared to the intensive heating, no matter what the concrete mechanism is. Quantitatively, this means that radiation power per unit volume $P_{\mathrm{rad}} \ll nTU/\lambda$, where n and T are the number density and temperature of the corona, λ is the width of the transition region, and U is the velocity of the solar wind. The influence of gravitational field with acceleration $g_\odot$ is also negligible for the temperature gradient in the TR $Mg_\odot\lambda \ll T$; cf. Ref. [20]

2 MHD equations and energy fluxes

First of all, we have to emphasize than MHD is not a model but adequate theory applicable for the dense plasma of TR. Let us now recall the main MHD equations.

The time derivative $\partial_t\mathbf{B}$ which implicitly participates in the energy conservation Eq. (2) obeys the equation cf. Ref. [22]

$$\begin{aligned} \mathrm{d}_t\mathbf{B} &= \mathbf{B}\cdot\nabla\mathbf{v} - \mathbf{B}\,\mathrm{div}\,\mathbf{v} - \nu_{\mathrm{m}}\,\Delta\,\mathbf{B} + \mathrm{grad}\,\nu_{\mathrm{m}}\times\mathrm{rot}\,\mathbf{B}, \\ \mathrm{d}_t &\equiv \partial_t + \mathbf{v}\cdot\nabla\,. \end{aligned} \tag{14}$$

Analogously the momentum equation Eq. (3) can be rewritten by the substantial derivative

$$\rho\,\mathrm{d}_t v_i = -\partial_i P + \partial_k\left\{\eta\left(\partial_k v_i + \partial_i v_k - \frac{2}{3}\delta_{ik}\partial_j v_j\right)\right\} + \partial_i\left(\zeta\partial_j v_j\right) - \frac{1}{\mu_0}\left(\mathbf{B}\times\mathrm{rot}\,\mathbf{B}\right)_i\,. \tag{15}$$

We analyze AW propagating along magnetic field lines $\mathbf{B}_0$. We focus our attention on the narrow TR, where the static magnetic field is almost homogeneous and the waves are within acceptable accuracy one dimensional. For the velocity and magnetic fields we assume

$$\begin{aligned} \mathbf{v}(t,x) &= U(x)\,\mathbf{e}_x + u(t,x)\,\mathbf{e}_z, \\ \mathbf{B}(t,x) &= B_0\,\mathbf{e}_x + B_0 b(t,x)\,\mathbf{e}_z, \end{aligned} \tag{16}$$

with vertical homogeneous magnetic field $B_0\mathbf{e}_x$ perpendicular to the surface of the Sun. The transverse wave amplitudes of the velocity $u(t, x)$ and magnetic field $b(t, x)$ we represent with the Fourier integrals

$$u(t,x) = \int_{-\infty}^{\infty} \tilde{u}(\omega, x)\mathrm{e}^{-\mathrm{i}\omega t}\frac{\mathrm{d}\omega}{2\pi}, \tag{17}$$

$$b(t,x) = \int_{-\infty}^{\infty} \tilde{b}(\omega, x)\mathrm{e}^{-\mathrm{i}\omega t}\frac{\mathrm{d}\omega}{2\pi}. \tag{18}$$

For illustrative purposes it is convenient to consider monochromatic AW with $u(t, x) = \hat{u}(x)\mathrm{e}^{-\mathrm{i}\omega t}$ and $b(t, x) = B_0\,\hat{b}(x)\mathrm{e}^{-\mathrm{i}\omega t}$. And later on to restore the summation on different frequencies which finally gives Fourier integration.

2.1 Wave equations

For transversal waves the general MHD equations Eq. (15) and Eq. (14) give the following exact system of linear for $\hat{u}(x)$ and $\hat{b}(x)$

$$(-\mathrm{i}\omega + U\mathrm{d}_x)\,\hat{u} = V_\mathrm{A}^2\mathrm{d}_x\hat{b} + \frac{1}{\rho}\mathrm{d}_x\,(\eta\mathrm{d}_x\hat{u}) - \nu_\mathrm{k}k_y^2\hat{u}, \tag{19}$$

$$-\mathrm{i}\omega\hat{b} = \mathrm{d}_x\hat{u} - \mathrm{d}_x(U\hat{b}) + \mathrm{d}_x(\nu_\mathrm{m}\mathrm{d}_x\hat{b}) - \nu_\mathrm{m}k_y^2\hat{b}, \tag{20}$$

where

$$V_\mathrm{A}(x) = B_0/\sqrt{\mu_0\rho(x)} \tag{21}$$

is the Alfvén velocity. For illustrative purpose and complete lack of observational data in the present work we analyze only the case of zero transversal to the magnetic field wave vector $k_y = 0$. This system can be Hamiltonized. In our numerical analysis we solve the first order linear system of equations

$$-\mathrm{i}\mathrm{d}_x\Psi = \mathsf{K}\Psi, \quad \Psi \equiv \begin{pmatrix} \hat{b} \\ \hat{u} \\ \hat{\gamma} \\ \hat{w} \end{pmatrix}, \quad \Psi^\dagger \equiv \left(\hat{b}^*, \hat{u}^*, \hat{\gamma}^*, \hat{w}^*\right), \quad \mathsf{K} \equiv \frac{\mathrm{i}\overline{\mathsf{M}}}{\nu_\mathrm{m}\nu_\mathrm{k}}, \tag{22}$$

where $\hat{\gamma} \equiv \mathrm{d}_x\hat{b}$, $\hat{w} \equiv \mathrm{d}_x\hat{u}$ and

$$\overline{\mathsf{M}} \equiv \begin{pmatrix} 0 & 0 & -\nu_\mathrm{m}\nu_\mathrm{k} & 0 \\ 0 & 0 & 0 & -\nu_\mathrm{m}\nu_\mathrm{k} \\ (\mathrm{i}\omega - \mathrm{d}_xU)\nu_\mathrm{k} & 0 & -(U + \mathrm{d}_x\nu_\mathrm{m})\nu_\mathrm{k} & \nu_\mathrm{k} \\ 0 & \mathrm{i}\omega\nu_\mathrm{m} & V_\mathrm{A}^2\nu_\mathrm{m} & -\left(U - \frac{\mathrm{d}_x(\eta)}{\rho}\right)\nu_\mathrm{m} \end{pmatrix}.$$

For homogeneous medium with constant η, ν_m, ρ, V_A, and U, in short for constant wave-vector matrix K, the exponential substitution $\Psi \propto \exp(\mathrm{i}kx)$ in Eq. (22) or equivalently Eq. (19) and Eq. (20) gives the secular equation $\det(\mathsf{K} - k\mathbb{1}) = 0$, which after some algebra gives the dispersion equation of AW

$$\left(\omega_\mathrm{D} + \mathrm{i}\nu_\mathrm{m}k^2\right)\left(\omega_\mathrm{D} + \mathrm{i}\nu_\mathrm{k}k^2\right) = V_\mathrm{A}^2k^2, \tag{23}$$

where $\omega_\mathrm{D} \equiv \omega - kU$ is the Doppler shifted frequency. Usually either viscosity coefficient is significantly bigger than the other. For example, for hot plasma $\nu_\mathrm{k} \gg \nu_\mathrm{m}$ and with an acceptable approximation we can neglect the smaller one and to work with matrix 3×3. Approximation of ideal fluid is reduced to both negligible viscosity coefficients and corresponding 2×2 matrix gives the well-known AW dispersion in moving fluid $\omega_{\mathrm{D}}^2 = V_{\mathrm{A}}^2 k^2$. A perturbative treatment of dispersion equation $\omega(k) = \omega' + i\omega''$ Eq. (23) gives [22] $\omega'' = -\frac{1}{2}(\nu_{\mathrm{m}} + \nu_{\mathrm{k}})k^2$. If WKB approximation is applicable for wave amplitude damping we have to use the eigenvalue of K matrix giving smallest damping for right running wave.

In order to make model evaluation of the influence of magnetic field on viscosity for the waves we can use the the Braginskii result [25]

$$\nu_{\mathrm{k},2} = \frac{\frac{6}{5}\tilde{x}^2 + \frac{7}{3}}{\tilde{x}^4 + 4.03\tilde{x}^2 + \frac{7}{3}}\nu_{\mathrm{k}}, \tag{24}$$

where $\tilde{x} \equiv \omega_{Be}\tau_{pp}$.

2.2 Wind variables

We solve the wave equation Eq. (22) from some height at which hydrogen plasma is approximately completely ionized $x = 0$ to some distance large enough $x = l$, where the short wavelength AW are almost absorbed. As we pointed out, this distance is much bigger than the *width of the transition region* λ, but much smaller than solar radius. That is why the problem is effectively one dimensional. The horizontal wavelength of the waves responsible for heating is much smaller than the range of area for which we are calculating the temperature profile.

The considered one-dimensional $0 < x < l$ time independent problem has three integrals corresponding to the three conservation laws related to mass, energy and momentum. The mass conservation law Eq. (1) gives the constant flow

$$j = \rho(x)U(x) = \rho_0 U_0 = \rho_l U_l = \mathrm{const}, \tag{25}$$

where $\rho_0 = \rho(0)$, $\rho_l = \rho(l)$, $U_0 = U(0)$, and $U_l = U(l)$. The energy conservation law reduces to a constant flux along the x-axis

$$q = q_{\mathrm{wind}}^{\mathrm{ideal}}(x) + \tilde{q}(x) = \mathrm{const}, \tag{26}$$

$$q_{\mathrm{wind}}^{\mathrm{ideal}} = q_{\scriptscriptstyle U} + q_{\scriptscriptstyle P}, \quad q_{\scriptscriptstyle U} \equiv \left(\frac{1}{2}U^2\right)\rho U, \quad q_{\scriptscriptstyle P} \equiv h\rho U.$$

Here the first term describes the energy of the ideal wind, i.e. a wind from an ideal (inviscid) fluid. The second term $\tilde{q}(x)$ includes all other energy fluxes; in our notations tilde will denote sum of the non-ideal (dissipative) terms of the wind and wave terms. In detail the non-ideal part of the energy flux $\tilde{q}(x)$ consists of: the wave kinetic energy $\propto |\hat{u}|^2$, viscosity (wind $\propto \frac{4}{3}\eta + \zeta$ and wave $\propto \eta$ components), heat conductivity $\propto \varkappa$, and Poynting vector $\propto \hat{b}^*$,

$$\tilde{q}(x) \equiv q_\xi + q_\varkappa + q_{\mathrm{wave}}, \tag{27}$$

$$q_\xi \equiv -\xi U \mathrm{d}_x U, \quad q_\varkappa \equiv -\varkappa\,\mathrm{d}_x T, \quad q_{\mathrm{wave}} = q_{\mathrm{wave}}^{\mathrm{ideal}} + q_{\mathrm{wave}}^{\mathrm{diss}},$$

$$q_{\mathrm{wave}}^{\mathrm{ideal}} = q_u + q_b + q_{ub}, \quad q_{\mathrm{wave}}^{\mathrm{diss}} = q_\eta + q_\varrho,$$

$$q_u \equiv \frac{j}{4}\sum_{\mathrm{waves}} |\hat{u}|^2, \quad q_b \equiv \frac{B_0^2}{2\mu_0}\sum_{\mathrm{waves}} U\left|\hat{b}\right|^2, \quad q_{ub} \equiv -\frac{B_0^2}{2\mu_0}\sum_{\mathrm{waves}} \mathrm{Re}(\hat{b}^*\hat{u}),$$

$$q_\eta = -\frac{1}{4}\eta\sum_{\mathrm{waves}} \mathrm{d}_x|\hat{u}|^2, \quad q_\varrho = -\frac{B_0^2}{2\mu_0}\nu_{\mathrm{m}}\sum_{\mathrm{waves}} \mathrm{Re}(\hat{b}^*\hat{\gamma}),$$

$$q_{\rm wave}^{\rm ideal} = \sum_{\rm waves} \left\{ \frac{j}{4} |\hat{u}|^2 + \frac{B_0^2}{2\mu_0} \left[U \left|\hat{b}\right|^2 - \mathrm{Re}(\hat{b}^*\hat{u}) \right] \right\}, \quad q_{\rm wave}^{\rm diss} = \sum_{\rm waves} \left\{ -\frac{1}{4}\eta \mathrm{d}_x |\hat{u}|^2 - \frac{B_0^2}{2\mu_0} \nu_{\rm m} \mathrm{Re}(\hat{b}^*\hat{\gamma}) \right\},$$

$$q_{\rm wave} = \sum_{\rm waves} \left\{ \frac{j}{4} |\hat{u}|^2 - \frac{1}{4}\eta\, \mathrm{d}_x |\hat{u}|^2 + \frac{B_0^2}{2\mu_0} \left(U \left|\hat{b}\right|^2 - \mathrm{Re}(\hat{b}^*\hat{u}) - \nu_{\rm m}\mathrm{Re}(\hat{b}^*\hat{\gamma}) \right) \right\},$$

where $\xi \equiv \frac{4}{3}\eta + \zeta$. In order to alleviate the detail analysis depicted in Fig. 3, we have introduced convenient notations for all terms with different nature. Here time averaged energy flux is represented by the amplitudes of the monochromatic oscillations, this is a standard procedure for alternating current processes. In our case we have, for example, $\left\langle \hat{u}^2 \right\rangle_t = \left\langle (\mathrm{Re}\,\hat{u})^2 \right\rangle_t = \left\langle \frac{1}{4}(\hat{u} + \hat{u}^*)^2 \right\rangle_t = \frac{1}{2} |\hat{u}|^2$. The other terms from Eq. (4) are averaged in a similar way like in the equation above. In short, we substituted the formulas for AW Eq. (16) propagating along the magnetic field in x-direction in the general formula for the energy flux Eq. (4) and made the time averaging and summation over waves with different frequencies. We will repeat the same approach to the momentum flux Eq. (6)

For constant spectrum of incoming waves the momentum conservation law Eq. (6) gives constant flux $\Pi = \Pi_{xx}$

$$\Pi = \Pi_{\rm wind}^{\rm ideal}(x) + \tilde{\Pi}(x) = \mathrm{const}, \qquad (28)$$
$$\Pi_{\rm wind}^{\rm ideal} = \Pi_U + \Pi_P, \quad \Pi_U \equiv \rho U U, \quad \Pi_P \equiv P,$$

the sum of the ideal wind fluid and the other terms

$$\tilde{\Pi}(x) \equiv \Pi_\xi + \Pi_{\rm wave}, \quad \Pi_\xi \equiv -\xi \mathrm{d}_x U, \quad \Pi_{\rm wave} \equiv \sum_{\rm waves} \frac{1}{4\mu_0} \left|\hat{b}\right|^2, \qquad (29)$$

which take into account the wave part of the Maxwell stress tensor $\propto |\hat{b}|^2$ and viscosity of the wind $\propto \xi$. The sum here and in Eq. (27) is actually integral over the spectral density $\mathcal{W}(\omega)$ of incoming AW

$$\Pi_{\rm wave} \equiv \sum_{\rm waves} \frac{1}{4\mu_0} \left|\hat{b}\right|^2 = \int_0^\infty \frac{\mathrm{d}\omega}{2\pi} \mathcal{W} \frac{1}{4\mu_0} \left|\hat{b}\right|^2. \qquad (30)$$

Let us mention that all components of the wave $\Psi(\omega, x)$ have frequency dependence. One can suppose that the spectral density $\mathcal{W}$ of the waves coming from the chromosphere has the same power law dependence, *i.e.*

$$\mathcal{W}_{\overline{\omega}} = C/\omega^\alpha. \qquad (31)$$

The parameter α is between 1.5 and 2: $\frac{3}{2}$ [6], 1.59 [5], 2 [26]. C is the unknown parameter of the theory, which we vary for fixed ξ in order to reproduce the temperature increase in the TR. In our earlier works [27] simulations of plasma heating with power law spectral density is given with up to 8 AW with different frequencies.

We have to solve the hydrodynamic problem for calculation of wind velocity and temperature at known energy and momentum fluxes. The problem is formally reduced to analogous one for a jet engine, cf. Ref. [28]. We approximate the corona as completely ionized hydrogen plasma, i.e. electrically neutral mixture of electrons and protons. The experimental data tells us that proton temperature T_p is higher than electron one T_e. This is an important hint that heating goes through the viscosity determined mainly by protons. However for illustration purpose and simplicity we assume proton and electron temperatures to be equal $T_e = T_p = T$. For such an ideal (in thermodynamic sense) gas the local sound velocity is

$$c_s^2(x) = \frac{c_p}{c_v}\frac{P}{\rho} = \gamma\frac{T}{\langle m\rangle}, \quad \gamma = \frac{c_p}{c_v} = \frac{5}{3}, \quad \langle m\rangle = \frac{n_p M + n_e m}{n_p + n_e} \approx \frac{1}{2}M,$$
$$n_e = n_p = \frac{1}{2}n_{\rm tot}, \quad P = n_{\rm tot}T = \frac{\rho T}{\langle m\rangle} = \frac{j}{U}\frac{T}{\langle m\rangle}, \quad h = c_p\frac{T}{\langle m\rangle} = \frac{\varepsilon + P}{\rho},$$

where, as we mentioned earlier, h is the enthalpy per unit mass and ε is the density of internal energy. Although there are some hints for different values of the adiabatic index γ [29], we will use the traditional value of 5/3 for our calculations since this choice will not change the essence of our presentation. Hydrogen plasma is well theoretically investigated and the 5/3 value of gamma is used in hundreds of works. There are no theoretical hints for significant deviations from this value obtained by ideal gas approximation. Actually a small variation of the value of gamma does not change qualitatively the temperature profile of the TR.

In order to alleviate the final formulae we introduce two dimensionless variables χ and τ which represent the non-ideal part of the energy and momentum flux respectively

$$\chi(x) \equiv \left.\frac{\tilde{q}(x)}{\rho_0 U_0^3}\right|_x^0, \quad \tau(x) \equiv \left.\frac{\tilde{\Pi}(x)}{\rho_0 U_0^2}\right|_x^0. \tag{32}$$

Here we could include gravitational field, Bremsstrahlung or other accessories, which are negligible for the narrow TR. Analogously, for the wind velocity and temperature we have

$$\overline{U}(x) \equiv \frac{U(x)}{U_0}, \quad \Theta(x) \equiv \frac{T(x)}{\langle m\rangle U_0^2}, \quad \Theta_0 = \Theta(0), \tag{33}$$

where $U_0 = U(0)$. The energy and momentum constant fluxes Eq. (26) and Eq. (28) in the new notation take the form

$$\frac{q - \tilde{q}(0)}{\rho_0 U_0^3} = \frac{1}{2}\overline{U}^2 + c_p\Theta - \chi = \frac{1}{2} + c_p\Theta_0, \tag{34}$$

$$\frac{\Pi - \tilde{\Pi}(0)}{\rho_0 U_0^2} = \overline{U} + \Theta/\overline{U} - \tau = 1 + \Theta_0. \tag{35}$$

From the second equation we express the dimensionless temperature Θ and substitute in the first one. Solving the quadratic equation for the wind velocity U we derive

$$U = U_0\overline{U}, \quad \overline{U}(x) = \frac{1}{\gamma+1}\left(\gamma + s^2 + \gamma\tau(x) - \sqrt{\mathcal{D}(x)}\right), \tag{36}$$

where for the discriminant we have

$$\mathcal{D} = (s^2 - 1)^2 - 2\chi(\gamma^2 - 1) + \gamma\tau\left[\gamma\tau + 2\left(\gamma + s^2\right)\right],$$
$$s^2 \equiv \frac{c_s^2(0)}{U_0^2} = \gamma\Theta_0, \quad c_s^2(x) = \left(\frac{\partial P}{\partial \rho}\right)_S = \frac{\gamma T(x)}{\langle m\rangle}. \tag{37}$$

If $\chi = 0$ and $\tau = 0$, we get the initial condition $\overline{U}|_{\chi=0,\tau=0} = 1$. This condition determines the sign in front of $\sqrt{\mathcal{D}}$ in Eq. (36). Here γ is the constant ratio of the heat capacities, and $s \equiv c_s(0)/U_0$ is the ratio of the sound and wind velocity at $x = 0$. We suppose that initial wind velocity is very small $U(0) \ll c_s(0)$. The velocity distribution Eq. (36) can be substituted in Eq. (35) and we derive the dimensionless equation for the temperature distribution expressed by the wind velocity profile

$$T(x) = \langle m \rangle\, U_0^2 \Theta(x), \tag{38}$$

$$\Theta(x) = \overline{U}(x)\left[1 + \Theta_0 + \tau(x) - \overline{U}(x)\right], \quad \Theta_0 = \frac{s^2}{\gamma}. \tag{39}$$

The solutions for velocity $\overline{U}(x)$ Eq. (36) and temperature $\overline{T}(x)$ Eq. (38) distributions are important ingredients in our analysis and derivation of the self-consistent picture of the solar wind. We use a one dimensional approximation and in addition the constant flux of mass, energy and momentum give 3 integrals of motion. This enables us to solve exactly the nonlinear part the MHD equations for the solar corona heating problem analytically. As the energy and momentum flux density equations, Eq. (26) and Eq. (28) respectively, contain space derivatives, these nonlinear equations are differential equations, in which the variables are explicitly expressed but not the derivatives, as it is usual.

The intuitive interpretation is unnecessary, but nevertheless rewriting Eq. (34) and Eq. (35) as

$$\frac{1}{2}\overline{U}^2 + c_p\Theta = \frac{1}{2} + c_p\Theta_0 + \chi, \tag{40}$$

$$\overline{U} + \Theta/\overline{U} = 1 + \Theta_0 + \tau, \tag{41}$$

one can say that $\chi(x)$ is the energy given from all non-ideal terms to the initial energy flux of the ideal wind $\frac{1}{2} + c_p\Theta_0$ in order ideal wind flux to reach his value $\frac{1}{2}\overline{U}^2 + c_p\Theta$ at coordinate x. In the same dimensionless units $\rho_0 = 1$ and $U_0 = 1$ one can consider the solar wind as a wind of ideal fluid with initial momentum density $1 + \Theta_0$ in $x = 0$. In the space interval $(0, x)$ this ideal wind obtains from non-ideal processes momentum flux $\tau(x)$ and finally the momentum flux at coordinate x is $\overline{U} + \Theta/\overline{U}$. Such a chain of notions could be useful to trace influence of every process to the wind velocity $U(x)$ and temperature $T(x)$ profiles.

2.3 Boundary conditions for the waves

At known background wind variables $U(x)$ and $T(x)$ we can solve the wave equation Eq. (22) for run-away AW at $x = l$. As we will see later the run-away boundary condition Eq. (55) corresponds to right propagating AW at the right boundary of the interval. The wave equation Eq. (22) is extremely stiff at small viscosity, and numerical solution is possible to be obtained only downstream from $x = 0$ to $x = l$. We have to find the linear combination of left and right propagating waves at $x = 0$, which gives the run-away condition at $x = l$.

The solution of wave equation according to Eq. (26) determines the energy flux related to the propagation of AW

$$\tilde{q}_{\mathrm{wave}}(\Psi(x)) \equiv \Psi^\dagger g \Psi = \frac{j}{4}|\hat{u}|^2 - \frac{1}{2}\eta\,\mathrm{Re}\,(\hat{u}^*\hat{w}) - \frac{\nu_{\mathrm{m}} B_0^2}{2\mu_0}\mathrm{Re}\left(\hat{b}^*\mathrm{d}_x\hat{b}\right) + \frac{B_0^2}{2\mu_0}\left(U\left|\hat{b}\right|^2 - \mathrm{Re}\left(\hat{b}^*\hat{u}\right)\right), \tag{42}$$

where

$$g(x) \equiv \begin{pmatrix} \frac{UB_0^2}{2\mu_0} & -\frac{B_0^2}{4\mu_0} & -\frac{\nu_{\mathrm{m}}}{4\mu_0}B_0^2 & 0 \\ -\frac{B_0^2}{4\mu_0} & \frac{1}{4}j & 0 & -\frac{1}{4}\eta \\ -\frac{\nu_{\mathrm{m}}}{4\mu_0}B_0^2 & 0 & 0 & 0 \\ 0 & -\frac{1}{4}\eta & 0 & 0 \end{pmatrix}. \tag{43}$$

Here j-term represents kinetic energy of the wave, η-term comes from the viscous part of the wave energy flux, and B_0-terms describe the Poynting vector of the wave.

For hot plasmas the Ohmic dissipation is negligible and we can use the approximation $\nu_m = 0$. In this case it is not necessary to use dynamic variable $\hat{y}$ and we can work with 3×3 matrix. For zero viscosity and Ohmic ressisitivity, equations for AW amplitudes can be reduced to 1D Schrödinger equation [30]. For the illustration of the wave boundary values we will use this approach.

In order to take into account the boundary condition at $x = l$, we calculate the eigenvectors of the matrix K, which according to Eq. (22) determine the wave propagation in a homogeneous fluid with amplitude $\propto \exp(ikx)$. Then the eigenvalues of K give the complex wave-vectors

$$k = k' + ik'' = \text{eigenvalue}(\mathsf{K}), \qquad \text{i.e.} \quad \det(\mathsf{K} - k\mathbb{1}) = 0. \tag{44}$$

The three eigenvectors L, D and R are ordered by spatial decrements of their eigenvalues

$$k''_L < 0 < k''_R < k''_D, \tag{45}$$

and are normalized by the conditions

$$-\mathrm{L}^\dagger g\,\mathrm{L} = \mathrm{R}^\dagger g\,\mathrm{R} = \mathrm{D}^\dagger g\,\mathrm{D} = 1, \tag{46}$$

where the sign corresponds to the direction of wave propagation. Notation L corresponds to left propagating wave, R to right propagating wave, and D for an overdamped at small viscosity mode.

For technical purposes we introduce the matrix notations

$$\mathrm{L} = \begin{pmatrix} \mathrm{L}_u(x) \\ \mathrm{L}_b(x) \\ \mathrm{L}_w(x) \end{pmatrix}, \quad \mathrm{R} = \begin{pmatrix} \mathrm{R}_u(x) \\ \mathrm{R}_b(x) \\ \mathrm{R}_w(x) \end{pmatrix}, \quad \mathrm{D} = \begin{pmatrix} \mathrm{D}_u(x) \\ \mathrm{D}_b(x) \\ \mathrm{D}_w(x) \end{pmatrix}. \tag{47}$$

For low enough frequencies $\omega \to 0$ and wind velocities the modes describe: 1) right-propagating AW with $k'_R \approx \omega/V_A$and small $k''_R \approx \nu_k\omega^2/2V_A^3 \ll k'_R$, 2) left propagating wave $k_L = -k_R$, and a diffusion overdamped mode $k''_D \approx V_A^2/\nu_k U \gg k'_D$ which describes the drag of a static perturbation by the slow wind $U \ll V_A$ in a fluid with small viscosity. In this low frequency and long wavelength limit the stiffness ratio of the eigenvalues is very large

$$r_{DR} = \frac{|k_D|}{|k_R|} \approx \frac{k''_D}{k'_R} \approx \frac{V_A^3}{\nu_k U\omega} \gg 1. \tag{48}$$

As we emphasized the wave equations Eq. (22) form a very stiff system and indispensably has to be solved downstream from the chromosphere $x = 0$ to the corona $x = l$ using algorithms for stiff systems. Let

$$\psi_L(x) = \begin{pmatrix} u_L(x) \\ b_L(x) \\ w_L(x) \end{pmatrix}, \quad \psi_R(x) = \begin{pmatrix} u_R(x) \\ b_R(x) \\ w_R(x) \end{pmatrix} \tag{49}$$

are the solutions of the wave equation Eq. (22) with boundary conditions

$$\psi_L(0) = \mathrm{L}(0), \quad \psi_R(0) = \mathrm{R}(0). \tag{50}$$

We look for a solution as a linear combination

$$\psi(x) = \psi_R(x) + r\,\psi_L(x), \tag{51}$$

in other words we suppose that from the low viscosity chromosphere plasma do not come overdamped diffusion modes. The strong decay rate make them negligible at $x = 0$. Physically this means that AW (R-modes) are coming from the Sun and some of them are reflected from the TR (L-modes)

$$\psi(0) = \mathrm{R}(0) + r\,\mathrm{L}(0). \tag{52}$$

Analogously for the configuration of open corona we have to take into account the run-away boundary condition for which we suppose zero amplitude for the wave coming from infinity

$$\psi(l) = \tilde{t}\,\mathrm{R}(l) + \tilde{c}\,\mathrm{D}(l). \tag{53}$$

Written by components

$$\begin{pmatrix} u_{\mathrm{R}}(l) \\ b_{\mathrm{R}}(l) \\ w_{\mathrm{R}}(l) \end{pmatrix} + r \begin{pmatrix} u_{\mathrm{L}}(l) \\ b_{\mathrm{L}}(l) \\ w_{\mathrm{L}}(l) \end{pmatrix} = \tilde{t} \begin{pmatrix} \mathrm{R}_u(l) \\ \mathrm{R}_b(l) \\ \mathrm{R}_w(l) \end{pmatrix} + \tilde{c} \begin{pmatrix} \mathrm{D}_u(l) \\ \mathrm{D}_b(l) \\ \mathrm{D}_w(l) \end{pmatrix}. \tag{54}$$

This boundary condition gives a linear system of equation for the reflection coefficient r, transmission coefficient $\tilde{t}$ and the mode-conversion coefficient $\tilde{c}$.

For $l \to \infty$ when $\exp[-k''_{\mathrm{D}}(l)l] \ll 1 \exp[-k''_{\mathrm{R}}(l)l]$ the amplitude of D-mode is negligible and the run-away boundary condition reads

$$\psi(l) = \psi_{\mathrm{R}}(l) + r\psi_{\mathrm{L}}(l) \approx \tilde{t}\mathrm{R}(l), \tag{55}$$

or by components

$$\begin{pmatrix} u_{\mathrm{R}}(l) \\ b_{\mathrm{R}}(l) \end{pmatrix} + r \begin{pmatrix} u_{\mathrm{L}}(l) \\ b_{\mathrm{L}}(l) \end{pmatrix} = \tilde{t} \begin{pmatrix} \mathrm{R}_u(l) \\ \mathrm{R}_b(l) \end{pmatrix}. \tag{56}$$

These systems give the amplitudes of the reflected wave r and transmitted wave $\tilde{t}$ in the solution Eq. (51). For this solution we have the energy flux of transmitted $\mathcal{T}$ and reflected $\mathcal{R}$ waves

$$\begin{aligned} \mathcal{T} &\equiv \psi^{\dagger}(l)\,g(l)\,\psi(l) = |\tilde{t}|^2 + |\tilde{c}|^2 + \left(\tilde{t}\,\tilde{c}^{*}\mathrm{D}^{\dagger}(l)\,g(l)\,\mathrm{R}(l) + \mathrm{c.c.}\right), \\ 1 - \mathcal{R} &\equiv \psi^{\dagger}(0)\,g(0)\,\psi(0) = 1 - |r|^2 + \left(r^{*}\mathrm{L}^{\dagger}(0)\,g(0)\,\mathrm{R}(0) + \mathrm{c.c.}\right). \end{aligned} \tag{57}$$

Then we introduce the absorption coefficient

$$\mathcal{A} \equiv -\psi^{\dagger}(x)\,g(x)\,\psi(x)\Big|_0^l = 1 - \mathcal{R} - \mathcal{T}. \tag{58}$$

The described solution is normalized by unit energy flux of the R-wave. If we wish to fix energy flux of the right propagating wave to be $q_{\mathrm{wave}}(0)$ we have to make the renormalization

$$\Psi(x) = A_{\mathrm{wave}}\psi(x), \tag{59}$$

using a parameter A_{wave}. In this section we have described Absorbing Boundary Conditions (ABC)[31] well known from radar calculations, but realization for AW is more complicated and require eigenvector analysis. Now using $\Psi(x)$ we can calculate the wave part of the energy flux Eq. (42) and the wave part of the momentum flux

$$\tilde{\Pi}_{\mathrm{wave}}(x) \equiv \frac{1}{4\mu_0}\left|\hat{b}(x)\right|^2. \tag{60}$$

If we work with 4×4 matrix the generalization is obvious, in this case we have to operate with one additional diffusion mode.

3 Self-consistent procedure and results

We have to say few words about the numerical methods. For high frequency waves we can use WKB approximation, which for our problem reads

$$\Psi(\omega, x) = \sqrt{W(\omega)}\,\mathrm{R}(x)\exp\left(\mathrm{i}\int_0^x k_\mathrm{R}(x')\,\mathrm{d}x'\right). \tag{61}$$

As the equations for wind variables are stiff, we can use the implicit Euler method for $T(x)$ and $U(x)$

$$\frac{\mathrm{d}y}{\mathrm{d}x} = f(x,y), \quad y(x+h) = y(x) + f(x+h, y(x+h)). \tag{62}$$

For the realization of this method we use the explicit formulas Eq. (36) and Eq. (38) for U and T respectively. As dissipative fluxes contains the derivatives $-\xi U \mathrm{d}_x U - \varkappa\,\mathrm{d}_x T$ for the energy flux Eq. (27) and $\tilde{\Pi}_\xi(x) \equiv -\xi \mathrm{d}_x U$ in (29), the explicit expression for the wind variables Eq. (36) and Eq. (38) is actually a system of first order ordinary differential equations. In the WKB approximation complete system for wind and wave variables can be solved simultaneously.

One alternative is to use explicit expressions for derivatives $\mathrm{d}_x U$ and $\mathrm{d}_x T$ which age given from the constant energy Eq. (26) and momentum Eq. (28) fluxes. The limit case of ideal fluid with small ξ and $\varkappa$ in the denominators gives the singular perturbation theory. Care is therefore needed for the choice of the numerical method.

For exact wave variables the implicit Euler method applied for Eq. (22) gives

$$\Psi(x+h) = \left[\mathbb{1} - \mathrm{i}h\mathsf{K}(x+h)\right]^{-1}\Psi(x). \tag{63}$$

For this case the WKB approximation is only a starting point. Then for fixed wind profiles T and U we solve the wave equation with the corresponding boundary conditions. And using wave fluxes of energy and momentum we calculate the next approximation for the wind variables T and U. The procedure has to be repeated to reach self-consistency of wind and wave variables.

The simple Euler method can be elaborated by incorporation of predictor-corrector method using for small interpolation the Eitken formula for polynomial interpolation. Sequential polynomial interpolations can be rearranged as Padé interpolation using well-known epsilon algorithm by Wynn and Backer; some details are given in the Appendix II.

Following the described numerical method we have calculated the temperature and wind velocity profiles and the results are depicted in Fig. 1 and Fig. 2.

The energy and momentum flux density of the calculated temperature and wind profiles are respectively given in Fig. 3 and Fig. 4.

Finally, we present a comparison between theory and observation in Fig. 5: the calculated temperature profile from Fig. 1 is placed on the presented observational data from Ref. [13].

4 Conclusion

4.1 Discussions

In spite that iron was suspicious from the very beginning the problem of Coronium was a 70 year standing mystery until unambiguous identification as Fe^{13+} by Grotrian and Edlen [33] in 1937. The same 70 year time quantum was repeated. In 1947 Alfvén [2] advocated the idea that absorption of AW is the mechanism of heating of solar corona. Unfortunately the idea by Swedish iconoclast [34] was never realized in its original form: what can be calculated, what is measured, what is explained

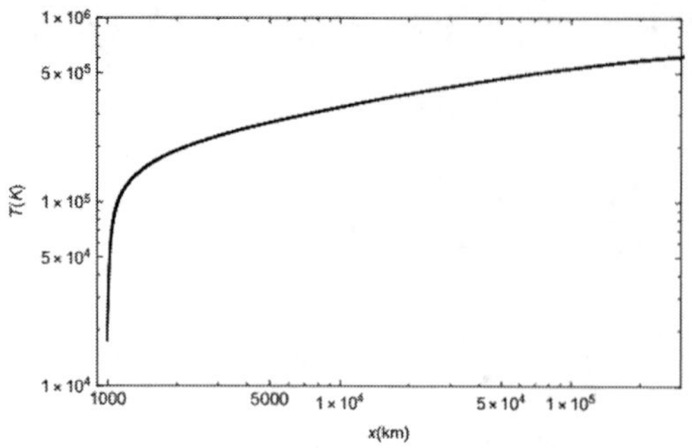

Figure 1: Temperature profile. Height dependence of the temperature T(x). MHD calculations qualitatively reproduce the well-known observational data [12, 13, 32]. This step-like behaviour confirms the scenario of self-induced opacity with respect to Alfvén waves.

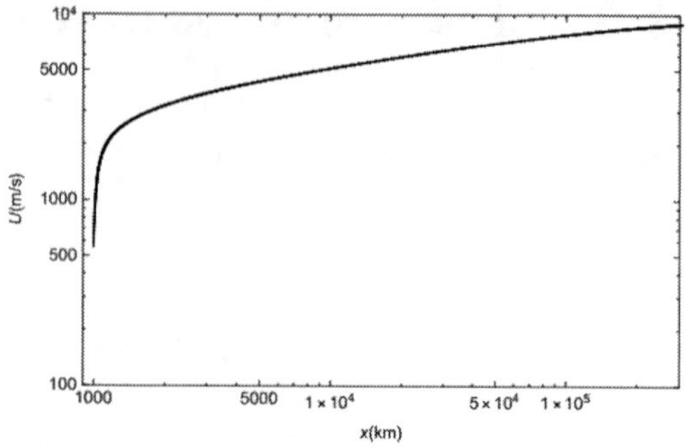

Figure 2: MHD calculation of height dependence of the solar wind velocity profile $U(x)$ in logarithmic scale. The wind profile resembles the temperature profile because Θ_0 is the dominating term in the brackets in Eq. (39), while τ is of the order of one and $\tau \ll \overline{U} \ll \Theta_0$.

and what is predicted. That is why there is a calamity of ideas still on the arena, for a contemporary review see the SOHO proceedings ([35]). From qualitative point of view, the narrow width of the TR $\lambda = \min |\mathrm{d}x/\mathrm{d}\ln T(x)|$ is the main property which should be compared against the predictions of other scenarios. For example, in order for the nanoflare hypothesis to be vindicated [36], such reconnections are needed to explain the narrow width of the TR at the same boundary conditions of wind velocity and temperature. Moreover, electric fields of the reconnections heats mainly the electron component of the plasma. How then proton temperature in the corona is higher? Launching of Hinode gave a lot of hints for the existence of AW in the corona [37], see also Ref. [38]. However, most of the those research was in UV region when high frequency AW which heat are already absorbed. All observations are for low frequency (mHz range) AW for which hot corona is transparent. The best can be done is to extract low frequency behavior of the spectral density of AW and to extrapolate to higher frequencies responsible for heating. So observed AW are irrelevant for the heating. In order to identify AW responsible for the heating, it is necessary to investigate high frequency (1 Hz range) AW in the cold chromosphere using optical, not UV spectral lines. We are unaware whether such type of experiments are planned. One of the purposes of the present work is to focus the attention of experimentalists on the 1 Hz range AW in the chromosphere, which we predict on the basis of our MHD analysis. For such purposes we suggest a special attention to be paid to the behavior of oscillations [39] and sunspot waves [40] above active regions. Another possibility is provided by Doppler tomography [41] of Hα or Ca lines. Doppler tomography was successfully used for investigation of rotating objects, such as accretion disks [42] and solar tornados [43]. Here we wish to mark also the Doppler tomography by Coronal Multi-channel Polarimeter build by Tomczyk [6]. For investigation of AW by Doppler tomography we suggest development of frequency dependent Doppler tomography operating as a lock-in voltmeter. The date from every space pixel should be multiplied by $\sin \omega t$ and integrated for many wave periods. Finally one can observe time averaged distribution of the AW amplitude. Systematic investigation of such frequency dependent Doppler tomograms will reveal that Swedish iconoclast [34] is again right that AW heat the solar corona, after another 70 years of dramatic launching of vast variety of ideas. In this article we wish to emphasize that the concept of self-induced opacity with respect to MHD waves

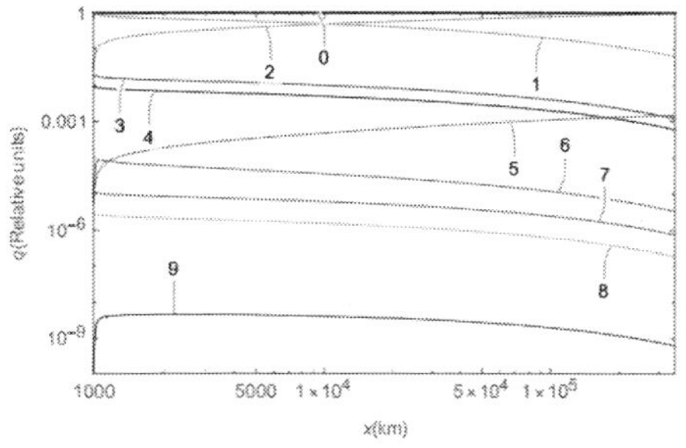

Figure 3: Detailed analysis of all energy flux density terms defined in Eq. (26) and Eq. (27). Height x dependence of the logarithms of the modulus of different components of the energy flux. Labels of curves correspond to: 0) the total constant energy flux density q, which is integral of motion of MHD equations, 1) the ideal wave energy flux term q_{ub} containing both wave velocity $\hat{u}$ and magnetic field $\hat{b}$, 2) the wind pressure energy flux term q_P, 3) the ideal wave energy flux term q_b; for this "aether" term a magnetic energy $\propto |\hat{b}|^2$ is moving with the velocity of the wind U, 4) the ideal wave velocity energy flux term q_u, 5) the ideal wind energy flux acceleration term q_U, 6) the wind thermal conductivity dissipative energy flux term $q_\varkappa$, 7) the wave dissipative viscous energy flux term q_η, 8) the wave dissipative Ohmic resistivity flux term q_ϱ and 9) the wind viscous dissipative energy flux term q_ξ. Both wind dissipative terms q_ξ and $q_\varkappa$ are negative and describe leeward energy flux. All other terms describe windward energy fluxes. Despite being small, $q_\varkappa$ and q_ξ contain space derivatives and determine the shape of temperature $T(x)$ and wind velocity $U(x)$ profiles.

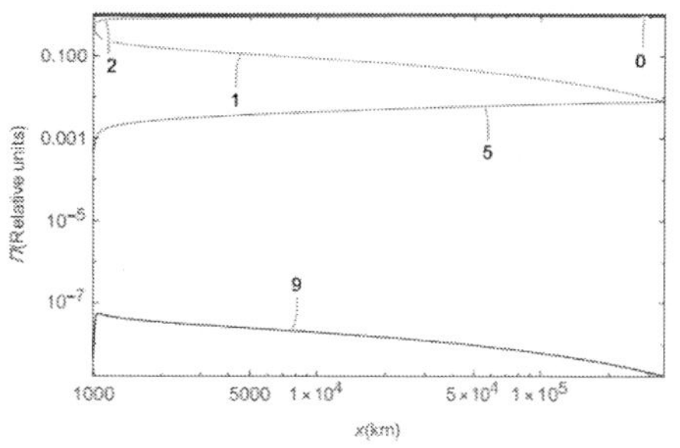

Figure 4: Detailed analysis of all momentum flux density terms defined in Eq. (28) and Eq. (29). Labels of curves correspond to: 0) the total constant momentum flux density Π, which is integral of motion, 1) the wave momentum flux term Π_{wave}, 2) the wind pressure momentum flux term Π_P, 5) the ideal wind momentum flux acceleration term Π_U and 9) the dissipative viscous momentum flux term Π_ξ. As for the energy flux, this small term contains space derivative $\mathrm{d}/\mathrm{d}x$ and we have singular perturbation theory for a system of differential equations for the wind variables $T(x)$ Eq. (38) and $U(x)$ Eq. (36).

was qualitatively proposed by us several months before launching the Hinode mission; now we are presenting only the detailed MHD analysis confirming our qualitative consideration ten years ago.

4.2 Plasma heating by AW – a historical perspective

What have we learned from the one-dimensional static MHD problem? We have demonstrated that qualitatively predicted self-induced opacity of plasma is an intrinsic property. Absorption of AW causes viscous heating of ions and that is why the proton temperature is higher than the electron one. In this way we have revealed an effective method for ion heating which can be applied to many

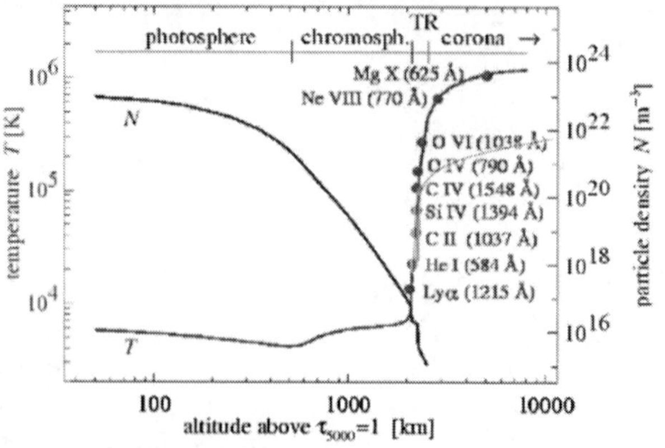

Figure 5: The thin increasing line is the temperature profile $T(x)$ from MHD calculation presented in Fig. 1. This illustrative MHD example is imposed together with observational data of Ref. [13]. The thick line with mark 'T' is $T(x)$ from Ref. [13], spectral lines of elements reveal how this observational curve is derived; the thick line with mark 'N' is the particle density $n(x)$ also in logarithmic scale in right ordinate. In the lower part of the TR between 10^4 K and 10^5 K MHD calculation almost perfectly matches the observational data. For higher temperatures the MHD calculation which contains only one monochromatic AW gives lower temperatures than the experimental profile. Optimistically we believe that MHD calculations with several AW could fit to whole observational $T(x)$.

plasma problems. Actually plasma heating by MHD waves is used in the MIT alcator [44]. We suggest however that the toroidal geometry can be replaced by Budker *probkotron* geometry [45], in which the energy of the AW will be focused in a narrow jet with a hundred times increased temperature. A de Laval nozzle will be realized by strong magnetic fields. We do believe that this will be an effective method for navigation in the Solar System (cf. [46]). Electric power from a nuclear reactor will create a fast electron-proton jet and this will dramatically decrease the initial mass of the rocket. For large-scale Earth-based installations such a jet of high-temperature deuterium will inject a fresh idea in nuclear fusion physics. For this purposes, however, we need of detailed function of temperature and magnetic field dependence of plasma kinetic coefficients. Such investigations started seventy years ago by R. Landshoff [47], E. S. Fradkin in 1951 [48], I. E. Tamm [49] and S. I. Braginskii in 1952 [50]. The main final results are listed in the Kinetics by Lifshitz and Pitaevskii [23]. Unfortunately we are unaware of recent reviews elaborating the old one by Braginskii [25].

Other unresolved problems are the second viscosity, kinetic coefficients and enthalpy of partially ionized hydrogen plasma. Having approximated experimental data with our calculation of temperature $T(x)$ and wind velocity $U(x)$ profiles, our work can be extended to the chromosphere. In such a way, solving the problem of kinetics and magnetohydrodynamics of solar plasma we can concentrate our attention to solve the inverse problem of revealing the spectral density of MHD waves coming from the invisible layers of the sun.

Our MHD analysis confirmed that the narrow transition region is almost a permanent explosion of chromospheric plasma by self-induced opacity with respect of MHD waves. There is no other scenario able to explain the width of the transition region. Except the thermal expansion of completely ionized plasma, a significant contribution for launching of the solar wind gives intensive low frequency MHD waves reflected from the TR as Fresnel reflection by a jump of the mass density $\rho(x) = Mn(x)$ and of the corresponding refraction coefficient $N_r = V_A(0)/V_A(x) = \sqrt{U(0)/U(x)}$. The local Alfvén speed

$V_A(x) = B_0/\sqrt{\mu_0\rho(x)}$ also has step-like behavior and the jump of the effective refraction coefficient creates strong reflection too, giving momentum to the TR. Unfortunately, up to now there is no reliable observational data for the height profile the solar wind $U(x)$. Using our routines we will continue MHD analysis in this direction but it will be nice to have alternative considerations. The contribution of low-frequency AW to the launching of solar wind is the next problem which has to be set in the agenda of the astrophysics.

Existence of intensive Alfénic modes in TR can be confirmed by systematic investigation of non-thermal broadening of spectral lines of Mg, and other elements both in the chromosphere and transition region. Moreover, taking into account that magnetic field is more or less vertical with respect to the Sun surface, the transversal AW create horizontal oscillations of the velocity and the transversal Doppler effect will give maximal broadening close to periphery of the solar disk and minimal at the center.

In the present work we concentrated our attention to the transversal to the magnetic field AW. In the chromosphere however, ionization-recombination processes will create a significant second viscosity ζ and high-frequency slow magnetosonic waves should be absorbed in the chromosphere. For heating of the chromosphere we need to know temperature dependence $\zeta(T)$ and the enthalpy of hydrogen plasma. We are unaware of such papers and will put this task in the agenda for further research.

For our approach, incoming AW are simply boundary condition. AW propagate along the magnetic force lines as through wave-guide. It is now subject of speculations where AW are created by perturbation of the magnetic pipes by turbulent convection or MHD waves are created in the solar tachocline [51] by self-sustained-wave turbulence in shear flow. The last scenario is perhaps working for the jets accompanying accretion [52]. In accretion disks the the effective viscosity and heating is created by self-sustained wave turbulence. Small part of accreting plasma following magnetic force lines is heated and accelerated in jets analogously to the solar wind; this will be a task of space MHD in the next decades.

One dimensional wave-guide approximation can be extended far from the TR to the distances of order of solar radius and even the size of solar system. In this case it is necessary to take into account slow adiabatic increasing of the area of the plasma stream and corresponding decrease of the constant longitudinal magnetic field at approximately constant magnetic flux.

Such type of contribution for the propulsion can be realized in a *probkotron* used simultaneously as a resonator. In such a way, hydrodynamic approach puts in the agenda many problems which can be easily solved knowing the detailed magnetic and temperature dependence of kinetic coefficients of solar plasma.

The main defect of the present research is that in frequency ω representation we solve static equations which require sophisticated numerical methods. All those results have to be repeated in time t dependent representation using the brute force of Monte Carlo averaging of standard MHD calculations.

A review of various mechanisms for solar corona heating could be easily made – most surely there is no plasma physical process that has not been used as the long-sought mechanism for solar corona heating. Nevertheless, we will point some alternative scenarios, which according to us even qualitatively contradict to the observations.

The longitudinal magnetosonic waves are intensely absorbed in the chromosphere through the large second viscosity and in practice do not reach the corona. And in reality, the intensity of the longitudinal waves in the solar wind is significantly lower than the Alfvén waves intensity.

If nanoflares and magnetic reconnections are the main mechanism for heating, then plasma energy absorption will be by Ohmic heating of the electrons. However, the proton temperature is higher than the electron one in the solar wind [26]. For more thorough review, please see one of the most recent works [32].

According the best we know there are no other attempts for calculation of the temperature profile of TR. In such a way the acceptable agreement between the observational data and our MHD analysis allows to conclude that after 70 years we have reliably returned to Alfvén [2] idea that heating of solar corona is predominantly by AW and the problem of set by Grotrian and Edlen: what heats the solar corona is already solved – 80 years perhaps the most long living problem of the science.

Authors are thankful to Yana Maneva [53–56] and Martin Stoev for the collaboration in the early stages of the present research [14] when the idea of self-induced opacity was advocated and consideration of many problems related to physics of solar corona. Fruitful comments and discussions with Tsvetan Sariisky are highly appreciated.

Authors are also thankful to Ramesh Chandra and Reetika Joshi for discussions of the non-thermal broadening of Mg II lines, Vassil Vassilev for his interest in our work, Ivan Zhelyazkov, Yana Maneva, Steven R. Cranmer, Eckart Marsch, Leon Ofman, Stefaan Poedts, Jaime Araneda, Plamen Angelov, Ramesh Chandra and Yurii Dumin, for short discussions and comments.

References

[1] P. Swings, Astrophys. J. **98**, 116–128 (1943); B. Edlén, Arkiv f. Matem. Astr. och Fys. **28**, B No. 1 (1941); B. Edlén, Zs. f. Ap. **22**, 30 (1942).

[2] H. Alfvén, MNRAS **107**, 211–219 (1947).

[3] H. Alfvén, Nature **150**, 405–406 (1942).

[4] J. Heyvaerts and E. R. Priest, A&A **117**, 220 (1983).

[5] L. P. Chitta, A. A. Ballegooijen, L. Rouppe van der Voort, E. E. DeLuca, and R. Kariyappa, Astrophys. J. **752**, 48, (2012); http://arxiv.org/abs/1204.4362.

[6] S. Tomczyk and S. W. McIntosh, Asprophys. J; http://arxiv.org/abs/0903.2002.

[7] R. Erdélyi and V. Fedun, *Are There Alfvén Waves in the Solar Atmosphere?,* Science **318**, 1572-1574 (2007);

[8] S. Tomczyk, S. W. McIntosh, S. L. Keil, P. G. Judge, T. Schad, D. H. Seeley, and J. Edmondson, Science **317**, 1192 (2007).

[9] G. L. Withbroe and R. W. Noyes, Annual review of astronomy and astrophysics **15**, 363–387, (1997), Figure 3.

[10] J. A. Eddy, *A New Sun: The Solar Results from Skylab,* ed. by R. Ise, prep. by George C. Marshall Space Flight Center, NASA, Washington, D.C., (1979), Fig on page 2, Fig on page 36, https://history.nasa.gov/SP-402/p2.htm.

[11] M. J. Aschwanden, *Physics of the Solar Corona. An Introduction with Problems and Solutions* (Springer, New York, 2005).

[12] E. H. Avrett and R. Loeser, The Astrophysical Journal Supplement Series **175**, Issue 1 p. 229–276, (2008).

[13] H. Tian, E. Marsch, C. Tu, W. Curdt, J. He, New Astronomy Reviews **54**, p. 13-30, (2010), Fig. 1 Lower.

[14] T. M. Mishonov, M. V. Stoev, and Y. G. Maneva, Eur. Phys. J. D **44**, 533–536 (2007); arXiv:astro-ph/0701908; arXiv:astro-ph/0701554; arXiv:astro-ph/0609609.

[15] T. K. Suzuki, Nonlinear Processes in Geophysics **15**, 2, 295–204 (2008).

[16] L. Ofman, Journal of Geophysical Research **115**, A4 (2010).

[17] L. Ofman, Space Science Reviews **120**, 1–2, 67–94 (2008).

[18] V. M. Nakariakov, L. Ofman, T. D. Arber, Astronomy and Astrophysics **353**, 741–748 (2000).

[19] L. Ofman, J. M. Davila, Journal of Geophysical Research **103**, A10, 23677–23690 (1998).

[20] L. Ofman, J. M. Davila, Journal of Geophysical Research **100**, A12, 23413–23426 (1995).

[21] S. Vasheghani Farahani *et al.,* Astronomy and Astrophysics **544**, A127, 5 (2012).

[22] L. D. Landau and E. M. Lifshitz, *Course of Theoretical Physics*, Vol. 8, *Electrodynamics of Continuous Media*, 6th ed. (Pergamon, New York, 1984), Chap. 8 "Magnetohydrodynamics" Eqs. (63.6-8), (65.7-15), (66.2-5,9), sec. 69, Problem, Eq. (69.8).

[23] L. D. Landau and E. M. Lifshitz, *Course of Theoretical Physics*, Vol. 10, *Physical Kinetics*, 1st ed. (Nauka, Moscow, 1979),
Sec. 13 "Kinetic phenomena in gases in external field" Eq. (13.18),
Sec. 42 "Energy transfer between electrons and ions" Eqs. (42.5,6),
Sec. 43 "Particle mean free path in plasma" Eqs. (43.4-10),
Sec. 58 "Hydrodynamic equations of magnetoactive plasma" Eq. (58.16),
Sec. 59 "Plasma kinetic coefficients in strong magnetic field" Eq. (59.38).

[24] E. Landi and M. Landini, Astronomy and Astrophysics **347**, 401–408, (1999).

[25] S. I. Braginskii, *Transport Phenomena in Plasma,* Vol. 1 (Atomizdat, Moscow, 1963) (in Russian);
Transport Processes in Plasma, Reviews of Plasma Physics Vol. 1 205, (1965), Eq. (4.44);
E. M. Epperlein and M. G. Haines, "Plasma transport coefficients in a magnetic field by direct numerical solution of Fokker-Planck equation," Physics of Fluids **29**, p. 1029 (1986).

[26] L. F. Burlaga and W. H. Mish, J. Gephys. Res., 92, 1261 (1987); E. Marsch, in Physics and Chemistry in Space – Space and Solar Physics, Vol. 21, Series Editors: M. C. E. Huber et al., Physics of the Inner Heliosphere, Vol. 2, Editors: R. Schwenn and E. Marsch, Springer-Verlag Berlin (1991); Fig. 10.4.

[27] R. V. Topchiyska, N. I. Zahariev, T. M. Mishonov, Bulg. J. Phys. **40**, 56–77 (2013);
N. I. Zahariev, T. M. Mishonov, AIP Conf. Proc. **1356**, 123 (2011);
T. M. Mishonov, N. I. Zahariev, R. V. Topchiyska, B. V. Lazov, S. B. Mladenov, in Proceedings of the Bulgarian National Conference on Physics 2014, Plovdiv, Bulgaria, Bulgarian Chemical Communications, Vol. 47, Special Issue B, pp. 368–379 (2015);
T. M. Mishonov, N. I. Zahariev, R. V. Topchiyska, B. V. Lazov, S. B. Mladenov, A. M. Varonov, https://arxiv.org/abs/1103.2233.

[28] R. P. Feynman, R. B. Leighton, and M. Sands, *The Feynman Lectures on Physics. Exercises* (Addison-Wesley, London, 1965), Problems 40.4-5.

[29] T. Van Doorsselaere *et al.,* The Astrophysical Journal Letters **727**, 2, L32, 4 (2011).

[30] L. D. Landau and E. M. Lifshitz, *Course of Theoretical Physics*, Vol. 3, *Quantum Mechanics*, (3 ed., Pergamon, New York, 1977),
Sec. 21, "General properties of one dimensional Schödinger equation.

[31] https://www.encyclopediaofmath.org/index.php/Absorbing_boundary_conditions
J. P. Berenger, "A perfectly matched layer for the absorption of electromagnetic waves" J. Comp. Phys. **114**, 185–200 (1994); C. Cerjan, D. Kosloff, R. Kosloff, M. Reshef, Geophysics **50**, 705–708 (1985) ; R. W. Clayton, B. Engquist, Bull. Seis. Soc. Amer. **67**, 1529–1540 (1977); B. Engquist, A. Majda, L. Halpern, L. N. Trefethen, J. Acoust. Soc. Amer. **84**, 1397–1404 (1988); E. L. Lindman, J. Comp. Phys. **18**, 66–78 (1975); IEEE Trans. Electromagn. Compat. **23**, 377–382 (1981); R. A. Renaut, J. Comp. Phys. **102**, 236–251; R. A. Renaut, J. Fröhlich, J. Comp. Phys. **124**, 324–336 (1996); R. A. Renaut, J. Peterson, Geophysics **54**, 1153–1163 (1989); A. C. Reynolds, Geophysics **43**, 1099–1110 (1978); P.A. Tirkas, C.A. Balanis, R.A. Renaut, IEEE Trans. Antennas and Propagation **40**: 10, 1215–1222 (1992).

[32] T. Sakurai, Proc. Jpn. Acad., Ser. B **93**, pp. 87–97 (2017).
[33] W. Grotrian, Naturwissenschaften **27**, 214, (1939);
B. Edlén, Z. Physik **104**, 407 (1937).
[34] A. J. Dessler, Science **170**, 604 - 606 (1970);
http://www.sciencemag.org/content/170/3958/604.
[35] *Proceedings of the SOHO 15 Workshop – Coronal Heating,* 6-9 September 2004, St. Andrews, Scotland (ESA SP-575, December 2004) Editors: R. W. Walsh, J. Ireland, D. Danesy, and B. Fleck.
[36] Ch. Day, Phys. Today **62**, May 18–21 (2009).
[37] B. De Pontieu, *et al.,* Science **318**, 1574–1577 (2007); T.J. Okamoto *et al.,* Science **318**, 1577-1580 (2007); Y. Katsukawa *et al.*, Science **318**, 1594-1597 (2007).
[38] D. Jess *et al.,* Science **323**, 1582–1585 (2009).
[39] D. Y. Kolobov, N. I. Kobanov, A. A. Chelpanov, A. A. Kochanov, S. A. Anfinogentov, S. A. Chupin, I. I. Myshyakov, V. E. Tomin, Advances in Space Research **56**, pp. 2760–2768 (2015).
[40] R. Chandra, G. R. Gupta, Sargam Mulay and Durgesh Tripathi, MNRAS **446**, pp. 3741–3748 (2015).
[41] T. Marsh, Europhysicsnews **36** No. 4, 133-138 (2005).
[42] T. R. Marsh and K. Horn, MNRAS **235**, 269–286 (1988).
[43] http://www.mps.mpg.de/en/forschung/sonne/.
[44] J. A. Snipes *et al.,* Phys. Plasmas **12**, 056102 (2005); http://dx.doi.org/10.1063/1.1865012.
[45] N. A. Krall and A. W. Trivelpiece, *Principles of Plasma Physics,* (McGraw-Hill, New York, 1973); Figs. 92, 101, 102, 104, 107.
[46] E. Y. Choueiri, Scientific American **300**, 58-65 (2009); doi:10.1038/scientificamerican0209-58.
[47] R. Landshoff, Phys. Rev. **76**, p. 904 (1949).
[48] E. S. Fradkin, JETP **5,** No. 5, p. 956 (1957);
Report FIAN, unpublished (1950–1951) (in Russian).
[49] I. E. Tamm, A. D. Sakharov, Physica Institute of the Academy of Sciences of U.S.S.R. Otchety (Russian title: Otchety o magnitnom termoyadernykh reaktorakh) (1951) (in Russian).
[50] S. I. Braginskii, ZhETF **33**, No. 3, p. 645, (1957) (in Russian);
JETP **6**, No. 3, p. 494 (1958) (English translation);
p. 494, Footnote: "This work was performed in 1952."
[51] https://en.wikipedia.org/wiki/Tachocline;
E. A. Spiegel, J.-P. Zahn, Astronomy and Astrophysics **265**, no 1, 106–114;
P. Charbonneau et al, The Astrophys. J. **527**, Issue 1, pp. 445–460 (1999).
[52] https://en.wikipedia.org/wiki/Astrophysical_jet;
I. Ciufolini and J. Wheeler, *Gravitation and Inertia*, (Princeton University Press, Princeton, New Jersey, 1995), Sec. 6.3, Pictures 4.5, 6.1-4;
Speical issue of the *Astrophysical Journal Letters* dedicated to the *Hubble Space Telescope* **435**;
H. C. Ford *et al.*, Astrophys. J. Lett. **435**, L27-L30 (1994);
R. J. Harms *et al.*, Astrophys. J. Lett. **435**, L35-L38 (1994);
R. D. Blandford, R. L. Znajek, MNRAS, **179**, Issue 3, p. 433–456, (1977).
[53] J. A. Araneda, Y. Maneva, and E. Marsch, Phys. Rev. Lett. **102**, 175001 (2009).
[54] Y. G. Maneva, J. A. Araneda, and E. Marsch, ed. by I. Zhelyazkov, AIP Conference Proceedings, vol. 1121, pp. 122–126 (2009).
[55] Y. G. Maneva, J. A. Araneda, and E. Marsch, Twelfth International Solar Wind Conference, eds. M. Maksimovic, K. Issautier, N. Meyer-Vernet, M. Moncuquet and F. Pantellini, AIP Conference

Proceedings, vol. 1216, pp. 227–230, (2010).
[56] Y. G. Maneva, Ph.D. Thesis, (Göttingen, uni-edition, 2010), ISBN 978-3-942171-39-7.
[57] G. A. Baker Jr. and P. Graves-Morris, *Padé Approximants,* Second edition, (Cambridge Univ. Press, Cambridge 1996).
[58] L. D. Landau and E. M. Lifshitz, *Course of Theoretical Physics*, Vol. 5, *Statistical Physics, Part 1*, (3 ed., Pergamon, New York, 1977),
Eq. (11.4), Eq. (15.12) "Theorem of small corrections", Sec. 78 "Thermodynamic variables of classical plasma", Eqs. (78.11-14), Sec. 79 "Method of correlation functions" Sec. 104 "Ionization equilibrium" Eqs. (104.4,5).
[59] A. B. Migdal, *Qualitative methods in quantum mechanics* (Moskow, Nauka, 1975) (in Russian), Chapter 1, "Dimensional and model evaluations".
[60] L. D. Landau and E. M. Lifshitz, *Course of Theoretical Physics*, Vol. 2, *The classical Theory of field*, 7th ed. (Pergamon, New York, 1989),
Sec. 33 "Tensor energy-momentum of electromagnetic field" Eqs. (33.3), (33.9),
Sec. 68 "Dipole emission at scattering", Eq (68.1), Sec. 70 "Radiation in the case of Coulomb interaction", Eq. (70.22), Prob. 2 and 4.
[61] L. D. Landau and E. M. Lifshitz, *Course of Theoretical Physics*, Vol. 4, V. B. Beresteckii, L. D. Landau and E. M. Lifshitz *Quantum electrodynamics*, (3 ed., Pergamon, New York, 1990), Sec. 92, "Electron-nucleus bremsstrahlung. The non-relativistic case ", Eq. (92.23), Problems 2 and 5.
[62] L. D. Landau and E. M. Lifshitz, *Course of Theoretical Physics*, Vol. 6, *Hydrodynamics*, 4th ed. (Pergamon, New York, 1985), Chap. 8 "Viscous liquid" Sec. 15 "Equation of motion of viscous liquid" Eq. (15.6).
[63] L. D. Landau and E. M. Lifshitz, *Course of Theoretical Physics*, E. M. Lifshitz and L. P. Pitaevskii Vol. 9, *Statistical Physics, Part 2*, Chap. VIII, "Electromagnetic fluctuations" Sec. 85, "Degenerated plasma", Eq. (85.23-24) (Pergamon, New York, 1979),
[64] L. D. Landau and E. M. Lifshitz, *Course of Theoretical Physics*, Vol. 1, *Mechanics*, (Pergamon, New York, 1989),
Sec. 40, "Hamilton equations".
[65] L. D. Landau and E. M. Lifshitz, *Course of Theoretical Physics*, Vol. 7, *Theory of elasticity*, (Pergamon, New York, 1989),
Sec. "Strength tensor" Eq. (2.1).

A Elementary statistics and kinetics of hydrogen plasma

In this section following the kinetics by Lifshitz and Pitaevskii [23], we will repeat the gas kinetic formulas applied to the plasma. Let us begin with the thermal velocities of the electrons and the protons

$$v_{T_e} \equiv \sqrt{T_e/m}, \qquad v_{T_p} \equiv \sqrt{T_p/M},$$

and introduce the variable $r_{_T} \equiv e^2/T$ with length dimension. The electron and proton transport sections for the Coulomb interaction expressed with the new variable have the form

$$\sigma_{ee} = \frac{\Lambda}{0.6} r_{_T}^2, \qquad \sigma_{pp} = \frac{\Lambda}{0.4} r_{_T}^2, \qquad \Lambda = \ln \frac{r_{_D}}{r_{_T}}. \tag{64}$$

For Rutherford scattering the cross-section is infinite and in the precise theory this infinity is substituted by the Coulomb logarithm from a big number: the ratio of the Debye radius $r_{_D}$ and the smallest distance r_T between two protons with thermal energy T. The coefficients 0.6 and 0.4 are added to the final exact formulas for respectively the electric conductivity and viscosity coefficients of completely ionized hydrogen plasma.

In a gaseous approximation the reciprocal mean free path for one electron is additive

$$\frac{1}{l_e} = \sigma_{ee} n_e + \sigma_{ea} n_a, \tag{65}$$

where σ_{ea} is the electron-atom cross section and n_a is the atom density. The mean free path l_e is connected with the mean free time

$$\tau_e = \frac{l_e}{\upsilon_{T_e}} = \frac{0.6}{\Lambda} \frac{m^{1/2}\, T^{3/2}}{e^4\, n_e}, \tag{66}$$

and the collision rate $\nu_e = 1/\tau_e$. The elementary kinetic estimate for the electric conductivity ς and the ohmic resistivity ϱ [23] connected with it

$$\varsigma = \frac{1}{\varrho} = n_e q_e^2 \frac{\tau_e}{m} \tag{67}$$

gives for the magnetic field diffusion coefficient

$$\nu_{\mathrm{m}} = \varepsilon_0 c^2 \varrho = \frac{c^2}{4\pi} \frac{m}{e^2 n_e \tau_e} = \frac{c^2}{4\pi} \frac{\sqrt{m T_e}}{e^2 n_e} \left(n_e \frac{\Lambda}{0.6} r_T^2 + n_a \sigma_{ea} \right). \tag{68}$$

At high temperatures $n_a \to 0$ this formula gives the well known result for the Ohmic resistance and electric conductivity of fully ionized hydrogen plasma

$$\varrho = \frac{1}{\varsigma} = \frac{1}{4\pi\epsilon_0} \frac{\Lambda}{0.6} \frac{e^2 m^{1/2}}{T^{3/2}}, \qquad \frac{\varsigma}{4\pi\varepsilon_0} = 0.6 \frac{T^{3/2}}{e^2 m^{1/2} \Lambda}. \tag{69}$$

In such a way, the approximate estimation for the transport section Eq. (64) exactly reproduces the well-known result [23].

Analogously, the heat conductivity coefficient is determined by the heat capacity per one electron $c_e = \frac{3}{2}$ and the elementary kinetic estimate

$$\varkappa = c_e n_e l_e \upsilon_{T_e} \tag{70}$$

gives

$$\varkappa = \frac{3}{2} n_e \sqrt{\frac{T_e}{m}} \left(n_e \frac{\Lambda}{0.6} r_T^2 + n_a \sigma_{ea} \right)^{-1}. \tag{71}$$

For completely ionized hydrogen plasma the heat conductivity coefficient becomes

$$\varkappa = \frac{0.9}{\Lambda} \frac{T^{5/2}}{e^4 m^{1/2}}. \tag{72}$$

Similarly, for the reciprocal proton mean free path we have

$$\frac{1}{l_p} = n_p \sigma_{pp} + n_a \sigma_{pa}, \tag{73}$$

where σ_{pa} is the proton-atom cross section. The elementary evaluation for the kinetic viscosity gives

$$\nu_k = l_p v_{T_p} = \sqrt{\frac{T_p}{M}}\left(n_p \frac{\Lambda_p}{0.4} r_T^2 + n_a \sigma_{pa}\right)^{-1}, \tag{74}$$

and for fully ionized plasma we arrive at the well known result for the viscosity Eq. (43.10) of Ref. [23]

$$\eta = \rho \nu_k = \frac{0.4}{\Lambda} \frac{T_p^{5/2} M^{1/2}}{e^4} \tag{75}$$

Let us also mention the following relations between the kinetic coefficients

$$\varkappa \varrho = 1.5\, T_e / q_e^2, \tag{76}$$

$$\eta/\varkappa \approx \frac{4}{9} \sqrt{mM}. \tag{77}$$

For the temperature dependent magnetic Prandtl coefficient $\mathrm{P_m} \equiv \nu_k/\nu_m$ we have

$$\mathrm{Pr_m} = \frac{4\pi e^2 n_e}{c^2 \sqrt{mM}} \left[\left(n_e \frac{\Lambda r_T^2}{0.6} + n_a \sigma_{ea}\right)\left(n_p \frac{\Lambda r_T^2}{0.4} + n_a \sigma_{pa}\right)\right]^{-1}, \tag{78}$$

and for high temperatures when $n_a \to 0$ we have

$$\mathrm{Pr_m} = \frac{0.96\pi}{\Lambda^2} \frac{T^4}{\sqrt{mM} c^2 e^6 n} \gg 1; \tag{79}$$

i.e. the electric resistance is negligible. While for the cold chromosphere the magnetic Prandtl number is of order of one, for the hot corona with a 100 times higher T it increases by 8 orders of magnitude. For this reason, in the whole interval where we consider MHD heating, the coefficient ν_m is negligible. In other words, the formal inclusion of Ohmic heating in the MHD equations will not change the profile $T(x)$ in the range of highest temperatures which is exactly the purpose of our consideration. Therefore, for illustrative purposes, we can take $\nu_m = 0$. While for low temperatures in the chromosphere, it is possible to neglect the viscosity and the heating could be purely ohmic.

Let us give an elementary estimate for ratios of the corresponding times and collision rates for electron, proton and electron-proton collisions for energy exchange τ_{ep}^{ε} for completely ionized hydrogen plasma

$$\left(\tau_e = \frac{1}{\nu_e}\right) : \left(\tau_p = \frac{1}{\nu_p}\right) : \left(\tau_{ep}^{\varepsilon} = \frac{1}{\nu_{ep}^{\varepsilon}}\right) = 1 : \sqrt{\frac{M}{m}} : \frac{M}{m}.$$

In the last proportion, it is taken into account that for elastic electron proton collision the exchange of energy is of order of m/M; it can be derived considering the scattering in the system of center of the mass. Our elementary consideration $\tau_{ep}^{\varepsilon} = \tau_e(M/m)$ gives

$$\tau_{ep}^{\varepsilon} = \frac{1}{8\sqrt{2\pi}} \frac{T^{3/2} M}{e^4 n \Lambda m^{1/2}}, \tag{80}$$

where the numerical coefficient $8\sqrt{2\pi} \approx 20$ in Eq. (42.5) of Ref. [23] requires state-of-the-art consideration. The kinetic of heat exchange between protons and electrons

$$\frac{\mathrm{d}T_e}{\mathrm{d}t} = -\frac{T_e - T_p}{\tau_{ep}^{\varepsilon}} \tag{81}$$

shows that in the transition region with width λ we can consider equal temperatures $T_e = T_p = T$ if the flight time through the transition region with wind velocity U is much longer than the time for which the protons heat up the electrons

$$\frac{\lambda}{U} \gg \tau^{\varepsilon}_{ep}. \tag{82}$$

In this way, for hydrogen plasma with known density ρ and temperature T, we can determine the pressure P and the kinetic coefficients $\varsigma, \varkappa, \eta$ necessary for our hydrodynamic analysis. For applicability of usual hydrodynamic approach the inequality

$$\nu_p = \frac{\Lambda}{0.4}\frac{e^4 n}{T^{3/2}M^{1/2}} \gg \omega_{Bp} = \frac{q_e B}{M}, \qquad B = H/c \tag{83}$$

holds perfectly for the chromosphere and the lower boundary of the TR. In the upper boundary of transition layer, *i.e.* in the beginning of the corona, this inequality is already broken because the density n decreases hundred times and simultaneously $T^{3/2}$ increases thousand times.

In order to take into account the influence of the magnetic field on the viscosity for transversal equations for the wave amplitudes in the present paper we suggest the Padé approximant

$$\eta_2 = \frac{1}{1 + C(\omega_{Bp}\tau_{pp})^2}, \qquad C \approx 0.9, \tag{84}$$

which has the same magnetic field dependence as longitudinal electric conductivity in perpendicular magnetic field in τ-approximation. For strong magnetic fields the suggested Padé approximant reproduces the well-known result

$$\eta_2(B \longrightarrow \infty) \approx \frac{8\pi^{1/2}e^4\Lambda n_p^2}{5(MT)^{1/2}\omega_{Bp}^2} \tag{85}$$

for $\omega_{Bp} \gg \nu_p$ case Eq. (X.59.38) of Ref. [23]. For all other coefficients η_1-η_4 we can suggest analogous formulas but the simplest Padé approximation [57] even quantitatively reproduces the main effect of the magnetic field at high temperatures and small densities – viscous heating stops in the hot corona.

B Ionization degree

The ionization degree $\alpha_{\imath} \equiv n_p/(n_p + n_a)$ satisfies the Saha equation [58]

$$\alpha_{\imath} = 1/\sqrt{1 + (1 + \alpha_{\imath})R_n(T)}, \tag{86}$$

where

$$R_n = \frac{n_p}{\mathrm{e}^{-I/T} n_q}, \quad n_q = \left(\frac{mT}{2\pi\hbar^2}\right)^{3/2}, \quad I = \frac{1}{2}mc^2\alpha_s^2, \quad \alpha_s = \frac{e^2}{\hbar c}.$$

Numerical calculation of the Saha equation requires several Newton iterations

$$\alpha_{\imath}^{(\mathrm{new})} = \alpha_{\imath} - \frac{F(\alpha_{\imath})}{\mathrm{d}F/\mathrm{d}\alpha_{\imath}} \tag{87}$$

for the function

$$F(\alpha_{\imath}) \equiv \alpha_{\imath}^2 - \frac{1}{1 + (1 + \alpha_{\imath})R_n(T)} = 0 \tag{88}$$

starting from $\alpha_I = 1/2$.

We also note that protons determine the mass density

$$n_\rho = n_p + n_a = n_p/\alpha_I, \tag{89}$$

and then for the pressure we have

$$P = n_{\rm tot}T, \qquad n_{\rm tot} = n_e + n_p + n_a = (1 + \alpha_I)n_\rho.$$

Ionization degree is important ingredient for the physics of the chromosphere. However, for the temperature of 10 kK and the low density of the upper chromosphere below the TR, the hydrogen is almost completely ionized. Since our calculation starting point is just before the TR, it is justified to assume full ionization $\alpha_I = 1$.

Knowing total cross-section of ionization by electron impact upon atom $\sigma_{\rm ion}$ we can evaluate the ionization rate and make elementary evaluation of second viscous coefficient of weakly ionized hydrogen plasma in the chromosphere. Large second viscosity in the chromosphere leads to significant damping of the MHD waves having pressure oscillations. That is why in the TR we concentrate our attention on AW.

C Bremsstrahlung

In this section we will make a model evaluation of the order of the *bremsstrahlung* in the TR. We follow the style of model evaluations by Migdal [59] Our staring point is the total radiation energy. The total radiation for the collision of electron and proton is [60], cf. also [61].

In the chromospheric boundary of the TR the temperature is low and electron thermal velocity $v_{T_e} = \sqrt{T/m}$ is much lower than the Bohr velocity $v_{\rm B} = e^2/\hbar$. For small relative velocities of the charged particles $v_0 \ll v_{\rm B}$ we have to use classical electrodynamics for calculation of the energy $\Delta\mathcal{E}$ of radiation emission. For the hydrogen plasma this means that temperature should be lower than atomic energy 1 Ry=13.6 eV=158 kK. Let us recall the well-known result for the back scattering of two charged particles [60]

$$\Delta\mathcal{E}(0) \equiv \frac{8\mu^3 v^5}{45c^3e^2}\left(\frac{e}{m} - \frac{e}{M}\right)^2 \approx \frac{8}{45}(mv^2)\left(\frac{v}{c}\right)^3, \tag{90}$$

where $\mu = mM/(m + M)$ is the reduced mass. We are applying this formula for electrons and protons with equal charges and $m \ll M$. With electron energy and the constant of Coulomb interaction we can construct a parameter with dimension of length $r_0 = e^2/\mu v_0^2$ and to make a model evaluation for the cross-section for energy loss by radiation (*effective radiation*)

$$\varkappa = \int_0^\infty \Delta\mathcal{E}(\rho)\cdot 2\pi\rho \mathrm{d}\rho \sim \pi r_0^2 \cdot \mathcal{E}(0) = C_{\rm BS}\frac{e^4 v_0}{c^3 m}, \tag{91}$$

where $C_{\rm BS}$ is a dimensionless constant of the order of one. This evaluation agrees with the exact result for the repulsing particles

$$\varkappa = \frac{8\pi}{9}\frac{|e_1e_2|\mu v_0}{c^3}\left(\frac{e_1}{m_1} - \frac{e_2}{m_2}\right)^2, \quad \mu = \frac{m_1m_2}{m_1 + m_2}. \tag{92}$$

For the hot Corona where $v_{T_e} \gg v_{\rm B}$ we have to use the results of the quantum electrodynamics [61]. For two electrons, for example, the dipole emission is zero and the quadrupole emission gives

$$\varkappa = \hbar\int_0^\varepsilon \omega\,\mathrm{d}\omega = 8.1\frac{e^2}{\hbar c}r_e^2\varepsilon, \quad r_e = \frac{e^2}{mc^2}, \quad \varepsilon = \frac{1}{2}mv_0^2, \quad v_0 \ll c.$$

We are giving also the result emission of radiation by elastic scattering of a slow $v \ll v_{\mathrm{B}}$ electron by a neutral atom [61]

$$\varkappa = \int \hbar\omega\, \mathrm{d}\omega = \frac{32}{45\pi}\frac{e^2}{\hbar c}\sigma_{\mathrm{elast}}\left(\frac{v}{c}\right)^2\left(\frac{1}{2}mv^2\right). \tag{93}$$

In our model qualitative consideration we will substitute the relative velocity by thermal velocity $v_0 \to v_{T_e}$. We hope that real integration of the corresponding statistical problem gives only multipliers of the order of one. For one scattering center the power of the emitted radiation can be evaluated as averaged flux of particles times the energy loss for scattering $v_{T_e} n\sigma\Delta\mathcal{E} = v_{T_e} n\varkappa$. For the total volume density of the emitted bremsstrahlung radiation $\tilde{Q}_{\mathrm{BS}}$ we have to multiply to the density of scattering centers

$$\tilde{Q}_{\mathrm{BS}} = (nv)(\Delta\mathcal{E}\sigma_{ep})n = n^2 v_{_{\mathrm{Te}}}\varkappa. \tag{94}$$

Substituting here

$$\sigma_{ep} = \pi r_{_T}^2 = \pi\left(\frac{e^2}{T}\right)^2, \qquad v = v_{T_e} \tag{95}$$

or directly the model evaluation of classical bremsstrahlung evaluation Eq. (91) we arrive at

$$\tilde{Q}_{\mathrm{BS}}(x) = \frac{8\pi}{45}\left(\frac{e^2}{mc^2}\right)^2 n^2(x)cT(x). \tag{96}$$

Let us follow a fluid particle moved by the wind with velocity $U(x)$. Integrating the power in time integral $\mathrm{d}t = \mathrm{d}x/U(x)$ we conclude that total energy loss per unit area in interval $(0, x)$ is

$$\int_0^x \tilde{Q}_{\mathrm{BS}}\frac{\mathrm{d}x'}{U(x')}$$

This means that energy flux is already not a constant and for the decreasing of the energy flux we obtain

$$\Delta q_{\mathrm{BS}}(x) = -U(x)\int_0^x \tilde{Q}_{\mathrm{BS}}\frac{\mathrm{d}x'}{U(x')}. \tag{97}$$

Finally, the dimensionless energy flux in the units used for the velocity Eq. (36) and temperature Eq. (38) we derive the term corresponding to the bremsstrahlung

$$\chi_{\mathrm{BS}} = \frac{\Delta q_{\mathrm{BS}}(x)}{\rho_0 U_0^3} = -\frac{U(x)}{\rho_0 U_0^3}\int_0^x \tilde{Q}_{\mathrm{BS}}\frac{\mathrm{d}x'}{U(x')}. \tag{98}$$

The negative sign corresponds to energy losses. For electron temperatures of order of 1 Ry and higher, the quantum electrodynamics result has a fine constant α_{s} [61] multiplier and as a result Bremsstrahlung is negligible for a cooling mechanisms in TR.

D Gravity

Since our calculation does not extend beyond 1 solar radius, for the gravitational energy flux we can use the approximate form of the gravitational potential

$$\mathbf{q}^{\mathrm{grav}} = \rho\mathbf{v}\Phi, \qquad \Phi = g_{\odot}x + \mathrm{const}, \quad g_{\odot} > 0, \tag{99}$$

where $g_{\odot}$ is the solar gravitational acceleration. The gravitational energy flux in our notations becomes

$$\tilde{q}(x)^{\mathrm{grav}} = jg_{\odot}x, \tag{100}$$

and according to definitions Eq. (32) can be easily expressed in our dimensionless variables

$$\chi_{\mathrm{grav}} = -\frac{j g_{\odot} x}{\rho_0 U_0^3}. \tag{101}$$

Analogously for the momentum the integrating in one dimensional case the density of gravitational force $f_g = -\rho(x)\,\mathrm{d}_x\Phi(x) = -\rho(x)g_{\odot}$ according to momentum conservation law Eq. (3) gives

$$\tilde{\Pi}_{xx}^{(\mathrm{grav})}(x) = -\int_0^x f_g(x')dx' = \int_0^x \rho(x')g_{\odot}\mathrm{d}x'. \tag{102}$$

Similarly, the momentum flux in dimensionless units becomes

$$\tau_{\mathrm{grav}}(x) = \left.\frac{\tilde{\Pi}^{\mathrm{grav}}(x)}{\rho_0 U_0^2}\right|_x^0 = -\frac{g_{\odot}}{\rho_0 U_0^2}\int_0^x \rho(x')\mathrm{d}x'. \tag{103}$$

The negative sign of τ_{grav} means that gravitational forces decrease the momentum density of solar wind.

The derived dimensionless energy and momentum gravitational fluxes could be easily added to the calculation, or can be evaluated as a perturbation to the calculated solution. Because both of these fluxes have negligible contribution in our region of interest [20], we have not included them in the numerical example analyzed in the present work.

Moreover, taking into account the TR width $\lambda \sim 100$ km and the temperature in its chromospheric end at $T = 10$ kK for the worst case scenario, we get $Mg_{\odot}\lambda/k_{\mathrm{B}}T' \approx 3\times10^{-2} \ll 1$, i.e. gravity is a negligible perturbation in the TR.

E Momentum equation and distribution of the temperature

After some algebra following Eq. (15.5) of Ref. [62],
Eq. (33.9) of Ref. [60] and Eq. (66.2) of Ref. [22] Eq. (3) can be rewritten as

$$\rho(\partial_t+\mathbf{v}\cdot\nabla)\,\mathbf{v}=-\nabla p+\eta\Delta\mathbf{v}+\left(\zeta+\frac{\eta}{3}\right)\mathrm{grad\ div}\,\mathbf{v}+(\nabla\eta)\cdot\nabla\mathbf{v}+(\nabla\mathbf{v})\cdot\nabla\eta+(\mathrm{div}\,\mathbf{v})\nabla\left(\zeta-\frac{2}{3}\eta\right)-\frac{\mathrm{rot}\,\mathbf{B}}{\mu_0}\times\mathbf{B}. \tag{104}$$

This equation can be derived as x derivative of the constant momentum flux Eq. (28) and different terms give the different densities of the external force.

Quite analogously differentiation of the energy flux Eq. (26) can be derived as the consequence of the equation of heat transport

$$\rho T\,(\partial_t+\mathbf{v}\cdot\nabla)\,s = \mathrm{div}\,(\varkappa\nabla T) + \sum_{i,k}\left[\eta\left(\partial_k v_i + \partial_i v_k - \frac{2}{3}\delta_{ik}\,\mathrm{div}\,\mathbf{v}\right) + \zeta\delta_{ik}\,\mathrm{div}\,\mathbf{v}\right]\partial_k v_i. \tag{105}$$

For an ideal fluid with negligible kinetic coefficients η, ζ and ϱ all terms of this equation are zero.

It is possible to check that upper equation for entropy transport and production is mathematically equivalent to the equation for the conservation of energy Eq. (2). In this sense Eq. (2) can be considered as a proof the evolution equation of the magnetic field Eq. (14). That is why one can say the the complete set of MHD equations Eq. (14) and Eq. (15) can be considered as consequence of mass Eq. (1), energy Eq. (2), and momentum Eq. (3) conservation equations, and the equation of the temperature field $T(\mathbf{r},t)$ is is also consequence of the conservation equations.

F Numerical methods

The differential equations we related to MHD analysis of solar corona are stiff and for them we have to apply the implicit Euler equation in the simplest case

$$\mathrm{d}_x y = f(x,y), \quad \frac{y(x+h)-y(x)}{h} = f(x+h, y(x+h)),$$
$$y = G(x, \mathrm{d}_x y), \quad y(x+h) = G\left(x+h, \frac{y(x+h)-y(x)}{h}\right), \quad \mathrm{d}_x y = \frac{\mathrm{d}y}{\mathrm{d}x}. \tag{106}$$

The wind variables $y = (T, U)$, and the conservation laws Eq. (26) and Eq. (28) give the possibility to express their derivatives explicitly. We however used the explicit formulas for wind variables $(T(x), U(x))$: Eq. (36) and Eq. (38) expressed by the derivatives $(\mathrm{d}_x T(x), \mathrm{d}_x U(x))$.

For the wave variables the corresponding system of equations Eq. (22) is linear, formally in the equation above $f(x,y) = \mathsf{A}(x)y$, and the realization of the Euler method reads

$$y(x+h) = (\mathbb{1} - h\mathsf{A})^{-1}\, y(h). \tag{107}$$

Let $\alpha_{\mathrm{L}}(x)$, and $\alpha_{\mathrm{R}}(x)$, are the smallest eigenvalues of matrix A describing left- and right propagating waves giving the solution we need. And $\alpha_{\mathrm{D}}(x)$ and $\alpha_{\mathrm{D'}}(x)$ are the big in modulus eigenvalues of the parasitic diffusion modes. The implicit Euler method can work in the case of

$$h|\alpha_{\mathrm{L}}|,\ h|\alpha_{\mathrm{R}}| \ll 1 \ll h|\alpha_{\mathrm{D}}|,\ h|\alpha_{\mathrm{D'}}|. \tag{108}$$

A few words we have to say for the numerical determination of limes. Let s_0, s_1, s_2, ...are the values of y(x+h) calculated by steps h, $h/2^1$, $h/2^2$, $h/2^3$, ... We can apply ε-algorithm for the calculation of $y(x+h)$ as the limes of the sequence $\{s_0, s_1, s_2, s_3, \dots\}$. In such a way, from the Euler method we can build a high-order method using Padé approximants. Analogously, let the sequence y_0, y_1, y_2, ... is the Aitken polynomial extrapolation of $y(x+h)$ calculated using 0,1,2,3, ... interpolation points. Applying again the epsilon algorithm we arrive at N-point Padé approximant predictor method for solving system of differential equation; according to the best we know such an opportunity is not used up to now. The ε-algorithm is fast convergent and the optimal order is determined by the error of truncation, i.e. our method has an adaptive order. In such a way, in spite of time consuming, we have a precision method for solving system of ordinary differential equations, which utilizes all numerical resources at a fixed machine precision. Details of the numerical realization of ε-algorithm and will be given in a separate publication.

G Influence of the magnetic field on the kinetic coefficients

The influence of magnetic field on the kinetic coefficients is significant for the solar corona, but for the TR it is also important to be taken into account. Unfortunately it is not standard for the MHD considerations of the solar corona and that is why the full set of equations is given only in this appendix. Let us start our consideration with the tensor of magnetic viscosity in the Drude τ-approximation for the electron drift velocity $\mathbf{V}_{\mathrm{dr}}$

$$m\mathrm{d}_t\mathbf{V}_{\mathrm{dr}} = -\frac{m}{\tau_{ee}}\mathbf{V}_{\mathrm{dr}} + q_e\left(\mathbf{E} + \mathbf{V}_{\mathrm{dr}} \times \mathbf{B}\right) \tag{109}$$

and corresponding tensors for the conductivity $\hat{\sigma}$, resistivity $\hat{\varrho}_\Omega$ and magnetic diffusivity $\hat{\nu}_{\mathrm{m}}$

$$\mathbf{j} = n_e q_e \mathbf{V}_{\mathrm{dr}} = \hat{\sigma}\mathbf{E}, \quad \mathbf{E} = \hat{\varrho}_\Omega \mathbf{j}, \quad \hat{\nu}_{\mathrm{m}} \equiv \varepsilon_0 c^2 \hat{\varrho}_\Omega. \tag{110}$$

In the static limit for frequencies $\omega \ll \omega_{Be}$ much lower than the electron cyclotron frequency

$$\omega_{Be} = \frac{q_e B}{m} \tag{111}$$

For the magnetic diffusivity tensor in the plane (y, z) perpendicular to the constant magnetic field $\mathbf{B}_0 = B_0 \mathbf{e}_x$ we have

$$\hat{\nu}_{\mathrm{m}} = \begin{pmatrix} 1 & -\omega_{Be}\tau_{ee} \\ +\omega_{Be}\tau_{ee} & 1 \end{pmatrix} \nu_{\mathrm{m}}, \tag{112}$$

and the equation for the magnetic field of the AW oscillations Eq. (20) now is reduced to the system

$$-\mathrm{i}\omega\hat{b}_z = \mathrm{d}_x\hat{u}_z - \mathrm{d}_x(U\hat{b}_z) + \mathrm{d}_x(\nu_{\mathrm{m}}\mathrm{d}_x\hat{b}_z) + \mathrm{d}_x(\omega_{Be}\tau_{ee}\nu_{\mathrm{m}}\mathrm{d}_x\hat{b}_y), \tag{113}$$

$$-\mathrm{i}\omega\hat{b}_y = \mathrm{d}_x\hat{u}_y - \mathrm{d}_x(U\hat{b}_y) + \mathrm{d}_x(\nu_{\mathrm{m}}\mathrm{d}_y\hat{b}_y) + \mathrm{d}_x(\omega_{Be}\tau_{ee}\nu_{\mathrm{m}}\mathrm{d}_x\hat{b}_z). \tag{114}$$

The generalization of the equations for the velocity oscillations Eq. (19) according the notations of Ref. [23] now reads

$$(-\mathrm{i}\omega + U\mathrm{d}_x)\,\hat{u}_z = V_{\mathrm{A}}^2\mathrm{d}_x\hat{b}_z + \frac{1}{\rho}\mathrm{d}_x\left(\eta_2\mathrm{d}_x\hat{u}_z\right) - \frac{1}{\rho}\mathrm{d}_x\left(\eta_4\mathrm{d}_x\hat{u}_y\right), \tag{115}$$

$$(-\mathrm{i}\omega + U\mathrm{d}_x)\,\hat{u}_y = V_{\mathrm{A}}^2\mathrm{d}_x\hat{b}_y + \frac{1}{\rho}\mathrm{d}_x\left(\eta_2\mathrm{d}_x\hat{u}_y\right) + \frac{1}{\rho}\mathrm{d}_x\left(\eta_4\mathrm{d}_x\hat{u}_z\right). \tag{116}$$

And the criterion to use approximation of viscous plasma with infinite conductivity becomes

$$\varepsilon_0 c^2 \varrho_{\mathrm{H}} = \omega_{Be}\tau_{ee}\nu_{\mathrm{m}} \ll \nu_{\mathrm{k}} = \frac{\eta_2}{\rho}, \tag{117}$$

for $\omega_{Be}\tau_{ee}$. Substituting here for the Hall resistivity the $\varrho_{\mathrm{H}} = q_e n_p/B$ we obtain the criterion for infinite conductivity

$$B \ll B_{\mathrm{H}} = H_{\mathrm{H}}/c \equiv \frac{q_e \eta_2(x)}{\varepsilon_0 c^2 M}. \tag{118}$$

The magnetic field dependence of the reduced viscous coefficients η_2/η and η_4/η are

$$\frac{\eta_2}{\eta} \approx \frac{\frac{6}{5}\tilde{x}^2 + \frac{7}{3}}{\tilde{x}^4 + 4.03\tilde{x}^2 + \frac{7}{3}}, \tag{119}$$

$$\frac{\eta_4}{\eta} \approx \frac{\left(\tilde{x}^2 + \frac{7}{3}\right)\tilde{x}}{\tilde{x}^4 + 4.03\tilde{x}^2 + \frac{7}{3}}. \tag{120}$$

If necessary, the frequency dependence of the kinetic coefficients can be also taken into account in the described system of equations describing the propagation of AW.

As the wind is along the magnetic field, the corresponding viscosity η multiplying $\mathrm{d}_x U$ is not changed. The influence of the magnetic field on the kinetic coefficient creates coupling between the AW with different polarization and double the range of the matrix. Nevertheless the work with 8×8 matrices is a routine doable task after development of the system of notions and notations and qualitative analysis of the oversimplified examples. Qualitatively the dissipative kinetic coefficients η and ϱ transform the wave energy of waves propagating into an ideal fluid into wind energy of an ideal gas, and the contribution of the other components is negligible.

In short, the influence of the magnetic field on the profiles of temperature $T(x)$ and wind velocity $U(x)$ is reduced to decreasing of the slopes $\mathrm{d}_x T(x)$ and $\mathrm{d}_x U(x)$ beyond the TR in the beginning of the corona.

H Correlation energy

For completely ionized plasma the correlation energy and pressure reads [58, 63, 64]

$$E_{\mathrm{corr}} = -e^3 \sqrt{\frac{\pi}{T}}\, n^{3/2} V, \quad N = Vn, \quad n = n_e + n_p, \quad n_e = n_p, \quad M = N\langle m\rangle, \tag{121}$$

$$P = P_{\mathrm{id}} + P_{\mathrm{corr}}, \quad P_{\mathrm{id}} = nT, \quad P_{\mathrm{corr}} = -\frac{1}{3}\sqrt{\frac{\pi}{T}}\, e^3 n^{3/2}.$$

Then for the correlation part of the enthalpy per unit mass we have

$$h = \frac{E + PV}{nV\langle m\rangle} = h_{\mathrm{id}} + h_{\mathrm{corr}}, \quad h_{\mathrm{id}} = c_p \frac{T}{\langle m\rangle}, \quad h_{\mathrm{corr}} = -\frac{4\sqrt{\pi}}{3}\frac{e^3}{\langle m\rangle}\sqrt{\frac{n}{T}},$$

and for the dimensionless energy and momentum fluxes we obtain

$$\chi_{\mathrm{corr}}(x) = \frac{U(0)h_{\mathrm{corr}}(0) - U(x)h_{\mathrm{corr}}(x)}{\rho_0 U_0^3}, \tag{122}$$

$$\tau_{\mathrm{corr}}(x) = \frac{p_{\mathrm{corr}}(0) - p_{\mathrm{corr}}(x)}{\rho_0 U_0^2}. \tag{123}$$

In such a way the Debye-Hückel correlation effects in solar plasma are incorporated in the developed scheme. To every term of the momentum flux we can ascribe corresponding volume density of a force $f_i = \partial_k \sigma_{ki}$ [65] and the correlation correction of the pressure is not an exception $f_{\mathrm{corr}} = -\nabla p_{\mathrm{corr}}$.

MATEC Web of Conferences **145**, 03010 (2018) https://doi.org/10.1051/matecconf/201814503010
NCTAM 2017

Aerodynamic characteristics and flow field of delta wings with the canard

Saya Mochizuki[1,*] , and *Gouji* Yamada[1]

[1]Tokai University, Department of Mechanical Engineering, 4-4-1 Kita-kaname, Hiratsuka-shi, Kanagawa, Japan

Abstract. Now, many kinds of explorations for outer planets have been proposed around the world. Among them Mars attracts much attention for future exploration. Orbiters and landers have been used for Mars exploration. Recently as a new exploration method, the usage of an airplane has been seriously considered and there are some development projects for Mars airplane. However, the airplane flying on the Earth atmosphere cannot fly on the Mars atmosphere, because atmospheric conditions are much different each other. Therefore, we focused on the usage of the airplane with unfolding wings for Mars exploration. These unfolding wings are designed as delta wings. However, delta wings do not have enough aerodynamics characteristics in a low speed region. In this study, to improve the aerodynamic characteristics of delta wings, we have proposed the usage of canard wings. The purpose of this study is to examine the effectiveness of canard wings to improve aerodynamic characteristics in a low speed region. CFD analysis is performed using four wing models with different canard shapes. The result shows that the usage of canards is effective to improve aerodynamic characteristics of delta wings in a low speed region. In addition, increasing lift coefficient is possible by changing the shape of canards.

1 Introduction

Now, many kinds of explorations for outer planets have been proposed around the world. Among them Mars attracts much attention for future exploration. [1], [2] Orbiters and landers have been used for Mars exploration. Although landers can obtain high resolution data, it cannot explore over a wide range. On the other hand, orbiters can explore over a wide range, but the resolution of the observed data is low. Recently as a new exploration method, there are some development projects for Mars airplane. [3] However, atmospheric conditions are much different between Mars and Earth. The flying conditions are low Reynolds number and high Mach number due to the decrease of atmospheric density and sound velocity. Additionally, the airplane must correspond to flight from low speed to high speed. Because, when it flies immediately after the discharge from a capsule, the flight condition is high speed, and as it approaches Mars, the flight condition is low speed. Therefore, it is necessary that new airplane which can fly on the Mars is developed.

* Corresponding author: 7BEMM084@cc.u-tokai.ac.jp

We focused on the airplane with unfolding wings by gliding for Mars exploration. Figure 1 shows a flight profile of the airplane with unfolding wings. The shape at a supersonic speed is streamlined-lifting-body which reduces the projected area. It is suitable for flight at a high speed. The lifting-body with unfolding wings that is expected to improve aerodynamic characteristics is used from transonic speed region to low speed region. These unfolding wings are designed as delta wings which are suitable for flight at a high speed. When the projected area expands, it is expected to improve lift coefficient and expand exploration range. Aerodynamic characteristics of delta wings were investigated from high subsonic to supersonic speed, and it has been clarified that the unfolding wings can improve aerodynamic characteristics. [4] However, delta wings do not have enough aerodynamic characteristics in a low speed region.

In this study, to improve the aerodynamic characteristics of delta wings, we have proposed the usage of canard wings. The purpose of this study is to examine the effectiveness of canard wings to improve aerodynamic characteristics in a low speed region by the computational fluid dynamics (CFD).

Fig. 1. Outline of the flight profile of the airplane with unfolding wings.

2 Analysis Overview

2.1 Condition

This Analysis use a calculator of usual personal computer level. Table 1 shows the specification of hardware and software. The simulation was conducted using OpenFOAM.

Table 1. The specification of hardware and software.

Hardware	
CPU	Intel Xeon E5-2640 (Core10, 2.30GHz)
Software	
CFD	OpenFOAM-4.0
Mesher	blockMesh snappyHexMesh
Postprocessor	paraView

2.2 Models

Figure 2 shows four wing models with different canard shapes. In Fig.2 (a), base (body) is an isosceles triangle with the vertical angle of 30 degrees. In Fig.2 (b), delta45 is composed of the body and an unfolding wing, and the wing is designed as delta wing with the sweep-back angle of 45 degrees. In Fig.2 (c), canard1 is composed of the body, unfolding wing and canard wing. The canard wing with the sweep-back angle of 45 degrees is positioned forward of the unfolding wing, and its wing area is 1/3 of the unfolding wings. In Fig.2 (d), canard2 is the same shape as canard1. Its wing area is 1/4 of the unfolding wing. All of these four models are flat plate with the total length of 300mm and the thickness of 2mm.

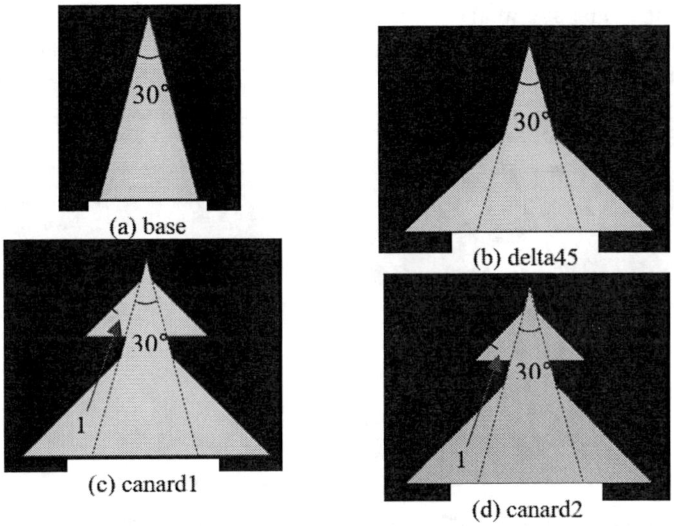

Fig. 2. Four wing models with different canard shapes.

2.3 Flow Field

The analysis fluid was air at standard atmosphere (1013hPa, 15°C), density ρ =1.225kg/m^3 and kinematic viscosity ν=1.47×10^{-5}m^2/s, and the airflow speeds were 15m/s, 20m/s, 25m/s and 30m/s. CFD analysis was conducted as the angle of attack of from 0 degree to 30 degrees in 3 degree intervals. Mach number is ranging from 0.044 to 0.088, satisfying a condition of incompressible fluid ($M_\infty < 0.14$). Therefore, flow field is incompressible fluid. The total length of models and the airflow speed gave the Reynolds numbers of 3.0×10^5~6.2×10^5. The flow field was assumed to be turbulent flow, because it cannot be judged as laminar flow or turbulent flow by the Reynolds number.

Numerical simulation of airflow around the four types of models was carried out in three dimensional steady states condition. The solver was simpleFoam which was standard solver in OpenFOAM based on SIMPLE method, because the airflow field was incompressible turbulent flow. The equation of the motion was the Reynolds averaged Navier-Stokes (RANS) in which the k-ω SST (Shear-Stress Transport) model was adopted as a turbulent model.

2.4 Numerical grid

Numerical grids are created by blockMesh and snappyHexMesh which are standard Meshers in OpenFOAM. Figure 3 shows numerical grids for the base. The number of cells of the case is about 2 million cells. Previously, CFD analysis was carried out using the case of 3.5 million cells, and it was confirmed that the suggested grid has no grid dependency on calculated results. The numerical grids are decided with the consideration of the calculation cost. The size of the case is set as 20 times the total length of models in X-axis direction (the direction of the total length of models), 20 times the wing span in Y-axis direction (the direction of the wing span), and 100 times the wing thickness in Z-axis direction (the direction of the wing thickness). Owing to the symmetry of the suggested delta wing because of no-roll and no-yaw angles, a half span model of the wing was analysed. As the boundary condition of the case, all the boundary surfaces of the front, rear, left, right and up of the space of Fig.3 (a) were freestream (the distant boundary), and the boundary surfaces of the down of the space of Fig.3 (a) was symmetryPlane (symmetry boundary).

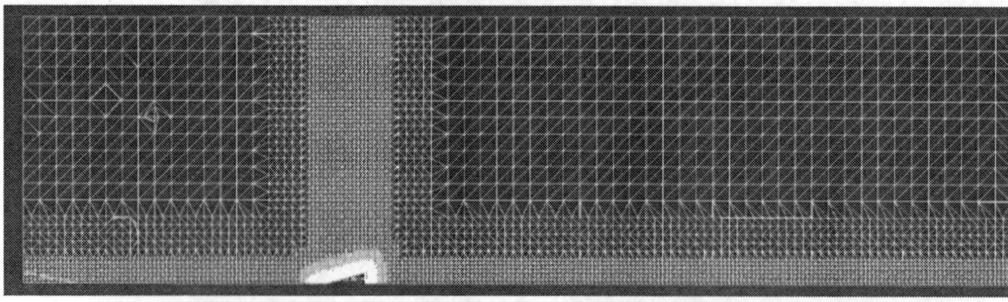

(a) the whole of numerical grids

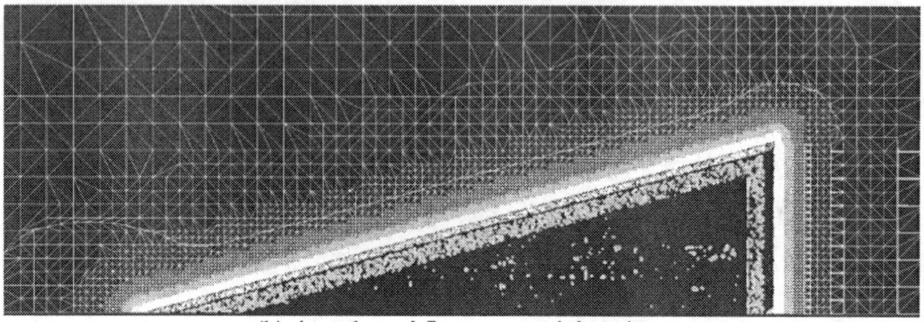

(b) the enlarged figure around the wing.

Fig. 3. Numerical grids for the base

3 Results and Discussion

3.1 Aerodynamic characteristics by CFD analysis

Aerodynamic characteristics of the suggested four types of models with flow velocity of 15m/s are shown in Fig.4. Aerodynamic characteristics of the base are lower than those of other models. It is found that aerodynamic characteristics are improved by the unfolding wings. The lift coefficients C_L compared between the suggested delta45 and canard1 or canard2 with flow velocity of 15m/s are shown in Fig.5. The lift slope of the delta45 become gentle from the angle of attack of 16 degrees, and a stall occurs at angle of attack of 23 degrees. The lift coefficient of the canard1 rises untill the angle of attack of 21 degrees and stagnates after that. On the other hand, although the lift coefficient of the canard2 stagnates from the angle of attack of 20 degrees that is lower than canard1, it rises again and stalls at the angle of attack of 26 degrees. Although the maximum lift coefficient is almost the same between the delta45 and canard1 as seen in Fig.5 (a), the maximum lift coefficient of the canard2 is larger than that of delta45. It is found that attaching the canard wings could delay the angle of attack where stall occurs. Figure 6 shows the lift coefficient compared between canard1 and canard2. The canard2 gave the best result of aerodynamic charasteristics in comparison with two types of the delta wings with canard. It is found that the smaller wing area of the canard wings insreases lift coefficient.

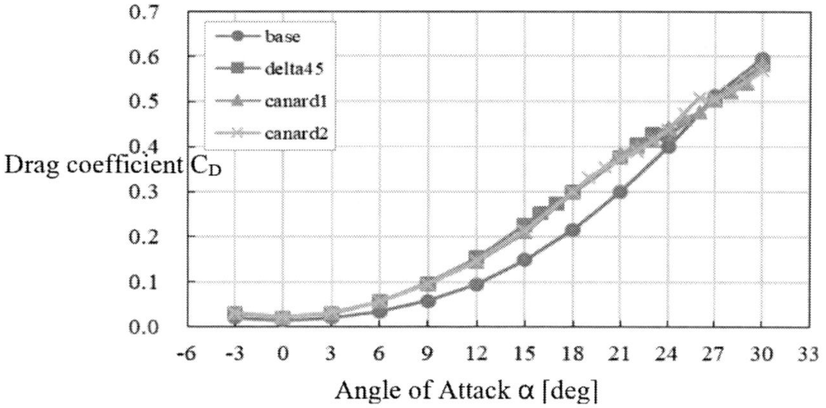

(a) Drag coefficient C_D

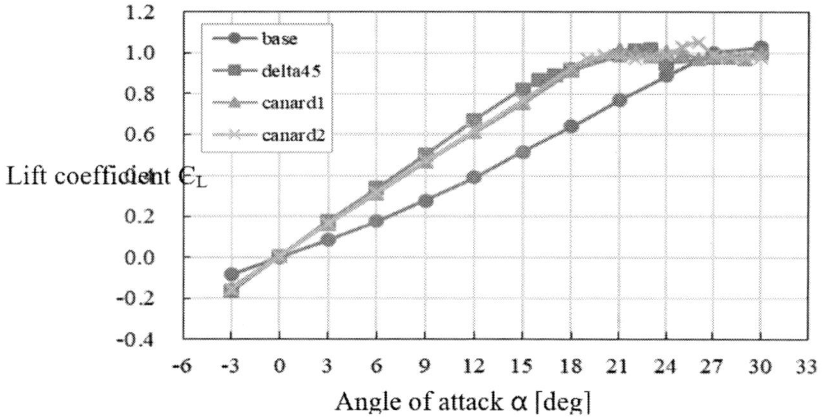

(b) Lift coefficient C_L

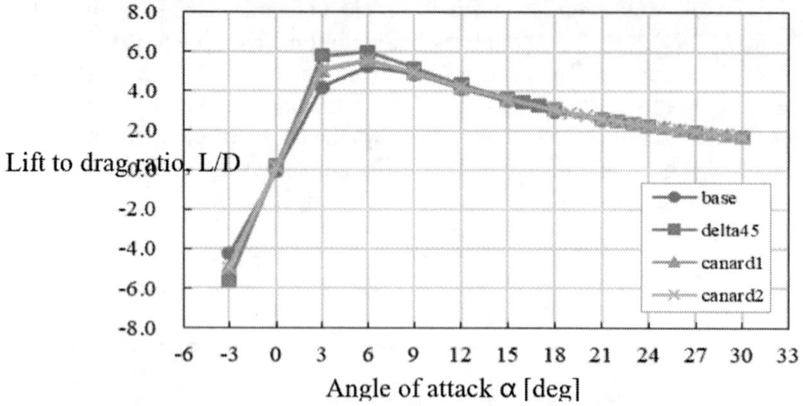

(c) Lift to drag ratio L/D

Fig. 4. Aerodynamic characteristics of the suggested four types of models with flow velocity of 15m/s.

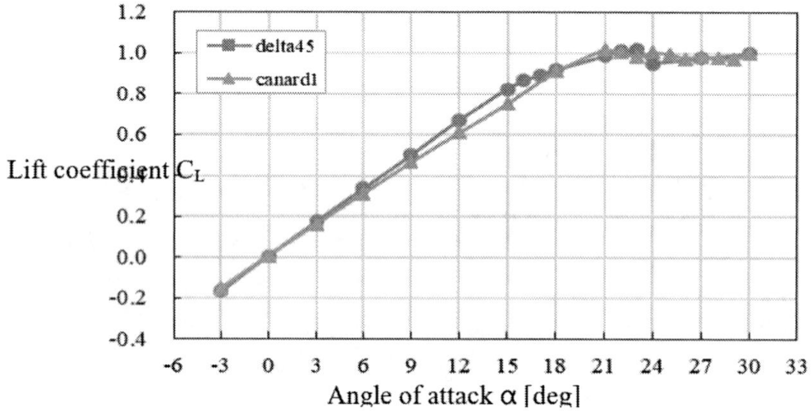

(a) Comparison between the suggested delta45 and canard1

(b) Comparison between the suggested delta45 and canard2

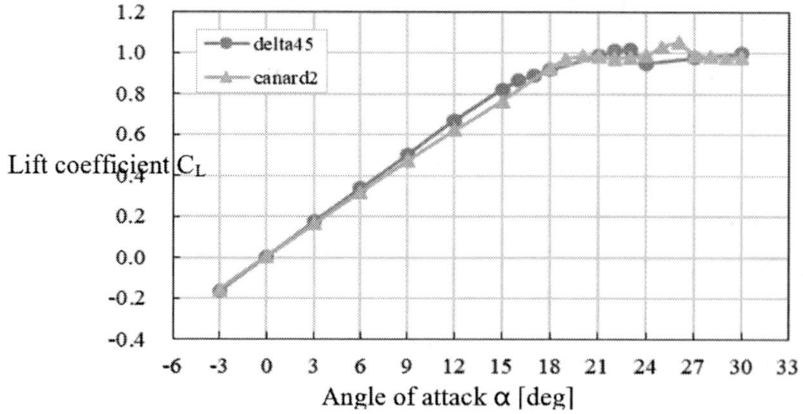

Fig. 5. The lift coefficient C_L compared between delta45 and canard1 or canard2 with flow velocity of 15m/s

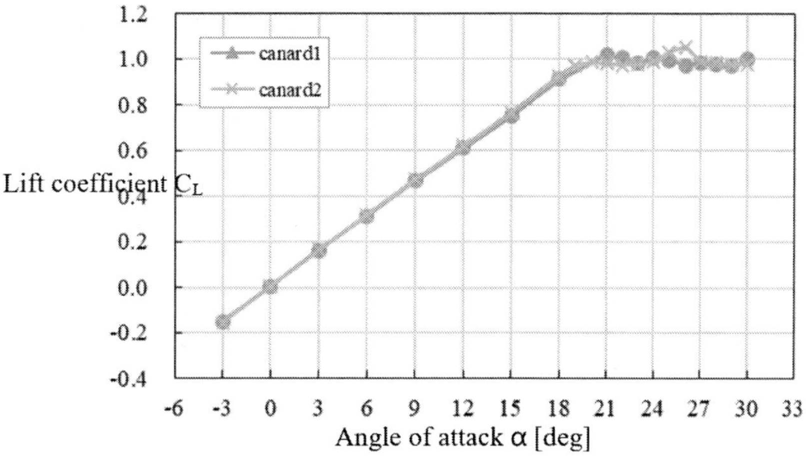

Fig. 6. the lift coefficient compared between canard1 and canard2.

3.2 Flow field

Figure 7 shows the pressure distribution of the delta45 at angles of attack of 23, 24 and 25 degrees. The pressure at α=24 degrees is higher than that at α=23 degrees in the rear region of the wings. It is considered that vortex breakdown occurs, causing a stall. On the other hands, it is found that leading edge separation vortex grow. That is the reason why lift coefficient increases again after the stall. Figure 8 shows the pressure distribution of the canard1 and canard2 at the angles of attack of 25, 26 and 27 degrees. The low-pressure area of the canard1 is decreasing with increasing of the angle of attack. The low-pressure area of the canard2 expands on the canard wing at the angle of attack of 26 degrees comparison with that of 25 degrees. Figure 9 shows the cross section of the pressure distribution of one fifth root chord point, x/c=0.2, and three fifth root chord point, x/c=0.6, at the angles of attack of 26 and 27 degrees of the canard2. It is found that vortex breakdown occurs on the canard and unfolding wing as the angle of attack increases. Figure 10 shows the cross section of the pressure distribution of three fifth root chord point, x/c=0.6, at the angle of attack of 18 degrees for the canard1 and canard2. The vortex from the tip of the canard2 exists inside of the main wings, and the pressure is lower in comparison with the canard1. The quantity of the vortex from the tip of the canard2, that reached the main wings without conjunction with the vortex of the canard wing, was larger than that of the canrd1, because the canard wing area of the canard2 is smaller than that of canard1.

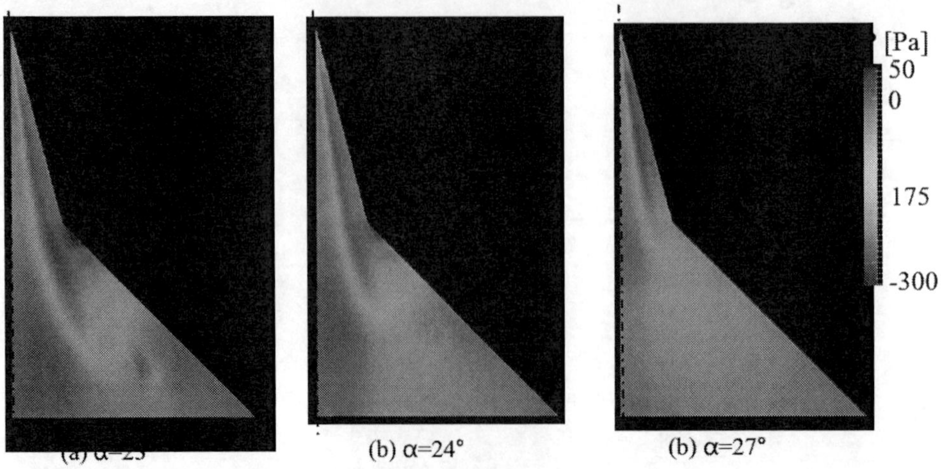

(a) α=25 (b) α=24° (b) α=27°

Fig. 7. the pressure distribution of the suggested delta45

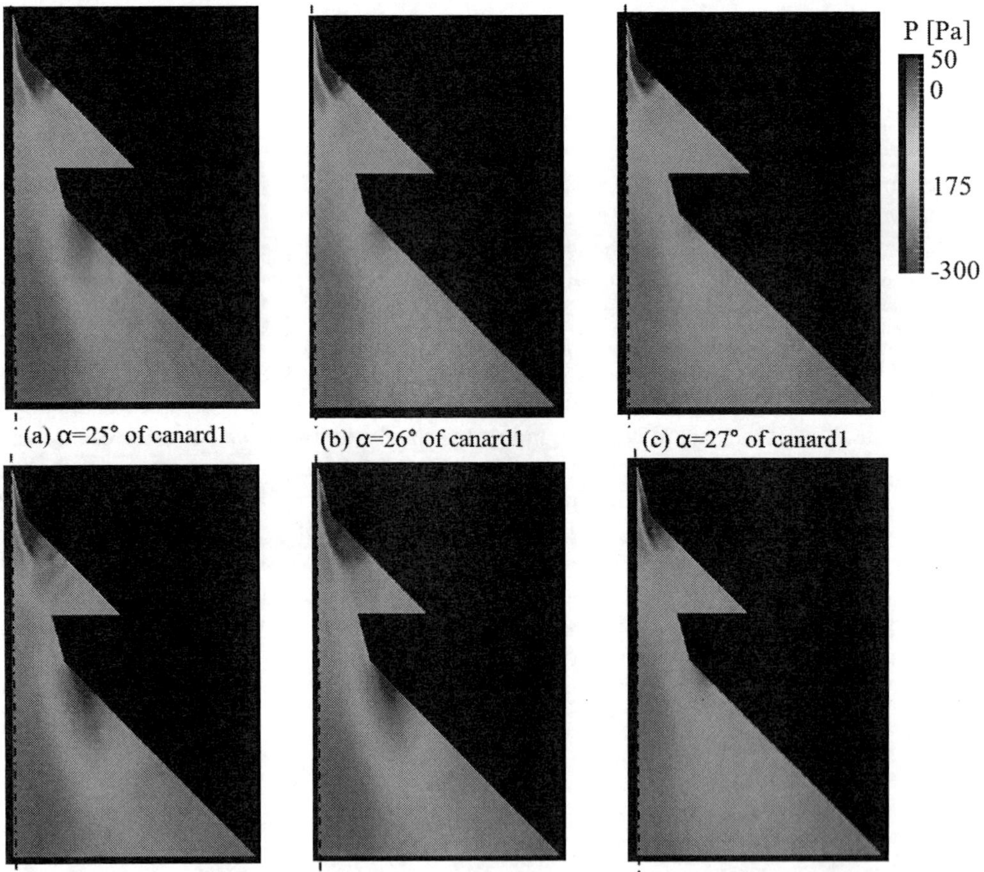

(a) α=25° of canard1 (b) α=26° of canard1 (c) α=27° of canard1

(d) α=25° of canard2 (e) α=26° of canard2 (f) α=27° of canard2

Fig. 8. the pressure distribution of the suggested canard1 and canard2

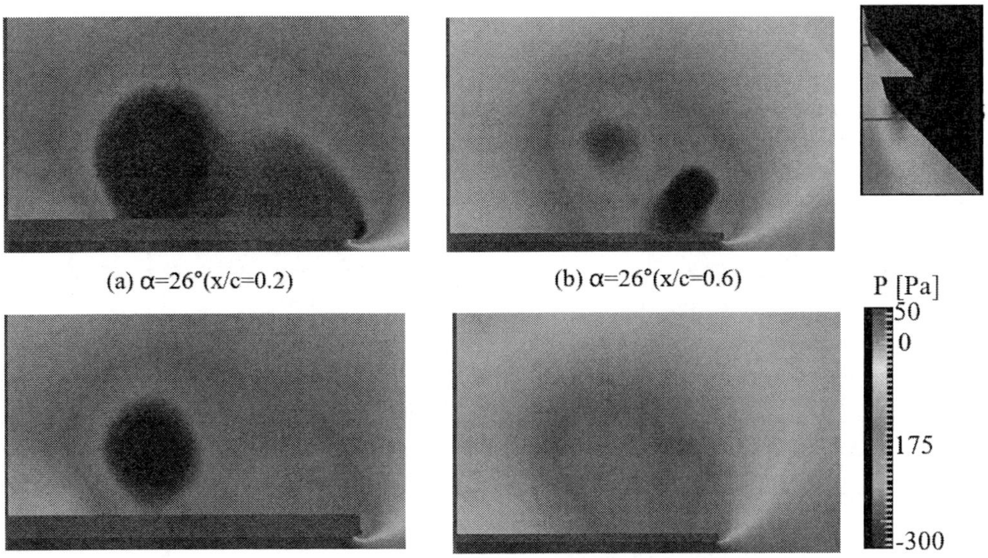

(a) α=26°(x/c=0.2) (b) α=26°(x/c=0.6)

(c) α=27°(x/c=0.2) (d) α=27°(x/c=0.6)

Fig. 9. the cross section of the pressure distribution of the canard2

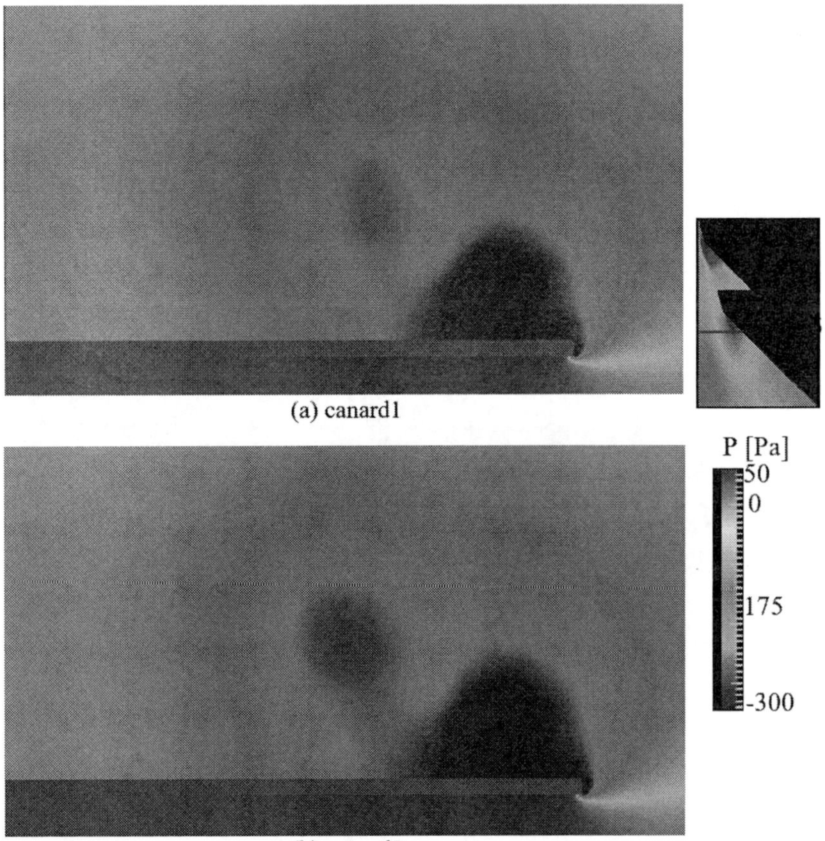

(a) canard1

(b) canard2

Fig. 10. the cross section of the pressure distribution at the attack of angle of 18 degrees (x/c=0.6)

4 Conclusions

In this study, to improve the aerodynamic characteristics of delta wings in a low speed region, we have proposed the usage of canard wings. The purpose of this study is to examine the effectiveness of canard wings to improve aerodynamic characteristics in a low speed region. CFD analysis are performed using four wing models with different canard shapes. The result shows that the usage of canards is effective to improve aerodynamic characteristics of delta wings in a low speed region. Lift coefficient of the canard2 is highest in all of four types of models. In addition, increasing lift coefficient is possible by changing the shape of canards.

References

1. D. Vaughn, H. C. Miller, B. F. James, M. M. Munk, *AIAA-2005-4110*, (2005)
2. P. Messina, D. Vennemann, Acta Astronautica, **57**, 156-160, (2005)
3. D. G. Mark, A. C. Mark, *AIAA-2003-6578*, (2003)
4. G. Yamada, S. Miyazaki, H. Kawazoe, Aerospace technology Japan, the Japan Society for Aeronautical and Space Sciences, **15**, 9-14, (2016)

Kamchia watershed groundwater recharge assessment by the CLM3 model

Olga Nitcheva[1,*], *Borislav* Milev[2] , *Tanya* Trenkova[3], Nina Philipova[1] and Polya Dobreva[1]

[1]Institute of Mechanics at the BAS, Acad. G. Bonchev 4, 1311 Sofia, Bulgaria
[2]European Politechnical University, St.St. Kiril i Methody 23, Pernik, Bulgaria
[3]National Institute of Geophysics, Geodesy and Geography at the BAS, Acad.G. Bonchev 3, 1113 Sofia, Bulgaria

Abstract. Estimating groundwater recharge is an important part of the water resources evaluation. In spite of the numerous existing methods it continues to be not easy value to quantify. This is due to its dependence on many meteorological, hydrogeological, soil type and cover conditions and the impossibility for direct measurement. Employment of hydrological models in fact directly calculates the influence of the above cited natural factors. The Community Land Model (CLM3) being loaded with all land featuring data in global scale, including an adequate soil filtration process simulation by the Richards equation, together with the possibility for input of NCEP/NCAR Reanalyses database, featuring the meteorological effect, gives an opportunity to avoid to great extent the difficulties in groundwater (GW) recharge estimation. The paper presents the results from an experiment concerning GW recharge monthly estimation during 2013, worked out for the Kamchia river watershed in Bulgaria. The computed monthly and annual values are presented on GIS maps and are compared with existing assessments made by other methods. It is proved the good approach and the applicability of the method.

Introduction

The GW is a part of the regions total runoff and the knowledge of their quantity enables their more rational utilisation from resource and environmental aspect.

The target of the paper is to analyse the possibilities of application of present-day advanced hydrology models, in particular the model CLM3, to assessment of the groundwater recharge.

* Corresponding author: olgani@imbm.bas.bg

1 Methodological aspects and methods for groundwater recharge evaluation

The groundwater bodies (GWB) in fact are underground reservoirs which are mainly recharged through the surface by precipitation or other artificially created sources as irrigation losses and others. The re-infiltration through the banks of rivers, occurring in periods of high water levels, is significant only for GWB in the terraces along big rivers with sustained high water levels, not typical for mountainous countries like Bulgaria. This doesn't relate to the East Kamchia lowland GWB either.

The annual runoff from a catchment is composed of the surface component of the runoff occurred in a short period after the rainfall and the groundwater one formed by the GWB discharge known as base flow. Regarded in a multiannual period the volume of the base flow and that of the abstracted water should be equal to the volume of the recharge during that time interval.

The advantage of GW use compared to the surface running waters is their regulated character, i.e. the abstraction rate does not depend from the recharge rate, due to the water accumulation in underground reservoirs. Besides, they are broadly available away from running waters. In addition, groundwater resources are less subjected to pollution.

The currently used methods for GW recharge assessment apply two possible approaches for solution of this problem. The first one consists in calculation procedures for evaluation of the GW actual recharge rate based on the records of the long term variation of the GWB regime characteristics as their water table fluctuation or springs discharge, as well as of the streamflow regime of the linked with them rivers [1]. The ground water level fluctuation and the rivers streamflow separation methods use this approach. Its advantage consists in the lack of necessity for possessing data concerning the soil properties, land cover and relief and climatic effects. That can be considered as conventional methods. They have been developed and used broadly since the second half of the last century.

In the last two decades an often practice for GW recharge assessment became the employment of models accounting for the meteorological loading [2-6]. Here considered is an improved hydrological model, which includes adequate simulation not only of the surface runoff formation and evaporation but also of the processes of water infiltration and transport through the soil unsaturated zone. Thus it became possible the direct imitation of the entire process of water penetration into the soil depth forming the potential recharge through further percolation to the GWB. Using these models gives a new alternative instrument for GW recharge assessment. Their application doesn't need input of GWB related information but requires accurate knowledge about all factors forming the GW like the soil unsaturated zone water retention and filtration properties, land surface topographic and vegetation cover data and detailed long term records of the registered climatic factors. This information relates to the entire GWB watershed area. All these data apart from assessment of the soil water problems are commonly used in different fields of human activity wherefore they are expected to be easily available. Soil properties can be provided from soil maps used in the agriculture.

The model used in this study is the Community Land Model version 3.0 (CLM3). It is the land surface model of the Community Earth System Model and the Community Atmosphere Model (http://www.cgd.ucar.edu/tss/clm/) [2-3]. It is a spatially distributed model developed in the USA National Center for Atmospheric Research (NCAR). The model computes the water balance between the precipitation, runoff, evapotranspiration and infiltration through the soil layer simulating the processes of water and heat transport through the soil profile and energy exchange between the soil surface and the atmosphere.

Regarded is a soil layer of given thickness made of joining to each other vertical columns with cross section equal to the area of the assumed spatial grid elements.

The main advantage of the assumed in this study model compared to other hydrological models is the detailed and realistic description of the water movement processes through the unsaturated zone by accurate mathematical dependencies. They account not only for the gravity but as well the porous and capillary uplift forces driving the soil moisture down or up the soil profile. The uplift forces depend on the degree of the wetness.

Numerical simulation of water transport in the soil zone is presented by the Richards equation [7], where the water transport in the soil profile depends on the wetness of the environment. This is the following nonlinear equation:

$$C(\psi)\frac{\partial \psi}{\partial t}=\frac{\partial}{\partial z}\left[K(\psi)\left(\frac{\partial \psi}{\partial z}+1\right)\right]-S(\psi,z) \qquad (1)$$

In (1) $C(\psi)=d\theta/d\psi$ is the differential water capacity (water retention), θ is the volumetric soil moisture (m^3/m^3), $K(\psi)$ is the hydraulic conductivity (m/day), ψ is the soil water pressure head (m) (soil suction), t is the time (days), $S(\psi,z)$ is the function of root water uptake (day^{-1}).

The Darcy-Richards equation (1) can be computed for the current time step $n+1$ after determining the moisture retention $\theta(\psi)$ and the hydraulic conductivity $K(\psi)$. It means for every time step n after determination of ψ, the values of $C(\psi)$ and $K(\psi)$ are evaluated. So for the time step $n+1$, the partial differential equation of second order becomes a linear equation.

In the CLM3 model the soil hydraulic properties $C(\psi)$ and $K(\psi)$ are estimated by the empirical functions of Clapp and Hornberger [8].

2 Kamchia River watershed description

Bulgaria is a mountainous country with not abundant but regulated in many reservoirs surface waters (Fig. 1). They satisfy the needs of the irrigation and a greater part of the domestic and industrial water supply. Yet the groundwater serves as basic water source in flat regions with scarce possibilities for water accumulation reservoirs. The groundwater is important for meeting the water demand of the population in the Kamchia River region too.

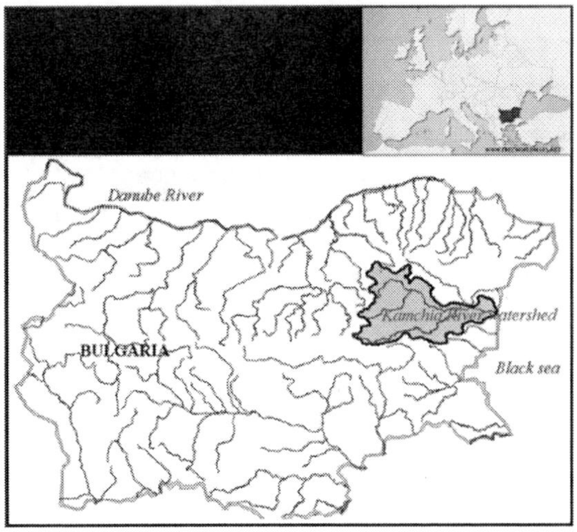

Fig.1. Map of Bulgarian territory and location of the Kamchia River watershed.

The Kamchia River is 244.5 km long situated in eastern Bulgaria and is the longest river on the Balkan Peninsula flowing directly into the Black Sea [9-10].

Its watershed occupies the middle east part of the country, from Eastern Balkan to the Black Sea. Its area is 5358 km^2 [10]. The west part is hilly with elevation from 700 to 300 m. Towards the east to the Black sea the relief becomes flatter. In the west part the average annual precipitation is 750-800 mm/a, in the east part - about 550 mm/a. The climate is continental, the maximum river runoff occurs usually in February [9-10].

Kamchia River has considerable discharge contribution into the Bulgarian Black Sea, ranging between 179,3 x 10^6 m^3 yr^{-1} and 1475,3 x 10^6 m3 yr^{-1} [11]. It is the total amount of surface and groundwater resource.

The Kamchia River hydrogeology is presented by porous, carst (carst-ctack) and crack waters, which are differentiated by 12 groundwater bodies located in 6 geological layers – Quaternary, Neogene, Paleogene, Upper Gretaceous, Hotriv Barem Gretaceous and Malm Valanginian (Fig.2) [12].

The porous groundwater occupy the first and second geological layers, while the carst, carst crack and crack groundwater are located in the third to sixth geological layers. The groundwater infiltration in some areas is very high, up to 30-40% of the precipitation. [13].

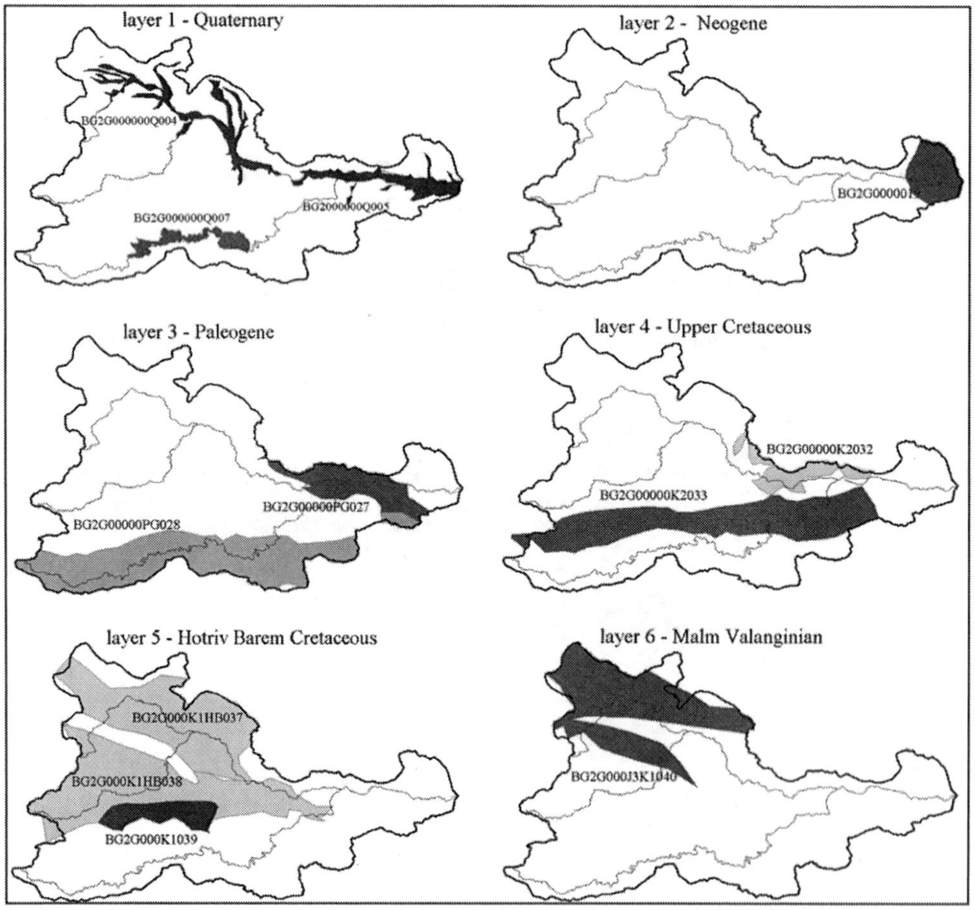

Fig. 2. Groundwater bodies location in 6 hydrogeological layers for the Kamchia River watershed.

On the basis of the published in the Ministry of Environment and Water (MEW) site long term (for the period 1981-2013) [14] GWB annual recharge values the authors has prepared a GIS map showing the mean recharge intensity on all GWB watersheds in mm. It is calculated through dividing the calculated by MEW mean annual GWB recharge in l/s to the watershed area. The intensity on areas with overlaying GWB is obtained by superposition.

Given the complexity of the GWB multilayer location and scarcity of observation data records because of insufficient observation wells and springs and river discharge measurement stations as well, the elaborated map hardly can give reliable detailed information about the areal distribution of the recharge intensity. The following facts contribute to the approximate nature of the map. The plotted intensity is averaged over the GWB watersheds varying from several dozens to several thousand km^2.

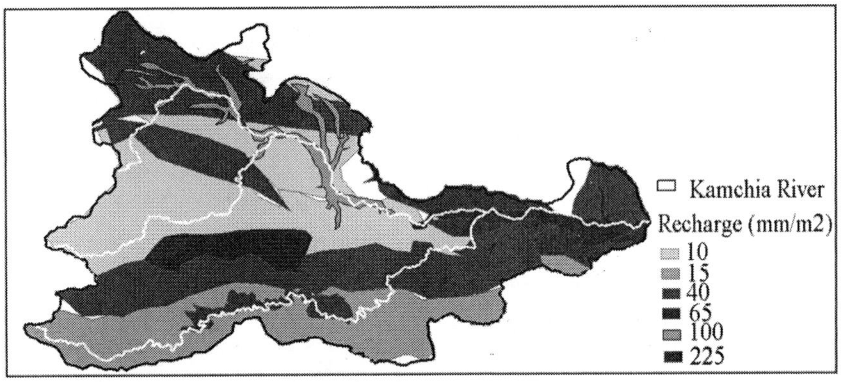

Fig. 3. Mapping the observed multi-annual groundwater recharge (by conventional methods) for the Kamchia River watershed through the period 1981-2013.

The shown on the map intensities in average are lower than real ones because the actual recharge areas of many GWB are less than their watersheds.

Nevertheless, the map is based on estimations using actual field data and regardless the above pointed out discrepancies with the lack of more trustworthy documents it can be used for various GWB related assessments.

3 Groundwater recharge estimation by CLM3 model

The CLM3 model is applied to evaluation of the GW recharge of the Kamchia River watershed during the year 2013. The year 2013 is characteristic with having country annual precipitation 638 mm nearly the same as the average precipitation for the period 1981-2013 [15]. For the studied region available are mean precipitation data measure in 12 meteorological stations – Omurtag (714 mm), Targovishe (646 mm), Shumen (598 mm), Koutel (794 mm), Preslav (641 mm), Smiadovo (678 mm), Varbitza (708 mm), Sindel (562 mm), Dalgopol (646 mm), St. Oriahovo (578 mm), G. Chiflik (534 mm) and Obzor (594 mm) [9,10], the average precipitation is 641,1 mm.

The land surface grid in the study is assumed with resolution (cells) 5 x 5 km. Every cell input contains specific for its water movement and transformation related properties data according to the topography, land cover, soil profile and vegetation [16].

The needed by the model meteorological input is supplied in this study with NASA atmosphere NCEP/NCAR Reanalysis data gathered by weather observations from ships, planes, RAOBS, stations, satellite observations and other ways [17]. They are freely available and easy to use. The atmospheric forcing data as air temperature, air pressure and humidity, solar radiation, wind speed, precipitation with 1 day time step are applied to the nodes of computational grid with cells 200 x 200 km. In the cell they are continuously distributed. The mean annual precipitation on the Kamchia River watershed for 2013 computed by the model meteorological input is 681 mm (Table 1), very close to the MEW estimation as mean of the period 1981-2013.

The necessary input related to the country territory soils characteristics and land cover for every grid cell with resolution 5 x 5 km are taken from the International Geosphere - Biosphere Programme (IGBP) data base [16].

The simulation period started two months before the beginning of the study period - on 1.11.2012, in order the soil moisture distribution to reach the state of real initial field conditions. The latter is important because the actual soil wetness is the main factor for the soil unsaturated zone water transport considered in the model. The soil infiltration capacity is controlled by soil properties and soil moisture within the top soil layers.

The regarded in the model soil layer thickness is equal to 3.43 m divided into 10 sub-layers. The computation is executed in 3 hour time step. The water amount calculated as filtrated below the vegetation root zone, with assumed maximum thickness 1 m, is considered as potential GW recharge. The model obtained potential recharge intensity monthly values for 2013, average for the whole district area, are shown in mm on Table 1. In the table the potential recharge intensity is named infiltration because it transforms into GW recharge only on a part of the watershed.

Table 1. Calculated monthly and annual hydrological parameters.

Months	**I**	**II**	**III**	**IV**	**V**	**VI**	**VII**	**VIII**	**IX**	**X**	**XI**	**XII**	**2013**
Rain (mm)	23	57	28	44	55	159	82	22	26	58	53	8	**615**
Snow (mm)	46	5									3	12	**66**
Prec. (mm)	69	62	28	44	55	159	82	22	26	58	56	20	**681**
Infiltr. (mm)	20	51	3	7	-17	53	-14	-27	2	24	36	13	**151**
Infiltr./ prec. (%)	29	82	11	16		33			8	41	64	65	**22**

From the results shown on Table 1 the following conclusions are possible to draw. The estimate of the district territory mean annual infiltration intensity in percent of the total annual precipitation is 22%. The monthly distribution of the infiltration quantities in response of the precipitation is rather logical. In cold but not frosty months in the period October-February the infiltration share of the precipitation is high while in the warm and hot period March – September the prevailing movement of soil water is upward due to capillary uplift forces. An exception happens in June when the precipitation is very high. The annual infiltration is 151 mm.

The shown infiltration turns into potential GW recharge only on the 2 000 km^2 of the whole district estimated by the National Hydrogeological Map [19]. If so the total annual recharge for 2013 on the Kamchia River Watershed can be evaluated as equal to 2000 x 10^6 x 0,151= 302 x 10^6 m^3.

The groundwater recharge of the country makes 30-50% of the total river runoff. It is observed that for Bulgaria for the period 1981-2013 the average annual total runoff amounts to 15 495 x 10^6 m^3 while the average annual groundwater recharge is 5 421 x 10^6 m^3 [15] the ratio being 35%. There is no information about the GW recharge to total runoff ratio of the Kamchia River watershed. Assessed by hydrometric observations the Kamchia River mean annual runoff is approximately 829 x 10^6 m^3 [9-10]. The modelled groundwater annual recharge, estimated for an year with precipitation close to the average one for the watershed, is equal to 36,4 % of the mean annual runoff. It is close to the total country one 35%. This proves to a great measure the realism of the model recharge estimations.

The CLM3 model calculates the average infiltration intensity on every grid cell with 25 km^2 area. This enables the authors to plot a GIS map [20] of the territorial distribution of the annual infiltration intensity of the Kamchia River watershed for the regarded year shown on Fig. 4. The map could be matched to the analogous map on Fig. 3. Their similarity consists in the fact that they both show the recharge intensity distribution on the district territory under approximately equal annual precipitation. On the other hand, there are substantial differences between them having in mind the described at the end of the previous paragraph way of obtaining the map on Fig. 3. The information on the map on Fig. 3 is a result of averaging and approximate assessments based on lump figures taken from the MEW published data [14], while the map on Fig 4 is directly based on computation values. This relates above all to the map resolution and infiltration intensity values. While the map on Fig. 3 gives more or less a rough picture of the district GW recharge territorial distribution the one on Fig. 4 shows the computed values on every land spots with equal to 25 km^2 area. The numbers on the map on Fig. 3 are more or less fictitious values, obtained by dividing the real GWB recharge volume to the GWB entire watershed area, while the map on Fig. 4 shows the value of the computed spot infiltration intensity. Therefore, a detailed comparison between them is not relevant.

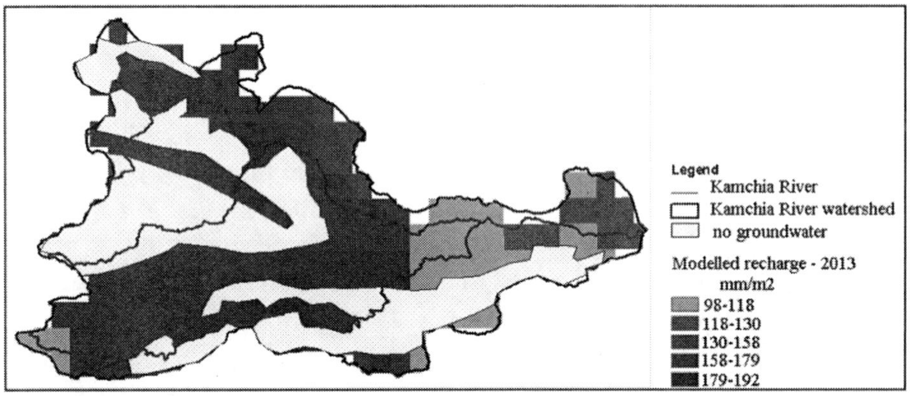

Fig. 4. Mapping the modelled annual groundwater recharge for the Kamchia River watershed in 2013.

The results of the comparison between the two maps obtained by completely different methods and data sources lead to two basic conclusions. The first is that the model CLM3 rather well reflects quantitatively the soil moisture infiltration processes and the assumption of the water infiltrated below the 1 m root zone as potential recharge is realistic. The second very important conclusion is that the huge amount of the model input related to the land surface, soil characteristics and climatic effects is correct and trustworthy as a whole. There is always possibility with peer analysis of the results from the model application to single GWB recharge simulation to achieve further calibration of the soil and land cover related data sets concerning some of the grid cells.

4 Conclusions

So far no hydrogeological modeling of the Kamchia River watershed has been done. The executed study is an innovative one. The results from the modeling have been compared to the scarce existing observation data.

A simulation has been done for the year 2013 with precipitation close to the average value for the period 1981-2013. The obtained GW recharge can be assumed as average value for the last 32 years. The total annual recharge amounts to 151 mm equal to 22% of the annual precipitation quantity.

GW recharge intensity GIS maps have been prepared based on the available data drawn from the MEW database as well as on the results from the modeling. The results from their comparison are commented.

The model obtained annual GW recharge of the Kamchia River watershed equal to 302 x 10^6 m^3 makes 36,4 % of the mean annual river runoff amounting to 829 x 10^6 m^3. The percentage is rather close to the same value valid for the whole country and equal to 35%.

The estimated GW recharge intensity varies from 98 mm to 192 mm on the areas with possibility for infiltration. It is established by observation that on some areas the GW recharge reaches up to 30-40% of the precipitation. This confirms the high values obtained by the modeling.

The accomplished study can serve as basis for prognostic assessments of the GW recharge response to climate changes under given scenarios.

References

1. NIMH, *Methodology for assessment of the groundwater body resources with consideration of the climatic factors changes and the necessary water quantity monitoring for its accomplishment*, 2012
2. K. Oleson et al, *NCAR Tech. Note NCAR/TN-461+STR, Natl. Cent. for Atmos. Res., Boulder* (2004)
3. K. Oleson, G. Niu, Z. Yang, D. Lawrence, J. Geophys. Res. (113) 2008
4. L. Pagliero, F. Bouraoui, P. Willems, J. Diels, Journal of Env. Quality **43**, 145 (2014)
5. A. Pistocchi, H. Bek, B. Bisselink, E. Gelati, C. Lavalle, J. Feher, *Water scenarios for the Danube River Basin*, JRC Technical Report (2015)
6. I. Trichakis, P. Burek, A. deRoo, A. Pistocchi, Environ Process J. **15**, 1 (2017)
7. L. Richards, Journal of Applied Physics **1**, 318 (1931)
8. B. Cosby, M. Hornberger, B. Clapp, T. Ginn, Water Resources Research J. **20 (6)**, 682 (1984)
9. O. Santurdjian et al, *General schemes for water use in the districts with river basin management of Bulgaria*, IWP BAS (2000)
10. O. Santurdjian et al, *Waters in the Kamchia River Basin, IWP* (2006)
11. S. Dineva, *International Symposium on Outfall Systems*, Mar del Plata, Argentina (2011)
12. Ministry of Environment and Water of Bulgaria, River Basin Management Plans for Black Sea Region, 2010-2015, http://www.bsbd.org/UserFiles/File/RAZDEL_1(1).pdf
13. Environmental Assessment Report „*Preparation of a general transport master plan of R. Bulgaria*“, Ministry of transport, information technologies and communications of Republic of Bulgaria (2010)

14. Ministry of Environment and Water of Bulgaria, *Registry of the resources and balans of the groundwater of the Black Sea basin Directorate* (2015)
15. Ministry of Environment and Water of Bulgaria, *Multiannual fresh water resources of Bulgaria up to 2013*, http://www5.moew.government.bg/?page_id=24269
16. G. Bonan, S. Levis, L. Kergoat, K. Oleson, Global Biogeochem. Cycles J. **16 (2)**, 5 (2002)
17. E. Kalney et al, Bull. Amer. Meteor. Soc J. **77**, 437 (1996)
18. O. Nitcheva, B. Milev, Journal of Theoretical and Applied Mechanics 47 (1), 85 (2017)
19. Maps, *Atlas of Bulgaria*, published by the Geographycal Institute of Bulgarian Academy of Sciences (1973)
20. A. De Jager, J. Vogt, Hydrol. Sci. J. **55 (5)**, 661 (2010)

Numerical investigation of high temperature synthesis gas pre-mixed combustion via ANSYS Fluent

Dmitry Pashchenko[1]*

[1]*Samara State Technical University, Russia, 443100 Samara, Molodogvardiyskaya str. 244*

Abstract. A numerical model of the synthesis gas pre-mixed combustion is developed. The research was carried out via ANSYS Fluent software. Verification of the numerical results was carried out using experimental data. A visual comparison of the flame contours that obtained by the synthesis gas combustion for Re = 600; 800; 1000 was performed. A comparison of the wall temperature of the combustion chamber, obtained with the help of the developed model, with the results of a physical experiment was also presented. For all cases, good convergence of the results is observed. It is established that a change in the temperature of the syngas/air mixture at the inlet to the combustion chamber does not significantly affect the temperature of the combustion products due to the dissipation of the H_2O and CO_2 molecules. The obtained results are of practical importance for the design of heat engineering plants with thermochemical heat recovery.

1 Introduction

Energy and environmental problems are the reasons for the constant perfection of energy installations, the search for new fuels, the development of innovative technologies for the use of traditional hydrocarbon fuels, etc. In recent decades, specialists and engineers from different countries have paid much attention to the use of hydrogen as a fuel source of energy. Thus, according to the Hydrogen Scenario of the International Energy Agency (IEA), large-scale hydrogen production on a global scale will increase by 20-25% annually after 2030 [1]. The main consumers of hydrogen should be transport and power plants using this gas as fuel. Known problems of storage and transportation of hydrogen to consumers stipulate the use of technologies in the transitional stage, in which fuel hydrogen is produced directly on the fuel-consuming installation (on-board hydrogen production) [2].

The production of hydrogen-containing gas immediately before consumption has found application in high-temperature plants (HTP) with thermochemical recuperation of the flue gases heat[4, 5].

The main concept of thermochemical waste-heat recuperation is transfor- mation of flue gases heat into chemical energy of a new synthetic fuel that has higher calorimetric properties such as low heating value. If in the conventional high-temperature plants, for example the industrial furnaces, conversion of chemical energy of fuel into heat energy occurs by fuel combustion, then in the HTP with TCR the conversion of chemical energy of fuel is divided into two stages. The first stage is the increasing of the low heat value of initial fuel by transformation of enthalpy waste-heat into chemical

*e-mail: pashchenkodmitry@mail.ru

energy of the new synthetic fuel. The second stage is combustion of the new synthetic fuel that has low heat value greater than the low heat value of primary fuel.

In HTP with thermochemical regeneration the hydrogen-containing gas is utilized as fuel at the same HTP [6, 7]. Combustion of synthesis gas of various compositions is a fairly well studied process both experimentally and theoretically. However, the publications in the literature do not cover the entire spectrum of practically interesting questions, in particular, the combustion of high-temperature hydrogen-containing gases in high-temperature plants with thermochemical heat recovery.

The present work is devoted to a numerical study of the characteristics of combustion of synthesis gas that produced in a system of thermochemical heat recuperation. The study was carried out in the software ANSYS Fluent (Academic Research).

2 Numerical setup

2.1 Computational domain

The schematic diagram of the high-temperature plant with thermochemical heat recuperation of waste flue gases by steam methane reforming is presented in Fig.1.

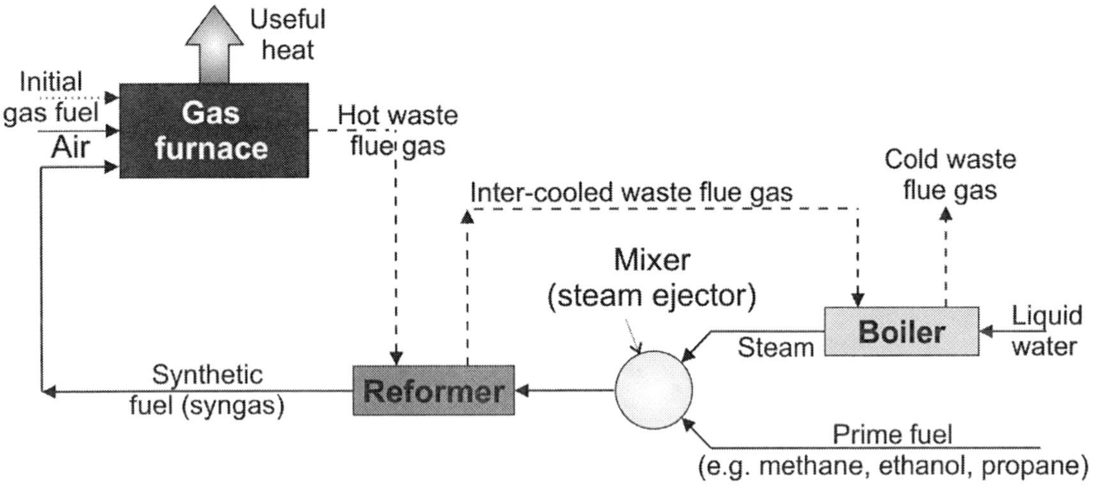

Figure 1. The schematic diagram of high-temperature plant with thermochemical waste-heat recuperation

The overall reaction in steam methane reforming process to form H_2, C, CO, CO_2, H_2O and other components may include the different equations, but the biggest contribution for energy and material balance are follow reactions:

$$CH_4 + H_2O = CO + 3H_2 - 206.2kJ/mol; \quad (1)$$

$$CO + H_2O = CO_2 + H_2 + 41.1kJ/mol. \quad (2)$$

The composition of the synthesis gas after conversion depends both on the technological parameters of the operation of the thermochemical heat recovery system: the temperature of the flue gases, the composition of the initial reaction mixture, the pressure, and the design characteristics of the reformer: the volume of the reaction space, the catalyst and its amount,

The composition of the synthesis gas after steam methane reforming process depends both on the technological parameters: the temperature of the flue gases, the composition of the initial reaction mixture, the pressure, and the design characteristics of the reformer: the volume of the reaction space, the catalyst and its amount. The equilibrium composition of the synthesis gas for the H_2O/CH_4 ratio in the initial reaction mixture of 2 is shown in the Table 1 .

Table 1. Synthesis gas composition

Temperature	H_2	CO	H_2O	CO_2	CH_4
700 K	33.04	4.07	32.34	6.52	23.63
800 K	58.11	9.86	18.94	7.23	5.86
900 K	61.34	12.89	16.51	6.08	3.18
1000 K	63.40	14.38	16.12	5.08	1.02
1100 K	63.65	16.14	16.15	3.98	0.08

A two dimensional cylindrical combustor model was built and numerically solved by using ANSYS code (Fluent solver). The model combustor, its dimensions and mesh structure are presented in Fig. 2. All dimensions are in millimeters. A wall material is steel.

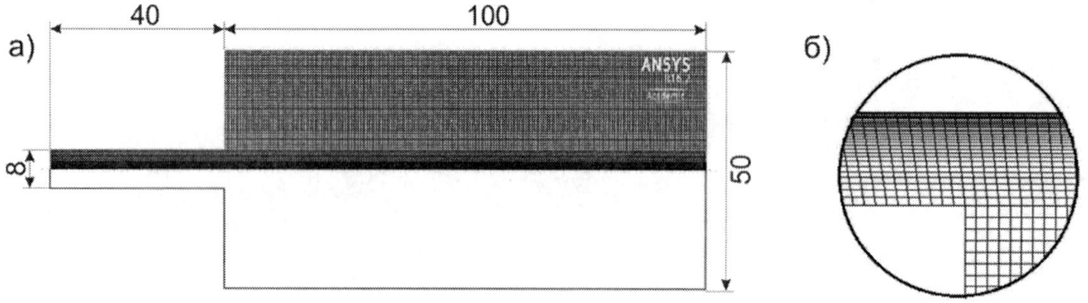

Figure 2. The computational domain and mesh structure of combustor model

2.2 Mathematical model

In this paper, syngas with composition according Table 1 used as fuel at stoichiometric conditions and to model combustion process in the combustor, steady state-three-dimensional continuity, energy, momentum and species equations are solved. Following assumptions are made for numerical simulations: steady-state, no energy flux due to mass concentration gradients, no work done by pressure and viscous forces, no gas radiation, no surface reactions and no slip condition at the combustor wall. Combustion reaction scheme was modeled using Eddy Dissipation Concept model.

Continuity equation for steady-state conditions [8]:

$$\nabla(\rho \cdot u) = 0. \tag{3}$$

where u is velocity; ρ is density.

$$\rho(u \cdot \nabla u) = -\nabla p + \nabla\left(\mu \cdot \left[\nabla u + (\nabla u)^T - \frac{2}{3}\nabla u \cdot I\right]\right) + \rho \cdot g + F, \tag{4}$$

where p - pressure; I - unit tensor; $(\nabla u)^T$ - the effect of volume dilation; $\rho \cdot g + F$ are the gravitational body force and external body forces.

Energy equation for combustion products:

$$\nabla u \cdot (\rho \cdot E_f + p) = \nabla \left[k_{eff} \nabla T - \sum h_j \cdot J_j + (\tau_{eff} \cdot u) \right] + S_h, \tag{5}$$

where k_{eff} is the effective conductivity; J_j is the diffusion flux of species j; left side of the equation is sum of energy transfer due to conduction, species diffusion, and viscous dissipation, respectively; S_h includes the heat of chemical reaction.

Initial and boundary conditions:

- syngas/air inlet: $x = 0$; $T_g = T_{g(in)}$; $g_i = g_{i(in)}$;
- combustion products outlet: $y = 0$; $\frac{\partial g_i}{\partial y} = 0$; $\frac{\partial T_g}{\partial y} = 0$;
- combustor wall: $\frac{k_{eff}\partial^2 T}{\partial x^2} = 0$; $\frac{k_{eff}\partial^2 T}{\partial y^2} = 0$.

Here T_g is temperature of reaction mixture; g_i is mass fraction of i element.

For developed model the standard k-ϵ model is used. For inlet and outlet of the combustor, mass flow inlet and pressure outlet boundary conditions are employed, where gauge pressure and turbulent intensity of the inlet and outlet are the same (0 Pa and 5%, respectively) [9].

Calculation method for specific heat and density are incompressible-ideal gas law and mixing law, respectively. Viscosity and gas thermal conductivity of the mixture are calculated by averaging mass fraction of all species. Convergence criteria for momentum, continuity and species are a value of residual less than 10^{-3}; for energy is 10^{-6}. All of the equations are discretized by a second-order upwind scheme.

3 Results and discussion

3.1 Model validation

For verification of the developed model, the experimental combustion data of the fuel mixture of the following composition was used: CH_4=68%; CO_2=22%; H_2=10% with complete premixing at equivalence air ratio 1.0.

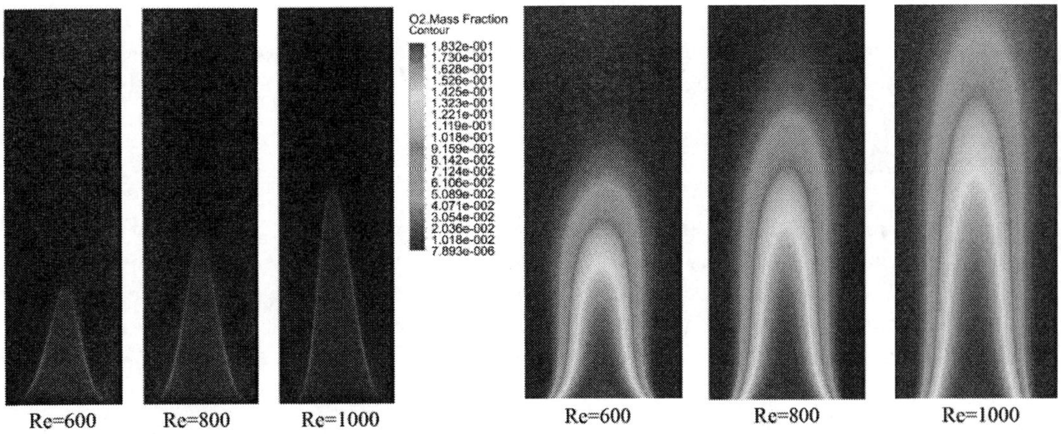

Figure 3. The contours of flame for various Reynolds number

In Fig. 3 the results of the experiment and the contour of the mass fraction of oxygen obtained as a result of calculations for different values of the Reynolds number are shown. The kinematic viscosity of the gas mixture for determining the Reynolds number is calculated by the Sutherland rule. Visual comparison of flame shows good convergence of the results obtained with the help of the model with experimental data. For simulation results, one can see the characteristic Bunsen cone - the combustion front.

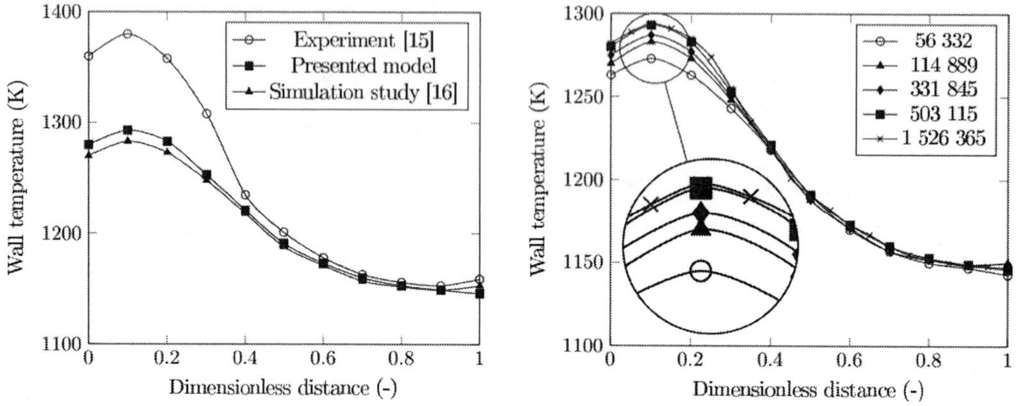

Figure 4. The profile of wall temperature for combustion chamber (left figure) and the Grid independency study (right figure)

To validate developed model additionally, calculated wall temperature were compared with experimental results of Yang and co-workers [59] and modeling investigation of Jiaqiang and co-authors [53]. As presented in Fig. 4, the agreement between results is well. Discrepancies between experimental data and simulation results may be seen at the zone close to inlet of the combustor. These discrepancies can be as the results of errors of the experimental setup and assumptions for the mathematical model. However, these discrepancies slow down towards the midsection of the combustor chamber and calculated wall temperature profiles show a good agreement with experimental results in terms of trend and value. Towards the outlet of the combustor, discrepancies become apparent again. Boundary conditions: Hydrogen flow rate and equivalence ratio are 5.25×10^{-7}kg/s and 1.0, respectively, the mass flow rate of the inlet is 1.8551×10^{-5} kg/s; both side walls are adiabatic.

3.2 Grid independency study

In numerical modeling of fluid dynamics, the accuracy of the results is determined not only by the solver setup and the specified convergence criteria, but also by the number of unit cells and the quality of the mesh [17, 18]. In this study, the minimum number of cells for the computational domain shown in Fig. 2 is estimated. Calculations are made for the following initial data: fuel - 100% hydrogen; Hydrogen flow rate and equivalence ratio are 5.25×10^{-7}kg/s and 1.0, respectively so mass flow rate of the inlet is 1.8551×10^{-5} kg/s; both side walls are adiabatic. The number of cells varied from 56 332 to 1 526 365 with the constant convergence criteria mentioned above.

From Fig. 4 (right side) it is seen that with increasing number of cells, the accuracy of calculation increases. However, with an increase in the number of cells greater than 500,000, the results change insignificantly. Therefore, for further calculations, the number of cells is recommended to be selected in the range from 500 to 600 thousand. In this case, a satisfactory quality of the results is observed with a minimum of processing power.

3.3 Combustion products temperature and contours

In Fig. 5 shows the distribution of the combustion products temperature for synthesis gas of various compositions and the temperature given in the Table 1. The number of cells for computational domain is 503 115. The calculations were carried out for the equivalence air ratio 1.0. The obtained dependences make it possible to state that the composition and temperature of the synthesis gas of the parameters under study do not have a significant effect on the temperature of the combustion products. This can be explained by the dissociation of molecules of water and carbon dioxide, which is accompanied by the absorption of heat generated as a result of combustion of fuel. Such a strong effect of dissociation on the heat balance can be explained by the large amount of physical heat introduced with the fuel-air mixture and the minimum excess pressure in the combustion chamber.

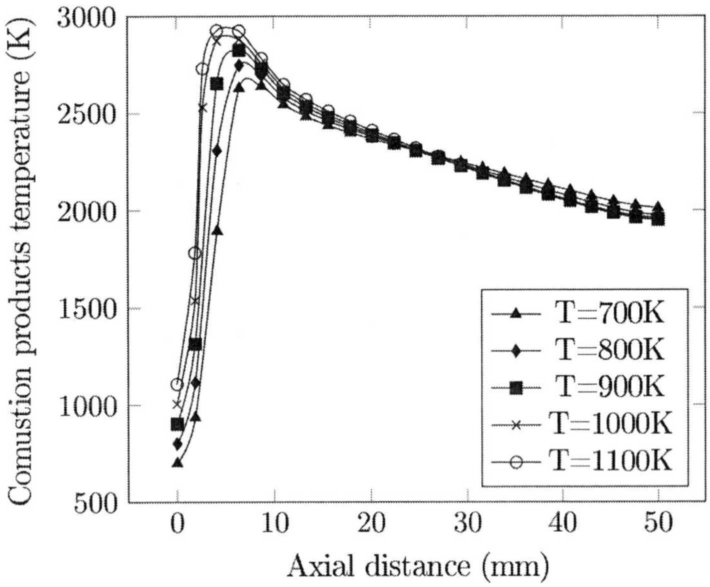

Figure 5. The profile of combustion products temperature along axial distance of combustion chamber

The results in Fig. 5 can be used to design combustion chambers for high-temperature plants with thermochemical heat recovery. Also, the results show that when the HTP is transferred for thermochemical recuperation of heat, it is necessary to take into account the change in the temperature regime of such an installation. The temperature in the combustion chamber can be controlled by recirculating the flue gases after the steam generator (Fig. 1).

In Fig. 6 shows the contours of temperature (a), velocity (b), turbulent viscosity (c), and mass fraction of oxygen (d) in the combustion chamber. The composition of synthesis gas obtained at the methane conversion process temperature of 900 K and at an equivalence air ratio of 1.0. The presented contours give a graphic representation of the processes occurring during the combustion of the fuel-air mixture.

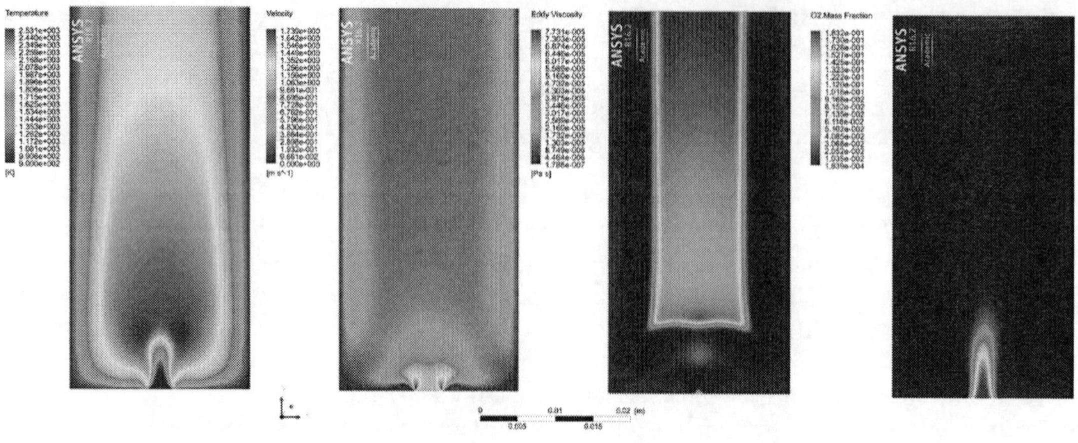

Figure 6. The profile of combustion products temperature along axial distance of combustion chamber

4 Conclusion

The developed numerical model of the syngas/air combustion makes it possible to determine the thermal characteristics of the process at each point of the combustor. With the help of the model it is established that an increase in the temperature of the fuel-air mixture before combustion does not lead to a significant change in the temperature regime in the combustion chamber. For all investigated fuel compositions, the temperature of the combustion products at the combustor outlet is about 2000K. Numerical results have good convergence with experimental data. Thus, the root-mean-square deviation of the results of the wall temperature of the combustion chamber, obtained with the help of the model, in comparison with the experimental data for all computational domain does not exceed 0.036. In addition, for the verification of the developed model, experimental combustion data of the fuel mixture of the following composition were used: CH_4 = 68%; CO_2 = 22%; H_2 = 10% with complete premixing at $\alpha = 1.0$. Visual comparison of physical and numerical flames shows good convergence of the results obtained with the help of the model with experimental data. For simulation results, a cone of Bunsen - the combustion front, is seen.

It has been established that an increase in the number of elementary cells for the investigated computational domain is more than 500-600 thousand does not lead to a noticeable increase in the accuracy of the results. A commercial software ANSYS Fluent with a particular choice of several physical models (regarding turbulence, combustion, radiation) is used for this analysis. The obtained data can be used in the design of heat engineering plants with thermochemical regeneration of the heat of the exhaust flue gases due to steam methane conversion.

References

[1] Soria A., Szabo L., Russ P., Suwala W., *World Energy Technology Outlook 2050* (European Commission. Joint Research Center. 2006) 221.
[2] Chakravarthy V. K., Daw C.S., Pihl J.A., Conklin J.C., Energy&Fuels **24(3)**, 1529-1537 (2010).
[3] Popov S. K., Svistunov I.N., Garyaev A.B., Serikov E.A., Energy **127**, 44-51 (2017).
[4] Pashchenko D. I., Thermal Engineering **60(6)**,438-443 (2013).
[5] Verkhivker G., Kravchenko V., Energy **29(3)**, 379-388 (2004).

[6] Pashchenko D., International Journal of Hydrogen Energy **42(22)**, 14926-14935 (2017).
[7] Zarov E. A., Kul'kov M. G., Filippov I. V., Water Resources. **44(2)**, 267-275 (2017).
[8] Tsynaeva A. A., Nikitin M. N., Procedia Engineering. **150**, 2340-2344 (2017).
[9] Tsynaeva A., Nikitin M., MATEC Web of Conferences. **86**, 04-018 (2017).

Technical project for a new water purification solution

Adina Toma[1,*], *Zheleva* Ivanka[2] , *Cristian* Puşcaşu[1], *Alexandru* Paraschiv[1] and *Mihaela* Grigorescu[1]

[1] COMOTI – Romanian Research and Development Institute for Gas Turbines, Gas Turbines Special Components. Physics and Mechanic Testing Department, 220D Iuliu Maniu, 061126 Bucharest, Romania
[2] University of Ruse „Angel Kanichev", 8 Studentska, 7017, Russe, Bulgaria

Abstract. This research is part of the RO-BG Cross-Border Cooperation Program, project "CLEANDANUBE", MIS-ETC 653, which has finalised by providing a common strategy to prevent the Danube's pollution technological risks with oil and oil products. This paper presents a new sustainable water purification solution. A short introduction will be offered and an overview regarding the research and new methods to greening the waste is provided. The theoretical aspects of the centrifugal separation phenomenon are studied and the preliminary project bases were established. The paper conveys the possible constructive variations and the technological implications of those. Ultimately, the technical project for a new water purification solution and conclusions with critical points encountered during the designing phase are presented.

1 Introduction

All EU countries use a common approach for river basin management conveyed by the adoption of the EU Water Framework Directive (2000/60/EC) [1], and based on this; a Management Plan of Danube River Basin was elaborated. The plan focuses on key border issues, so-called significant management issues (which may directly or indirectly affect the quality of rivers, lakes and groundwater bodies across borders): Organic pollution; Pollution from nutrients; Pollution with hazardous substances; and Hydro morphological alterations.

From all forms of pollution, the most severe impact on water has pollution with oil during transportation, although it represents only 24% of the pollution causes.

The presence of oil waste is a major problem in terms of environmental factors, as well as economic. These wastes are hazardous for fauna and flora and further investment development is needed to greening the affected areas. Waste oil composition is diverse, depending on the types of waste and stock of origin and the nature of oil spilled.

Analyzing the risk factors of shipping accidents that cause oil pollution, we can highlight the categories: Risks arising from physical and chemical properties of certain

* Corresponding author: adina.toma@comoti.ro

dangerous goods carried, including oil and petroleum products; Risks of hidden defects caused by the vessel; and Risks caused by human mistakes.

Waste oil has a composition consisting basically of three compounds phase's types:

- The first part consists of phase oil and contains various species of heavy hydrocarbon compounds and resins or asphalt category.
- The second phase consists of water in various minerals that are dissolved.
- The third phase consists of mineral compounds that are in solid phase. Separation of the three phases is a difficult technical problem and is realized with high costs.

The novelty of the solutions presented in this paper is an entirely different technology (centrifugation), that will complement and improve the existing ones, in order to minimize the effects of pollution, the response time and the intervention time in case of pollution, as well as to increase in the same time the capability and efficiency of the separation process for those three phases.

2 Theoretical aspects of the centrifugal separation phenomenon

The new water purification solution proposes the use of the most effective and used multiphase mixtures separation method based on the difference between the different specific weights of the components. In most cases the system is non-invasive (does not require chemical treatment than in cases of catastrophic spills).

Centrifugal separation is the division of the constituents of a heterogeneous mixture, different based on their specific weight, using centrifugal force created during rotation.

Decanter centrifuge separates by specific gravity liquids, spilling the top, centre, easy fluid, and one hard on the periphery. In the same manner can be employed the separation of a liquid which has solid particles in suspension. Speed centrifuges can reach up to 60,000 revolutions per minute (ultracentrifuge). Separation is facilitated by dividing the conical liquid layer by means of conical plates.

For components with different densities the separator works only by sedimentation, a process that produces the separation of liquids and solids in suspension due to the difference in density. If the density difference is large, then gravity can provide enough force for separation to occur within a reasonable time - as is the case with large tanks or blades separators. If the density difference is small, then it would take too much gravitational separation, and separation force must be increased by adding the centrifugal forces several times higher than that of gravity. Centrifugal force can be created either by the flow of the mixture as a hydro cyclone or mechanically driven rotation, as in sedimentation centrifuges. Centrifugal separator can be used for most types of liquid/solid, given its ability to handle a variety of different mixtures and concentrations. The separator can be used for three phase separation in which liquid is composed of water and oil. It can be operated so as to give a high degree of separation.

In order to design such solution, first of all we must understand the centrifugal forces.

Centrifugal acceleration developed inside the cylinder radius r and angular rotational speed "ω" is:

$$a = r\omega^2 \tag{1}$$

Centrifugal force is:

$$G = \frac{r\omega^2}{g} \tag{2}$$

$$G = \frac{r}{g}\left(\frac{2\pi n}{60}\right)^2 \tag{3}$$

$$G = \frac{n^2 \emptyset}{1800} \tag{4}$$

Where: r = container inner radius [m]
ω = angular speed [rad/sec^2]
g = gravitational acceleration = 9.81 [m/sec^2]
n = container speed [rpm]
360 ° = 6.287 radian
S, the slippage force (figure 1) is:

$$S = \sin\alpha\ G \tag{5}$$

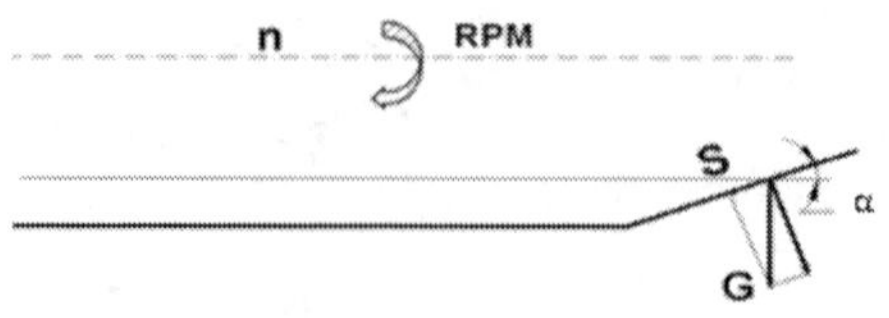

Fig. 1. Slippage force

The equivalent surface (sigma value) is:

$$\Sigma = \frac{1}{160} n^2 L_{cyl} r^2 \tag{6}$$

Where;
Σ = sigma value [m^2]
n = Container speed [rpm]
Lcyl = barrel length [m]
r = maximum radius of cylinder (half diameter) [m]
Sigma value is the equivalent surface in sq. m of a static flotation tank to produce the same solid/liquid separation results.

Dewatering surface

Dewatering surface [m^2] is the surface of the cylinder section (the cylindrical portion):

$$A_{c,n} = \pi D L_{cyl} = \frac{\Sigma}{G} \tag{7}$$

Where:
$A_{c,n}$ =Decanter surface settling [m^2]
Σ = sigma value [m^2]
G = centrifugal force
D = internal diameter [m]
L_{cyl}= length of the longitudinal section [m].

The dewatering volume

Centrifugal separator dewatering volume is obtained assuming the total volume of liquid in the cylinder of the container. This volume is a function of surfaces diameter (figure 2).

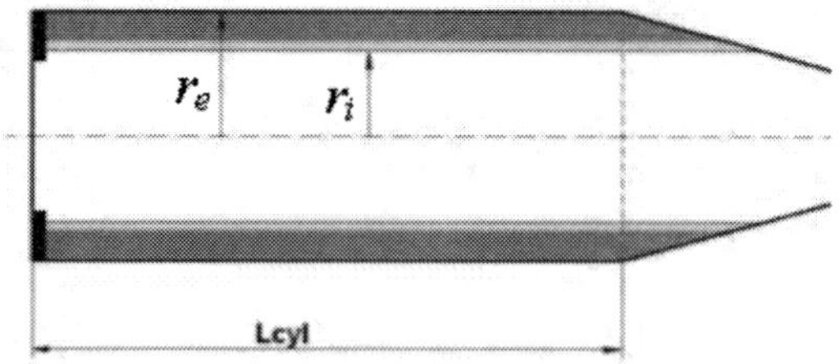

Fig. 2. Centrifugal separator dewatering volume scheme

To calculate the dewatering volume cylindrical one must take into a consideration the cylindrical length L_{cyl}, the container inner radius r_e[m] and the r_i [m], where r_i is approx. = 60% r_e.

Dewatering volume is:

$$D_v = (r_e^2 - r_i^2)\pi L_{cyl}\ [\text{m}^3] \tag{8}$$

Retention time:

$$R_t = \frac{3600V_{td}}{Q_{input}} \quad (9)$$

Where:
R_t = retention time
V_{td} = container total dewatering volume
Q_{input} =Slurry feed rate [l/h]
The retention time is the time in which the liquid remains in the container before being discharged. Retention time depends on characteristics of the feed slurry like this:
- specific weight of solid;
- viscosity of liquid phase
- amount of substance treatment.

Longer retention time will result in more efficient solid/liquid separation.
Depending on water quality by introducing a speed differences between roller and separator conveyor within walking distance increases and hence the quality of the separation process.
Efficiency of separation is achieved by changing Δ_n, where Δ_n is the differential between the cylinder and conveyor speed.

The power required to drive the centrifuge

- For training centrifuges power consumption involved the following:
- N_a power required to bring the masses in motion, from sleep mode to speed;
- N_b power required to bring the liquid to the centrifuge at the speed of system;
- N_c power to overcome friction in the spinning shaft bearings;
- N_d power to overcome friction between the centrifugal and ambient air;
- N_e power for other resistance to overcome, eg scrapers which remove sediment friction (in centrifuges with continuous operation), the friction in the flow of fluid through the centrifuge, etc.

The starting power is:

$$N_p = N_a + N_b + N_c + N_d + N_e \quad (10)$$

Nominal power for discontinuous centrifuges is:

$$N_{r,c} = N_c + N_d \quad (11)$$

Nominal power for continuous centrifuges is:

$$N_{r,c} = N_b + N_c + N_d + N_e \quad (12)$$

Nominal power system is therefore lower than the starting power. The starting power is even greater as the starting time (up to the speed of the system) is smaller.
More details regarding the theoretical aspects of the centrifugal separation phenomenon can be found in the "***Study on the designation of technological parameters and constructive to be attained***" [2].

3 Possible constructive variations and the technological implications

Technologically, the water purification solution must meet the following conditions: to be provided with a floating system to maintain it over the level of water; to create a barrier that will not allow the crossing of polluted waters downstream; to be provided with a admission system for the polluted waters with a rough preliminary filtration of the collected products.

The filtration system must be dimensioned in such a way to allow the crossing of designed flow in the quantity and quality needed for the suction pump to function.

The suction pump must ensure the designed flow for the centrifugal installation. For a medium capacity installation, as it results from the study began for stage 3 of the project (15÷25 m³/h), it must ensure an adjustable output pressure of 4÷7 bar.

The dimensioning of the building elements of the separator that will match the functional requirements (designed flow for processing polluted water).

The achievement of some variable speeds n_c (speed of the separator housing) and n_i (speed from the entry in the differential gearbox) so the speed difference Δn between cylinder and transporter can process the separation for every features of the polluted element. One solution that was studied for the research theme was sedimentation centrifuge with horizontal screw (figure 3) [3].

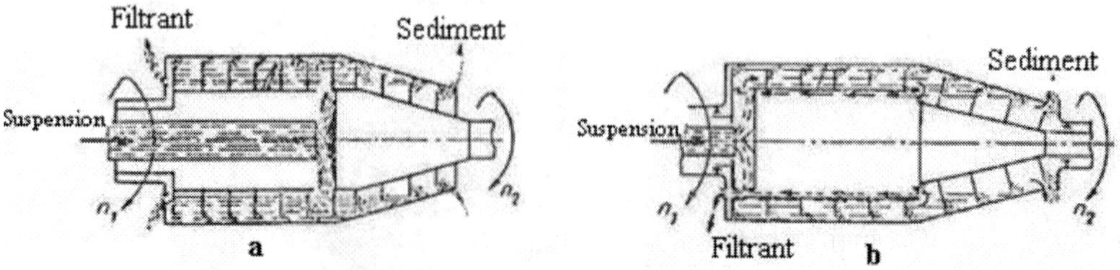

Fig. 3. Screw centrifuge functioning sketch a) in counter-current; b) in echi- current

Other solutions, as a result of the numerous working conditions, imposed by the raw material and the resulted products, were several types of centrifuges (table 1).

Table 1- Centrifuge classification

Classification criteria	Classification groups
Separation power (efficiency factor z)	Normal Centrifuge ($z<3000$) Super centrifuge ($z>3000$) Ultracentrifuge ($z>5\cdot 10^5$)
Centrifugation purpose	Filtering the granular or crystalline suspensions / Suspensions sedimentation / Emulsion separation
Function	Discontinuous (in batch) / Continuous
Download (precipitate evacuation)	Manual / By gravitation, at low speed / With pulsed pistons / With spiral conveyer / With knives and scrapers / By centrifugal force / Hydraulics
Axis position	Vertical / Horizontal / Oblique
The support of the centrifuge	Suspended / Supported
Operation	Upper / Lower

These centrifuges are characterized by a perforated brake drum, lining with mesh or cloth. They serve to filter suspensions, separation of crystals from initial solution, drying the textiles and are indicated for separating suspensions that needs o long period for centrifugation and washing. Developing several possible constructive variants, analysing concomitant the technologic implications was very complex to be performed because it has a multidisciplinary character (chemistry, material sciences, fluid mechanics, fluid flows in machines with rotating blades, oppositions of materials, etc.). This activity concluded with a final project theme, to achieve the technical project for a new water purification solution.

3 Preliminary project basis

The centrifugal separation installation must meet the next general requirements:
- to be efficient and adaptive to the goal; to be the most simple in construction; and to be not so expensive in exploitation, regarding the energy consumption and the raw materials and materials needed. For the designation of the technological and constructive parameters needed to develop the preliminary project theme, first it must be defined the final technological scheme and then the adopted constructive solution. The most appropriate technological scheme for the proposed application is the one from figure 4. In order to

estimate in a correct way the dimensions of the storage capacity for the solid components, an analysis over the solid phase concentration of Danube's water both in the normal and in flooding period must be carried out.

Following the analysis over the constructive solutions adaptable to the goal of the CLEANDANUBE project, the solution of a centrifugal separator in a horizontal position is the most convenient, both constructive and functional. In figure 5 is represented the general scheme of a centrifugal separator. The centrifugal separator is in principle a quite simple device, although is far from being easy to build. Has in his composition a rotating cylinder and an interior transport screw, the separate liquids getting out at one end and the solids at the opposite end. The first quality of the separator is its capacity to remove large amounts of suspended solids from a mixture of liquids, with a reasonable low level of retained liquid in removed solids. This apparent simplicity of the separator is complicated by the diverse range of projects variations.

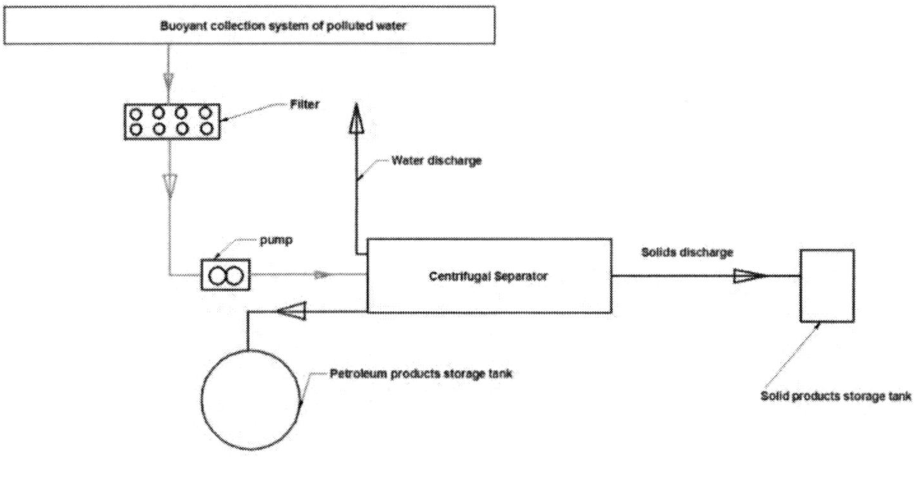

Fig. 4. The general scheme for the petroleum wastes separation system

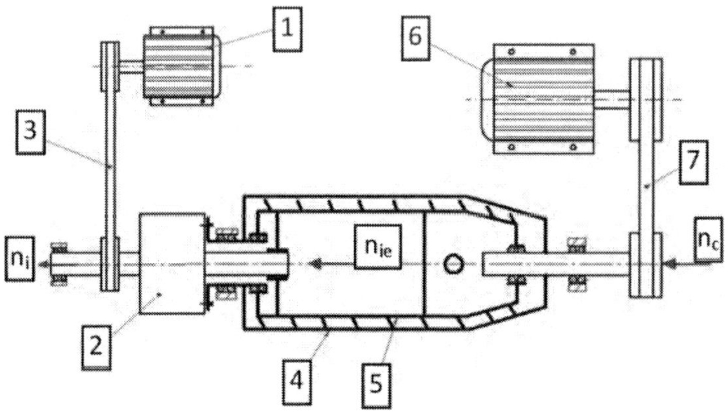

Fig. 5. General scheme of a centrifugal separator (1-Differential gear drive motor; 2-Differential gear; 3-Belt drive differential gear input shaft; 4 - Container; 5-Conveyor ans.; 6- Centrifugal container drive motor; 7- Centrifugal container belt drive; n_c- rotational container speed [rpm]; n_i - rotational speed of a differential gear input shaft [rpm]; n_e - rotational conveyor speed [rpm])

5. Technical project for a new water purification solution

Following previous presented aspects on the development of technological and constructive parameters and effective technological schemes to prevent and remove the effects of pollution the Danube, were determined technical conditions for the design and construction of treatment plants based mainly on centrifugal separation technology.
In figure 6 is represented overall design of the separation plant.

The centrifugal separator designed in the context of this research can be used both as solid-liquid phase separator, and is part of decanters (used for clarification of liquids) or three-phase separator consisting of two immiscible liquids and solid and are known as the tricanters.

Materials of construction of the centrifugal separator are many and varied. It is more usual to make the contact parts, particularly in the rotating assembly, of some form of stainless steel. This is to avoid assembly problems and misalignment due to corrosion on mating surfaces. This has to be avoided with high speed rotating equipment.

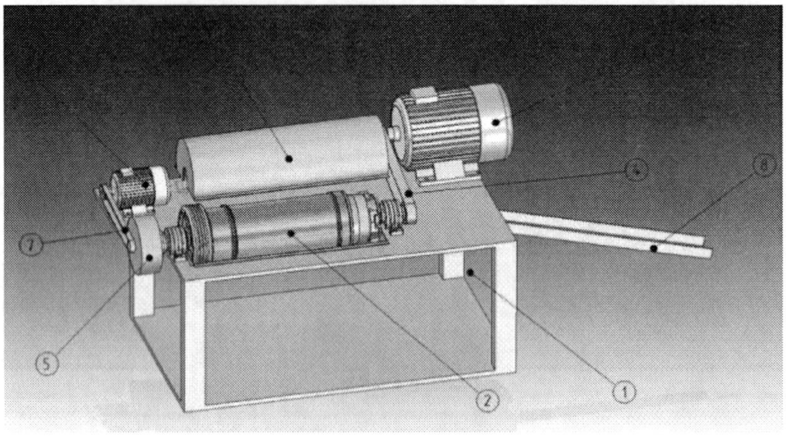

Fig. 6. General configuration of the installation of centrifugal separation (1-Frame; 2-Centrifugal separator; 3-Main drive; 4-Belt drive centrifuge container; 5-Gearbox; 6-Gearbox drive motor; 7-Belt drive gearbox; 8- Guide for solid discharge; 9-Casing)

Nevertheless, it must be said that there are many centrifugal separators in operation with bowls of carbon steel, where their manufacturer claims to be able to overcome, or avoid, corrosion. For stationary contact components there is no need for a high grade of stainless steel. When the process used is non-arduous, simple neoprene seals and gaskets will suffice. Supporting framework will be in ordinary or even cast steel. For parts that goes into housing and screw austenitic steel 316L is chosen.

The research comprises also the standard and auxiliary elements that go into water infested with petroleum products treatment plants and the documentation of detailed execution. Auxiliary devices that go into the system are the following (figure 7): Protective-barrier systems (dams); Infested water collection system; Drives a centrifugal separator plant; Storage system of processed products; Instrumentation and control of technological process.

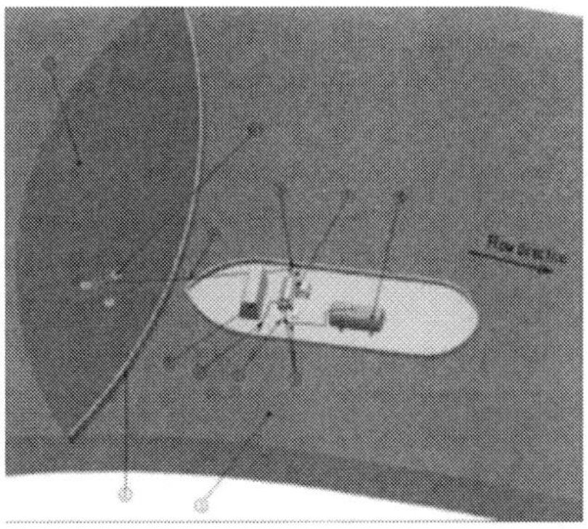

Fig. 7. Overview of an ship-mounted installation for the recovery of petroleum products spilled in the waters of the Danube (1-Danube river; 2-Protective-barrier systems; 3-Wastewater; 4-Infested water collection system; 5-Centrifugal separator assembly; 6- Infested water pipe; 7-Oil pump; 8-Recovered oil pipeline; 9-Water discharge pipe; 10-Oil tank; 11- Solids collection tank; 12-Control panel)

In principle, the technological flow for treating water contaminated with petroleum products in marine accidents includes the following operations:

a) Information, Warning.
b) Measures to stop pollution and limit the polluted area.
c) Monitoring. The initial phase of monitoring is to determine pollutants by analysis of samples, analyses that will be resumed in the terminal phase to confirm the completion of remediation and the possibility of recovering the water and the solid part without the oil component, which is stored separately. Depending on the appearance and colour of the oil slick on the water surface, we can determine the thickness (volume) of the pollutant. Surveillance of the area will continue after the completion of works by checking for anything that may have been overlooked. According to the initial observations a method of intervention is defined in order to be applied (minor or major intervention).
d) Isolating the contaminated area.
e) Recovering the water polluted with oil products.
f) Processing and separating the composing phases of the polluting mixture.
g) Controlling and appreciating the decontamination.

6. Conclusions

The water and solids must be restored in the environment and petroleum stored for transport. According to oil waste centrifuging systems, three steady phases result that must be verified and enclosed within legal limits of use restoration or source discharge.
Water resulted from centrifuging must be verified to abide legal requirements for used water discharge by means of quality indicators for waste water provided by legislation in force. The proposed solution involves processing, on-site contaminated water with oil products and reintegration into the natural circuit of water and solid components, oil obtained being taken into special tanks, total and immediate problem solving. This is possible using special ***centrifugal processing equipment of water polluted with oil in one step and the separation into organic products, solid water and oil.***

During CLEANDANUBE implementation, critical points were encountered, both technological and administrative. From technological view the challenging was given by the particularity of the Danube river, because the solution is perfectly adapted to be used in Romania-Bulgaria cross border. From administrative point of view, the bureaucracy of the entire process, from signing the contract to implement him and even now, to assure his sustainability it is a burdensome process. The most challenging is to find founds in order to implement the solution along the Danube cross-border.

References

1. Directive 2000/60/EC of the European Parliament and of the Council establishing a framework for the Community action in the field of water policy, modified by 2008/32/CE, 2008/105/CE and 2009/31/CE Directives and 2455/2001/CE Decision
2. C. Puscasu, I. Zheleva et all, Study 2, Study on the designation of technological parameters and constructive to be attained, Common strategy to prevent the Danube's pollution technological risks with oil and oil products – CLEANDANUBE, Operation: no.2(2i)-2.2-5, code MIS-ETC 653, 2010

   http://www.comoti.ro/en/participarea_la_proiecte_europene_2_11.htm
3. C.Puscasu, I. Zheleva, et all, Study 3, Study on possible constructional versions, technological implications, Common strategy to prevent the Danube's pollution technological risks with oil and oil products – CLEANDANUBE, Operation: no.2(2i)-2.2-5, code MIS-ETC 653, (2011)

MATEC Web of Conferences **145**, 03014 (2018) https://doi.org/10.1051/matecconf/201814503014
NCTAM 2017

Computational fluid dynamics calculus and analysis for gas and water turbines

Valeriu Vilag[1,*], *Ivanka* Zheleva[2] , *Jenia* Popescu[1] and *Krasimir* Tujarov[2]

[1]COMOTI Romanian Research and Development Institute for Gas Turbines, 061126 Bucharest, Romania
[2]"Anghel Kanchev" University of Ruse, 7017 Ruse, Bulgaria

Abstract. The paper presents the utilization of Computational Fluid Dynamics for calculating the flow through turbines. The first and most extended part of the paper is focused on gas turbines where the simulations are very precise and can be successfully used even for optimization of blade geometry. Flow details and results for an axial turbine are presented along with a proposal of optimization algorithm. The second part of the paper is dedicated to water turbines and there is presented the calculus realized for a kinetic water turbine. I this case, the flow around the turbine blades is presented and some data about the predicted performances along with many ways for improving the simulations. The conclusions of the paper are related to similarities and differences between the two types of simulations and to the many ways of using these simulations for practical applications.

1 Introduction

Energetic machines like gas and water turbine are in general use from many decades but their performances are still a research subject since new and improved methods for designing them can be applied. A very useful tool is in our days Computational Fluid Dynamics (CFD) thanks to the great advance in computing capabilities. Hence, any proposed design can be evaluated through numerical simulations and a decision between several ones is taken with way lower costs in comparison to when doing real experiments for all of them. The results are in very good correlation to the experimental results giving more and more confidence to the designers to use this evaluation tool even for optimization.

2 Gas turbine simulations

For gas turbines as a specific type of turbomachinery, there are many commercial CFD codes that give satisfactory results and pretty quick. A typical case study is presented below where an axial turbine having a stator blade row a rotor blade row is analyzed using ANSYS CFX.

* Corresponding author: valeriu.vilag@comoti.ro

The full geometry of the axial gas turbine stage is presented in Fig. 1 where there can be observed 30 stator blades and 61 rotor blades.

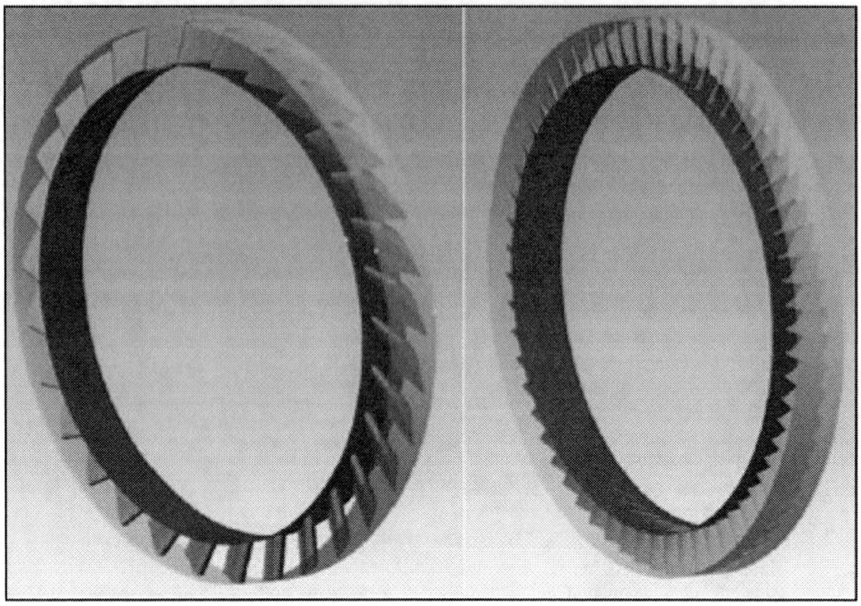

Fig. 1. Full axial gas turbine stage

Since the corresponding calculus domain for such a complex geometry would lead to very large computing time there is introduced a simplification which is very close to realty. So, it is considered that the blades belonging to the same row are identical and therefore the flows between them are also identical. When applying this hypothesis, also called periodicity of flow [1], for both stator and rotor we can reduce the calculus domain to only one stator blade and one rotor blade as shown in Fig. 2.

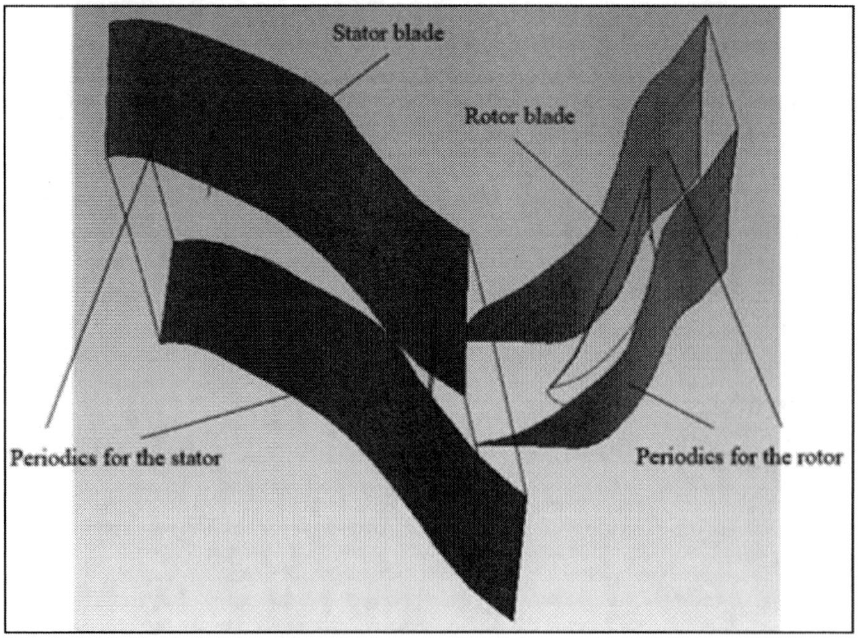

Fig. 2. Simplified domain using the so-called periodicity

Once we have established the calculus domain, the next step consists in defining the grid or mesh. Many meshing tools exist also as commercial codes the main purpose being to obtain a discretization fine enough to capture all the geometrical features and coarse enough to reduce the computing time. We have to mention here that unlike for compressor simulations where mesh quality may affect the obtained results in a very complex way, for gas turbines the mesh quality is important for capturing the shape of the blade. The difference can be explained by the different flow characteristics as for the compressors the pressure must rises being a forced or unnatural process whereas for turbines the pressure naturally falls.

The case study continues with the boundary conditions where the flow is fully or partially known. In Table 1 there are presented the boundary conditions and their names as they appear in ANSYS CFX together with the imposed flow characteristics.

Table 1. Imposed flow characteristics on the boundaries

<table>
<tr><th>No.</th><th>Boundary name</th><th>Flow characteristics</th></tr>
<tr><td>1</td><td>Stator Inlet</td><td>Total temperature; Total Pressure</td></tr>
<tr><td>2</td><td>Stator hub</td><td rowspan="3">No slip adiabatique wall, stationary</td></tr>
<tr><td>3</td><td>Stator shroud</td></tr>
<tr><td>4</td><td>Stator blade</td></tr>
<tr><td>5</td><td>Periodics for the stator</td><td>Periodic interface</td></tr>
<tr><td>6</td><td>Stator Outlet</td><td rowspan="2">Fluid-to-fluid frozen rotor interface</td></tr>
<tr><td>7</td><td>Rotor Inlet</td></tr>
<tr><td>8</td><td>Rotor hub</td><td rowspan="3">No slip adiabatic wall rotating with 22000 rpm</td></tr>
<tr><td>9</td><td>Rotor shroud</td></tr>
<tr><td>10</td><td>Rotor blade</td></tr>
<tr><td>11</td><td>Periodics for the rotor</td><td>Periodic interface</td></tr>
<tr><td>12</td><td>Rotor outlet</td><td>Mass flow rate (per one blade passage)</td></tr>
</table>

The fluid-to-fluid frozen rotor interface is a frame change model which is used to predict the transition from the stationary frame of the stator to the rotational frame of the rotor. It is one of the many possible choices to pass between frames and in combination with the periodic interfaces, for the stator and the rotor, gives satisfactory results even for multiple stage turbomachinery.

The solver requests also imposing the turbulence model, which in the present study is "k-ε" and some other details of the computational model including different numerical methods coefficients. We don't discuss them in the current paper since the default settings from ANSYS CFX when entering "Turbo-mode" are very appropriate.

After the solver computes several iterations, usually around 500 for one turbine stage are enough, the results are generated and they can be post-processed to find the predicted performances [2]. A report can be automatically generated showing several useful parameters of the turbine stage, the most important ones being displayed in Table 2 for the presented case study.

Table 2. Computed results

No.	Name	Value
1	Rotor blade torque	9.9884 [Nm]
2	Total power	1403 [kW]
3	Total-to-total isentropic efficiency	87.654 %
4	Expansion ratio	2.4581

Apart from the computed values the user can display contours plots of flow variables like pressure and temperature, vectors, streamlines and other parameters showing the turbulence, energy loss etc. Pressure variation, Fig. 3, is used to evaluate the losses that are characteristics in turbomachinery like profile losses and secondary losses, while the temperatures. Fig. 4, are used when choosing the materials from which the turbine will be constructed.

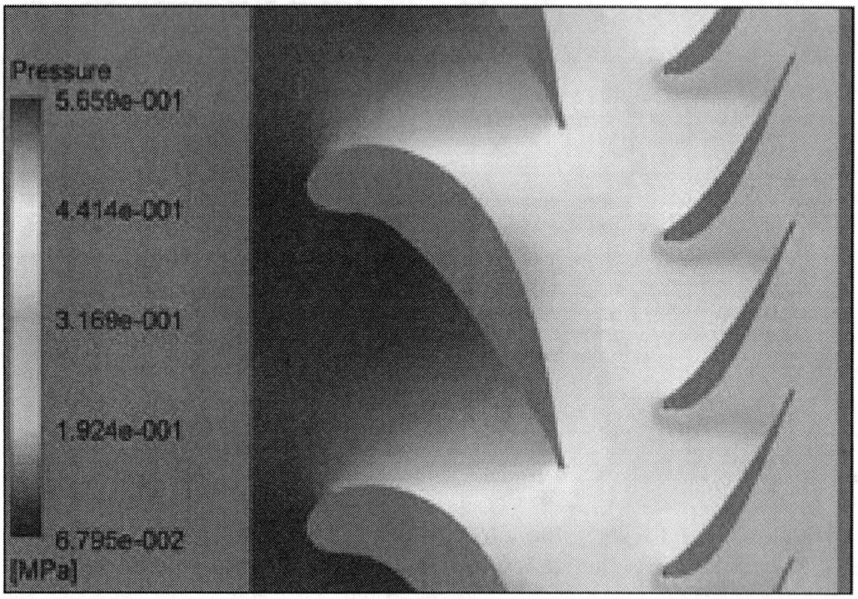

Fig. 3. Pressure variation at mid span

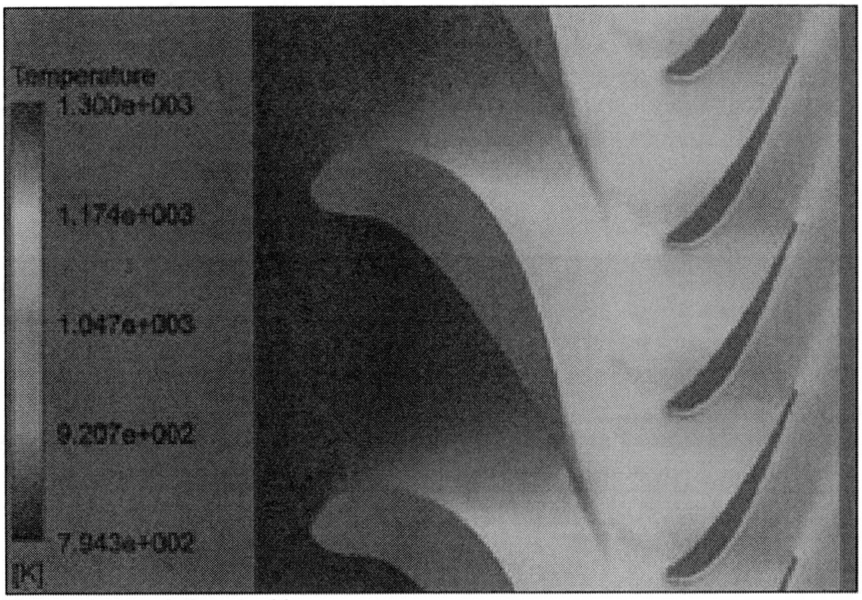

Fig. 4. Temperature variation at mid span

3 Gas turbine optimization

Optimization in the gas turbine field is seen as the quest for the optimum shape giving the desired performances and it is performed also using computers. Since it is always about shape, it all starts with defining the aerodynamic profile into the computer. There are many ways of defining a turbine row but the one presented in [4] is the most used since it contains all necessary and sufficient parameters for turbine design. In Fig. 5 we present these parameters, some being purely geometrical and some depicting the flow.

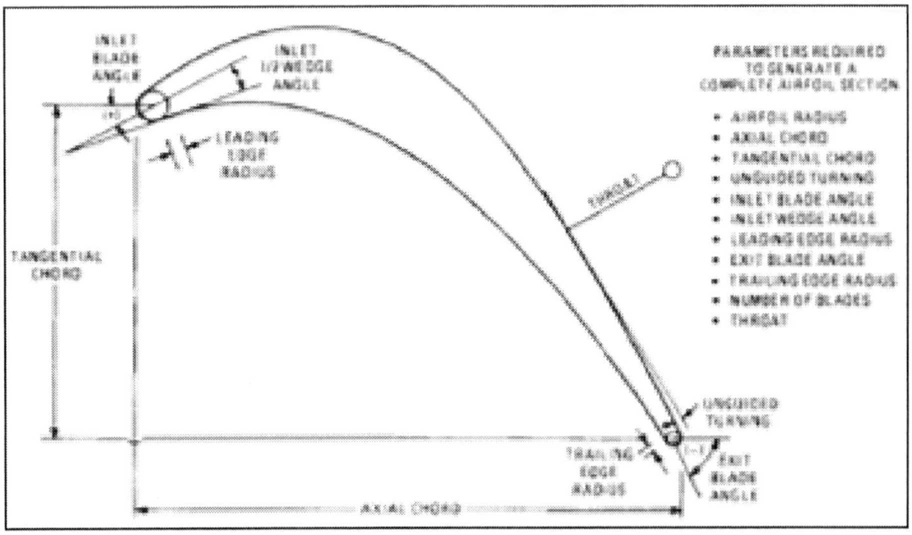

Fig. 5. Parameters defining a turbine row [3].

By imposing some limits for each of the parameters one can obtain a very wide design space consisting in many turbine geometries to be evaluated.
In Fig. 6 there is presented the shape evolution for a stator blade [4] when optimized using the algorithm presented in Fig. 7 where an initial population is created and then, via some operations inspired from the nature, new individuals are obtained.

Fig. 6. Stator blade shape evolution

The presented algorithm is an evolutionary one meaning that it creates individuals at each step and then CFD evaluation tools can be used. Its convergence can be demonstrated but still, the finding of the optimum can consume a lot of computing effort.

```
t := 0;
Create initial population B_0 = (b_{1,0}, ..., b_{m,0});

WHILE stopping condition not fulfilled DO
BEGIN

(* proportional selection *)

   FOR i := 1 TO m DO
   BEGIN
      x := Random[0, 1];

      k := 1;
      WHILE k < m & x < Σ_{j=1}^{k} f(b_{j,t}) / Σ_{j=1}^{m} f(b_{j,t}) DO
         k := k + 1;

      b_{i,t+1} := b_{k,t}
   END

(* one-point crossover *)

   FOR i := 1 TO m - 1 STEP 2 DO
   BEGIN
      IF Random[0, 1] ≤ p_C THEN
      BEGIN
         pos := Random{1, ..., n - 1};

         FOR k := pos + 1 TO n DO
         BEGIN
            aux := b_{i,t+1}[k];
            b_{i,t+1}[k] := b_{i+1,t+1}[k];
            b_{i+1,t+1}[k] := aux
         END
      END
   END

(* mutation *)

   FOR i := 1 TO m DO
      FOR k := 1 TO n DO
         IF Random[0, 1] < p_M THEN
            invert b_{i,t+1}[k];

   t := t + 1
END
```

Fig. 7. Optimization algorithm [5]

4 Water turbine simulations

This section is dedicated to kinetic water turbine simulations and it contains some insight view of the simulations performed in a Romanian-Bulgarian collaboration project. Unlike the gas turbine case, the kinetic water turbine is a specific type of turbomachinery which is very difficult to evaluate using numerical simulations and the obtained results are in many cases unsatisfactory [6]. Still, the steps are very similar to the ones for the gas turbine case and are briefly presented in Fig. 8 and 9.

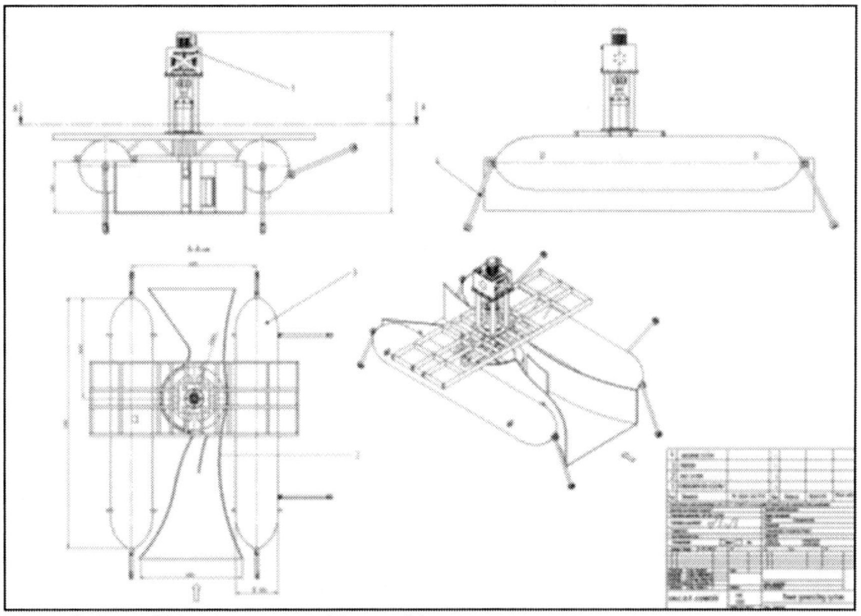

Fig. 8. General view of kinetic water turbine architecture

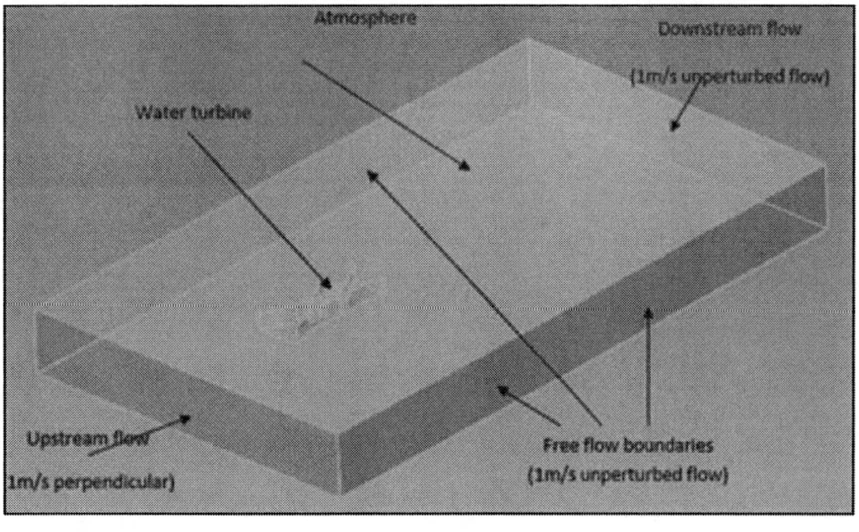

Fig. 9. Calculus domain and boundary conditions

Many simplifications where also employed, the most important being the one saying that the rotor is fully under water and it doesn't create waves. This hypothesis might not be valid for all the working regimes of the presented water turbine, but we consider that the numerical simulation still is able to give qualitative information.

We present below some results in the form of contour plots at midspan of the kinetic turbine, Fig. 10 velocity variation.

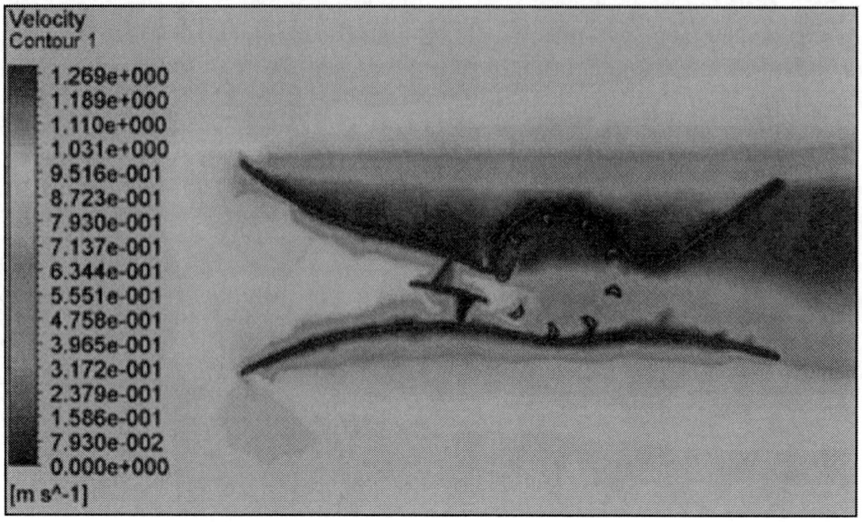

Fig. 10. Velocity magnitude at mid span

Velocity plots can be used to determine the efficiency of the kinetic water turbine highlighting also the areas where the flow has big gradients creating losses.

An indication about the waves can be seen by showing velocity vectors at the surface of the water, where the atmosphere begins, Fig. 11, but good results may be obtained only when modelling two phase flow water / air.

Fig. 11. Velocity vectors at the surface of the water.

Theoretical performance curves for kinetic water turbines are presented in Fig. 13, the points being obtained for tip speed ratios up to the value of 2.

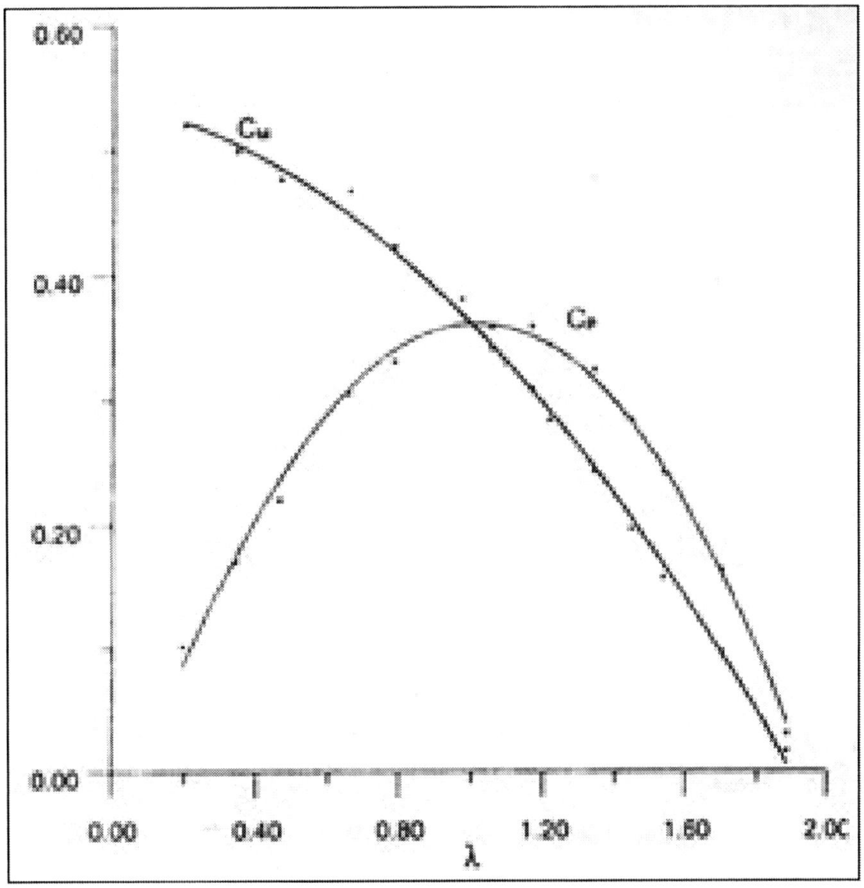

Fig. 12. Theoretical performance curves of kinetic water turbine.

Conclusions

When speaking about turbomachinery, CFD simulations are currently extensively used to conduct virtual experiments providing a good overview of the performances. For gas turbines, some specific hypothesis are applied in order to reduce the computational domain and hence to reduce the computational effort. Thus, they can be used even in optimization problems where a lot of evaluations must be performed in order to obtain optimum shapes. An evolutionary algorithm was presented and successfully used for stator blade optimization showing the wide design space a parametric model can offer.

A special type of turbomachinery is represented by turbines that extract the energy from natural flows like the wind turbines and kinetic water turbines are. For kinetic water turbines, although some results were presented, the complexity of the phenomena makes mandatory the improvement of the simulations in such way to be able to predict the water/air interface (waves) which is an important source of losses.

In conclusion, similar way of treating gas and kinetic water turbines is useful to some extent but, the later must be treated with much more precaution to obtain satisfactory results.

Acknowledgment

The work presented herein was funded by the Romanian National Research Program Nucleu TURBO 2020 and the Romania - Bulgaria CBC Programme 2007-2013 (project "Joint study regarding an electro-generator system powered by water turbine for cross-border ecological electrical transport system – ElectroRiver", MIS ETC Code 128).

References

1. ANSYS CFX user guide;
2. Vilag V., Popescu J., Petcu R., Bogoi A., Silivestru V., Numerical study on an axial gas-turbine stage, Annals of DAAAM for 2008 & Proceedings of the 19th International DAAAM Symposium, 22-25th October 2008, Trnava, Slovakia, ISSN 1726-9679, ISBN 978-3-901509-68-1, Katalinic, B. (Ed.), pp. 1397-1398, Published by DAAAM International Vienna, Vienna;
3. Pritchard, L. J., An Eleven Parameter Axial Turbine Airfoil Geometry Model, ASME 1985 International Gas Turbine Conference and Exhibit, Vol. 1: Aircraft Engine; Marine; Turbomachinery; Microturbines and Small Turbomachinery, Houston, Texas, USA, 1985;
4. Vilag V., Contributions regarding optimization of axial turbines using evolutionary algorithms and computational gasdynamics, Doctoral Thesis, Politehnica University of Bucharest, Faculty of Aerospace Engineering, July 2013;
5. Bodenhofer U., Genetic Algorithms: Theory and Applications, Lecture Notes Second Edition – WS 2001/2002, Institut für Algebra, Stochastik und wissensbasierte mathematische Systeme Johannes Kepler Universität, A-4040 Linz, Austria;
6. Dediu G., Vilag V., Popescu J., System instrumentation and data acquisition for TurboRiver-RB7, Annual Conference, "Angel Kanchev"University of Ruse, 26-27.10.2012, Ruse, Bulgaria

Design of short low speed Göttingen type wind tunnel: CFD simulation

Hardy Weisweiler[1], *Jasmina* Kojouharova[1,*], and *Roland* Dückershoff[1]

[1]Department of Mechanical Engineering, Mechatronics and Materials Technology, THM University of Applied Sciences, 61169 Friedberg, Germany

Abstract. The paper is focused on design simulations by the means of commercial software for a low speed circuit wind tunnel with an open test section, recently built at the Department of Mechanical Engineering, Mechatronics and Materials Technology at THM University of Applied Sciences in Friedberg. The proposed wind tunnel has a test section with cross sectional area of 1 X 1 m^2 and a length of 1.6 m. The maximum achievable speed is about 50 m/s with empty test section. The simulation had the target to propose and to verify various geometries - test chamber, tunnel contraction, diffuser etc. as well the tunnel corner including the vanes. In addition, the influence of the vanes number and their shape on the ventilator power needed and the flow velocity uniformity as well, have been subjects of the investigation. Also the ventilator type (rotational segment) impact on the flow pattern within the operational area has been scrutinized.

1 Introduction

Wind tunnels generate nearly uniform air flows, with low turbulence intensity and have the advantage to test the flow properties under well controlled conditions compared to experiments in the open environment. These devices have been in operation since the end of the 19th century and could be built by two different manners: closed loop Gottingen type respectively open Eiffel type wind tunnel (s. Fig. 1). They can be further divided into open and close test section types.

The wind tunnel can be classified also by flow speed into four groups: (i) subsonic or low-speed; (ii) transonic; (iii) supersonic; (iv) hypersonic.

At low flow speed, due to the dynamic similarity, the same Reynolds numbers as for the real application are required in order to enhance the reliability of the test data. The scale of the test models is predefined by the test chamber size and since the almost all wind tunnels operate at the air atmospheric pressure, the only option left is to increase the flow velocity. Therefore, achieving of high flow velocity has been sought.

When talking about computational engineering, it is meant that we attempt to "mimic" the nature, to describe the real processes by means of conservation equations or in other words to fix the mathematical model and then playing with the parameters to learn more about the processes there.

* Corresponding author: jasmina.kojouharova@m.thm.de

Here the questions are: first - how realistic we describe the physical processes and second - how reliable the computations are. Certainly, the experimental studies should run, especially at high Reynolds number, for full identification of the flow characteristics and thus to validate the numerical simulations. In this aspect even if the velocity of interest is out of the wind-tunnel operation range, the fluid flow simulations still can be validated if no other physical effects are met. Therefore, the wind-tunnel experiments and CFD simulations might be considered as complementary ones.

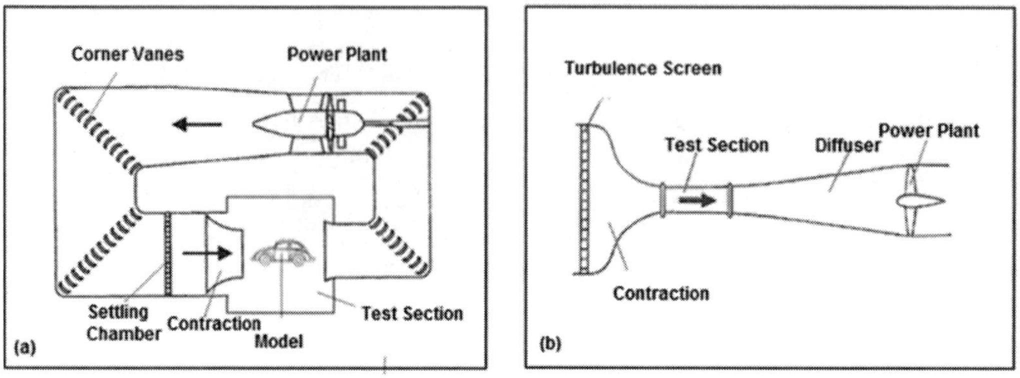

Fig. 1. Schematic of (a) closed loop Göttingen type wind tunnel (b) open Eiffel type wind tunnel

A wind tunnel is characterised by the flow quality inside the test chamber. The drawback of that devices is that the flow field could not be well developed alongside the tunnel, especially in the case of short one. Moreover, the side's effects could not be completely excluded and thus also they affect the flow characteristics.

A compromise has to be made and a wind tunnel with a flow uniformity within the range 0.5 – 2 %, the turbulence intensity of 0.5 – 2 % is considered as a good one. It should provide temperature uniformity within 0.1 – 0.5 % at the inlet of the test section.

The design of the wind tunnels is a complex task involving fluid mechanics issues and typical engineering aspects. Nevertheless, the topic is well discussed in the literature (for instance, the already classical publication [1-2] or the recent reports [3-4]). However, some concerns still remain to be clarified due to the fact that the design depends mainly on the wind tunnel final purpose. In addition some restrictions, like for instance the limited space available for the wind tunnel, makes difficulties to achieve the flow quality desired. It requires an innovative design to be studied in order to keep the test section within the specification: a cross sectional area of 1 X 1 m^2 and a length of 1.6 m.

Due to specific infrastructure reason we have faced also severe electric power restriction. So that the accurate assessment of the losses along the tunnel is compulsory.

2 Wind tunnel design considerations

At first, the decision has been taken in favour of a closed circuit wind tunnel. Besides of the infrastructural circumstances in this case, due to kinetic energy conservation of the airflow, typically less power is needed for reaching the velocity required, although the general design complexity evokes more difficulties.

The overall layout of the proposed wind tunnel is shown in Fig. 2. The airflow circulates counter clockwise (s. Fig. 2) from the suction zone at the end of the test camera through diffuser, settling chambers and contraction being driven through the power plant.

Based on long lasting experience, the E.T.S.I. group at the Madrid Technical University [4] elaborates a methodology focused on the reduction of construction and operation cost of low speed wind tunnel (LSWT) for a given performance and quality requirement. Thereafter they implemented it in an Excel spreadsheet (web page htts://www.aero.upm.es/LSLCWT) as an assistance of new wind tunnel designers.

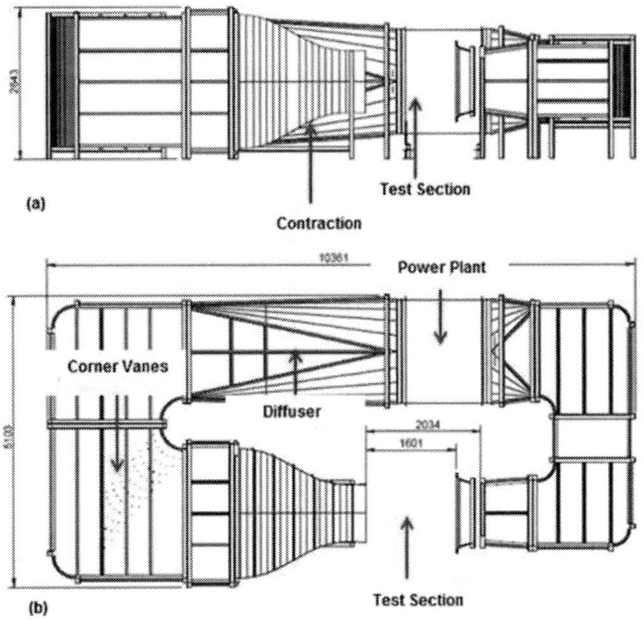

Fig. 2. Layout of the THM wind tunnel: (a) vertical intersection (b) longitudinal intersection.

Following the recommendations with the specification given: maximum achievable velocity of 50 m/s and the prescribed test section dimensions, we got the LSWT length of 13.1 m and width of 5.4 m far above the specification of 10. 5 m length and 4.5 m width.

The possible way to reduce the overall size is to decrease the contraction ratio, i.e. the ratio between the largest cross section area, found in the stagnation chamber and the smallest cross section area found in the test section. However, a high contraction ratio is a key factor in achieving a good quality flow. Another way is to use the wide angle diffusers which may lead to flow detachment. It is crucial to avoid flow separation close to the wall of the contraction zone. Separations may cause vibrations which are also undesirable.

The corners with the incorporate turning vanes could sufficiently improve the flow quality. The parameters to be considered in the design are the radius, the spacing between vanes and the flow impact angle. The use of stagnation chamber with honeycomb and screen layer leads to decrease the turbulence intensity.

At the stage of design, the most adequate method to verify that design meets those criteria is computational fluid dynamics (CFD).

3 Physical model

Numerical simulations have been performed for the 3D turbulent incompressible flow built on a Reynolds Average Navier-Stokes (RANS) steady state calculations by κ-ε turbulence model. The development in the time the flow is not a matter of discussion in this paper.

Due to the complex geometry of the construction (s. Fig. 2) some simplification have been implemented in order to keep the calculation time at some reasonable limits. In the subdomain, representing the impeller, a set of virtual forces is applied, which replace the modelling of the axial fan blades and produce similar effects in the outer flow space.

3D fan zones, considered as fluid cell zones, simulate the effect of an axial fan by applying a distributed momentum source in a toroid-shape fluid volume. The axial and the swirl momentum sources are expressed as:

$$S_{ax}[kg\, m^{-2}s^{-2}] = \frac{\Delta p}{\Delta x}$$

$$S_{swirl}[kg\, m^{-2}s^{-2}] = \frac{\dot{m}}{A_{fan}} \cdot \frac{1}{\Delta x} \cdot r \cdot \omega$$

The parameters needed are the pressure jump Δp in the flow direction and the swirl components, both expressed in relation to the flow rate. In most of the cases the swirl components are not included: the fan characteristics are then solely represented by the pressure jump. This value is typically obtained from experimentally generated fan blades curve. In the formulae above Δx stays for the fan thickness, A_{fan} for the fan cross sectional area, $\dot{m}$ for the mass flow rate, r is the fan radius and ω the angular velocity.

The settling chamber is located just before the stream entering the nozzle and thus has a major influence on the flow quality in the test section. The THM wind tunnel is equipped with one honeycomb and three screens.

The honeycomb layer has a thickness of 18 mm and the cylindrical shaped cells are with a diameter of 2.5 mm and a wall thickness is 0.2 mm. Two of the screens have the thickness of 140 μm and one of them is with the thickness of 70 μm. Their porosity is 68 %, 49 % and 58 % respectively.

To minimise the computational power needed, the topology of the settling chamber has been neglected and the simulations have been performed under the assumptions that the volume there is empty. While the honeycombs essentially serve very efficiently to reduce the lateral turbulence and the screens attenuate the longitudinal one, the simulated velocity uniformity is to be considered as a low limit. In addition the pressure losses through this chamber has been analytically evaluated.

4 Results and discussion

First and foremost, the pressure losses through all the wind tunnel components have been analytical evaluated, however it is beyond the scope of this paper and are not to be discussed now.

To find correlations between the geometry proposed and the pressure losses, only the axial driving forces have been considered initially. This is reasonable because the fan could be equipped with additional grids in order to reduce the swirl momentum forces.

At first, the primary CAD model has been simulated, critically studied and the impact of the particular components has been figured out. Different layouts have been proven to keep the outer dimensions - length, width and high, within the specification given.

Two main simulations targets - to keep the total pressure loss minimal and to look for the best fluid flow performance within the test section, are in some sense contradictory, though the quality criteria are considered as prevailing ones.

The contraction ratio between 4 and 6 is considered as sufficient in the case of wind tunnels for civil or industrial application. In our design the contraction ratio was set to 5.1.

The flow quality is strongly dependent on the geometry of the diffuser just after the power plant. Its length was fixed to 2.9 m keeping its semi-open angle less than 3.5°.

The outer dimensions of the corner chamber and the corner radius as well also have been selected. The corner radius is proportional to the chamber entrance so that it was set to be 0.3 m for the front und 0.4 m for the back chambers.

The corner with the incorporated turning vanes are accountable for flow structure and for the major part of the total pressure loss, often more than 50 % [4]. Thus, the main simulations issue targets the defining of the inner corner topology.

In order to keep the construction expenses at reasonable level, the vanes have been defined as simple curved plates and a special attention has been draft on the vanes spacing, arc length and their orientation as well.

Two different scenarios have been simulated: distance of 0.2 m respectively 0.15 m within the front chambers and 0.25 m respectively 0.18 m within the back chambers. It has been proven that the increase of vanes number proposed, results in an increase of the pressure loss their within 10 % which is still tolerable.

Initially, very short vanes, less than 0.1 m, have been considered, which turn out to be ineffective. Then different type of arcs (length and attack angle) have been simulated and the result, relevant to the arc with an angle of 96 o, has been found to be the most efficient one. The arc lengths have been fixed to 0.51 m for front chambers and 0.67 m for the back chambers.

The variety of design layouts has been examined. Here we are not to follow all of them and further the result only for the design selected, which might be identified in Fig. 4, are presented.

The pressure losses as a function of the volume flow rate have been studied and the results are summarised in Table 1.

Table 1. Pressure losses within the wind tunnel in dependence on the flow velocity

Volume Flow Rate [m³/s]	30	40	50
Pressure losses [Pa]	155	272	434

The total pressure loss is understood as a sum of pressure losses through all the construction components which are defined by the use of the relation:

$$\Delta p = \sum_i \frac{1}{2} \cdot \rho \cdot V_i^2 \cdot \zeta_i$$

Here ρ is the gas density, taken to be 1.185 kg/m³, V_i is the gas velocity and ζ_i is the pressure loss coefficient within the construction components respectively.

Based on that, the total pressure loss coefficient ζ can be defined as

$$\Delta p = \frac{1}{2} \cdot \rho \cdot V^2 \cdot \zeta$$

where V is the operating velocity. Eventually, the THM wind tunnel is characterized with a total pressure loss coefficient of about 0.29.

Once the total pressure loss has been fixed, the main power P of the plant, needed to maintain the flow running inside the wind tunnel at a constant speed, may be assessed as

$$\mathrm{P} = \Delta p \cdot \frac{\dot{\mathrm{Q}}}{\eta}.$$

Here $\dot{Q}$ stays for the volumetric flow rate and η is the fan efficiency. At a typical fan efficiency value of about 50 %, the fan power needed is specified to be of about 43.5 kW. One has to mention that the figure given should be increase by 10 % to account the pressure loss within the settling chamber and the dissipations as well as.

The velocity distribution within the cross section of the nozzle exit is presented in Fig. 3 at the volume flow rate of 50 m/s. The velocity uniformity is estimated to be of about 3 % within the core.

The flow structure depending on the swirl momentum imposed have been also studied. The rotational motor speed of 1000 rpm has been assumed. So that three different scenarios have been simulated: 8 rev/s, 16 rev/s and 32 rev/s respectively. While the swirl impact affects only little the total pressure losses, it significantly modifies the flow structure within the test section.

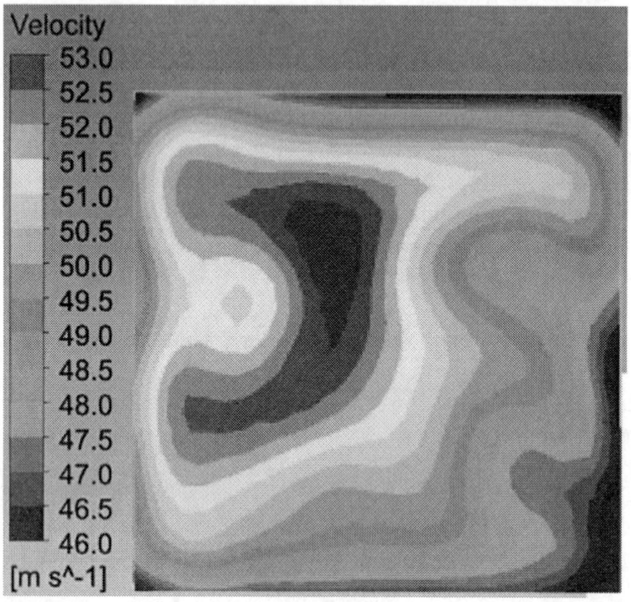

Fig. 3. Velocity distribution within the nozzle exit at volume flow rate of 50 m^3/s

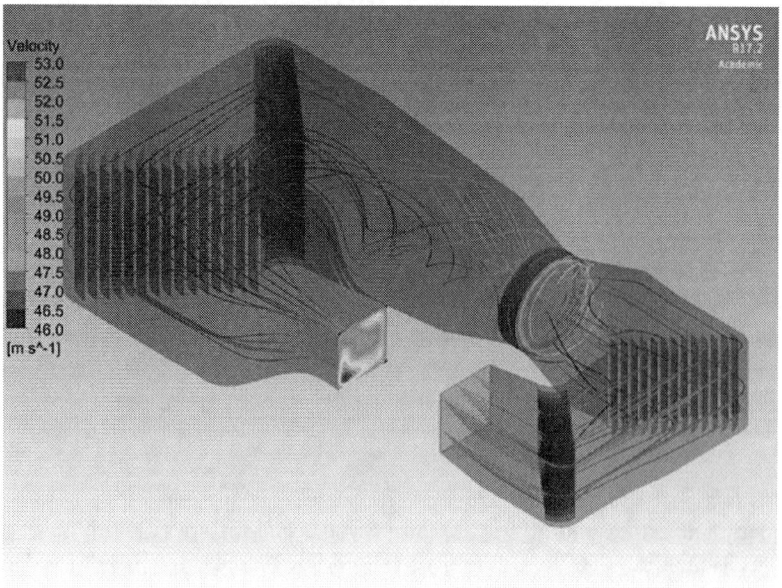

Fig. 4. Streamlines and velocity distribution within the wind tunnel at volume flow rate of 50 m^3/s and 16 rev/s rotational impeller speed.

To fully understand the flow structure, the velocity components in horizontal and vertical direction of the nozzle exit also have been studied. The streamlines and the velocity distribution within the wind tunnel at volume flow rate of 50 m³/s and 16 rev/s rotational impeller speed are given in Fig. 4, while the velocity components distribution within the cross section of the nozzle exit are shown in Fig. 5.

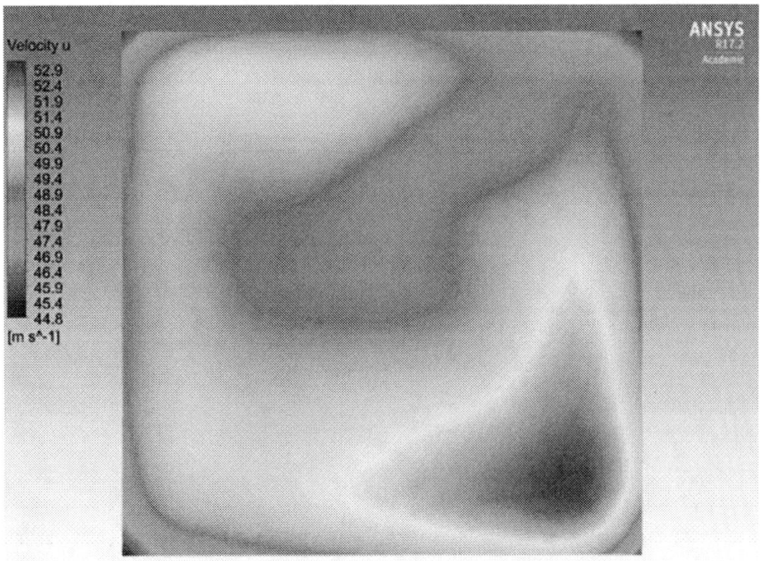

(a)

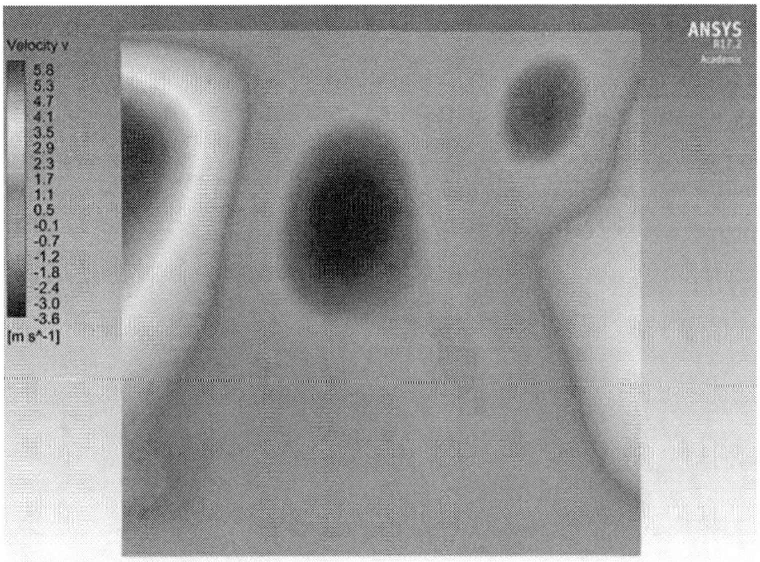

(b)

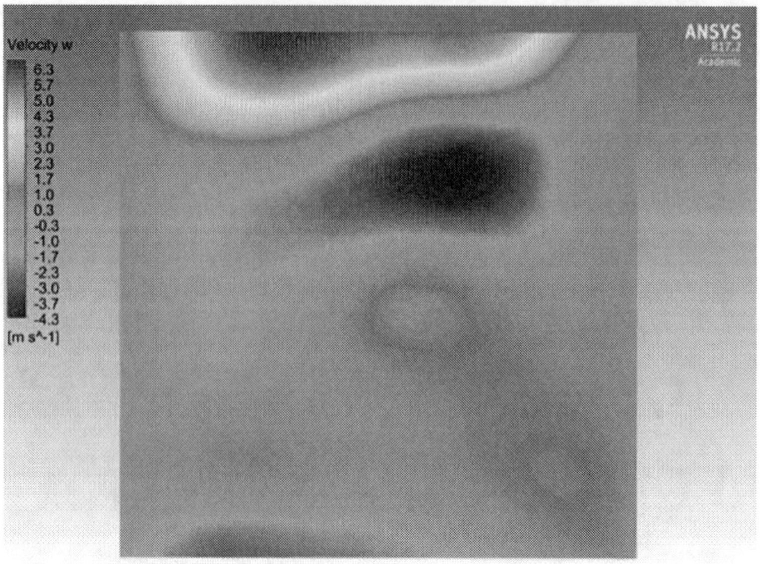

(c)

Fig. 5. Velocity components distribution within the cross section of the nozzle exit at volume flow rate of 50 m³/s and 16 rev/s rotational impeller speed: (a) velocity component in the direction perpendicular to the cross section exit; (b) velocity components in the horizontal direction; (c) velocity components in the vertical direction.

5 Wind tunnel validation

After wind tunnel has been built at the THM in Friedberg the specification has been validated by L. Feucht, D. Krön and S. Neumann [6]. The measurements have been made at three cross

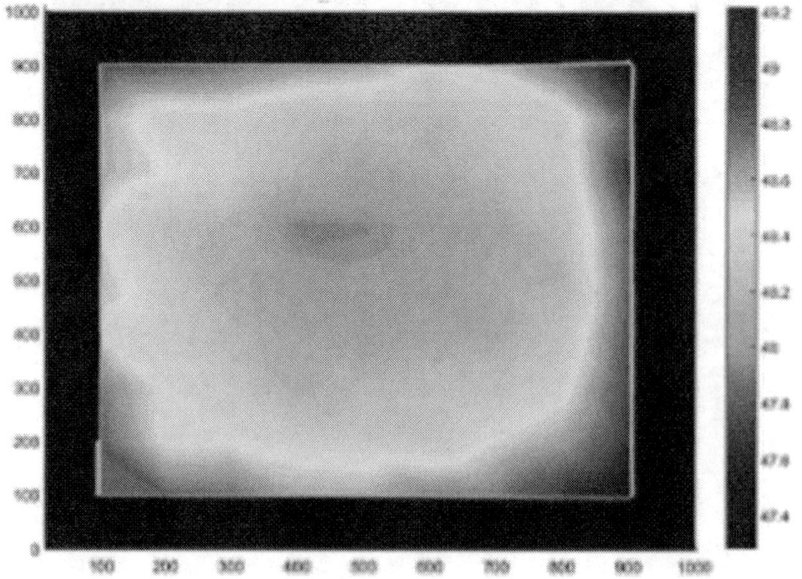

Fig. 6. Velocity distribution at the cross section with a distance of 0.4 m from the nozzle exit and a volume flow rate of 50 m³/s

sections of the test section with distances to the nozzle exit of 0.4 m, 0.8 m and 1.2 m respectively. The specified maximal velocity of 50 m/s was reached and the flow characteristics fulfilled the specification given (s. Fig. 6).

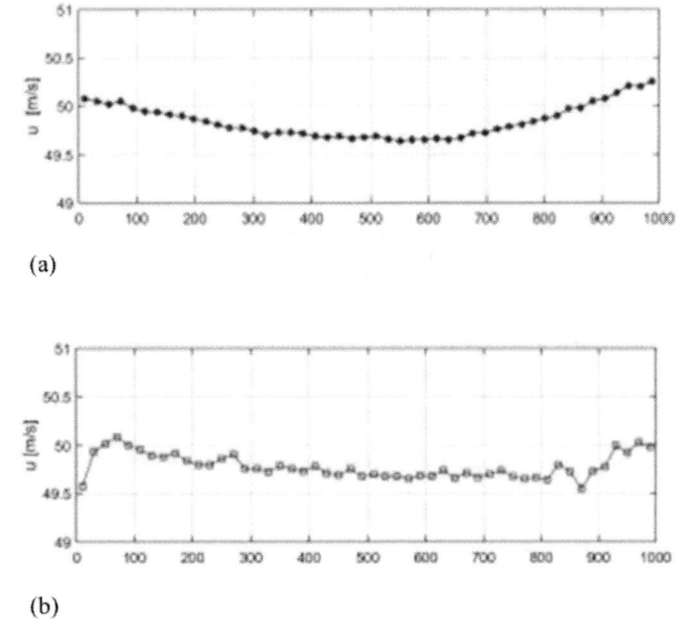

Fig. 7. Measured velocity at the middle of cross section of the test section at a distance of 0.4 from the nozzle exit as a function of the distance to the nozzle wall and volume flow rate of 50 m^3/s: (a) in horizontal direction; (b) in vertical direction.

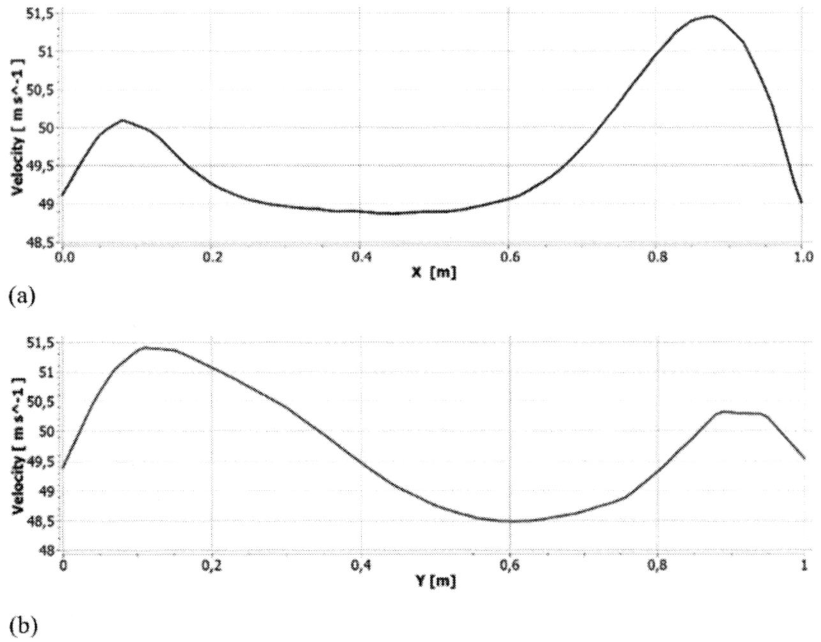

Fig. 8. Velocity at the middle of the nozzle cross section as a function of the distance to the nozzle wall and volume flow rate of 50 m^3/s: (a) in horizontal direction; (b) in vertical direction.

The data predicted have been compared with the experimental data and a good correlation has been found (s. Fig. 7 and Fig.8 where the velocity at the middle of the nozzle cross section as a function of a distance to the nozzle wall is given). The discrepancies observed can be attributed to the specifics of the measurement. The velocity of the flow has been calculated at the exit of the tunnel taking into account the wall effects, however, the measurements have been carried out somewhere in the middle of the test section. The wall effects, affecting the flow in the tunnel, are there already negligible.

References

1. P. Branshaw, R. C. Pankhurst, Progress in aeronautical sciences, **6**, 1 (1964)
2. W. H. Rae, A. POPE, *A. Low-speed wind tunnel testing* (2nd edn. John Wiley & sons. 1984)
3. B. Lindgren, A.V. Johansson, *Design and evaluation of a low-speed wind-tunnel with expanding corners* (Technical Reports from Royal Institute of Technology, Departments of Mechanics, SE-100, 44 Stockholm, Sweden 2002)
4. M.A.G. Hernández, A.I.M. López., A.A. Jarzabek, J.M.P. Perales, Y., Wu, S. Xiaxiao, *Design methodology for a quick and low-cost wind tunnel, Chapter 1 in Wind Tunnel Designs and Their Diverse Engineering Applications* (Edited by N. A. Ahmed, ISBN 978-953-51-1047-7 (Open Acces) 2013)
5. *ANSYS CFX-Solver Theory Guide, Inc. Release 14.5*, Southpointe October (2012)
6. D. Krön, *Messung und Bewertung der Strömungsturbulenz im Freistrahl des 1x1 m^2 Windkanals Göttinger Bauart GWK der THM* (Projektarbeit 2015)

Identification of the influence of the heating upon the heat transfer during pyrolysis process used for End-of-Life tires treatment

Ivanka Zheleva [1,*], *Ivan* Georgiev [2], and *Margaritka* Filipova[1]

[1] Department Thermoingeneering, Hydraulics, Ecology, Ruse University, 7017, Ruse, 8 Studentska Str., BULGARIA
[2] Department Applied Mathematics and Statistics, Ruse University, 7017, Ruse, 8 Studentska Str., BULGARIA

Abstract. End-of-Life tires (EOLT) are waste that does not practically decompose in nature and because of this their sound treatment is needed for environment protection. One of the possible methods for such a treatment is pyrolysis process. Globally, around 23% of all EOLT are processed through pyrolysis, whereas in the Republic of Bulgaria only 5% are processed by pyrolysis. This shows that for Bulgaria this method pyrolysis still has a good potential for development and further research. In our previous paper [1], an adequate mathematical model of non-stationary heat transfer during pyrolysis used for the treatment of end-of-life tires has been developed. A numerical algorithm has been also developed in MATLAB to solve the respective mathematical initial and boundary value problems. In this study on the base of the previously developed model we examine the temperature regimes in the pyrolysis station with three cameras and their strong dependence upon the characteristics of the initial heating. Results for the temperature fields, temperature isolines and gradients for different kind of initial heating of the process are presented and commented.

1 Introduction

End-of-Life tires (EOLT) are waste that practically does not decompose in nature and because of this their sound treatment is needed for environment protection. One of the possible methods for such a treatment is pyrolysis process. Thus pyrolysis is an opportunity to solve the ecological problems related to the accumulation of large quantities of end-of-life tires. Pyrolysis is a chemical process that takes place without the presence of oxygen. Pyrolysis differs from the combustion process due to the lack of waste products. Globally, around 23% of all EOLT are processed through pyrolysis, whereas in the Republic of Bulgaria this percentage is much smaller - only 5%. This shows that for Bulgaria this method still has a good potential for development and further research. In Republic of Bulgaria there are several pyrolysis stations working for treatment of end-of-life tires but their management is predominantly manual and therefore not efficient. Because of this the

* Corresponding author: izheleva@uni-ruse.bg

theoretical examination of the temperature regimes in that kind of stations is needed. In our previous paper an adequate mathematical model of non-stationary heat transfer during pyrolysis used for the treatment of end-of-life tires has been developed. A numerical algorithm has been also developed in MATLAB to solve the respective mathematical initial and boundary value problems [1].

In this study we examine the temperature regimes in the pyrolysis station with three cameras and their strong dependence upon the characteristics of the initial heating. Results for the temperature fields, temperature isolines and gradients for different kind of initial heating of the process are presented and commented.

2 Mathematical posing of the problem

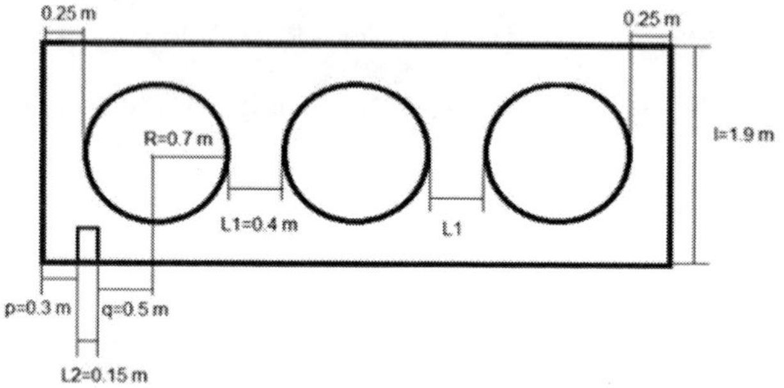

Fig. 1. Horizontal cross section of the pyrolysis station with three cameras and one heater

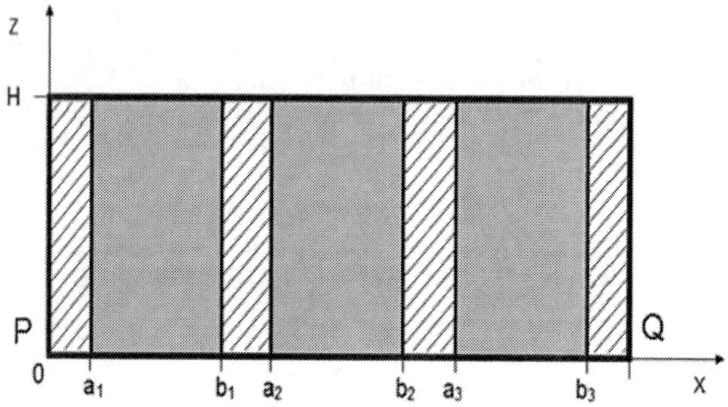

Fig. 2. Vertical cross section of the pyrolysis station with three cameras in the plane Oxz, a_1b_1, a_2b_2, a_3b_3 – diameters of pyrolysis cameras, 1.4 m, OH – 1.45 m, PQ = 5.5 m

Fig. 3. 3D figure of the pyrolysis station with three cameras

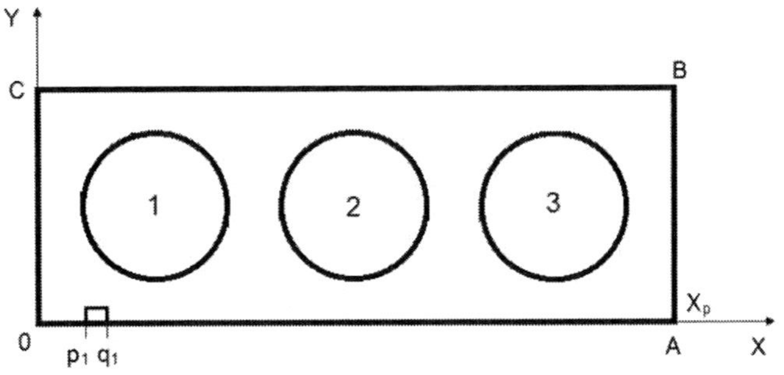

Fig. 4. Geometrical area for the first model in the plane Oxy, p_1q_1- burner, p_1=0.3m, q_1=0,45m, 1, 2, 3 – circles - bottoms of pyrolysis cameras, OA= 5,5 m, OC = 1,9 m, diameters of circles – 1,4 m

Here, following [1] we will write the mathematical model for studying non-stationary heat transfer in pyrolysis chemical reactor (station) with three cameras and one heating burner, which are schematically presented in Fig. 1 – Fig.4. So the model for Oxy cross section (Fig.4) consists of the equation

$$\frac{\partial T}{\partial t} = k\left(\frac{\partial^2 T}{\partial x^2} + \frac{\partial^2 T}{\partial y^2}\right) \tag{1}$$

Here T is the temperature in Kelvin T [K], and k is the coefficient of thermal conductivity, [k] = W / (m.K).

This equation has to be solved in the double-bonded geometric area of Fig. 4 under the following initial and boundary conditions:

At $t = t_0 \quad T = T_0 = 20\,^{\circ}C$ for $\forall\ x, y$ (2)

For $y = 0, \forall\ t\ x \in (0, p_1) \cup (q_1, X_p)$ $\quad \frac{\partial T}{\partial y} = k_1(T - T_0)$ (3)

For $y = H \quad \forall\ t, \forall\ x.$ $\quad \frac{\partial T}{\partial y} = k_1(T - T_0)$ (4)

For $x = 0 \quad \forall\ t, \forall\ y.$ $\quad \frac{\partial T}{\partial x} = k_1(T - T_0)$ (5)

For $x = L \quad \forall\ t, \forall\ y.$ $\quad \frac{\partial T}{\partial x} = k_1(T - T_0)$ (6)

For $y = 0, \forall\ t\ x \in (p_1, q_1)$ $\quad$ T = T*(t, x) from Fig. 5 (7)

The thermal conductivity coefficients are respectively for refractory bricks $k_1 = 0{,}29\ W/mK$, for steel (material of pyrolysis chamber, circles 1, 2, 3 Fig. 1) $k_2 = 45\ W/mK$,

The model for Oxz cross section (Fig.2) consists of the equation:

$$\frac{\partial T}{\partial t} = k\left(\frac{\partial^2 T}{\partial x^2} + \frac{\partial^2 T}{\partial z^2}\right). \tag{8}$$

This equation has to be solved in the geometrical area, shown in Fig.2 with the following initial and boundary conditions

At $t = 0, T = T_0 = 20\,^0C = 293\ K$ (11)

At $z = 0$ and $x \in [0, Q]$ $T = T_{01}(t)$ (12)

At $z = H$ and $x \in [0, Q]$ $\frac{\partial T}{\partial z} = k_*(T - T_0)$ (13)

At x=0 and x=Q and $z = 0$ and $z = H$ $\frac{\partial T}{\partial x} = k_1(T - T_0)$ (14)

Coefficient k_* is equal to k_1 if $x \in [0, a_1] \cup [b_1, a_2] \cup [b_2, a_3] \cup [b_3, Q]$ and k_* is equal to k_3 if $x \in [a_1, b_1] \cup [a_2, b_2] \cup [a_3, b_3]$ The function $T_{01}(t)$ is calculated after the first stage of modeling. Here $k_3 = 0{,}16 W/mK$ is for the material of the tires.

From the mathematical point of view the problem is to fine the temperature T as a function of space and time $T(t, x, y, z)$ in the geometrical area, shown at Fig.1 – Fig.4, by solving the initial and boundary value problems for equation (1) and (8) with initial and boundary conditions respectively (2) – (7) and (9) – (14). To find solutions of these problems is very complicated because the geometrical areas are very complex (it is not a convex polytope) and also the coefficients of thermal conductivity are different in the different subareas of the geometric area.

To solve these mathematical problems we will use the numerical algorithm, developed by using MATLAB software and proposed in our paper [1].

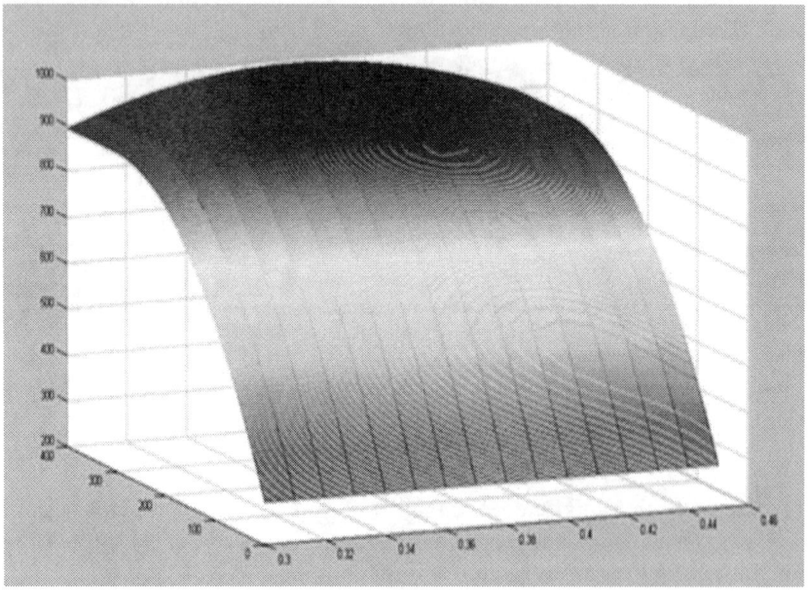

Fig. 5. 3D Graphic of the temperature $T^*(t, x)$ of the burner as a function of the space $0.3m < x < 0.45m$ and time and $T_{max} = 900\ K$.

3 Modelling of the heating process

The heating of the pyrolysis stations is carried out by one burner placed on the bottom of the station (see p_1q_1 on the fig.3). The used fuel can be different. To describe the heating process of the pyrolysis station we will create a function $T^*(t,x)$ for $p_1 \leq x \leq q_1$ and for $\forall\, t$. We suppose that $T^*(t,x)$ is symmetric quadratic function with T_{max} in the middle of the interval for $p_1 \leq x \leq q_1$ and for $\forall\, t$

Using also MATLAB software we develop $T^*(t,x)$. The Fig. 5 shows such a function for $p_1 = 0.3\ m,\ q_1 = 0.45\ m$ *and* $T_{max} = 900\ K$.

In this paper we study the influence of the heating on the whole pyrolysis process. We look at two heating regimes – with $T_{max} = 2773\ K$ and $T_{max} = 32730\ K$.

4 Results and Discussion

First we examine the heating regime with $T_{max} = 2773K$. Figures 4, 5 and 6 show the calculated temperatures in Oxy plane at t = 500 s, t = 3600 s and t =15000 s. Figures 7, 8 and 9 show the temperatures in Oxz plane at the same moments of time t = 500 s, t = 3600 s and t =15000 s. Figure 10 shows the temperature as a function of time (at t = 500 s, t = 3600 s, t = 15000 s) in the Oxy plane in the PQ section, which is the initial and boundary condition of the mathematical problem of studying the heat exchange of the pyrolysis process in the Oxz plane.

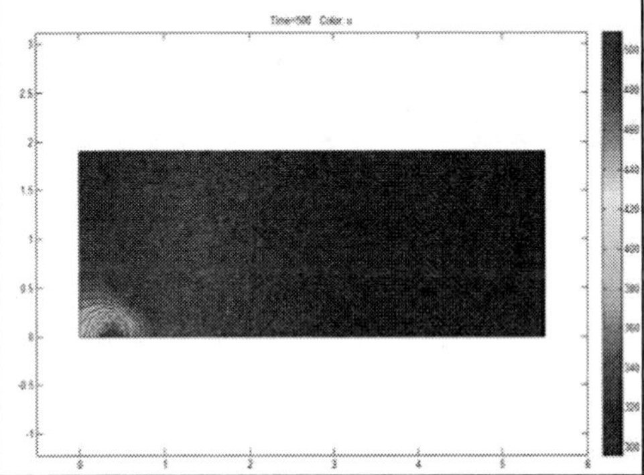

Fig. 4. Temperature field with temperature isolines in the plane Oxy at t = 500 s and Tmax= 2773K

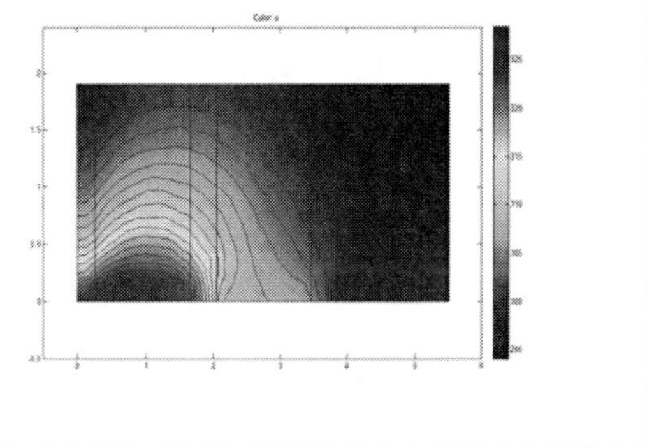

Fig. 7. Temperature field with temperature isolines in the Oxz plane at t = 500 s and Tmax= 2773K

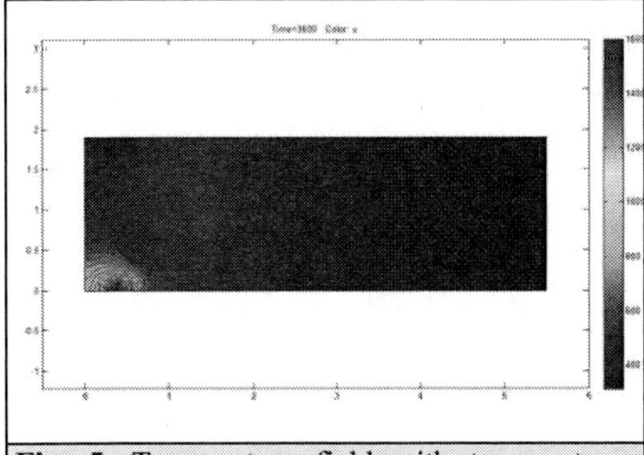

Fig. 5. Temperature field with temperature isolines in the plane Oxy at t = 3600 s and Tmax= 2773K

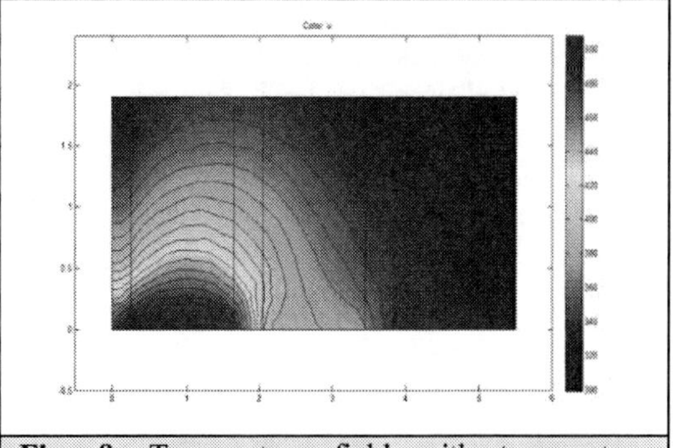

Fig. 8. Temperature field with temperature isolines in the Oxz plane at t = 3600 s and Tmax= 2773K

https://doi.org/10.1051/matecconf/201814503016

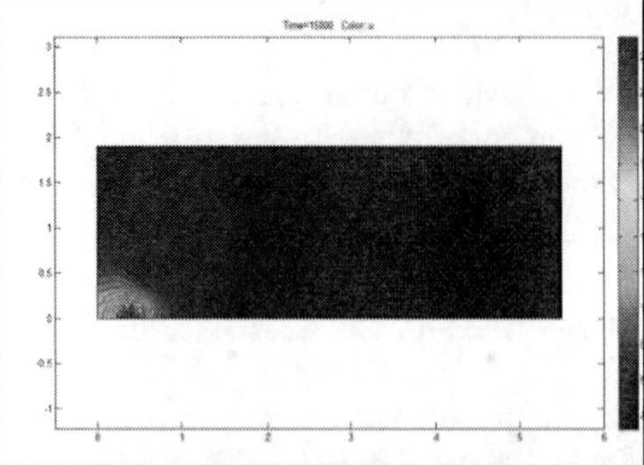

Fig. 6. Temperature field with temperature isolines in the plane Oxy at t =15000 s and Tmax= 2773K

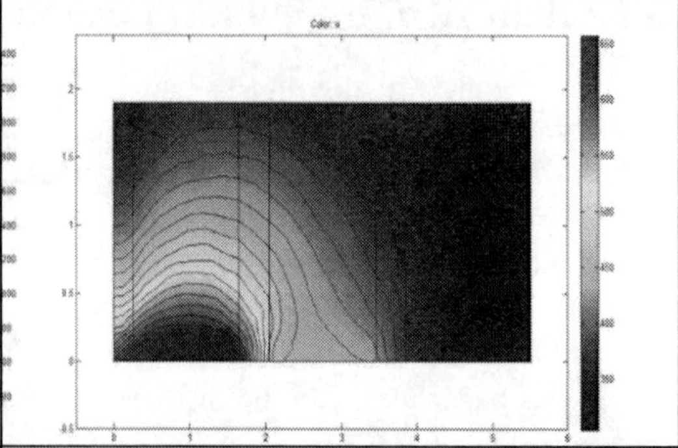

Fig. 9. Temperature field with temperature isolines in the Oxz plane at t = 15000 s and Tmax= 2773K

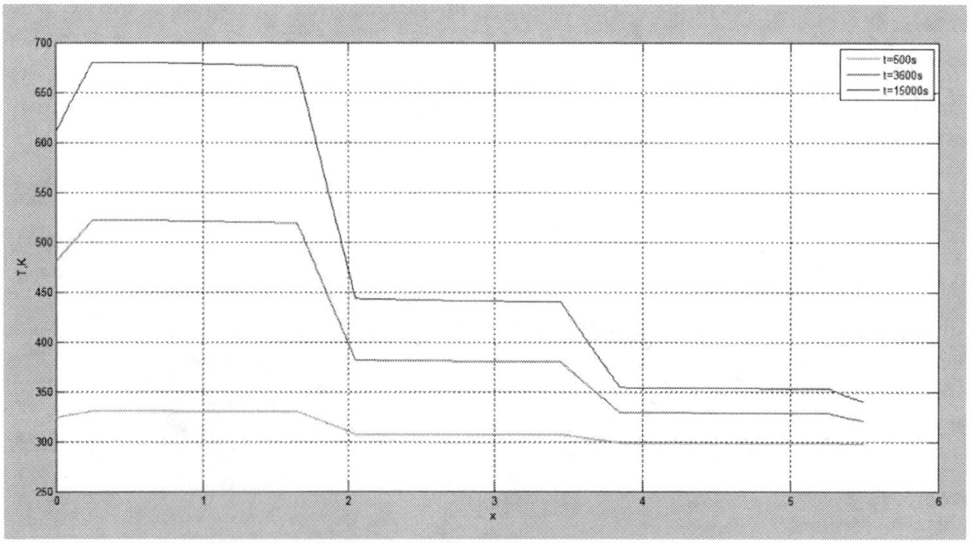

Fig. 10. Graph of the temperature as a function of time T*(t, x) (at t = 500 s, t = 3600 s, t = 15000 s) in the Oxy plane in the PQ section, which is the initial and boundary condition of the mathematical problem of studying the heat exchange of the pyrolysis process in the Oxz plane

Second we examine the heating regime with $T_{max}= 3273K$. Figures 11, 12 and 13 show the calculated temperatures in Oxy plane at t = 500 s, t = 3600 s and t =15000 s. Figures 14, 15 and 16 show the temperatures in Oxz plane at the same moments of time t = 500 s, t = 3600 s and t =15000 s. Figure 17 shows the temperature as a function of time (at t = 500 s, t = 3600 s, t = 15000 s) in the Oxy plane in the PQ section, which is the initial and boundary condition of the mathematical problem of studying the heat exchange of the pyrolysis process in the Oxz plane.

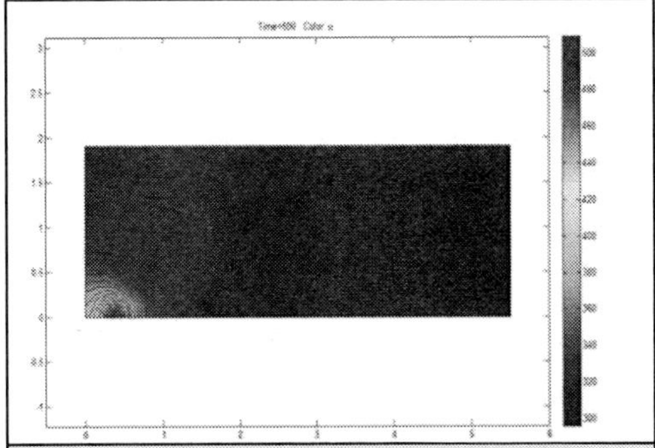

Fig. 11. Temperature field with temperature isolines in the plane Oxy at t = 500 s and Tmax= 3273K

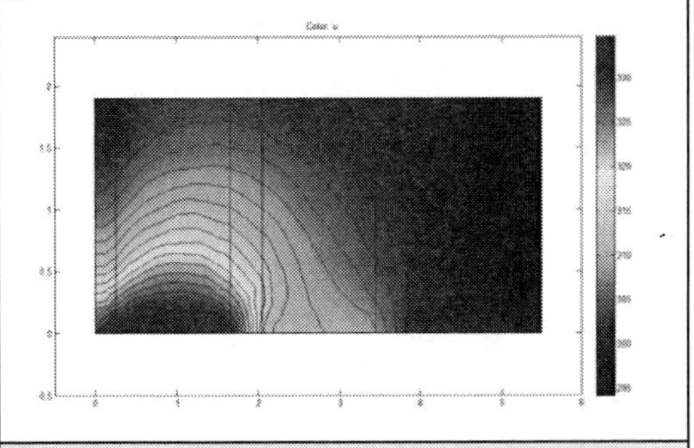

Fig. 14. Temperature field with temperature isolines in the Oxz plane at t = 500 s and Tmax= 3273K

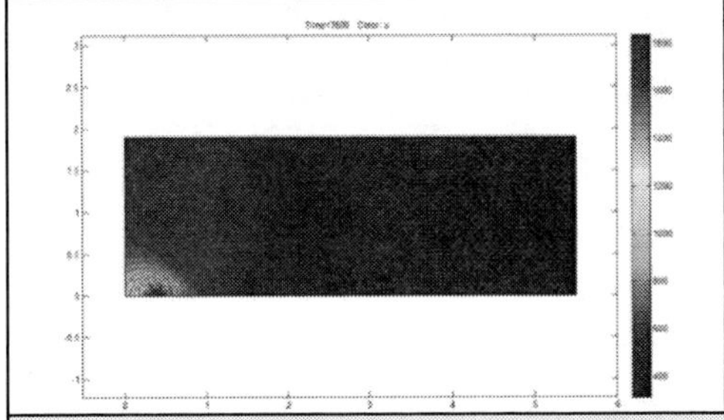

Fig. 12. Temperature field with temperature isolines in the plane Oxy at t = 3600 s and Tmax= 3273K

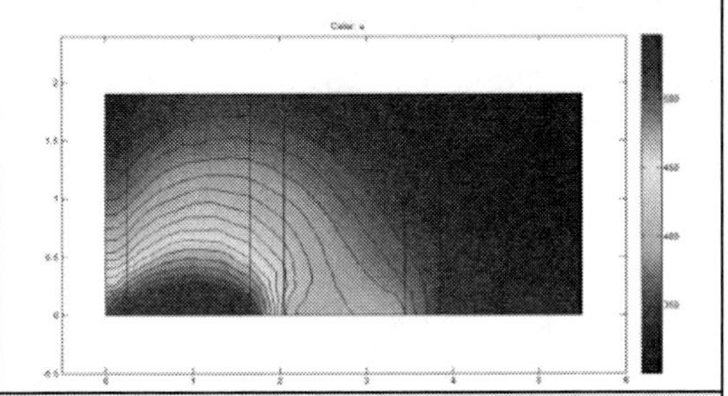

Fig. 15. Temperature field with temperature isolines in the Oxz plane at t = 3600 s and Tmax= 3273K

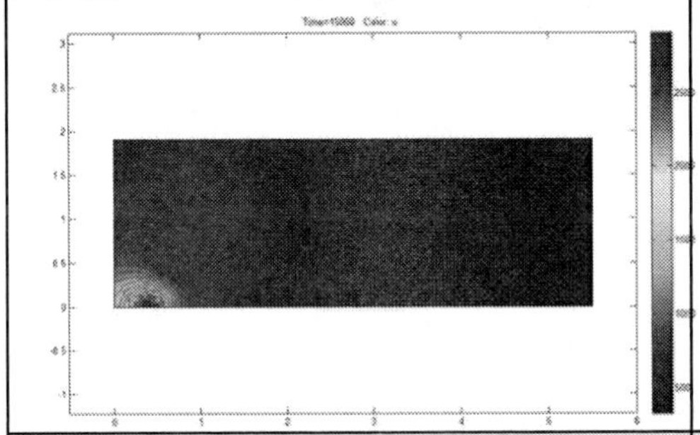

Fig. 13. Temperature field with temperature isolines in the plane Oxy at t =15000 s and Tmax= 3273K

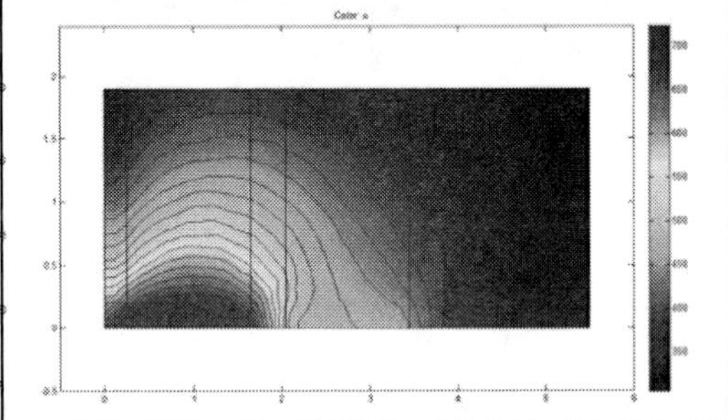

Fig. 16 Temperature field with temperature isolines in the Oxz plane at t = 15000 s and Tmax= 3273K

Figure 17 shows the temperature as a function of time (at t = 300 s, t = 3600 s, t = 15000 s) in the Oxy plane in the PQ section, which is the initial and boundary condition of the mathematical problem of studying the heat exchange of the pyrolysis process in the Oxz plane if Tmax= 3273K .

If we compare the maximal temperatures at the bottom of the most heated camera (the left one) from Fig.10 and Fig.17 for t = 15000 s one will see that in the first case T* = 680^{0}C (Fig.10) while in the second case T* = 750^{0}C (Fig.17).

 https://doi.org/10.1051/matecconf/201814503016

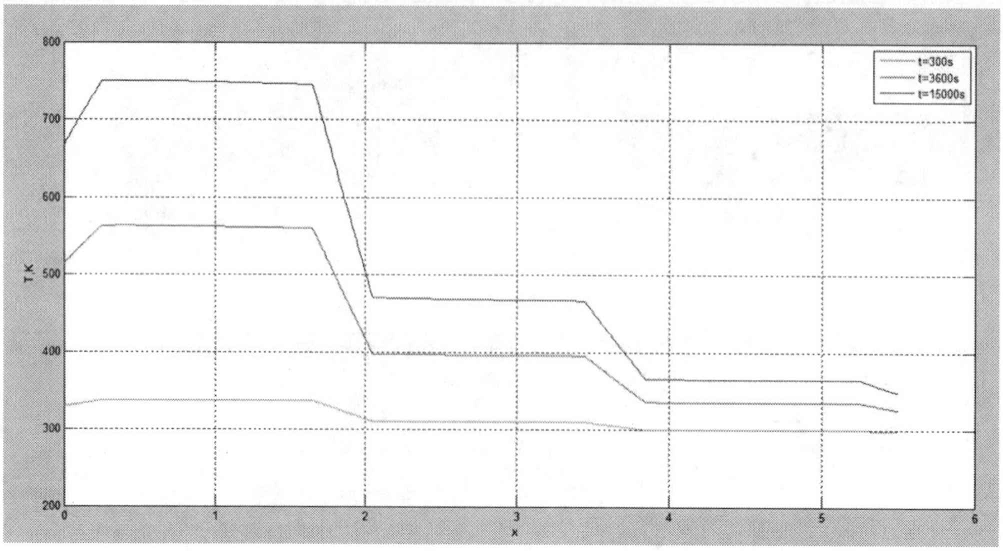

Fig. 17. Graph of the temperature as a function of time T*(t, x) (at t = 500 s, t = 3600 s, t = 15000 s) in the Oxz plane in the PQ section, which is the initial and boundary condition of the mathematical problem of studying the heat exchange of the pyrolysis process in the Oxz plane

Figure 18 shows the temperature at t = 15000 s on the upper end of the cameras (z = H) for the two heating regimes with T_{max}= *2773 K* – red line and with T_{max}= *3273 K* – blue line. The maximum temperature reached for z = H in the first case is T = 370^{0}C and in the second case is T* = 388^{0}C.

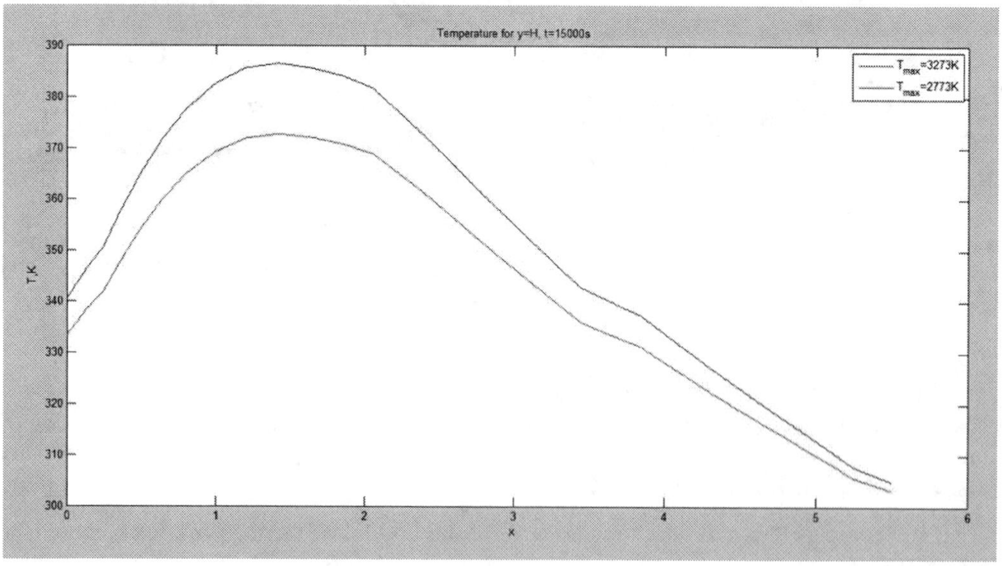

Fig. 18 Temperatures for z = H, t = 15000 s for the two heating regimes with T_{max}= *2773 K* – red line and with T_{max}= *3273 K* – blue line.

5 Conclusions

An opportunity to solve the ecological problems related to the accumulation of large quantities of end-of-life tires is pyrolysis which is a complex process of thermal destruction. It is conducted usually in chemical reactors where the tires are heated to a certain temperature under which they are decomposed into several fractions - liquid, solid and gaseous. After the initial heating, the process is continued for a certain period of time, after which the reactor is allowed to cool. The liquid and solid products of pyrolysis are then removed and thus a pyrolysis cycle is completed. The gaseous fraction is a fuel mixture and usually is returned in the pyrolysis reactor for its heating.Within this technology a significant economic effect can be achieved because solid, liquid and gaseous substances received by pyrolysis are useful and the amount of the EOLT is reduced.

The management of pyrolysis stations for treatment of end-of-life tires in Republic of Bulgaria is predominantly manual and therefore not efficient. This necessitates the theoretical examination of the temperature regimes in that kind of stations, which determines the topic of the present paper.

Using an adequate mathematical model and a numerical algorithm for studding non-stationary heat transfer during pyrolysis used for the treatment of end-of-life tires [1] we explore the heating of pyrolysis process. Two different heating ways were described and tested. First was with $T_{max}= 2773\ K$ and second with $T_{max}= 3273\ K$. Results for temperature fields were obtained and commented. They can be used to more accurately choice of the temperature measurement position in the existing pyrolysis station and thus to improve its management.

This paper contains results of the work on project No 2017 – AIF - 3 financed by „Scientific Research Fund" of Ruse University.

References

1. I. Zheleva, I. Georgiev, M. Filipova, D. Menseidov, Mathematical Modelling of the Heat Transfer during Pyrolysis Process Used for End-of-Life Tires Treatment, AMiTaNS AIP Conference Proceedings (2017) (in print).

Efficient control of mechatronic systems in dynamic motion tasks

Desislava Despotova, *Petko* Kiriazov*, and *Stefan* Karastanev

Institute of Mechanics, Bulgarian Academy of Sciences, Acad. G. Bonchev Str., bl.4, BG-1113 Sofia, Bulgaria

Abstract. Robots and powered exoskeletons have often complex and non-linear dynamics due to friction, elasticity, and changing load. The proposed study addresses various-type robots that have to perform dynamic point-to-point motion tasks (PTPMT). The performance demands are for faster motion, higher positioning accuracy, and lower energy consumption. With given motion task, it is of primary importance to study the structure and controllability of the corresponding controlled system. The following natural decentralized controllability condition is assumed: the signs of any control input and the corresponding output (the acceleration) are the same, at least when the control input is at its maximum absolute value. Then we find explicit necessary and sufficient conditions on the control transfer matrix that can guarantee robust controllability in the face of arbitrary, but bounded disturbances. Further on, we propose a generic optimisation approach for control learning synthesis of various type robotic systems in PTPMT. Our procedure for iterative learning control (LC) has the following main steps: (1) choose a set of appropriate test control functions; (2) define the most relevant input-output pairs; and (3) solve shooting equations and perform control parameter optimisation. We will give several examples to explain our controllability and optimisation concepts.

1 Introduction

Controlled mechanical systems (CMS) are various-type robotic systems [1,2], vehicles/platforms with active suspension [3,4], and other engineering structures with active vibration/shape control. CMS have to be considered as functionally directed compositions of mutually influencing subsystems: control, actuator, structural, and sensor subsystems. Such systems have highly complex dynamics and advanced conceptual frameworks are needed that consider at the same time the problems of full dynamic modelling, optimal system design, accurate parameter identification, and optimal robust control. There are various design tasks for controlled mechanical systems (CMS), where continuously increasing demands for higher speed, better motion accuracy, and reduced energy consumption are to be satisfied. In order to achieve such complicated performance optimization, it is very important to study their controllability and find efficient solutions for the control-related, design optimization problems. Our intention is to present novel concepts and criteria for design optimization of CMS having, as usual, decentralized controllers at the lowest (joint) level. The following natural decentralized controllability

* Corresponding author: kiriazov@imbm.bas.bg

condition is assumed: the signs of any control input and the corresponding output (the acceleration) are the same, at least when the control input is at its maximum absolute value.

In optimal feedback control problems, one has in principle to optimize between error and energy for finding the most appropriate control gains. In the well-known control methods using linear-quadratic or H-infinity optimization functionals [4], the weight matrices are critical towards the required system performance. Choosing their coefficients only by trial-and-error, however, may not guarantee achievement of the required motion accuracy within the existing control power limits. Moreover, optimality and robustness cannot be guaranteed if a control design method does not take into account a realistic estimate for modelling errors and system disturbances, [5].

In this paper we present a generic design and control optimization approach, which has the potential to overcome any of the above problems in an efficient way. CMS of various-type can be considered with the premise that they can be equipped with a sufficiently large number of actuators. Our concept is that system performance that is optimal and robust to arbitrary-in-place external disturbances is hard to provide if the actuators' number is smaller than the number of the controlled performance variables.

The first step is to find a dynamic model relating the control inputs and the controlled outputs which is suitable for both purposes: accurate identification and robust control design. To do that, we can apply the so-called multibody system (MBS) approach: the mechanical structure of CMS can be approximated by a composition of rigid bodies connected by joints, actuators, springs, and dampers [6]. Then we find explicit necessary and sufficient conditions on the control transfer matrix that can guarantee robust controllability in the face of arbitrary, but bounded disturbances. Thus the design optimization process has to involve, besides the basic strength/load capacity criterion, additional design relations for optimal trade-off between the bounds of disturbances and the control force limits [7,8]. Further on, we propose a generic optimisation approach for control learning synthesis of various type robotic systems in dynamic point-to-point motion tasks (PTPMT) Our procedure for iterative learning control (LC) has the following main steps: (1) choose a set of appropriate test control functions; (2) define the most relevant input-output pairs; and (3) solve shooting equations and perform control parameter optimisation. We will give several examples to explain our controllability and optimisation concepts.

2 Controllability of multi-joint dynamics

First, it is very important to properly define the controls, the performance variables to be controlled, and the structure of the input-output relations, [9]. We have to find nominal dynamic models that are appropriate for both purposes, accurate parameter identification [10] and robust control design, [11]. As said above, any mechanical system can be approximated, in general, by a composition of rigid bodies connected by joints, springs, dampers, and driving forces. Applying the Lagrange formalism, the dynamic performance of such a MBS can be described by

$$\ddot{\mathbf{q}} = \mathbf{M}^{-1}(\mathbf{q})(\mathbf{Bu} - \mathbf{C}(\mathbf{q}, \dot{\mathbf{q}}) + \mathbf{g}(\mathbf{q})), \tag{1}$$

where $\mathbf{q}$ is the vector of performance variables/controlled outputs, $\mathbf{M}$ is the inertia matrix, $\mathbf{C}$ is the vector of velocity forces, $\mathbf{g}$ stands for friction and gravitation forces, matrix $\mathbf{B}$ represents the actuator locations/gains, and $\mathbf{u}$ is the vector of control inputs (forces, torques).

Thus for MBS with elastic elements, dynamic model structures can be described by a relatively small number of parameters. All coefficients in (1) can be identified in a systematic way as proposed in [10].

In principle, using both feedforward and feedback control functions is needed for optimal performance to be achieved. With a feedforward control, we can compensate, to some extent, for the gravitation, stiffness and other forces. As usual, we have decentralized controllers at the lowest (joint) level and the control magnitudes $|\mathbf{u}_i|$ should fulfil the following natural condition for independent joint controllability:

$$\operatorname{sign} \ddot{\mathbf{q}}_i = \operatorname{sign} \mathbf{u}_i, \quad \text{if} \quad |\mathbf{u}_i| = \bar{\mathbf{u}}_i \ , \ i = 1,...,n \ . \tag{2}$$

For the purpose of feedback control design, the following simple, but sufficiently general, model for the error dynamics can be used:

$$\ddot{\mathbf{e}} = \mathbf{A}(\mathbf{q})\mathbf{u} + \mathbf{d} \ , \tag{3}$$

where: $\ddot{\mathbf{e}} = \mathbf{q} - \mathbf{q}^{ref}$; vector $\mathbf{d}$ stands for the model imprecision and the uncompensated terms as well as for the measurement and environment noises; $\mathbf{A} = \mathbf{M}^{-1}(\mathbf{q})\mathbf{B}$ is the control transfer matrix which will play a central role in the next control design considerations.

A necessary and sufficient condition for the dynamic system (1) to be robustly controlled by a decentralized controller is that matrix **A** be generalized diagonally dominant (GDD), [11]. It can be easily verified that any mechanical system with two degrees of freedom has control transfer matrix which is GDD.

$$c_i \mathbf{A}_{ij} > \sum_{j \neq i} c_j |\mathbf{A}_{ij}| \ , \text{ for some } c_i > 0, i = 1,...,n \ . \tag{4}$$

With **A** being GDD, the non-negative matrix theory [12] states that there always exists a positive vector $\bar{\mathbf{u}}$ of control magnitudes that solve the following system of equations

$$\mathbf{A}_{ii}\bar{\mathbf{u}}_i - \sum_{j \neq i} |\mathbf{A}_{ij}| \bar{\mathbf{u}}_j = \bar{\mathbf{d}}_i \ , \tag{5}$$

where $\bar{\mathbf{d}}_i$ are some upper bounds for the disturbances.

Eqs. (5) present optimal trade-off relations between the bounds of model uncertainties and the control force limits. The greater the determinant Δ of this system of linear equations, the less control forces required to overcome the disturbances. Therefore, the index Δ can be used to analyze, measure, and optimize the independent joint controllability.

3 Examples on how controllability can be improved

3.1 Two-link manipulator/exoskeleton

Consider a two-degree-of-freedom manipulator and assume, for simplicity, that the masses m_1 and m_2 of the links are concentrated at their ends, Fig. 1.

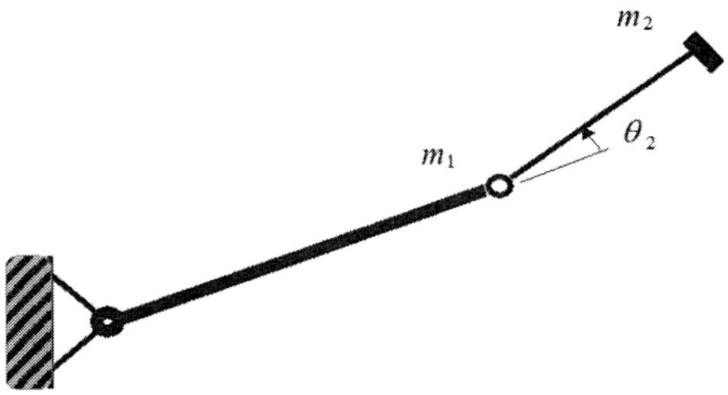

Fig. 1. Manipulator with rotational joints

The inertia matrix coefficients read as follows

$$\mathbf{M}_{11} = (m_1 + m_2)l_1^2 + m_2 l_2^2 + 2m_2 l_1 l_2 c \tag{6}$$

$$\mathbf{M}_{12} = m_2 l_2^2 + m_2 l_1 l_2 c, \mathbf{M}_{22} = m_2 l_2^2 \ , \tag{7}$$

where l_1 and l_2 are the links' lengths and $c = \cos\theta_2$. Then,

$$\Delta = (l_1^2 l_2^2 (m_1 m_2 + m_2^2 s^2))^{-1}, s = \sin\theta_2 . \tag{8}$$

We can see that, with given mass of the first link m_1, the less the second link's mass m_2, the more capable for robust decentralized controllability of such manipulation systems are.

3.2 Planar manipulator with three rotational joints

The GDD-conditions, though not so restrictive, may not be satisfied in this case. Considering this linkage system in the worst (as regards the inertia interaction) case of stretched-out configuration, we have found that

(a) when the manipulator links are of equal length, the GDD condition for the inverse inertia matrix will be satisfied if the link masses strongly decrease in progressing from the base to the tip:

$$m_1 / m_3 - 16 m_3 / m_1 > 16 \ ;$$

(b) if the link lengths decrease from the base to the tip, e.g., $l_1 = 2l_2 = 4l_3$ Fig. 2 , the decrease of link masses is not so strong as in case (a):

$$4m_1 / m_3 - 8m_3 / m_1 - 9m_2 / m_1 > 14 .$$

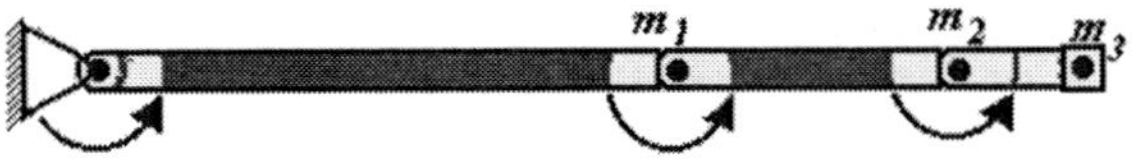

Fig. 2. Three-link manipulator

Thus the stronger the decrease of link masses from the base to the tip, the better the capability of the manipulation systems (with either two or three joints) for independent joint controllability.

3.3 Upright posture stabilisation

This case-study is an extension of the study presented in [13], and now we consider a human body with assistive exoskeleton/exosuit. Unlike the usual adoption for ankle and hip control strategy, our proposal, from the controllability point of view, is that the main controlled joints are to be the ankle and the lumbar spine ones, Fig. 3a. By fixing (freezing) the hip joints and considering the pelvis and the legs as one (first) segment, the so-considered two-link manipulator system (hinged to the ground), Fig. 3b, [14], has much better controllability.

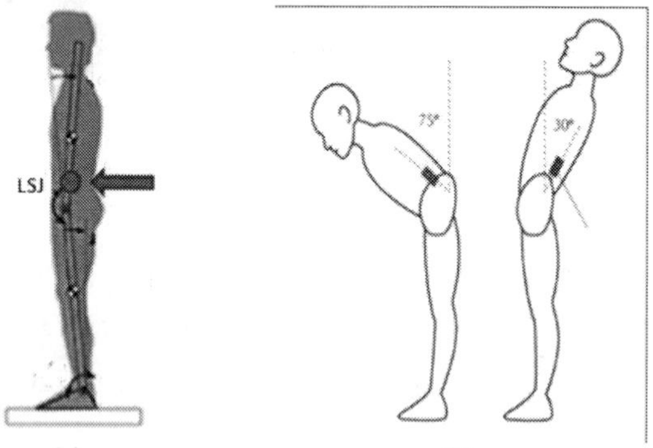

(a) (b)

Fig. 3. Upright posture stabilization (LSJ is lumbar spine joint)

Thus, in upright postural control, human (with assistive device) has to take a strategy change from ankle to lumbar spine in response to increasing perturbation. Further on, our study can help designing robust balancing controller mimicking human control adaptation.

4 Learning control and its optimisation

The usual performance criteria for goal-directed movements are: the positioning accuracy, the movement execution time and the energy consumption. Once the control structure is defined, the next main stage in the control design is how to learn and optimize the specific control functions with a given point-to-point motion task. The main steps in our procedure for iterative motor control learning and optimization are as follows

1. Choose a set of appropriate test control functions: The term "appropriate" concerns the structure, shape, and bounds of the test control functions. Simple linear-spline control functions of "bang-pause-bang" or "bang-slope-bang" shape can be used for a first-order optimization. Parameters, describing the test control functions, like switch times, slopes, or pause lengths, mostly influence the reached position and the performance criteria.
2. Define the most relevant pairs of control parameters and control outputs: It means that for each controlled output we have to assign a control input, which mostly influences it. Regarding the controllability in multijoint tasks, the input-output pairs will be most relevant if couplings between these single-input single-output subsystems are to be as small as possible.

3. Solve shooting equations and perform control parameter optimization: With the above input-output pairing, the given goal-directed motion task is transformed into systems of learning parameter equations with respect to the switch times. They are solved applying natural bisection algorithms and this is the first level of our control synthesis procedure. At the next level, the rest of the control parameters are varied to optimize the movement execution time and the energy consumption.

Following the above approach, we can find satisfactory suboptimal solutions with minimum number of control parameters to be learnt. Existence of feasible solutions and convergence of the bisection algorithms are guaranteed, which ensures feasibility and efficiency of our control learning procedure, [15]. This approach has been successfully applied to dynamics models of various-type CMS, e.g., planar manipulators, Fig. 4 and Fig. 5.

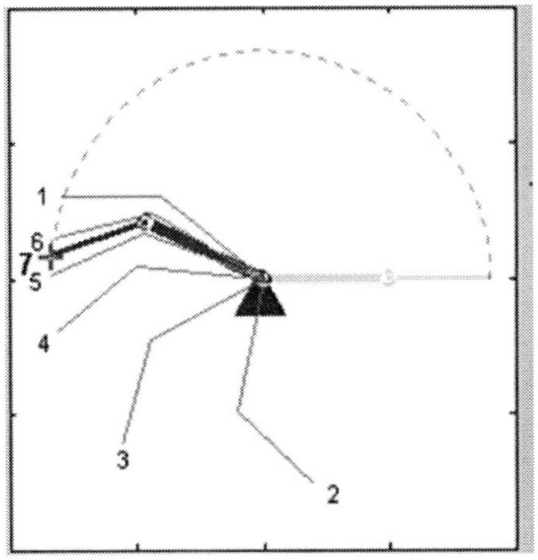

Fig. 4. Fast learning to reach a target

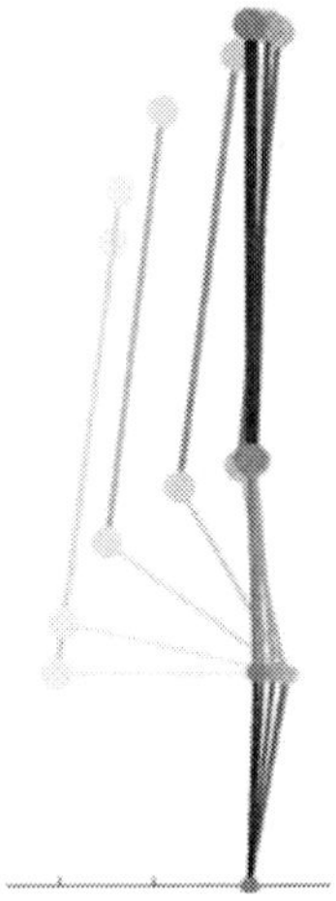

Fig. 5. Standing-up motion task

In case of wearable robotics, such movements can be generated by the corresponding muscles and/or by robotic orthoses (exoskeletons). The excitation signals naturally come from the brain and they may be processed via brain-computer interface (BCI) and/or electro-myographical (EMG) devices.

5 Conclusion

We have proposed a generic conceptual framework for structure identification, analysis and design of control functions to achieve the best possible human/robot motion. It is based on underlying principles and advanced concepts from optimal control theory, neurophysiology, and controlled multi-segmental dynamics.

Recent studies at leading R&D centres for neurological rehabilitation say that three important principles must be observed: assistance-as-needed, embodied motor control and learning, and motivate patients for self-motor learning. Our control concepts and methodology for control design and optimal control learning are in accordance with these principles! The proposed concepts and algorithms could be very useful in designing and controlling arm/leg prostheses as well as in neuro-muscular rehabilitation of disabled people. Our current research intention is to extend the proposed approach to meet the next challenge: optimal control synthesis of 3D humanoid locomotion to produce efficient and naturally looking walking.

References

1. J. Slotine, S. Shastry, Tracking control of nonlinear systems using sliding surfaces with application to robotic manipulators, I J Control, **2,** 465-492 (1983)
2. P. Kiriazov, Robust Decentralized Control of Mechanical Systems, In: *Solid Mechanics and its Applications*, **52**, (Ed. D. van Campen, Kluwer Acad. Publ., 175-182, 1997)
3. S. Riebe, H. Ulbrich, Model-based vibration isolation of a hexapod-system using a combined feedforward-feedback control concept, *Proc. IEEE Conf. on Mechatronics and Robotics*, 400-405 (2004)
4. R. Amirifar, N. Sadati, A low-order H-infinity controller design for an active suspension system via linear matrix inequalities, J. Vibr. and Cont, **10(8),** 1181-1197 (2004)
5. L. H. Keel, S. P. Bhattacharyya, Robust, fragile, or optimal? IEEE Trans. on Autom. Cont., **42(8)** 1098-1105 (1997)
6. W. Schiehlen, *Multibody Systems Handbook* (Berlin, Springer, 1990)
7. P. Kiriazov, Necessary and sufficient condition for robust decentralized controllability of robotic manipulators, *American Control Conference*, 2285-2287 (1994)
8. P. Kiriazov, Optimal robust controllers for multibody systems, *Solid Mech. and its Applications*, **30**, (Ed. H. Ulbrich and W. Günthner, Springer-Verlag GmbH, 183-192, 2005)
9. S. Skogestad, I. Postlethwaithe, *Multivariable Feedback Control: Analysis and Design*, (Wiley, 1996)
10. P. Kiriazov, Efficient approach for dynamic parameter identification and control design of structronic systems, *Solid Mechanics and its Applications*, **89** (Eds. U. Gabbert and H. Tzou, Kluwer Publ, 323-330, 2001)
11. P. Kiriazov, Optimal robust controllers for multibody systems, *Solid Mechanics and its Applications*, **130**, (Ed. by H. Ulbrich and W. Günthner, Springer-Verlag GmbH, 183-192, 2005)
12. J. Lunze, *Feedback Control of Large-Scale Systems* (Prentice Hall, UK, 1992)
13. P. Kiriazov, Humanoid robots: how to achieve human-like motion, *Humanoid 2003*, 1-15, (2003)
14. https://design.tutsplus.com/articles/human-anatomy-fundamentals-flexibility-and-joint-limitations--vector-25401
15. K. Kiryazov , P. Kiriazov, Efficient learning approach for optimal control of human and robot motion, *Emerging Trends in Mobile Robotics* (Eds: H. Fujimoto, M O Tokhi, H. Mochiyama, G. S. Virk, 1219-1227, 2010)

A multiphase flow in the antroduodenum: some results of the mathematical modelling and computational simulation

Marat Kamaltdinov[1,*], *Peter* Trusov[2] , and *Nina* Zaitseva[1]

[1]Federal Budget Scientific Institution "Federal Scientific Center for Medical and Preventive Health Risk Management Technologies", Department of Mathematical Simulation of Systems and Processes, 82 Monastyrskaya street, 614000, Perm, Russia
[2] Perm National Research Polytechnic University, Department of Mathematical Simulation of Systems and Processes, 29 Komsomolsky Prospect, 614000, Perm, Russia

Abstract. This work is devoted to a problem of creation of a multiphase flow model in an antroduodenum that includes the lower part of the stomach and the upper part of the small intestine (duodenum). The mathematical model is developed for the next purposes: description of the process of digestion in normal physiological state and with functional disorders, prediction of the flow characteristics in distinct conditions. The antroduodenum is considered as a complex form canal with moving boundaries. The mathematical statement of the problem includes mass and momentum conservation equations for phases and components as well as initial and boundary conditions with peristaltic movement. To describe secretion and absorption processes we use mass sources/sinks in a layer adjacent to the tract wall. Using computational simulation we got the digestion processes characteristics in dynamic: pH of the medium, components and phases velocities and fractions. In some scenarios with secretory and motor disorders we identified zones of abnormal acidity. The results of the simulation showed that motor functionality of the antrum and pyloric sphincter affect the stomach evacuation. Additionally, the chemical absorption rate is analyzed in scenarios with distinct food parameters (size, viscosity, density).

1 Introduction

Diseases of the gastrointestinal tract (GIT) hold high position in the structure of digestive system pathologies. Note there was a percentage increase of global incidence of gastritis and duodenitis (18.3%) and peptic ulcer diseases (4.8%) between 2005 and 2015 [1]. Although the prevalence of atrophy in the stomach and the incidence of peptic ulcer disease decrease in some countries [2, 3], we observe reports that the prevalence of atrophic corpus gastritis and duodenitis is still high. It is about from 5 to 12% for distinct countries and age groups [4-6]. Additionally, it is well known that the chemicals found in drinking water and food can increase health risk [7, 8].

* Corresponding author: kmr@fcrisk.ru

For a deeper understanding the pathological processes we are working on a multilevel mathematical model which describes accumulation of damages in a human body, associated with environmental exposure [9]. The upper level of the model deals with averaged interaction of body systems. For more detailed description we considers "meso-level" models of the cardiovascular, respiratory [10], digestive, neuro-endocrine [11], and other systems.

This work is devoted to the problems of development of the antroduodenum model, that details damages to antroduodenum caused by acid and other chemicals from the tract cavity. CFD (Computation Fluid Dynamics) models meet these purposes, allowing to assess and predict flow characteristics at any point of simulation domain. Nowadays models of fluid flow in the esophagus [12, 13], stomach [14, 15] and bowel [16-18] are developing rapidly. Available models of the stomach or duodenum mainly account for the peristaltic waves in the antrum [19, 20] and gut [21]. Researchers mostly consider single-phase flow without biochemical reactions, secretion of the digestive glands and functional disorders of the tract. Only separate interesting aspects are considered in some cases, for example, diffusion of mixture component (pepsin) in the stomach lumen [22], transfer of the mixture through the pyloric sphincter [23], pouch emptying of a patient after gastric surgery [24]. Xue et. al showed that particle phase had a sufficient impact on the flow characteristics [25], but the mass exchange between phases due to the food dissolution were not considered. Also, available literature does not address CFD models describing acidity in the tract. Thus, it is necessary to continue development of the flow models in the gastrointestinal tract. Particularly, we should pay attention to individual patient parameters and diseases.

For a more detailed description we presented a multiphase flow model in the 3D antroduodenum allowing for the mass sources/sinks due to particle dissolution, reactions between components, mixture components secretion and absorption, peristaltic movement, functional disorders of the tract [26]. The antroduodenum is a section of the gastrointestinal tract that includes the lower part of the stomach (antrum), the upper part of the small intestine (duodenum) and the pyloric outlet that divides antrum and duodenum. The mathematical model is developed for the next purposes: description of the process of digestion in a normal physiological state and with functional disorders, prediction of the flow characteristics in the distinct conditions. In this paper we consider some variants of the model application and results of computational simulation.

2 Conceptual statement

We consider a digestion stage when stomach evacuation begins and the temperature of the multiphase media in the stomach is equal to the body temperature. At first approximation all processes are considered to be isothermal. The first phase is a multicomponent liquid with components dissolved at the molecular level: water, hydrochloric acid, sodium bicarbonate, the reaction product with hydrochloric acid and sodium bicarbonate (carbon dioxide and sodium chloride), dissolved proteins, fats, carbohydrates, chemicals, polypeptides, and pepsin. Food particles of different sizes are considered as separate liquid phases and dissolved under the action of hydrochloride acid. Besides, the model takes into consideration an acid neutralization reaction with sodium hydrogen carbonate and enzymatic reaction of pepsin and protein.

The antroduodenum is considered as a complex form canal with moving boundaries. Antroduodenum motility is assumed to be periodical in the model; it allows for peristaltic waves in the antrum and duodenum as well as for pyloric sphincter contraction. Wave processes parameters are thought to be independent of tract contents properties. To describe the secretion and absorption processes we use mass sources in a layer adjacent to the tract wall. When describing sodium bicarbonate secretion, we allowed for functionality of control signalling through neuro-endocrine mechanisms. Receptors are assumed to determine acidity level in the stomach body and give signals to change intensity of sodium bicarbonate mass source in the antroduodenum.

We consider only absorption of chemicals in the antroduodenum because nutrients absorption is not particular for this area of the tract. The amount of chemicals absorbed into blood is proportional to the difference between concentration at the border of the tract cavity and the blood concentrations. Absorption rate does not depend on the flow characteristics and it is defined by physicochemical properties of the toxicant.

We divided all modelling area into four sections: acid and alkaline secretion zones in the stomach, pyloric outlet with duodenum and area of the inflow of the pancreatic juice. Each l-th section of the tract wall has its own damage $D_{(l)(m)} \in [0;1]$ in terms of accomplishing m-th function. Functionality of the tract wall section is related to the damage by the following formula: $F_{(l)(m)} = 1 - D_{(l)(m)}$, functionality that is different from 1, is shown by a quantitative reduction of the secretory $F_{(l)(1)}$, absorption $F_{(l)(2)}$ and motor $F_{(l)(3)}$ tract functions. We assume that when disorders in secretory function of the stomach body occur it results in a decrease of the intensity of hydrochloric acid mass source. From the medical point of view, hyposecretion can be a sign of atrophic gastritis with various severity degree [27]. We don't deal with acid hypersecretion as it can rather be caused by regulation disorders while decreased secretion can result from damage to secretory cells. Similarly, the model considers only decrease in the intensity of sodium bicarbonate mass source of the antrum, duodenum, pancreas. Disorders in motor function become apparent via decrease in peristaltic waves amplitude in the antrum and duodenum, or in weaker contraction of the pyloric sphincter which can be caused by weakness of muscles in the circular layers of the tract walls. Other motility disorders, for example, changes in contractions periodicity are not considered. Disorders in absorption function reduce income rate to the blood because chemicals can change wall properties of the tract. It is one of the tolerance mechanisms.

3 Mathematical statement

The mathematical statement of the problem includes mass and momentum conservation equations for the phases and components as well as initial and boundary conditions with peristaltic movement.

Generally advection-diffusion equations for the first phase components can be given as:

$$\frac{\partial}{\partial t}(\rho_{(1)}\alpha_{(1)}Y_{(i)}) + \nabla \cdot (\rho_{(1)}\alpha_{(1)}\mathbf{v}_{(1)}Y_{(i)}) = -\nabla \cdot \mathbf{J}_{(i)} + R_{(i)} + S_{(i)} + \sum_j m'_{(j)(i)}, \qquad (1)$$

$$\mathbf{r} \in \overline{\Omega},\ t \in [0;T),\ i = \overline{0,I},\ j = \overline{2,J},$$

where i is a component index of the first phase; j is a phase index, index $j = 1$ corresponds to the first phase which is a multicomponent liquid, index $j > 1$ corresponds to phases of food particles with different sizes, which in first approximation are described with a model of Newtonian liquid with different viscosity; $\rho_{(j)}$ is a density of the j-phase, kg/m^3; $\alpha_{(j)}$ is a volume fraction of the j-phase; $Y_{(i)}$ is a mass fraction of the i-component of the first phase; ∇ is a nabla operator; $\nabla\cdot$ is a divergence operator; $\mathbf{v}_{(j)}$ is a velocity of the j-phase, m/s; $\mathbf{J}_{(i)}$ is an intensity of mass flow of the first phase i-component due to diffusion processes, *kg*/(m^2·s); $R_{(i)}$ is an intensity of mass sources of the first phase i-component due to reactions between components, *kg*/(m^3·s); $S_{(i)}$ is a mass sources intensity of the i-component due to secretion/absorption, *kg*/(m^3·s); $m'_{(j)(i)}$ is an augend in mass balance equation, determining the transition intensity from j-phase into the first phase i-component, *kg*/(m^3·s); t is an independent variable (time), s; Ω is an interior of the whole area; $\partial\Omega$ is the area boundary; $\overline{\Omega} = \Omega \cup \partial\Omega$ is the area closure (area interior and its boundary); $\partial\Omega_{(l)}$ is l- boundary of area, $l = \overline{1, L}$; we introduce secretion as mass sources in layer adjacent to the tract wall, so we define $\Omega_{(l)}$ as internal layer adjacent to the l-boundary, $l = \overline{1, \mathrm{L}}$.

The mass conservation equations for the liquid phases of the food particles in general terms can be expressed as:

$$\frac{\partial}{\partial t}(\rho_{(j)}\alpha_{(j)}) + \nabla \cdot (\rho_{(j)}\alpha_{(j)}\mathbf{v}_{(j)}) = -\sum_i m'_{(j)(i)} + m''_{(j-1)(j)} - m''_{(j)(j+1)}, \tag{2}$$

$$\mathbf{r} \in \overline{\Omega},\ t \in [0;T),\ i = \overline{0, I},\ j = \overline{2, J},\ m''_{(1)(2)} = m''_{(J)(J+1)} = 0,$$

where $m''_{(j-1)(j)}$ is an augend in mass balance equation, determining the transition intensity from (j-1)-phase into j-phase, *kg*/(m^3·s).

The system of mass conservation equations is complemented by the following 2 equations:

$$\sum_i Y_{(i)} = 1, \sum_j \alpha_{(j)} = 1,\ i = \overline{0, I},\ j = \overline{1, J}. \tag{3}$$

The momentum-conservation equations can be written as:

$$\begin{aligned} &\frac{\partial}{\partial t}(\alpha_{(1)}\rho_{(1)}\mathbf{v}_{(1)}) + \nabla \cdot (\alpha_{(1)}\rho_{(1)}\mathbf{v}_{(1)}\mathbf{v}_{(1)}) = -\alpha_{(1)}\nabla p + \nabla \cdot \boldsymbol{\tau}_{(1)} + \\ &+\alpha_{(1)}\rho_{(1)}\mathbf{g} + \sum_j K_{(j)(1)}(\mathbf{v}_{(j)} - \mathbf{v}_{(1)}) + \sum_j (\sum_{i=3,4,5,6,9} m'_{(j)(i)}\mathbf{v}_{(j)}), \end{aligned} \tag{4}$$

$$\begin{aligned} &\frac{\partial}{\partial t}(\alpha_{(j)}\rho_{(j)}\mathbf{v}_{(j)}) + \nabla \cdot (\alpha_{(j)}\rho_{(j)}\mathbf{v}_{(j)}\mathbf{v}_{(j)}) = -\alpha_{(j)}\nabla p + \nabla \cdot \boldsymbol{\tau}_{(j)} + \alpha_{(j)}\rho_{(j)}\mathbf{g} + \\ &+\sum_q K_{(q)(j)}(\mathbf{v}_{(q)} - \mathbf{v}_{(j)}) - (\sum_{i=3,4,5,6,9} m'_{(j)(i)})\mathbf{v}_{(j)} + m''_{(j-1)(j)}\mathbf{v}_{(j-1)} - m''_{(j)(j+1)}\mathbf{v}_{(j)}, \end{aligned} \tag{5}$$

$$\mathbf{r} \in \Omega, t \in [0;T), j = \overline{2, J}, q = \overline{1, J}, m''_{(1)(2)} = m''_{(J)(J+1)} = 0,$$

$$\boldsymbol{\tau}_{(j)} = \alpha_{(j)}\eta_{(j)}[\nabla\mathbf{v}_{(j)} + (\nabla\mathbf{v}_{(j)})^T],\ j = \overline{1, J},\ \mathbf{r} \in \overline{\Omega}, \tag{6}$$

where p is a pressure, Pa; $\boldsymbol{\tau}_{(j)}$ is a deviatoric part of Cauchy stress tensor of the j-phase, Pa; $\mathbf{g}$ is a vector characterizing mass force impact, m/s^2; $K_{(j)(q)}$ is a coefficient of interphase interaction j and q-phase, *kg*/(m^3·s), $\eta_{(j)}$ is a shear viscosity of the j-phase, Pa·s.

We pass by detailed description of mass sources terms in order not to overload the paper. Kinematic boundary conditions on the antroduodenum walls and conditions of components zero flow through the boundary are set allowing for peristaltic motion (no slip condition):

$$\mathbf{v}_{(j)}(t,\mathbf{r}) = \mathrm{d}\mathbf{r}(t)/\mathrm{d}t, \quad \frac{\partial Y_{(i)}}{\partial n} = 0,$$
$$i = \overline{0,I}, \quad j = \overline{1,J}, \quad t \in [0;T), \quad \mathbf{r}(t) \in \partial\Omega_{(l)}, \quad l = \overline{2,5}, \tag{7}$$

where $\mathbf{r}(t) \in \partial\Omega_{(l)}$ is a radius vector of material point on the antroduodenum wall in the gastrointestinal tract. The peristaltic wave amplitude is proportional to the functionality $F_{(l)(3)}$. Values for mass fractions of the first phase components, pressure and conditions for tangent components of stress vector being equal to zero are set on the rest of the examined area boundaries (inlet/outlet sections):

$$Y_{(i)} = Y_{(i)}^{\partial\Omega}, \ \mathbf{n}\cdot\boldsymbol{\sigma}_{(j)}\cdot\mathbf{n} = p, \ \mathbf{t}_{(j)} - (\mathbf{n}\cdot\boldsymbol{\sigma}_{(j)}\cdot\mathbf{n})\mathbf{n} = \mathbf{0}, \ \mathbf{t}_{(j)} = \mathbf{n}\cdot\boldsymbol{\sigma}_{(j)}, \tag{8}$$

$$t \in [0;T), \ \mathbf{r} \in \partial\Omega_{(l)}, \ l = 1,6.$$

Equation system is completed with initial conditions:

$$\alpha_{(j)}(t,\mathbf{r}) = \alpha_{(j)}^{0}, \ Y_{(i)}(t,\mathbf{r}) = Y_{(i)}^{0}, \ \mathbf{v}_{(j)}(t,\mathbf{r}) = \mathbf{v}_{(j)}^{0}, \ \mathbf{r} \in \overline{\Omega}, \ t = 0. \tag{9}$$

Major difficulties related to parameters identification are caused by determining spatially distributed flow characteristics of the gastric content at various digestion stages. At given stage, most of the parameters are determined on the basis of literature data [28-31]. The results discussed in given paper are obtained on the example of an antroduodenal GIT section which is average statistical as per its geometric properties (size and shape). As a result of an ultrasonic study, stomach images were obtained and analyzed. The reconstruction algorithm of three-dimensional area is based on approximation of a real geometrical shape by a finite number of ellipses. Dynamic reconstruction of the computational grid during peristaltic activity is carried out with Fluent solver using the script in C programming language. Motility baseline parameters in the antrum (provided there are no functional disorders) are set as follows: periodicity is 18 s, wave width is 0.02 m, amplitude is 0.011 m. The wave starts in the middle part of stomach body and spreads towards the pyloric outlet at the velocity equal to 0.0022 m/s for 38 s. Waves parameters in the duodenum are as follows: periodicity is 9 s, wave width is 0.04 m, amplitude is 0.0035 m; the wave starts near the pyloric outlet and spreads at the velocity equal to 0.005 m/s for 36 s. The pyloric opening opens and closes for 2 s each 18 s. So, under the set parameters, motility of the antrum, pyloric opening and duodenum is synchronized. The calculated area geometry occasionally changes.

4 Results and discussion

Using computational simulation we got the digestion processes characteristics in dynamic: pH of the medium, components and phases velocities and fractions. We analyzed zones of high acidity, processes of stomach evacuation and chemical absorption in some scenarios.

4.1 Acidity in the tract

We identified zones of abnormal acidity in some scenarios with initially increased acidity (for example, due to acidic drink intake), secretory [32] and motor disorders. On the whole we can note that if there are no functional disorders, sodium hydrogen carbonate neutralizes excessive acidity in antroduodenum more efficiently when a drink in a set volume with pH > 3.5 is consumed. An area with small acidity (from 5 to 7 pH) evolves in near-wall layers of the antrum and duodenum cavity to protect from negative influence. Drinks with pH=2.3 have much greater damaging potential and are much more harmful for the mucous coat of the GIT walls. In this case abnormally low pH levels are observed in the duodenal cap area (pH=3). But if any disorders in alkaline secretion appear the situation becomes substantially worse as an area of potential damage to the tract walls grows. In this case high acidity of a mixture (pH=2.5) is detected near both the concave and convex walls of the duodenum. If there are disorders in the antrum motor function, acidic contents of the stomach transfer into the bowels slower and the duodenum neutralizes excessive acidity.

4.2 Stomach evacuation

We considered five phases of food particles with different sizes. Initial placement of particles phases is in the stomach only. Food particle distribution in the process of crashing in oral cavity is obtained on the basis of Rosin–Rammler equation modification considering functional disorders of the dentofacial system. Parameters of the distribution function were set for carrots (20 chewing cycles, functionality of the chewing equals 1) based on available literature [33]. In case with functional disorders particles sizes are much greater (Table 1). At the beginning of the numerical simulation, the first phase contains only a water component ($\rho_{(1)} = 1000$ kg/m^3, $\eta_{(j)} = 10^{-3}$ Pa·s). The density of food particles equals to 1005 kg/m^3, viscosity is $2{\cdot}10^{-3}$ Pa·s.

Table 1. Sizes of the particles phases.

No.	Medium size, mm	Fraction with dentition functionality = 1, $\alpha^0_{(j)}$	Fraction with chewing functionality = 0.5, $\alpha^0_{(j)}$
j=1	–	0.1	0.1
j=2	3.6	0.066	0.510
j=3	2.8	0.165	0.160
j=4	1.7	0.516	0.195
j=5	0.7	0.132	0.031
j=6	0.2	0.021	0.005

The results of the simulation showed that motor functionality of the antrum and pyloric sphincter affect the stomach evacuation (Table 2). Functional disorders in the antrum are considered at two levels (1 and 0.75) and maximal wave amplitudes are 11 mm and 8.25 mm respectively. Functional disorders in the pyloric outlet are considered at four levels (1, 0.9, 0.5 and 0) which correspond to distinct levels of incomplete closure. In 324 s some mass of the dissolved food components and food particles are observed in the duodenum due to the stomach evacuation process. The particles sized ≥1.7 mm do not evacuate into the bowel, but settle into the lower part of the stomach. This particles need to be dissolved for evacuation to the gut. The obtained results correspond to the available experimental data, some studies have shown that the food particles sized < 2 mm are evacuated during fed motility [34].

Table 2. Results in scenarios with disorders of GIT motor function, t=324 s.

No.	Functional disorders in the antrum motility	Functional disorders in the pyloric outlet motility	Mass of the dissolved food components in the duodenum, *g*	Mass of the particles with average size 0.7 mm in the duodenum, *g*	Mass of the particles with average size 0.2 mm in the duodenum, *g*
1	1	1	0.064	0.762	0.524
2	1	0.9	0.064	0.792	0.538
3	1	0.5	0.073	0.414	0.419
4	1	0	0.072	0.393	0.378
5	0.75	1	0.046	0.077	0.205
6	0.75	0.9	0.045	0.081	0.209
7	0.75	0.5	0.046	0.081	0.215
8	0.75	0	0.046	0.095	0.221

It is important to note that food mass in the duodenum are much lower in scenarios with antrum disorders. First of all, mass of the particles with size 0.7 mm is approximately ten times lower (0.762 and 0.077 *g* in scenarios 1 and 5 respectevily). There is a weak relation of the pyloric outlet motility to the stomach evacuation in scenarios 5–8. Thus, movement of the large particles is more dependent on the antrum motor functionality. When the gastric outlet is opening, the particles sized 0.7 mm transit into the bowel at 0.026 m/s and 0.016 m/s in scenarios 1 and 5 respectevily (Fig. 1).

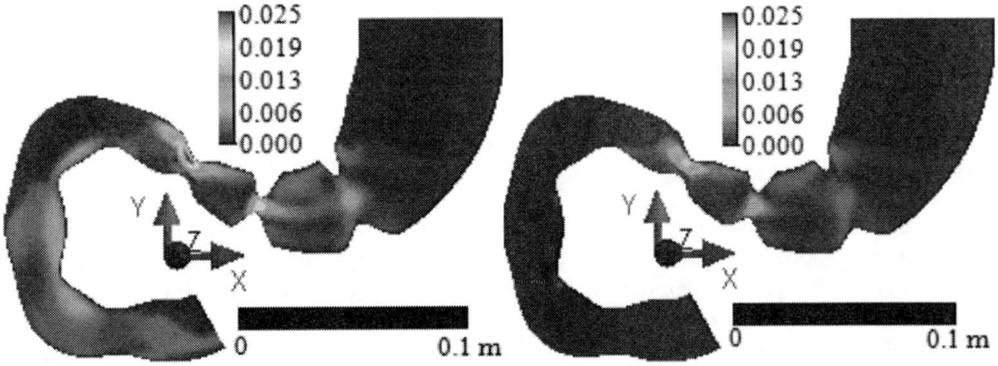

Fig. 1. Absolute values of velocities in the antroduodenum (scenario 1 is on the left, scenario 2 is on the right)

Results showed that mass of the dissolved food components in scenarios 3 and 4 is a little more than in scenarios 1 and 2. Components easily pass through incomplete closure of the pyloric outlet. But mass of the particles is in 1.5–2 times lower in scenarios 3 and 4. Probably, co-operative work of the antrum and pyloruc outlet muscles is needed for succesful particles transition to the duodenum.

4.3 Chemical absorption

The chemical absorption rate is analyzed in scenarios with functional disorders and distinct food parameters (size, viscosity, density). For example, we used lead parameters, the diffusion coefficient is taken as $5.12 \cdot 10^{-10}$ m^2/s, and the permeability coefficient is taken as $3.34 \cdot 10^{-3}$ s^{-1}. Another model parameters are presented in Table 3.

Table 3. Model parameters for analysis of the chemical absorption rate.

No.	Functional disorders in the alkaline secretion	Functional disorders in the antrum motility	Density of particles phases, *kg*/m3	Viscosity of particles phases, Pa·s	Functional disorders in the chewing function
1	1	1	1020	$2 \cdot 10^{-3}$	1
2	1	1	1005	$2 \cdot 10^{-3}$	1
3	1	1	1005	10^{-1}	1
4	0.5	1	1005	$2 \cdot 10^{-3}$	1
5	1	0.75	1005	$2 \cdot 10^{-3}$	1
6	1	1	1005	$2 \cdot 10^{-3}$	0.5

Modelling results showed that maximal absorption rate in the antroduodenum are observed in scenario 4 (up to $4.38 \cdot 10^{-14}$ *kg*/s for a moment time equals 324 s). In this case high acidity in the tract enhance particles dissolution and chemical concentration increases faster. We observed that chemical absorption rate reached $3.94 \cdot 10^{-14}$ *kg*/s in scenario 2 without functional disorders. Functional disorders in the antrum motility, high particles viscosity and high density lead to weaker food mixing and dissolution, therefore the absorption rate reduces. We observed that absorption rate values are $3.16 \cdot 10^{-14}$, $2.83 \cdot 10^{-14}$ and $2.63 \cdot 10^{-14}$ *kg*/s in scenarios 5, 3 and 1 respectively. In scenario 6 we saw minimal values ($1.29 \cdot 10^{-14}$ *kg*/s) for absorption process. Mass fraction of the largest particles in the stomach are very high in this case (more than 50%). These particles need more time to dissolute, therefore lead concentration is relatively low.

5 Conclusion

Thus, given approaches allows to describe the process of the multiphase flow in the antroduodenum both as normal and as under pathological disorders. In this paper we proposed some variants of model application. On the one hand, mathematical modelling can improve our understanding of the digestive processes. On the other hand, usage of individual patient parameters may lead to medical and prophylactic recommendations based on modelling results as in the case of cardiovascular system. This is a future model perspective.

Further identification of the model parameters is the next important step. It is necessary to carry out experimental research on determination of the rheological characteristics of the contents of the digestive tract, also viscosity and density of the media are not constant and must change in the process of the digestion.

The reported study was funded by RFBR according to the research project № 16-01-00126 A

References

1. GBD 2015 Disease and Injury Incidence and Prevalence Collaborators, Lancet **388(10053)**, 1545–1602 (2016)
2. T. Kamada, K. Haruma, M. Ito, K. Inoue, N. Manabe, H. Matsumoto, H. Kusunoki, J. Hata, M. Yoshihara, K. Sumii, T. Akiyama, S. Tanaka, A. Shiotani, D.Y. Graham, Helicobacter **20**(3), 192–198 (2015)
3. H. Malmi, H. Kautiainen, L.J. Virta, N. Färkkilä, J. Koskenpato, M.A. Färkkilä, Aliment Pharmacol Ther **39**(5), 496–506 (2014)
4. L.D. Roman, R. Lukyanchuk, O.A. Sablin, E.I. Araslanova, C. Eklund, P. Hendolin, L. Paloheimo, K. Syrjänen, Anticancer Research **36(8)**, 4129–4138 (2016)
5. A. Alper, S. Hardee, D. Rojas-velasquez , S. Escalera, R.A. Morotti, D.S. Pashankar, J Pediatr Gastroenterol Nutr **62**(2), 314–316 (2016)
6. H. Song, M. Held, S. Sandin, H. Rautelin, M. Eliasson, S. Söderberg, G. Hallmans, L. Engstrand, O. Nyrén, W. Ye, Clin Gastroenterol Hepatol **13**(9), 1592–1600 (2015)
7. R. Kroes, D. Muller, J. Lambe, M. Lowik, J. van Klaveren, J. Kleiner, R. Massey, S. Mayer, I. Urieta, P. Verger, A. Visconti, Food Chemistry and Toxicology **40**, 327–385 (2002)
8. A. Singh, R.K. Sharma, A. Agrawal, F.M. Marshall, Food and Chemical Toxicology **48**(2), 611–619 (2010)
9. P.V. Trusov, N.V. Zaitseva, D.A. Kiryanov, M.R. Kamaltdinov, M.Yu. Tsinker, V.M. Chigvintsev, D.V. Lanin, Mathematical Biology and Bioinformatics **7**(2), 589–610 (2012)
10. P.V. Trusov, N.V. Zaitseva, M.Yu. Tsinker, Mathematical Biology and Bioinformatics **11**(1), 64–80 (2016)
11. N.V. Zaitseva, D.A. Kiryanov, D.V. Lanin, V.M. Chigvintsev, Computational and Mathematical Methods in Medicine **2014** (2014)
12. H. Gregersen, D. Liao, J.G. Brasseur, Ann N Y Acad Sci. **1380**(1), 6–18 (2016)
13. W. Kou, A.P. Bhalla, B.E. Griffith, J.E. Pandolfino, P.J. Kahrilas, N.A. Patankar, Journal of Computational Physics **298**, 446–465 (2015)
14. M.J. Ferrua, R.P. Singh, Current Opinion in Food Science **4**, 116–123 (2015)
15. S. Hao, B. Wang, Y. Wang, European Journal of Pharmaceutical Sciences **70**, 72–81 (2015)
16. L.A. Fullard, W.J. Lammers, M.J. Ferrua, Journal of Food Engineering **160**, 1–10 (2015)
17. Y.F. Lim, C. de Loubens, R.J. Love, R.G. Lentle, P.W.M. Janssen, Food and Function **6**, 1787–1795 (2015).
18. M.D. Sinnott, P.W. Cleary, S.M. Harrison, Applied Mathematical Modelling **44**, 143–159 (2017)
19. M.J. Ferrua, Z. Xue, R.P. Singh Journal of Biomechanics **47(15)**, 3664–3673 (2014)
20. Y. Imai, I. Kobayashi, S. Ishida, T. Ishikawa, M. Buist, T. Yamaguchi, American Journal of Physiology – Gastrointestinal and Liver Physiology **304**(5), G536–G542 (2013)
21. S. De Loubens, R.G. Lentle, R.J. Love, C. Hulls, P.W.M. Janssen, Journal of the Royal Society Interface **10** (2013)

22. H. Kozu, I. Kobayashi, M. Nakajima, K. Uemura, S. Sato, S. Ichikawa, Food Biophysics **5**(4), 330–336 (2010)
23. S. Dillard, S. Krishnan, H. S. Udaykumar, World Journal of Gastroenterology **13**(9), 1365–1371 (2007)
24. F. Gao, D. Liao, J. Zhao, A.M. Drewes, H. Gregersen, Obesity Surgery **18**, 243–250 (2008)
25. Z. Xue, M. J. Ferrua, R.P. Singh, Alimentos Hoy **21**(27), 3–13 (2012)
26. P.V. Trusov, N.V. Zaitseva, M.R. Kamaltdinov, Computational and Mathematical Methods in Medicine **2016** (2016)
27. P.H. Katelaris, F. Seow, B.P. Lin, J. Napoli, M.C. Ngu, D.B. Jones, Gut **34**, 1032–1037 (1993)
28. A.J. Cornish-Bowden, J.R. Knowles, Biochemical Journal **113**(2), 369–375 (1969)
29. N. Karmakar, G. Jayaraman, IMA Journal of Mathematics Applied in Medicine & Biology **5(1)**, 33-43 (1988)
30. F. Kong, R.P. Singh, Food Biophysics **6**, 84–93 (2011)
31. R.B. Voegborlo, A.M. El-Methnani, M.Z Abedin, Food Chemistry **67**(4), 341–345 (1999)
32. M.R. Kamaltdinov, N.V. Zaitseva, P.Z. Shur, Health Risk Analysis, **1**, 38–46 (2017)
33. M. A. Peyron, A. Mishellany, A. Woda, Journal of Dental Research, **83**(7), 578–582 (2004)
34. F. Kong, R.P. Singh, Journal of Food Science, **73**(5), R67–R80 (2008)

Effect of post type on the fracture resistance of endodontically treated premolars

Ekaterina Karteva[1,*], *Neshka* Manchorova[1], *Dessislava* Pashkouleva[2], *Donka* Keskinova[3], and *Stoyan* Vladimirov[1]

[1]Department of Operative Dentistry and Endodontics, Faculty of Dental Medicine, Medical University – Plovdiv, 4002 Plovdiv, Bulgaria
[2]Institute of Mechanics, Bulgarian Academy of Sciences, 1113 Sofia, Bulgaria
[3]Department of Applied and Institutional Sociology, University of Plovdiv "Paisii Hilendarski", 4000 Plovdiv, Bulgaria

Abstract. Endodontically treated teeth (ETT) are more susceptible to fracture during mastication. The aim of our study was to investigate the fracture resistance of premolars, restored with different techniques. Eighteen human premolars, extracted for orthodontic or periodontal reasons, were used. Mesio-occlusal cavities were prepared on all of them. They were divided into 3 groups according to the restoration method: with a cemented fiber post (FP), with a cemented metal post (MP) and without a post, but restored with a fiber reinforced composite core (FRC). After thermocycling, their fracture resistance was evaluated in a standard testing machine. The results were analyzed with one-way ANOVA and Tukey Honestly Significant Difference (HSD) tests ($p<0.05$). The fracture modes were also determined. The MP group showed the lowest fracture resistance rates and the highest number of unrestorable root fractures. In conclusion, teeth restored with FRC cores or FP show better survival rates than those restored with metal posts.

1 Introduction

Endodontic treatment is one of the most demanding dental procedures. Each tooth can present various challenges to the clinician, such as severe pain, curved canals and difficult access because of transposition [1]. In addition, endodontically treated teeth (ETT) are more susceptible to fracture during mastication [2]. They are reported to have shorter survival rates than vital teeth. One of the main reasons for this phenomenon is the reduction in tooth structure, which is a result of caries, trauma and cavity preparations. Other factors that influence the survival rate of these teeth are abrasion, erosion, noncarious lesions and the age of the patient [3]. Nevertheless, the primary reasons for the occurrence of partial or complete cusp fractures are the excessive loss of hard dental tissues and endodontic therapy. In many cases, the failure of the restorations results in vertical root fracture that cannot be restored, leading to the extraction of the tooth.

* Corresponding author: katya.kk@gmail.com

In cases with substantial hard tissue loss, the retention of the restorative material is usually provided by a post [4]. According to the method of fabrication, there are two main groups of posts available: cast post and core systems, and prefabricated posts, upon which a direct core is placed. The materials used for the fabrication of a post also vary widely. Initially, they were made of stainless steel or precious gold alloys. Later, posts made of titanium alloys were introduced. Due to the recent developments in the field of dental materials and the advances in adhesive dentistry, posts made of fiber-reinforced resin composite or ceramic have become widely popular. In particular, the zirconium dioxide and glass fiber-reinforced posts offer a reliable bonding to the hard dental tissues, as well as a highly aesthetic result [5].

There are a large number of studies on the restoration of severely with various post systems [6, 7]. However, the influence of post type on the fracture resistance of teeth with minor hard dental tissue loss is not extensively studied [8].

In recent years, the use of new type of core material – fiber-reinforced composite (FRC), has been widely advocated. Compared to the commercial particulate filler composite resin (PFC), it has superior mechanical properties, combined with significantly lower polymerization shrinkage [9].

With respect to the type of teeth most susceptible to fractures, the premolars show the highest failure rates [10]. This motivated us to investigate the fracture resistance of premolars with mesio-occlusal (MO) cavities, restored with either metal posts, fiber posts or FRC only. The hypotheses were:

1. Premolars with MO cavities, restored with prefabricated fiber posts and a direct composite, have a higher fracture resistance than premolars restored with metal posts;
2. Premolars, restored with prefabricated metal posts show more severe fracture modes than premolars restored with prefabricated fiber posts.
3. Premolars with MO cavities, restored with no post, but strengthened with a FRC core and a direct composite, have higher fracture resistance than premolars, restored with prefabricated posts.

2 Materials and methods

Eighteen human, single root premolars, free of cracks and defects were extracted for orthodontic or periodontal reasons. They were stored in 0.2% thymol solution for no longer than three months. The bucco-lingual and mesio-distal widths of the crowns were measured and only teeth of similar sizes were selected for the experiment. They were divided in 3 groups (n=6): teeth restored with a fiber post (FP), teeth restored with a metal post (MP) and teeth restored with a FRC core and no post. MO cavities were prepared on all of the samples.

The teeth were then thermocycled for 5000 cycles between 5±5°C and 55±5°C (LTC 100, LAM Technologies, Italy). Subsequently, their roots were embedded in self-curing resin to a level 2 mm apical to the cementoenamel junction (CEJ) using a modified technique, proposed by Soares et al. [11]. The periodontal ligament was simulated using a polyether-based impression material (Impregum Garant L Duo Soft, 3M ESPE). First, the roots were dipped into melted wax to a point 2 mm apical to the CEJ. Then, a radiographic film with a circular hole was used to stabilize the tooth for the embedding step. The tooth was thus positioned in a plastic cylinder (25 mm in diameter and 35 mm in height) and the self-curing resin was inserted. After the polymerization of the resin, the teeth were removed and the wax was cleaned from the roots. The polyether was placed in the resin and the teeth were re-inserted until the setting of the material.

The specimens were then subjected to the static fracture resistance test by using a universal testing machine (Fu1000e, Germany). They were loaded in compression at a constant speed of 4 mm/min along the tooth axis until failure (Fig. 1). The mechanical testing is performed at a room temperature. A computer recorded the data on the forces and the displacements. The digitally recorded values are used to draw the load-strain curves and to define the fracture resistances. The results were analyzed with One-way Analysis of Variance (ANOVA). Differences between groups were identified with Tukey Honestly Significant Difference (HSD) tests. A significance level of 0.05 was used for all the tests.

Fig. 1. Specimen in a testing machine.

The fractured specimens were visually analyzed and divided into 4 groups according to the modes of fracture. A modified classification system based on that used by Burke et al. [12] was used: 1. Fractures involving a small part of the coronal hard dental tissues (HDT); 2. Fractures involving a small part of the HDT and a cohesive fracture of the obturation; 3. Fractures involving the HDT, cohesive and/or adhesive fracture of the obturation with root involvement (that can be restored). 4. Severe fractures of the crown and root that cannot be restored (Fig. 2).

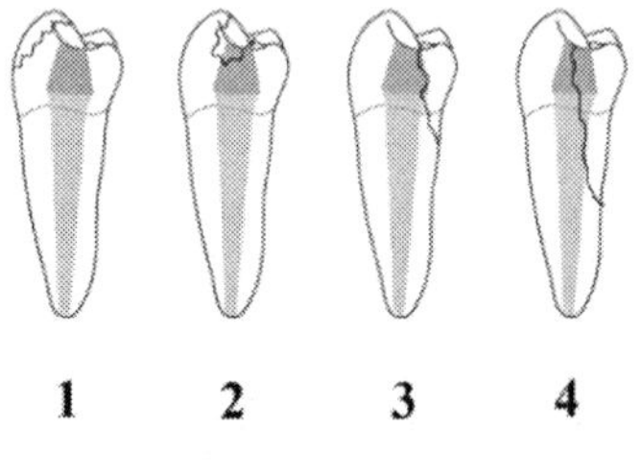

Fig. 2. Classification of the modes of fracture.

3 Results

Obtained typical diagrams of the load of premolars are showed in Fig. 3. The mean fracture resistance and standard deviation are illustrated in Table 1.The one-way ANOVA showed that there were significant differences among the three groups with respect to their resistance to fracture ($p<0.05$) (Table 1). The FRC group presented higher fracture resistance values, compared to the other groups. The fracture resistance between the FP and MP groups was also significantly different, with the MP showing the lowest values.

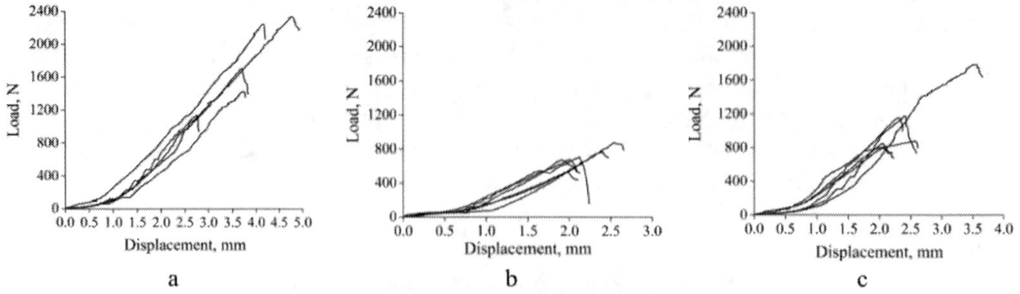

Fig. 3. Load-strain diagrams of the groups: FRC (a), MP (b) and FP (c).

Table 1. Mean fracture resistance of ETT.

Group	Teeth number *n*	Fracture resistance [N]	Standard deviation
FRC	6	1770.5	517.8
MP	6	707.6	113.9
FP	6	1108.9	369.6

p values < 0.05

Fracture mode distribution according to Fig. 2 is presented in Table 2. The MP group presented the highest incidence of fracture types 4. The teeth in the FRC and FP group showed more favourable fracture modes.

Table 2. Fracture mode distribution among groups.

Group	Fracture type			
	No of teeth, assigned to group 1	No of teeth, assigned to group 2	No of teeth, assigned to group 3	No of teeth, assigned to group 4
FRC	-	1	5	-
MP	-	-	1	5
FP	-	2	4	-

4 Discussion

The results confirmed the first hypothesis – that teeth, restored with fiber posts show greater fracture resistance than teeth restored with metal posts. A possible reason for this is the difference in the mechanical properties of the two materials, namely – their modulus of elasticity. The mean modulus of elasticity of dentin is between 16.5 and 18.6 GPa [13], while the values for metal and fiber posts are 210 and 39, respectively [14, 15]. The much lower modulus of elasticity of fiber posts means that they should cause less stress to the root dentin than metal ones. Thus, by lowering the stress that occurs during mastication, the risk of root fracture should be reduced. De Castro et al. examined the influence of the post material (stainless steel, titanium or carbon/Bis-GMA posts) and post configuration on the stress distribution in ETT [16]. The results showed that the shape of the post had a relatively small effect on stress concentration. However, the stainless steel posts showed the highest levels of stress, compared to the other post types.

The hypothesis that premolars restored with metal posts would show more severe fractures than those restored with fiber posts was also confirmed. Our results showed the highest number of unrestorable fractures in teeth from the MP group. Compared to the other groups, they also presented the lowest fractures resistance modes. This is in agreement with the results from previous studies [17]. However, there is scientific evidence that the post material used does not significantly influence the survival rates of ETT. In a clinical long-term pilot trial, Sterzenbach et al. tested the biommetic concept that posts with low modulus of elasticity are more successful in restoring non-vital teeth [18]. The results for both materials used (fiber posts and titanium posts) were favourable after 7 years, with similar failure rates. The reasons behind this contradictory data could be several: the Young's modulus of titanium posts is 103 GPa, which is two times lower than that of the stainless steel posts used in our study [19]; all of the teeth in the clinical trial were restored with metal-fused ceramic crowns, which increased their fracture resistance. Therefore, the materials and techniques used for determining the fracture resistance of ETT vary considerably and should be taken into account when comparing the results.

The third hypothesis was also confirmed – the teeth restored with FRC used for core build-up with no post, showed higher fracture resistance compared to the other groups. FRC resin shows better flexural strength due to the introduction of short, variously orientated fibers [20]. This ensures that the toughness of the material remains the same, irrespective of the direction of the masticatory forces. The presence of the fibers also contributes towards the reduction of the polymerization shrinkage of the material. This in turn decreases the formation of micro-gaps, which are responsible for stress concentration within the restoration, micro-leakage and the formation of cracks.

5 Conclusion

The results of our study suggested that teeth restored with either a fiber-reinforced composite core or fiber post show better survival rates that teeth restored with stainless steel posts. Premolars, restored with no post but strengthened with a fiber-reinforced composite core have highest fracture resistance because it reinforces the material by filling the cracks. Further studies are needed to determine whether and how the extent of tissue loss can affect these results.

References

1. G. Yordanova, M. Dinkova, Compt. Rend. Acad. Bulg. Sci. **68**, 107-114 (2015)
2. D. Assif, C. Gorfil, J. Prosthet Dent. **71**, 565-567 (1994)
3. S.C. Khera, Z. Askarieh, J. Jakobsen, Dent. Mater., **6**, 5-9 (1990).
4. I. Peroz, F. Blankenstein, K.P. Lange, M. Naumann, Quintessence Int., **36**, 737-746 (2005)
5. A. Almaroof, et al, J. Dent. **43**, 1308-1315 (2015)
6. M. Naumann, A. Preuss, R. Frankenberger, Dent. Mater. **23**, 138-144 (2007)
7. G. Heydecke, F. Butz, A. Hussein, J.R. Strub, J. Prosthet Dent. **87**, 438-445 (2002)
8. G. Heydecke, F. Butz, J.R.Strub, J. Dent. **29**, 427–433 (2001)
9. S. Garoushi, P. K. Vallittu, D. C. Watts, L. V. J. Lassila, Dent. Mater. **24**, 211–215 (2008)
10. A. Tamse, I. Zilburg, J. Halpern, Int. Endod. J. **31**, 127-132 (1998)
11. P. V. Soares, et al., J. Prosthet. Dent. **99**, 30-37 (2008)

12. F. J. Burke, J. Dent. **20**, 131-139 (1992)
13. R. G. Craig, F. A. Peyton, J. Dent. Res. **52**, 710-718 (1958)
14. A. Lanza, R. Aversa, S. Rengo, D. Apicella, A. Apicella, Dent. Mater. **21**, 709-715 (2005)
15. T. Nakamura, et al., Dent. Mater. J. **24**, 275-279 (2005)
16. R. de Castro Albuquerque, et al., J. Oral Rehab. **30**, 936-943 (2003)
17. C. J. Goodacre, J. Evid. Based Dent. Pract. **10**, 32-34 (2010)
18. G. Sterzenbach, Al. Franke, M. Naumann, J. Endod. **38**, 1557-1563 (2012)
19. C. S. M. Coelho, et al., Dent. Mater. J. **28**, 671-678 (2009)
20. S. Garoushi, P. K. Vallittu, L. V. J. Lassila, Dent. Mater. **23**, 1356-1362 (2007)

Biomechanical modelling of bile flow in the biliary system

Alex Kuchumov[1,*]

[1]Perm National Research Polytechnic University, Department of Theoretical Mechanics and Biomechanics, 614990 Perm, Russia

Abstract. The biliary system consists of the biliary tree, gallbladder and major duodenal papilla. Soft tissues compliance plays important role in the bio-fluids transport. Particularly, bile flow disturbances due to bile duct wall motor function changes in the extra-hepatic ducts, from medicine point of view are called dyscinesia of biliary tract. Fluid motion in the elastic and compliant ducts can be described by different models (for example, Windkessel model, peristaltic fluid motion, FSI algorithm). Our approach is decomposition of the biliary system into three compartments (extra-hepatic biliary tree, gallbladder, major duodenal papilla). Bile flow in the extra-hepatic ducts is simulated using FSI algorithm. Bile flow in the gallbladder can be described as flow in the reservoir with compliant ducts using Windkessel model. Bile flow in the major duodenal papilla is considered as peristaltic fluid motion, because the wall contraction is really important factor of fluid motion in that segment. The coupling of these compartments is performed by boundary conditions. The biliary system geometry was obtained using MRI patient-specific data. It was confirmed that normal bile can be modeled as Newtonian fluid and lithogenic bile can be modeled as non-Newtonian fluid (Carreau fluid). Bile ducts were modeled as hyperelastic material.

1 Introduction

The biliary system is responsible for bile producing, storing and transport to the duodenum in order to emulgate fats. Despite several biomechanical models of bile flow in different segments, there is no entire biomechanical model. Our approach is related to considering of the biliary system as three compartments. Each of compartment is modelled using different models depending on mechanical performance of this compartment based on physiological data. Computational models of bile flow dynamics were proposed by Luo et al [1], [2], [3], Agrawal et al [4], [5], Kuchumov et al [6], [7], Al-Atabi et al [8], Ro-Chin Lo et al [9]. Peristaltic bile flow models were proposed by Maiti and Misra [10] and Kuchumov et al [11], [12]. There is no entire biomechanical model of the bile flow in the biliary system and the purpose of this paper is present a such model.

Our approach is decomposition of the biliary system into three compartments (extra-hepatic biliary tree, gallbladder, major duodenal papilla) using several models and

* Corresponding author: kychymov@inbox.ru

approaches. The compartments coupling into the entire model is performed using boundary conditions.

• Bile flow in the gallbladder.

We used a modified Windkessel model to evaluate bile flow in the patient-specific gallbladder model. Using ultrasound data we can evaluate Windkessel model parameters for the patient and non-invasively determine pressure changes on time during the gallbladder refilling and emptying. The pressure on time dependence is applied as a boundary condition when we investigate bile flow in the biliary tree.

• Bile flow in the biliary tree.

We considered 4 cases: healthy bile flow (considered as Newtonian fluid), lithogenic bile (considered as non-Newtonian fluid (Carreau fluid)) flow, lithogenic bile flow in the case of gallstone presence; bile flow after cholecystectomy (the gallbladder removal). The velocity and pressure distributions during the gallbladder emptying were presented. Velocities were found to have lower magnitude in the case of lithogenic bile, but the pressures are higher in this case. Stress state of bile ducts was also evaluated. It was shown that von Mises stress distribution is located in the cystic duct and hepatic ducts mostly; whereas, the shear stress distribution is mostly prevailing in the common hepatic duct. It is believed that duct wall shear stress can be considered as indicator of gallstones formation and von Mises stress is related to biliary pain. The developed model can be applied in the medical practice to evaluate the circumstances of surgical interventions.

• Bile flow in the major duodenal papilla.

Choledochopancreatic reflux (i.e. the flow of the gallbladder bile coming out the common bile duct into the pancreatic duct instead of the duodenum) is known to be one of the reasons of the pancreatitis (inflammation of the pancreas). Understanding of the reasons of the reflux from physiological, hydrodynamic, biomechanical points of view is still a challenging task. The current paper aims at developing mathematical model of the peristaltic bile transport flow through the duct at papillary stenosis as a tapered finite-length tube. It allows evaluating velocities and pressure distribution along the tube, and detecting choledochopancreatic reflux occurrence conditions. Adopting the perturbation method, the analytical solutions for velocities and pressures are obtained. Pressure distribution versus axial coordinate at different time instants are plotted for various values of Weissenberg number and amplitude ratio. It revealed that the amplitude ratio has more effect on the pressure distribution along the tube compared to the Weissenberg number. The values of the pressure gradient corresponding to reflux occurring are obtained. Moreover, it is reported that the pressure drop value corresponding to average flow rate equal to zero may serve as reflux occurrence criterion.

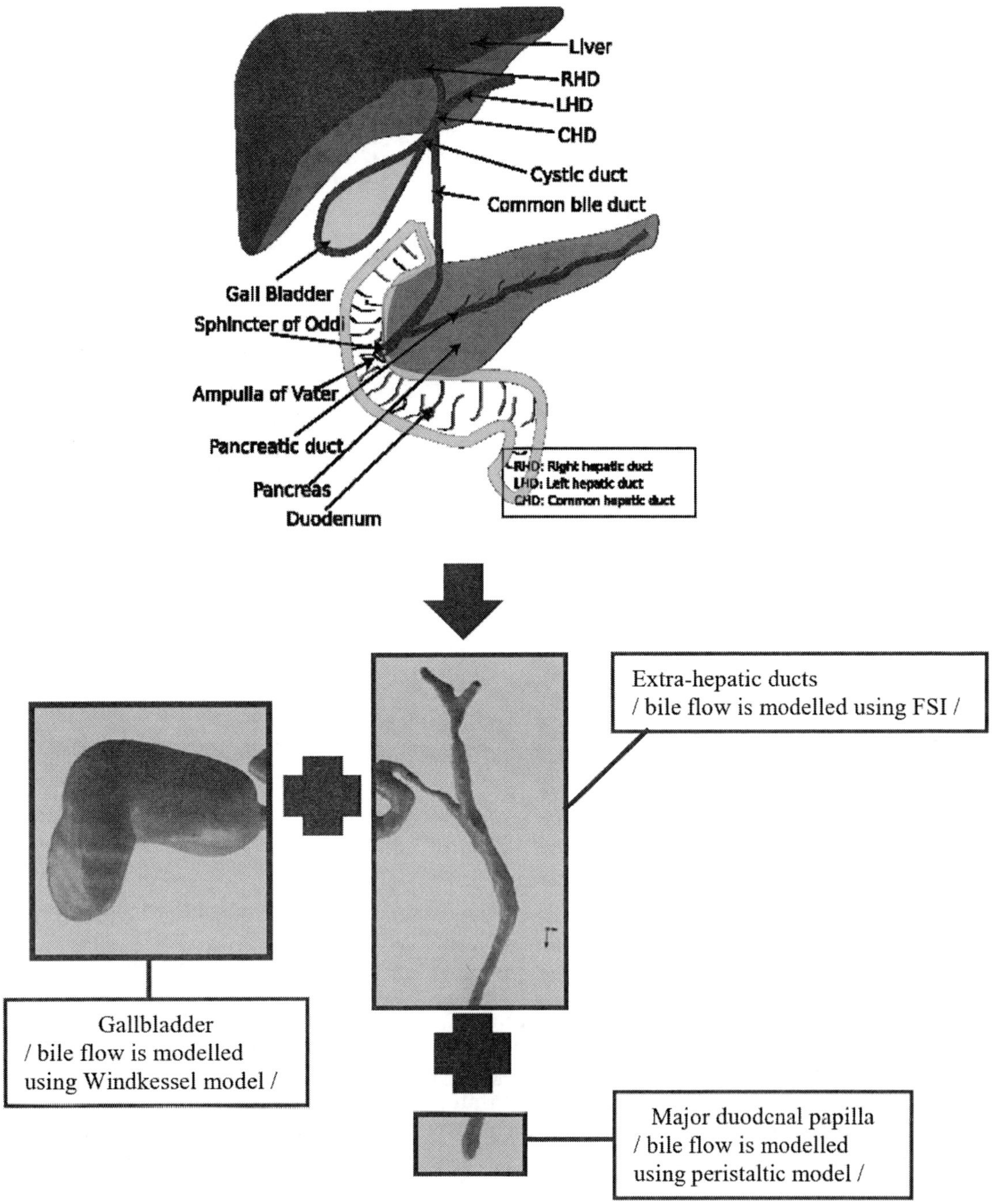

Fig. 1. Biliary system scheme and compartment models.

2 Biomechanical modelling of the biliary system

2.1 Bile flow in the gallbladder

As was mentioned above the bile flow in the gallbladder during the emptying is performed due to pressure drop and can be modelled by modified Windkessel model [13]. Figs 2 and 3 present schemes of gallbladder volume changes during refilling and emptying and bile flow scheme, respectively.

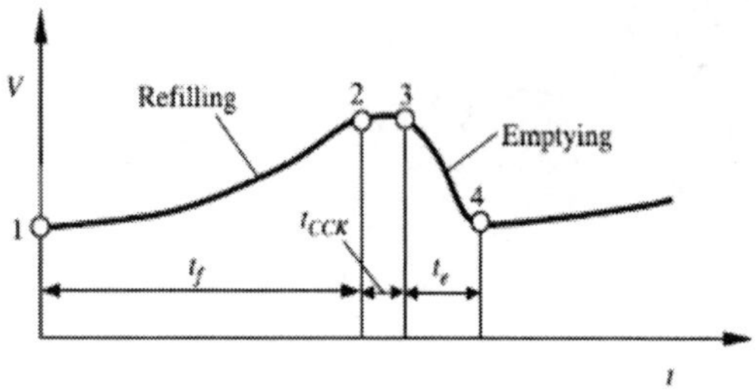

Fig. 2. Dependence of the gallbladder volume vs. time. during the refilling and emptying phases.

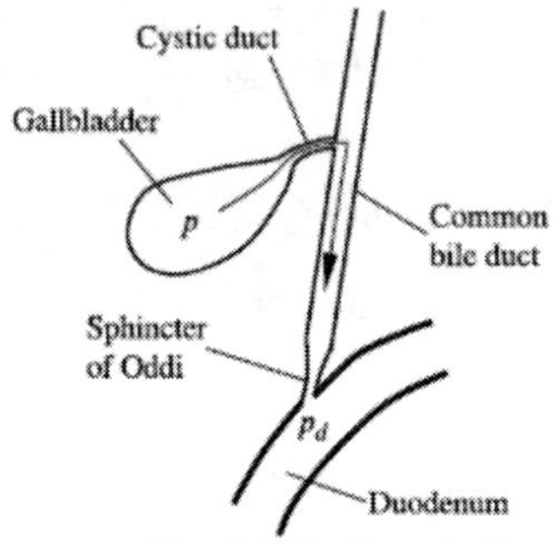

Fig. 3. Bile flow in the gallbladder during emptying scheme .

2.1.1 Modified Windkessel model

The flow scheme due to pressure gradient and interaction with elastic walls can be described by Windkessel model [14]:

$$-\frac{dV}{dt} = Q, \tag{1}$$

$$dV = d(C \cdot P), \tag{2}$$

$$Q = \frac{p - p_d}{R}, \tag{3}$$

where V is the gallbladder volume, C is a gallbladder wall compliance, Q is a bile flow rate, p is a pressure inside the gallbladder, p_d is a pressure in the duodenum ($p_d = 6$ mm Hg) [15], R is a hydraulic resistance. Substituting (3) in (1) with account of (2), considering $C = \text{const}$, one can obtain:

$$p = p_d + (p_e - p_d)\, e^{\frac{t_e - t}{RC}}, \tag{4}$$

where p_e is the pressure when gallbladder is empty ($p_e = 11$ mm Hg) [12], t_e is the emptying time. Volume change dependence :

$$V(t) = B + C(p_e - p_d)\, e^{\frac{t_e - t}{RC}}, \tag{5}$$

where $B = V_e - C(p_e - p_d)$.

2.1.2 Ultrasound gallbladder performance

In order to find patient-specific parameters of Windkessel model, the experiment was performed. The ultrasound measurements of volunteer's full gallbladder volume were made initially. The typical ultrasound image of the gallbladder is presented in Fig. 4. After that, the volunteer had a meal in order to force the gallbladder emptying. After 20 minutes when the meal was taken with 5 minute interval, the volume measurements were made. The measurements were made from 20 to 240 minutes after taking a meal. After that using least square method, we obtained Windkessel model parameters can be found by (6) (Fig. 5). Using the obtained parameters, we can substitute them into the pressure on time dependence and use it as a boundary condition, when bile flow modelling in the biliary tree is performing.

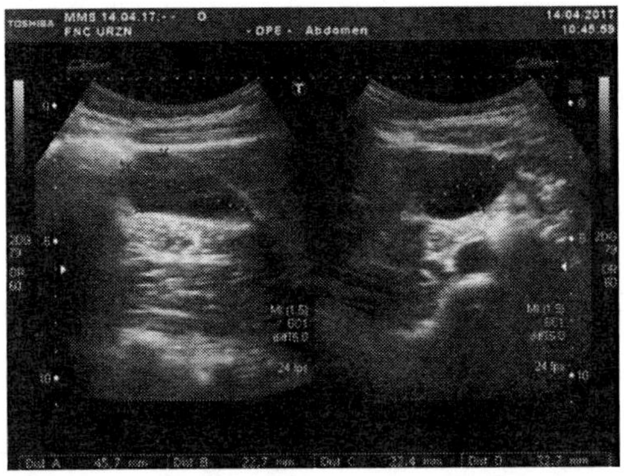

Fig. 4. Ultrasound images of the gallbladder.

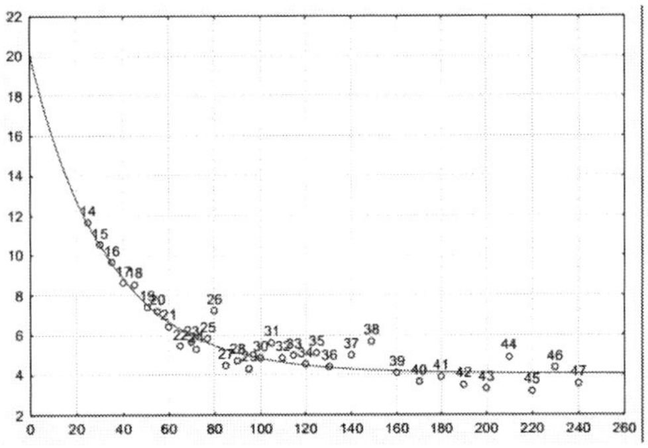

Fig. 5. Experimental dots and simulated *V*(*t*) curve.

2.2 Bile flow in the biliary tree

The second stage of simulation of bile flow in the biliary system is FSI modeling of the bile flow in the biliary tree. To make a patient-specific FSI modeling you need to create 3D patient-specific geometry; to evaluate fluid and solid properties Symmetric finite element model was created using ANSYS APDL Mechanical (ANSYS, Inc., Canonsburg, PA, USA). The dimensions of the model were taken from experiment images processing using ImageTool software (UTHSCSA, USA). The thickness was taken as 10^{-3} m.
The obtained parameters were exported to the ANSYS Workbench 13.0 (ANSYS, Inc., Canonsburg, PA, USA) (Engineering Data), which allowed to recompute stress–strain curves for uniaxial, biaxial and shear experiments for the hyperelastic material described by Mooney-Rivlin 2 parameter model.

2.2.1 Fluid-Structure Interaction Problem Statement

Mass and momentum conservation equations for an incompressible fluid can be expressed as

$$\nabla \cdot u = 0 , \tag{6}$$

$$\rho_f \left(\frac{\partial u}{\partial t} + ((u - u_g) \cdot \nabla) u \right) = -\nabla p + \nabla \cdot \tau , \tag{7}$$

where is ρ_f is the fluid density, p is the pressure, u is the fluid velocity vector and u_g is the moving coordinate velocity. In the Arbitrary Lagrangian – Eulerian formulation, $(u-u_g)$ is the relative velocity of the fluid with respect to the moving coordinate velocity. Here τ is the deviatoric stress tensor. This tensor is related to the strain rate tensor; if the fluid is incompressible and viscosity is constant across the fluid, this equation can be written in terms of an arbitrary coordinate system as

$$\tau_{ij} = \mu\left(\frac{\partial v_i}{\partial x_j} + \frac{\partial v_j}{\partial x_i}\right), \tag{8}$$

where x_j is the th spatial coordinate, v_i is the fluid's velocity in the direction of axis i, τ_{ij} is the jth component of the stress acting on the faces of the fluid element perpendicular to axis.

For healthy bile, which is considered as Newtonian Fluid

$$\eta = \text{const},$$

The Carreau's equation, which is used for simulation of lithogenic bile, is written as follows

$$\eta = \frac{\eta_0 - \eta_\infty}{(1+(a\cdot\dot{\gamma})^2)^k} + \eta_\infty, \tag{9}$$

The momentum conservation equation for solid body can be written as

$$\nabla \cdot \sigma_s = \rho_s \,[\text{illegible}], \tag{10}$$

where ρ_s, σ_s, and [illegible] are solid density, solid tensor, and local acceleration of the solid, respectively.

It is known that blood vessels can be described as hyperelastic materials [16]. Due to similar anatomical composition, the bile ducts can also be considered as hyperelastic materials. For hyperelastic materials, the stress-strain relation is written as follows

$$\sigma_s = \frac{\partial W}{\partial \varepsilon}, \tag{11}$$

The Mooney-Rivlin hyperelastic potential takes on the form

$$W = c_{10}\left[\bar{I}_1 - 3\right] + c_{01}\left[\bar{I}_2 - 3\right] + \frac{1}{d}[J-1]^2, \tag{12}$$

The FSI interface should satisfy the following conditions:

1. The displacements of the fluid and solid domain should be compatible i.e.

$$\delta_s = \delta_f, \tag{13}$$

2. The tractions at this boundary must be at equilibrium:

$$\sigma_s \cdot \hat{n}_s = \sigma_f \cdot \hat{n}_f, \tag{14}$$

3. and the no-slip condition for the fluid should satisfy

$$u = u_g. \tag{15}$$

In the above conditions, δ, σ and $\hat{n}$ are respectively, displacement, stress tensor and boundary normal with the subscripts '*f*' and '*s*' indicating a property of the fluid and solid.

2.2.2 Boundary conditions for gallbladder refilling

Solid Domain

As shown in Fig. 6, the ends of the extra-hepatic biliary tree were constrained by specifying zero-displacements in all directions and prohibition of the rotation about all axes (boundary condition type: fixed support).

Fluid domain

During the gallbladder refilling, the bile comes out from the the liver via right and left hepatic ducts and flows into the gallbladder. Bile pressures and initial velocity were applied to the inlets and the outlet (Fig. 6). Following the work of Howard et al [17], who measured the mean flow rate of the bile from the gallbladder (2.0–3.0 ml/min) after meal. Using relation

$$Q = \int_S v dS, \tag{16}$$

where v is velocity, S is the cross section area, which can be evaluated for the given patient-specific model, we expressed the velocity from this equation and obtained the mean value of velocity 6 mm/s and used this value as inlet velocity. Outlet pressure corresponds to $p(t)$ obtained in the previous section. Pressure in the end of the biliary tree was taken equal to the pressure in the duodenum i.e. 0.96 kPa [18].

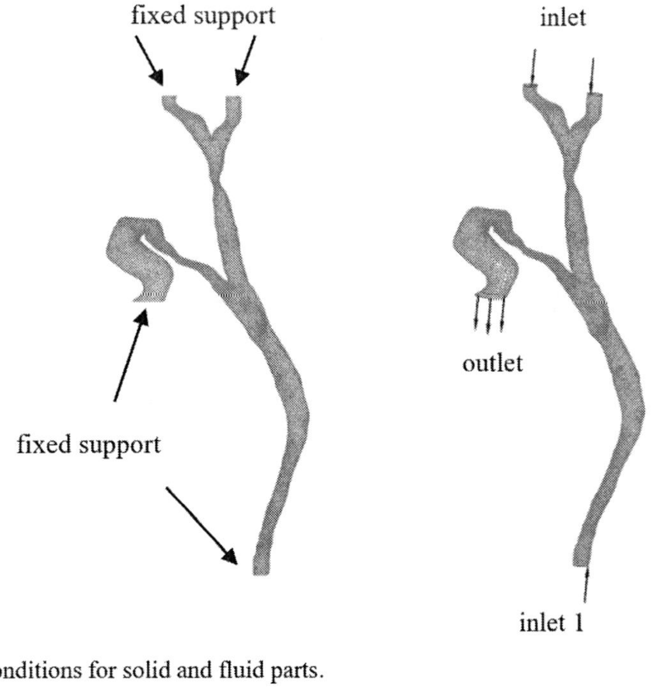

Fig. 6. Boundary conditions for solid and fluid parts.

2.2.3 Numerical Procedure

Fluid and solid meshes consisted of approximately 610,000 and 79,000 elements, respectively. Solid mesh was mapped with tetrahedral element size 0.3 mm. The fluid mesh was free with tetrahedral elements also. One-way FSI algorithm was adopted and the simulations were performed using ANSYS Workbench 13.0 (ANSYS, Inc., Canonsburg, PA, USA). The two solvers (ANSYS Mechanical and ANSYS CFX) were coupled and solved iteratively. The time step used for this simulation was 0.04 s. Initially, the fluid field is solved until the convergence criteria are reached. The calculated forces at the structure boundaries are then transferred to the structure side. The structure side is calculated until the convergence criterion is reached. The solution is finished when the maximum number of time steps (100) is reached.

2.2.4 Velocity distributions

Velocity distributions for two cases (flow of lithogenic bile in the biliary tree and bile flow in the extra-hepatic tree after cholecystectomy) are presented in Fig. 7. When gallbladder empties, the bile flows out of the gallbladder via cystic duct and common bile duct. At the same time bile flows out of the liver via hepatic ducts and common bile duct. We simulated these physiological phenomena in the present investigation. From Fig. 7 one can see that the peak of velocity is observed in the cystic duct, where bile accelarates towards the common bile duct. In our opinion, it is related with the fact that cystic duct has very complex inner geometry (spiral valves of Heister), which function is to damp such high velocities. In this case only external geometry was created without account of inner structure. As it was mentioned previously, the bile flow velocities play role in the stone formation. Lower velocity fields are potential zones for stone occurence. The minimal velocities are observed in the case of bile flow in the biliary tree after cholecystectomy. It is related with the fact that the gallbladder and cystic duct (where commonly the maximal velocity is observed) are absent. The bile amount is less and thus the velocity are lower than in the other cases. It also stated that lower velocities may lead to the stone formation. Our results showed that after cholecystectomy bile velocity significantly decreases and stone formation is probable for the current situation. The comfirmed fact of repeated stone formation after cholecystectomy can be also found in the medical literature [19].

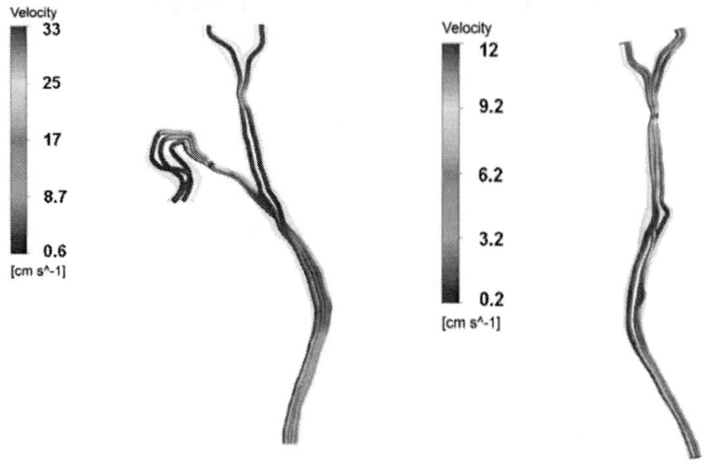

Fig. 7. Velocity distributions in case of lithogenic bile flow and bile flow in the extra-hepatic tree after cholecystectomy.

2.3 Bile flow in the major duodenal papilla

We consider the peristaltic flow of a non-Newtonian fluid (Carreau's fluid) in a tapered finite-length tube. Sinusoidal waves of constant speed propagate along the channel boundaries. The wavelength is comparable with the channel length ($L \approx \lambda$), thus the wave number is very small, and Reynolds number is negligible.

A wall geometry is described by the relation

$$\bar{H}(\bar{Z},\bar{t}) = a - l\bar{Z} + b\sin\frac{2\pi}{\lambda}(\bar{Z} - c\bar{t}), \tag{17}$$

where a is an inlet radius; l is a slope coefficient; c is a wave velocity; b is an amplitude of peristaltic wave; $\bar{Z}$ is a longitudinal coordinate.

The equations of motion are

$$\frac{1}{\bar{R}}\frac{\partial(\bar{R}\bar{U})}{\partial\bar{R}} + \frac{\partial\bar{W}}{\partial\bar{Z}} = 0, \tag{18}$$

$$\rho\left(\frac{\partial\bar{W}}{\partial\bar{t}} + \bar{U}\frac{\partial\bar{W}}{\partial\bar{R}} + \bar{W}\frac{\partial\bar{W}}{\partial\bar{Z}}\right) = -\frac{\partial\bar{P}}{\partial\bar{Z}} + \frac{1}{\bar{R}}\frac{\partial}{\partial\bar{R}}\left(\bar{R}\bar{S}_{13}\right) + \frac{\partial\bar{S}_{33}}{\partial\bar{Z}}, \tag{19}$$

$$\rho\left(\frac{\partial\bar{U}}{\partial\bar{t}} + \bar{W}\frac{\partial\bar{U}}{\partial\bar{Z}} + \bar{U}\frac{\partial\bar{U}}{\partial\bar{R}}\right) = -\frac{\partial\bar{P}}{\partial\bar{R}} + \frac{1}{\bar{R}}\frac{\partial}{\partial\bar{R}}\left(\bar{R}\bar{S}_{11}\right) + \frac{\partial\bar{S}_{31}}{\partial\bar{Z}}, \tag{20}$$

where $\bar{W}$, $\bar{U}$ are axial and radial velocities. Let us introduce non-dimensional variables:

$$r = \frac{\bar{R}}{a},\quad z = \frac{\bar{Z}}{\lambda},\quad w = \frac{\bar{W}}{c},\quad u = \frac{\lambda\bar{U}}{ac},\quad t = \frac{c\bar{t}}{\lambda},\quad \delta = \frac{a}{\lambda},\quad p = \frac{a^2\bar{p}}{c\lambda\eta_0},\quad \mathrm{We} = \frac{c\Gamma}{a},$$

$$S_{11} = \frac{\lambda\bar{S}_{11}}{c\eta_0},\quad S_{13} = \frac{a\bar{S}_{13}}{c\eta_0},\quad \dot{\gamma} = \frac{a\bar{\dot{\gamma}}}{c}\quad \mathrm{Re} = \frac{ca\rho}{\eta_0},$$

$$h = \frac{\bar{h}}{a} = 1 - \frac{\lambda l z}{a} + \varphi\sin 2\pi(z - t)$$

where $\varphi = \frac{b}{a}$ is an amplitude ratio, $\mathrm{We} = \frac{c\Gamma}{a}$ is the Weissenberg number, $\mathrm{Re} = \frac{ca\rho}{\eta_0}$ is the Reynolds number.

After non-dimensionalization, the equations are as follows

$$\frac{1}{r}\frac{\partial(ru)}{\partial r} + \frac{\partial w}{\partial z} = 0, \tag{21}$$

$$\mathrm{Re}\cdot\delta\left[u\frac{\partial w}{\partial r} + w\frac{\partial w}{\partial z}\right] = -\frac{\partial p}{\partial z} + \frac{1}{r}\frac{\partial}{\partial r}\left(rS_{13}\right) + \delta\frac{\partial}{\partial z}\left(S_{33}\right), \tag{22}$$

$$\mathrm{Re}\cdot\delta^3\left[w\frac{\partial u}{\partial z} + u\frac{\partial u}{\partial r}\right] = -\frac{\partial p}{\partial r} + \delta^2\frac{1}{r}\frac{\partial}{\partial r}\left(rS_{11}\right) + \delta^2\frac{\partial}{\partial z}\left(S_{31}\right), \tag{23}$$

To reduce the problem, we will use the lubrication theory approximations of infinitesimally small wall curvature ($\delta \to 0$) and small Reynolds number ($\mathrm{Re} \to 0$). The approximations assume that inertial effects are negligible and that the dominant axial scale is much larger than the dominant radial scale [20].

Thus,

$$\frac{1}{r}\frac{\partial(ru)}{\partial r}+\frac{\partial w}{\partial z}=0, \tag{24}$$

$$\frac{\partial p}{\partial z}=\frac{1}{r}\frac{\partial}{\partial r}\left(r\left[1+\left(\frac{m-1}{2}\right)\mathrm{We}^2\left(\frac{\partial w}{\partial r}\right)^2\right]\left(\frac{\partial w}{\partial r}\right)\right), \tag{25}$$

$$\frac{\partial p}{\partial r}=0, \tag{26}$$

where $S_{31}=S_{13}=\left(1+\frac{m-1}{2}\mathrm{We}^2\left(\frac{\partial w}{\partial r}\right)^2\right)\left(\frac{\partial w}{\partial r}\right)$.

Boundary conditions are written as follows:

---- for axial velocity

$$\frac{\partial w}{\partial r}=0 \quad \text{at} \quad r=0, \tag{27}$$

$$w=1 \quad \text{at} \quad r=h, \tag{28}$$

---- for radial velocity

$$u=0 \quad \text{at} \quad r=0, \tag{29}$$

$$u=\frac{\partial H}{\partial t} \quad \text{at} \quad r=h. \tag{30}$$

The finite tube length requires the pressures at two ends of the tube

$$p=p_1 \quad \text{at} \quad z=0, \tag{31}$$

$$p=p_L \quad \text{at} \quad z=L. \tag{32}$$

In order to solve equations (24)–(26) and account for boundary conditions (27)–(32), the perturbation method is adopted and the solution is represented as an expansion over Weissenberg number (We):

$$w=w_0+(\mathrm{We})^2 w_1+O(\mathrm{We}^4), \tag{33}$$

$$u=u_0+(\mathrm{We})^2 u_1+O(\mathrm{We}^4), \tag{34}$$

$$p=p_0+(\mathrm{We})^2 p_1+O(\mathrm{We}^4). \tag{35}$$

Solving the system (24)–(26) subjected to the boundary conditions, the axial and radial velocities are obtained in the following form:

$$w_0(r,z) = \frac{1}{4}\left(r^2 - h^2\right)\frac{\partial p_0}{\partial z}, \tag{36}$$

$$u_0(r,z) = \frac{r}{4}\left[h\frac{\partial h}{\partial z}\frac{\partial p_0}{\partial z} - \frac{1}{4}\left(r^2 - 2h^2\right)\frac{\partial^2 p_0}{\partial z^2}\right]. \tag{37}$$

$$w_1(r,z) = \frac{1}{4}\left(r^2 - h^2\right)\frac{\partial p_1}{\partial z} - \frac{m-1}{64}\mathrm{We}^2\left(\frac{\partial p_0}{\partial z}\right)^3\left[r^4 - h^4\right], \tag{38}$$

$$u_1(r,z) = \frac{1}{4}hr\frac{\partial h}{\partial z}\frac{\partial p_1}{\partial z} + \frac{r}{8}\left(h^2 - \frac{r^2}{2}\right)\frac{\partial^2 p_1}{\partial z^2} - \frac{3(m-1)}{128}\mathrm{We}^2\left(\frac{\partial p_0}{\partial z}\right)^2\left(r\left[\frac{r^4}{3} - h^4\right]\frac{\partial^2 p_0}{\partial z^2}\right) - \frac{(m-1)h^3 r}{32}\mathrm{We}^2\frac{\partial h}{\partial z}\left(\frac{\partial p_0}{\partial z}\right)^3. \tag{39}$$

$$p_0(Z,t) - p_0(0,t) = G_0(t)\int_0^z \frac{d\xi_1}{h^4(\xi_1,t)} + 16\int_0^z \frac{1}{h^4(\xi_1,t)}\left(\int_0^{\xi_1} h(\xi_2,t)\frac{\partial h(\xi_2,t)}{\partial t}d\xi_2\right)d\xi_1, \tag{40}$$

where $G_0(t)$ is equal to

$$G_0(t) = \frac{\left(p_L - p_1\right) - 16\int_0^L \frac{1}{h^4(\xi_1,t)}\left(\int_0^{\xi_1} h(\xi_2,t)\frac{\partial h(\xi_2,t)}{\partial t}d\xi_2\right)d\xi_1}{\int_0^L \frac{d\xi_1}{h^4(\xi_1,t)}}. \tag{41}$$

$$p_1(Z,t) - p_1(0,t) = \left[\frac{(m-1)}{12}\mathrm{We}^2\int_0^Z h^2(\xi,t)\left(\frac{\partial p_0}{\partial z}\right)^3 d\xi + G_1(t)\int_0^Z \frac{d\xi}{h^4(\xi,t)}\right], \tag{42}$$

where is equal to

$$G_1(t) = -\frac{\frac{m-1}{12}\mathrm{We}^2\int_0^Z h^2(\xi,t)\left(\frac{\partial p_0}{\partial z}\right)^3 d\xi}{\int_0^Z \frac{d\xi}{h^4(\xi,t)}}. \tag{43}$$

2.3.1 Influence of Weissenberg number on the pressure distribution

This section is devoted to the analysis of Weissenberg number (We) and amplitude ratio (φ) influence on the pressure distribution $[p(\xi, t) - p(0, t)]$ along the tube, and dependence of averaged flow rate (Q) versus pressure to analyze the reflux occurring conditions. Figures are plotted by using *MatLab* software.

The parameters were taken as follows: $a = 0.006$, $\lambda = 0.5$, $l = 1$, $m = 0.56$. Inlet and outlet pressures correspond to common bile duct pressure (1.12 kPa) and duodenum pressure (0.98 kPa) values.

Pressure distributions at various Weissenberg numbers (We = 0.1; 0.5) are shown in Figs. 8–9. The pressure distribution evolution along the tapered tube length at Weissenberg number We = 0.5 is displayed in Figs. 9, *a–f*. Fig. 9, *a* ($t = 0$) shows that at first the pressure decreases slowly and finally rises sharply to meet the leading end of the bolus. After one half of the periodic cycle (Fig. 9, *d*), pressure value reaches its minimum. Finally, at the time $t = 1$ (Fig. 9, *f*), which represents the completion of one period, the pressure distribution resembles the unity at $t = 0$; it indicates that a new cycle is about to begin (Fig. 9, *a*). The same behavior can be seen in Fig. 5 plotted for Weissenberg number We = 0.9. It is observed that pressure difference in different segments of the tube increases with the increase of Weissenberg number. It can be physically interpreted that if the fluid viscosity is less, then less pressure is required for the Carreau's fluid propagation through the tapered tube.

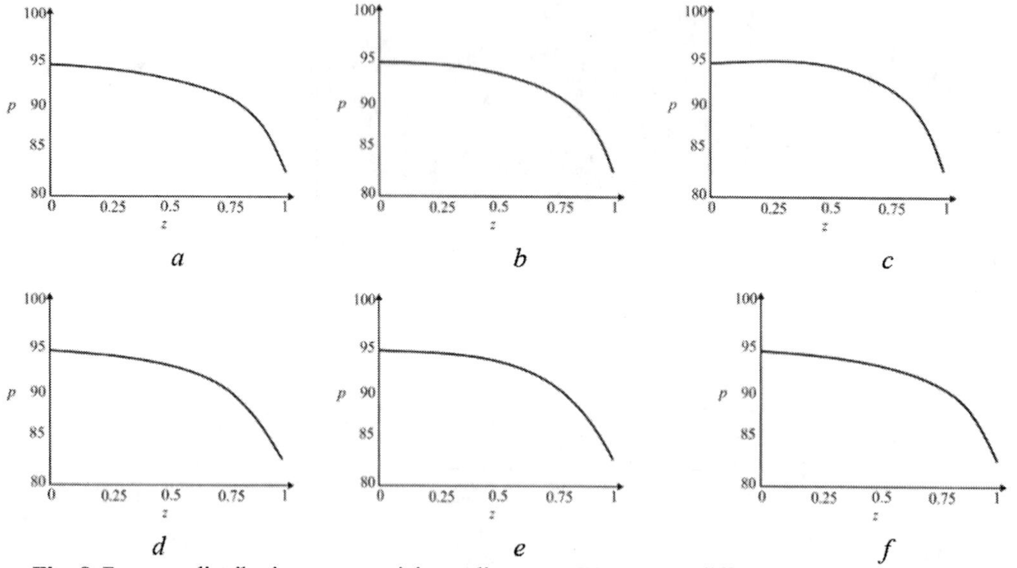

Fig. 8. Pressure distribution versus axial coordinate z at We = 0.1 at different time instants: $a - t = 0$, $b - t = 0.2$, $c - t = 0.4$, $d - t = 0.6$, $e - t = 0.8$, $f - t = 1$.

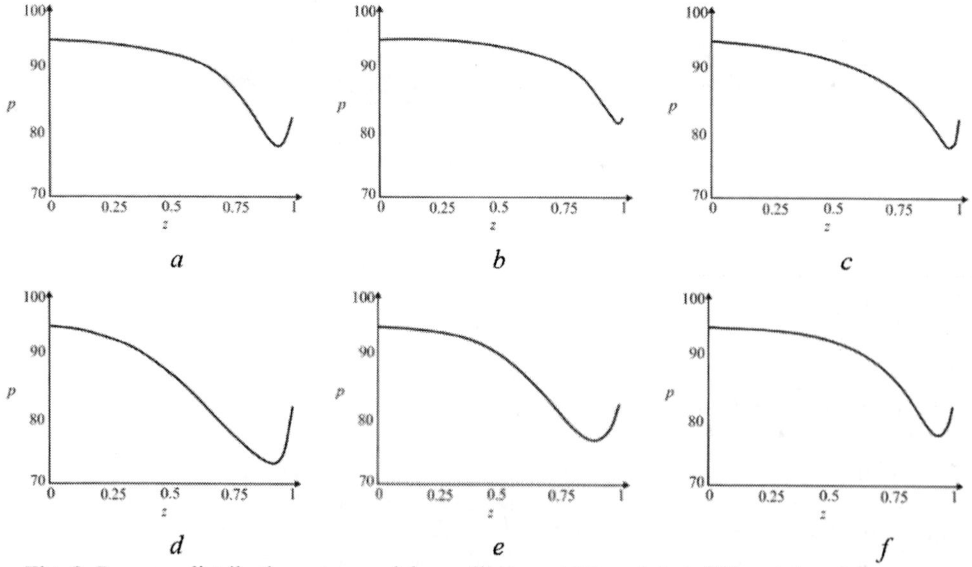

Fig. 9. Pressure distribution versus axial coordinate z at We = 0.5 at different time instants: $a - t = 0$, $b - t = 0.2$, $c - t = 0.4$, $d - t = 0.6$, $e - t = 0.8$, $f - t = 1$.

2.3.2 Influence of amplitude ratio on the pressure distribution

Influence of amplitude ratio on the pressure distribution along the tube is shown in Figs. 10–11.

The calculations were made for three amplitude ratios: φ = 0.1; 0.2. Weissenberg number value was taken equal to 0.5. The increase of amplitude ratio decreases cross-section during bolus propagation and simultaneously increases the pressure value difference. It can be seen also that amplitude ratio has more influence on the pressure distribution compared with the Weissenberg number.

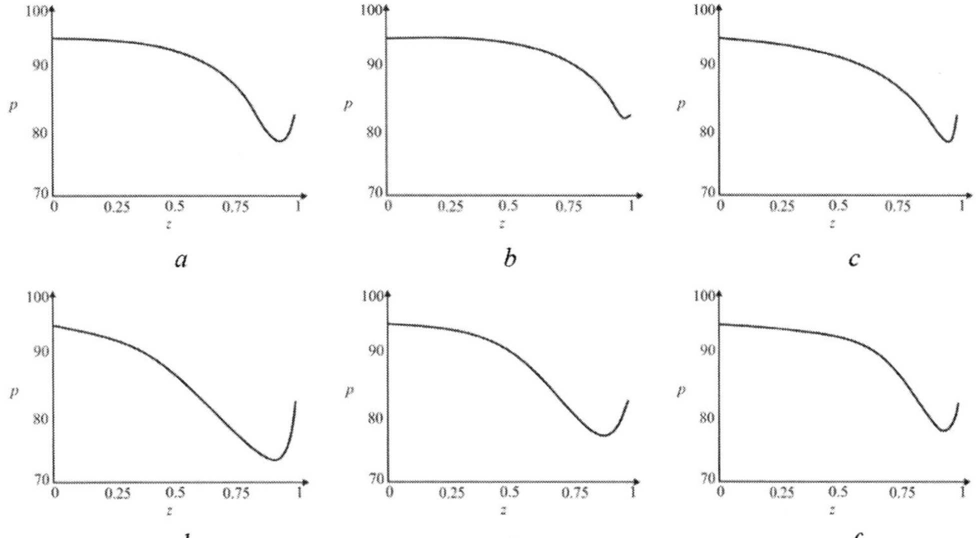

Fig. 10. Pressure distribution versus axial coordinate z at φ = 0.1 at different time instants: a – t = 0, b – t = 0.2, c – t = 0.4, d – t = 0.6, e – t = 0.8, f – t = 1.

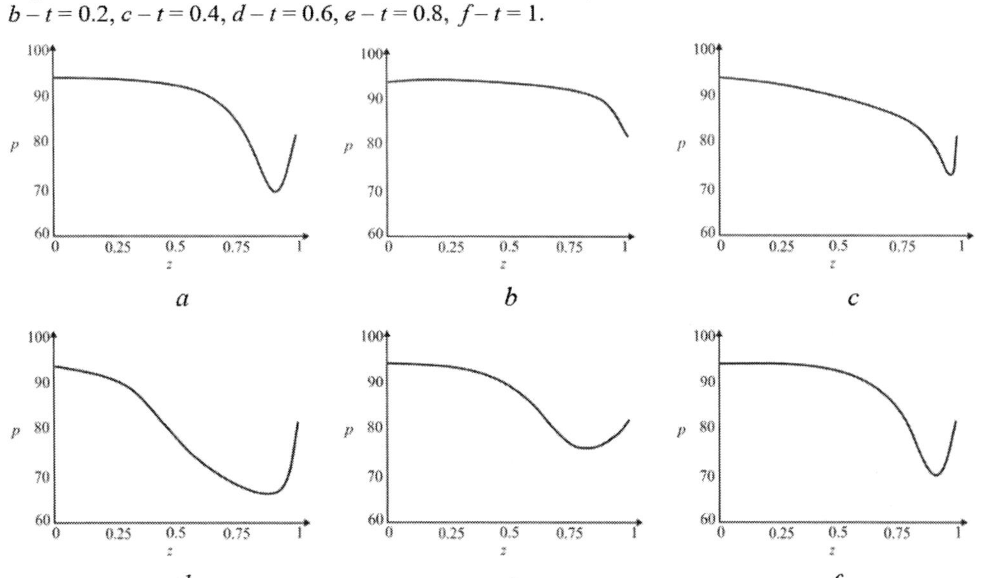

Fig. 11. Pressure distribution versus axial coordinate z at φ=0.2 at different time instants:a – t = 0, b – t = 0.2, c – t = 0.4, d – t = 0.6, e – t = 0.8, f – t = 1.

Conclusions

According to [21], the use of elective cholecystectomy has increased dramatically following the widespread adoption of laparoscopic cholecystectomy (Fig. 6). The number of complications is still high. One of the reasons is the absence of patient-specific approach for patient's treatment. The surgeons use their own experience, but do not use the results of biomechanics to evaluate the intervention outcomes. Our research will help surgeons to decrease the number of complications in population and will shorten the recovery time after surgery.

The reported study was funded by RFBR according to the research project № 16-08-00718 A and Russian Government Assignment grant 19.7286.2017/8.9.

References

1. R.C. Ooi, X.Y. Luo, S.B. Chin, A.G. Johnson, N.C. Bird, J. Biomech. **37** (2004)
2. W.G. Li, X.Y. Luo, S.B. Chin, N.A. Hill, A.G. Johnson, N.C. Bird, Ann. Biomed. Eng. **36** (2008)
3. W.G. Li, X.Y. Luo, A.G. Johnson, N.A. Hill, J. Biomech. Eng. **129** (2007)
4. S. Agarwal, A.K. Sinha, S.P. Singh, Int. J. Appl. Math Mech. **8** (2012)
5. S. Agarwal, S.P. Singh, Int. J. Eng. Trends Technol. **32** (2016)
6. A.G. Kuchumov, V. Gilev, V. Popov, V. Samartsev, V. Gavrilov, Korea Aust. Rheol. J. **26** (2014)
7. A.G. Kuchumov, Y.I. Nyashin, V.A. Samarcev, V.A. Gavrilov, Acta Bioeng. Biomech. **15** (2013)
8. M. Al-Atabi, R.C. Ooi, X.Y. Luo, S.B. Chin, Med. Eng. Phys. **34** (2012)
9. R.C. Lo, W.L Huang Y.M. Fan, Comput. Biol. Med. **60** (2015)
10. S. Maiti, J.C. Misra, Int. J. Eng. Sci. **49** (2011)
11. A.G. Kuchumov, Y.I. Nyashin, V.A. Samartsev, *Proceedings of 7th WACBE World Congress on Bioengineering* (Springer, 2015)
12. A.G. Kuchumov, Russ. J. Biomech. **20** (2016)
13. W.G. Li, X.Y. Luo, N.A. Hill, R.W. Ogden, A. Smythe, A. Majeed, N. Bird, Ann. Biomed. Eng. **39** (2011)
14. W.G. Li, X.Y. Luo, N.A. Hill, A. Smythe, S.B. Chin, A.G. Johnson, N. Bird, Comput. Math. Methods Med. **9** (2008)
15. V.A. Samarcev, *Paths of the improvement of the cholelithiasis treatment at the high operative risk groups: optimization of the diagnostics methods, landmark endoscopic and less-invasive tratment, predicting and prophylactics of the complications* (Doctor Thesis, Academician EA Wagner Perm State Medical Academy, 2005)
16. Ã. Avril, P. Badel, A. Duprey, J. Biomech. **43** (2010)
17. P.J. Howard, G.M. Murphy, R.H Dowling, Gut **32** (1991)
18. A.G. Kuchumov, Y.I. Nyashin, V.A. Samartsev, V.A. Gavrilov, M. Mesnard, Russ. J. Biomech. **15** (2011)
19. A. Sharma, V. Soni, Y. Afaque, M. Baijal, P. Chowbey, A. Goswami, K. Najma, R. Khullar, J. Minim. Access Surg. **11** (2015)

20. Y. Abd El Maboud, Kh.S. Mekheimer, S.I. Abd El Salam, Applied Bionics and Biomechanics **11** (2011)
21. D.R.Urbach, T.A. Stukel, Candadian Medical Association Journal **172** (2012)

Human upper limb manipulator mass center motion and mass moments of inertia variation

Gergana Nikolova[*], *Daniel* Dantchev, and *Alexander* Kazakoff

Institute of Mechanics, Bulgarian Academy of Sciences, Acad. G. Bonchev Str., Building 4, Sofia 1113, Bulgaria

Abstract. Motion control is complicated for people having traumas or neurological diseases. An underlying assumption in our work is that the motion of healthy people is optimal with respect to positioning accuracy, movement response, and energy expenditure. In this paper, a new approach for determination of the human upper limb mass-inertial characteristics is presented by using the 3D geometrical mathematical modeling analysis approach. Two examples will be given to illustrate the main features and advantages of the proposed design concepts. The objective of the work presented in this paper is a determination of the mass properties of a two joints human upper limb manipulator. Results are aimed to have application in an exoskeleton design, the design of manipulation system and external manipulation system, serving people with some motion difficulties, as well as in sport and rehabilitation.

1 Introduction

A new modified biomechanical geometrical model of the human upper limb is presented, with the help of three-dimensional geometrical mathematical modeling – 3D. This model represents a modification of the ones suggested in [1, 2] and is based on our own anthropometric measurements of 50 men aged between 30–40 years. Basic anthropometric dimensions were reported before [3, 4]. The investigation on the human body geometric and mass-inertial characteristics is of key importance in human motion analysis. This investigation is a natural continuation of the work, presented in [5]. The objective of the work presented in this paper is a definition of the mass properties of a two joints human upper limb manipulator. Two examples of two different positions of the upper limb are presented in order to illustrate the main features and advantages of the proposed design concept. The two positions are

[*] Corresponding author: gergana1973@gmail.com

depicted in Figs 2 and 3. We define a coordinate frame ***F*** with its origin at the midpoint of the shoulder joint with the z-axis along the body vertically downwards, the x-axis in the sagittal plane, and the y-axis is in the transversal plane. For any of the two positions we present the coordinates of the center of mass, the principal moments of inertia, the components of the unit vectors that define the coordinate frame in which the inertial tensor is diagonal, and the moments of inertia with respect to the ***F***, as well as its components with respect to a system that has its center at the center of mass but with axes aligned with ***F***. Finally, we present data how the center of mass of the upper limb changes when it moves so that its end reaches the area of the mouth of the human – one imagines a movement of a disabled person who shall be helped in order to drink a cap of water, eat a meal, etc.

The objectives of the work presented in this paper have been achieved by fulfilling the following main tasks:

1) building up a 3D geometrical mathematical model of the upper human limb manipulator, using the statistical data presented in [3, 4];
2) Secondly, verification of the proposed model via comparison of the data obtained from the determination of the human upper limb manipulator mass properties by using the computer code and those derived with the analytical means from the previous studies. The comparison demonstrated a nice agreement between these data.
3) Investigation on how the center of mass of the proposed manipulator changes with respect to angle variations in two planes: *i)* one defined by the plane of the manipulation system and *ii)* the second one, motion of this particular plane with respect to the origin of the coordinate system, situated in the human shoulder, i.e., with respect to the laboratory coordinate frame.
4) Visualization of the mass center motion in 3D space, which is expected to be a partial saddle like surface;
5) Derivation of data that can be used in the design of devices aimed to help disabled people, having problems with the movement of their upper limbs.

We consider two joints model of the upper limb. We take that the upper arm can move only in the (z, x) plane with an angle α that characterizes its position. The lower arm is taken to be able to move in the plane formed by the upper and the lower arm with angle β determining the position of one with respect to the other one. The proposed topic seems to be very interesting and attracting the attention of many scientists worldwide. For example, investigations on a human-arm-like mechanical manipulator are reported before [5, 6, 8, 9, 10]. Reference [7] can be used as a general source of knowledge on the subject. In addition, upper limb portable motion analysis system, based on inertial technology for neuro-rehabilitation purposes is revealed in [11]. Reference [12] deals with control of hand equilibrium trajectories in multi-joint arm movements. In [13] one investigates the coordination of arm movements and present there an experimentally confirmed mathematical model. The effect of visuomotor displacements on arm movement paths is presented in [14]. Finally, the mechanics of multi-joint posture and movement control are revealed in Ref. [15].

2 Model building up

The ideology of the 3D geometrical mathematical modelling – 3D (GMM) analysis is a creation of the geometrical model of the investigated body and by applying on the geometry, the material density and the analytical properties of the 3D bodies used in the approximation to determine all properties of interest including the body mass properties and the mass moments of inertia, respectively. The model geometry of human upper limb presented in Fig. 1. Fig 1a) represents the cross-section of the upper part of the arm at its upper anthropometric point, Fig. 1b) – the corresponding cross-section where the upper arm meets the lower arm, and Fig. 1c) – the cross-section of the lower end of the lower arm. We recall that the upper arm, that extends between *acromion-radiale* anthropometric points, while the lower arm is taken to be, as usual, between the points *radiale-stylion*. The cross sections shown in Fig. 1 are known under the name stadium solids. Its cross-section consists of a rectangular + two parts of ellipsis on the left and right-hand sides. The radii of the ellipsis are such that one has the width and the proper circumference of the crosses section, which are measured for the set of individuals considered.

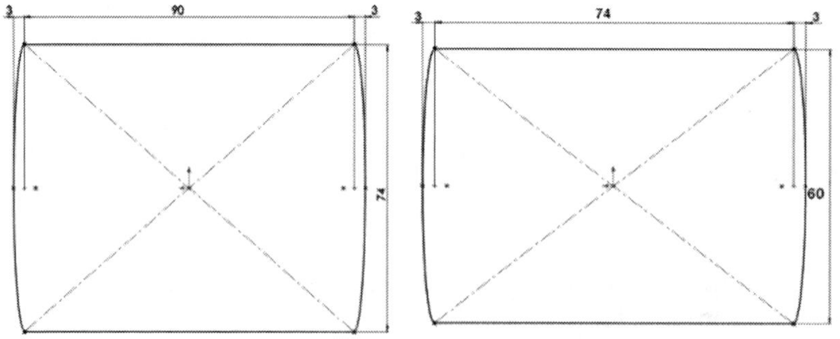

a) Upper Arm upper surface; b) Upper Arm lower surface;

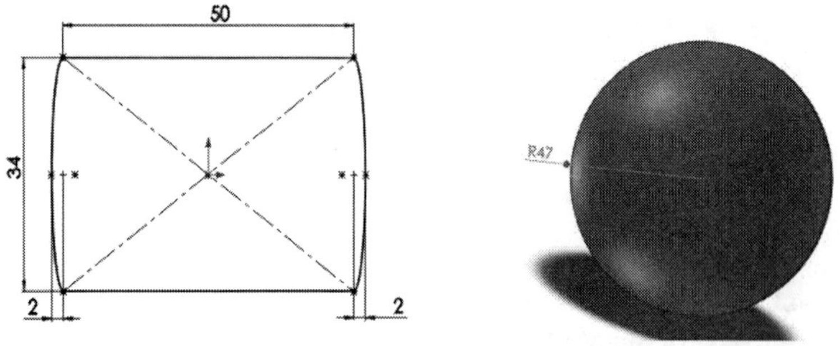

c) Lower Arm lower surface; d) Human hand modeled like sphere;

Fig. 1. Sketches for human upper limb and hand, modeled as a sphere - basic dimensions in mm.

The full model geometry, presented in Figs 2 and 3, can be achieved by extruding the plane figures 1a) and 1b) to obtain the upper arm and figures 1b) and 1c) to form the lower arm of the upper limb with lengths 309 mm and 247 mm, respectively [16]. The model mass properties and the corresponding mass moments of inertia can be computed applying the corresponding material density on the model geometry. The calculated results for the three human upper limb parts are compared with the analytically already calculated and reported in [5] mass moments of inertia. The agreement between the computer model and the analytical results for any of the segments separately gives us confidence that the model can be used for determining the mass inertial characteristic of the whole upper limb for any its special position.

3 Results from the model

The role of the angles α and β and the principal positions of the upper limb in our model are depicted in Figs. 2 and 3. The Fig. 3 demonstrates the role of the angle β for the movement of the lower arm when the upper arm is simply kept fixed aligned with the body along the z-axis. Axes 1-5 in Figs. 2 and 3 are automatically created by the computer system and represent three axes in the basic co-ordinate system X, Y, Z with origin in the shoulder and the other two are oriented longitudinally along the upper and the lower arm, respectively.

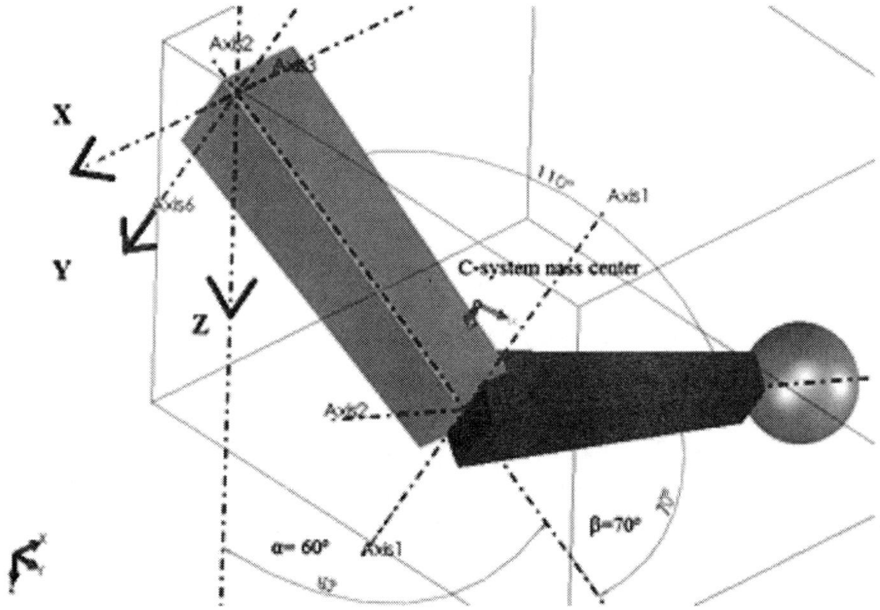

Fig. 2. Manipulator with two joints, position α= 60°, β=70°.

Before reporting the mass inertial parameters for the upper limb in the above positions let us state that our computer model generated in CAD media SolidWorks, has been verified against analytical calculation of the mass inertial parameters for

any of the components of the upper limb which includes the volume, mass, position of center of mass and the inertial moments of the components [5]. That is why we have a confidence that it reliably works for a general position of the upper limb. In principle, our computer model can provide data for mass, volume, surface area, center of mass-coordinates, principal axes of inertia and principal moments of inertia, moments of inertia taken in a coordinate frame at the center of mass and axes aligned with the output coordinate system and moments of inertia, taken at the output coordinate system. For example, this data for the upper limb position shown in Fig. 2 are:
Mass = 3248.13 g, Volume = 3015170.96 mm^3 , Surface area = 193039.49 mm^2, Center of mass coordinates (in mm): $C_x = -67.97, C_y = -205.80, C_z = -205.80$. The unit vectors defining the principal axes of inertia are with components $\vec{e}_1 = (0.59, 0.70, -0.40)$, $\vec{e}_2 = (-0.81, 0.51, -0.30)$, and $\vec{e}_3 = (0.00, 0.50, 0.87)$, and the principal moments of inertia: ($kg \times dm^2$) are

$$P_x = 1.25, P_y = 8.3, P_z = 9.2 \tag{1}$$

Moments of inertia taken at the center of mass and aligned with the output coordinate system are:

$$\begin{matrix} Lxx = 5.8 & Lxy = 2.9 & Lxz = -1.7 \\ Lyx = 2.9 & Lyy = 5.1 & Lyz = -2.4 \\ Lzx = -1.7 & Lzy = -2.4 & Lzz = 7.9 \end{matrix} \tag{2}$$

while moments of inertia taken at the output coordinate system are:

$$\begin{matrix} Ixx = 24.2 & Ixy = 7.5 & Ixz = -4.3 \\ Iyx = 7.5 & Iyy = 11.2 & Iyz = -10.3 \\ Izx = -4.3 & Izy = -10.3 & Izz = 23.1 \end{matrix} \tag{3}$$

The point C represents the center of mass of the manipulator in the configuration given – see Figs 2 and 3. It also is the origin of the coordinate system shown in Figs 2 and 3, in which the inertial moments tensor is in its diagonal form. Similar detailed data can be reported for any position of the upper limb. For example, for the configuration shown in Fig. 3 one obtains:

$$C_x = -36.16, C_y = 0, C_z = 150.26 \tag{4}$$

Using the possibilities offered by the model we have determined also how the center of mass of the upper limb changes when it moves in a way that the hand reaches near the mouth of the human. Fig. 4 shows the C-mass center-coordinates in [mm]: α varies with seven values from zero to 60°, while β varies with eleven values from 70° to 170°. A saddle like surface of the positions of the center of mass is defined in this way.

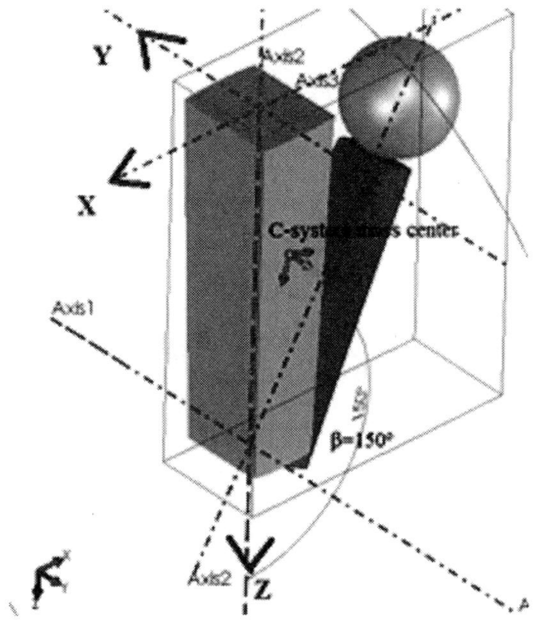

Fig. 3. Manipulator with two joints-α=0° and β=150°.

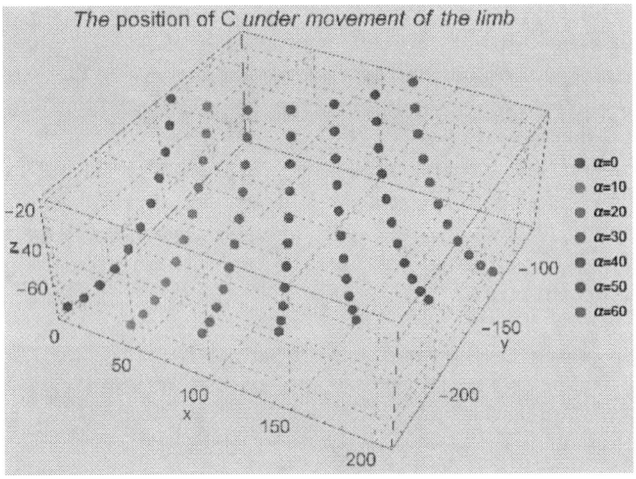

Fig. 4. Graphics of mass center of motion-coordinates in [mm].

4 Conclusion

In this paper, a new approach for determination of the human upper limb mass-inertial characteristics is presented by using the 3D geometrical mathematical modeling analysis approach. Results are presented, derived from the built-up 3D geometrical mathematical analysis, demonstrating few of the options offered by our approach. Analysing the results, defined by the new approach, the following conclusions can be stated along the z-axis.

1. A new 3D geometrical mathematical model of the human upper limb is build up, utilizing a computer code.
2. Mass properties of the human upper limb manipulator are determined and partially illustrated in the current paper.
3. A comparison is performed with the mass properties, defined by the analytical analysis and reported before in previous studies. The achieved precision is perfect and this fact allows us future calculations, using this 3D model to be performed with confidence.
4. Visualisation of the manipulation mass center motion is performed under the variation of both investigated angles. A partial saddle like surface is generated, defining the motion in the original laboratory coordinate system.
5. Data are generated, that can be used in design of devices aimed to help disabled people having problems with the movement of their upper limbs.

5 Future work

An obvious direction of future activities is the improvement of the proposed model so, that the geometrical modelling becomes as close as possible to the real shape of the human upper limb. In the current study we used the geometrical parameters directly as measured from a set of individuals. The approximation with specific 3D bodies of the segments of the limb leads to a given mistake in reproducing its mass. If one insists of using the physically measured density one way around that problem is to modify with, say, 5% precision the actual measure data in such a way that the obtained mathematical model reproduces exactly the mass of segments of the limb as follows from the corresponding regression equations. Since they do depend on the gender of the corresponding person in such a way our model will also become, to a given extend, gender specific.

Authors believe that the presented study can be really helpful in the design of manipulation systems with application in sport and rehabilitation providing the handy opportunity to calculate and determine the mass-inertial parameters of the upper limb of a given individual provided the anthropometric easily measurable data for that individual are known.

The financial support by the Bulgarian National Science Fund: Contract DN-07/5 "Study of anthropometric and mass-inertial characteristics of the Bulgarian men and women via mathematical models of the human body" is gratefully acknowledged.

References

1. G. Nikolova, Y. Toshev, J. Biomech., **40**, 3700–3707 (2007).
2. G. Nikolova, J. Vibroeng., **11**, 2, 274-278 (2009).
3. G. Nikolova, J. Theoretical Appl. Mech., **40**, 3, 101–116 (2010).
4. G. Nikolova, J. Theoretical Appl. Mech., **40**, 4, 73–80 (2010).
5. G. Nikolova, D. Dantchev, Al. Kazakoff, Proc. of the 3-rd World Congress on New Technologies (NewTech'17) Rome, Italy, 2017, (to be published).
6. V. Grecu, N. Dumitru, L. Grecu, Proc. of the International Multi Conference of Engineers and Computer Scientists, Vol **II**, IMECS, Hong Kong (2009).

7. Ph. Ringrose, M. Bemtley, Springer Dordrecht Heidelberg New York London, ISBN 978-94-007-5496-6, DOI 10.1007/978-94-007-5497-3(eBOOK).
8. F. J. Valero-Cuevas, DOI 10.1007/978-1-4471-6747-1_2, Springer-Verlag London, 2016.
9. M. Z. Al-Faiz, A. F. Shanta, Intelligent Control and Automation, **6**, 29-37 (2015).
10. A. Hemami, Robotica, **5**, 23-28 (1987).
11. R. Perez et all, Sensors, **10**, 1, 0733-10751 (2010).
12. T. Flash, Biol. Cybern, **57**, 257-274 (1987).
13. T. Flash, N. Hogan, The Journal of Neuroscience, **5**, 7, 1688-1703 (1985).
14. S. J. Goodbody, D. M. Wolpert, Exp Brain Res, **127**, 213–223, (1999).
15. N. Hogan, Biol. Cybern, **52**, 315-331 (1985).
16. Y. Yordanov et al. Anthropology of the Bulgarian population at the end of the 20th century (30-40 years old persons), Prof. M. Drinov Publishing House, Sofia, (2006).

MATEC Web of Conferences **145**, 04006 (2018) https://doi.org/10.1051/matecconf/201814504006
NCTAM 2017

CAD Design of Human Male Body for Mass–Inertial Characteristics Studies

Gergana Nikolova[*], *Vladimir* Kotev, and *Daniel* Dantchev

Institute of Mechanics, Bulgarian Academy of Sciences, Acad. G. Bonchev Str., Building 4, Sofia 1113, Bulgaria

Abstract. The aim of the present research is to present a 16-segmental biomechanical model of the Bulgarian male to determine the mass-inertial characteristics of the body of the Bulgarian male based on parameters available in the literature and its 3D generation within SolidWorks software. The motivation of the research is to support mainly sport, rehabilitation, wearable robots and furniture design users. The proposed CAD model of the human body of men is verified against the analytical results from our previous investigation, as well as through comparison with data available in the provided references. In this paper we model two basic human body positions: standing position and sitting with thighs elevated. The comparison performed between our model results and data reported in literature gives us confidence that this model can be reliably used to calculate the mass-inertial characteristics of male body at any postures of the body that is of interest. Therefore, our model can be used to obtain data for positions which the human body has to take in everyday live, in sport, leisure, including space exploration, for investigating criminology cases – body fall, car crash, etc. The model is suitable for performing computer simulation in robotics, medicine, sport and other areas.

1 Introduction

The human motion is a topic of analysis since ancient times. To make this analysis scientifically sound one obviously needs the knowledge of the geometric and mass-inertial characteristics of the body as total, as well as of the different human body segments. In modern times such studies were related to topics like investigations oriented in the design of air or space transport systems – they have been conducted in the 70-s on elderly male cadavers [1]. In the same period the anthropometric measurements and data for moments of inertia and centers of gravity have been also obtained for living male individuals of the Air Force population. They were collected for the inertial parameters about three mutually perpendicular axes for various body positions [2, 3]. In the early 21st century the same positions were examined in the scale-up study, see [4]. In paper [5] authors present a study for modeling and simulation of human body by considering the skeleton's bones. Body segment inertial parameters (BSIPs) are obtained by cameras and motion analysis methods

---

[*] Corresponding author: gergana1973@gmail.com

[6]. Authors computed individual BSIPs from scan images. Body segment parameters of young Chinese men are determined with magnetic resonance imaging in Ref. [7]. Mass, center of mass, and moment of inertia estimates for infant limb segments are studied in [8]. The analysis of lower limb movements has been well established in biomechanics research and clinical applications for a long time. For these studies, powerful and very advanced tools have been developed to measure movement parameters and reaction forces. The main focus of interest aims towards gait movements where the understanding of the basic concepts is supported by numerous models. Definitions of physiological ranges and detection of pathological changes in movements open an increasingly valuable clinical field of application. If, however, the primary function of the upper extremities as highly variable and adaptive organ for manipulating tasks is the subject of interest, the situation becomes considerably more complex [9]. It is necessarily to determine mass-inertial characteristics in order to study human movement and its mechanics [10].

During the past decade, a lot of research of wearable robotics has been done [11, 12]. In order to design and develop exoskeletons and wearable robots it is again necessarily to determine mass-inertial characteristics of each user. There is a relationship between the wearable devices and the dimensions of mechanical construction which are specific for each individual and thus the power of the motors (electric or pneumatic) is different. Generally, in order to study the human dynamics of different postures, one unavoidably has to determine mass inertial characteristics of human body in different positions. These characteristics are, obviously, individual and depend on the size of body segments and their mass [5-12].

In the current article we propose an approach to the problem based on a mathematical model realized in a suitable computer media. The design of 3D model of the human body requires knowledge of anthropometry and biomechanics. Moreover, one should successively solve specific problems, related to: i) appropriate body decomposition; ii) choice of necessary anthropometric points; iii) organization of data that give the geometrical and mass-inertial characteristics of the human body and its segments; iv) model generation within a CAD system (e.g. SolidWorks); v) verification of the model via analytical results. The 3D model should enable to perform fast individualization using a small range of parameters that could be easily determined.

In the current study, we suggest such a model and provide data for the mass-inertial parameters of the human body in two basic positions, and compare our results with those available from the literature. In order to do that we are going to use a realization of the model within a CAD system - Solid Works.

2 Model and method

We use a mathematical model of the human body described in details in [13]. There this model has been used to determine the mass inertial parameters of the different segments of the body. Here we only present some basic facts for the model used in the current study - see Fig. 1.

The model consists of 16 segments representing: head + neck, upper, middle and lower part of torso, thigh, shank, foot, upper arm, lower arm and hand, assumed to be relatively simple geometrical bodies. We assume full body symmetry with respect to the sagittal plane. The geometrical data needed is taken from a detailed anthropological investigation of the Bulgarian population [14]. A total of 2435 males were measured. We take the average values and design a model, which represents the so defined average Bulgarian male. After determining the mass inertial parameters of the segments, one can also study the corresponding characteristics of the total body assuming the body to be in a given position of interest.

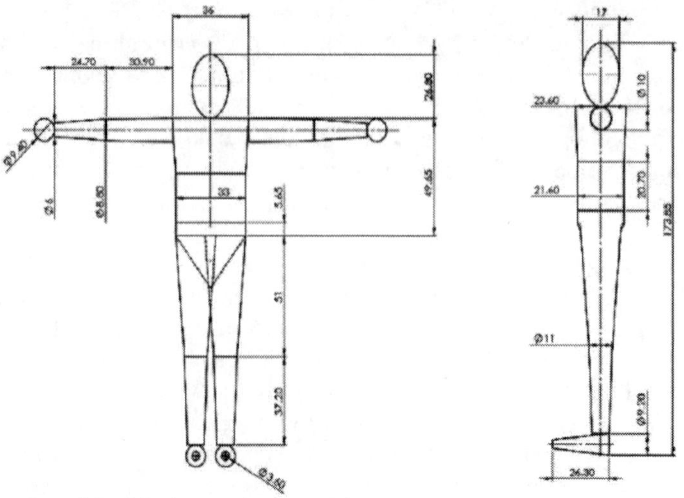

Fig. 1. 16-segemental model of the human body and the corresponding dimensions.

In order to achieve this goal, we have performed a realization of the model in CAD system – SolidWorks. We have *verified* the computer realization by comparing the results it delivers for the mass-inertial parameters of the segments of the body with those reported in [13]. The basic positions of the body have been classified long ago in the literature – see, e.g., [1, 2, 3]. One normally considers eight principal body positions. Here, we will present data for two of these positions: the so-called "standing position" and the so-called "sitting, thighs elevated position" – see Fig. 2.

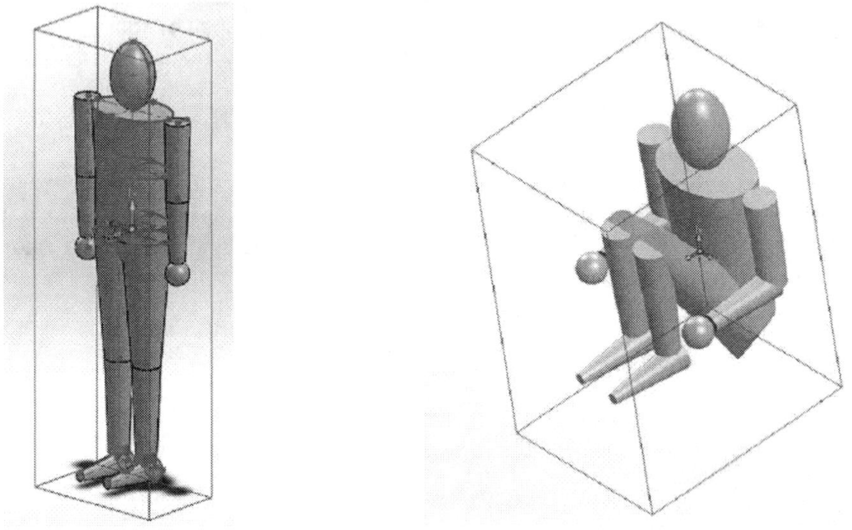

Fig. 2. Standing position (left); Sitting, thighs elevated position (right)

3 Determination of mass-inertial characteristics using SolidWorks media

The 3D CAD Design software SolidWorks, is used to generate the model.

The computer-generated model is verified by comparing its results with those reported in [13] for the different segments of the body. The program procreates segment-by-segment data about volume, mass, center of mass and moments of inertia. That gives us assurance that this model could be used to calculate these characteristics at different body positions of the body. The data for the mass-inertial parameters of the human body in positions shown in Fig. 2 are presented in Tables 1 and 2. Table 1 contains the data for the so-called standing position, Table 2 - data for the principal moments of inertia of the body in so-called sitting, thighs elevated position as well as the coordinates of whole center of mass of the body.

Table 1. Standing position.

	Ref. [4]		**Ref. [1]**	**Ref. [2]**	**Ref. [3]**		**Our data**
	50%	95%			50%	95%	
I_{XX} [kg.cm^{2}x10^3]	14.4	18.5	17.2	12.7	9.1	14.1	9.7
I_{YY} [kg.cm^{2}x10^3]	129.2	163.4	118.9	116.0	116.2	161.9	105.3
I_{ZZ} [kg.cm^{2}x10^3]	144.5	182.3	134.0	129.5	122.3	171.1	112.0
Center of mass [cm]	80.2	84.7	72.3	78.7	80.0	83.8	74.6
Total mass [kg]	82.2	98.5	65.2	75.5	73.4	90.9	72.5
Height [cm]	179.9	190.1	172.1	176.3	175.5	185.7	171.5

Following the definition of [3] in the "standing position" the individual stands erect with head oriented in the so-called Frankfort plane and with arms hanging naturally at the sides - see Fig. 2 (left). In the "sitting, thighs elevated position" thighs and forearms are placed parallel to the X-axis, the upper arms, shanks and spine are parallel to the X-axis; the soles are parallel to X-Y plane; wrist axes are parallel to Z-axis, and the head lies in Frankfort plane - see Fig. 2 (right).

Table 2. Moments of inertia in standing position and the whole body center of mass location in sitting, thighs elevated position.

Characteristics	**Ref. [4]**		**Ref. [1]**	**Our data**
	50%	95%		
I_{XX} [kg.cm^{2}x10^3]	13.1	15.2	17.9	18.3
I_{YY} [kg.cm^{2}x10^3]	48.6	55.8	42.8	33.3
I_{ZZ} [kg.cm^{2}x10^3]	48.7	59.8	44.0	31.4
L (x)	59.4	61.5	58.7	53.6
L (y)	0	0	0	0
L (z)	9.7	10.5	11.7	7.3

The coordinate frame used is with axes which coincide with the approximate body axes: the sagittal (y), the frontal (z) and the longitudinal (x) ones. The Tables contain also data from the available literature found by using other methods. The coordinate system for the results reported, as well as the units used there, have been converted to the ones used in present investigation. The 50% and 95% marks mean that the corresponding percent of measured data are below the value reported.

The analysis shows a reasonably good agreement between our results and those previously reported in the literature.

4 Conclusion

We proposed a 16-segment model of the human body of the average Bulgarian male and its 3D model realization in SolidWorks environment. Data for the mass-inertial characteristics of the body in two of its basic positions are obtained and compared with those reported in the literature. The comparison between our model results and data reported in literature – see Tables 2 and 3 - gives us confidence that this model could be used to calculate the mass inertial characteristics at any specific posture of the male body.

The motivation of the research is to support mainly sport, rehabilitation, wearable robots and ergonomics (furniture design users). The model may be applied also in other areas such as simulation of the human behavior in space, medicine (orthopedics, orthotics and prosthetics design), criminology, etc.

To reduce the differences between the results obtained and those available in the literature, we plan to improve segment modeling by using geometric bodies closer to the real shape of the human body segments. For that aim, one needs to collect data for additional anthropometric parameters. This is currently underway. Naturally, for the new set of geometric bodies one will need again to derive analytical expressions for their mass-inertial parameters and to compare with those obtained through the corresponding realization of the modified model within a suitable CAD/CAM system.

The financial support by the Bulgarian National Science Fund: Contract DN-07/5 "Study of anthropometric and mass-inertial characteristics of the Bulgarian men and women via mathematical models of the human body" is gratefully acknowledged.

References

1. C. E. Clauser, J. T. McConville, J. W. Young, Weight, volume, and center of mass of segments of the human body, TR AMRL-TR-69-70, Ohio, (1969).
2. W. R. Santschi, J. DuBois, C. Omoto, Moments of Inertia and Centers of Gravity of the Living Human Body, AMRL TR 63-36. Ohio, (1963).
3. E. P. Hanavan, A mathematical model of the human body, AMRL-TR-64-102, Ohio, (1964).
4. NASA-STD-3000, The man_system integration standards, Anthropometry and biomechanics at: https://msis.jsc.nasa.gov/sections/section03.htm#3.2.1.
5. D. Andreescu, Fl. Ionescu, H. Riehle, *Modelling and Simulation of Human Body by Considering the Skeleton's Bones*, (IASTED Intern Confer BioMechanics August 16-18 2007, Honolulu Hawaii, USA, 2007).
6. M. Rossi, A. Lyttle, N. Benjanuvatra, B. Blanksby, J Sports Sciences Med., **12**, 4 (2013).
7. Ho Wh., S. Ty, L. Cc, Y. Cheng, Med Sci Sports Exerc. **45**, 9 (2013).

8. K. Schneider, R. Zernicke, J Biomech., **25**, 2 (1992).
9. G. Rau, C. Disselhorst-Klug, R. Schmidt, J Biomech., **33**, 10 (2000).
10. A. Tözeren, *Human Body Dynamics Classical Mechanics and Human Movement* (Springer, 1999).
11. J. Pons, *Wearable Robots: Biomechatronic Exoskeletons* (John Wiley and sons, Ltd, England, 2008).
12. W. Huo, S. Mohammed, J. Moreno, Y. Amirat, IEEE Systems Journal, **10**, 3 (2016).
13. G. Nikolova, Y. Toshev, J Biomech., **40**, 16 (2007).
14. Y. Yordanov et al. *Anthropology of the Bulgarian population at the end of the 20th century (30-40 years old persons)* (Prof. Marin Drinov Academic Publishing House, Sofia, Bulgaria, 2006).

MATEC Web of Conferences **145**, 04007 (2018) https://doi.org/10.1051/matecconf/201814504007
NCTAM 2017

Biomechanics of the immediate impact of wearing a rigid thoracolumbar corset on gait kinematics and spatiotemporal parameters

Redha Taiar[1,*], *Alexandra Mitton*[2], *Julien Cambier*[1], *Islem Guenaoui*[1], *Sophie Michel*[3], *Ellie Abdi*[4], *and Guillaume Polidori*[1]

[1]Research Group in Engineering Sciences (GRESPI), University of Reims, France
[2]Masseur Kinésithérapeute, Doulaincourt Saucourt, France
[3]Institut de Formation en Masso-Kinésithérapie (IFMK), Reims, France
[4]Montclair State University, New Jersey, USA

Abstract. The corset support is a device classified as orthosis. It compensates a functional deficiency with means of protection, recovery, correction, maintenance, and support or contention. There are two types of orthosis 1) rest orthosis and 2) corrective orthosis. Rest orthosis maintains joints in a defined position to avoid deformities or to relieve a pain at joints. Corrective orthosis adjusts joint deformity either passively or actively. Corset is used in various pathological use, thoracic-lumbar fracture, scoliosis, Scheuermann's disease or spinal dystrophy. The purpose of this study was 1) to determine the immediate impact of wearing a semi-rigid thoracolumbar corset, the Lombax® Dorso on gait kinematics and 2) spatiotemporal parameters in 6 adults. These parameters were recorded using the optoelectronic system Vicon® on treadmill gait subjects with and without corset for the comparison. The results showed that wearing a corset significantly decrease the rotation amplitudes of the scapular and pelvic girdles ($p<0.05$) in the frontal plane. The movement of the pelvis and hip in this same plane was decreased also when comparing with and without a corset effects ($p<0.05$). The corset significantly increased the range of flexion-extension of the hip during the gait cycle. At the conclusion of this study the discriminate parameters of wearing a corset was quantified. The results and in association with manufacturer will help to improve materials for better optimization support. Comparable perspectives and after improvement of materials will aim to experiment with patients on real daily life situation.

1 Introduction

The purpose of this study was to determine immediate impact of wearing a semi-rigid thoracolumbar corset, the Lombax® Dorso on gait kinematics and spatiotemporal parameters in 6 healthy adults. The corset is an apparatus classified as orthosis. It

[*] Corresponding author: redha.taiar@univ-reims.fr

compensate for a functional deficiency with means of protection, recovery, correction, maintenance, and support or contention [1]. There are several categories of orthoses depending on the searched purpose. The corsets were comprised of two types, 1) rest orthosis and 2) corrective orthosis. Rest orthosis maintains joints in a defined position to avoid deformities or to relieve a pain at joints. Corrective orthosis corrects joint deformity either passively or actively. Corsets are used in various pathological aspects, thoracolumbar fracture [2-4], scoliosis [5], Scheuermann's disease or spinal dystrophy [6]. By its application, a corset limits the movement of several joints therefore; a corset could impede walking and increase its energy costs. It is noteworthy to evaluate effects of wearing a corset on gait spatiotemporals and kinematics parameters in order to collect elements to be worked during its removal.

Gait is sequenced in cycle beginning with the contact of one foot on the ground and ending with the next contact of this same foot. It alternates a support phase where the foot stays on the ground and an oscillating phase where the foot is not in contact with the ground [7-9]. Technological advances made it possible to develop new tools allowing a fine analysis of gait. It provided quantitative and objectives data. In this study, an optoelectonic system has been chosen for the measurements. Various studies on impact of wearing a corset on gait and kinematics parameters have been investigated [10-13]. The trends obtained in this study were validated.

2 Materials and methods

Six subjects (4 women's and 2 men's) participated in this study with 21 ± 1, 5 Age, 60 ± 5 kg and 171 ± 6, 5 cm size. The criteria of inclusion are the absence of old traumas, orthopedic disorders at the musculoskeletal system and rachis. Before the realization of the experiment, they were informed by modalities of the experimentation and gave by writing their informed consent.

The Lombax® Dorso, a corset of the Thuasne® brand (Fig.1), was used in this study. It is indicated in muscular and static insufficiencies, osteoporosis (vertebral compression fracture) and postoperative maintain. Depending on the manufacturer [14], the corset acts by a 3-point support of the recovery and stabilization of the thoracic spine. This is possible by the recoil of scapular belt, 1) on the discharge of the thoracic and lumbar vertebral corpus and 2) on the increased intra-abdominal pressure.

A treadmill was used to give a constant gait speed to the subjects. A 480 CX treadmill model of Proform® was utilized. In a study Stoquart, Detrembleur, and Lejeune [15] have shown that gait kinematics and spatiotemporals parameters were dependent of the speed. It is therefore necessary to develop a constant gait speed in order to avoid a bias. Several studies [15-18] concluded that the recorded parameters during gait on the treadmill were similar to those of gait on the ground.

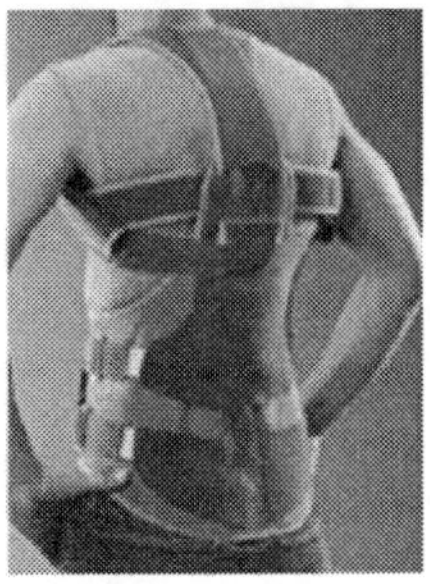

Fig. 1. Lombax® Dorso corset of the brand Thuasne® [24]

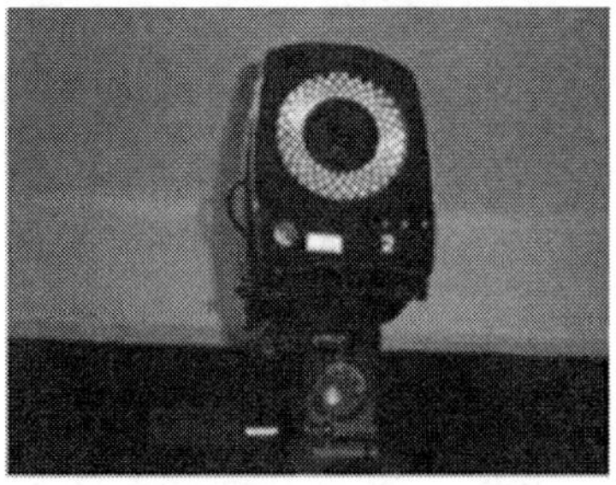

Fig. 2. Camera of Vicon® optoelectronic system

The Vicon® optoelectonic system was used to measure gait kinematics and spatiotemporal data. It was composed of 5 cameras (Fig.2) which produced infrared light. The subject was equipped with markers that reflected this infrared light. This allowed the cameras to capture and to record the markers' trajectory in the 3 dimensions with an accuracy order of a millimeter. Cameras were connected to a digital recorder (Oxford Metrics Vicon® Data station 3D motion capture interface). This digital recorder was connected to a central processing unit. Eighteen markers infrared light reflectors with a diameter of 14 mm were placed on the subjects to calculate the joints centers.

The experimental protocol was itemized into 9 steps:

1. Arranging the acquisition volume around the treadmill (Fig.3).
2. System calibration.
3. Placement of 18 reflective markers at 14 mm in diameter on the level of the lower limbs and the scapular belt (Fig.4). These are placed on anatomical point defined by the Plug-in-Gait model.
4. Measurement of anthropometrics parameters of each subject essential for the calculation of the joints centers.
5. Walking at 4 km/h for 3 minutes by the subject in order for familiarization with the treadmill to eliminate leaning bias.
6. Recording of kinematics and spatiotemporal data by the Vicon® optoelectronic system with a sampling frequency of 120 Hz. Three 30-seconds tests were performed during which the subject was asked to walk at 4 km/h while looking straight (Fig.5-a). In between each trial, the subject continued to walk.
7. Placement of Lombax® Dorso corset.
8. Recording of gait kinematics and spatiotemporal data with wearing of corset (Fig.5-b).
9. Data processing was filtered to eliminate noise. A biomechanical model OLGA (Optimized Lower-Limb Gait Analysis) was applied to calculate the joints centers by the positions of the markers combined with the anthropometrics parameters of the subject.

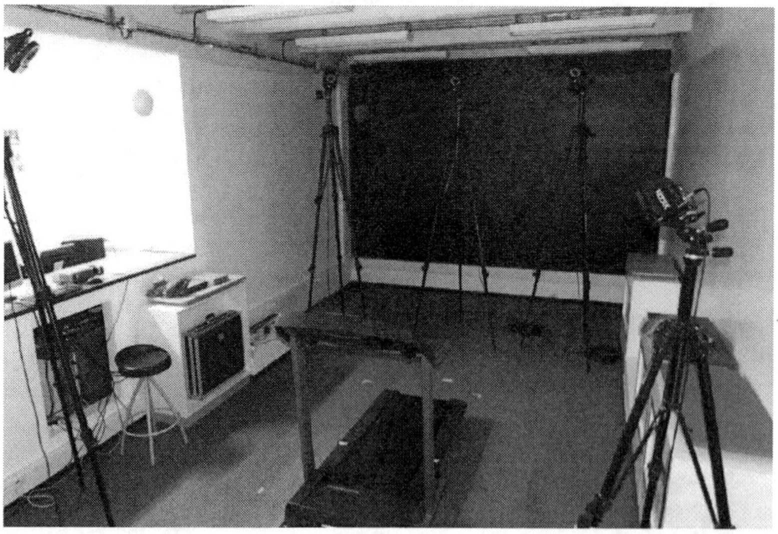

Fig. 3. Acquisition volume

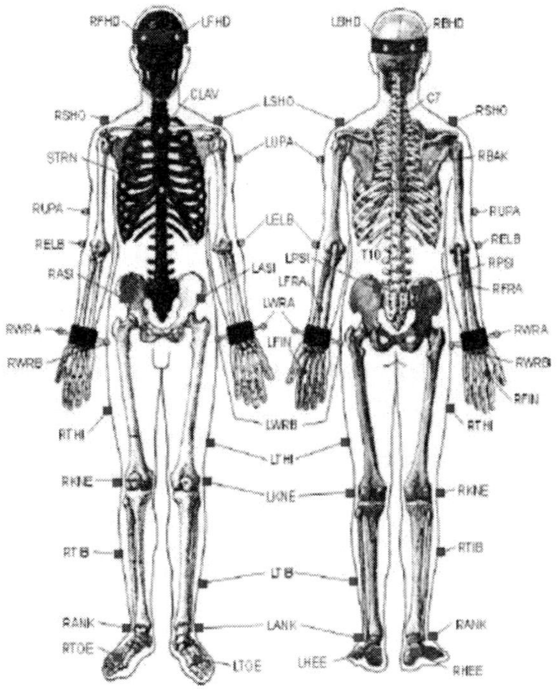

Fig. 4. Plug-in-Gait model [34]

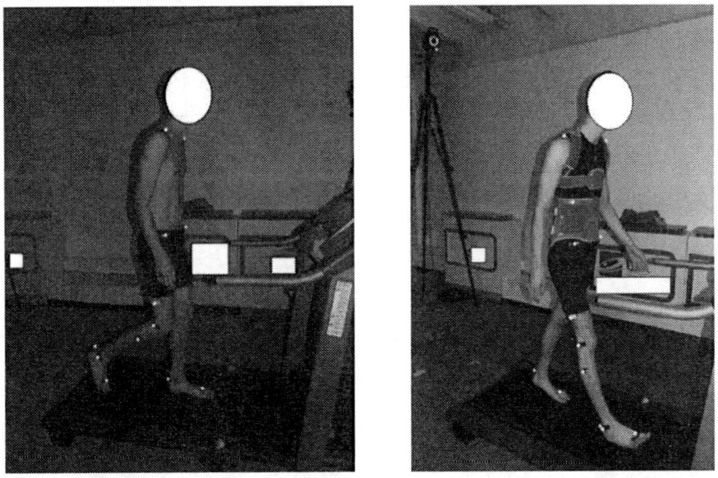

Fig. 5. Gait without (a) and with (b) corset at 4 km/h

3 Results and discussion

The results obtained for gait without a corset and gait with Lombax® Dorso corset on a treadmill at 4 km/h were compared (Table 1 and Fig. 6).

It is observable that wearing of the Lombax® Dorso corset has little effect on the spatial parameters analyzed (Step Length (ST) and duration of the support phase (DSP)) during a gait at 4 km/h, deviations not exceeding 3% in this study. Results on kinematics parameters at the scapular belt, pelvis, hip, knee, and ankle were more significant. At the level of scapular belt, the amplitude of rotation decreased by 41.25 % when corset was worn during gait. This decrease of 6° is significant, which confirms the hypothesis that this movement is limited due to the application of the corset in three points (sternal, spinal and pelvis). At the level of the pelvis in transverse plane, a significant difference was noted. The amplitude of pelvis rotation decreased by 30% between the two experimental conditions. The hypothesis which during the physiological gait, there was a dissociation of the scapular and pelvic belts thus was confirmed.

Table 1: Comparison of saptiotemporal and kinematic parameters between gait conditions with and without corset.

	Average and standard deviation		Difference	Variation %
	Without corset	With corset		
Step Length SL (m)	0,66 ± 0,05	0,66 ±0,06	0	0,55
Duration of Support phase DSP (s)	0,43 ± 0,05	0,44 ± 0,06	0,01	2,24
Scapular belt – Transverse plane SBTP (°)	14,15 ± 4,71	8,31 ± 2,64	-5,84	-41,25
Pelvis – Transverse plane PTP (°)	9,03 ± 4,73	6,26 ± 2,12	-2,78	-30,74
Pelvis – Frontal plane PFP (°)	9,10 ± 2,62	5,76 ± 2,97	-3,34	-36,7
Hip – Sagittal plane HSP (°)	40,33 ± 3,33	42,91 ± 4,45	2,57	6,38
Hip – Frontal plane HFP (°)	11,37 ± 2,28	8,5 ± 2,45	-2,86	-25,19
Knee – Sagittal plane KSP (°)	56,72 ± 3,90	58,16 ± 3,98	1,44	2,53
Ankle – Sagittal plane ASP (°)	29,47 ± 10,19	31,52 ± 10,71	2,05	6,95

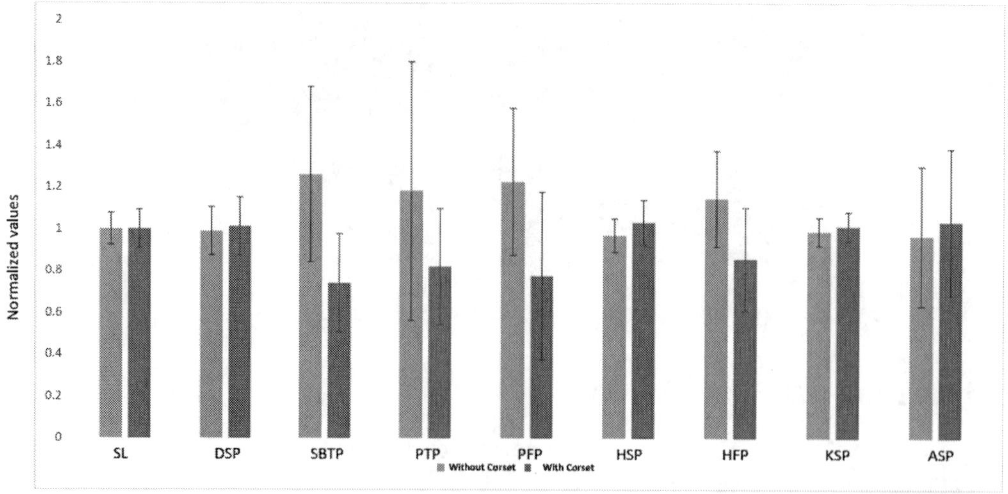

Fig. 6. Comparison of parameters without and with corset (values normalized: division by the average).

In the literature, the results diverge according to the studies. In the study of Mahaudens, Banse, and Detrembleur [10] along with Vogt, Pfeifer, Portscher, and Banzer [13] thermoformed rigid corset had no significant influence on this movement of the pelvis during gait. In addition, the experimentation of Konz, Fatone, and Gard [12] also shows that wearing a fiberglass corset significantly reduces the rotation of the pelvis during gait. In the frontal plane of the pelvis, all studies [10, 12, and 13] are in agreement with this study. A significant decrease in the amplitude of the pelvis in the frontal plane either for a thermoformed rigid corset, fiberglass or for a lumbar corset was found. At the level of the hip in the sagittal plane, while wearing the semi-rigid corset increased by approximately 6% amplitude of flexion-extension during the gait. This confirmed the hypothesis adopted beforehand.

The speed of walking was imposed at 4 km /h. With the intention of maintaining this rhythm, the amplitude of flexion-extension of the hip increased in order to compensate for the movements limited by the corset. The length of the step was not affected by the modification of these kinematic parameters which Konz et al., [12] confirmed as a found result. In the frontal plane of the hip, while the Lombax® Dorso corset was worn it resulted in a significant decrease in the amplitude of the adduction-abduction movements. Thus, the worn corset limiting the movement of the pelvis in the frontal plane led to a reduction in the amplitude. This reduction was observed at the adduction-abduction of the hip during walking, a result also noticed in [10, 12]. Concerning the influence on the knee and the ankle this study respected that there was no significant evolution yet again confirmed by [12].

4 Conclusion

This study was attentive in immediate impact of worn rigid thoracolumbar corset, the Lombax® Dorso, on gait kinematics and spatiotemporal parameters in 6 healthy adults. It followed those amplitudes of the rotations scapular and pelvic belts of the inclination of the pelvis in the frontal plane. In addition adduction-abduction of the hip which were significantly diminished by wearing the corset while amplitude of flexion-extension of hip increased significantly for the study population. For the latter, the results obtained made it possible to establish rehabilitation objectives and to propose physiotherapy methods in order to recover a physiological march.

Acknowledgement

The authors declare that there was no personal financial interest in conducting this research. The information on companies' names is integral to the research.

References

1. Y. Xhardez et collaborateurs, *Vade-mecum de kinésithérapie et de rééducation fonctionnelle*, 6e édition (Maloine, Paris, 2009), pp 96-97.
2. J. C. Dosch, *Traumatologie du rachis. Issy-les-Moulineaux* (Elsevier, Masson, 2012) p.103.
3. M. Freslon, D. Bouaka, P. Coipeau, G. Defossez, N. Leclercq, J. Nebout, E. Marteau, N. Poilbout, R. Prebet, Revue de chirurgie orthopédique et réparatrice de l'appareil locomoteur **94S**, S22-S35 (2008).
4. P. Guigui, B. Lassale, A. Deburge, *Fractures et luxations récentes du rachis dorsal et lombaire de l'adulte* (EMC, 15-829-A-10, 1998).
5. A. V. Bruyneel, P. Chavet, S. Mesure, Kinésithérapie la revue **8**, 23-29 (2008).

6. C. Palazzo, F. Sailhan, M. Revel, Revue de rhumatisme monographies **81**, 7-15 (2014).
7. P. A. Willems, B. Schepens, C. Detrembleur, *Marche normale* (EMC, 26-007-B-75, 2012).
8. S. Armand. Analyse Quantifiée de la Marche : extraction de connaissances à partir de données pour l'aide à l'interprétation clinique de la marche digitigrade, Thèse présentée à l'Université de Valenciennes et du Hainaut Cambrésis. Numéro d'ordre 05-07 (2005).
9. Viel E et collaborateurs, *La marche humaine, la course et le saut : biomécanique, explorations, normes et dysfonctionnements* (Masson, Paris, 2000) p.18.
10. P. Mahaudens, X. Banse, C. Detrembleur, Gait & Posture **28**, 703-707 (2008).
11. M. S. Wong, C. Y. Cheng, B. K. W Ng, Gait & Posture **28**, 189-195 (2008).
12. R. Konz, S. Fatone, S. Gard, J. Rehabil. Res. Dev. **43**, 161-170 (2006).
13. L. Vogt, K. Pfeifer, W. Portscher, W. Banzer, J. Rehabil. Res. Dev. **37**, 495-499 (2000).
14. W. H. Wu, X. C. Lin, O. G. Meijer et al., Human Movement Science **33**, 194-202 (2014).
15. G. Stoquart, C. Detrembleur, T. Lejeune, Clinical Neurophysiology, **38**, 105-116 (2008).
16. O. P. Riley, G. Paolini, U. Della Croce, K.W. Paylo, D.C. Kerrigan, Gait & Posture **26**, 17-24 (2006).
17. K. Parvataneni, L. Ploeg, S. J. Olney, B. Brouwer, Clinical Biomechanics **24**, 95-100 (2009).
18. D. Pradon, E. Hutin, S. Khadir, R. Taiar, F. Genet, N. Roche, Clinical Biomechanics **26**, 867-872 (2011).

Application of magnetoelastic effects to the evaluation of elastic and plastic strains in steel structural members

Eduard Gorkunov[1,*]

[1]Institute of Engineering Science, RAS (Ural Branch), Ekaterinburg, Russia

Abstract. The paper deals with different methods for evaluating macroscopic and microscopic stresses in ferromagnetic materials by magnetic characteristics. It demonstrates the possibility of reconstructing strain curves according to measured magnetic characteristics of steel products.

1 Introduction

Magnetic methods are successfully used to evaluate the mechanical properties, structural state and phase composition of steel and cast-iron products. Besides the structural state of members, their stress state is of great importance. Macrostresses and induced anisotropy (texture) can be easily revealed by magnetic methods. The evaluation of the level of microstresses is a challenge. The evaluation methods are based on the recording of parameters associated with the displacement of 90° domain walls, that is, with the redistribution of magnetoelastic energy in a ferromagnet.

Under magnetization reversal, the magnetization of ferromagnets changes discretely in the form of irreversible magnetization jumps of different amplitude (the Barkhausen effect). Barkhausen jumps (BJs) appear due to the interaction of domain walls (DWs) with defects of different nature (pores, inclusions, secondary phase precipitations, dislocations, grain boundaries, stress gradients, etc.). Therefore, the recorded BJ flux parameters represent the structural state of ferromagnets, and they are sensitive indicators of its changes, and they can be used both for nondestructive testing and for analysing the degradation of product properties under operation.

2 Evaluation of working macrostresses

Iron-based ferromagnets are characterised by the existence of two subsystems of 180° and 90° domain walls. Figure 1 shows a transition of a 180° domain wall through an inclusion defect in the longitudinal magnetization of an iron-silicon alloy crystal. DW separation from a defect (at a critical field H_{cr}) generates electromagnetic and elastic waves (fig. 1 b).

* Corresponding author: ges@imach.uran.ru

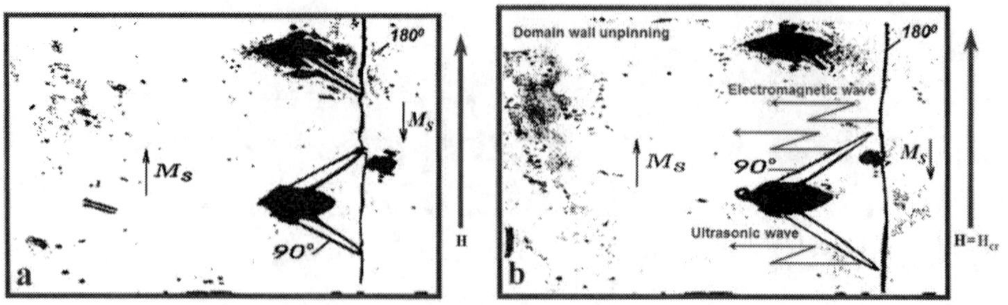

Fig. 1. Interaction of domain walls with an inclusion-type defect.

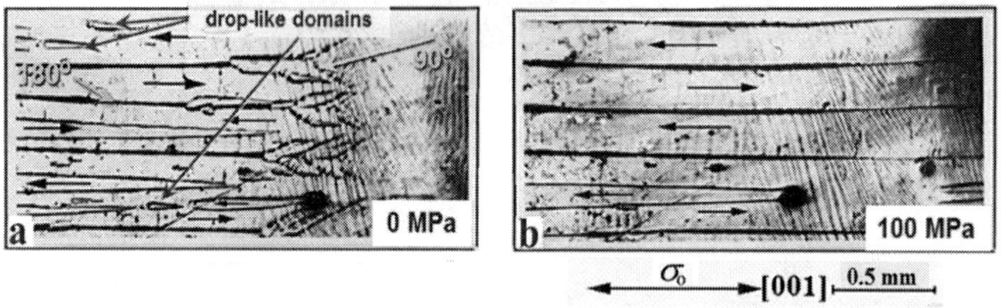

Fig. 2. Rearrangement of 90° domains into 180° ones in silicon iron (a) under 100 MPa tension (b).

A similar process takes place in crystal portions with internal residual stresses. The approach of DWs to these portions causes a jump-like rearrangement of the complex of 90° and 180° domains corresponding to the compression and tension zones [1].

For instance, applied stresses σ_0 change magnetic anisotropy thus reducing the volume of 90° magnetic phases (fig. 2 a, b). At small tensile strains, 90° DWs are substituted by 180° ones; therefore, a texture with magnetization vectors lying along the tetragonal axes closest to the direction of tension is formed in the crystals. The area of 180° DWs grows, and the processes of magnetization reversal become easier.

Thus, in steels with positive magnetostriction ($\lambda>0$) under tension ($\sigma>0$) (positive magnetoelastic effect, $\lambda\sigma>0$) the magnetization reversal processes become easier, whereas under compression (negative magnetoelastic effect, $\lambda\sigma<0$) they are impeded, and this can be illustrated by magnetic hysteresis loops (fig. 3 a) for steel 45.

For nickel, which has negative magnetostriction, the processes of magnetization reversal are more difficult under tension ($\lambda\sigma<0$ – negative magnetoelastic effect), and they are easier under compression (fig. 3 b).

This principle is a basis for the evaluation of working stresses in members made of ferromagnetic materials (fig. 4).

3 Evaluation of the level of microstresses

To evaluate the microstress level in ferromagnetic materials, three approaches are proposed, which involve magnetic energy redistribution in magnetization reversal during the motion of 90° DWs.

One approach involves creating a predominant displacement of 90° DWs, with applied stresses σ_0 in some regions of a ferromagnet being equal to internal ones σ_i. At $\sigma_0 \sim \sigma_i$, the magnetic permeability μ^{90} associated with the displacement of 90° walls is much greater

 https://doi.org/10.1051/matecconf/201814505001

than that ensuing from the mobility of 180° ones ($\mu^{90} \gg \mu^{180}$). Since the mobility of 90° DWs is governed by the microstress level, the separated contribution of 90° DWs to the process of magnetization reversal enables the microstress level to be evaluated.

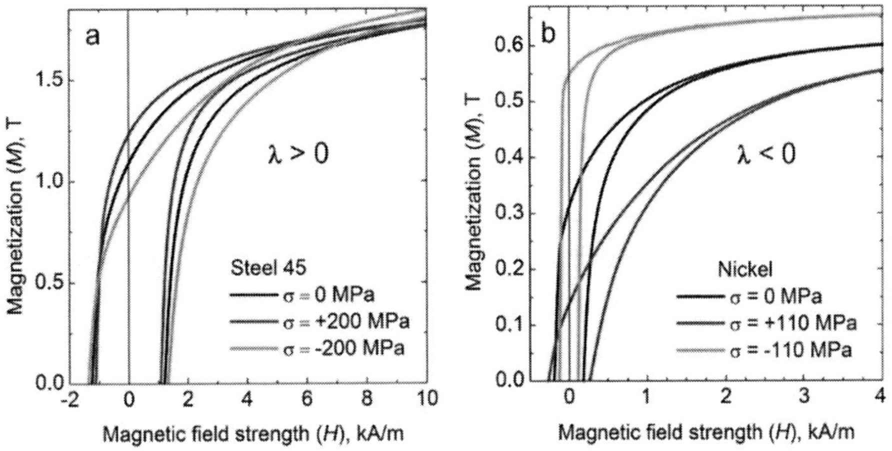

Fig. 3. Hysteresis half-loops for steel 45 (C~0.45%) (a) and nickel (b)

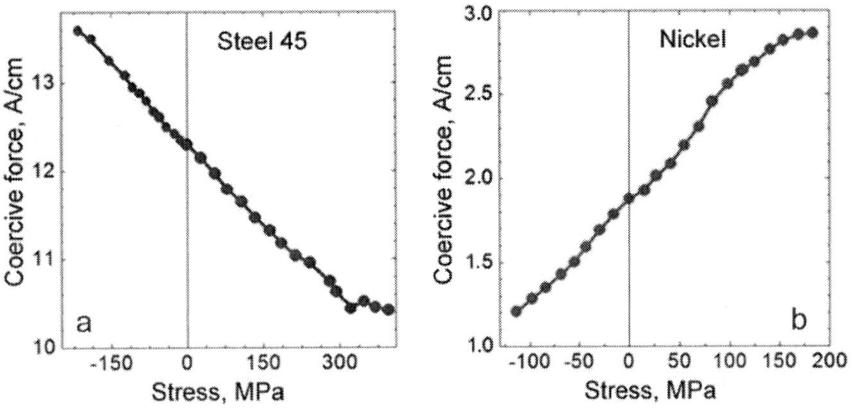

Fig. 4. The coercive force of steel 45 (a) and nickel (b) affected by tensile and compressive stresses

Magnetoelastic studies were conducted as follows (fig. 5): a demagnetized specimen was subjected to tensile stressing (σ_0), then a magnetic field (H) creating induction (B_0) was applied along the tension axis, whereupon low-frequency elastic dynamic vibration of small amplitude ($\Delta\sigma^{\sim}$) was excited in the specimen. The amplitude of magnetic induction ($\Delta B^{\Delta\sigma^{\sim}}$) generated in the specimen by dynamic vibration was recorded by an encircling coil. The measurements were repeated at different tensile loads ranging between 0 and $0.6\sigma_{0.2}$. The plots of measured induction versus tensile loading (fig. 6) have a characteristic form with a peak [2, 3].

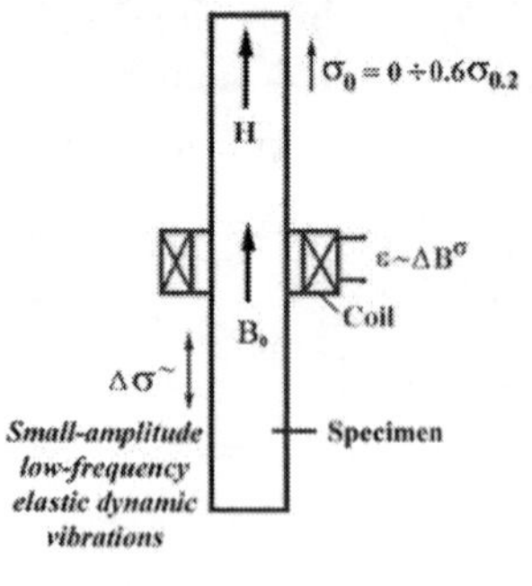

Fig. 5. Experimental scheme.

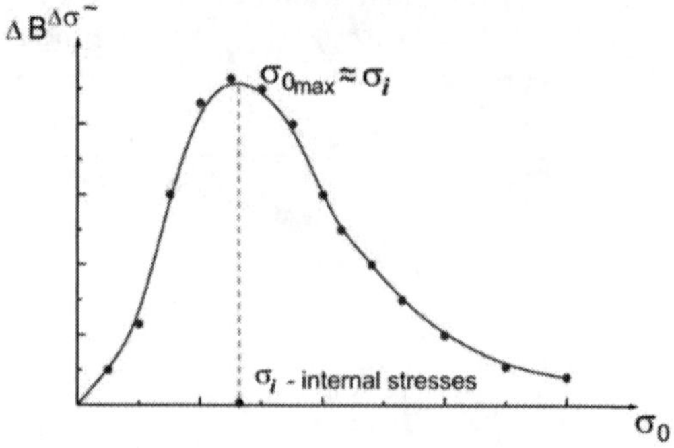

Fig. 6. The behaviour of magnetic induction caused in the specimen by elastic vibrations, $\Delta\sigma^{\sim}$, as a function of applied external stresses.

It follows from [2, 3] that the load value at which the peak is reached correlates with the mean amplitude of internal stresses in differently deformed steels. Thus, according to [2, 3], the value of magnetic permeability $\mu\sigma$ associated with the displacement of 90° DWs is determined as

$$\mu^{\sigma} \approx k \frac{\Delta B^{\Delta\sigma}}{\Delta\sigma B_o} \tag{1}$$

Experimental studies demonstrate (fig. 7) that the values of σ_{0max} and μ^{σ} correlate well with dislocation density in iron and steel St3 under plastic deformation.

Another parameter enabling the microstress level in ferromagnetic materials to be estimated is magnetoelastic acoustic emission (MAE). According to current concepts, magnetoelastic acoustic emission is based on the magnetostrictive mechanism of exciting signals produced by local sources of magnetostrictive strain at irreversible displacements of 90° DWs (fig. 1 b) and carrying information about changes in the magnetoelastic state of a ferromagnet.

When a ferromagnetic specimen is affected by an alternating magnetic field, it undergoes magnetization reversal along the magnetic hysteresis loop, and the irreversible displacement of 90° DWs causes elastic wave pulses, (fig. 1 b), which are recorded by a piezotransducer as electric signals. After amplification, the number of jumps of electrical acoustic emission signals is recorded, or the rms values of emf U_{mae} of the envelope of all the jumps are measured. Materials with high values of magnetostriction show high acoustic activity in magnetization reversal.

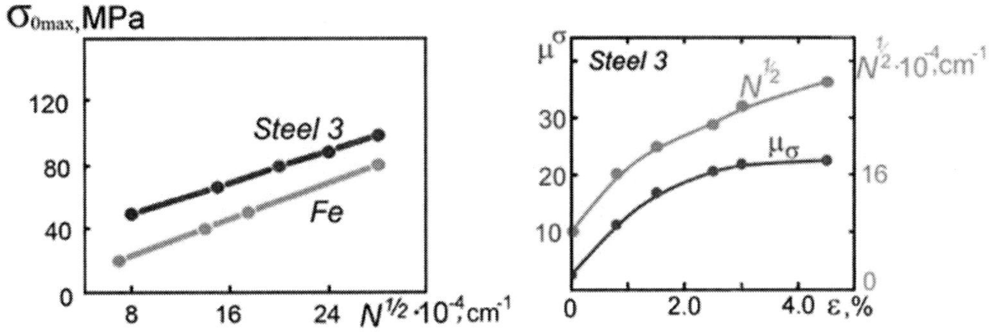

Fig. 7. σ_{0max} and magnetic permeability μ^{σ} caused by the displacement of 90° domain walls as dependent on dislocation density in iron and steel 3 (C~0.2%) suffering plastic deformation

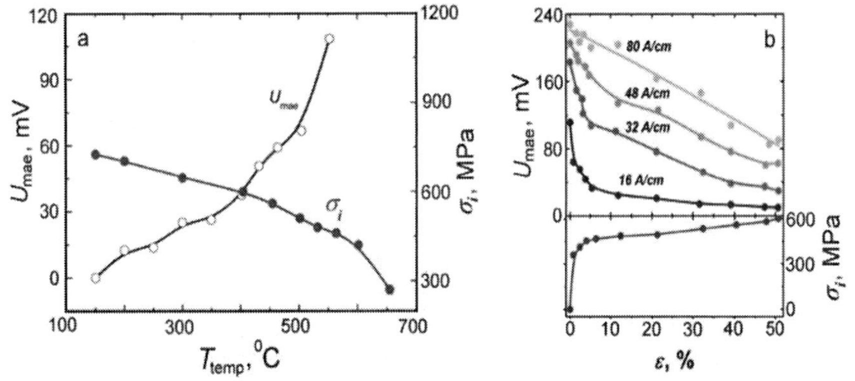

Fig. 8. The behaviour of the rms value of U_{mae} and microstresses as dependent on tempering temperature for steel 34KhN3M (a) and on the amount of plastic strain for steel 40 (b).

As there is a relation between the stress state and magnetostriction, the parameters of magnetoelastic acoustic emission must also be related to the elastic properties of a ferromagnet, i.e., to the microstress level.

Figure 8 illustrates the behaviours of the rms value of U_{mae} and the microstress level as dependent on the tempering temperature for steel 34KhN3M (a) and on the amount of plastic strain for steel 40 (b). It follows from the analysis of fig. 8 that U_{mae} correlates fairly well with the values of microstresses found by the X-ray technique. The MAE technique is employed to evaluate the microstress level in heat-treated and elasto-plastically deformed ferromagnetic materials.

The third approach is magnetostrictive mutual transformation of elastic and electromagnetic waves in ferromagnets, i.e., electromagnetic-acoustic transduction (EMAT). EMAT signal parameters are also sensitive to the microstress level in a ferromagnet.

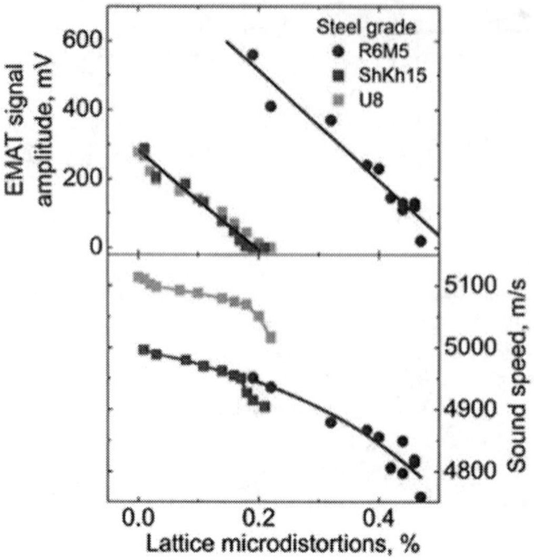

Fig. 9. A relation between the EMAT signal parameters and crystal lattice microdistortions for heat-treated steels.

EMAT in steels was studied in a strong polarizing magnetic field by means of exciting passage coils and recording by the resonance technique to generate a standing elastic wave in the specimen. The EMAT signal amplitude was determined at elastic vibration resonance, as well as the resonance frequency enabling elastic wave propagation velocity to be calculated.

It is obvious from fig. 9 that the EMAT signal amplitude and elastic wave propagation velocity decrease almost linearly with the growth of crystal lattice microdistortions caused by heat treatment (quenching followed by tempering) of specimens made of carbon steel U8 (C≅0.80%; Cr≅0.23%; Mn≅0.22%), low-alloy steel ShKh15 (C≅0.98%; Cr≅1.38%; Mn≅0.32%), high-alloy steel R6M5 (C≅0.84%; Cr≅3.30%; W≅5.92%; V≅1.80%; Mo≅4.83%), and this is suggestive of the applicability of EMAT parameters to the evaluation of the level of microstresses.

4 Reconstruction of strain curves

The determination of the uniaxial stress-strain state of metals by magnetic techniques enables one, using different models of the mechanics of solids, to approach the problem of predicting the behaviour of the mechanical properties of a metal under deformation and evaluating its residual life. It is proposed that magnetic methods of nondestructive testing [4] should be used for monitoring permanent strains in structural members. A technique has been developed for restoring strain diagrams by known relationships of standard mechanical properties of steel and measured magnetic characteristics.

The hardening curve is approximated by the equation

$$\sigma = \sigma_Y + A \cdot \Lambda^n \tag{2}$$

where σ and σ_Y are current working stress and yield stress, $A = (\sigma_U - \sigma_Y)/\Lambda_{sh}^n$, $n = \sigma_U \cdot \Lambda_{sh} / (\sigma_U - \sigma_Y)$, $\Lambda_{sh} = 2 \cdot \ln(1+\delta)/\sqrt{3}$ is shear strain at the end of uniform tension.

If we know correlation between standard mechanical characteristics and magnetic ones, like Y = a+bX, then, after substituting the values obtained from the correlations in eqn. (2), we arrive at a new strain curve equation taking into consideration measured magnetic characteristics, eqn. (3). Thus, by the data obtained in [4], after substituting the correlation expressions $\sigma_Y \approx H_c$, $\sigma_U \approx H_c$ into eqn. (2), we have

$$\sigma = \sigma_Y(H_c) + A(H_c) \cdot \Lambda^{n(H_c)}. \tag{3}$$

The results are presented in fig. 10.

The mechanisms of the strain-induced behaviour of the magnetic characteristics of low- and medium-carbon steels have been experimentally studied on all the portions of the stress-strain diagram up to necking under plane stress conditions in elastic-plastic tensile deformation.

In order to analyse the behaviour of the coercive force in tension, the H_c versus strain curves were superposed on the corresponding stress-strain diagrams (fig. 11). Three characteristic portions are obvious in the behaviour of the coercive force; namely, the region of elastic strain, the yield plateau and/or drop (sharp yield point) and the region of developed plastic strain.

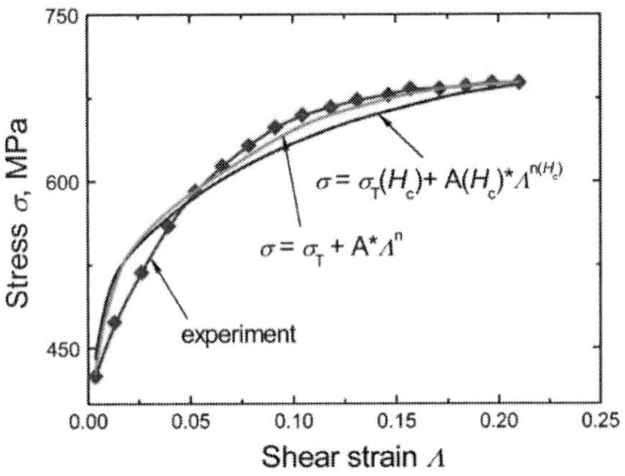

Fig. 10. Experimentally and analytically obtained tension diagrams, eqn. (3).

The effect of stresses reaching and exceeding $\sigma_{0.2}$ leads to a collapse of the magnetic texture of stresses in the elastic strain region, and the coercive force in the plastic strain region is mainly affected by an increase in the density of dislocations and dislocation clusters ($H_c \sim N^{1/2}$, where N is dislocation density [5]) and the formation of a crystallographic strain texture.

As the stress $\sigma_{0.2}$ is approached, the coercive force of all the tested specimens grows significantly, the most active growth of H_c being observed prior to $\sigma_{0.2}$ rather than at $\sigma_{0.2}$, and this agrees well with the results obtained in [6]. In steel St3 the coercive force stops growing as soon as the upper yield stress is reached, H_c remains constant or slightly decreases, and then, after passing the lower yield stress, it resumes growing, though less intensively (see fig. 11 a).

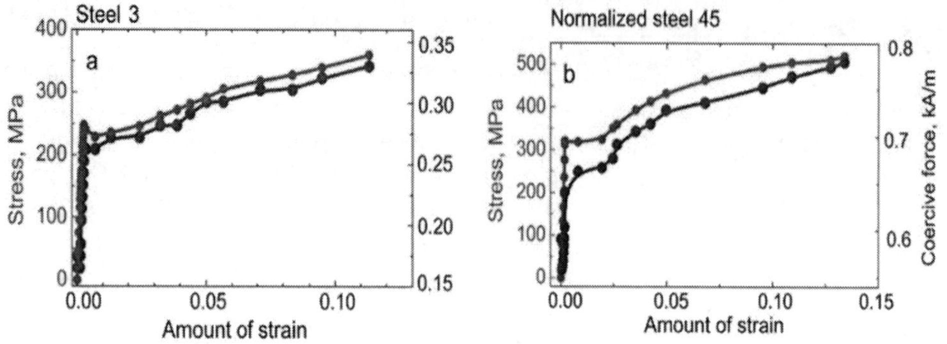

Fig. 11. Tensile stress and coercive force as dependent on the amount of strain for steels.

The investigation has yielded a technique for the restoration of stress-strain diagrams for homogenous steel products by coercive force, and this is a basis for magnetic monitoring of the stress-strain state of loaded steel members.

5 Deformation processes analysed with the use of differential magnetic permeability

To test non-destructively the condition of the constituents of a multilayer product and to assess their residual life, it is suggested that, as was proposed earlier in [7], the value of the field of maximum differential magnetic permeability $H_{\mu dmax}$ of the magnetically soft and magnetically hard constituents of a compound product should be used as an informative parameter. A method for calculating the stress-strain diagram of a flat two-layer steel product under uniaxial tension from the strain-dependences of the magnetic characteristics of the constituents was proposed in [8], the dependences being obtained directly in the course of loading.

Figure 12 a shows the major hysteresis loops of a compound specimen (No. 1 in the table) and its constituents No. 2 and No. 3, both measured separately, as well as the corresponding field dependences of maximum magnetic permeability in the initial state (before loading). The hysteresis loop of the compound specimen (curve 1 in fig. 12 a) has a form characteristic of two-layer ferromagnets [7], and it differs from the hysteresis loops for homogeneous materials (curves 2 and 3), among other things, in that they have two bends. One of them, localized in the region of smaller fields, corresponds to the magnetically soft constituent, and the other corresponds to the magnetically hard one. The mechanism of the formation of these bends was discussed earlier in [5] and [7]. On the field dependence of differential magnetic permeability, bends of the kind manifest themselves as maximums (peaks) of differential magnetic permeability (fig. 12 b). The number of peaks corresponds to the number of layers with different magnetic hardness, the peak height proportion being determined, in particular, by the proportion of the thickness of the respective layers [5, 7].

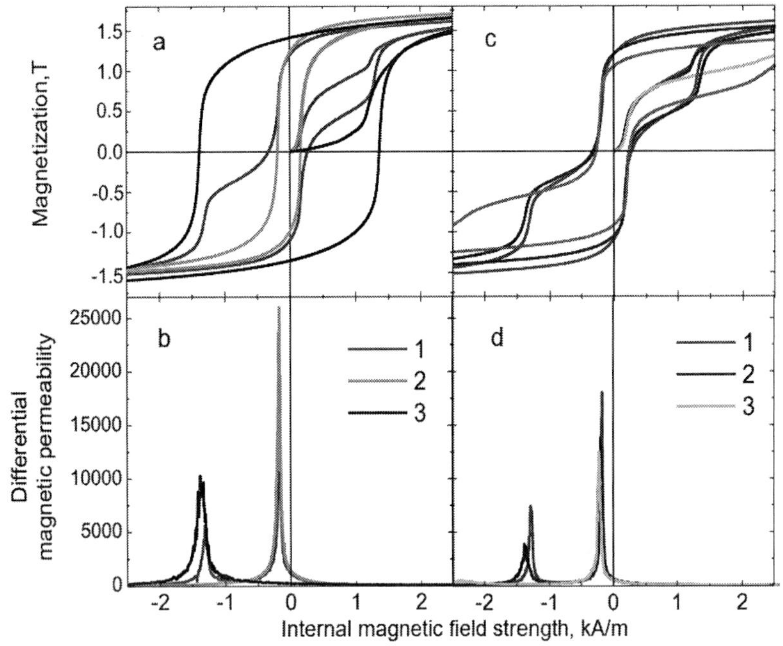

Fig. 12. Major magnetic hysteresis loops for the compound specimen (curve 1) and its constituents (curves 2 and 3) (a) and the corresponding field dependences of differential magnetic permeability (b) in the initial state; for the compound specimen in the initial state (curve 1) (c), under external stresses amounting to 0.5 of the fracture stress in the magnetically hard constituent (2) and immediately before the fracture of the magnetically hard constituent (3).

Table 1. The geometric dimensions of the test parts of the specimens and their magnetic characteristics in the initial state.

No.	Specimen	Dimensions, mm	H_c, kA/m	B_r, T
1	compound specimen: St3+45 (quenching+tempering)	90 × 20 × (2.14+1.10)	0.30	1.14
2	St3 (annealing): 0.03% C, the rest Fe	90 × 20 × 2.04	0.18	1.11
3	45 (quenching+tempering): 0.45% C; the rest Fe	90 × 20 × 1.20	1.36	1.37

Figures 12 c and d show a change in the form of the field dependences of differential magnetic permeability and the evolution of the major hysteresis loops of the compound specimen (No 1). To avoid complication of the figures, we show the curves fitting only three points of the stress-strain diagram; namely, curves 1 correspond to the initial state, curves 2 were taken under external stresses amounting to 0.5 of the fracture stress in the magnetically hard constituent, and curves 3 were obtained immediately before the fracture of the magnetically hard constituent. It follows from fig. 12 c, d that, in the course of deformation, the peak of the magnetically hard constituent on the field dependence of the differential magnetic permeability shifts into the region of stronger fields and, besides, it becomes less intensive and more widened. The peak of the magnetically soft constituent also shifts into the region of stronger fields, due to the small absolute shift magnitude (as compared to that for the magnetically hard constituent), this shift is less pronounced on the field dependence. The bends of the major loop also shift into the region of stronger fields (see fig. 12 c).

Thus, the field dependences of differential magnetic permeability determined by differentiating the descending branches of the corresponding hysteresis loops with respect to the field reveal the maximums (peaks) of differential magnetic permeability of the individual layers in two-layer products.

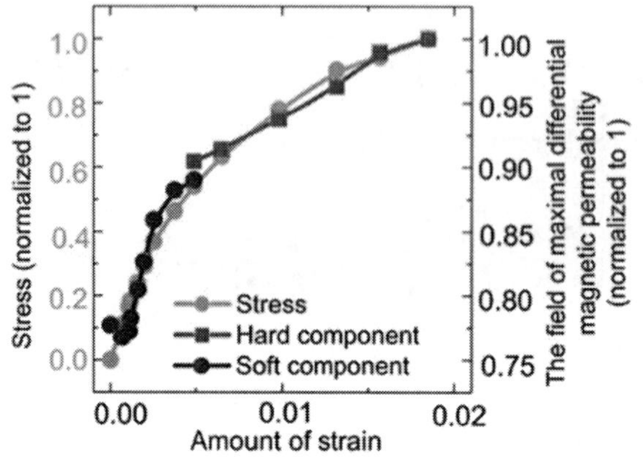

Fig. 13. Tensile stress and the field of differential magnetic permeability as dependent on the amount of strain for a two-layer product.

The characteristics of the peaks of differential magnetic permeability (peak field, height and width) are shown to vary regularly under elastic and plastic strains; therefore, they can serve as a tool for testing the condition of multilayer products. Figure 13 shows a procedure of restoring the stress-strain diagram (green curve) of a two-layer product by its magnetic characteristics. The procedure is based on the following mechanism: when a two-layer product is tensioned, a sharp increase in the value of the field of maximal differential permeability for each constituent occurs under the effect of strains corresponding to the yield stress of the constituent. This enables the strain curve for a two-layer product to be restored – the field $H_{\mu dmax}$ of the magnetically soft constituent is the more informative in the earlier stages of deformation (fig. 13 (o)), whereas $H_{\mu dmax}$ of the magnetically hard constituent is the more informative at higher amounts of strain (fig. 13 (□)).

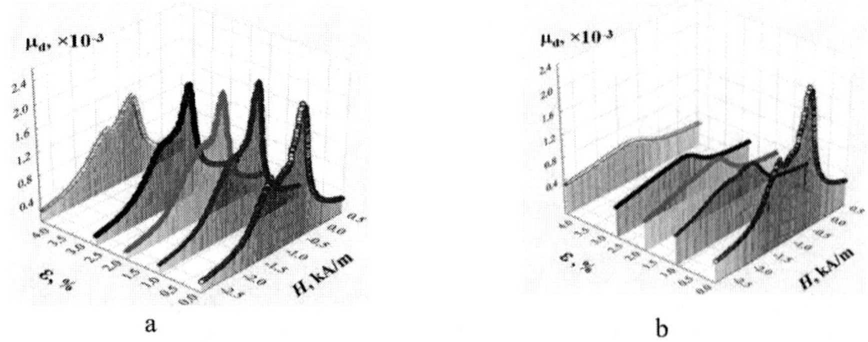

Fig. 14. Evolution of the field dependence of differential magnetic permeability of a laser heat-treated specimen as a function of specific elongation: measurements made under loading (a); measurements made after unloading (b)

Similar results on the change in the field dependence of differential magnetic permeability for carbon steel surface-hardened by laser radiation were obtained for plastic tensile strain (fig. 14).

Conclusion

Magnetic methods are applicable to the evaluation of working macrostresses in structural members. Three techniques for the correlative determination of the level of microstresses in ferromagnetic materials have been discussed. All of them involve the recording of 90° domain wall motion parameters.

A technique for the restoration of stress-strain diagrams of homogeneous steel products by coercive force and/or residual induction has been proposed, which is a basis for magnetic monitoring of the stress-strain state of loaded steel members.

It has been demonstrated that the characteristics of the peaks of differential magnetic permeability (peak field, height and width) vary regularly under the effect of elastic and plastic strains and, therefore they can be an efficient tool for testing the condition of multilayer products. A procedure of restoring the stress-strain diagram of a two-layer product by its magnetic behaviour has been developed. The procedure is based on a discovered regularity, which is as follows: when a two-layer product is tensioned, a sharp increase in the values of the field of maximum differential permeability for each constituent is induced by strains corresponding to the yield stress of the constituent. This enables one to restore the strain curve for a two-constituent product, i.e., $H_{\mu dmax}$ of the magnetically soft constituents is the more informative parameter in the early stages of deformation, whereas $H_{\mu dmax}$ of the magnetically hard constituent is the more informative at higher amounts of strain.

The study was partially financed by the RFBR, project No 16-08-01077.

References

1. E.S. Gorkunov, Y.N. Dragoshanskii, Russ. J. Nondestr. Test. **35,** 6, 409-426 (1999)
2. A.P. Nichipuruk, E.S. Gorkunov, N.I. Noskova, E.G. Ponomareva, FizMM **12**, 81-87 (1992)
3. A.P. Nichipuruk, E.S. Gorkunov, L.D. Gavrilova, L.V. Atangulova, M.V. Degtyarev, B.I. Kamenetskii, L.S. Davydova, T.I. Chashchukhina, L.M. Voronova, Russ. J. Nondestr. Test. **34,** 4, 273-279 (1998)
4. E.S. Gorkunov, A.B. Bukhvalov, A.Z. Kaganovich, S.S. Rodionova, V.N. Durnitskii, Russ. J. Nondestr. Test. **32,** 6, 469-477 (1996)
5. M.N. Mikheev, E.S. Gorkunov, *Magnetic methods of structural analysis and nondestructive testing*. (Nauka, Moscow, 1993)
6. J.M. Makar, B.K. Tanner, J Magn Magn Mater. **184,** 2, 193-208 (1998)
7. E.S. Gorkunov, B.M. Lapidus, A.V. Zagainov, S.A. Voronov, G.Y. Bushmeleva, Russ. J. Nondestr. Test. **7**, 441–446 (1988)
8. E.S.Gorkunov, S.M. Zadvorkin, I.G. Yemelianov, S.Yu. Mitropolskaya, Phys.metals metallog **103**, 6, 624-632 (2007)

Magnetic flux distribution in a ferromagnetic material magnetized by U-shaped electromagnets of different geometric dimensions types

Anna Povolotskaya[1,*], *Eduard* Gorkunov[1], *Pavel* Nosov[1], and *Sergey* Zadvorkin[1]

[1]Institute of Engineering Science, RAS (Ural Branch), Ekaterinburg, Russia

Abstract. A model experiment has been carried out on studying the type of changes in the magnetic induction in a homogeneous isotropic sample that is locally magnetized with attached U-type electromagnets of different geometrical dimensions. The study was aimed at finding out magnetic flux distribution at different locations within the sample and determining the effect that the geometry of attached electromagnets has on this distribution.

1 Introduction

Magnetic methods of non-destructive testing and evaluation are widely used for assessing the structural state, determining the strength characteristics, and analyzing the phase composition of products made of ferromagnetic materials [1–4]. In non-destructive testing of products under industrial conditions, magnetic characteristics are most often measured locally by using attached magnetic devices (AMD) that include magnetization facilities (coils, permanent magnets, electromagnets) and gaging elements (pickup coils, flux gates, Hall probes, magnetoresistors, etc.) [5]. The geometrical sizes of test products in most cases by far exceed the AMD sizes. Therefore, it is important that the magnetization facility should ensure magnetization or magnetization reversal of the evaluated area of the test product along a close-to-major hysteresis loop. In order to choose the optimum AMD configuration and adequately interpret the results of local magnetic measurements during in-production control of the structure and mechanical properties of metallurgical products (roll stock, forged pieces, rolls, etc.) without using costly "classical" testing methods, it is important to know magnetic flux distribution in near-surface and internal layers of test objects.

When evaluating surface-hardened products [6, 7], a more complicated problem arises. To assess the characteristics of the surface-hardened layer, it is necessary to concentrate magnetic flux in the hardened layer, whereas when evaluating properties of a viscous bulk of the product, one needs to ensure significantly deeper magnetic-flux penetration [8].

Thus, depending on the local magnetic evaluation task, the topography of magnetic-flux distribution in the product as well as the conditions of magnetization and magnetization

* Corresponding author:us@imach.uran.ru

reversal in separate tubes of product's magnetic flux should be known [9].

Some experimental studies of magnetic-flux topography in ferromagnetic products that are locally magnetized by an attached electromagnet have been reported [10–14]. Theoretical calculations of the steadystate magnetic-flux distribution that are based on the system of Maxwell equations and the fundamental integral equation of magnetostatics have been performed only for some particular cases. Some works on the numerical modeling of flux and field distributions inside ferromagnets have been published recently [15–18]. In most experimental studies, methodological obstacles that are related to imperfection of measuring transducers impose certain restrictions on the shape of test products, and the shape, in its turn, affects the magnetic-flux distribution in these samples. For example, in [10] the width of the plate coincides with the width of electromagnet poles; this considerably reduces lateral magnetic-flux straying and increases magnetization depth. The same is confirmed by modeling results in [15]. Miniature high-sensitivity measuring transducers are currently used to take measurements of magnetic-flux parameters without the above restrictions on the sample shape.

The aim of this work is to study experimentally the type of changes in magnetic induction in a homogeneous isotropic sample with allowance for lateral straying when the sample is locally magnetized by attached U-type electromagnets of different sizes in order to examine the spatial magnetic-flux distribution in the sample and to determine the effect that the geometry of attached electromagnets has on this distribution.

2 Experimental procedure and material

A compound model sample of 40Kh chromium steel, which is extensively used in machine building, was prepared for the experiments. The sample consisted of two parts A and B with dimensions 100 × 70 × 90 and 85 × 70 × 90 mm, respectively. A schematic drawing of the sample is presented in Fig. 1.

A 10 × 10-mm groove was milled on the inner surface of part A that is in contact with part B. A 10 × 10 × 10-mm measuring insert made of steel of the same grade could be displaced along the entire groove length. Three Hall probes were attached to each of the three mutually orthogonal facets of the measuring insert. The insert is schematically shown in Fig. 1*b*.

Auxiliary 40Kh-steel 10 × 10-mm inserts with lengths of 2, 5, 10, 20, 40, and 50 mm were manufactured to ensure the homogeneity of the magnetic medium that was tested. These inserts were used to fill the entire length of the groove. During experiments, parts A and B were fayed together, with the measuring insert being fixed in the groove. To avoid distortions of magnetic flux in the model sample, all the above parts were subjected to identical thermal treatment (quenching at a temperature of 860 °C followed by tempering at a temperature of 300 °C in a vacuum furnace) so as to ensure identical magnetic and mechanical properties.

The geometrical dimensions and the number of turns of electromagnets that were used to magnetize the sample are listed in Table 1.

Table 1. Geometrical dimensions of used U-type electromagnets.

AMD no.	Pole thickness a, mm	Pole width b, mm	Distance between pole centers d, mm	$\sqrt{S_e}$, mm
1	11.5	28	45.5	17.9
2	5.5	15	18.5	9.08
3	10	25	23	15.8

The electromagnet current varied within a range of −2.6 to +2.6 A using a regulated power supply unit. The origin of a Cartesian coordinate system was put at the center of electromagnet's neutral cross section, as show in Figs. 2*a* and 2*b*. The Hall probes that registered induction along *x*-, *y*-, and *z*-axes are denoted as P_x, P_y, and P_z, accordingly. At the beginning of measurements, the electromagnet was placed on the preliminarily demagnetized sample so that the Hall probe P_x be at the center of the neutral cross section of the electromagnet at a depth of 5 mm [at a point with coordinates (0; 0; 5)].

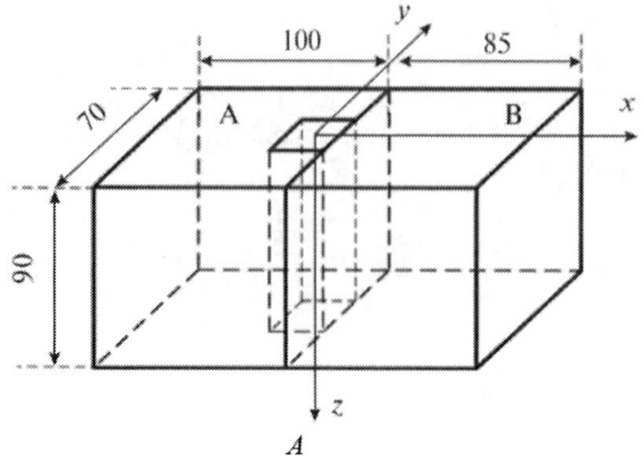

Fig. 1. Schematic drawings of (*a*) sample and (*b*) measuring insert.

Measurements with electromagnets were carried out in the following manner. The measuring insert was installed so that its upper facet should coincide with the surface of the model sample (see Fig. 2*a*). Given such an arrangement, the Hall probes had the following coordinates: P_x ($x = 0, y = 0, z = 5$); P_y ($x = -5, y = 5, z = 5$); and P_z ($x = -5, y = 0, z = 10$). A cycle of measurement of magnetic properties was executed in this initial position (see Fig. 2*b*). The entire system was demagnetized in all cases prior to the measurements, with the complete demagnetization performed after each measurement cycle.

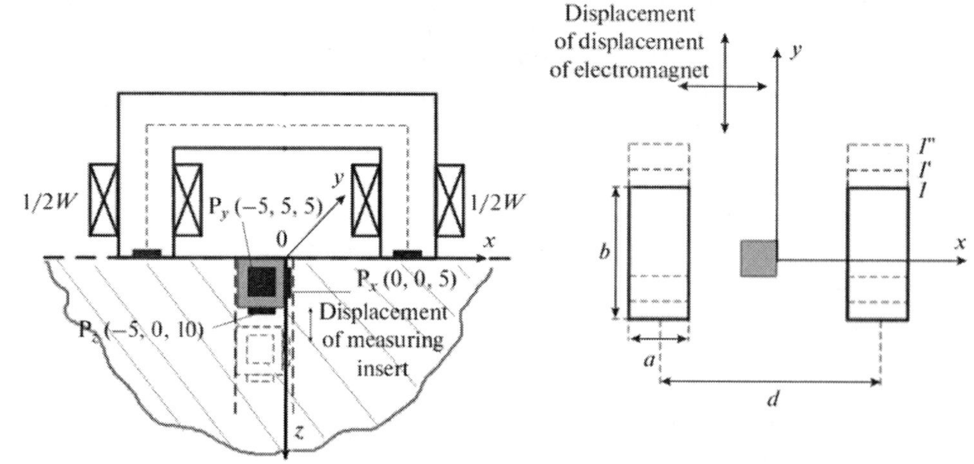

Fig. 2. Disposition of electromagnet on the sample and orientation of Hall probes (*a*); electromagnet displacement (view from above) (*b*).

In order to study lateral straying, the attached electromagnet was sequentially displaced along the y-axis in 5-mm steps into positions 1', 1" and so on (see Fig. 2*b*), with the entire measurement cycle of hysteresis loops by the Hall probes repeated every time. Then, the electromagnet was returned to the original position, and similar measurements were taken, with the electromagnet displaced in 5-mm steps along the x-axis. After measurements for different prescribed values of x had been taken, the electromagnet returned to the initial position.

The next series of measurements were taken after sequential 5-mm displacements of the measuring insert, with the free space filled every time with the auxiliary inserts.

3 Results and discussion

Figure 3 shows how the value of the x-component of magnetic induction B_x as measured with AMD 1 and AMD 2 for the maximum applied field depends on the distance x to the plane of AMD's neutral cross section for different positions along the y-axis at a depth of z = 5 mm from the sample surface.

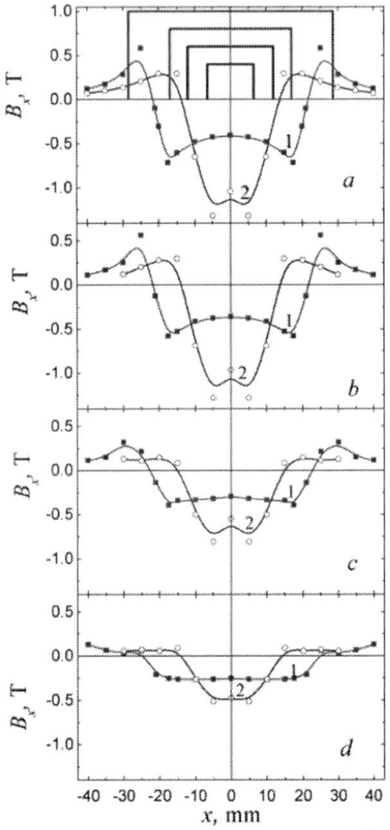

Fig. 3. Schematic arrangement of AMD 1 and AMD 2 on the sample and values of magnetic induction components B_x at different distances from the center of AMD neutral cross section at depth z = 5 mm from the sample surface at (*a*) y = 0; (*b*) 5; (*c*) 15; and (*d*) 20 mm. Curves 1 and 2 are obtained with AMD 1 and AMD 2, respectively.

For the sake of clearness, Fig. 3*a* schematically presents attached magnetic devices. It can be seen that the x-component of magnetic induction B_x – both for AMD 1 and for AMD

2 – increases when approaching the electromagnet pole starting at half the interpolar gap, attains its maximum approximately at the border of the inner pole edge, and then decreases to zero. When the electromagnet moves still further away from the measuring insert, B_x changes sign, reaches its maximum near the outer pole edge, and then rapidly drops down to zero. When the AMD is displaced from the measuring insert along the y-axis, the x-component of magnetic induction B_x decreases, the maxima near the borders of the poles become less pronounced, and at $y = 20$ mm the x-component of the magnetic-flux density only slightly changes in the interpolar zone.

The x-component of the magnetic induction B_x changes in such a way only at a relatively small depth of $z = 5$ mm. At $z = 10$ mm (Fig. 4*a*), the maxima near the borders become less pronounced (for any displacement along the y-axis), while the sign-reversal position of B_x is displaced farther from the center. At $z = 20$ mm (Fig. 4*b*), the x-component of magnetic induction B_x does not change sign for any of the values of the x-coordinate that are indicated in the figures.

Figure 5 shows dependences of the z-component of the induction B_z for the maximum applied field for different positions along the y-axis ($y = 0, 5, 10, 15, 20$ mm) at the depth of $z = 5$ mm. For clearness, the figures schematically indicate AMDs. The magnetic-induction component B_z at the center of the interpolar gap equals zero for any displacement along the y-axis, increases when approaching the pole, reaches its maximum value beneath the pole center, and decreases to zero when moving further away from the measuring insert. For deeper locations of the measuring insert, the values of B_z are smaller, but the overall type of the dependences is retained.

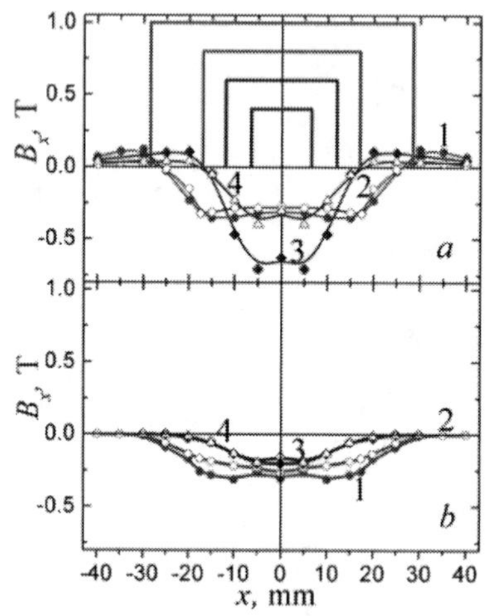

Fig. 4. Schematic arrangement of AMD 1 and AMD 2 on the sample and values of magnetic-induction components B_x for different positions of the plane x, y at depths z = (*a*) 10 and (*b*) 20 mm from the sample surface for $y = 5$ mm — curves 1 (•), and 3 (♦), for $y = 20$ mm — curves 2 (o) and 4 (Δ). Curves 1 (•), 2 (o) and curves 3 (♦), 4 (Δ) were obtained from measurements with AMD 1 and AMD 2, respectively.

For the maximum applied field, the magnetic-induction component B_y (Fig. 6) depends on the displacement along the y-axis in the following way. At $y = 0$ mm, B_y is equal to zero everywhere; then it grows when approaching the pole, reaches its maximum value near the border of the inner pole edge, and decreases to zero when the AMD moves further away from the measuring insert.

Displacement along the y-axis alters the type of the dependence of B_y in the following manner. The maximum position is shifted closer to the pole center, whereas the very maximum value of B_y first increases, reaches its maximum at y =10–15 mm, and then reduces (Fig. 7). The lack of monotony and the presence of a maximum on the $B_y(y)$ curve can be explained by the fact that in this case two "concurrent" processes occur. On the one hand, magnetic-flux density should in toto decrease to zero when the AMD is displaced along the y-axis, but, on the other hand, the ratio $B_y/|B|$ monotonically increases from zero and does not exceed unity, given such a displacement.

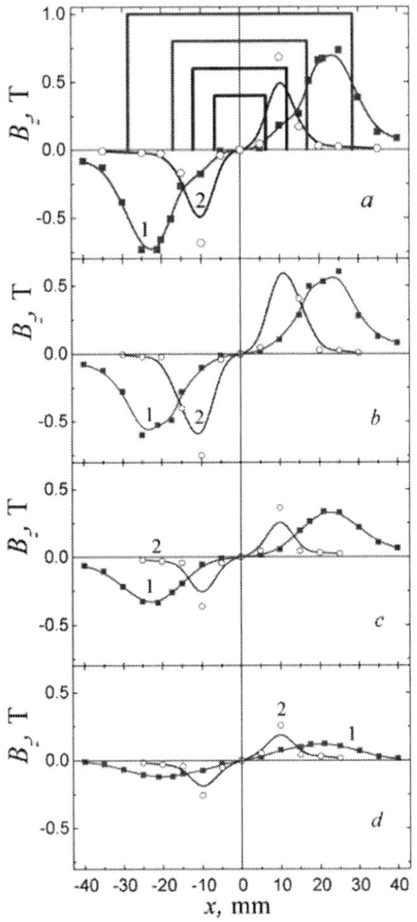

Fig. 5. Schematic arrangement of AMD 1 and AMD 2 on the sample and values of magnetic-induction components B_z at different distances from the center of AMD neutral cross section at depth z = 5 mm from the sample surface for (*a*) y = 0; (*b*) 5; (*c*) 15; and (*d*) 20 mm. Curves 1 and 2 were obtained with AMD 1 and AMD 2, respectively.

Figure 8 presents dependences of the modulus of the total magnetic-induction vector $|B|$ for the maximum applied field for different positions along the y-axis (y = 0, 5, 10, 15, 20 mm) at depths of z = 5, 10, 20 mm, as measured with AMD 1. It can be seen that the modulus of the total magnetic-induction vector $|B|$ increases when approaching the pole and reaches its maximum value at the inner pole edge (x = 17 mm). At the pole center, the value of $|B|$ decreases and then reaches its second maximum approximately 2–3 mm away from the outer pole edge (x = 28 mm). As the AMD moves further away from the measuring insert, the modulus of the total magnetic-induction vector $|B|$ reduces and tends to zero. Such a type of changes in $|B|$ takes place only at a relatively small depth of z = 5 mm. At z =

10 and 20 mm, the first maximum (near the inner pole edge) stops being very pronounced, while the second maximum (near the pouter pole edge) disappears. Displacement of the AMD along the y-axis also diminishes the maxima of $|B|$ near the edges of the poles.

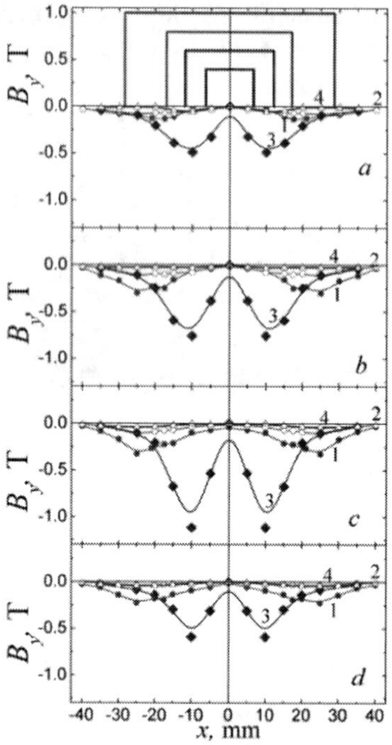

Fig. 6. Schematic arrangement of AMD 1 and AMD 2 on the sample and values of magnetic-induction components B_y for different positions of the plane x, y at y = (*a*) 5; (*b*) 10; (*c*) 15; and (*d*) 20 mm at depths z = 5 mm [curves 1 (•) and 3 (♦)] and z = 20 mm [curves 2 (o) and 4 (Δ)] from the sample surface. Curves 1 (•), 2 (o) and curves 3 (♦), 4 (Δ) were obtained from measurements with AMD 1 and AMD 2, respectively.

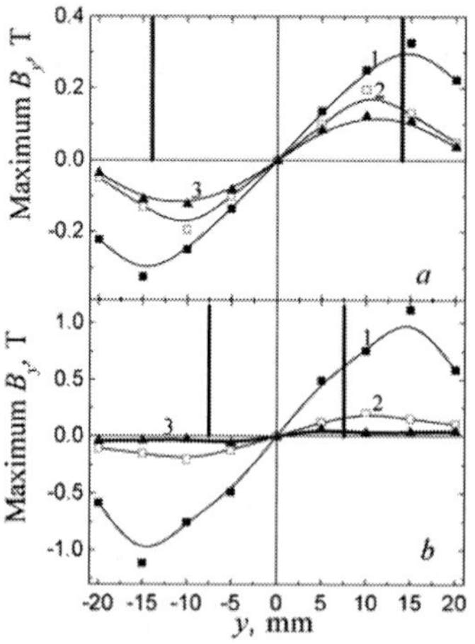

Fig. 7. Maximum values of magnetic-induction component B_y depending on the displacement along the y-axis: (*a*) measurements with AMD 1, (*b*) measurements with AMD 2. Curves 1, 2, and 3 correspond to z = 5; 10; and 20 mm, respectively.

The fact that the maxima are located near pole edges can be explained as follows. The main contribution to the value of the modulus of the total magnetic-induction vector is rendered by the B_x and B_z components (the B_y component contributes much less). However, the magnetic-induction component B_x turns to zero approximately beneath the center of a pole of the attached magnetic device (see Figs. 3 and 4), thereby reducing the modulus $|B|$. Therefore, a small decrease in the modulus of magnetic-induction vector is observed beneath the pole center as compared with the pole edges. At $z = 20$ mm, the component B_x no longer demonstrates sign reversal beneath the pole, which is manifested in the lack of double maxima of $|B|$. It should be noted that the sign reversal of B_x beneath the pole center is accounted for by the presence of stray fluxes in the transducer–sample circuit.

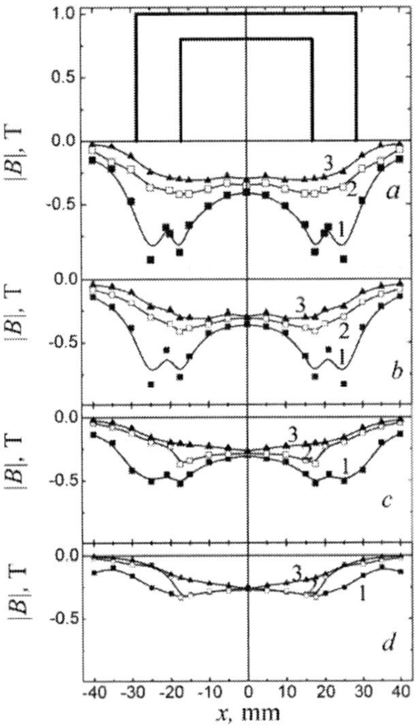

Fig. 8. Magnetic-induction modulus $|B|$ at different points of the sample for depths $z = 5$ mm (curves 1), 10 mm (curves 2), and 20 mm (curves 3) at $y =$ (*a*) 0; (*b*) 5; (*c*) 10; and (*d*) 20 mm. Measurements are taken with AMD 1.

When magnetizing products, it is important to know the depth to which they are magnetized. Figure 9 presents dependences of the modulus of the total magnetic-induction vector $|B|$ at the center of the neutral cross section ($x = 0$, $y = 0$) on the distance z to the sample surface (Fig. 9*a*) and on $z/\sqrt{S_e}$ (Fig. 9*b*). The dependences were measured when the sample was magnetized by three electromagnets with different geometrical dimensions, where S_e is the cross section area of the relevant AMD. The values of currents in measurements with the electromagnets of AMD 2 and AMD 3 were set so as to ensure a field of 4.2 kA/m beneath their poles, a field which is created beneath the pole in case of magnetization with AMD 1 operating at the maximum current. It can be seen from Fig. 9*a* that the value of the total magnetic-induction vector $|B|$ monotonically decreases with the depth z, with its value dependent, to a considerable degree, on the geometrical dimensions of the attached magnetic device (the cross section area of the poles of the electromagnet and the interpolar distance). It should be noted that at small depths, in near-surface layers of the neutral section, the smaller the gap, the greater the values of $|B|$; whereas the greater the cross section area of the AMD poles, the larger the magnetization depth. The latter statement is illustrated more visually in Fig. 9*b*. At small depths of less than approximately $1.3\sqrt{S_e}$ mm, the value of $|B|$ depends, to a considerable degree, on the distance d between the centers of electromagnet poles, while at larger depths it only depends on the cross section area of AMD pole. Reducing the pole section area, in particular, its width, leads to increasing lateral straying and, accordingly, decreasing the magnetization depth and flux density in the test object.

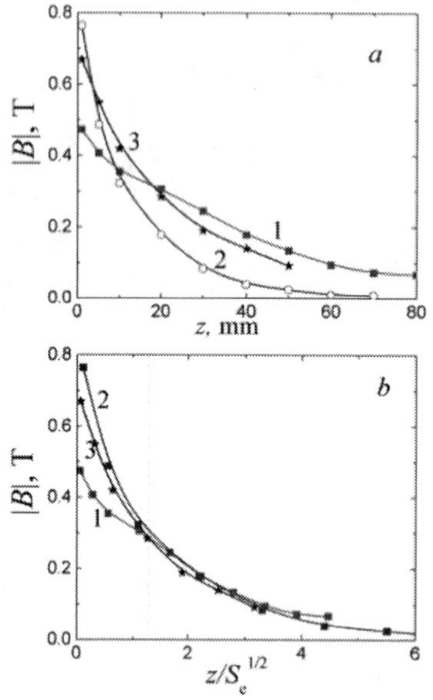

Fig. 9. Dependence of the modulus of total magnetic-induction vector $|B|$ at the center of the neutral cross section (*a*) on the distance z to the sample surface and (*b*) on the value of $z/\sqrt{S_e}$. Curves 1, 2, and 3 were obtained with AMD 1, AMD 2, and AMD 3, respectively.

Kostin et al. [17] performed numerical modeling and obtained quantitative estimates of the lateral straying and percentage of magnetic flux that goes beyond the profile of electromagnet poles from the overall magnetic flux in the magnetic circuit. They demonstrated that, first of all, the greatest lateral straying of magnetic flux is observed in the neutral cross section, and, secondly, the lateral straying depends, to a considerable degree, on both the ratio of the width of the magnetized plate to the pole width and the interpolar distance, provided the cross sections of the poles are equal. Based on the experimental data for all the AMDs, the fraction of magnetic flux that leaves the boundaries of the poles to the overall magnetic flux that passes through the neutral cross section was calculated. The magnetic flux BS was calculated by dividing the neutral cross section into square 5 × 5-mm cells and assuming the induction to be the same within any cell (Table 2).

Table 2. Fraction of the lateral straying of magnetic flux in the neutral cross section of the plate when it is magnetized by AMDs with different geometrical dimensions.

AMD no.	Ratio of plate width to AMD pole width	Ratio of distance between AMD pole centers to pole width	Lateral straying, %
1	2.5	1.6	57
2	4.7	1.2	65
3	2.8	0.92	60

As compared to AMD 3, AMD 1 has almost a twice larger distance between the pole centers, with the pole width of AMD 1 exceeding that of AMD 3 by only 1.12 times. It can be seen from the results provided in the table that reduction in lateral straying of magnetic flux in the neutral cross section and, hence, growth of the magnetization depth are affected to a much greater degree by reduction in the ratio of the width of the magnetized plate to the AMD pole width than by decrease in the ratio of the interpolar gap to the pole width.

4 Conclusion

A technique has been developed for determining the topography of magnetic induction in a ferromagnet that is magnetized by attached magnetic device. The distributions of all the magnetic-induction components over the volume of a bulky homogeneous isotropic sample of steel 40Kh that is locally magnetized by attached U-type electromagnets with different geometries.

It has been established that if the width of the magnetized plate is greater than the electromagnet-pole width, the distribution of magnetic flux in the plate at a distance to the plate surface of less than $1.3\sqrt{S_e}$ is affected by both the cross section area of AMD poles and the distance between the centers of the poles of the electromagnet.

It has been demonstrated experimentally that the main factor that significantly affects the growth of the magnetization depth is reduction of the ratio of the magnetized plate width to the AMD pole width due to reduction of the fraction of strayed magnetic flux that occurs in this case.

The above results can prove useful in magnetic nondestructive testing when choosing the most expedient disposition of measuring transducers within the space between electromagnet poles, magnetic structural analysis when adjusting the depth of magnetization of a test product as well as verifying the adequacy of numerical modeling of processes of magnetization and magnetization reversal of products made of steel 40Kh.

The study was partially supported by UB RAS project No 15-10-1-22.

References

1. M. N. Mikheev, E. S. Gorkunov, *Magnetic Methods of Structural Analysis and Nondestructive Testing* (Nauka, Moscow, 1993)
2. V. V. Klyuev, V. N. Filinov, V. I. Matveev, *Nondestructive Testing. A Handbook. In 8 Vols. Magnetic Methods of Testing, Vol. 6, Book 1* (Mashinostroenie, Moscow, 2006)
3. E. S. Gorkunov, Diagn. Resour. Mech. Mater. Struct. **1**, 6-40 (2015)
4. E. S. Gorkunov, Diagn. Resour. Mech. Mater. Struct. **3**, 6-50 (2015)
5. E. S. Gorkunov, V. A. Zakharov, Russ. J. Nondestr. Test. **31**, 8, 625-641 (1995)
6. E. S. Gorkunov, V. A. Zakharov, Soviet J. Nondestr. Test. -USSR, **27**, 1, 1-18 (1991)
7. L. Jian, K. T. Chau, Progress in Electromagnetics Research **92**, 1-16, (2009)
8. V. E. Shcherbinin, E. S. Gorkunov, *Magnetic Testing of Quality of Metals* (Ural Branch, Russ. Acad. Sci., Ekaterinburg, 1996)
9. V. A. Zakharov, *Magnetostatics of Systems with Ferromagnets* (Ural Sci. Center, USSR Acad. Sci., Sverdlovsk, 1986)
10. M. N. Mikheev, Izv. Akad. Nauk SSSR, **3**, 68-77 (1948)
11. G. V. Bida, Russ. J. Nondestr. Test. **35**, 9, 702-711 (1999)

12. G. V. Bida, E. Y. Sazhina, Russ. J. Nondestr. Test. **32,** 5, 411-417 (1996)
13. L. A. Fridman, V. P. Tabachnik, G. S. Chernova, Soviet J. Nondestr. Test.-USSR, **13** , 4, 449-455 (1977)
14. M. N. Mikheev, Tr. Inst. Phys. Metals, **21** (1959)
15. O. N. Vasilenko, V. N. Kostin, Russ. J. Nondestr. Test. **49** (9), 510-518 (2013)
16. V. R. Skal'skii, B. P. Klim, E. P. Pochapskii, Russ. J. Nondestr. Test. **48** , 1, 23-34 (2012)
17. V. N. Kostin, O. N. Lukinykh, Y. G. Smorodinskii, K. V. Kostin, Russ. J. Nondestr. Test. **46,** 6, 403-410 (2010)
18. V. N. Kostin, O. N. Vasilenko, A. M. Porseva, A. A. Kabakova, A. I. Melchakova, Diagn. Resour. Mech. Mater. Struct. **5**, 8-15 (2015)

Studying the effect of elastic-plastic strain and hydrogen sulphide on the magnetic behaviour of pipe steels as applied to their testing

Anna Povolotskaya[1], *Eduard* Gorkunov[1,*], *Sergey* Zadvorkin[1], and *Igor* Veselov[2]

[1]Institute of Engineering Science, RAS (Ural Branch), Ekaterinburg, Russia
[2]Ekaterinburg branch of the Russian Research Institute for the Tube and Pipe Industries, Ekaterinburg, Russia

Abstract. The paper reports results of magnetic measurements made on samples of the 12GB pipe steel (strength group X42SS) designed for producing pipes to be used in media with high hydrogen sulphide content, both in the initial state and after exposure to hydrogen sulphide, for 96, 192 and 384 hours under uniaxial elastic-plastic tension. At the stage of elastic deformation there is a unique correlation between the coercive force measured on a minor hysteresis loop in weak fields and tensile stress, which enables this parameter to be used for the evaluation of elastic stresses in pipes made of the 12 GB pipe steel under different conditions, including a hydrogen sulphide containing medium. The effect of the value of preliminary plastic strain, viewed as the initial stress-strain state, on the magnetic behaviour of X70 pipe steels under elastic tension and compression is studied. Plastic strain history affects the magnetic behaviour of the material during subsequent elastic deformation since plastic strain induces various residual stresses, and this necessitates taking into account the initial stress-strain state of products when developing magnetic techniques for the determination of their stress-strain parameters during operation.

1 Introduction

Problems of estimating the degradation of the strength properties of pipe steels, especially when they operate in corrosion-active environments containing hydrogen sulphide, and diagnosing the stress-strain state of pipes are currently urgent in oil and gas industry. In hydrogen sulphide containing media, intensive hydrogenation of metal takes place, which causes pores filled with hydrogen, laminations and cracks. Note also that the effect of a corrosion-active medium on metals enhances under high residual stresses in products [1, 2]. However, such important parameter as the level of residual stresses in pipes is tested only in production, without regard for the fact that a pipe may suffer additional uncontrollable plastic deformation while being transported to the installation site and installed. As a result, the condition of pipes after pipeline laying may differ greatly from their condition at the

* Corresponding author: ges@imach.uran.ru

moment they leave the production plant. This may result in higher material damage rates, higher probability of emergencies and a shorter service life of pipelines.

Nondestructive magnetic techniques are currently finding increasing application in diagnosing the stress-strain state of pipes. The development of these techniques was discussed in numerous studies, e.g. in [3-6], but practically all of them ignored the history of the preliminary plastic strain of the material.

This paper studies the effect of elastic-plastic tensile deformation and the effect of hydrogen sulphide on the magnetic characteristics of steel pipe specimens in order to substantiate the choice of informative parameters for diagnosing the condition of pipes in production and use. The history of the specimens was modelled by uniaxial plastic tension and holding in a hydrogen sulphide containing medium.

2 Experimental procedure and material

Test specimens of control-rolled pipe steel of strength class X70 (according to the API classification), cut out from a large pipe, and steel 12GB designed for operation in media with high hydrogen sulphide content are studied.

The effect of preliminary plastic strain on the magnetic behaviour of the specimens during subsequent elastic deformation is studied on X70 steel specimens cut out from a 1420 mm diameter pipe along the direction of rolling. In stage I, the specimens were subjected to uniaxial tension to various values of plastic strain δ, namely, 0.08, 0.23, 0.49 and 1.65%, with the determination of their magnetic characteristics. The loading–unloading diagrams for each specimen are presented in fig. 1. In stage II, the specimens plastically pre-deformed in stage I underwent elastic tension and compression, the magnetic characteristics being simultaneously measured in a closed magnetic circuit and with attached magnetic devices with the aid of the Remagraph C500 hysteresisgraph, the specimens being magnetised along the direction of applied loading. The specimens were deformed at room temperature.

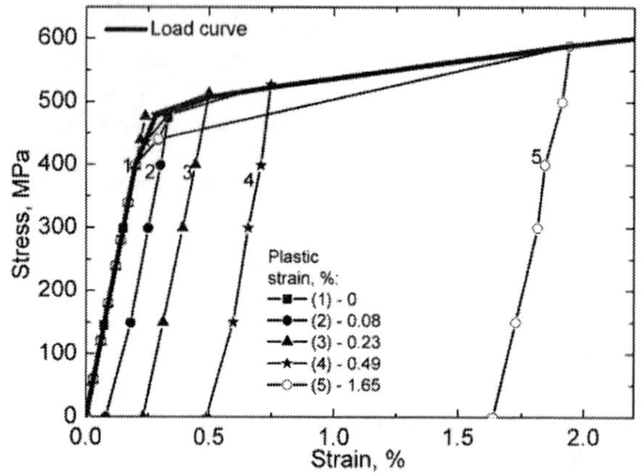

Fig. 1. Loading–unloading diagrams for the test specimens.

The effect of hydrogen sulphide on the magnetic properties of the 12GB steel was studied on specimens cut out lengthwise from a 114 mm diameter pipe, under conditions of elastic and plastic deformation by uniaxial tension at room temperature. Four sets of specimens were examined in the initial state and under the action of hydrogen sulphide lasting 96, 192 and 384 hours. The specimens were saturated with hydrogen sulphide at 20 °C in the A model medium according to the NACE TM0177-2005 standard.

Magnetic measurements were made in a closed permeameter-type magnetic circuit by means of the Remagraph C500 hysteresisgraph.

3 Results and discussion

3.1 Effect of preliminary plastic strain on the magnetic characteristics of the X70 steel

Figure 2 shows magnetic characteristics (coercive force H_c, residual induction B_r and maximum magnetic permeability μ_{max}) as dependent on tensile stresses, the dependences being obtained in the stage of plastic deformation to a certain level and under subsequent unloading. The magnetic behaviour of the specimens under tension agrees with the current ideas of the processes occurring in the magnetic structure of steels under force action. Similar results of studying the effect of elastic-plastic tension on magnetic behaviour were obtained earlier and discussed in [7, 8].

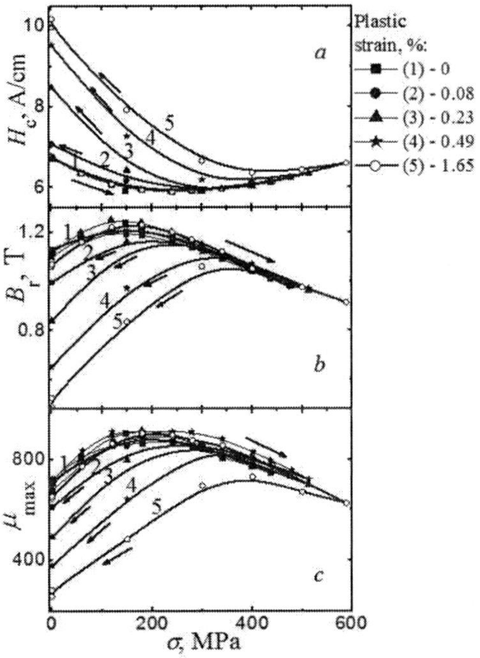

Fig. 2. Coercive force H_c (*a*), residual induction B_r (*b*) and maximum magnetic permeability μ_{max} (*c*) as dependent on tensile stresses under loading to different values of plastic strain followed by load removal: 0 (curve 1), 0.08% (2), 0.23% (3), 0.49% (4), 1.65% (5).

It follows from fig. 2 that, when the specimens are unloaded after plastic strain, the magnetic characteristics change irreversibly; the higher the value of plastic strain, the greater these irreversible changes. The higher values of the coercive force in the no-load state than those in the loaded state can be explained as follows: in the course of unloading, significant residual compressive elastic stresses arise along the direction of tension in a considerable number of grains; herewith, prerequisites appear for the formation of the "easy magnetization plane" magnetic texture, when it is more energetically advantageous for the spontaneous magnetization vectors to line up in the plane normal to the tension axis and hence to the magnetization and magnetization reversal fields; as a result, the magnetization reversal is hampered, and this increases the values of the coercive force and decreases the values of $B_r(\sigma)$ and $\mu_{max}(\sigma)$. These results agree with those obtained in [9], where they also reported a significant difference in the coercive force values for loaded and unloaded steel specimens under plastic tension.

For a vivid demonstration of the effect of plastic strain on the magnetic behaviour of the specimens, fig. 3 shows the values of magnetic characteristics after load removal, reduced to the values in the no-load state, as dependent on the value of plastic strain. The monotonic behaviour of the magnetic characteristics with growing plastic strain enables these parameters or their combination to be used to evaluate the strain state of a pipe which results from pipe manufacture, transportation and installation and which must be considered later on, when testing the stress-strain state during operation.

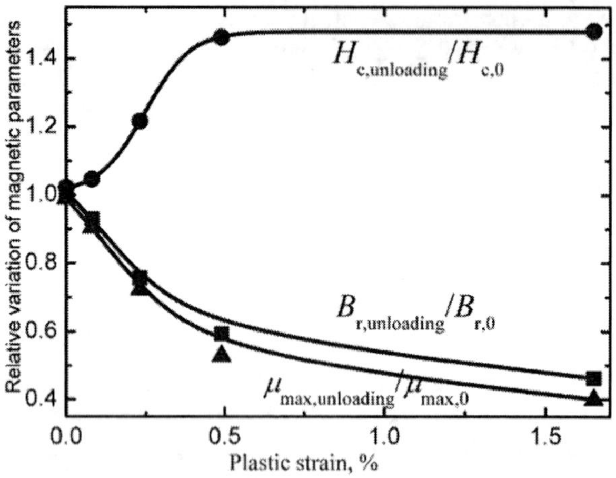

Fig. 3. Relative variation of magnetic parameters after load removal from the value of plastic strain.

3.2 Effect of elastic tensile/compressive strain on the magnetic characteristics of the X70 steel after preliminary plastic deformation

Figure 4 shows the magnetic characteristics of the specimens as dependent on applied stresses measured in a closed magnetic circuit under elastic tension/compression after previous plastic tension to various values.

Under elastic tension, the magnetic behaviour of specimens with different initial stress-strain states is identical; namely, when the tensile stress values increase, the magnetic characteristics change nonmonotonically, with extreme points. It follows from fig. 4 that the higher the value of plastic pre-strain, the higher the values of stresses matched by these extrema. The values of stresses at which there are minimum coercive force values (and, correspondingly, peaks on the dependences $B_r(\sigma)$ and $\mu_{max}(\sigma)$) are attributed to a partial compensation of a part of residual internal compressive stresses by external elastic tensile stresses [10, 11]. It is obvious that the higher the value of plastic pre-strain, the higher the level of residual internal compressive stresses; consequently, the higher the value of stresses at which the coercive force is minimum and the wider the range of applied elastic stresses where the magnetic behaviour is monotonic. Note that the applied stress dependences of the magnetic characteristics measured under elastic tension coincide, within the measurement error with similar dependences obtained under load removal. The reversibility of the changes in the magnetic parameters is characteristic of all the specimens, regardless of the value of their plastic pre-strain. These results agree with those obtained in [10].

Under elastic compression (see fig. 4), the dependences of the magnetic characteristics of the metal plastically pre-strained by tension to small values, below 0.23% (curves 1 and 2 in fig. 4), remain monotonic over the entire range of compressive stresses. For the compression of the specimens with plastic pre-strain of 0.23% and higher and, correspondingly, with a higher level of residual compressive stresses (curves 3 to 5 in fig. 4) there are extrema on the dependences $H_c(\sigma)$, $B_r(\sigma)$ and $\mu_{max}(\sigma)$. This is uncharacteristic of the magnetic behaviour of α-iron-based magnetic materials under elastic compression. Besides, irreversible changes of the magnetic parameters are observed for these specimens when the load is removed after elastic compression. The irreversible changes in the magnetic characteristics of the specimens with high values of plastic pre-strain may be due to the transition into the region of developed plastic strain of in metal regions with external compressive stresses imposed upon residual compressive stresses; as a result, the total stresses overshoot the yield stress. Besides, a decrease in the yield stress must be taken into consideration in the compression tests of the specimens subjected to plastic pre-tension (the Bauschinger effect [12, 13]). Under plastic compression, the "easy magnetization plane" magnetic texture collapses and, correspondingly, the processes of magnetization reversal along the direction of compression become easier. The latter, in turn, decreases the values of the coercive force; as a consequence, the curve $H_c(\sigma)$ corresponding to unloading after compression lies below the similar curve for loading. On the other hand, increased dislocation density under plastic strain must result in increased coercive force. Obviously, this circumstance is of secondary importance in the formation of the level of the hysteretic characteristics of the studied materials under compression.

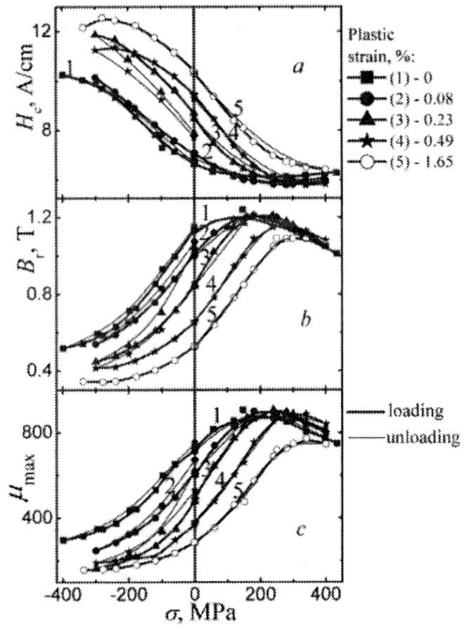

Fig. 4. Applied external stress dependences of coercive force H_c (*a*), residual induction B_r (*b*) and maximum magnetic permeability μ_{max} (*c*) measured in a closed magnetic circuit. The plastic strain values are 0 (1), 0.08% (2), 0.23% (3), 0.49% (4) and 1.65% (5).

In practice, it is easier to measure the magnetic parameters with the use of attached transducers, particularly, for testing large objects, such as large-diameter pipes. Test results obtained with the use of attached transducers are depicted in fig. 5 showing the values of the coercive force H_{ce} and the rms values of magnetic Barkhausen noise voltage U

(measured when the transducer is placed on the specimens along the direction of applied loading) as dependent on applied stresses. It follows from a comparison between fig. 4 and fig. 5 that in the tension region there is a qualitative similarity between the dependences $H_c(\sigma)$ and $H_{ce}(\sigma)$ obtained, respectively, by measurements in a closed magnetic circuit and with an attached transducer.

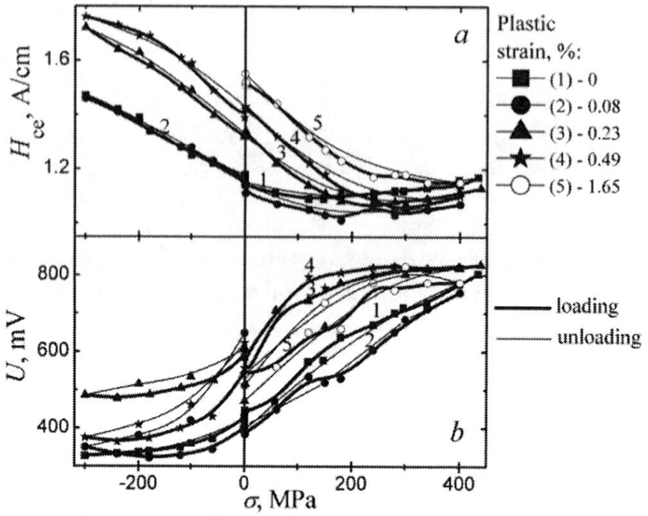

Fig. 5 Coercive force $H_{ce}(a)$ and rms voltage $U(b)$ measured by attached transducers as dependent on applied external stresses. The plastic strain values are 0 (1), 0.08% (2), 0.23% (3), 0.49% (4) and 1.65% (5).

It follows from figs. 4 and 5 that, at applied elastic stresses ranging approximately between -200 MPa and 120 MPa, the magnetic characteristics of all the specimens plastically pre-strained to different amounts vary monotonically and thus can be used to test working elastic stresses.

3.3 The effect of hydrogen sulphide on the magnetic parameters of the 12GB steel under uniaxial elastic-plastic tension

The values of ultimate strength σ_u, yield stress σ_y, percentage elongation at fracture δ, contraction ratio ψ and the magnetic characteristics of the steel under study in the initial state and after holding in hydrogen sulphide containing environment are given in the table. Here, $h_c^{0.4}$ and $h_c^{0.05}$ are coercive force values on minor hysteresis loops at a maximum magnetic induction of 0.4 and 0.05 T respectively. The values shown in the table are averaged experimental data for three specimens of each set. The deviations of individual characteristics from the mean values do not exceed 5%.

Table. The mechanical and magnetic properties of the 12GB pipe steel.

Duration of holding in H_2S, h	σ_u, MPa	σ_y, MPa	δ, %	ψ, %	H_c, A/cm	B_r, T	μ_{max}	$h_c^{0.4}$, A/cm	$h_c^{0.05}$, A/cm
0	474	374	23	77	3.81	1.47	1485	3.18	0.60
96	458	361	22	76	3.70	1.47	1496	2.86	0.58
192	498	426	14	70	3.95	1.52	1446	3.16	0.63
384	500	437	12	74	4.05	1.51	1515	3.19	0.71

Figure 6 shows stress-strain diagrams and strain (ε) dependences of the magnetic parameters of the specimens for specimens with mechanical and magnetic characteristics closest to the mean values. For the other specimens, the stress-strain diagrams and the strain dependences of the magnetic characteristics are similar to those shown in fig. 6. It follows from the table and fig. 6 that the 96-hour holding has little effect on the mechanical properties. At the same time, an increase in holding in a hydrogen sulphide containing media to 384 hours is accompanied by some increase in the strength characteristics and an almost twofold decrease in percentage elongation at fracture; it also results in numerical differences on the dependences $H_c(\varepsilon)$ and $h_c^{0.4}(\varepsilon)$ in the region of high strains; namely, the longer the holding in H_2S, the higher lie the curves corresponding to these dependences.

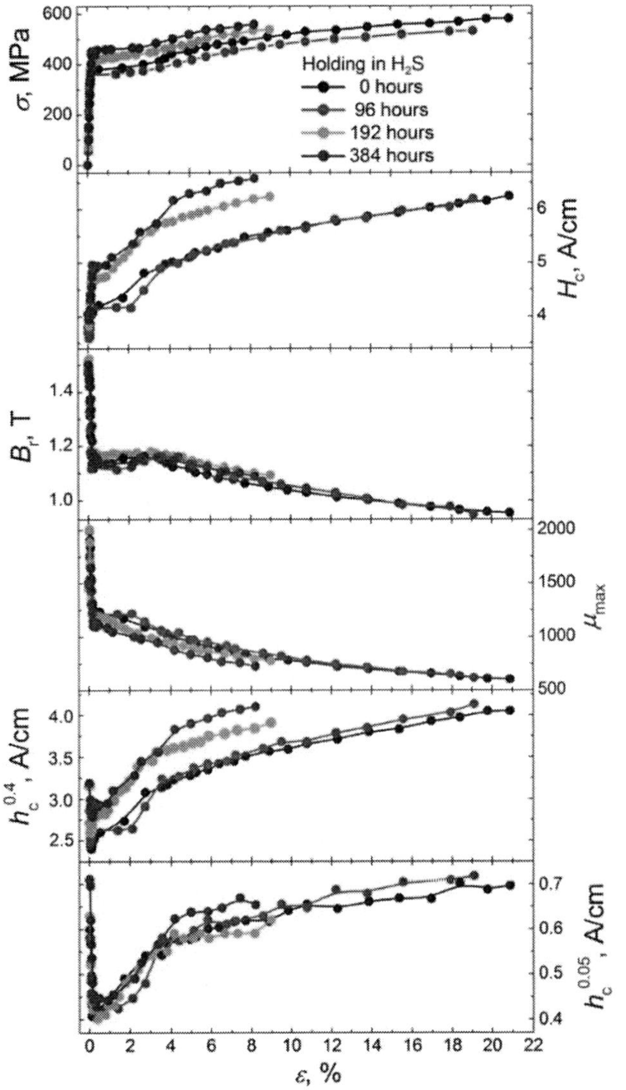

Fig. 6. Stress-elongation diagrams and elongation dependences of magnetic characteristics for the 12GB pipe steel specimens.

The action exerted by a hydrogen sulphide containing medium on the 12GB steel for

96 hours has practically no effect on magnetic characteristics. The parameters H_c and $h_c^{0.05}$ are seen to tend to increase with the duration of holding in hydrogen sulphide.

As with the carbon steels St3 and 45 [14], for the alloy steel 12GB, except in the initial stage of loading, the strain dependences of the coercive force are qualitatively similar to the stress-strain diagrams, whereas the dependences of maximum magnetic permeability and residual induction are inverse. In the initial portion of the tension diagram (see fig. 7) the values of the coercive force slightly decrease and then increase. The nonmonotonic behaviour of the coercive force as a function of elastic strain is explainable by the appearance of a magnetic texture due to a positive magnetoelastic effect.

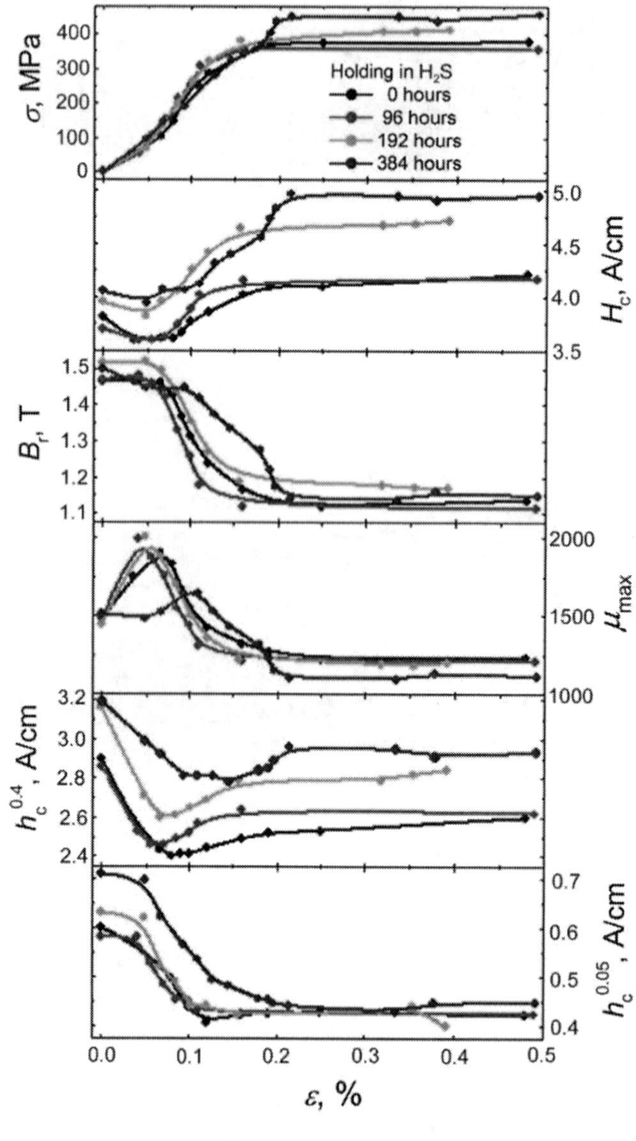

Fig. 7. The initial portions of loading diagrams for the 12GB steel specimens and the strain dependences of the magnetic characteristics of the steel.

Figure 8 shows coercive force, residual induction and maximum magnetic permeability as dependent on the magnitude of tensile stresses σ in relative coordinates. The magnetic characteristics are normalized to their values measured at $\sigma = 0$, the stresses being normalized to yield stress σ_y. As is obvious, the dependences $B_r(\sigma)$ and $h_c^{0.05}(\sigma)$ are unique in the practically important region of elastic stresses; namely, when the stresses grow from 0 to σ_y, the values of $h_c^{0.05}$ decrease by 30 to 40%, and the values of B_r decrease approximately by 30%, residual induction remaining practically unchanged in the initial stage of loading, when $\sigma < 100$ MPa. Thus, the value of the coercive force measured under magnetization reversal in weak fields can be used as a parameter for the evaluation of elastic strains in the 12GB pipe steel under conditions including the action of a hydrogen sulphide containing medium. Besides, it is obvious from fig. 8 that the minimum values of $h_c^{0.05}$ correspond to the yield stress of the specimens. In this connection, the value of $h_c^{0.05}$ can be used as a parameter for the evaluation yield stress for the 12GB steel.

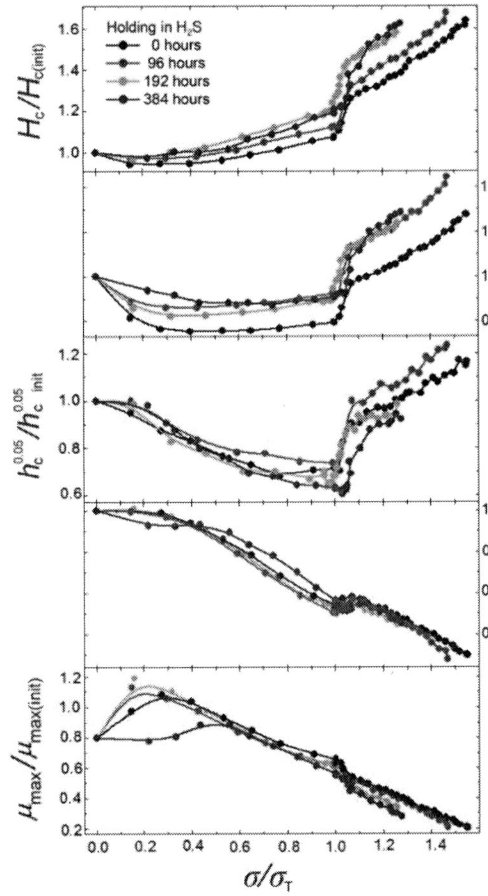

Fig. 8. Magnetic characteristics of the 12GB pipe steel as dependent on applied tensile stresses in relative coordinates.

4 Conclusion

It has been revealed that the behaviour of the magnetic characteristics of the steel (coercive force, residual induction and maximum magnetic permeability) with increasing plastic strain is monotonic. Therefore, these parameters or their combination can be used to evaluate the strain state resulting from pipe making, transportation and installation.

Due to residual stresses induced by plastic strain, the prehistory in the form of plastic strain affects the magnetic behaviour of the material during its subsequent elastic deformation, and this necessitates considering the initial stress-strain state of a metallic structure when developing magnetic techniques for the determination of its stress-strain parameters in operation.

For the steel studied, the range of elastic stresses (between -200 and 120 MPa) has been determined where the magnetic characteristics measured in the longitudinal direction vary uniquely and hence are usable for the evaluation of working elastic stresses.

The 96-hour action of H_2S has practically no effect on the mechanical and magnetic characteristics of the 12GB pipe steel. A longer holding in hydrogen sulphide causes notably lower plasticity, higher values of ultimate strength and yield stress and slightly increased coercive force.

The strain dependences of ultimate coercive force are qualitatively similar to the stress-strain diagrams for the 12GB steel, whereas those of maximum magnetic permeability and residual induction are inverse, except in the initial portion of loading, where the action of induced magnetoelastic anisotropy is evident.

In the stage of elastic strain of the 12GB steel, both in the initial state and after holding in hydrogen sulphide, a unique correlation was found between the coercive force measured on a minor magnetic hysteresis loop in weak fields and tensile stresses, and this enables this parameter to be used for the evaluation of elastic stresses in products made of the 12GB pipe steel. The minimum on the dependence $h_c^{0.05}(\sigma)$ corresponds to the yield stress of this steel both in the initial state and after holding in hydrogen sulphide.

In this study, to measure the magnetic and mechanical characteristics, we used the equipment of the Plastometriya collective use center affiliated to IES UB RAS. The study was partially supported by UB RAS project No 15-10-1-22.

References

1. L. L. Shreir, R. A. Jarman, G. T. Burstein, *Corrosion* (3rd ed., 2 Vols., Newnes-Butterworths, London, 1993)
2. I. Yu. Pyshmintsev, I. N. Veselov, B. A. Yerekhinsky, V. I. Chernukhin, A. G. Shiriaev, Truboprovodnyi Transport: Teoriya i Praktika **5,** 57, 26-31 (2016)
3. R. A. Sadrtdinov, V. B. Geitsan, Yu. P. Surkov, V. G. Rybalko, D. V. Novgorodov, A. Yu. Surkov, Russ. J. Nondestr. Test. **46,** 2, 84-91 (2010)
4. A. P. Nichipuruk, A. N. Stashkov, V. N. Kostin, M. K. Korkh, Russ. J. Nondestr. Test. **45,** 9, 616-622 (2009)
5. E. S. Gorkunov, S. Yu. Mitropolskaya, S. M. Zadvorkin, L. S. Shershneva, E. A. Tueva, Tyazheloe Mashinostroenie **10**, 2-6 (2010)
6. V. F. Muzhitskii, B. E. Popov, G. Y. Bezlyudko, V. V. Zarudnyi, E. A. Levin, Russ. J. Nondestr. Test. **32,** 2, 97-102 (1996)
7. E. S. Gorkunov, A. M. Povoltskaya, K. E. Solov'ev, S. M. Zadvorkin, Russ. J. Nondestr. Test. **46,** 9, 638-644 (2010)
8. E. S. Gorkunov, Yu. V. Subachev, A. M. Povolotskaya, S. M. Zadvorkin. Russ. J. Nondestr. Test. **49,** 10, 584-594 (2013)
9. V. G. Kuleev, T. P. Tsarkova, A. P. Nichipuruk, V. I. Voronin, I. F. Berger, Phys.metals metallog **103**, 2, 131-141 (2007)
10. V. G. Kuleev, E. S. Gorkunov, Russ. J. Nondestr. Test. **33,** 11, 741-753 (1997)

11. S. V. Vonsovsky, JETP **17,** 12, 1094-1105 (1947)
12. Johann Bauschinger, Dingler's Polytechnisches Journal **224,** 129-134 (1877)
13. J. F. Bell, *Experimental Principles of the Mechanics of Solids. Part 2. Finite Deformations* (Nauka, Moscow, 1984)
14. E. S. Gorkunov, S. M. Zadvorkin, S. V. Smirnov, S. Yu. Mitropolskaya, D. I. Vichuzhanin, Phys.metals metallog. **103,** 3, 311-316 (2007)

Investigation of the structure, phase composition and physical-mechanical characteristics of high-strength laminated materials which contain corrosion-resistant steels with high nitrogen content

Evgeniia Putilova [1,*], *Eduard* Gorkunov, *Sergey* Zadvorkin

[1]Institute of Engineering Science, RAS (Ural Branch), Ekaterinburg, Russia

Abstract. In this work, we studied the structure, phase composition, mechanical and magnetic properties of structural steel clad with corrosion-resistant steels with varying degrees of austenite stability in order to determine the possibility of using magnetic methods for diagnostics the current state of such materials.

1 Introduction

The development of modern industry is closely connected with the need to create and introduce new technologies and especially promising materials that are resistant to aggressive external influences and ensure the durability, reliability and cost-effectiveness of the used structures [1]. The analysis of the trends in the development of modern production shows that one of the effective solutions to this problem is the development and creation of a new generation of materials with a hierarchically organized structure, in particular multicomponent or multilayered materials. The layers or components of such materials individually have different sets of required properties, but when they are joined, it is possible to obtain a material that will combine the advantages of each layer.

So, for example, to ensure high corrosion-erosion resistance of hulls of ice navigation vessels, it is advantageous to use steels resistant to action of sea water and ice fields [2, 3]. Since corrosion wear affects only the outer surface of the hull, it is economically sound to increase the corrosion resistance of not the full hull's wall thickness but to provide it in a thin clad outer layer of corrosion-resistant steel [4]. However, the phase transformations taking place in many austenitic corrosion-resistant steels can lead to formation of new constituents in the structure of the material, which can cause deterioration in the operational properties of the product as a whole, reduce the durability and, subsequently, lead to the fracture. To ensure safety during the operation of products from laminated materials, as well as to assess the residual life of parts and structural elements, it is necessary to detect changes occurring during operational loads in the laminate as a whole as well as in its components. At present, there are only a little amount of works concerning the methods for

* Corresponding author: tuevaevgenya@mail.ru

nondestructive testing of the current state of individual layers in bimetallic materials under operating stress conditions [5-7].

In this paper, we present the results of investigations of various laminated materials, consisting of corrosion-resistant steels with different austenite stability are present and structural steels with different strength characteristics. Also, the article shows the possibility of using magnetic parameters for estimating changes in the structure, phase composition and stress-strain state of products from the investigated materials.

2 Experimental procedure and material

The objects of research were various laminates based on structural steels clad with corrosion-resistant steels with varying degrees of austenite stability (manufactured by hot-rolling technology "steel 11CrN4CuMoV - steel 08Cr18N10" and " steel 11CrN4CuMoV - steel 04Cr20Ni6Mn11Mo2NVNb"), and also their separate components. Both the laminates themselves and their individual components were subjected to various kinds of deformation effects: cold rolling, hot rolling, uniaxial tension.

Hot rolling, both bimetals and individual components, was carried out by several passes until the specified value of the total reduction was achieved. The value of the degrees of deformation during rolling for all materials is indicated in Table 1.

Table 1. Rolling deformation degree

Material	Rolling deformation degree				
steel 11CrN4CuMoV	50	38	28	20	
steel 08Cr18N10	50	40	32	25	20
steel 04Cr20Ni6Mn11Mo2NVNb	49	41	30	20	
«steel 11CrN4CuMoV – steel 04Cr20Ni6Mn11Mo2NVNb»	49	40	33	21	
«steel 11CrN4CuMoV – steel 08Cr18N10»	47	45	30	20	

After rolling the samples were heat-treated in the following manner: quenching in water from a temperature of 950 °C and a high tempering at 600 °C during 7 hours and subsequent cooling in water. This heat treatment is usually used for the main layer - steel steel 11CrN4CuMoV.

Magnetic parameters were determined by means of the Remagraph C-500 magnetic measuring system. The magnetic measurements were made in a closed permeameter circuit. A magnetic field of strength up to 500 A/cm was applied along the axis of the sample. The measurement error of the field and induction did not exceed 3%. Before and after each measurement of magnetic properties the specimen was demagnetized. The differential magnetic permeability was determined by differentiating the descending branches of the magnetic hysteresis loop (hereinafter the figures present only those portion of the field dependences of differential permeability where peaks are formed).

The strength of the bonding of the layers in bimetals was determined by shear test of a corrosion-resistant layer by means of a Tinius Olsen machine.

Using an X-ray energy dispersive microanalysis system, Oxford INCA Energy 450 XT with a non-gas detector ADD from Oxford Instruments, the distribution of alloying elements along a line passing through two steels and an intermediate zone between them has been obtained.

X-ray phase analysis was performed on a Shimadzu XRD-7000 X-ray diffractometer.

3 Results and discussion

Rolling with different degrees was carried out in order to determine the minimum necessary rolling deformation to obtain a sufficiently high strength of joint. The strength of joint of the layers in bimetals was determined by the shear test of a corrosion-resistant layer. It is established that at a 20% deformation degree for both materials the shear strength of the layers is low. With an increase of deformation degree, the strength of the compound increases and it can be said that when the deformation is greater than 30% for the bimetal «steel 11CrN4CuMoV – steel 08Cr18N10» and more than 35% for the bimetal «steel 11CrN4CuMoV – steel 04Cr20Ni6Mn11Mo2NVNb» the strength does not change significantly. The maximum shear strength was determined for the material «steel 11CrN4CuMoV – steel 04Cr20Ni6Mn11Mo2NVNb» and is 600 MPa.

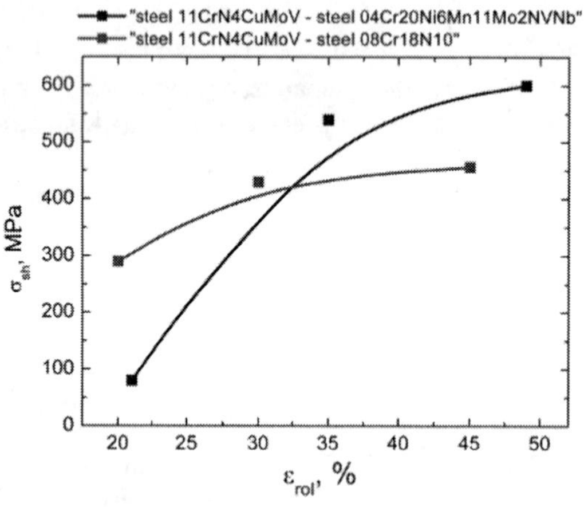

Fig. 1. Dependence of the strength joint of layers in bimetals depending on the degree of deformation by rolling

A measurement of the hardness distribution over the cross-section of the resulting bilayer materials was carried out. For the bimetal «steel 11CrN4CuMoV – steel 04Cr20Ni6Mn11Mo2NVNb» the hardness values of both layers are approximately on the same level, however, in the transition zone (bond zone) there is the decrease of the values of hardness approximately on 20-30%.

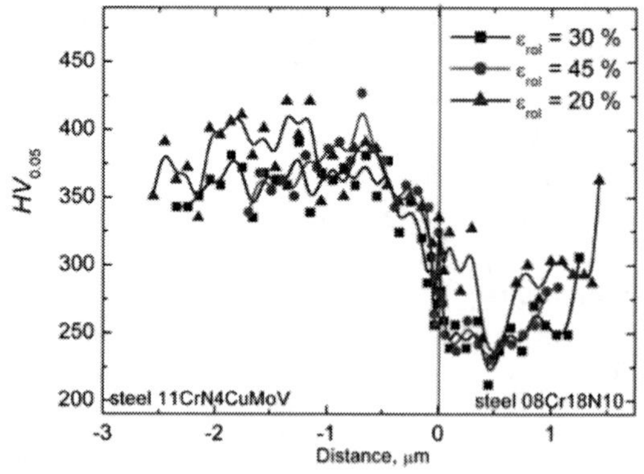

a)

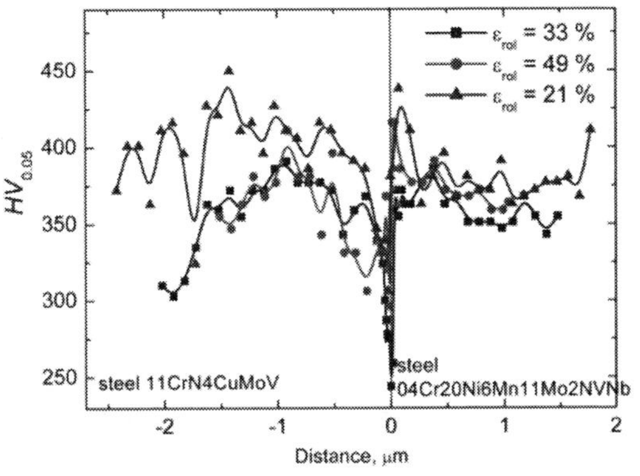

b)

Fig. 2. The hardness distribution over the cross-section of the resulting bilayer materials

The distribution of alloying elements along a line passing through two steels and an intermediate zone between them is obtained. It can be seen that in the transition region there is a decrease in such elements as chromium, nickel, manganese in the direction from corrosion-resistant steel to the main steel 11CrN4CuMoV. I.e. in the process of hot rolling the diffusion of chromium, nickel and manganese from corrosion-resistant steels into less doped steel 11CrN4CuMoV to a depth of up to 10 microns occurs. Accordingly, corrosion-resistant steels are depleted by same elements at approximately the same distance.

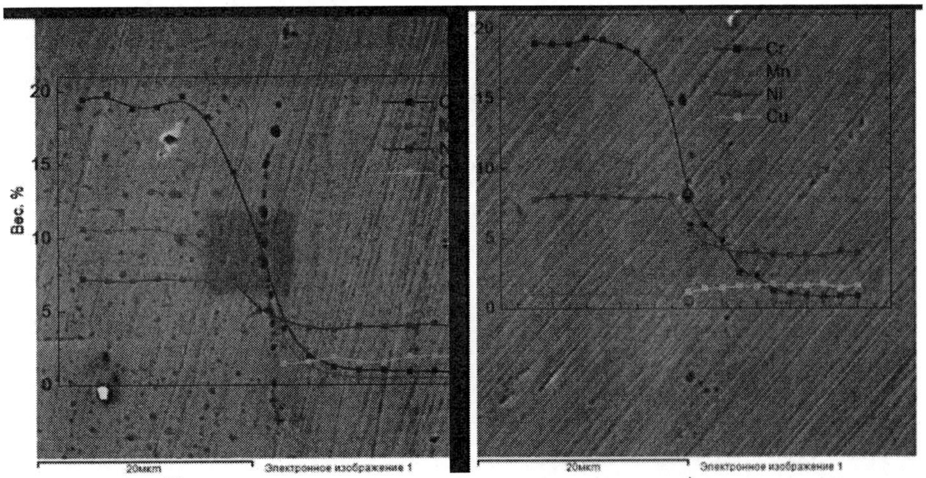

b) *a*)

Fig. 3. The distribution of alloying elements along a line passing through two steels and an intermediate zone between them

Fig. 4 for comparison shows the dependences μ_{dif}(H) for corrosion-resistant steel 08Cr18N10T [8], and 08Cr18N10, studied in this work. Figure 1b shows the dependences of μ_{dif}(H) for corrosion-resistant steel 08Cr18Ni10Ti. In the initial state of 08Cr18Ni10Ti steel (Fig.4b) there is no if maximum on the field dependence due to the insufficient amount of a ferromagnetic phase in the material that is also confirmed by the X-ray diffraction analysis. However, when the strain increases, the α'-phase is formed [8, 9] and at some value of ε_{rol} a peak begins to appear on the field dependence of the differential magnetic permeability. The dependence of the amount of strain induced α′-martensite on ε_{rol} for steel 08Cr18Ni10Ti is practically linear (Fig. 4c), and at $\varepsilon_{rol} = 0.19$ it is about 20%, therefore the formation of the peak at the field dependence becomes almost immediately visible (Fig. 4b). The behaviour of the peak field for all materials is similar to that of its coercive force.

For steel 08Cr18Ni10 the peak on the field dependence is formed only for the initial state sample. According to X-ray diffraction analysis the non-deformed sample contains about 10% of the ferromagnetic phase. Then this steel was subjected to hot rolling with cooling at the room temperature. Hot rolling was carried out at a temperature of 1050 °C with several passes until the specified values of the total accumulated strain been achieved. The ferromagnetic phase was absent in all the deformed samples, because of the fact that during the hot rolling the ferromagnetic phase dissolved and upon further cooling it did not have time to form. Also, in contrast to cold rolling, where the anisotropy of properties is clearly distinguished with respect to the rolling direction, it is not observed during hot rolling. The lack of anisotropy is also confirmed by the fact that the values of the residual stresses in 08Cr18Ni10 steel, determined by the X-ray diffraction analysis, are practically the same level in mutually perpendicular directions.

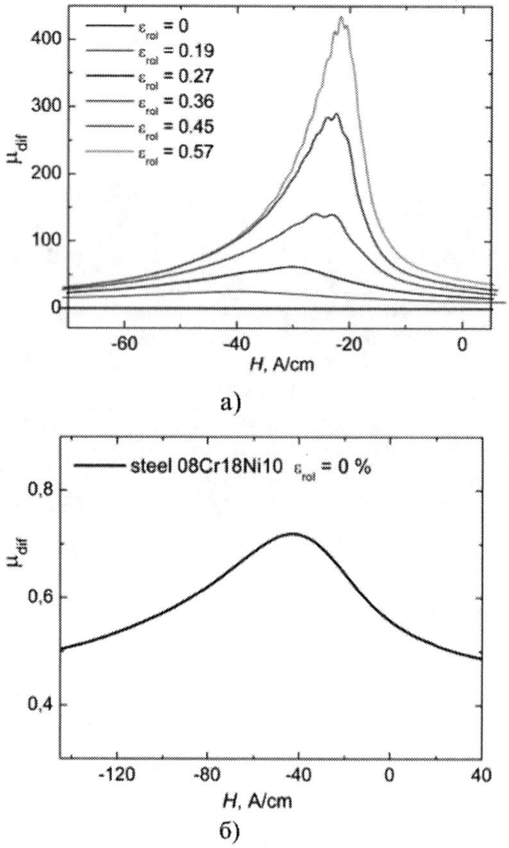

Fig. 4. The field dependences of the differential magnetic permeability of 08Cr18Ni10Ti (a) and 08Cr18Ni10 (b) steels

The only one peak on the field dependence corresponding to ferrite is observed for structural steel 11CrNi4CuMoV (Fig. 5a) as well as for the above mentioned steel St3.

However in comparison with early studied steel St3 [8], the peak field is shifted toward higher fields, what could be explained by the more alloyed 11CrNi4CuMoV steel. The peak fields (Fig. 5a) differ insignificantly, and the higher the strain, the stronger is the field the peak is located at. Figure 5b shows the field dependences of the laminate "steel 11CrN4CuMoV - steel 08Cr18N10" and its individual components. The main fig. 5b shows the curves for all materials deformed at strain about 40%. It can be seen that the peak field for a laminate practically coincides with that of for 11CrNi4CuMoV steel, and the field dependence for a corrosion-resistant steel does not show a peak. Thus exactly the constructional steel makes the main contribution to the formation of the magnetic properties of this laminate. As for the steel 08Cr18N10, then Fig. 5b shows the peak in the field dependence of the differential magnetic permeability of this steel in the initial state. As already mentioned above, in addition to the austenite there is ferromagnetic phase in the initial state of the steel 08Cr18N10. However, with further hot rolling at 1050 °C, these phases decompose, so one can not observe any peaks in the field dependence after rolling.

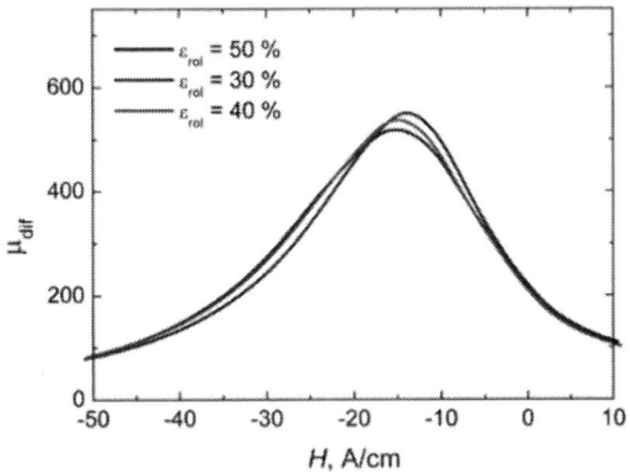

a)

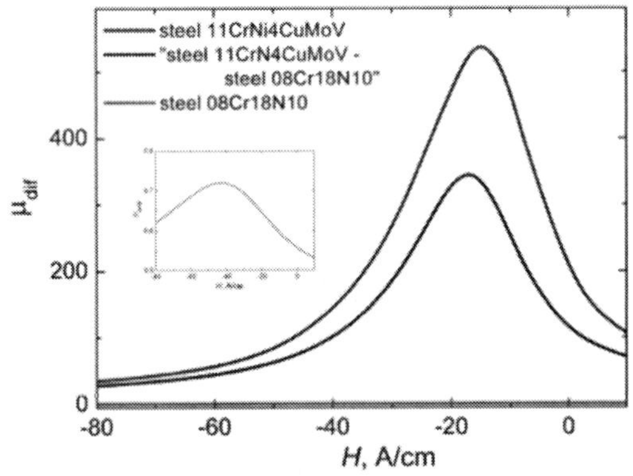

б)

Fig. 5. The dependences of μ_{dif}(H) for the steel 11CrN4CuMoV with different rolling strains (a); the steels 11CrN4CuMoV, 08Cr18Ni10 and the laminated material made from these steels (b)

Figure 6 shows the field dependencies of differential magnetic permeability for specimens of 11CrNi4CuMoV steel and laminates "steel 11CrN4CuMoV - steel 08Cr18N10" and " steel 11CrN4CuMoV - steel 04Cr20Ni6Mn11Mo2NVNb" at the maximum strain accumulates after hot rolling. All the curves clearly show one peak corresponding to the ferrite in the structural steel 11CrN4CuMoV.

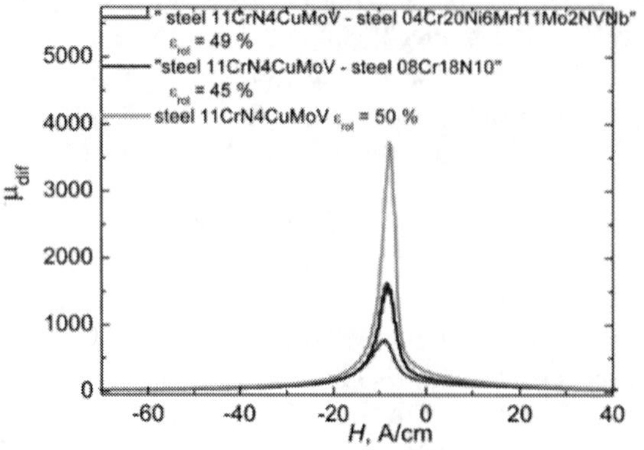

Fig. 6. The dependences of μ_{dif}(H) for the steel 11CrN4CuMoV and laminates "steel 11CrN4CuMoV - steel 08Cr18N10", " steel 11CrN4CuMoV - steel 04Cr20Ni6Mn11Mo2NVNb" under maximal deformation

4 Conclusion

The microstructure, phase composition, mechanical and magnetic properties of two-layered joints of high-strength steel and corrosion-resistant steels with different degrees of austenite stability obtained by hot rolling are studied. The strength joint of the layers was estimated by means of shear tests. It is shown that at higher strain degrees the strength of the joint increases. The possibility of diagnostics of the formation of a magnetically ordered phase in austenitic steels by the maximum magnetization in the maximum applied field (saturation magnetization) and the maximum magnetic permeability, during hot rolling and further heat treatment is established. It is shown that the coercive force and the maximum magnetic permeability can be used as parameters of non-destructive testing to assess the quality of such heat treatment operations as quenching and tempering of the materials under study.

It is shown that the value of the maximum differential permeability can be used for the controlling the strain-induced phase transformations in corrosion-resistant austenitic steels constituting a macroscopically inhomogeneous material. The values of the maxima on the field dependences of the differential magnetic permeability can be used as parameters to detect the formation of a new magnetically ordered phase during plastic and elastoplastic deformation, as well as its amount. The field values that correspond to the peaks on the field dependences of the differential magnetic permeability can be used to estimate the strain accumulated by rolling and uniaxial tension of corrosion-resistant metastable steels.

The work was partially supported by the RFBR, grant No 16-38-60027 and performed within the state task, No 01201354598. The researchers used the equipment the "Plastometriya" collective use center.

References

1. M.V. Kostina, O. A. Bannykh, V.M. Blinov, E. V. Blinov, V. N. Karpov, Almanac of Clinical Medicine **17**, 2, 92-95 (2008).
2. V.V. Sagaradze, N.V. Kataeva, S.Y. Mushnikova, O.A. Khar'Kov, G.Y. Kalinin, V.D. Yampol'Skii, Phys.metals metallog. **2** ,115, 202-210 (2014).

3. N. Nesterov, Morskoy flot **5**, 24-28 (2005).
4. S.Y. Mushnikova, V.A. Malyshevskii, G.Y. Kalinin, S.K. Kostin, V.V. Sagaradze, Y.I. Filippov, N.V. Kataeva, V.A. Zavalishin, Phys.metals metallog. **6** ,116, 626-635 (2015).
5. M.B. Rigmant, A.P. Nichipuruk, M.K. Korkh, Russ. J. Nondestr. Test. **9**, 511-521 (2012).
6. M.B. Rigmant, A.P. Nichipuruk, B.A. Khudyakov, M.K. Korkh, S.A. Zinchenko, A.V. Zagainov, Russ. J. Nondestr. Test. **10**, 583-587 (2016).
7. A.P. Nichipuruk, B.A. Khudyakov, M.K. Korkh, M.B. Rigmant, V.G. Kuleev, Russ. J. Nondestr. Test. **6**, 405-409, (2009).
8. E.S. Gorkunov, S.M. Zadvorkin, E.A. Putilova, A.M. Povolotskaya, L.S. Goruleva, I.A. Veretennikova, I.S. Kamantsev, Russ. J. Nondestr. Test. **6**, 346-356 (2012).
9. E.S. Gorkunov, S.M. Zadvorkin, E.A. Kokovikhin, E.A. Tueva, Yu.V. Subachev, L.S. Goruleva, A.V. Podkorytova, Russ. J. Nondestr. Test. **66**, 369-380 (2011).

The interrelationships of magnetic and magneto acoustic-emission characteristics of heat-treated steels of various chemical composition

Vladimir Kostin[1,2*], *Evgeniy* Serbin[1,2], and *Olga* Vasilenko[1,2]

[1]M.N. Mikheev Institute of Metal Physics of Ural Branch of Russian Academy of Sciences, 620990 Ekaterinburg, Russia
[2]Ural Federal University n.a. B.N. Yeltsin, 620002 Ekaterinburg, Russia

Abstract. The relationship between the amplitude of magnetoacoustic emission and the residual magnetic induction of the materials is shown. The dependence of magnetoacoustic emission (MAE) on the frequency of magnetization reversal of a group of steels of different chemical composition was investigated. In all the investigated cases, the MAE amplitude maximum corresponds to a field frequency of 3-5 Hz. The magnetic properties of two groups of steels subjected to various thermal treatments have been studied. It is shown that the residual magnetic induction of steels is a sufficiently universal parameter for controlling softening thermal treatments and that the amplitude of magnetoacoustic emission can be used as a testing parameter in scanning systems of structurescopy of ferromagnetic steels.

1 Introduction

At the core of magnetic structurecopy of steel products lies the presence of a correlation between magnetic and hardness properties. However, the use of magnetic testing parameters alone is often not sufficient for complete determination of the structural-phase state of steel products. Additional information can be acquired by parameters related to interaction of moving domain walls with defects of the crystal structure. One of such effects is magnetoacoustic emission (MAE), which is understood as a totality of elastic vibrations that arise in a ferromagnet upon its magnetization reversal.

In the case of low-frequency magnetization reversal of ferromagnets, the main mechanism of MAE is magnetostriction, which manifests itself in a change of the sample size (magnetostriction), as well as in the appearance of local elastic perturbations, mainly caused by irreversible displacements of 90-degree domain walls (magnetic noise). Thus, magnetoacoustic emission makes it possible to register jumps of domain boundaries not only near the surface, but also in the volume of the material [1, 2].

The rearrangement of the domain structure upon magnetization reversal can be characterized by various dimensional (area or mean path of the domain boundary), quantitative (the number of simultaneously shifting boundaries), and temporal (time of Barkhausen jump) parameters. The

* Corresponding author: kostin@imp.uran.ru

MAE signal also depends on the reflection of the elastic vibrations from the boundaries of the ferromagnet and, consequently, the MAE parameters depend on the shape and size of the magnetized object. These features should be reflected in the MAE spectrum, which makes it advisable to use broadband piezoelectric transducers for detection of the MAE instead of the previously used narrow-band [2, 3].

The dependence of the MAE on the frequency of the magnetization reversal field is no less important in the analysis of the MAE. The use of the MAE phenomenon in structurescopy implies carrying out a measurement cycle with averaging the amplitude readings over several periods. In addition, the noise nature of the MAE signal and the high sensitivity to noise of an acoustic nature and vibrations requires rapid measurement. This means that the optimal magnetization reversal frequency of a ferromagnet should not be too low, since for a long duration of measuring the field distribution of the MAE amplitude, the amount of interference during the period will increase dramatically [2].

The aim of this work is to establish the relationships between magnetic and magnetoacoustic emission parameters, to establish the correlation of the residual magnetic induction and amplitude of the MAE, to find new testing parameters of ferromagnetic objects in structurescopy, and to determine the dependence of the amplitude of the MAE on the frequency of the remagnetizing field.

2 Materials and experimental methods

To study the dependence of magnetoacoustic emission on the frequency of magnetization reversal, samples were made of three grades of steel in the form of plates 4.6 × 40.1 × 90 mm in size, differing in structural-phase composition and, accordingly, in magnetic properties. The plates were quenched and tempered at various temperatures (*Tque* and *Ttem*, respectively). (see Table 1).

Table 1. Characteristics of heat-treated steel plates

Material	*Tque*, °C	*Ttem*, °C	*Br*, T	*Hc*, A/cm	HB
Steel 10	930	400	1,1	8,9	269
Steel 10	930	500	1,3	7,9	248
Steel 10	930	600	1,5	6,0	192
65G	870	200	0,9	32,3	601
65G	870	400	1,45	14,4	505
65G	870	600	1,55	13,6	331
95Cr18	1070	200	0,5	74,5	601
95Cr18	1070	550	0,8	28,0	534

The magnetic properties of the samples were measured using the REMAGRAPH C-500 measuring system manufactured by Magnet-Physik Steingroever GmbH, Germany. The error in magnetization measuring did not exceed 2%, and the error in field measuring was 1%. The hardness of the samples was measured with a hardness tester TK14-250 with an error of no more than 0.5 HB (Brinell scale).

The effect of the shape and geometric dimensions of ferromagnets on the magnitude and nature of magnetoacoustic emission was studied on samples from annealed permendur (50% Fe, 50% Co), from which three samples were made: a short rectangular plate 4.1 × 40 × 89 mm (sample P1); a long plate with dimensions of 2 × 8.9 × 295 mm (sample P2) and a shortened plate with dimensions of 2 × 9 × 161 mm (Sample P3).

To evaluate the influence of the stress-strain state on the MAE signal, samples of 20G and 70G steels were chosen, the stress-strain state of which was varied over a wide range by means of cold plastic deformation by rolling up to respectively 40% and 63% (by changing the cross-section) and then annealing at various temperatures in the range from 20 to 800 °C for 1 hour, followed by cooling in air. After the samples were then grinded to remove the dross and the decarburized layer. The final dimensions of the 20G steel samples were 4x10.2x69 mm, and the 70G steel was 6x9.5x88 mm. The coercive force of all the samples studied after the grinding has changed insignificantly, which indicates the absence of appreciable plastic deformation during grinding [4].

For excitation of MAE two solenoids with different degrees of field homogeneity were used. The solenoid 1 had a length of 154 mm, a rectangular hole of 30 × 60 mm, and the field in it was inhomogeneous: at a distance of a quarter of the length from the center, the maximum field amplitude decreased by more than 15% of the value at the center of the solenoid. The solenoid 2 had a length of 750 mm, a round hole 48 mm in diameter, and a region of uniform (with an accuracy of 2%) field in it was 120 mm. The maximum amplitude of the field in solenoids 1 and 2 reached 160 and 100 A/cm, respectively. To eliminate extraneous vibrations, the samples in the solenoids were placed either on the foam rubber (solenoid 1) or on the suspension (solenoid 2).

Due to wideness of the spectrum of magnetoacoustic emission (from a few hertz to several megahertz) tested methods and means of acoustic emission testing were used for the measurements. Two broadband acoustic emission sensors LD-11 and A-9 ELTEST (3) with working area diameters of 17.5 mm (transducer 1) and 21 mm (transducer 2) were used to record acoustic vibrations. The stability of the acoustic contact was ensured by using Lythol-24 couplant and by fixation of the piezoelectric sensor with an elastic band. In more detail, the setup used for excitation and detection of the MAE is described in [2, 3].

3 Results and discussion

To determine the character of the dependence of the amplitude of the MAE on the frequency of the magnetization reversal U_{MAE} (f_{rm}) for a group of samples with different structural-phase states (see Table 1), magnetoacoustic measurements were made on the described setup with solenoid 1 (Fig. 1).

As can be seen from Fig. 1, the presence of a maximum on the dependence of the amplitude of the MAE on the frequency of the alternating field is characteristic for all the samples studied. When the field frequency is reduced below 0.5 Hz, the signal decreases sharply to noise level. As the frequency increases, the signal rapidly decreases and at a frequency of 15 Hz or more it also becomes commensurate with the noise level. The nonmonotonicity of the dependence of the amplitude of the MAE on the frequency of the alternating field and the "resonant" frequency of the MAE occurring at frequencies of 3-4 Hz are observed for all samples of the group, despite the differences in their physical properties.

The measurement of magnetoacoustic emission in an inhomogeneous field has practical importance, since in the future a local testing of the material properties is assumed. However, in a nonuniform magnetic field, the ponderomotive force is additionally affects ferromagnetic samples.

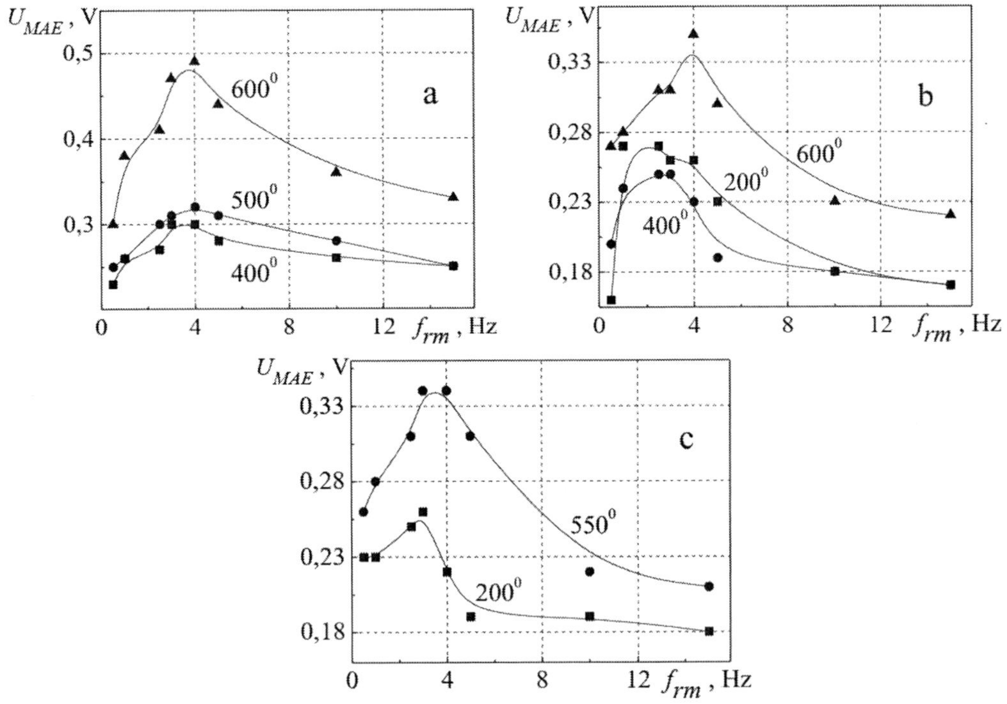

Fig. 1. Dependences of MAE amplitude of quenched and tempered at different temperatures samples of (a) steel 10, (b) steel 65G and (c) steel 95Cr18 on frequency of the field f_{rm}. Tempering temperatures of particular sample shown on the figure.

It is known that the coercive force *Hc* decreases monotonically with increasing tempering temperature of more than 300 °C only for hardened steels with a carbon content of less than 0.3% (see the properties of steel 10 in Table 1). With a higher carbon content, the coercive force at medium and high temperature tempering varies little or even ambiguously. However, with an increase in tempering temperature from 300 °C and higher for steels with different carbon content, monotonic growth of the residual magnetic induction of substance *Br* is noticeable (by a factor of 1.5 - 2), which is also evident from Table. 1.

A decrease in the coercive force *Hc* and an increase in the residual magnetic induction *Br* also take place during softening of the cold-deformed steels as a result of their annealing. Dependences of the residual magnetic induction and coercive force on the annealing temperature of 20G and 70G steels are shown in Fig. 2.

The coercive force of these steels decreasing, starting at *Tann* = 100 °C. The drop in *Hc* in steel 20G occurs over the entire annealing temperature range of the samples.

The *Hc*(*Tann*) dependence for steel 70G has a local maximum at *Tann* ~ 600 °C, the presence of which is explained by the coagulation of carbides, when the size of the inclusions becomes commensurate with the thickness of the interdomain boundary. A weak and ambiguous change in the coercive force at high annealing temperatures does not allow the use of its value as a quality testing parameter for the annealing of the investigated steels.

The residual magnetic induction *Br* is monotonically increasing to ~ 700 °C. The change in the magnetic parameters at temperatures below (450-500) °C is associated with the removal of distortions of the crystal lattice and the yield of carbon from the lattice of the supersaturated α-solid solution, and at higher temperatures - recrystallization processes occur. The decrease in *Br* at *Tann* above 700 °C is due to the deterioration of the structure and the increase in the level of internal stresses due to rapid cooling on air [3].

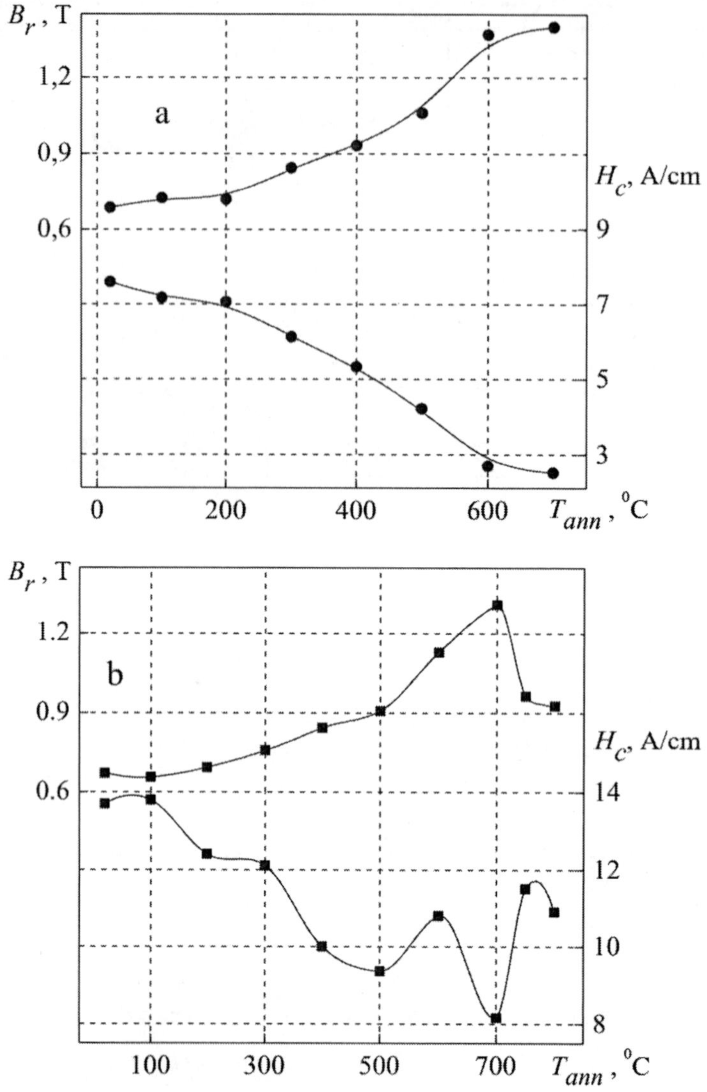

Fig. 2. Dependences of residual magnetic induction and coercive force on annealing temperature of steels 20G (a) and 70G (b)

Dependences of the residual magnetic induction *Br* on the hardness of cold-deformed and annealed at different temperatures samples from steels 20G and 70G are shown in Fig. 3.

With an increase of the annealing temperature of 20G steel in the temperature range from 300 to 700 °C, an increasing drop in hardness occurs, while the residual magnetic induction on the whole range not changing so intensively, especially at high annealing temperatures, where it is practically constant (Fig. 3a). This leads to a nonlinear dependence of the residual magnetic induction on the hardness and, accordingly, to a loss of sensitivity of *Br* to hardness as a testing parameter. Thus, for steel 20G, it is necessary to search for new hardness testing parameters. For steel 70G, the residual magnetic induction decreases linearly with increasing hardness and the value can be a hardness testing parameter.

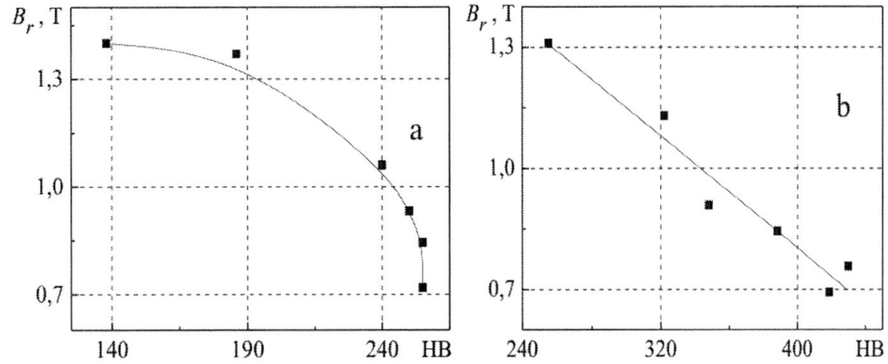

Fig. 3. Dependences of residual magnetic induction on hardness of samples of cold-deformed and annealed steels 20G (a) and 70G (b)

From the foregoing it follows that it is relevant to search for rapidly measured parameters of structurescopy suitable for use in scanning diagnostic systems. To this end, the parameters of magnetoacoustic emission were investigated on samples of these steels 20G and 70G.

On Fig. 4 shown the dependences of the amplitude of the MAE (U_{MAE}) and the residual magnetic induction on the annealing temperature of samples from cold-deformed and annealed 20G and 70G steels. In Fig. 4 is clearly seen correlation between the residual magnetic induction and the amplitude of the MAE.

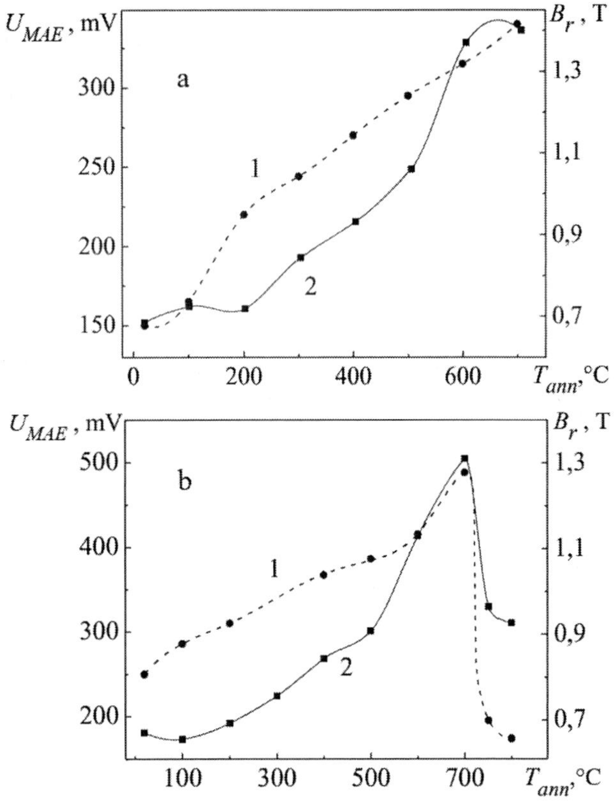

Fig. 4. Dependences of MAE amplitude (1) and residual magnetic induction (2) on temperature of annealing of samples of cold-deformed and annealed steels 20G (a) and 70G (b)

In particular, the dependence of the amplitude of magnetoacoustic emission on the annealing temperature of steel 70G accurately reproduces the maximum of the dependence of the residual magnetic induction on the annealing temperature. With the subsequent increase in *Tann*, the MAE signal is rapidly damped due to the deterioration of the steel structure.

4 Conclusion

It is established that the dependence of the amplitude of magnetoacoustic emission on the frequency of the magnetizing field has nonmonotonic character. For a large group of different in physical properties and chemical composition steels, the field frequency corresponding to the maximum MAE amplitude is 3-5 Hz.

The existence of a relationship between the residual magnetic induction of matter and the amplitude of magnetoacoustic emission of annealed steels 20G and 70G is established, which consists in correlation of the dependences of these parameters on the annealing temperature of steels. The value of U_{MAE} along with the residual magnetic induction can be a parameter in the scanning systems of structurescopy of ferromagnetic steels.

The work is carried out under the program of fundamental research of UrB of RAS 2015-2017, Project 15-17-2-5.

References

1. V.N. Kostin., V.I. Pudov, E.D. Serbin, O.N. Vasilenko, J. Deform. Fract. Mater. **2**, 41 (2017) (on Russian)
2. V.N. Kostin, D.Yu. Filatenkov, Yu.A. Chekasina, O.N. Vasilenko, E.D. Serbin, Acoust. Phys. **2**, 237 (2017)
3. V.N. Kostin, M.A. Guriev, O.N. Vasilenko, D.Yu. Filatenkov, Y.G. Smorodinsky, Phys. Mesomech. **16**, 103 (2013)
4. V.N. Kostin, O.N. Vasilenko, D.Yu. Filatenkov, Yu.A. Chekasina, E.D. Serbin, Russ. J. Nondestr. Test. **51**, 624 (2015)

High-speed deformation of copper samples with the use of magnetic pulse method

Sergey Krivosheev[1,*], *Sergey* Magazinov, and *Dmitrii* Alekseev

[1]Peter the Great Saint-Petersburg Polytechnic University, 195251 Saint-Petersburg, Russia

Abstract. The possibility of using the magnetic pulse method for studying the high-speed deformation modes of metals with strain rates up to 100,000 1/s without a preliminary phase of material compression under loading in the microsecond range of durations is shown. 3D calculation of the magnetic field and deformation modes has shown the possibility of creating loading schemes that are free of induced currents in the sample in the zone of maximum mechanical stresses. The first experiments show the need for verification of the Jones-Cook model at high deformation rates.

1 Introduction

Magnetic pulse method for creating controlled pressure pulses of microsecond duration in the study of the pulse strength of brittle non-conductive materials revealed a number of common patterns of fracture process [1, 2]. Using of thermodynamic approach for analysis of the results of brittle non-conductive materials fracture process allowed to reveal specific relations among extreme destructive loading, the duration of this loading and specific material parameter – energy storage time [3]. This approach can be used to describe transitions between different states of the loaded system, not only the catastrophic transition caused by fracture, but also elastic-plastic transition associated with a change in shape or deformation under any strain rates. The special feature of this method is the ability to form stressed states in which there is no pre-stressed modes in material, typical for spall test schemes. Implementation of this feature is especially important for the test of metals, composite and laminate materials, with an explicit spatial anisotropy of the deformation characteristics.

2 Conductive samples with defect of type of crack

Typical magnetic pulse loading of samples with macro defects is shown in Fig. 1. Selection of parameters of the generator and the magnetic system for generating controlled pressure pulses of microsecond duration can be implemented on the recommendation of [4, 5].

The cross-section of magnetic pulse driver (MPD) busbars was selected with taking into account the integral of the action, at which the material of busbars remains in the solid

* Corresponding author: ksi.mgd@gmail.com

phase [6]. An insulating layer of fiberglass with a thickness of 0.1 mm was located between MPD and grooves to exclude the current flow through the sample. The breakdown electric strength of fiberglass is 25 – 30 kV/mm and provides electrical isolation of the MPD from the sample. Such parameters of the insulating layer ensure that a pressure corresponding to the magnetic pressure generated by the MPD is transmitted to the sample.

The tensile strength of fiberglass is approximately 500 MPa and significantly exceeds the elastic limit of the research material. To exclude the impact of possible deformation of the busbars MPD on the parameters passed in the pressure pulse the edges of the busbars are fixed by fiberglass clamps. In this case, the parameters of the magnetic pressure are determined only by the geometry of the MPD and the current flowing through its busbars.

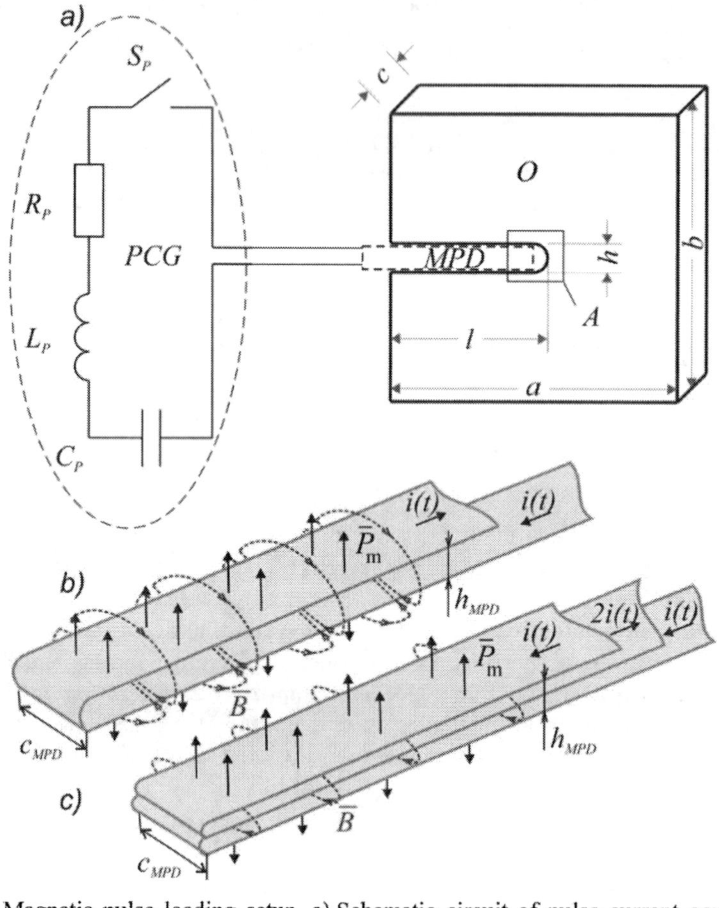

Fig. 1. Magnetic pulse loading setup. a) Schematic circuit of pulse current generator (PCG) device and scheme of loading for pulse testing. C_p, L_p - capacity and self-inductance of PCG; S_p – high-voltage commutation switch; R_P - resistor; O – sample under test; a, b, c – length, height and thickness of sample; *h* and *l* – height and length of groove. b) simple MPD; c) quasi-coaxial MPD; h_{MPD}, c_{MPD} - height and thickness of MPD; $\boldsymbol{P_M}$ – vector of magnetic pressure; $\boldsymbol{B}$ – vector of magnetic induction; *i(t)* – electric current.

The pressure amplitude can reach 2 - 10 GPa. Such pressure amplitude is sufficient for testing metallic materials. However, the transition to tests of conductive samples imposes a number of limitations related to the possible impact of induced current in the sample on the deformation curves of metal.

2.1 Induced currents structure

The simulation results of induced currents structure in metallic sample for the simple magnetic pulse driver (MPD) in the form of a system of flat bars (Fig. 1b) - simple MPD and the quasi-coaxial MPD (Fig. 1c), which are represented in [3], are shown in Fig. 2.

The calculation was performed in a 3D setting with the consideration of the induced currents impact on the sample and the MPD. The used material for the sample and the MPD was copper. Pulsed electrical current in MPD, with duration of 5 μs and amplitude of the current 40 kA, was set as the external impact.

Such impulse duration did not cause significant heating of the MPD busbars. The cross section of the MPD busbars was 2×0.2 mm.

Geometric dimensions, in accordance with the designations of Fig. 1 were: $a = b = 3$ cm, $c = c_{MPD} = 2$ mm, $l = 1.5$ cm, $h = 2$ mm, $h_{MPD} = 1.6$ mm.

The numerical simulation results showed that the use of a simple MPD induced current density in the sample can reach values up to 40 kA/mm2 or more in the zone of maximum mechanical stress (top of the groove – point M, position 1 in Fig. 2), and the sample heating due to a current flow is approximately 5°C. Calculated magnetic field between simple MPD busbars is 14 T.

According to the data, submitted in [7, 8] it is known that such current density causes the electroplastic effect (EPE), which can influence on the process of deformation. This effect is experimentally observed for low deformation rates and is revealed in the reduction of the yield stress. This reduction of the yield stress may reach 30-40 % for different metals [9]. A negligible amount of experimental data on the impact of EPE on the process of high-speed deformation, e.g. [10], does not exclude the possibility of its manifestation in pulsed load mode.

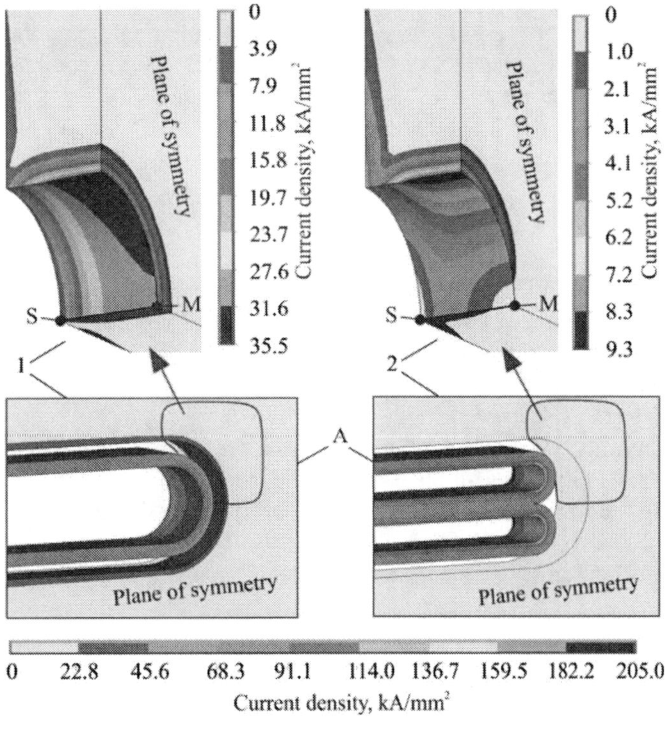

Fig. 2. Current density distribution near the groove top for different magnetic pulse drivers at the moment of maximum induced current. SM - line of symmetry

On the contrary, using of the quasi-coaxial MPD (position 2 in Fig. 2) leads to a significant current density reduction and changes the spatial distribution of current density. The quasi-coaxial driver provides the formation of such stressed state in which the local minimum of current density close to zero is in the apex of the groove in point M, thus we can say that in the apex of the groove the zone free from the EPE impact is formed. Calculated magnetic field between quasi-coaxial MPD busbars is 20 T.

The derivation and comparison of the relation between sample deformation and applied magnetic pressure for samples with the use of simple and quasi-coaxial MPD in the same mechanical stress conditions, from the point of view of formation of these fields, will allow to determine the EPE impact on the deformation under pulse loading.

2.2 Simulation of Elastic-Plastic deformation

The simulation of elastic–plastic deformation of a copper busbar with macro defect of type of crack in the three-dimension setting was carried out using the ANSYS Autodyn environment [11]. The Johnson–Cook (JC) plasticity model for OFHC copper [12] was selected as a calculation model of deformation. The reliability of mentioned model for various metals is supported by its implementation in different programming environments, e.g. [11]. According to JC model the elastic limit of material varies depending on plastic strain, strain rate v_p, which must be equal or more then 1/sec, and temperature is described by the next expression:

$$\sigma=[A+B\varepsilon^{n}{}_{p}][1+C\ln v^{*}{}_{p}][1-T^{*m}] \tag{1}$$

where ε_p is an effective plastic strain, $v^*_p = v_p/v_p0$ is a standardized effective plastic strain rate ($v_p0 = 1$ 1/s – effective plastic strain rate used for determination parameters of model – A, B, and n), $T^{*m} = (T - T_{room})/(T_{melt} - T_{room})$ is homologous temperature, $A = 90$ MPa, $B = 292$ MPa, $n = 0.31$, $C = 0.025$, $m = 1.09$, $T_{melt} = 1082.9$ °C are model parameters.

The JC fracture model is applied as the failure criterion:

$$D=\Sigma\Delta\varepsilon/\varepsilon^{f} \tag{2}$$

$$\varepsilon^{f}=[D_1+D_2e^{D_3\sigma^*}][1+D_4\ln v^{*}{}_{p}][1+D_5T^{*m}] \tag{3}$$

where ε_f is an effective fracture strain, $\Delta\varepsilon$ is the increment of equivalent plastic strain which occurs during an integration cycle, $\sigma^*=\sigma_m/\sigma_{av}$ is the dimensionless pressure-stress ratio (σ_{av} is the average of the three normal stresses and σ_m is the von Mises equivalent stress). T^{*m} and v^*_p are identical to those used in the equation (1). $D_1 = 0.54$, $D_2 = 4.89$, $D_3 = -3.03$, $D_4 = 0.014$, $D_5 = 1.12$ are parameters of fracture model. When the parameter D reaches the value 1, the material is destroyed.

The JC plasticity model describes deformation process up to fracture, and the validity of the use of the JC plasticity model is justified by the good correspondence between calculated and experimental deformation curves for different plastically deformable materials at strain rates up to ~ 2000 1/s. For instance, the applicability of JC model for description of the high-speed deformation process for steel 09G2S at strain rates up to 1500 1/s is shown in [13].

The deviation the deformation curve from a linear one due to plastic flow leads to a significant reduction in the achievable stress for the case of the wave loading (curves 1, 2, 3 in Fig. 3a). This leads to the failure to ensure a high strain rate, as it shown in Fig. 3.

Transition to a quasi- stationary mode allows to extend significantly the range of variation of achievable stress, and to provide a large deformation of the material at the top of the groove, until the fracture of the sample (Fig. 3, curve 4.).

The character of the stress state is defined by the characteristic size of the sample b (Fig. 1) and the loading wavelength $\lambda = c_1 \cdot T$, where c_1 - speed of sound in the material. If λ is much smaller than b wave mode is implemented, otherwise, the quasi-static is. Calculations were provided for two samples with different sizes: 1 – $a = 3$ cm, $b = 3$ cm, $c = 1$ mm, $h = 2$ mm, $l = 1.5$ cm; 2 – $a = 12$ cm, $b = 12$ cm, $c = 1$ mm, $h = 2$ mm, $l = 6$ cm (designations according to Fig. 1).

The quasi-static mode of loading was implemented in the first sample by impacting distributed pulse of pressure with duration $T = 50$ μs to groove shores. Wave mode was implemented in the second sample by the impact of the distributed pulse of pressure with duration $T = 8$ μs. The results of simulation show that in both cases the maximum stresses are formed in a zone located near the top of the groove (along line MS Fig. 2).

The analysis of numerical simulation results show that when the sample is impacted with an evenly distributed along the shores of the groove pressure created by magnetic pulse driver, it is possible to provide the deformation of the sample at a strain rate of up to 10^5 1/s.

The simulation was carried out at the supercomputer centre "Polytechnic" with heterogeneous cluster HPC Tornado. The parameters of one node of the cluster: RSC Tornado – 2 CPU with 14 cores (2xXeon E5-2697v3 2.6 GHz 64 GB RAM).

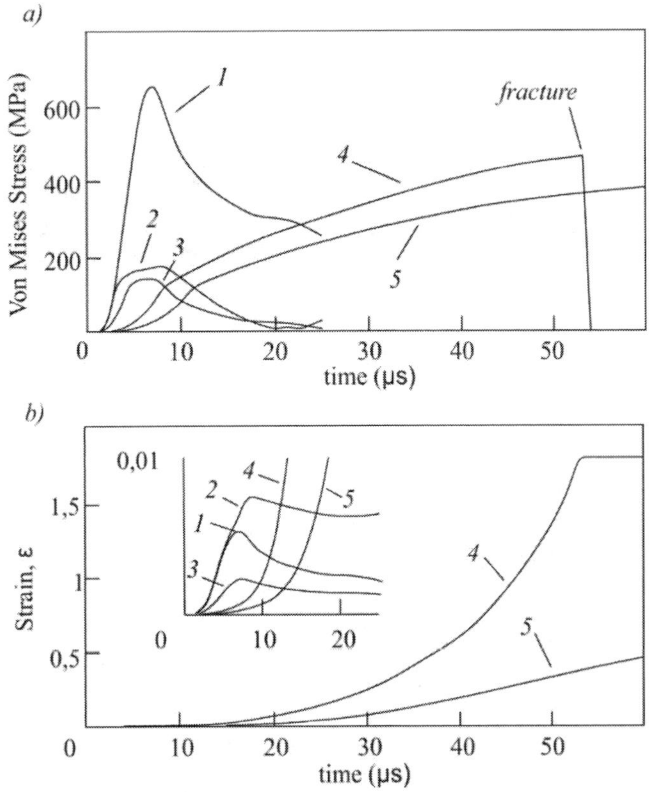

Fig. 3. a) Relation between stress on the groove top and time; b) Relation between relative deformation on the groove top and time.
1 – $P_m = 250$ MPa, $T = 8$ μs (excluding plastic deformation); 2 – $P_m = 250$ MPa, $T = 8$ μs, $v_p = 1600$ 1/s; 3 – $P_m = 85$ MPa, $T = 8$ μs, $v_p = 500$ 1/s; 4 – $P_m = 250$ MPa, $T = 50$ μs, $v_p = 90000$ 1/s; 5 – $P_m = 85$ MPa, $T = 50$ μs, $v_p = 14000$ 1/s.

2.3 Experimental setup and procedure

The experimental setup - the pulse current generator GIT-50/12 (storage capacity $C_p = 12$ μF, charging voltage $U_p = 50$ kV, self-inductance $L_p = 53$ nH) was used to provide initial test of samples according to the scheme shown in Fig. 1. The material of the sample was copper Cu ETP, sample sizes was – $a = 30$ mm, $b = 30$ mm, $c = 2$ mm, $h = 2$ mm, $l = 15$ mm. The hole with diameter $d_h = 10$ mm (see Fig. 4a) was made in the sample to increase the level of plastic deformation due to the localization of the mechanical stress in the zone between the groove and hole, which leads to a significant increase in stress compared to the sample without a hole.

After charging the generator up to 30 kV, GIT – 50/12 forms pulse current with amplitude ~ 360 kA in a simple magnetic pulse driver. Discharge current was measured by Rogowski coil and has a form of strongly damped sinusoid with the attitude of neighbouring amplitudes ~ 0.6 and period of 5 μs The corresponding to this current pressure pulse had an amplitude ~ 0.7 GPa in test with magnetic pulse driver with width $c_{MPD} = 11$ mm. The current lead busbars were fixed with dielectric gasket made of fiberglass to exclude any movement.

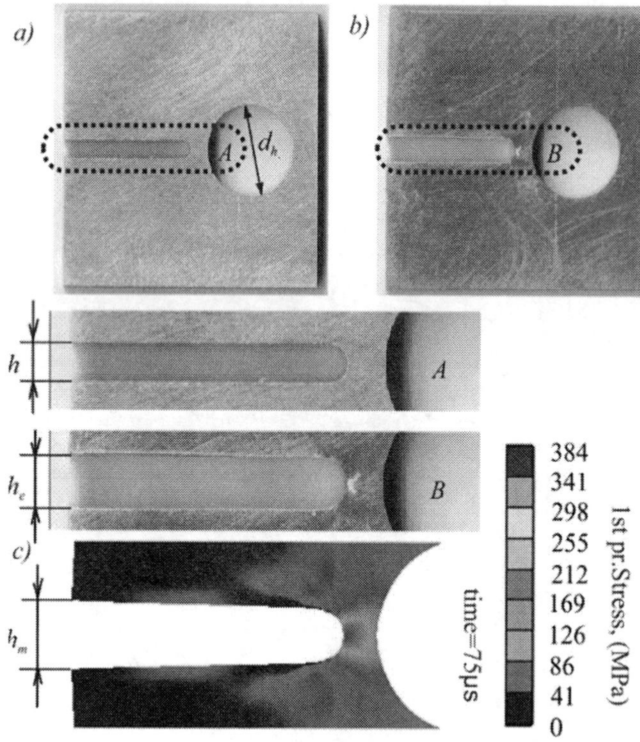

Fig. 4. Comparison of experimental data and numerical simulation. a) The sample before testing; b) The sample after testing; c) Numerical Simulation results – 1st principal stress (time = 75 μs). h, he, hm – height of groove before testing, after testing and from simulation; dh – hole diameter.

Comparison of experimentally registered permanent deformation with the results of numerical simulation is given in Fig. 4 and shows the inflated value of the groove deformation $h_m = 3.6$ mm, compared with the experimental value $h_e = 2.8$ mm. The mismatch of the calculated strain may be due to a variety of factors, including overestimated applied pressure, due to inaccurate recalculation of the current pulse into the magnetic pressure. According to the results of numerical simulation the highest mechanical stresses during deformation are concentrated in the top of the groove and plastic deformation strain rate reaches $v_p \sim 15000$ 1/s, however the JC model described in paragraph II.B may be insufficiently adequate for a given strain rate

3 Conclusion

Analysis of simulation results showed the possibility of using magnetic-pulse method of deformation of conductive materials at strain rates up to ~ 100 000 1/s. The proposed loading scheme allows to combine the process of deformation with a pulse current exposure and investigating the impact of the electroplastic effect. In addition, transition to higher strain rates will provide data for Johnson – Cook model verification in a wider range of influences.

References

1. S.I. Krivosheev, N.F. Morozov, Y.V. Petrov, and G.A. Shneerson, J. Mater. Sci. **32**, 3, 286-295 (1996)
2. S.I. Krivosheev and G.A. Shneerson, *MEGAGAUSS* (2008).
3. S.I. Krivosheev, N.V. Korovkin, V.K. Slastenko, and S.G. Magazinov, Int. J. Appl. Mech. **9**, 293 (2015)
4. S.I. Krivosheev, Tech. Physics **50**, 3, 334 (2005)
5. G.A. Shneerson, M.I. Dolotenko, S.I. Krivosheev, *Strong and superstrong pulsed magnetic fields generation* (Walter De Gruyter Incorp., **439,** 2014)
6. Heinz Knoepfel, *Pulsed high magnetic fields: physical effects and generation methods concerning pulsed fields up to the megaoersted level. North-Holland publishing company* (Amsterdam – London , 1970)
7. O.A. Troitskii, and L.G. Maistrenko, Soviet Mater. Sci. **8**, 6, 686 (1974)
8. H. Conrad, Mater. Sci. and Engineer. **A322**, 100-107 (2002)
9. A.F.Sprecher, S.I. Mannan, H. Conrad, Acta metal. **34**, 7, 1145 (1986)
10. B. Kinsey, G. Cullen, A. Jordan, S. Mates, CIRP Annals – Manufacturing Technology **62** (2013)
11. ANSYS Academic Research Customer #00420725 (from 2010), Comp Mech Lab
12. G.R. Johnson and W.H. Cook, J. Eng. Frac. Mech. **21**, 1, 31 (1985).
13. Vl. Vas. Balandin, Vl. Vl. Balandin, A. M. Bragov, L. A. Igumnov, A. Yu. Konstantinov, and A. K. Lomunov, Mech. solids **49**, 6, 666 (2014)

MATEC Web of Conferences **145**, 05007 (2018) https://doi.org/10.1051/matecconf/201814505007
NCTAM 2017

Non-destructive monitoring of tensile of mild steel samples by magnetic Barkhausen and ultrasonic methods

Yonka Ivanova [1,2,*]

[1]Sofia University "St.Kl.Ohridski", Faculty of Physics, Sofia, Bulgaria
[2]Institute of Mechanics-BAS, Solid Mechanics, Sofia, Bulgaria

Abstract. The paper presents an experimental study of deformations of mild steel samples by magnetic Barkhausen, magnetoacoustic and ultrasonic methods. The purpose of the study is to investigate and analyze the dependence between mechanical, magnetic and acoustic properties and to find appropriate parameters for stress state evaluation in the elastic and plastic regions.

1 Introduction

The working conditions of the materials lead to degradation of microstructure and properties of engineering materials. Among mechanical properties, static tensile properties are fundamental and have been used as a standard to represent mechanical properties. When materials deform, elastic deformation occurs in an early stage. When the deformation exceeds one critical value, named yield point, the deformation becomes plastic. The increasing of the mechanical stress beyond the yield stress has an undesirable effect because of subsequent plastic deformation and unpredictable failure. In such a manner, tensile properties have been crucial for discussion of plastic deformation process. The destructive methods such as tensile, compression or bending are traditionally used for obtaining the mechanical properties. Non-destructive techniques offer an alternative to traditional methods.

Non-destructive testing (NDT) methods enable easy, fast and "in situ" receiving the information for materials on the base on the correlations between NDT parameters and mechanical characteristics. Thus the researchers can study the alteration of microstructure and physical properties during exploitation (stress, deformation, heat treatments, neutron irradiation etc.). Many research papers concerning the stress effect on non-destructive testing parameters reveal results due to different materials and microstructure as well as NDT parameters [1-10, 11, 12, 15, 16, 33, 34]. The most used methods for stress (residual and applied stress) and structure evaluation are magnetic [1, 2, 15, 16, 22, 24, 35], magnetic Barkhausen Noise (MBN) [18-25], magneto acoustic emission (acoustic Barkhausen noise) (MAE) [9, 11, 12, 13, 14] and ultrasonic methods [4, 5, 26-29, 30, 31]. The first two

* Corresponding author: yonka@imbm.bas.bg

methods and effects are due to abrupt movement of magnetic domain walls during the magnetization of samples. They are sensible to microstructure (carbon content, grain size etc., chemical composition) and stress state of materials [10, 12, 14, 17, 21]. The working principle of MBN is based on Barkhausen noise when a ferromagnetic material is subjected to a varying magnetic field [14, 21]. The MBN parameters depend on penetration depth of magnetic field, frequency and physical properties such as specific electrical conductivity and magnetic permeability. The detected MBN signal is obtained from depths less than 0.2 mm below the surface, i.e. MBN is a surface technique [9, 11-13]. MAE is associated with the generation of acoustic signals as a result of continuous changes in magnetization and change of magnetostriction in separate areas of the material. The MAE parameters are obtained from the whole volume of material [8, 9, 11-13]. The third method is ultrasonic. It uses the variation of ultrasonic wave velocity as a function of stress level on the base of the acoustic-elastic theory. Internal stress measurement can be classified into groups according to the different type ultrasonic waves (longitudinal, transverse, longitudinally refracted and surface waves) [3-5, 26-30]. The choice of NDT information parameters is difficult because of complexity of many factors such as microstructure features and stresses. The complicated problem has a solution by applying several NDT methods based on different physical approaches [1-7].

The aim of the study is to perform monitoring of the deformation during uniaxial tension by three NDT methods. Comparing the results from surface techniques such as magnetic Barkhausen noise and ultrasonic surface wave methods with volumetric one magneto-acoustic emission gives the possibility to find out the appropriate non-destructive parameter for evaluation of applied stress. The combination of NDT methods for stress state evaluation in elastic and plastic regions the stress-strain diagram allows obtaining more information about materials and it is still topical. The variation of the Rayleigh wave velocity in the elastic and plastic regions of the stress-strain diagram of steels has to be obtained experimentally and compared with the results from another surface method. Another research goal is to establish the sensitivity and reliability of used methods to plastic deformation and damaging process of materials.

2 Experimental methodologies

Experimental procedure for investigation of deformation during tensile in low carbon steel specimens includes the following stages:

- Preparation of specimens;
- Tensile testing;
- MBN and MAE measurements during tensile testing.
- Ultrasonic surface waves measurements, carried out on-line during tensile testing

3 Materials and tensile testing

The specimens are obtained from a 3mm thick hot rolled mild steel plate. The dimensions of working part of specimens are 160mm (L) x 30mm (W) x 3mm (T) so that the specimen axis was parallel to the rolling direction.

Tensile tests are performed using a universal tensile machine, equipped with a load cell of 100 kN. The mechanical properties (yield strength, ultimate tensile strength and elongation after fracture) are preliminary determined from a stress-strain diagram after tension tests. The speed of the loading is 0,5 mm/s. The yield stress is about 300 MPa and tensile strength is about 410 MPa.

4 NDT measurements

The experimental system for in-line non-destructive monitoring is illustrated in the Fig.1. It consisted of tensile testing machine and NDT apparatus for MBN, MAE and ultrasonic surface wave setups. The NDT measurements are completed before reaching the maximum loading force and forming necks in the specimens. No measurements have been made after unloading.

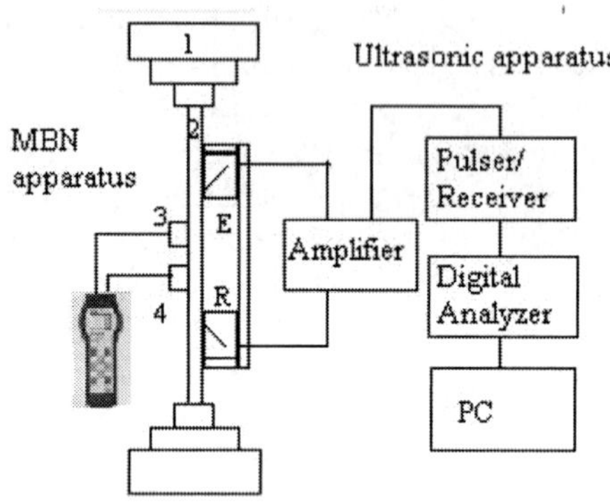

Fig.1.Experimental system for real time NDT monitoring using MBN, MAE (MULTITEST 010) and ultrasound. 1-Tensile machine, 2-tensile specimen, 3-MBN probe, 4-MAE probe

3.1 MBN and MAE measurement

The MBN apparatus consists of device "MULTITEST MC010', contact magnetic-noise probe and acoustic emission probe type [5, 32]. The principle of operation and methods for calibration of "MULTITEST-MC10" are presented in [32]. The device consists of signal generator, amplifiers of the magnetic and acoustic noise, microprocessor system and programmable software, for controlling the frequency and the magnitude of magnetic excitation. The excitation is by sine wave signal generator. The excitation frequency is set to 80 Hz. The MBN probe contains a magnetic exciter ("U" ferrite core magnet) and MBN pick-up sensor situated between the magnetic poles. The magnetic head was perpendicular to the surface of the specimens. The MBN is acquired using adjustable filters up to 100 kHz. To detect MAE signals an AE sensor (150 kHz, VS 150MVallen system) with a preamplifier included (40 dB) was positioned and fixed on the bottom surface of the specimens by specially made holder. The acoustic emission sensor was coupled to the surface of the specimens using silicon acoustic couplant that insures good surface contact between the two surfaces of the samples and facilitates the transmitting of acoustic waves.

The informative non-destructive parameters are Root Mean Square (RMS) values of magnetic noise voltage (U_m) and RMS values of magneto acoustic voltage (U_a).They are measured during the increase of the magnetizing current from 0 to 500 mA, saved in the device memory for signal processing and graphically presented. The processing and analysis is developed in a MATLAB environment. The applied magnetic field is perpendicular to the tensile stress direction. The parameters of magnetization and signal analysis are tunable in order to select the optimal ones as well as to obtain the maximum information of surface structure and stress state.

3.2 Ultrasonic measurement

When ultrasonic Rayleigh wave propagates in the same direction as the applied uniaxial stress, the dependence of the velocity on the tensile stress (σ) can be described in the form [26-28].

$$d\,C_R = \frac{C_{Rx} - C_{Ro}}{C_{Ro}} = \beta_R.\sigma \tag{1}$$

The relative change of surface wave velocity is proportional to the change in time of flight ($d\,t_R$) of the wave propagating along a distance L.

$$dC_R = -dt_R \tag{2}$$

$$dt_R = \frac{t_R - t_{Ro}}{t_{Ro}} = K.\sigma \tag{3}$$

where t_{Ro} is the time of flight of wave passing the distance L in an sample without stress, t_R denotes the time of flight during the applied stress in the same direction as the load, K is an acoustoelastic coefficient.

The setup for real time ultrasonic measurement is an ultrasonic US box, consisted of pulser and receiver, a 160 MHz A/D converter with 12-bit resolution and a notebook. For our investigation we choose surface Rayleigh ultrasonic waves. They were excited by an angular transducer with variable angle at 4 MHz. The propagation depth of these waves are of the order of 1.5 λ_R (Rayleigh wave length) and are about 1.5 mm. The signal was emitted by transducer E, received by transducer R and recorded as shown in Fig.3. Two transducers were mounted on the specimens in a pitch-catch arrangement by guiding device that allowed the movement of the receiver and ensured the precision of time of flight measurements. The measurements were carried out under same conditions. The time between loading steps is sufficient for recording the signals. The ultrasonic signals obtained from the stressed surface under increasing loading are stored in the computer for further signal processing and compared to reference signal from the stress-free sample.

During tensile testing, the ultrasonic measurements were carried out without stopping the tensile machine to ultimate stress strength. The distance between the transmitter and receiver was kept by the holder and does not change. The propagation of the wave is parallel to the stress direction. The signals corresponding to different loads were saved. The arrival times of flight are calculated by zero crossing technique. The measuring procedures were repeated in order to diminish problems of coupling or thermal changes. For analysis was used the relative difference in time of flight $d\,t_R$, defined as (3).

4 Results and discussion

4.1 MBN and MAE measurements

In order to carry out MBN analysis the test specimens were loaded with gradually increasing tensile force and after small retention the magnetic characteristics RMS values of MBN voltage (U_m) and MAE voltage (U_a) were measured and were used to characterize the microstructural and stress state of the materials.

Figure 2 presents calibration curves: RMS values of magnetic noise voltage (U_m) and magnetic acoustic emission (U_a) on the excitation current (I_n) and their dependences on loading tensile forces P. The magnetizing probe was placed in the middle of specimens, so the magnetic field was perpendicular to stress direction. The acoustic emission transducer was fixed on the other side of the specimens at the same position in all the measurements.

At the zero tensile stress, the MAE parameter is seen to be highest. With increased the loading to and after yield point, MAE parameter reduces values.

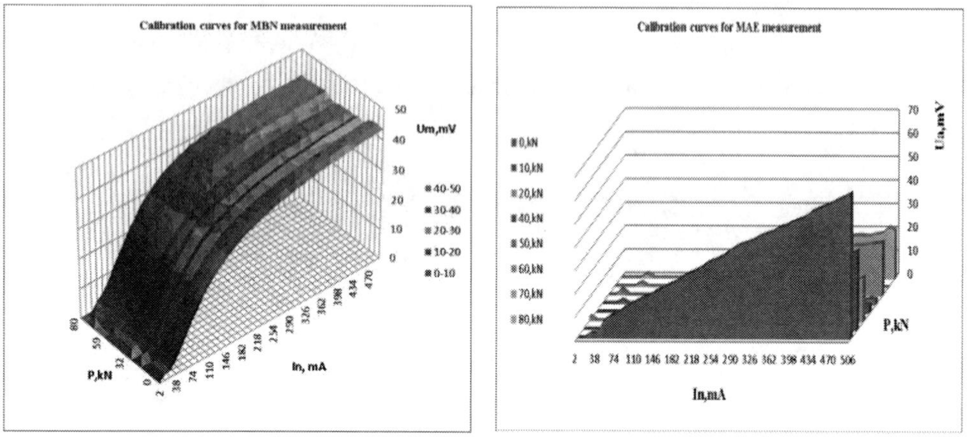

Fig.2. Calibration curves for MBN (a) and MAE measurement (b)

The stress-strain diagram and the variation of RMS value of MBN (U_m) are presented in Fig.3a. MBN signal is quite sensitive to stress during the elastic stage, and has a good linear relationship with stress in the plastic stage.

Figure 3b show the behaviors of MBN and MAE parameters with applied tension stress (σ). It can be seen from the curves: from 0 to 100 MPa, U_m rises with the stress. As the stress continues to increase from about 100 MPa to the yield stress (σ_y), the magnetic parameters start to decrease slowly. Upon the reach of yield limit the RMS value of MBN abruptly drops. The rising of the strain causes small growth of Um. At the beginning of plastic region U_m slowly increases.

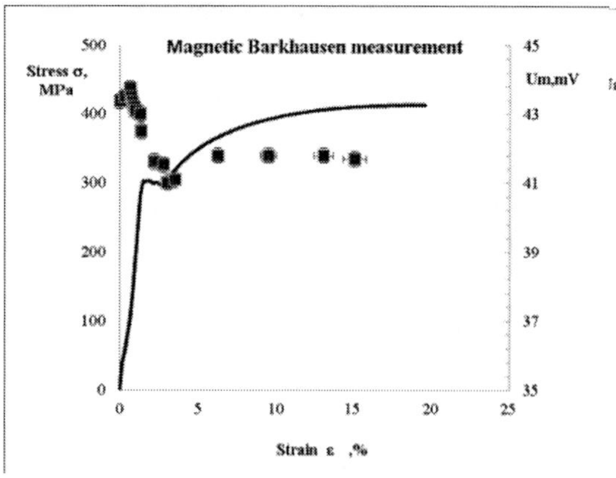

Fig.3a. Stress-strain curve and MBN

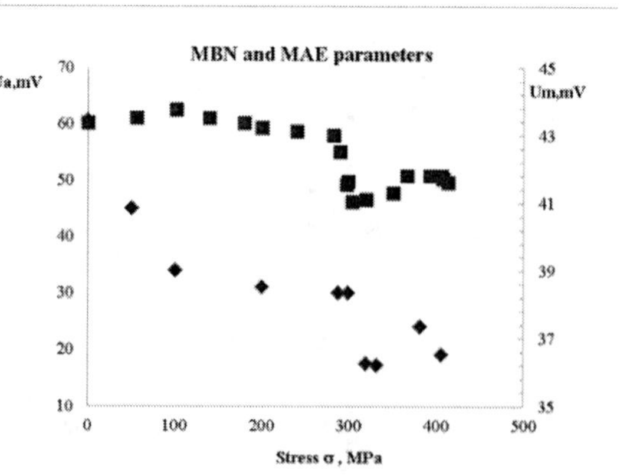

Fig.3b. Variations of RMS values of MBN and MAE parameters vs. tensile stresses

In the elastic region the behavior of MBN can be explained with changes of domain configuration as a response of variation of interatomic spacing. The domains within individual grains must reorient to minimize their energy. Stefanita and others [18,25] proposed two mechanisms for explanation of domain pattern. The first one is an orientation of 180° domains into the direction closest to the stress direction. The second mechanism involves the rising of domain wall numbers. The tensile stress creates more domain walls in the stress direction. Hence in this stage, the MBN signal rises with the increase of the applied stress [10,14]. Further increase of the stress causes the augmentation of the dislocation density. In the material, the regions with high dislocation density are separated by regions with low dislocation density, and then dislocation tangles start to form. The newly emerging dislocation tangles and cell boundaries increases the number of microscopic pinning sites. They act as obstacles to domain walls movement [14,18,25]. With alteration within the deformation of the microstructure MBN parameter decreases with the increasing of the stress. The theoretical analysis discussed above is in agreement with the experimental results.

Figure 4a,b show the variations of dimensionless parameters dU_m and dU_a versus parameter σ / σ_y. The relative change of MBN dU_m and MAE dU_a can be defined by:

$$dU_m = \frac{U_m - U_{mo}}{U_{m_o}} \tag{4}$$

$$dU_a = \frac{U_a - U_{ao}}{U_{a_o}} \tag{5}$$

Where U_m and U_a are the values of magnetic noise and MAE corresponding to current stress; U_{mo} and Ua_o are the values of magnetic parameters corresponding to unstressed sample.

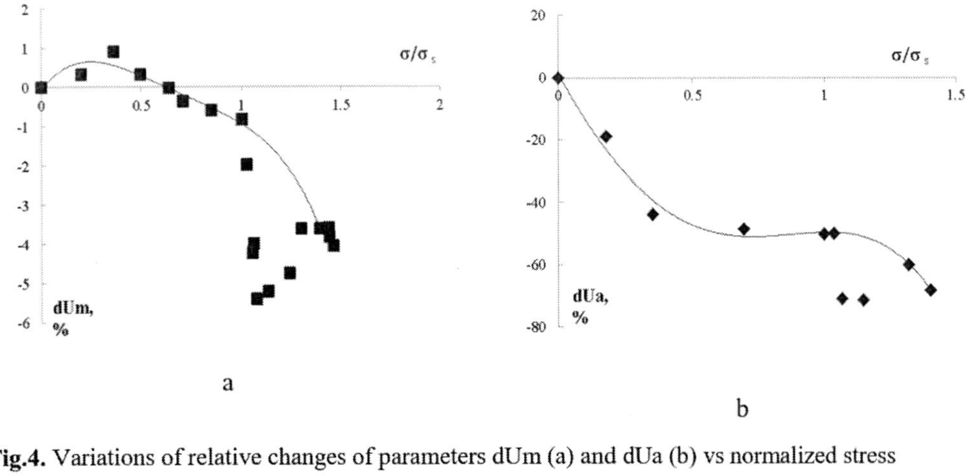

Fig.4. Variations of relative changes of parameters dUm (a) and dUa (b) vs normalized stress

With increasing load the material passes from elastic to plastic state and magnetic BN and MAE parameters decrease. The results show that the MBN and MAE parameters are sensitive to the initial stage of plastic deformation, especially for the stresses close to the yield stress.

4.2 Ultrasonic measurement

Figure 5 show the stress-strain curve and variation of ultrasonic parameter $d\,t_R$ during the tensile deformation. The strain and $d\,t_R$ are presented on x-axe. The values $d\,t_R$ are obtained from repeated experiments and closely distributed. It can be observed that the ultrasonic parameter grows with increasing of applied stress. In the elastic region the stress is linearly proportional to the strain and the ultrasonic parameter varies almost linearly with tensile stress.

The dependence between the time of flight $d\,t_R$ and stress in elastic region is expressed by linear regression:

$$d\,t_R = K_1.\sigma \tag{6}$$

where K_1 is a coefficient of proportionality between stress and ultrasonic parameter.

In the plastic region the variation of ultrasonic parameter can also be approximated by power low equation:

$$d\,t_R = K_2.(\sigma_{plastic})^p \tag{7}$$

Using least square fitting of the data, the coefficients are obtained: K_1=1,0845.10^{-5} MPa, K_2=2,10^{-18} MPa, p=6,1.

The presentation of the results in relative units allows comparing results obtained for different type of materials. Figure 6 presents the variation of dimensionless parameter $d\,t^*$ ($dt^* = d\,t_{R(\sigma)} / d\,t_{R(\sigma=\sigma_y)}$) upon the normalized stress σ^* ($\sigma^* = \sigma / \sigma_y$) .

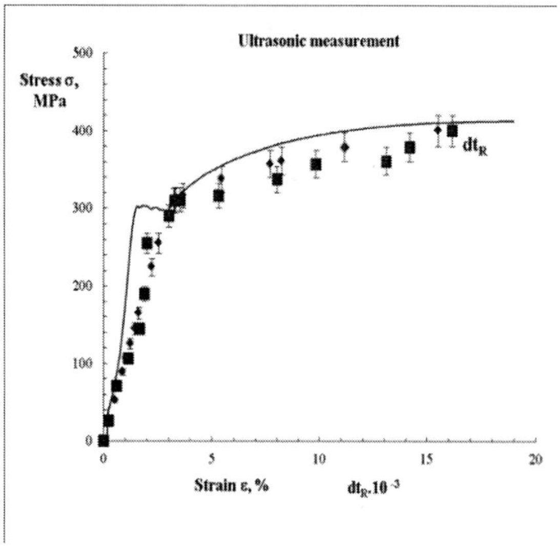

Fig.5. Stress-strain curve and variation of ultrasonic parameter $d\,t_R$ during tensile

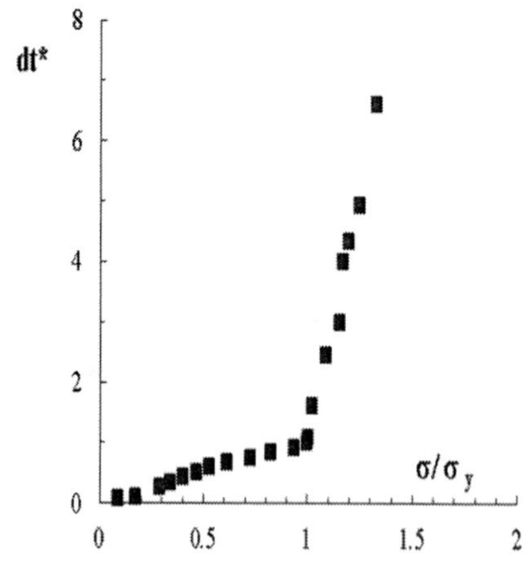

Fig.6. Variation of normalized ultrasonic parameter $d\,t^*$ vs. normalized stress

The dimensionless parameter dt^* is time of flight of ultrasonic Rayleigh waves referred to its value in the end of elastic region and the normalized stress σ^* is the current value of stress referred to the yield stress σ_y. In the plastic region the parameter dt^* sharply increases, i.e. the relative change of ultrasonic Rayleigh waves abruptly decreases.

Conclusion

The obtained results characterize the variations of NDT parameters: magnetic, magneto-acoustic and ultrasonic (velocities of ultrasonic Rayleigh waves) parameters in low carbon steels samples during tensile stress. They are in good agreement with the theory and experiments and can be explained by deformation processes in the elastic and plastic areas. Due to the relationship with applied tensile stress at different stages, the combination of the three techniques offers the potential for developing the methodology for the study of elastic-plastic behavior of ferromagnetic materials. The sharp change in non-destructive parameters at mechanical stresses close to the yield stress can be used as an indicator of plastic deformation.

References

1. E. Schneider, I. Altpeter, W.Theiner, *Nondestructive methods for material property determination* (Boston, Springer, 1984)
2. P. Holler, *Nondestructive characterization of materials* (New York: Springer, 1987)
3. C.H. Gur, I. Cam, Mater Charact. **58, 5,** 447–454 (2007).
4. P. Pereira, F. Armando, F. Grijalba, A Santos, J.Strain Analysis **51**, 8, 563–571(2016)
5. Y. Ivanova, T. Partalin, Russ J Nondestruct Test **48**, 2, 137–146 (2012)
6. I. Tomas, G. Vertesy, S. Barroso, S. Kobayashi, Nucl Eng Des. **265,** 201–209 (2013)
7. S. Takahashi, S. Kobayashi, I. Tomáš, L. Dupre, G. Vértesy, NDT and E Int. (to be published)
8. D.J. Buttle, C. B. Scruby, G.A.Briggs, J.P.Jakubovics, *Proc R Soc Lond A* 414, **1847**, 469-497 (1987)
9. M.Shibata, K.Ono, NDT&E Int. **14**, 5, 227-234 (1981)
10. J.L.Rodriguez, J.A. Perez-Benitez, J. Capo-Sanchez, Rev Mex Fis Journal S **54**, 2, 51–53 (2008).
11. V.N. Kostin, D.Y. Filatenkov, O.N. Vasilenko, Al.N. Stashkov, 11 ECNDT, Prague 2014 (2014).
12. V. N. Kostin, O.N.Vasilenko, D.Y. Filatenkov, Y.A. Chekasina, E.D. Serbin, Russ J. Nondestruct. Test. **51**, 10, 624-632 (2015)
13. V. N. Kostin, D.Y.Filatenkov, Y.A. Chekasina, O.N.Vasilenko, E.D.Serbin, Acoust. Phys. **63**, 2, 237-244 (2017)
14. E. S. Gorkunov, Yu. N. Dragoshanskii, V. A. Khamitov, Russ J Nondestruct Test. **37**, 12, 835–858 (2001)
15. E.S.Gorkunov, Yu.V. Subachev, A.M. Povolotskaya, S.M. Zadvorkin, Russ J Nondestruct Test. **49,** 10, 584-594 (2013).
16. E.S.Gorkunov, A.M. Povolotskaya, S.M. Zadvorkin, E.A. Putilova, AIP Conf. Proc, **1785** (2016)
17. D. M. Stewart, K.J. Stevens, A.B. Kaiser, Curr. Appl. Phys. **4**, 308–311 (2004)
18. A. Dhar, L.Clapham, D.L.Atherton, NDT &E Int. **34**, 507-514 (2001)
19. J. Pala, J. Bydžovský, AEEE **7** (2008)
20. T.Ren, G. Chen, W. Zhang , Z Qiu, Appl Mech Mater. **455** (2014)
21. Y. Huijuan and all, Telkomnika **12**, 3, 623-630 (2014)

22. J. Anglada-Rivera, L.R. Padovese, J. Capo-Sanchez, J Magn Magn Mater. **231**, 2, 299-306 (2001)
23. D. G. Hwang, H. C. Kim, J. Phys. D, Appl. Phys. **21**, 1807–1813 (1988)
24. H. Kikuchi and all, J Magn. **16,** 4, 427-430 (2011)
25. C. G. Stefanita, D. L. Atherton, L. Clapham, Acta Mater. **48,** 3545–3551 (2000)
26. A. N. Guz, F. G. Makhort, Int. Appl. Mech. **36** , 9 , 1119–1149 (2000)
27. V. V. Klyuev, V. N. Filinov, V. I. Matveev, *Nondestructive Testing. A Handbook. Ultrasonic Methods of Testing, Vol. 4, (Mashinostroenie, Moscow, 2007)*
28. A. A.Chernoochenko, F.G. Makhort, O.I. Gushcha, Int. Appl. Mech., **27,** 38–42 (1991)
29. E. Hu, Y. He, Y.Chen, Appl. Acoust. **70,** 2, 356–360 (2009)
30. M. Kobayashi, Int J Plasticity **26,** 107–125 (2010)
31. M. Kobayashi, Int J Solids Struct. **121** (2017)
32. G.Velev, Y. Ivanova, A. Markovski, B. Velev, *Emerging Technologies in Non-Destructive Testing V* (Taylor & Francis Group, London, 499-502, 2012)
33. L. Lyubimova, A.Tashlykov, R.Tabakaev, A.Levin, A.Popov, Eng. Fail. Anal. **81**, 45-56 (2017)
34. A.Belyaev, A. Semenov, V. Polyanskiy, D.Tretyakov, Y.Yakovlev, Procedia Structural Integrity **6**, 201-207 (2017)
35. V. Kostin, Y. Smorodinskii, Russ J. Nondestruct. Test. **53**, 493-504 (2017)

Methods for evaluation of mechanical stress condition of materials

Yordan Mirchev[1,*], *Pavel* Chukachev[2] , and *Mitko* Mihovski[1]

[1]Institute of Mechanics of Bulgarian Academy of Science, 1113 Sofia, Bulgaria
[2]Multitest Ltd., 9000 Varna, Bulgaria

Abstract. Primary attention is given to the following methods: method by drilling cylindrical holes (drill method) and integrated ultrasonic method using volume (longitudinal and transverse), surface, and sub-surface waves. Drill method allows determination of residual mechanical stress in small depth of material surfaces, assessing type, size, and orientation of principal stresses. For the first time, parallel studies are carried out of mechanical stress in materials using the electroacoustic effect of volume, surface and sub-surface waves on the one hand, and effective mechanical stresses on the other. The experimental results present electroacoustic coefficients for different types of waves in the material of gas pipeline tube of 243 mm diameter and 14 mm thickness. These are used to evaluate mechanical stresses in pipelines, according to active GOST standards.

1 Introduction

In recent years, scientific researches and their practical use has contributed to create a significant number of effective methods for assessing mechanical stresses based on dependencies between residual and effective stresses in material and certain physical characteristics (coercive force, magnetic memory, magnetic noise, propagation velocity of ultrasound waves, X-ray diffraction, etc.). There are also a number of semi-destructive methods, among which the so-called "drill method", is also successfully applied [7, 8].

In this study, complex investigations were carried out using volumetric, surface and subsurface ultrasonic waves for the evaluation of mechanical stress condition (MSC) in pipeline, and examinations by the semi-destructive "drill method". The main task in studying the MSC of pipelines is to determine axial σ_z and tangential (along the circumference) σ_h mechanical stresses.

The aim is to propose an algorithm to estimate the relative variation of MSC in surface and sub-surface layers of pipeline material, as well as the relative variation of MSC in material for the whole pipeline wall thickness.

[*] Corresponding author: mirchev@imbm.bas.bg

2 Method to investigated mechanical stress condition

2.1 "Drill method"

The hole drilling method is one of the widely used semi-destructive methods for determining the residual stresses in the surface layer of material. Damage caused to the material is a not deep small diameter cylindrical hole that is permissible or can be repaired [7, 8].

The method is performed by drilling a hole of up to *2 mm* depth using cylindrical drill (of up to *1.8 mm* diameter) in the investigated area of material. By means of properly attached strain gauges around the hole, induced deformations are determined at each hole drilling step. From deformations obtained, residual stresses are estimated using proper algorithms. Calibration constants are calculated for each strain gauge rosette used. Good knowledge of the modulus of elasticity and mechanical characteristics of the material is involved.

The method is mainly applied to isotropic materials. The recommended depth of the cylindrical hole is up to 0.5 times D, where D is the diameter of the drill used [7].

2.2 Ultrasonic method

Acoustic strain gauge measurements present the methods for determining the stressed-deformed state of the material by investigating the characteristics of the elastic waves propagating in material (amplitude, velocity, polarization, etc.). These dependencies are subject to study in non-linear acoustics and are related to the existence of the following phenomena:

- Nonlinear interaction of the elastic waves associated with the transmission of energy by interacting waves in the heterodyne frequency wave.
- Refraction of the sound wave in the presence of mechanical stresses where the vector describing the energy flow does not coincide with the wave propagation vector.
- Modulation of the sound wave in wave propagation in media of periodically varying parameters, resulting in wave phase, amplitude and frequency variations.
- Acoustic elasticity related to the dependence of elastic waves distribution velocity in solids on mechanical stresses (strains), type of wave polarization and the mutual orientation of the wave vector relative to the direction of mechanical effects.

A prerequisite for the application of acoustic-elastic effect methods is the variation in the elastic properties of the material and the propagation velocities of the ultrasound waves in a material in the absence or presence of mechanical stresses [1-6].

3 Technical implementation of the methods

3.1 "Drill method"

According to the operating instructions of the unit MTS 3000 RESTAN [8] made by SINT Technology- Italy it is recommended to work with strain gauge rosettes of type *A*, *B* or *C*. During the experiments carried out in this work, a type *A* rosette is used complying with the requirement for drill depth and recommended depth increments.

In fig. 1a, a schematic diagram is shown of fixing strain gauges in three-element rosette of type *A* used to drill a hole to evaluate residual mechanical stresses. Here, σ_{max}, σ_{min} are maximum and minimum residual stresses around the drilled hole.

Fig.1b shows block-scheme of typical hole drilling device. It presents an optical device for centering the tool holder and drilling tool. There are a number of requirements to the device, including:
- The hole shall be drilled concentric to the circle of strain gauges of diameter D with accuracy $\pm\ 0.004.D$.
- Drills or cylindrical hard-alloy mills driven by an electric motor are used with a rotational speed of up to *500 000 rpm*.
- Radial clearance angles of the tool cutting edges ensuring uniformity in depth up to *1 %* of tool diameter.
- Drilling of the hole is carried out under constant temperature conditions.
- The measuring instrument for recording the deformations shall have a resolution of $\pm\ 1.10^{-6}$ with the same repeatability of measurements.

After the tests, the residual stresses in the material are calculated, according to the requirements of ASTM E 837-08 [7].

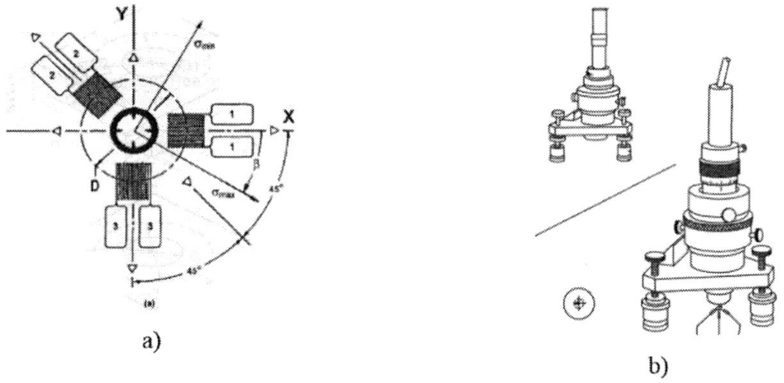

Fig. 1. a) Three grid rosette, type *A* and b) block-scheme of hole drilling device.

3.2 Ultrasonic method

MSC determination is performed by determining the relative variation of velocity or times of ultrasonic wave propagation in the material of the pipelinece. The relative variation is read against material in pipe of MSC assumed as base. The relation between the relative velocity variation and the mechanical stress variations is determined.

In the present work, three approaches to determine MSC are used - in the surface, subsurface, and the entire volume of the pipe material wall. The relative variations of the following are studied times of travel of:
- Transverse waves, polarization along and transverse to generant of the pipe and longitudinal wave propagating along the radius of the pipe.
- Surface Rayleigh wave.
- Sub-surface lateral wave.

Within the first approach, the MSC is determined over the entire wall volume of the pipe for the investigated area in which the waves propagate. The times of travel are determined by the echo method. With surface Rayleigh waves and sub-surface waves, MSC is investigated in surface and sub-surface layer respectively of the pipe wall, the times being determined by through transmission method.

GOST R 52890-2007 [5] regulates the basic requirements for the determination of a mechanical stress condition by volumetric ultrasound waves propagating along the radius of the controlled pipes. The standard is applicable to steel pipes (welded and seamless) where ratio of wall thickness to outside diameter is *20* and to pipes of diameter over *325mm*. The standard has been drawn up based on scientific researches in Mechanical Engineering Research Institute of the Russian Academy of Sciences in Nizhniy Novgorod conducted by Nikitina N. E. [1].

The investigations of the times of the ultrasound waves propagated in the material are carried out on pre-prepared reference samples of thickness from *13.817 mm* to 15.545 mm, cut out along the pipe away from the welded joint [9]. The samples are prepared by grinding machine Okamoto from pipe *X52N PSL2/L360* and marked by numbers *FE01*, *FE02*, *FE03* and *FE04*. Appearance of samples to determine times of wave travel in a material under tensile load is shown in fig. 2.

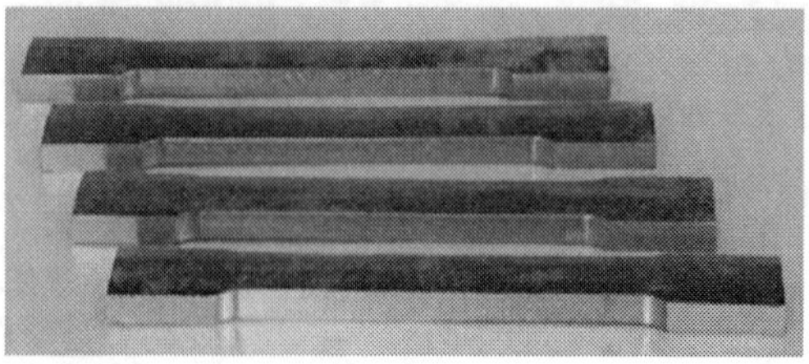

Fig. 2. Reference samples for evaluation of acoustic elastic factors.

Mechanical tests of material are carried out in advance in order to assess the yield point, creep and rupture limits. On this basis, the load and its velocity are regulated (to avoid the influence of the temperature effect when changing the velocity of the wave propagation). Stressing the tensile samples is carried out by EU100 machine. Loading mode is set so that it does not penetrate the metal creep area.

Determination of ultrasonic wave travel time in material is carried out by flaw detector Omniscan MX and probes for transverse waves of type V156-RM, for longitudinal of type V110-RM, for surface – type A574S, for sub-surface - type C567-SM.

Calipers, gauge blocks and clamping devices of high accuracy are used. Coupling the probes to the test samples is one of the most important tasks in order to achieve optimum acoustic contact and reduce the coupling effect on time determination. For each of the probes in use special clamping units are made equipped with magnets of clamping force *4kg*. Probes and clamping units are illustrated in fig.3 (attached to a pipe). In fig.4, probes are placed on a reference sample.

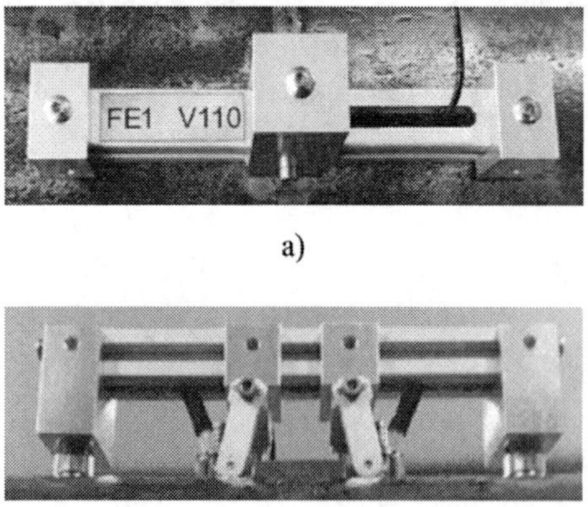

a)

b)

Fig. 3. Clamping units for (a) normal probe and (b) subsurface.

Fig.4. Setup of sensors attached to test sample and sample to the clamps of the test machine.

In the process of acoustic measurements shown in fig. 4, records of the load-time dependencies are made. With short interruptions of the mechanical tensile loader, using the ultrasonic unit, 3 to 5 independent measurements are made to determine the times of the waves travelling in the material. The number of measurements shall be repeated until the required uncertainty is reached.

On the basis of the acoustic paths data and the travel times for the respective load, the ultrasonic wave propagation velocities are calculated. The dependencies of the relative variation of velocities or travel times of ultrasound waves from load for surface and subsurface waves are plotted. When working with volumetric longitudinal and transverse waves, dependence of the calculated mechanical stresses is plotted using the theoretically-experimentally obtained acoustic-elastic factors.

4 Results of investigations of the reference samples to the ultrasonic method

The relative variations in the times for ultrasonic wave propagation in the material were determined. The time for the corresponding velocity in unstressed condition of the sample is accepted as a base time. The relative variation in % of times is calculated by the following equation

$$\Delta\tau,\% = \frac{\tau(P \neq 0MPa) - \tau(P = 0MPa)}{\tau(P = 0MPa)} * 100 \tag{1}$$

Relative variations of times calculated from tensile stresses applied of $P = 50MPa \div 300MPa$, by *50MPa* steps are presented in fig.5. The calculations are performed for values of the times determined by five independent measurements at a confidence level of *95%*. The relative variations of the transverse wave are marked by transverse 0 and transverse 90 with polarization respectively in the direction of the tensile stresses (along the pipe generant of which the reference samples are made) and perpendicular to the direction of the tension stresses.

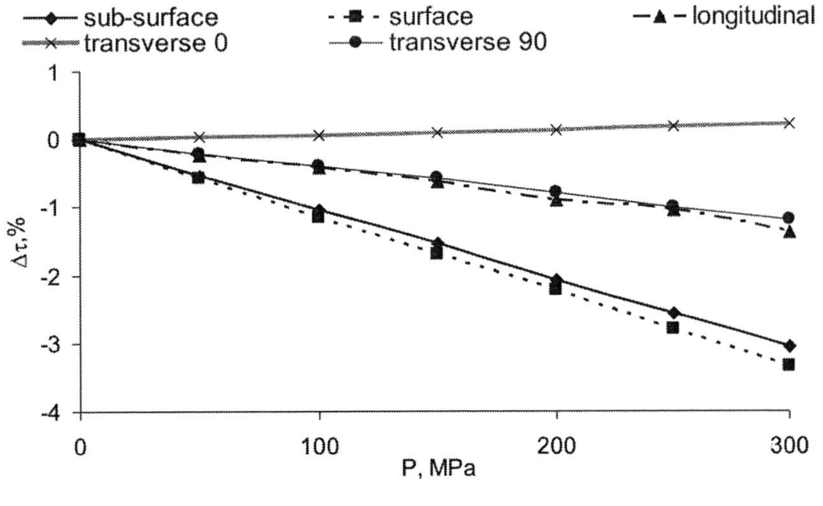

Fig. 5. Relative variation of travel times of ultrasonic waves in material of samples with mechanical tensile loading.

Dependencies in fig. 5 can be used as comparison curves for determination of the relative variation of mechanical stresses in the investigated area of the object.

In determining the mean variation of mechanical stresses across the wall thickness of the object under investigation, an algorithm is used specified in the standards GOST R 52890-2007 [5] and GOST R 52731-2007 [6]. The use of this algorithm disregards the wall thickness of the tested object in experimental determination of the travel times of longitudinal and two mutually polarized at *90* ° transverse waves. Their values are used to calculate acoustic-elastic factors. By the algorithm and the dependencies given in [1, 10] acoustic-elastic factors κ_1, κ_2, κ_3 are calculated. Factors in uniaxial mechanical stressed condition are calculated by the following equations

$$k_1 = \frac{\frac{L}{\tau_1}.\frac{\tau_{01}}{L_0} - 1}{\sigma_z}, \qquad k_2 = \frac{\frac{L}{\tau_2}.\frac{\tau_{02}}{L_0} - 1}{\sigma_z}, \qquad k_3 = \frac{\frac{L}{\tau_3}.\frac{\tau_{03}}{L_0} - 1}{\sigma_z}, \quad (2)$$

where L_0 and L are respectively, the paths of the propagating waves before and after the action of mechanical stresses, τ_{0i} and τ_i - the wave propagation times, respectively, before and after application of the tensile stress σ_z in longitudinal direction of reference samples. Here k_1 is related to propagation of transverse waves at *0*° polarization, k_2 – transverse wave at 90^0 polarization and k_3 – longitudinal wave. The polarization given for transverse wave at *90*° is in direction of the tensile stresses.

Using the experimental data obtained for the propagation times of the longitudinal and transverse polarized waves to equations (2), acoustic-elastic factors κ_1, κ_2, κ_3 are calculated. In Table 1, their values are listed as obtained for each load value P.

Table 1. Calculated acoustic-elastic factors k_1, k_2 and k_3

Load P, MPa	**k$_1$, 10^{-6}MPa^{-1}**	**k$_2$, 10^{-6}MPa^{-1}**	**k$_3$, 10^{-6}MPa^{-1}**
50	6.659	-40.336	-47.557
100	6.661	-40.225	-41.001
150	6.663	-39.435	-41.631
200	6.109	-39.541	-45.093
250	6.667	-40.013	-42.024
300	6.670	-39.933	-45.594

According to the values of acoustic-elastic factors k_1, k_2 and k_3 for each load value *P* the coefficients K_1 and K_2 are calculated to the equations given in [1, 10]

$$K_1 = \frac{k_1 - k_3}{(k_1 - k_3)^2 - (k_2 - k_3)^2}, \quad (3)$$

$$K_2 = \frac{k_2 - k_3}{(k_1 - k_3)^2 - (k_2 - k_3)^2}. \quad (4)$$

In Table 2, calculated values for K_1 and K_2 are listed for each load step.

Table 2. Calculated values for K_1 and K_2

Load P, MPa	K_1, 10^5MPa	K_2, 10^5MPa
50	0.1878	0.0250
100	0.2099	0.0033
150	0.2075	0.0094
200	0.1977	0.0214
250	0.2057	0.0085
300	0.1936	0.0210

The following mean values are obtained $K_1 = 0.2004.10^5$ *MPa* and $K_2 = 0.0148.10^5$ *MPa*.

5 Evaluation of residual mechanical stresses in pipe of pipeline

Distribution of mechanical stresses is studied along the length and circumference in submitted pipe of material *X52N PSL2/L360* from pipeline of diameter $D = 273$ *mm* and wall thickness *14 mm*. The pipe is not subjected to mechanical stresses. However, it is presumed that along the length and circumference of the pipe there are irregularly spread mechanical stresses.

Particularly significant changes are expected in the part of the pipe located in the near seam zone and in the welded joint itself.

5.1 "Drill method"

The measurements are carried out with the MTS 3000 RESTAN system, manufactured by company SINT Technology – Italy. The method is developed based on ASTM E 837-08 [7]. Software product is also used of this company.

Measurements are taken in 4 points. Experiments *E1* and *E2* are carried out with drilling holes in the inner surface of the pipeline, and *E3* and *E4* in the outside surface. Summarized obtained results of investigations in the inner and outside surface are listed in table 3 and table 4. Safety factor $n = \sigma_s / \sigma_e$ is higher for measurements in inner surface and lower – for outer surface. In all cases $n > 2$.

Table 3. Results of investigations of mechanical stresses in the inner surface of the pipe.

Experiment	$\sigma_{min} = \sigma_3$, N/mm^2	$\sigma_{max} = \sigma_1$, N/mm^2	σ_2, N/mm^2	β, deg	$\sigma_e = \sigma_1 - \sigma_3$, N/mm^2	n
E1	- 84.4	20.89	0	31.15	105.29	3.42
E2	69.29	121.50	0	79.75	121.5	2.96

Table 4. Results of investigations of mechanical stresses in the outer surface of the pipe.

Experiment	$\sigma_{min} = \sigma_2$, N/mm^2	$\sigma_{max} = \sigma_1$, N/mm^2	σ_3, N/mm^2	β, deg	$\sigma_e = \sigma_1 - \sigma_3$, N/mm^2	n
E3	46.8	166.4	0	- 24.49	166.44	2.16
E4	72.85	159.3	0	-12.46	159.3	2.26

The data analysis shows that σ_{min}, σ_{max} and σ_e are larger on the outer surface of the material. The residual stresses in material of pipeline are relatively high due to the method used to drill the pipeline and the presence of heterogeneities, non-metallic inclusions and different structure in different areas of pipeline.

5.2 Ultrasonic method

Measurements of initial data parameters (velocities of propagation of longitudinal, transverse, surface and subsurface waves) are carried out along the length of three generants of the pipe, located at *0°*, *90°* and *180°* to circumference. The zero reading along the generant coincides with central part of the welded joint. The weld reinforcement was removed by mechanical cleaning as they do in advance for many responsible welded joints, for example in Nuclear Power Plants.

Along the generant, measurements are presented from *0 mm* (in the center of the welded joint) to *350 mm* by *50 mm* steps. In some cases, additional measurements have also been carried out and immediately in the welded joint with a step of *5 mm*.

Fig.6a presents the obtained dependencies for changes in the propagation time of the surface Rayleigh waves τ_R to length L of the pipe. The obtained values are designated by *0°*, *90°* and *180°* with directions *0°*, *90°* and *180°* to circumference of the pipe. The spacing between fronts of the probes *C567SM* used is *10 mm*.

Fig.6b presents dependence τ_R on location of the middle of the distance between fronts of the probes to generant placed at *90°* to circumference of the pipe. Measurements were made at steps of *5 mm* from the center of the welded joint to L = *50 mm* to generant of the pipe.

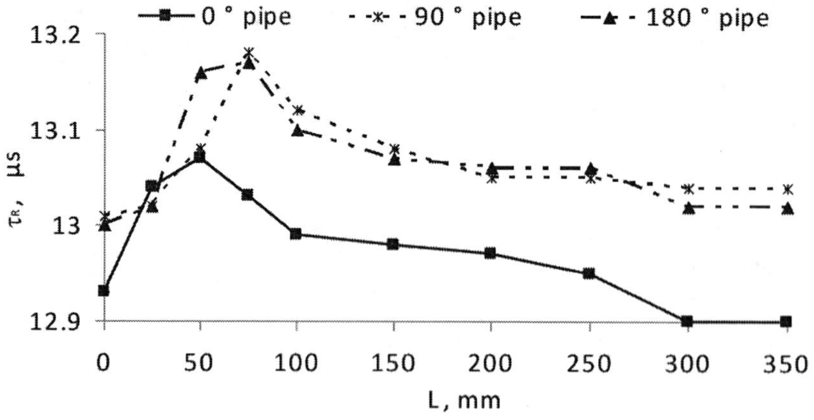

a)

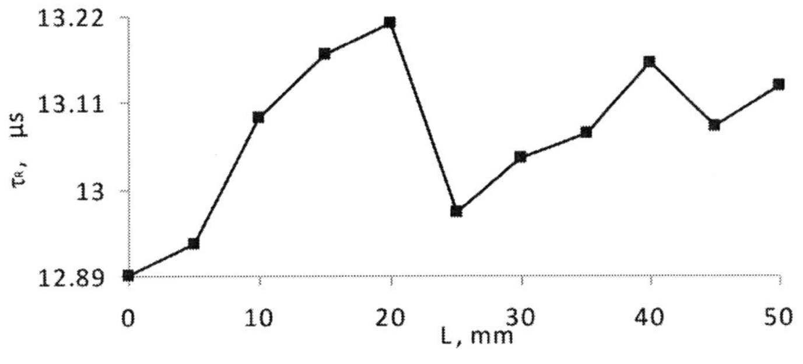

b)

Fig. 6. Dependencies of the travel time of surface wave $\tau_R = F$ (L)

Significant differences in travel times are observed for the area of the welded joint and the surrounding area.

Fig.7a presents the dependence of the travel time τ_{Π} of the subsurface wave measured along the generant *L*. Spacing between the probes for subsurface waves is *30 mm*.

Fig.7b, Fig.7c and Fig.7d present the dependencies for the travel times, respectively of transverse τ_t *(0°)*, (direction of polarization to generant), transverse τ_t *(90°)*, (direction of polarization transverse to generant) and longitudinal τ_l waves, for various points on the generant *L*, located at *0°*, *90°* and *180°* on the circumference of the pipe. It is worth noting that there is a difference in the thickness of the pipe at different points in which the times of distributed waves are determined.

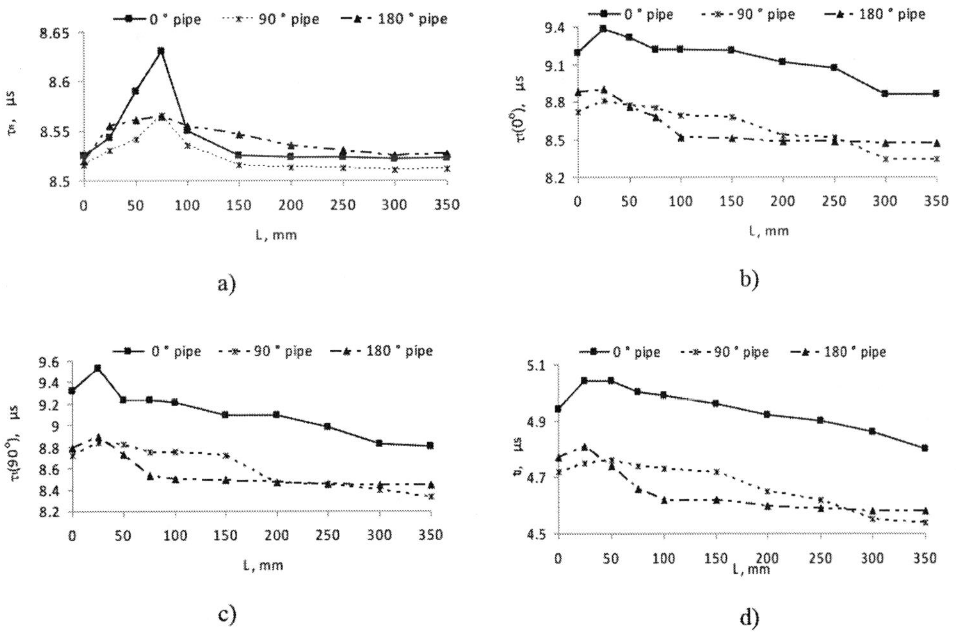

Fig.7. Dependencies $\tau_{\Pi} = F(L)$ a), $\tau_t(0°) = F(L)$ b), $\tau_t(90°) = F(L)$ c) and $\tau_l = F(L)$ d)

Using the dependencies for the times of surface and subsurface waves in material of reference samples and tested pipe distribution of mechanical stressed condition is determined in the points investigated of the surface and subsurface layer of the pipe.

Based on values obtained of times of travel of longitudinal τ_3 and transverse τ_1 (in direction of polarization of wave to generant) and τ_2 (in direction of polarization of wave transverse to generant) waves in reference samples distribution of residual equivalent stresses is calculated (to generant σ_z and transverse to generant along the circumference of the pipe σ_h) for a pipeline. In this case mechanical stresses are equivalent and related to the whole thickness of pipeline wall. Applying the mean values of theoretically-experimentally obtained acoustic-elastic factors K_1 and K_2 the relative variations of mechanical stresses are expressed by the following equations

$$\sigma_z = K_1.\Delta_1 - K_2.\Delta_2, \tag{5}$$

$$\sigma_h = K_1.\Delta_2 - K_2.\Delta_1, \tag{6}$$

where

$$\Delta_1 = \left(\frac{\tau_3.\tau_{01}}{\tau_1.\tau_{03}} - 1\right), \tag{7}$$

$$\Delta_2 = \left(\frac{\tau_3.\tau_{02}}{\tau_2.\tau_{03}} - 1\right), \tag{8}$$

τ_1, τ_2 and τ_3 are, respectively, times of travel of both transverse and longitudinal waves, τ_{01}, τ_{02} and τ_{03} – times of travel in absence of mechanical stresses.

The distribution of residual stresses in the pipe investigated is presented in fig.8. The stress distributions are observed along the generant L from the weld center. Generant is located at *90°* to circumference of the pipe. The values of residual mechanical stresses are compared with that located in the weld center. They are determined using ultrasonic methods and theoretically-experimentally obtained acoustic-elastic coefficients for the reference samples in the present work.

Fig. 8a gives the values of the residual stresses σ_z, acting in direction of the generant of pipeline. The residual stresses obtained from determination of times of travel in material of surface, subsurface and volumetric ultrasonic waves are designated respectively, by surface, subsurface and volumetric.

Fig.8b shows the values of the residual stresses σ_h, acting along the circumference of the pipe (perpendicular to generant). Here, the residual stresses are determined from the obtained times of travel of volumetric ultrasonic waves in material and acoustic-elastic coefficients obtained from the reference samples by equation (6).

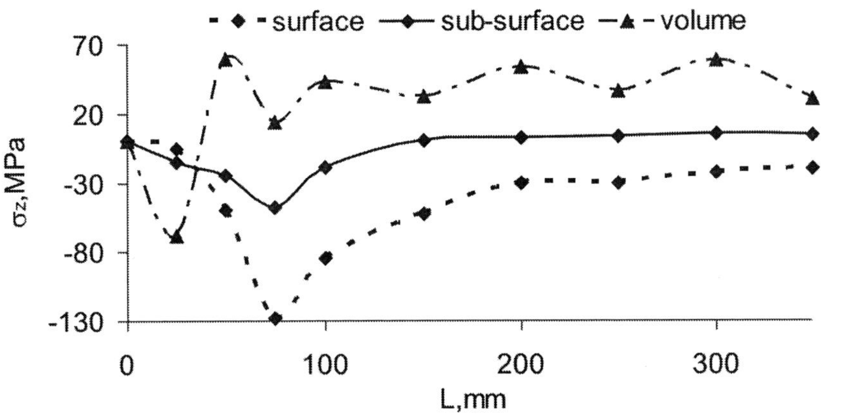

a)

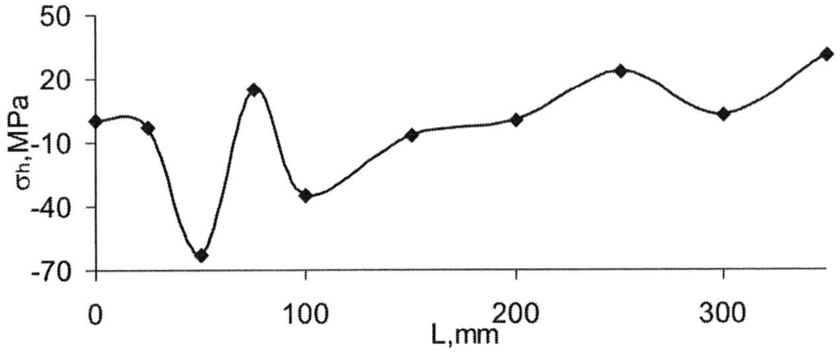

b)

Fig.8. Distribution of the residual stresses to generant L, a) acting in direction of generant and b) along the circumference of the pipe (perpendicular to generant).

The most significant are the variations of the residual stresses σ_z in surface layer by up to *128 MPa* and of equivalent residual stresses for the whole volume in radial direction in depth by up to *140 MPa* (Fig.8a). In subsurface layer the residual stresses obtained vary by up to *50 MPa*.

The dependence on the distribution of the equivalent volumetric stresses in fig.8a is non-monotonic and different from that for the surface and subsurface stresses, which is due to the fact that the residual stresses in wall depth vary non-monotonously from one surface to the other. This fact affects considerably the values of equivalent stresses determined by volumetric ultrasonic waves, especially in the welded joint and weld surrounding area of *75 mm* length.

The dependence for the residual stress distribution in fig.8b varies by up to *80 MPa*.

The complex use of the "drill method" with the acoustic strain gauge measurements allows to obtain true values of the residual stresses from their relative variations in the surface layer of material. This is done by determining the residual stresses using the drill method for one of the points investigated by acoustic strain gauge measurements in the surface for the same direction of action determined by acoustic strain gauge measurements.

Logically, the values obtained for the different measurement points and by different method used differ. The pipeline volume in which the propagation times for the various types of waves are obtained is different - the surface layer, the subsurface layer and the volume in depth.

Nevertheless, the obtained values show the applicability of the developed algorithm for the evaluation of the mechanical stresses in investigated material from an object.

Conclusion

Basic ultrasonic methods, as well as a semi-destructive method of MSC assessment were presented. Acoustic elastic coefficients for the test material and changes in velocities from mechanical stresses for surface and subsurface waves were obtained.

A MSC evaluation algorithm has been proposed together with producing and performing preliminary ultrasonic tests of reference samples to obtain acoustic elastic coefficients and comparative curves of times of travel of surface and subsurface ultrasound waves to determine relative variations of mechanical stresses in the investigated material.

The proposed algorithm is applied to a part of the pipeline for which the relative variations of the residual mechanical stresses along the generant, acting along the generant and the circumference of the pipe are determined

The complex use of "drill method" and ultrasonic strain measurements allows to determine the actual residual stresses in the surface layer of the investigated material.

References

1. N. E. Nikitina, *Acoustic-elastic. Application in practical experiance* (N. Novgorod: TALAM 2005)
2. V. M. Bobrenko, A.N. Kutsenko, I.P. Malahov, *Acoustical testing of mechanical stress* (Odesa: Astroprint 1997)
3. V. M. Bobrenko, M. S. Vangeli, A. N. Kutsenko, *Acoustical stress measurement* (Kishinev, Shtintsa, 1991)
4. A. N. Guz, F.G. Mahort, O. I. Gushta, V. K. Lebedev, *Basic of ultrasonic non-destructive testing for stress measurement in solid material* (Kiev, Naukova dumka. 1974)
5. GOST-R-52890-2007. Non-destructive testing. Acoustical method for stress testing in material of pipeline. General requeriments.
6. GOST-R-52731-2007. Non-destructive testing. Acoustical method for stress testing.
7. ASTM E 837-08. Standard Test Method for Determining Residual Stresses by the Hole-Drilling Strain-Gage Method.
8. Work instruction of Restan MTS3000 of SINT Technology srl., *System for measuring residual stress by the hole-drilling method.*
9. Y. Mirchev, P. Chukachev, M. Mihovski, *Proceedings of the XVII National Scientific-Technical Conference "Acoustic 2015"*, Sofia, Bulgaria **15**,164 -166 (2015)
10. M. Mihovski, Y. Mirchev, P. Chukachev, V. Sergienko, AIP Conf. Proc. **1785**, 040038 (2016)

A method of laser ultrasound tomography for solid surfaces mapping

Vasily Zarubin[1,2,*], *Anton* Bychkov[1,2], *Alexander* Karabutov[1,3], *Varvara* Simonova[4] and *Elena* Cherepetskaya[1]

[1]National University of Science and Technology "MISiS", 119049 Moscow, Russia
[2]Lomonosov Moscow State University, Faculty of Physics, 119234 Moscow, Russia
[3]Lomonosov Moscow State University, International Laser Centre, 119991 Moscow, Russia
[4]Institute on Laser and Information Technologies of the Russian Academy of Sciences, 140700 Shatura, Moscow Region, Russia

Abstract. Surface control is an important problem for production inspection in industry. Traditionally it is solved using mechanical and optical profilometry, providing high spatial resolution. However, such methods are inappropriate for bodies immersed in opaque liquids and require preliminary preparation of samples. Current work focuses on an application of a method of laser ultrasound tomography to mapping of solid surfaces. The suggested approach combines high acoustical resolution, high performance and allows profilometry of contaminated and immersed samples. An automated prototype for real-time surface profile measurements is constructed and tested on several bodies of revolution. Linear approximation of a cylinder profile obtained by the prototype shows maximum approximation inconsistency of 15 μm. The results are verified by those obtained by the conventional x-ray method.

1 Introduction

The problem of solid surface profilometry is important for a wide range of industrial applications. For instance, profilometry is used for an automatic detection of critical damages and defects on the surfaces of pipelines, reservoirs where a high resolution is not required, but the area of controlled surface is large [1]. On the other hand, there are problems requiring high resolution while inspecting small surface areas. These include such tasks as control of polymeric films with complex surface [2], parts for precision systems [3] and many others. In general, profilometry of surface is conducted using methods that can be divided into two groups - contact and non-contact methods.

The most commonly used contact method of profilometry is a mechanical stylus profilometry [4]. Despite the relative simplicity of the device design, the profilers of such type reliably provide a measurement of the solid surfaces with high resolution (up to fractions of nanometres) simultaneously providing a wide range of measurements. Difficulties in inspection of sensitive films and low operation performance are among the

* Corresponding author: zarubin.vasily@gmail.com

drawbacks of contact profilometers. This often forces technicians to use other methods of profilometry.

The mentioned problems can be solved by using non-contact profilometry methods, such as optical and ultrasonic. The most commonly used methods of optical profilometry are laser interferometry and laser triangulation [5, 6]. The advantages of these methods include high resolution, comparable with that provided by the stylus profilers, as well as high-speed performance. However, an expensive preliminary cleaning of samples is required for conduction of the optical profilometry. At the same time, industrial and factory control requires the study of contaminated objects, which may be immersed in opaque liquids (for example, oil tanks). In such situations it is convenient to use ultrasonic methods.

Acoustic microscopy is the most appropriate ultrasonic method for measurements of the surface [7]. It uses ultrasound of high frequency greater than 50 MHz, which provides a high submicronic longitudinal resolution. However, high frequencies are strongly attenuated in organic liquids, for example, oil, so the characteristic longitudinal resolution of ultrasonic profilometry is of the order of ~ 10 μm, which, nevertheless, is sufficient for many industries.

In the present work the laser ultrasound tomography method is used to perform profilometry of objects immersed in liquids. This approach has a number of advantages over traditional methods of profile imaging.

Laser ultrasound is the ultrasound induced by a specific way of generation. The laser beam is delivered to absorbing material, called an optoacoustic generator. Due to absorption in the near-surface layer and subsequent instant heating and short expansion of this substance [8] the ultrasonic waves are excited. The advantage of the laser method of ultrasound generation is the ability to generate powerful aperiodic pulses that are not available to conventional ultrasonic generators. The short submicrosecond duration and smooth shape of the probe pulse leads to an increased longitudinal resolution comparing to the traditional methods of ultrasound microscopy. At the same time, the use of the tomographic approach leads to an increased speed of operation.

2 Experimental method

Current work is focused on the development of principles, algorithms and a setup for high-speed laser-ultrasonic profilometry of solids.

The main principle of an experimental method is the following. The radiation from Nd:YAG laser (wavelength $\lambda = 1064$ nm, pulse duration $\tau = 10\ ns$, pulse repetition rate 20 Hz, Q-switched) is delivered by an optical fiber (Fig. 1) to the optoacoustic generator for laser excitation of ultrasound. Ultrasonic waves are focused by an acoustic lens on the surface of the object under study, which partially scatters them. The scattered radiation is concentrated by the same lens on the surface of the receiving elements of the acoustic antenna array. The acoustic antenna consists of 16 receiving piezoelectric PVDF transducers, which provide effective reception of a wideband ultrasonic signal in the frequency band of $1.6 - 9$ MHz [9]. Although the transducers are flat, they are glued to the surface of the acoustic lens, what makes the antenna effectively cylindrically focused.

Using a 3-axis positioning system, the sample is moved in the immersion liquid (water) to a focal plane of the acoustic antenna. After that, the sample is irradiated by ultrasound and the scattered signal is recorded by the receiving antenna. Analog electrical signals are fed to the preamplifier, get digitized by a high-speed multi-channel data acquisition and processing system. This system is based on FPGA NI FlexRIO architecture, has a built-in multichannel ADC with a sampling rate of 50 MHz. Its functions are digitizing analog signals, storing, averaging and transferring data to a PC for further processing of data and reconstruction of tomograms with a subsequent detection of the surface profile.

To operate the experimental setup a specialized software was developed. This software has a graphical interface that allows to control the sample movement. The software operates in real time (frame rate is 10 Hz) and displays tomograms of an object located in the focal plane of the antenna. For the reconstruction of tomogram the filtered back-projection algorithm is used [10]. The algorithm constructs an image over "arcs of probability", which have a shape of parabolas. A plane probing acoustic pulse is scattered by an inhomogeneity and then it's recorded by the receiver elements at a time moment $t = (z + d) / c$, where z is the distance travelled by the probe pulse from the generator to acoustic inhomogeneity, and d is the distance from the inhomogeneity to the receiving transducer.

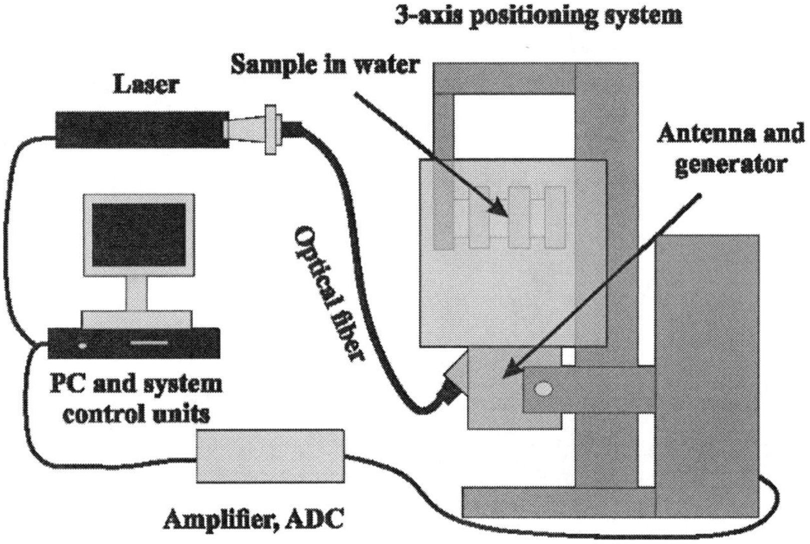

Fig. 1. Schematics of the experimental prototype for laser ultrasound profilometry.

Following this algorithm, the reconstructed acoustical pressure is defined by a formula

$$\epsilon(\mathbf{r}) = \frac{\sum_{i=1}^{N} \Delta\Omega_i \times b\left(\mathbf{d}_i, t = \frac{z + |\mathbf{d}_i - \mathbf{r}|}{c}\right)}{\sum_{i=1}^{N} \Delta\Omega_i} \quad (1)$$

where $\epsilon(\mathbf{r})$ denotes effective distribution of scatterers (tomogram values) in point with position defined by vector $\mathbf{r}$, $\mathbf{d}_\mathrm{i}$ – position of the ith transducer, N – number of transducers, $\Delta\Omega_i$ – solid angle of the ith transducer as it is seen from point defined by vector $\mathbf{r}$. A solid angle is introduced here to take into account a rate of acoustic energies, received by different transducers. And $b(\mathbf{d}_\mathrm{i}, t)$ is a backprojection value, calculated as

$$b(\mathbf{d}_\mathrm{i}, t) = -2\frac{\partial p(\mathbf{d}_\mathrm{i}, t)}{\partial t} \quad (2)$$

where $p(\mathbf{d}_\mathrm{i}, t)$ – pressure, detected by ith transducer.

After the reconstruction of tomograms, the profile of objects has to be detected on them. As it is seen from formula (1), the calculation of tomograms with sufficient number of pixels (typically 800×800 pixels for image of $20\,\mathrm{mm} \times 20\,\mathrm{mm}$ size) is the time consuming operation even not taking into account profile recognition. Therefore, the algorithms were implemented using parallel programming technology on graphical processing units NVIDIA CUDA [11].

For the spectral filtering of raw signals a bandpass filter was used. This filter is defined by a formula

$$K(f) = K_0 \exp\left(-\left(\frac{f_l}{f}\right)^8 - \left(\frac{f_l}{f}\right)^2 - \left(\frac{f}{f_h}\right)^4\right) \tag{3}$$

where f_l and f_h - cut-off lower and higher frequencies respectively, which were set to 0.1 MHz and 10 MHz in experiments. Both frequencies in (4) are determined by spectral band effectively acquired by PVDF transducers as well as by noise spectrum.

Several bodies of rotation were used as reference samples: a cylinder, a cylinder with irregular grooves and a cylinder with regular grooves. A photograph of the samples is shown in Fig. 2, and the tomograms obtained using the experimental setup are shown in Fig. 4. The main features of the samples are clearly revealed by these tomograms. Therefore, the next step is the implementation of algorithms for automatized profile detection from previously reconstructed tomograms.

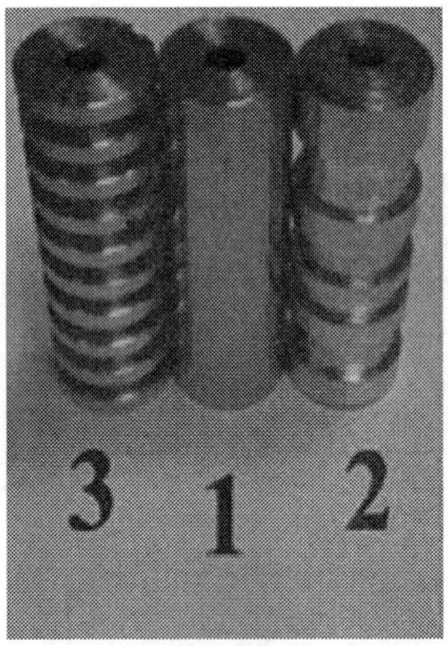

Fig. 2. Reference samples. 1 – cylinder, 2 – cylinder with irregular grooves, 3 – cylinder with regular grooves.

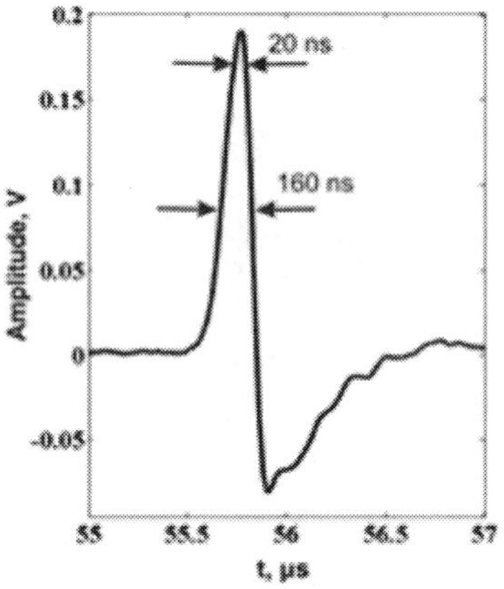

Fig. 3. Scattered from the cylinder surface signal.

A two-step algorithm was proposed for this purpose.

1. Determination of the position of the tomogram peaks in each column (parallel to the vertical z-axis direction, see Fig. 4). Theoretically this operation as follows from Nyquist-Shannon theorem can be carried out with arbitrary precision if the sufficient sampling rate is provided. Within the current experiment the condition of the theorem is fulfilled: the sampling rate of 50 MHz is more than 2 times greater than the significant upper frequency of the signal (15 MHz). The definition of the maximum position was conducted in two stages. Initially, an approximate position of the maximum was determined, after which a parabolic interpolation of the tomogram in a vicinity of the maximum was used to refine maximum position (Fig. 3).
2. After a set of the tomogram maxima is determined, the Hough transform [9] and the least-squares method were used to determine linear sections on the profile, their equations, and line approximation errors. As it follows from the experimental results, the

approximation inconsistency of the cylinder surface is 15 μm. This confirms the possibility of the reliable detection of solid surfaces profile with high precision.

3 Results and discussion

To verify the results of the profiler prototype measurements, a comparison with a measurements carried out by X-ray tomograph XD7600NT was done for cylindrical samples with regular and irregular grooves.

The equations of the outer and inner surfaces of the regular grooves were obtained

$$y = k \cdot x + b \tag{4}$$

$$k_1 = (-2.24 \pm 0.03) \cdot 10^{-2}, b_1 = (0.930 \pm 0.002)\ \text{mm},$$

$$k_2 = -2.14 \pm 0.03, b_2 = (2.902 \pm 0.001)\text{mm},$$

where coefficients errors are standard errors calculated following linear least-squares method.

From equations it follows that the depth of the grooves defined by the prototype is $1.972 \pm 0.005\ mm$. Measured with an X-ray tomograph, the depth of the grooves is found to be equal to $1.968 \pm 0.005\ mm$. Therefore, the longitudinal measurements coincide within the instrumental error (Fig. 5).

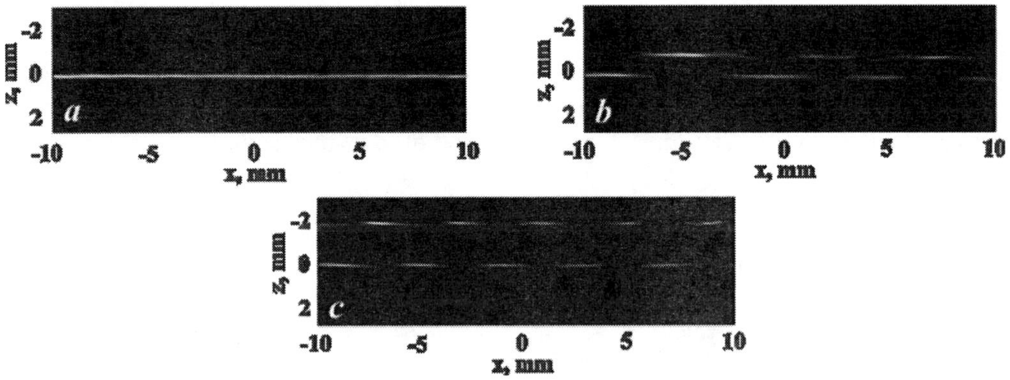

Fig. 4. Tomograms obtained using automatized prototype. a) Cylinder, b) cylinder with irregular grooves, в) cylinder with regular grooves.

The next inspected reference sample is the cylinder with irregular grooves. Measured with the developed laser-ultrasonic profilometer prorotype depth is 0.988 ± 0.004 mm, while the X-ray tomograph obtained 0.950 ± 0.005 mm. The discrepancy of the results is due to the fact, that the algorithm for calculation of the straight lines gives an average value of the depth over all grooves, and X-ray tomograph gives a depth measurement of a single groove.

One of the issues related to the profilometer prototype is its low transversal resolution. The combination of a receiving elements large width (1 mm) and of the small number of transducers leads to the high errors in determination of the positions of the grooves boundaries (up to 100-200 μm). It's possible to improve prototype transverse resolution in several ways. For instance, an antenna of modified configuration (e.g. toroidal with increased number of transducers) could be used. Another way is the development of techniques for reconstruction of composite images obtained from a set of tomograms with a translational or rotational shift.

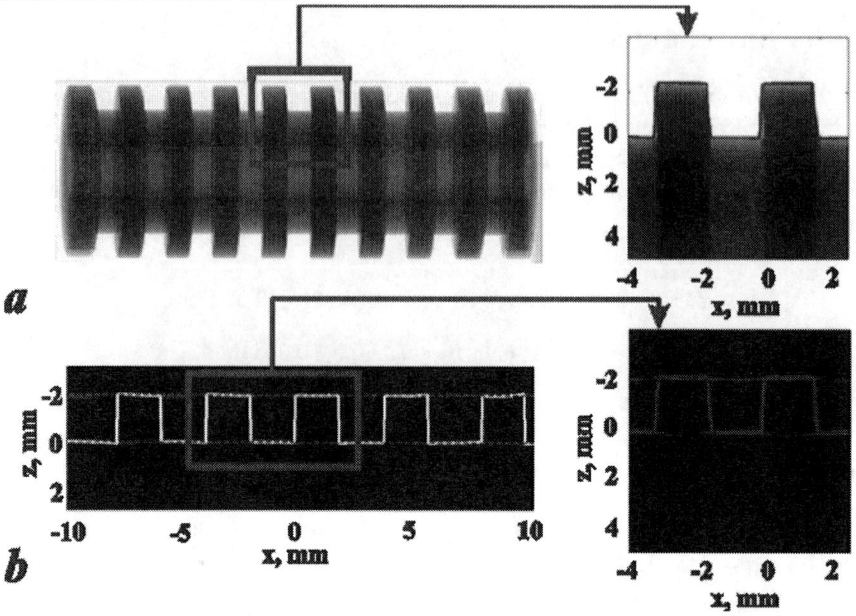

Fig. 5. a) X-ray tomogram of the sample with regular grooves, b) laser ultrasound tomogram of the sample with regular grooves.

The task of laser-ultrasonic profilometry with high precision is one of the important steps on the way to the development of the method of laser-ultrasonic tomography of solid objects with a curved surface. One of the applications of the laser-ultrasonic tomography method could be the flaw detection in specimens with curved surfaces, which could simplify the control of such samples, at the same time maintaining high spatial resolution.

Among the factors affecting the resolution of profilometry, there are several that require a particularly detailed study [12]. Here, the influence of the finite dimensions of the receiving elements is especially important. The integration of the acoustic signal over the surface of the piezoelectric element leads to the possibility to obtain only a locally averaged position of the surface of the object. This effect, leading to the decreased resolution, however, can be used to calculate the local roughness of objects by the increased duration of the ultrasonic pulse caused by its prolongation due to reflection on inhomogeneities.

4 Conclusions

Within the framework of the current study, a prototype of a laser-ultrasonic profilometer operating in real time is designed and tested. Filtered back-projection algorithm and the group of algorithms for tomogram processing and profile detection are parallelized using NVIDIA CUDA parallel programming framework and implemented on graphics processors. The best achieved inconsistency of profile approximation is 15 μm for a cylindrical surface. The measurements of the depth of the sample grooves performed by the laser-ultrasound prototype and the X-ray tomograph are compared. In general, the results of comparison showed the coincidence of measurements within the instrumentation error of measurements. According to the results of the current investigation, laser ultrasound profilometry is a perspective method for high-speed profilometry, which could be used for inspection of bodies immersed in opaque liquids.

This work was carried out with financial support from the Ministry of Education and Science of the Russian Federation in the framework of the NUST MISiS Competitiveness Program (No. K1-2015-025), and by the Russian Science Foundation (Grant No. 16-17-10181).

References

1. L.C. Lynnworth, *Ultrasonic measurements for process control: theory, techniques, application* (Academic press, 2013).
2. L.T. de Haan, V. Gimenez-Pinto, et. al. Adv. Funct. Mater. **24**, 9 (2014).
3. B.H. Lu, J.L. Zhang, B.G. Liu, Y.H. Sun, G.D. Liu, Proc. SPIE, Selected Proceedings of the Photoelectronic Technology Committee Conferences **1**, (2014)
4. D.H. Lee, N.G. Cho, Meas. Sci. Technol. **23,** 10 (2012)
5. J. Schmit, J. Reed, E. Novak, J.K. Gimzewski, J. Opt. A : Pure Appl. Opt. **10**, 6 (2008)
6. M.F. Costa, J. Indian Inst. Sci. **76**, 2 (2013)
7. A. Briggs, O. Kolosov, *Acoustic microscopy* (Oxford University Press, 2009)
8. V.E. Gusev, A.A. Karabutov, *Laser Optoacoustics* (American Institute of Physics, 1993)
9. V.A. Simonova, E.V. Savateeva, A.A. Karabutov, Moscow Univ. Phys. Bull. **64**, 4 (2009)
10. G. Wurzinger G., R. Nuster, N. Schmitner, S. Gratt et al., Biomed. Opt. Express. **4**, 8 (2013)
11. R. Szeliski, *Computer vision: algorithms and applications* (Springer Science & Business Media, 2010)
12. I.M. Pelivanov, V.A. Simonova, T.D. Khohlova, A.A. Karabutov, Proc. SPIE, Photons Plus Ultrasound: Imaging and Sensing, **756428** (2010)

Application of Synthetic Aperture Focusing Technique for inspection of plate-like structures using EMAT generated Lamb waves

Yordan Mirchev[1,*], *Krasimir* Staykov[2], and *Damyan* Ganchev[2]

[1] Institute of Mechanics, Bulgarian Academy of Sciences, Sofia, Bulgaria
[2] Technical University of Sofia, Sofia, Bulgaria

Abstract. The main challenge for guided wave inspection is exact defect characterization and sizing. EMAT generated Lamb waves usually have low signal-to-noise ratio which reduces the defect detection, characterization and sizing capabilities. That's why in most cases the method is used only as a screening tool. The Synthetic Aperture Focusing Technique is a process that increases the signal-to-noise ratio by numerically focusing the acoustic fields. In this paper the application of SAFT is tested over EMAT generated Lamb waves. The improvement of lateral resolution and signal-to-noise ratio is evaluated. Results are presented as a comparison between standard B-scan and SAFT processed data.

1 Introduction

Application of ultrasonic guided waves in non-destructive testing is growing rapidly for a variety of reasons. Quite often this is the only way to solve a problem because of limited access to a structure and their ability to inspect large area from a single position [1]. Recently, a lot of industrial solutions based on EMAT generated guided waves have been developed. EMAT is contactless technology, which makes it very well suited for automated inspection. Also EMAT can inspect through coatings and are not affected by pollutants, oxidation, or roughness. The most common and simplest configurations for generation of guided waves in plates and pipes are the so-called meander coil EMAT. A meander coil in conjunction with a normal magnetic field causes a series of alternating tractions parallel to the surface of the sample. As a result, guided waves, with a wavelength equal to double the spacing between the wires, are excited [2]. This configuration has very good mode selectivity due to its meander-coil design and tone-burst excitation. The technique can cover shorter distances (0.1-3 m), compared to long range UT and can be used for detection of corrosion, cracks and discontinuities on exposed tubes, gas lines, oil pipelines and storage tanks. It typically works with frequencies ranging from 100 kHz to 1 MHz and has up to 10 times better resolution than long range UT systems with minimal or no dead zone.

* Corresponding author: mirchev@imbm.bas.bg

The equipment usually includes a hand-held scanner which provides axial and circumferential scanning for inspection of pipes and plates using guided waves. This allows real time images (B-scans) to be generated [3].

The main challenge in guided wave inspection is the exact defect characterization and sizing. That's why in most cases the method is used only as a screening tool. One of the signal processing techniques used in conventional ultrasonic testing to improve understanding of detected signals and increase defect sizing potential is Synthetic Aperture Focusing Technique (SAFT) [4].

The SAFT analysis is an imaging process that increases the signal-to-noise ratio and improves lateral resolution by numerically focusing the acoustic fields. More recently, this imaging technique has been applied to imaging defects in plates using guided waves generated using piezoelectric transducers [5-6].

In this paper an offline post-processing SAFT technique is described which uses the complete B-scan data recorded from hand-held scanning of steel plates using EMAT generated Lamb waves.

2. Data post-processing algorithm

The movement of the EMAT transducer simulates a multi element synthetic aperture, where every element transmits and receives simultaneously. Schematic representation of the synthetic aperture is presented on fig. 1. For each position of the transducer n at regular intervals raw A-scans are recorded and no time correction is applied at that time.

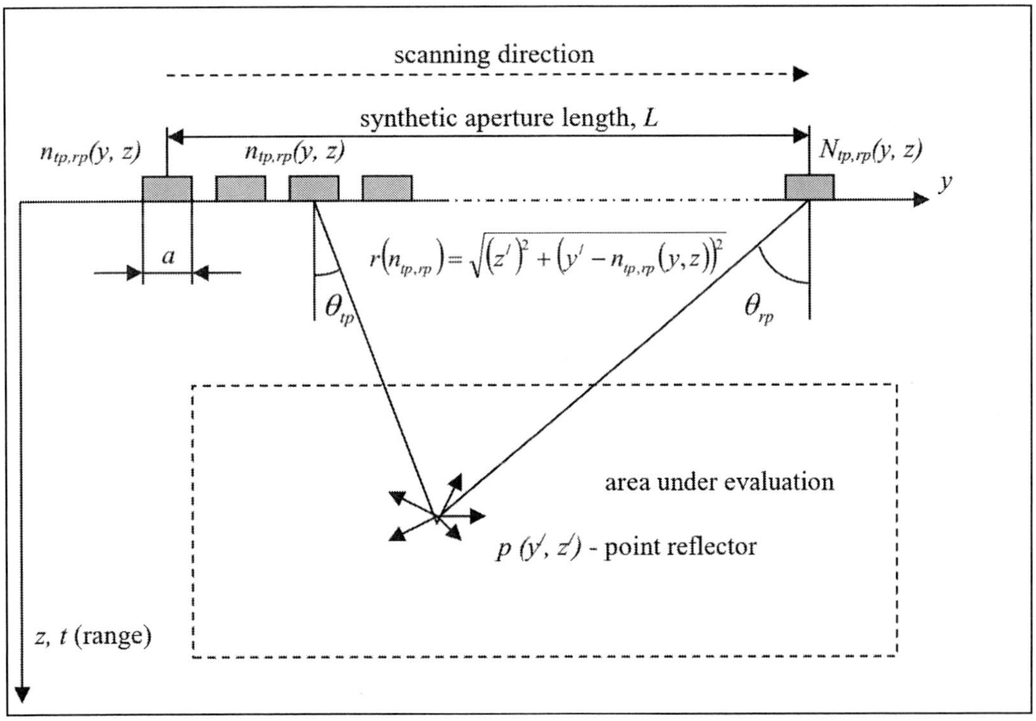

Fig. 1. Schematic representation of the synthetic aperture obtained by mechanical scanning

Flow chart of the data post-processing algorithm is presented on fig. 2.

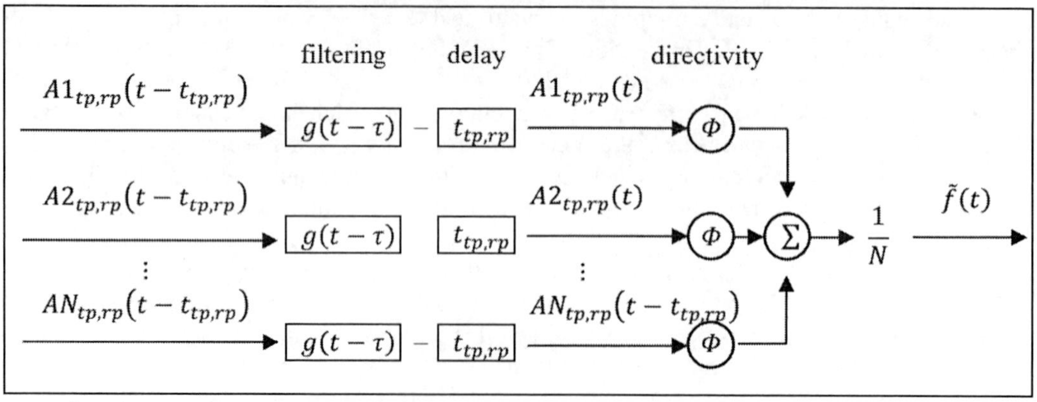

Fig. 2. Post-processing algorithm

The information signals $An_{tp,rp}(t - t_{tp,rp})$ are represented in a matrix giving an unprocessed B-scan image. Each signal is processed by a bell shaped correlation function $g(t-\tau)$ using the following equation:

$$An_{tp,rp} = A_0 e^{i\omega_0(t-t_{tp,rp})-\beta^2\omega_0^2(t-t_{tp,rp})^2}, \tag{1}$$

where A_0 is the maximum value of the transmitted impulse (in all cases in the present work $A_0 = 1$), e – Euler's number, $\omega_0 = 2\pi f_0$ – nominal angular frequency, f_0 – nominal linear frequency, t – time scale or range, $\beta = \frac{0.265}{m}$ – pulse shape and length coefficient, m – number of oscillations in the pulse at -6 dB level, $t_{tp,rp} = \frac{d_{tp,rp}}{C}$ – pulse time-of-flight for distance $d_{tp,rp}$ and velocity of ultrasonic waves in the test environment C, j – imaginary unit, k – wave number.

To reduce the noise in the received signal, a short-time Fourier transform is used. It is given by the following equation:

$$xn_{tp,rp}(\tau,\omega) = \int_{-\infty}^{\infty} An_{tp,rp}(t) gn_{tp,rp}(t-\tau) e^{-j\omega t} dt. \tag{2}$$

The filtered signal is represented by dependence (3) for the nominal angular frequency ω_0 of the probe. The Fourier spectrum of the processed signal is calculated with the following equation:

$$Fn_{tp,rp}(\omega) = \int_{-\infty}^{\infty} xn_{tp,rp}(\tau,\omega) d\tau. \tag{3}$$

Compensation of the delay time in the signal's spectrum for every individual position of the probe is given with the following dependence:

$$Gn_{tp,rp}(\omega) = Fn_{tp,rp}(\omega) e^{jk(\omega)(d_{tp,rp}-z')}, \tag{4}$$

where $d_{tp,rp}$ is the distance traveled by the wave from the transmitter to the reflecting point $p(z',y')$ and back to the receiver. z' - is the distance along the Z axis perpendicular to the scanning axis Y.

For the SAFT algorithm the distance $d_{tp,rp}$ is calculated as follows [7]:

$$d_{tp,rp} = 2r(n_{tp,rp}) = 2\sqrt{(z')^2 + \left(y' - n_{tp,rp}(y,z)\right)^2}, \tag{5}$$

where z' and y' – the coordinates of the point reflector, tp and rp – refer to the operation of the probe as emitter and receiver for the point reflector p in the area under consideration, n – position of the probe during mechanical scanning.

The response spectrum is calculated after the time delay is compensated. The directional characteristics of the ultrasonic probe and the amplitude drop due to beam divergence are also taken into account. This is achieved with the following dependence:

$$Hn_{tp,rp}(\omega) = \frac{\Phi n_{tp,rp} G n_{tp,rp}(\omega)}{\sqrt{r(n_{tp}) r(n_{rp})}}, \tag{6}$$

$$\Phi n_{tp,rp} = \frac{\sin\left(\frac{a}{2} k \sin(\Theta_{tp,rp})\right)}{\frac{a}{2} k \sin(\Theta_{tp,rp})}, \tag{7}$$

where a is the length of the probe, $\Theta_{tp,rp}$ – the angle between the Z axis of the probe and the point reflector $p(z', y')$ given on fig. 1, k – wave number.

From the dependence (6) signals $Xn_{tp,rp}(t)$ are obtained by a reverse Fourier transformation for each point of the area under evaluation using the following equation:

$$Xn_{tp,rp}(t) = \int_{-\infty}^{\infty} Hn_{tp,rp}(\omega) e^{j\omega t} d\omega. \tag{8}$$

The final transformed B-scan is calculated using:

$$h(z,y) = \sum_{n} Xn_{tp,rp}\left(\frac{d_{tp,rp}}{C}\right). \tag{9}$$

3. Used equipment, samples and experimental procedure

Experiment are conducted to evaluate the suitability of proposed SAFT post-processing for application in inspection of plate-like structures using EMAT generated Lamb waves. During the experiments an Innerspec Powerbox H ultrasonic instrument with permanent magnet sensor and temate® MRUT hand-held scanner is used. The sensor consists of "XXL" magnets and corresponding "hourglass" coils [3]. Evaluated samples are steel plates (grade S235JR according to EN 10025:2005) having dimensions 1000x200 mm and thickness of 4 mm. In the samples a through holes are drilled having diameter d = 10 mm, 15 mm and 20 mm. For the experimental work A0 Lamb wave mode at frequency 405 kHz is generated using meander coil with period 6.35 mm and 5 cycle excitation pulse. The EMAT transducer length is 50 mm, and the selected aperture length is 100mm. The samples are scanned according to the scheme shown on fig. 3. The distance between the EMAT and the through hole defect is 300 mm. During the scanning B-scans are recorded composed of A-scans at 1 mm steps in lateral direction. The sampling frequency was set to 10 MHz.

All B-scans are then post-processed offline using SAFT algorithm described above.

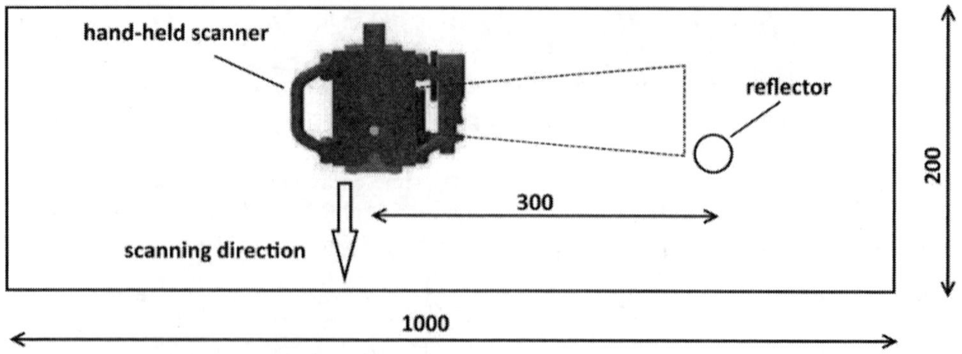

Fig. 3. Scheme showing the through hole reflector and the scanner during the experiments.

4. Results and discussion

On fig. 4 are presented B-scans from the experiments for different diameter reflectors and their corresponding SAFT processed images (eq. 9). On the B-scans are shown RF signals reflected from through holes having diameter d = 20 mm, 15 mm and 10 mm. Scanning direction is aligned with the Y-axis and the position changes from 0 to 100 mm. The Z-axis corresponds to the direction of sound propagation and is presented from 270 mm to 350 mm in front of the transducer position. The amplitude of the signal is encoded in gray scale.

After processing the initial B-scans with the proposed SAFT algorithm the width of the reflected ultrasonic beam in the scanning direction Y-axis is reduced. This can clearly be seen on fig. 4b, 4d and 4f. Considerable noise reduction can also be seen. This leads to better spatial resolution in the scanning direction.

In order to evaluate the ultrasonic beam width data is extracted from the recorded B-Scans at distance Z = 300 mm. This corresponds to the center line of the trough holes. The resulting graphs in the form of normalized amplitude at different lateral positions of the scanner are presented on fig. 5.

Data from scanning a 10 mm diameter through hole, presented on fig. 5c, shows that the ultrasonic beam width at -6dB is equal to 44 mm. After post-processing with the proposed algorithm the beam width is reduced to 30 mm. The improvement of spatial resolution in the scanning direction would be around 30%. Furthermore an improvement of signal-to-noise ratio around 18 dB was also observed. Similar results are observed from the scanning of other samples.

The results show the ability of SAFT to increase the inspection range and respectively the sensitivity of meander coil EMAT equipment. This post-processing additionally allows more precise characterization of discontinuities in plate-like structures tested using guided waves.

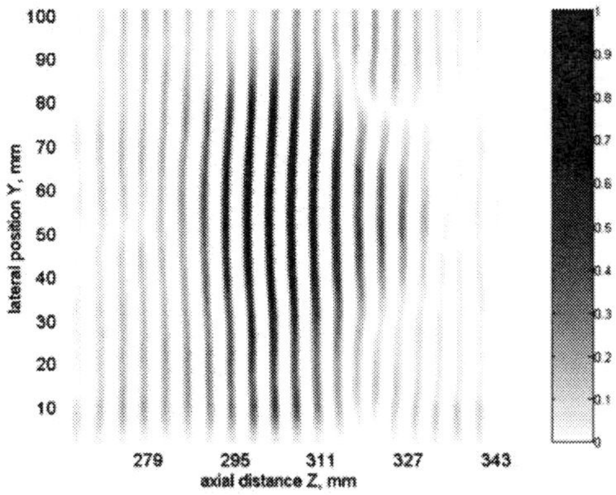

a) B-scan, $d = 20\ mm$

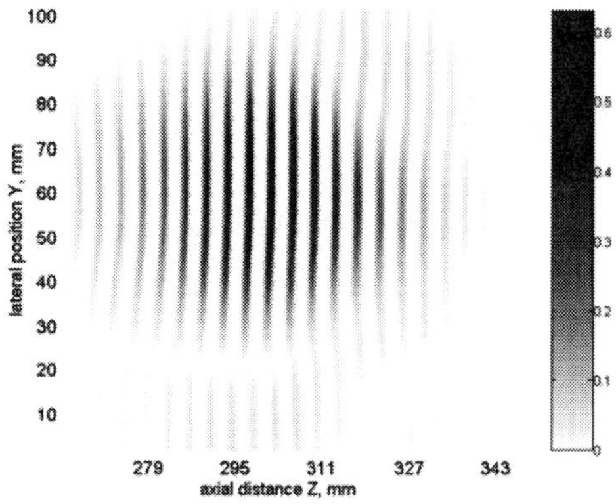

b) SAFT, $d = 20\ mm$

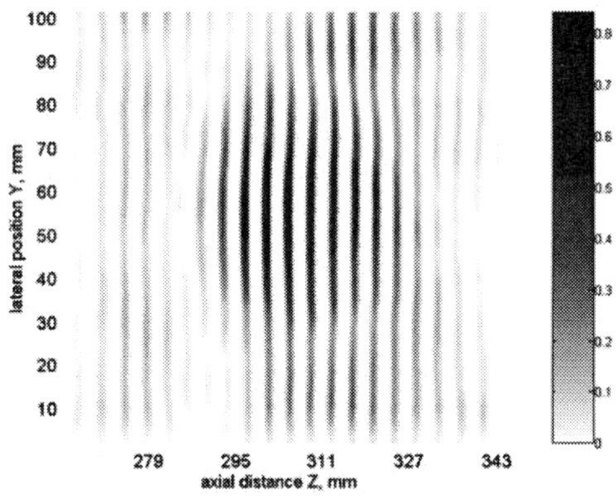

c) B-scan, $d = 15\ mm$

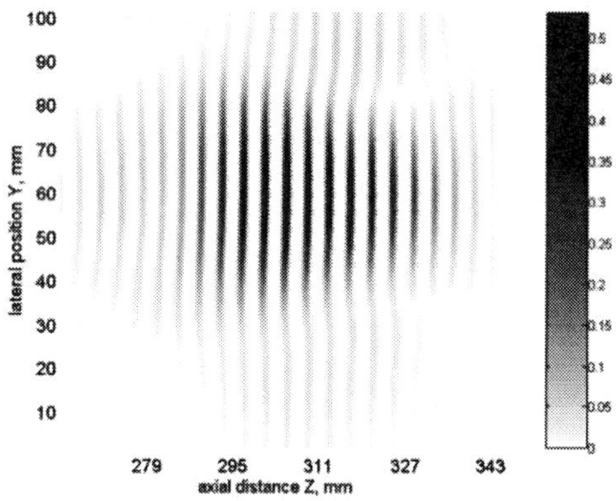

d) SAFT, $d = 15\ mm$

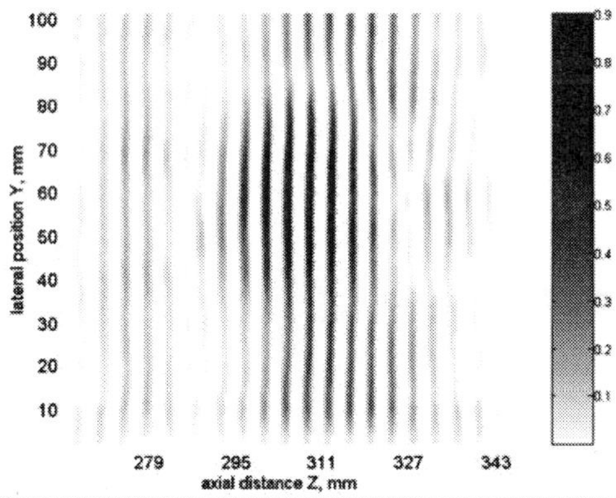

e) B-scan, $d = 10\ mm$

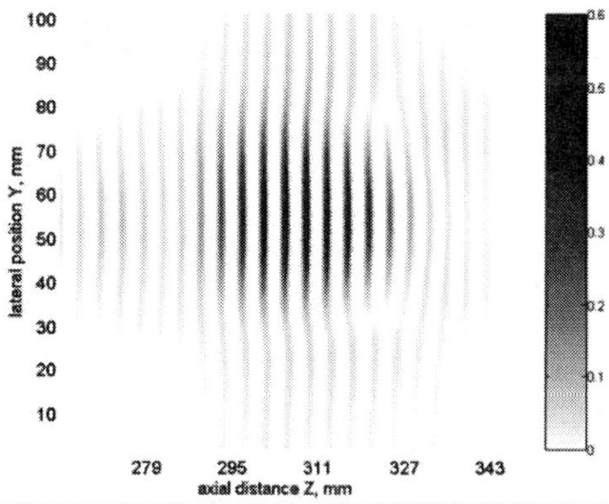

f) SAFT, $d = 10\ mm$

Fig. 4. B-scan images from the ultrasonic testing and their corresponding SAFT processed images

https://doi.org/10.1051/matecconf/201814505010

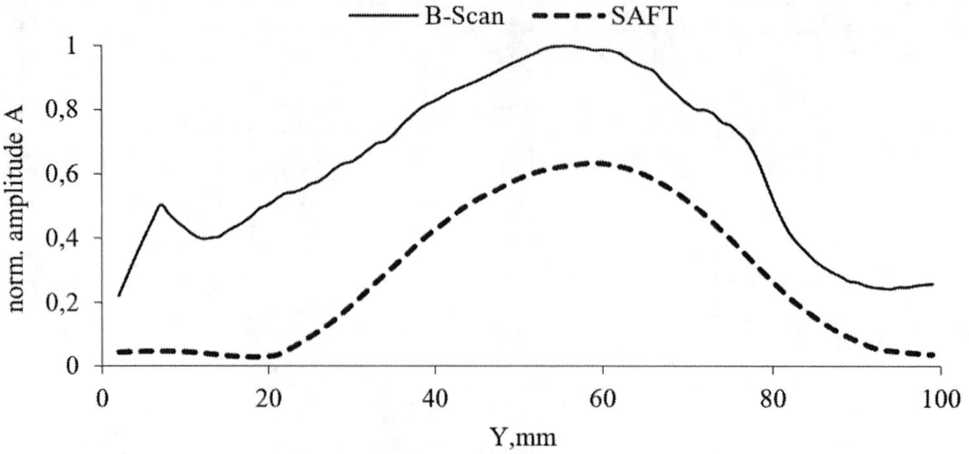

a) d = 20 mm

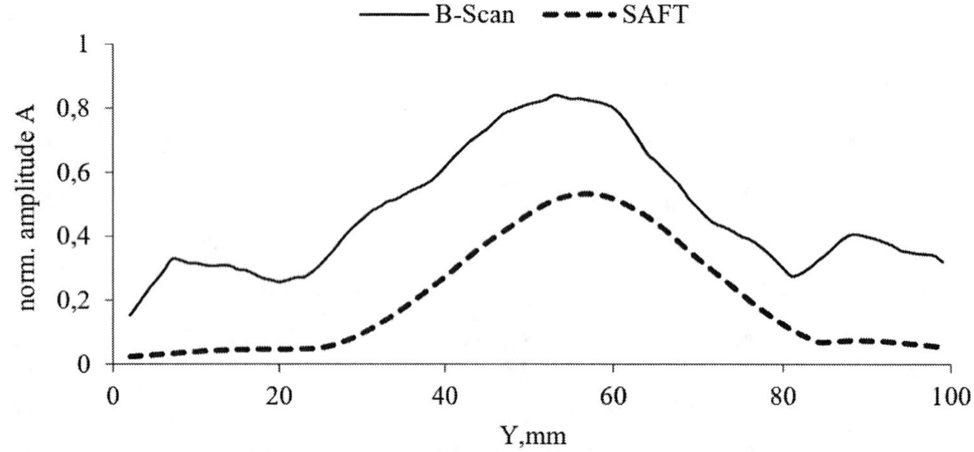

b) d = 15 mm

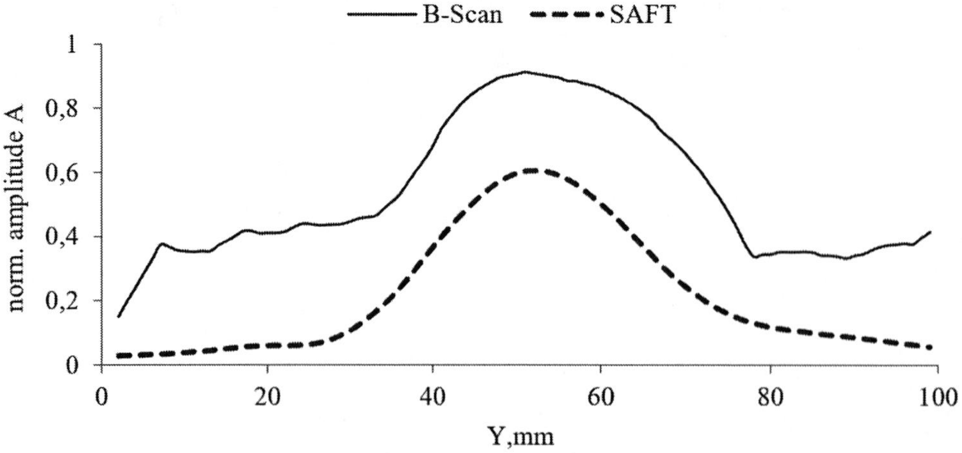

c) d = 10 mm

Fig. 5. Width of the ultrasonic beam at distance Z = 300 mm

5. Conclusion

The proposed algorithm for post-processing of ultrasonic signals received from EMAT generated Lamb waves is based on SAFT. The experimental results show around 30% improvement of the lateral resolution at -6 dB level. Improvement of the signal-to-noise ratio (more than 16 dB) is also observed which is of great importance in EMAT ultrasonic testing known with its low SNR. One advantage is that the focusing possibility is independent on the length of the sound path. Those results show the ability of proposed algorithm to improve defect sizing and characterization capabilities of conventional EMAT generated Lamb waves inspection techniques. The proposed off-line processing technique can be embedded into new inspection equipment as a real-time version that will enhance the defect detection and evaluation potential.

6. References

1. J. L. Rose, *Proc. SPIE* **7983**, 798302 (2011)
2. R. Ribichini, *Modelling of electromagnetic acoustic transducers*, PhD thesis, Imperial College London, London (2011)
3. Innerspec Technologies, Inc., http://innerspec.com/en/
4. J. Pitkänen, *ECNDT 2006*, Poster 211 (2006)
5. J. Davies, F. Simonetti, M. Lowe, P. Cawley, *ECNDT 2006*, Tu.4.1.3 (2006)
6. R. Sicard, A. Chahbaz. J. Goyette, IEEE Trans. Ultrason. Ferroelec. Freq. Cont. **51**, 1287-1297 (2004)
7. C. Holmes, B. Drinkwater, P. Wilcox, NDT & E Intern. **38**, 8, 701-711 (2005)

Preparation of sol-gel SiO_2 coatings on steel and their corrosion resistance

Irina Stambolova[1], *Stancho* Yordanov[2*], *Lyuben* Lakov[2], *Sasho* Vassilev[3], *Vladimir* Blaskov[1], and *Bojidar* Jivov[2]

[1]Institute of General and Inorganic Chemistry – Bulgarian Academy of Sciences, Acad. G. Bonchev St., bl.11, 1113 Sofia, Bulgaria
[2]Institute of Metal Science, Equipment and Technologies with Hydro- and Aerodynamics Centre “Acad. A. Balevski”, Bulgarian Academy of Sciences, Shipchenski Prohod Blvd. 67, 1574 Sofia, Bulgaria
[3]Institute of Electrochemistry and Energy Systems, Bulgarian Academy of Sciences, Acad. G. Bonchev St., bl. 10, 1113 Sofia, Bulgaria

Abstract. A protective SiO_2 coating was prepared by sol-gel method on stainless steel plates, using precursor solutions having 2 different concentrations: low (0.5M) and high (2M) concentration. The dip coating-drying cycles have been repeated three times, after that the coatings were thermally treated at 300°C or 500°C in air. Scanning electron microscopy (SEM), AFM and X-ray diffraction analyses have been applied. The corrosion resistances of SiO2 coated steels were examined in NaCl medium for 346 hours. Many microcracks and craters in the samples, prepared from solution of high concentration appear after corrosion attack. The coatings obtained from solution of low concentrations reveals almost unchanged surface structure without visible cracks and pits. The evaluated corrosion rate of this coating is lower than those of uncoated steel.

1 Introduction

The corrosion control of metals is of economical, technical and environmental importance. Corrosion processes are accompanied by a number of reactions that change the composition and properties of the metal surface and can be reduced with the selection of appropriate materials for particular application, by using new alloys, corrosion inhibitors or protective films and coatings deposited onto the metal surface, especially in aggressive environments. Oxide films and coatings such as SiO_2, ZrO_2, Al_2O_3 and TiO_2 can be deposited on metals to improve their surface and anticorrosion properties [1-6]. There are different methods for the obtaining of thin films and coatings on metal surface, such as electrochemical deposition [7], physical vapor deposition (PVD) [8], chemical vapor deposition (CVD) [9], plasma spraying [10] and sol–gel process [1, 5, 11-16]. The advantages of sol–gel technique are: low processing temperature, good homogeneity, use of compounds that do not introduce impurities into the end product as initial substances. Cost of precursor solutions has limited

*Corresponding author: stancho14@abv.bg

the use of sol-gel processing in many applications [17]. The sol–gel method is a wet-chemical technique and involving several stages: hydrolysis and polycondensation, gelation, aging, drying, densification and crystallization. The sol–gel technology is based on hydrolysis and condensation reactions [18, 19].

Typically the addition of catalysts increases the hydrolysis rate. The type of catalyst determines the pH of sols and affects the form of sol-gel materials, obtaining the films, powders or monoliths. The pH value of the reaction mixture, which is dependent on the used catalyst, influences the hydrolysis and condensation reaction and affects the form and structure of the received materials. One of the most technologically important aspects of sol-gel processing is the ability to produce coatings with sintering temperatures lower than 500◦C. Depending on heat treatment conditions, specially the final treatment temperature, the structure of sol-gel coatings may undergo changes. A sol–gel coating can be applied to a metal substrate through two techniques: dip-coating or spin-coating [11, 12]. The proper selection of synthesis parameters and the way of preparing materials is significant in successful sol–gel synthesis. These advantages make the sol–gel process one of the most appropriate technologies for preparation of thin, nanostructured films. The sol gel process gives possibilities to obtain coatings with unique diffusion, abrasion, dielectric etc. properties [17]. The preparation of coating by sol-gel method is an established method to produce homogeneous coatings with uniform thickness below 2μm [20]. Sol–gel technology has been significantly improved in the past 20 years. The sol–gel coatings are widely investigated in the area of optics [21], nanobiocomposites [22], biomaterials [23], self-cleaning materials [24] and others [25]. Different sol-gel coatings have been developed to increase the corrosion resistance of metals, although the pure inorganic films have limitations such as micro cracks, residual porosity and thickness limitations [26, 27]. Coatings with good passivity, insulating properties and good tribological properties show high corrosion resistance in aggressive eenvironment. Some papers [28-31] showed the feasibility of obtaining homogeneous and free of cracks sol-gel coatings on steel substrates. Other works [32] have demonstrated increases in the corrosion resistance by using zirconia or silica coatings.

The aim of this work is the obtaining of protective SiO_2 layers by sol-gel method (dip coating technique) on stainless steel plates and investigation their surface morphology, phase composition and chemical stability in salty solution of NaCl.

2 Experimental part

The preparation procedure of SiO_2 films by the sol– gel method was as follows: tetraethoxysilane (TEOS) wass dissolved into ethanol and stirred for 1 h. The mixture of ethanol, water and hydrochloric acid was added to the TEOS solution and also followed by stirring for 2 hours in order to obtain 0.5M and 2 M solutions. The mole ratio H_2O:TEOS was 3.7. The sample obtained from 0.5 M solution were denoted A, while the sample deposited from 2M solution were denoted B. The stainless plates AISI 316 were used as substrates. The substrates were cleaned ultrasonically in ethanol and acetone. The films were immersed in the corresponding solution and withdrawn at a constant rate of 3 cm/min and then were dried consequently at 60◦C and at 90°C. The samples A and B wre heated at 300°C. These steps were repeated 3 times. In order to study the effect of thermal treatment temperature on the protection properties, another sample denoted C produced from 2M solution was treated at 500°C for 1h. The phase composition of the samples was studied by X-ray diffraction (XRD) with CuKα-radiation (Philips PW 1050 apparatus). The average crystallite sizes of the films were estimated according to Scherrer's equation: The morphology of the coatings was investigated by means .of Scanning electron microscope PHILIPS SEM 515, (Belgium), operating at 20kV voltage. In order to study their chemical resistance the investigated samples as well as uncoated steel (reference sample) were exposed to the action of salty solution of 3.5% NaCl at 25°C (EN ISO10289/2006) in the course of 346 hours.The temperature of solution and the air temperature were controlled by calibrated thermometers. It was determined the mass weight loss after 256 and 346 hour.

3 Results and discussion

XRD analyses revealed that all samples are amorphous,since no sygnals of crystallized SiO_2 were observed. Several research groups were also obtained amorphous silica coatings thermally treated at 500°C and 600°C [33,34]

The SEM photographs of the substrate and silica coated steel before and after corrosion test ae shown in Figures 1-3.

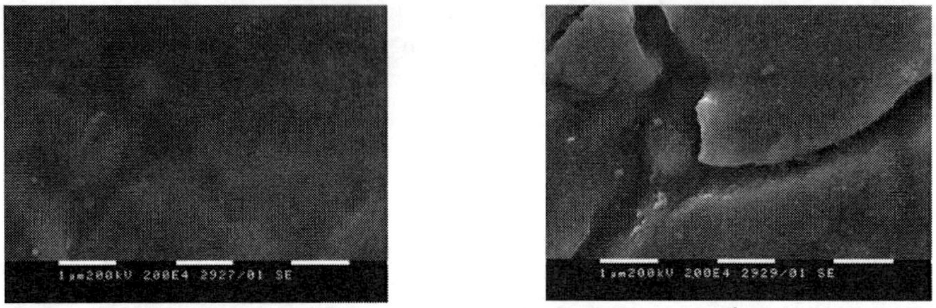

a   b

Fig.1. SEM photographs of bare steel substrate a) before and b) after corrosion test.

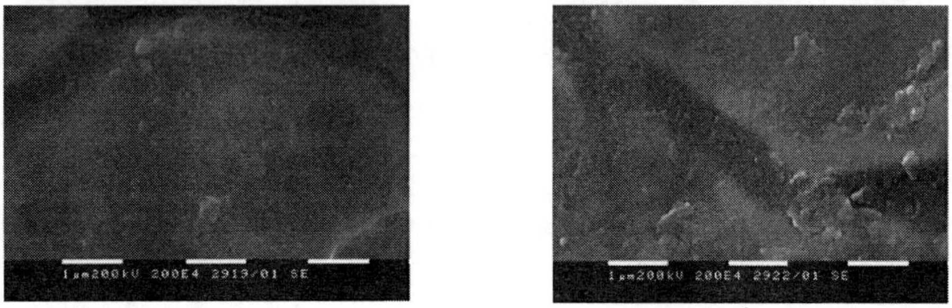

a   b

Fig.2. SEM photographs of sample A: a) before and b) after corrosion test.

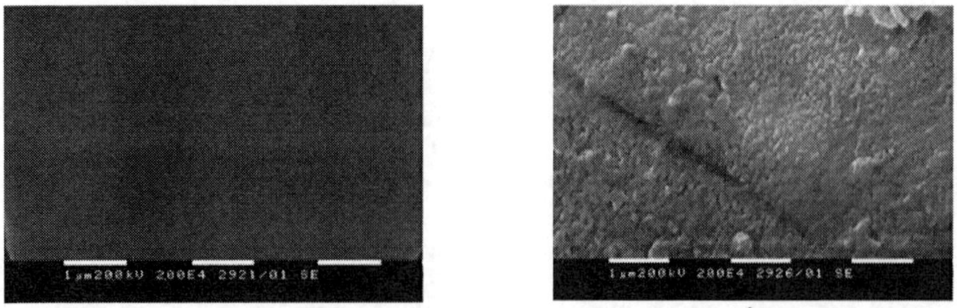

a   b

Fig.3. SEM photographs of sample C: a) before and b) after corrosion test.

Figure 1 reveals the SEM images of steel substrate – bare substrate and after corrosion test. The films structure has been undergoing considerable changes during the corrosion test. Some cracks are clearly visible on the microphotographs. The silica films surface, prepared from 0.5 M TEOS solution is relatively dense with some crystal imperfections (Fig 2-a). Sample A preserves the surface structure without visible cracks and pores after corrosive attack. (Fig. 2-b) TEOS solution of high concentration leads to more flat surface of the coatings after treatment at 300 and 500°C. After corrosion test there are many deep cracks in both samples B and C (Fig.3b). Sample C also reveals significant errosion of the surface.

Figure 4 represents the surface of SiO_2 coatings after corrosion test. On the surface structure of the sample A no visible cracks and pits are to be observed after corrosive test, but has some pitting effects (Fig 4-a). The roughness analysis gives the value Ra about 15 nm and Rq 24 nm for sample A after attack. The corrosive attack induces erosion of the surface and many cracks in the films, obtained from 2M TEOS solution. The film surface displays significant effect of corrosion. The roughness analysis of sample B gives the value Ra 20 nm and Rq 31 nm. There is appearance of microcracks and craters. (Fig 4-b). The erosion of the coatings becomes stronger after treatment at 500°C, leading to eruption of some areas of sample C. Duran et al. have proved that the SiO_2 coatings, deposited on stainless steel and galvanized steel treated at 550°C were more cracked because of the elimination of organic groups during the treatment reduces the plasticity of the coatings [33].

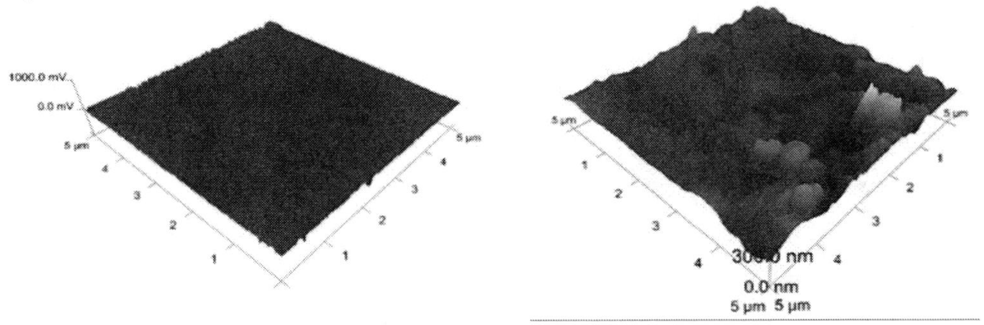

a b

Fig. 4. AFM images (3dimensional) of samples A (a) and C (b)

Acccording to DTA-TG analysis of the film A (Fig 5) exothermal and endothermal peaks are not registered.

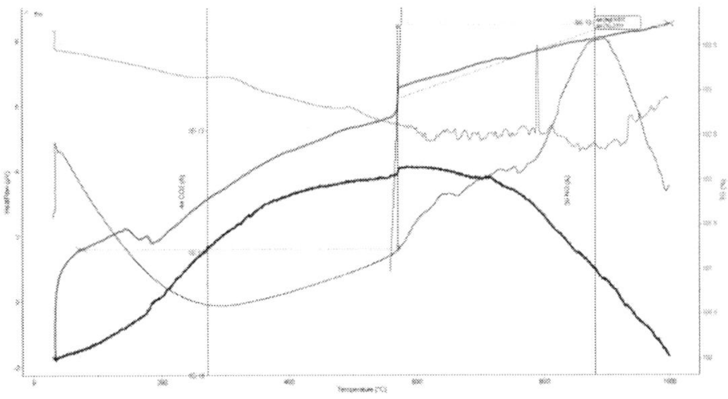

Fig. 5. DTA-TG analysis of sample A.

 https://doi.org/10.1051/matecconf/201814505011

The weight losses of the and the uncoated steel and SiO_2 coatings (sample A) after 250 in the NaCl solution are 1.81 and 1.06 mg/cm^2, respectively. No weight changes were observed after 346 hours. As can be seen from Fig. 6 the weight loss increases up to 250 hours of the test, after that remains constant. It can be seen that the sample A exhibit the better protection properties in comparison to the uncoated steel.

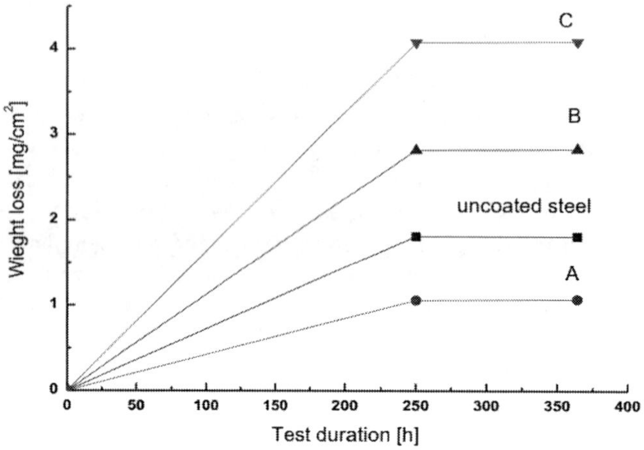

Fig. 6. Weight loss of the SiO_2 sol-gel coatings *vs* time.

The corrosion rate is based on the number of illimeters (thousandths of an inch) it penetrates each year. The corrosion rate was evaluated, according to the Standard - ASTMG31.

The corrosion rate of uncoated steel is 0.058mm/year, while for samples A, B and C are 0.035; 0.092 and 0.133mm/year, respectively. The obtained results revealed that the SiO_2 coatings, prepared from 0.5 M solution of TEOS exhibit the best protection properties. The formation of cracks is dependent on the internal stress in the coatings [18]. This effect is a result of the evaporation of large amount of solvents and organic residuals. Probably, in the samples produced from TEOS solution of high concentration (B and C) during the the thermal treatment the formation of cracks and pits are induced by the evaporation of the larger amount of intermediates (produced by hydrolysis). As a results the mechanical properties (hardness, plasticity etc) and the surface characteristics are deteriorated. This could be reason for decreased anti-corrosion properties. The application of TEOS solution of low concentration seemed more siutable for production iof SiO_2 coatings with enhanced corrosion resistance, due to their surface and mechanical characteristics.

4 Conclusions

Silica sol gel coatings, obtained from two different sol-gel solutions of low and high concentrations of TEOS and treated at 300°C and 500°C possess relatively dense surface. The protection efficiency of silica coatings decreases with the increasing of TEOS solution concentration. The coatings, deposited from 2M solution showed several microcracks and some erosion after immersing in corrosive medium regardless the treatment temperature. After corrosion attack, the sample, prepared from 0.5 M TEOS solution treated at 300°C reveals almost unchanged surface structure. The corrosion rate of this coating is lower than those of uncoated steel.

The authors are grateful to the financial support of Bulgarian National Science Fund at the Ministry of Education and Science, Contract No DN07/2 /14.12.2016/. This support is gratefully acknowledged.

References

1. G. Ruhi, O.P. Modi, I. B. Singh, Corros. Sci. 51, 3057 (2009).
2. C. X. Shan , X. Hou, K. L. Choy, Surf. Coat. Technol. **202**, 2399 (2008).
3. Z. F. Zhou, E. Chalkova, S.N. Lvov, P. Chou, R. Pathania, Corros. Sci. **49**, 830 (2007).
4. N. Padhy, U. Kamachi Mudali, V. Chawla, R. Chandra, B. Raj, Mater. Chem. Phys. **130**, 962 (2011).
5. J. Gallardo, A. Duran, J.J. de Damborenea, Corros. Sci. **46**, 795 (2004).
6. V.S. Saji, J. Thomas, Curr. Sci. **92**, 51 (2007).
7. E. Setare, K. Raeissi, M.A. Golozar, M.H. Fathi, Corros. Sci. **51**, 1802 (2009).
8. A. Perez, A. Billard, C. Rébéré, C. Berziou, S. Touzain, J. Creus, Corros. Sci. **74**, 240 (2013).
9. R. Hausbrand, B. Bolado-Escudero, A. Dhont, J. Wielant, Corros. Sci. **61**, 28 (2012).
10. Y. Wang, W. Tiana, T. Zhang, Y. Yang, Corros. Sci. **51**, 2924 (2009).
11. T.L.Metroke, R.L. Parhhill, E.T. Knobbe, Prog. Org. Coat. **41**, 233 (2001).
12. S.K.Tiwari, M. Tripathi, R. Singh, Corros. Sci. **63**, 334 (2012).
13. M.Atik, P.L. Neto, L.A. Avaca, M.A. Aegerter, Ceram. Int. **21**, 403 (1995).
14. R. Gheriani, R. Chtourou, J. Nano Res. **16**, 105 (2012).
15. S. Li, J. Fu, Corros. Sci. **68**, 101 (2013).
16. G. Ruhi, O.P. Modi, A.S.K. Sinha, I.B. Singh, Corros. Sci. **50**, 639 (2008).
17. B.D. Fabes, B.J.J. Zelinski, D.R. Uhlmann, in: J.B. Wachtman, R.A. Haber (Eds.), *Ceramic Films and Coatings* (Noyes, New Jersey, 1993).
18. C.J. Brinker, G.W. Scherer, *Sol–gel science. The physic and chemistry of sol–gel processing* (Academic Press, San Diego, 1990).
19. B. Babiarczuk, A. Szczurek, A. Donesz-Sikorska, I. Rutkowska, J. Krzak, Surf. Coat. Technol. **285**, 134 (2016).
20. P. Innocenzi, M.O. Abdirashid, Guglielmi M., J. Sol-Gel Sci Technol. **3**, 47 (1994).
21. C. M. Malengreauxa, G. M.-L. Léonarda, S. L. Pirarda, I. Cimierib, S.D. Lamberta, J.R. Bartlettc, B. Heinrichsa, Chem. Eng. J. **243**, 537 (2014).
22. P. E. Johnson, P. Muttil, D. MacKenzie, E. C. Carnes, J. Pelowitz, N.A. Mara, W.M. Mook, S.D. Jett, D.R. Dunphy, G.S. Timmins, C. J. Brinker, ACS Nano **9** (7), 6961–6977, (2015).
23. E. Nouria, M. Shahmiria, H. Reza Rezaiea, F. Talayianb, Surf. Coat. Technol. **205**, 5109 (2011).
24. S. R. Meher, L. Balakrishnan, Mater. Sci. Semicond. Process. **26**, 251 (2014).
25. E. Nouri, M. Shahmiri, H. Reza Rezaie, F. Talayian, Surf. Coat. Technol. **206**, 3809 (2012).
26. O. De Sanctis, L. Gomez, N. Pellegri, J. Non-Cryst Sol. **121**, 338 (1990).
27. J. De Damborenea, N. Pellegri, O. De Sanctis, A. Duran, J. Sol-Gel Sci Tech **4**, 239-244, (1995).
28. L.M. Seara, D.C. Leite, W.L. Vasconcelos, Proc.38th Braz. Cer. Cong., **1**, 62, (1994).

29. D.C. Leite, L.M. Seara, W.L. Vasconcelos, Proc.38th Braz.Cer.Cong, **1**, 138 (1994).
30. L.M. Seara, D.C. Leite, W.L. Vasconcelos, Proc.38th Braz.Cer.Cong., **3**, 1168 (1994).
31. M. M. Castro, C.H.S. Boucflas, W.L. Vasconcelos, Proc.53th Cong on Metallurgy and Materials, **12** (1998).
32. K. Izumi, N. Minami, Y. Uchida, Key Eng. Mater. **150**, 77 (1998).
33. A. Durán, Y. Castro, M. Aparicio, A. Conde, J. J. de Damborenea, Intern. Mater. Rev. **52**, 175 (2007).

Mechanical properties of composite coatings of chromium and nanodiamonds on aluminum

Nelly Gidikova[1], *Maciej* Sulowski[2], *Marcin* Madej[2], *Radoslav Valov*[1,*] and *Vladimir* Petkov[1]

[1] Institute of Metal Science, Equipment and Technologies with Hydro Aero Dynamics Center "Acad. Angel Balevski" at Bulgarian Academy of Sciences, 67 Shipchenski prohod, 1574 Sofia, Bulgaria
[2] AGH University of Science and Technology, Al. Mickiewicza 30, 30-059 Krakow, Poland

Abstract. Aluminum offers engineers weight saving advantages in their product design. However, aluminum has poor wear and friction properties. In addition, the surface oxide layer of this chemically active metal, which gives it the corrosion resistance, makes it a very difficult metal to plate [1]. Specific pre-treatment must be applied to remove the oxide layer from the aluminum surface. The nanodiamond particles additionally facilitates the process of chromium deposition. The object of this study is to evaluate the impact of nanodiamonds on the mechanical properties of the chromium coating plated on

1 Introduction

The aluminum has a lot of advantages in the design of engineering products because of its low weight. On the other hand it is relatively soft metal which means poor wear resistance and friction coefficient. New applications are possible when the light weight of aluminum is combined with the durability of the hard chrome.

The most widely commercially used now-a-days process for preparing an aluminum surface for plating is the zinc immersion process [2]. In this process a thin and adherent film of metallic zinc or alloy of zinc replaces the aluminum oxide film. The zinc provides a surface which is more readily plated with other metals than the aluminum. Not all metals, however, may then be simply deposited directly on the zinc surface.

The electrochemical chromium coatings have a wide practical application. They increase the hardness and the wear resistance of the matrix material and possess increased corrosion resistance. The modification of the chromium galvanic coatings with nanodiamond particles (ND) additionally increases these chemical and mechanical properties.

The main goal of this study is to investigate the influence of the ND on the readiness of chromium to be plated on the aluminum surface, the thickness and the continuity of the coating, the microhardness and wear resistance of the composite material comprised of electrolytic chromium coatings with nanodiamond particles and aluminum matrix material.

* Corresponding author: radoslav.valov@gmail.com

2 Experimental

The research conducted in this study is directed to evaluate the mechanical properties of composite chromium coating modified with nanodiamond particles on samples of aluminum alloy (dural AC 2017). The chromium was deposited on the surface of the aluminum items by electrolytic process with the traditional acidic electrolyte containing CrO_3 - 220 g/l and H_2SO_4 - 2.2 g/l. The parameters of the electrolytic process were: current density – 45 A/dm^2; duration of the process – 45 min and temperature of the electrolyte – 50°C. The diamond nanoparticles were added to the electrolyte as an aqueous suspension. Their concentration was 5, 10 and 25 g/l. The nanodiamond particles were produced by detonation synthesis [3].

The thickness of the coatings was determined by light microscopy. The microhardness tests were carried out according to Vickers standard at minimum 10 points. The applied load was 10 g. The dry friction wear resistance was studied by the pin-on-plate test unit (fig. 1). During the test a rectangular 15 x 5 x 10 mm wear sample (1) was mounted in a sample holder (4) equipped with a hemispherical insert (3) ensuring proper contact between the test sample and a steel ring (2), heat treated to 55 HRC, which was rotated at a constant speed of 136 rpm. The wear surface of the sample was perpendicular to the loading direction. Double lever system was used to force the sample towards the ring at 157 N ± 1% and 307N.

The loss of sample mass was measured after a sliding distance of 500m.

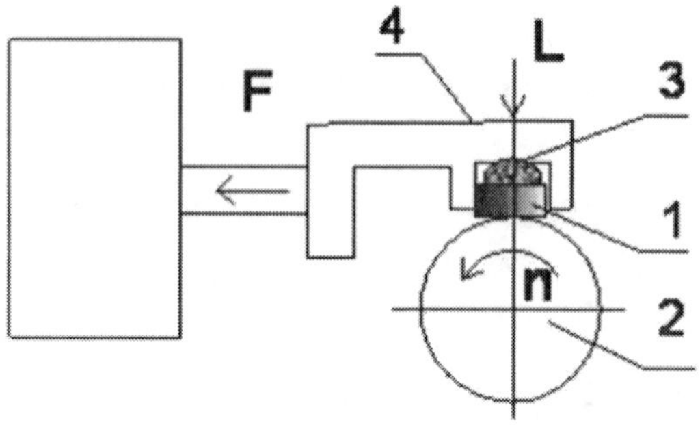

Fig. 1. Schematic view of a block-on-ring test unit

3 Results and discussion

Figures 2, 3 and 4 show the scanning electron image of coatings obtained from electrolytes with different concentrations of diamond nanoparticles.

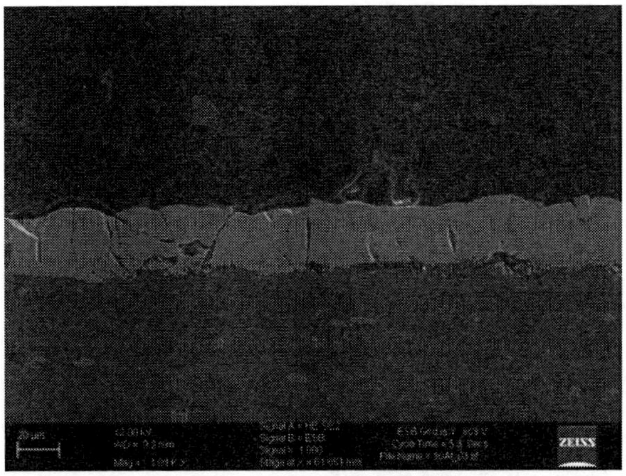

Fig. 2. SEM image of composite coating on aluminum, concentration of nanodiamonds – 5 g/l

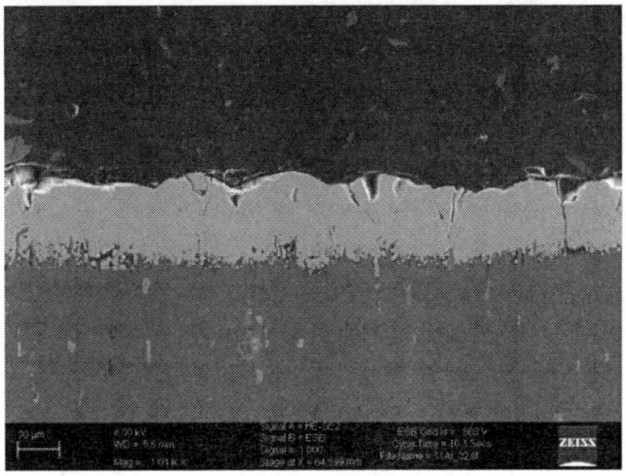

Fig. 3. SEM image of composite coating on aluminum, concentration of nanodiamonds – 10 g/l

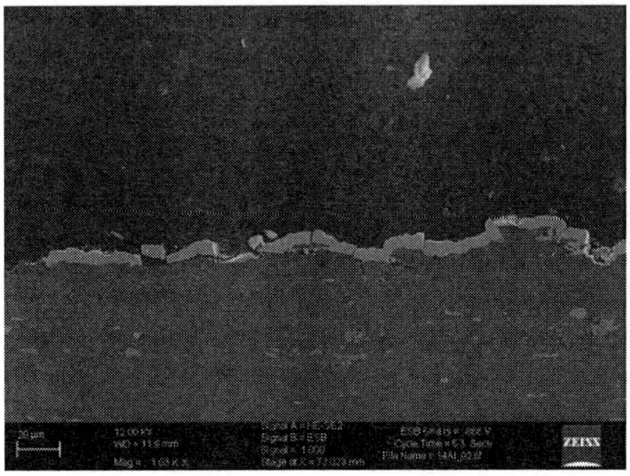

Fig. 4. SEM image of composite coating on aluminum, concentration of nanodiamonds – 25 g/l

It was found out that the yield of chromium on steel and the thickness of the coating is increased with the increase of the concentration of the diamond nanoparticles [2]. It is not the same with the chromium coating on aluminum and especially on this alloy. It is clearly seen from the figures that the best coating is achieved from electrolyte with concentration of nanodiamonds 10 g/l (fig.2). The same tendency is observed when testing the mechanical properties of the coating. The microhardness is the biggest at concentration of nanodiamonds 10 g/l (fig. 5). It goes from 260 kg/mm^2 (5 g/l) to 283 kg/mm^2 (10 g/l) and falls down to 230 kg/mm^2 (25 g/l). The matrix under the coating is also with increased microhardness (140 – 200 kg/mm^2) compared to 90 – 100 kg/mm^2 for the uncoated aluminum alloy.

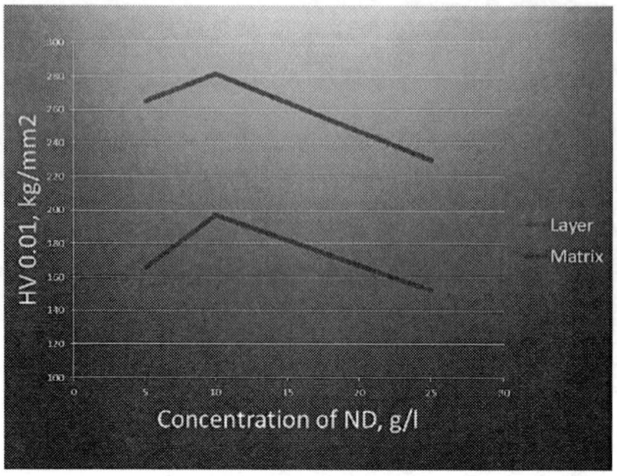

Fig. 5. Microhardness of the composite coating and the aluminum matrix at different concentrations of nanodiamonds

The wear resistance and the friction coefficient of the composite coating have similar behavior as the thickness of the coating at different concentrations of nanodiamonds (fig. 6 and 7). The best results are obtained at concentration of nanodiamonds in the electrolyte 10 g/l.

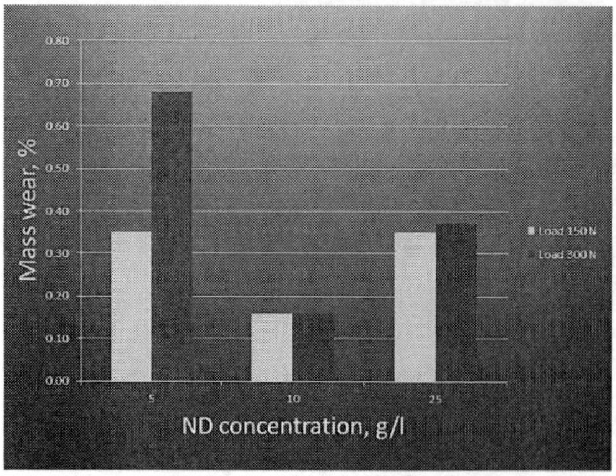

Fig. 6. Wear resistance of the composite coating expressed as mass wear at different concentrations of nanodiamonds

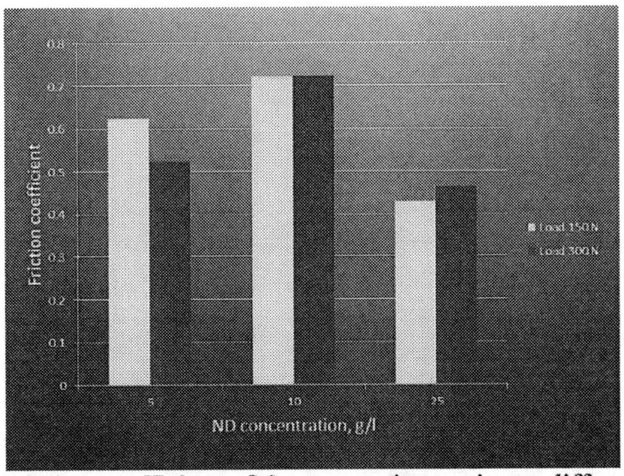

Fig. 7. Friction coefficient of the composite coating at different concentrations of nanodiamonds

The surface topography of the samples prepared at concentrations of nanodiamonds 5 and 10 g/l after the wear tests at sliding distance 500 m are shown in figures 8 and 9. The applied load is 150 N. Two types of wear mechanism can be seen, abrasive and corrosive wear. The dominate mechanism is corrosive wear. There are oxide regions on the surface after the tribological test.

Fig. 8. Surface topography of sample with 5 g/l ND after the wear test. Applied load 150 N and sliding distance 500 m.

Fig. 9. Surface topography of sample with 10 g/l ND after the wear test. Applied load 150 N and sliding distance 500 m.

The typical wear mechanism of the sample with 25 g/l ND after the wear test at sliding distance 500m is abrasive wear, but there is small amount of adhesive wear (fig. 10).

Fig. 10. Surface topography of sample with 25 g/l ND after the wear test. Applied load 150 N and sliding distance 500 m.

4 Conclusions

The presence of nanodiamond particles in the electrolyte facilitates the process of chromium deposition on aluminum items.

The chromium coating adheres very well on the substrate aluminum with thickness of about 20 μm.

The chromium coating increases the wear resistance and improves the friction of the aluminum products.

Two types of wear mechanism can be seen, abrasive and corrosive wear. The dominate mechanism is corrosive wear. There are oxide regions on the surface after the tribological test.

The financial support of the Fund "Scientific Researches" at the Ministry of Education and Science, Contract No DN 07/8/15.12.2016 is gratefully acknowledged.

References

1. U.S.Chrome Inc. Hard Chrome Plating, http://www.uschrome.com/chromium-on aluminum.
2. P.J. Topelian, US Pat **2**, 847, 371, Chromium Plating on Aluminum (2014)
3. G.K. Burkat, V.Y.Dolmatov, E.Osawa, E.A.Orlova, J.of Superhard Mater. **32**, 2, 98–111 (2010)

Airborne sound insulation of new composite wall structures

Yonka Ivanova [1,2*], *Todor* Partalin [3], *Luben* Lakov [4], and *Bojidar* Jivov [4]

[1]Sofia University, Faculty of Physics
[2]Institute of Mechanics-BAS, Sofia, Bulgaria
[3]Sofia University, Faculty of Mathematics and Informatics, Sofia, Bulgaria
[4]Institute of Metal Science, Equipment and Technologies with Hydroaerodynamics Centre "Acad. A. Balevski" – BAS, Sofia, Bulgaria

Abstract. Protection against noise is one of the essential requirements of the European Construction Product directive. In buildings, airborne sound insulation is used to define the acoustical quality between rooms. In order to develop wall structures with optimal sound insulation, an understanding of the physical origins of sound transmission is necessary. To develop a kind of knowledge that is applicable to the improvement of real walls and room barriers is the motive behind this study. The purpose of the work is to study the sound insulation of new composite wall structure.

1 Introduction

An important problem in the design, construction and operation of buildings is the achievement of energy efficiency and good sound insulation of housing and industrial constructions. One of the possible approaches to solving the problem is the use of composite sound and heat insulating materials to meet the requirements of the existing standards. At present, various insulation materials are found: cellular concrete products, foamed glass products, flammable and combustible polystyrene boards (EPS, XPS, etc.), mineral and glass wools with low water resistance, structural polystyrene, etc. [1]. Alternative composite insulation materials with good mechanical properties and low cost, obtained with recycled waste products are in process of developing.

2 Development of new composite materials based on glass-foam granules

A team of Institute of Metal Science, Equipment and Technologies "Acad. A. Balevski" Sofia, Bulgaria improved the technology for the production of foam glasses and developed a model for the production of composite heat and sound insulating materials made of glass-foam [2]. The developed technology is innovative as it recycles glass wastes. The

* Corresponding author: yonka@imbm.bas.bg

developed composite material meets European requirements for fire safety [3] and is claimed by patents [4, 5].

Experimental specimens are made of 5 to 20 mm glass-foam granules made from glass waste and a bonding agent in the form of an aqueous dispersion colloidal silicate based system of Portland cement, zeolite, air-entraining addition for concrete and water. The introduction of the zeolite modifying addition into the composition of the binder increases the physical-mechanical performance. Experimental quantities of modifying additive, granule sorting techniques, excess liquid phase separation, different drying temperatures, and composite block box were performed to provide a material with good physical and mechanical performance for non-bearing walls and barriers. Various monolithic composite blocks have been developed. Figure 1 presents a photo of the composite wall model, made from glass-foam granules.

The aim of the present work is to study the sound insulation properties of the developed innovative materials with and without glass foam, but with an identical ratio of the other components. A sandwich panel is also designed that is developed from an inner layer of the composite and front and outside layers of waterproof screed. The results are compared with the ytong building material. Figure 2 shows a photo of sandwich panel, made of glass-foam granulated composite with external waterproof screeds.

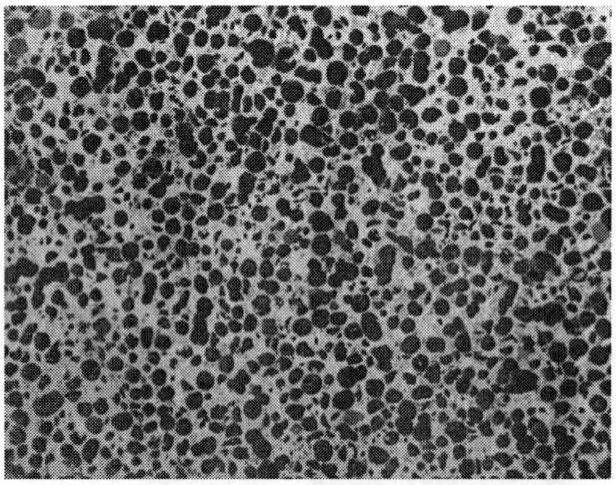

Fig. 1. Composite wall model with glass-granulated granules

Fig.2. Sandwich panel, made of glass-foam granulated composite with waterproof screeds

2 Methodology of acoustic studies

The acoustic measurements are carried out according to the procedures for field measurements of airborne sound insulation between two cameras (fig.3) [6-9]. The sound source camera contains one loudspeaker that creates a uniform sound field. The reference microphone measures the pressure of incident sound wave. Another microphone, placed in the receiving camera measures the pressure of the sound wave passed through the test specimen.

The sound pressure measurements L (dB) in the two cameras (rooms) are carried out in two modes:

1. Using a sinusoidal signal supplied to the single frequency speakers in octave bands 63, 125, 250, 500, 1000, 2000, 4000, 8000 Hz and sound pressure measurements L (dB) with microphones in both chambers at each detected frequency. The recording of the sound level in the cameras is done with the RFT 00024 sounder.

2. By using a broadband signal generated by a sweep generator in the 65 to 10000 Hz frequency band that is fed to the loudspeaker, and the sound pressure measurements L (dB) of the sounding waveform passing through the sample with microphones, subsequent recordings and signal processing (spectral analysis).

The coefficient of sound insulation R is determined by the dependence [6-9]:

$$R = L_1 - L_2, \text{ (dB)} \tag{1}$$

where L_1 and L_2 are the measured sound pressure levels respectively in the camera with the sound source and the receiving camera.

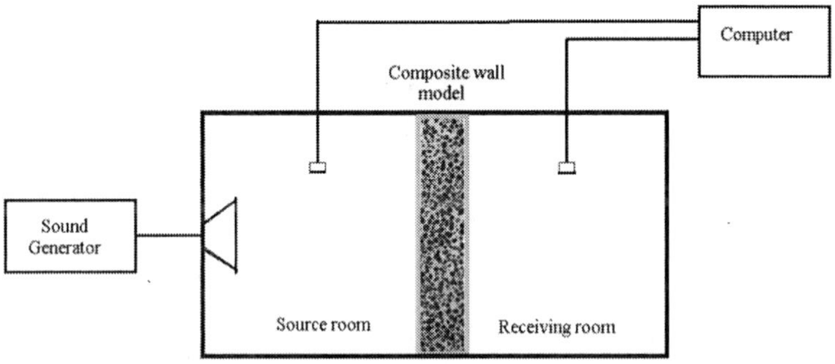

Fig.3 Schema of sound insulation measurements

4 Results and Discussion

Figures 4 and 5 show the results of the acoustic measurements for the developed composite panel with glass-foam granules and the non-granular composite material realized as described above in two modes: measured at individual frequencies and in the all frequency range. It can be seen that higher sound insulation properties possess glass-foam granules composite material.

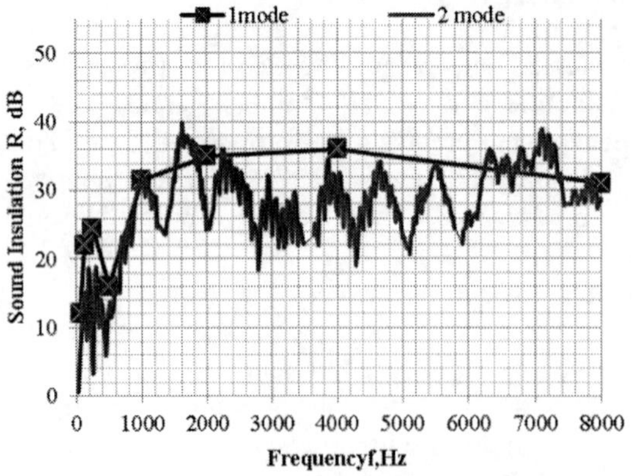

Fig.4. Dependence of sound insulation on frequency for composite materials based on glass-foam granules (GGCM)

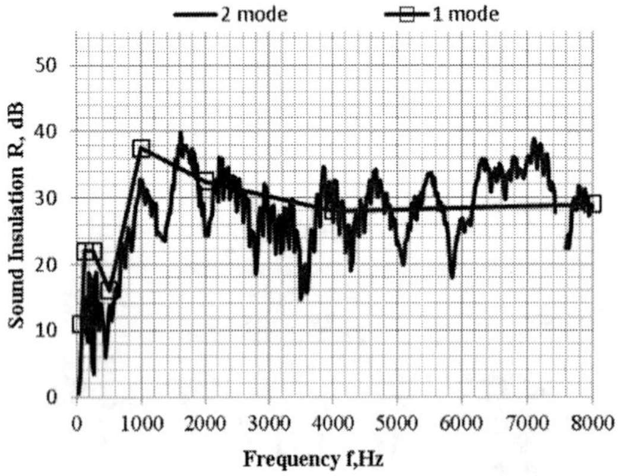

Fig. 5. Dependence of sound insulation on frequency for composite materials without glass-foam granules (NGCM)

Figure 6 compares the results for the composite wall model with glass granules (GGCM) and the reference panel without granules (NGCM) obtained at single frequency measurements. On the same figure are given the results for the sandwich composite panel with waterproof screeds (SP) and for ytong.

The mean value of sound insulation over the entire frequency range is R = 30 dB for the innovative composite material and R = 28 dB for the material without glass-foam granules, R = 36 dB for the sandwich panel with waterproof coatings and R = 33 dB for ytong. Comparing the results it can be seen that higher sound insulation properties possess sandwich panel consisted from glass granules composite material and waterproof coatings.

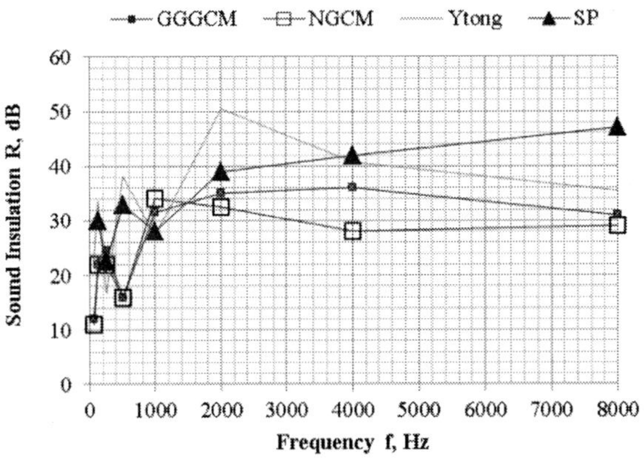

Fig. 6. Dependences of sound insulation on frequency for composite materials with (GGGCM) and without glass-foam granules (NGCM), composite panel with a waterproof coatings on their external sides (SP) and ytong.

4 Conclusion

The paper presents the results of sound insulation properties of innovative composite materials from glass foam granules and binder in the form of an aqueous dispersion colloidal system on a silicate basis as well as composite panels with waterproof coatings on their external sides.

The designed composite materials have suitable mechanical properties, good sound-insulation ability and could be used as sound insulated partitions and non-bearing walls in buildings.

The developed panels are considered to be potentially applicable for the construction of sound and thermal insulated non-porous interior partitions and external insulation of residential and industrial buildings and would extend the product scope to market for building materials.

References

1. D. Nazyrski, *Building insulations* (Sofia 2004).
2. L. Lakov, K.Toncheva, B. Jivov, S. Rafailov, S.Yordanov, Proc.of Intern.Conf. on Civil Eng.Des.and Const., 15-17 September 2016, Varna, Bulgaria (2016)
3. EU Regulation № 305/2011-2013
4. Patent № 111891/19.12.2014
5. Patent № 112126/ 27.10.2015
6. F. Fahy, *Foundation of Engineering Acoustics* (Elsevier Academic Press, San Diego, 2005)
7. Y. Ivanova, V. Vassilev, P. Djondjorov, Str. Djoumaliisky, Bulg.Chem. Commun. **42**, 1–8 (2015)
8. V. Hongisto, *Airborn sound insulation of wall structures-measurement and prediction methods* (Doct. Dissert, Helsinki University of Technology, Helsinki 2000)
9. M. Prascevic, D.Cvetkovic, D. Mihajlov, Facta Univers, Arch and Civ Engin, **10**, 2, 155-167 (2012)

Critical defect size assessment in pipelines on a nuclear power plant

Galya Dimova[1,*], *Vladimir* Yurukov[2], *Kiril* Djivdjanov[3]

[1]Kozloduy NPP EAD, ICC-DCC, 3321 Kozloduy, Bulgaria
[2]ATP – AtomToploProekt" LTD, Sofia, Bulgaria
[3]SENSATA , Sofia, Bulgaria

Abstract. In many energy industry structures, pipeline systems are subject to the impact of mechanical forces, moments of forces and fluid flows of high pressure and temperature. These load factors cause defects in the pipeline metal. As the years of operation increase, defects may occur and grow, which may lead to the destruction of pipeline walls. Special measures have been planned and implemented to ensure the safe operation of high-energy facilities. This study focused on pipelines and nozzles of nuclear power plant equipment with bimetal welded joints on which the size of critical defects was assessed. The base of assessment covers material properties, temperature and stress fields, fracture mechanics calculations. This study involves developing of finite element models and implementing simulations on them in order to obtain temperature fields and determine the stress-strain state of the component.

1 Introduction

In many energy industry structures, pipeline systems are subject to the impact of mechanical forces, moments of forces and fluid flows of high pressure $17 \div 18$ MPa and temperature330 $\div 350$ °C After years of operation under such conditions discontinuities (defects) can occur in the metal walls of the pipelines [1], as well production defects (from steel's melt) can appear. It is known that defects can remain stable without development, or, conversely, defects can grow further. Critical defect's size is a condition which produces a rapid and unexpected destruction of a component (fast fracture). The presence and size of defects is detected through nondestructive testing (NDT) methods, while strength analyses taking into account a load factors can forecast their locations with high probability - the so called critical areas in the facilities.

This paper considers a pipe with both cladding and thermal shirt (shield, jacket) made of austenitic stainless steel - a different steel type from that of the pipe itself. In some cases, the thermal shirt can be welded to the pipe with bimetallic welded joints. The two types of welded metals have different thermal characteristics. Under the impact of high temperature and pressure values of the fluid in the pipeline systems typical for the RPV (reactor pressure vessel), and following multiple load cycles, stress zones with higher strains can form in the boundary of welded joints. The difference in thermal properties of steels in bimetallic welds suggests that they will expand and contract at different rates during the plant modes of warming and cooling. Occurrence of discontinuities in the critical zones is expected as well as their further growth under the influence of strain values throughout the different operation conditions NOC (Normal Operating Conditions), DNOC (Deviation of Normal Operating Conditions) thus turning into defects of metal [2]. To ensure safe operation of NPP pipeline systems with bimetallic joints, it is important to specify/predict the place and the growth of discontinuities. Prediction of the defects behaviour is essential. Such information will enable the prioritised examination of these critical zones in full scope and using testing techniques ensuring high data resolution. The process of studying discontinuities and justifying their behaviour provides a modern approach to safe operation of the facilities with relatively low investments.

The purpose of this paper is to present the approach for detecting of critical zones in bimetallic pipe nozzles, where it is justified to presume that defects will appear, and assess the defects' critical dimensions. This study used a really existing object – nozzles of the reactor pressure vessel emergency core cooling systems. A 3D FEM (Finite Element model) model of the object of study was prepared in which commercial software product was used to made the geometry, the finite elements mesh and then to imitate the load factors (pressure, temperature, forces& moments) in the different operating modes of a nuclear power plant. The strength analyses identified the critical zones where series of discontinuities (defects) are postulated. These can be different types of defects, such as planar ones of the crack type, axially oriented as well as circumferentially – (postulated as) semi elliptical, located on the interface surface (i.e. sub clad ones) or elliptical internal ones (i.e. in the pipe wall, far from the cladding and outer surface). Growth of defects under the operating environment impact was imitated using software. A limiting condition for the calculations is that these defects should not lead to a fast fracture of the nozzle. Through solving the inverse problem, or reverse iteration, the maximum size defect is determined that, on the one hand, can be detectable via ultrasonic nondestructive testing and, on the other, the defect will not reach its critical size throughout the planned operational lifetime of the equipment.

The study is expected to assist in the development of the processes of repair, maintenance, and monitoring of pipeline systems as well as the projects for their long-term operation once their design lifetime has been reached. The study is also intended to support the systems qualification processes of nondestructive testing, and to ensure risk-informed control.

2 Methodology

2.1 Calculations of strains-stresses fields

Unstable condition of reactor unit equipment most often occurs under DNOC and EC – Postulated Accidents (Emergency Conditions). On the other hand, the defects develop due to the cycle loads of the nozzle, affected by different mechanical and thermal influences typical for the NOC, DNOC, and HT modes (HT – Hydrotest or Pressure Test). The number of the actual repeats of the operations transient modes was taken from the I&C (I&C - Instrumentation and Control System for technological data collection measured by sensors) and MDS System measurements (MDS – Monitoring and Diagnostic system for collection of supplementary data (temperature of the metal, stress and strain, etc.) The calculations of stress and strain fields cover all the operation modes of the nuclear power plant [2]. The calculations have considered specific load factors typical for Nuclear Power Plant, [3] a) inside / outside pressure in the pipeline system; b) pipeline weight; c) fluid weight; d) residual stress after welding; e) seismic loads; f) load originating from supports; g) dynamic forces of fluid pulsations, vibrations; h) stresses generated by temperature differences and differences between the materials thermal expansion coefficient (base metal, weld metal); j) stresses due to pressure and thermal fluctuations. To complete the structural analyses a $3D$ finite elements model (FEM) of the pipeline was created, using CAE module pack NISA (originally developed by EMRC co, now Crane co), [3].

The first stage of the structural calculations covers assessment of thermal transients and included simulations of the transients' heat transfer (taken into account processes of convection and conductivity) at different operational transient modes. The pipeline temperature distributions in the model of the tube and nozzle for the different moments of the analyzed temperature transients were produced by FEM simulations using NISA's the HEAT III module.

The second stage of the study considers the mechanical loads added to the thermal ones. Strain-stress conditions (SSC) for every characteristic situation of transients is defined by FEM simulations using module NISA II. The result represents the distribution of total stress due to both thermal and mechanical loads.

After the structural stress calculations, the pipeline nozzle sections with the greatest equivalent stresses were defined.

2.2 Size assessment of the critical defects

Following the structural calculations, in the critical zones thus identified, the presence of a family/series of defects are postulated, supposing every defect's semi-axis have a size of $a_0 = a + \Delta_a$, Fig. 1.

a – depth (minor semi-axis) of a schematized (semi-)elliptical defect;

Δ_a - an uncertainty value followed by ultrasonic testing UT procedure

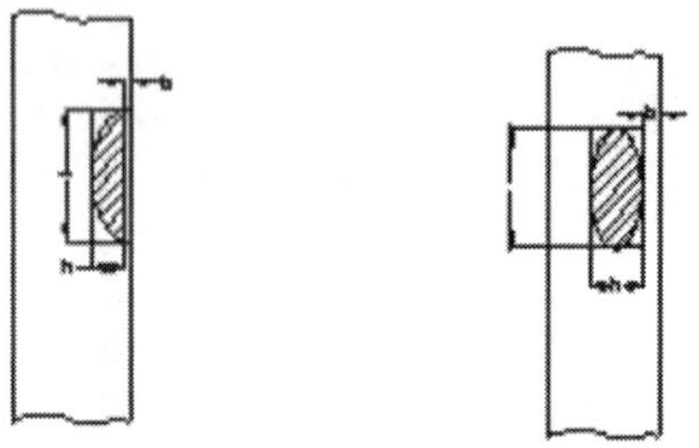

a. b.

Fig. 1. Defect schematization, $h = a_0$ - semielliptical, $h = 2a_0$ - elliptical, for both types - $l = 2c$; h – depth of a schematized (semi-) elliptical defect, l – length of a schematized (semi-) elliptical defect; b – size of a ligament. If the ligament, b, between the defect and outer surface is smaller than $a_0/9$, then the defect is considered a (conditional) surface one

(a) Semielliptical conditionally surface defect (b) Elliptical inner defect (embedded crack).

Under the different operating conditions of the component, the defect will be expected to develop by growing (however, it is also possible that the defect will not grow within the operational lifetime limits). The growth of the defect da can be detected by non-destructive testing of the pipeline zone, with an uncertainty value of Δ_a. In reality, on NPPs in operation, non-destructive tests on components are not performed continuously, but at specified time intervals. Considering the intervals between pipeline NDTs and the increment value da over one interval, the defect will reach a size of:

$$a_1 = a_0 + da \text{ ; or } a_1 = a + \Delta_a + da \tag{1}$$

Repeating this step, after n NDT intervals the defect will reach the size of:

$$a_n = a_{n-1} + \Delta_{a_{n-1}} + da_{n-1} \tag{2}$$

This step is reiterated until the end of the pipeline system operating lifetime (limiting condition), while the defect should not reach its critical size during this limiting timeframe. The target is to define that value of a_n, for which the defect will not reach its critical size a_{crit}, or $a_n \leq \lfloor a \rfloor$

$[a]$ - critical size of a schematized (semi-) elliptical defect

The growth rate of the defect da is defined according to the Methodology [1].

The defect growth of a component subjected to cyclic loads is calculated by the Paris model [1,2].

$$\frac{da}{dN} = C_0 \cdot \left(\frac{\Delta K}{\sqrt{1-R}} \right)^m \tag{3}$$

N – number of load cycles; $\Delta K = K_{max} - K_{min}$ - range of stress intensity factor

C_0, m - material properties; R - loading cycle asymmetry coefficient

The limit condition of a pipeline with a defect is determined as follows [3]:
- for ferritic components (SIF criterion or fracture toughness criterion):

$$K_I \leq [K_I] \tag{4}$$

- for ferritic and austenitic components (limit load failure (LLF) criterion)

$$R_F{}^T = 1.2.R_{P_{0.2}}{}^T \tag{5}$$

K_I - stress intensity factor , $MPa.\sqrt{m}$

$[K_I]$ - allowable value of stress intensity factor, after applying all safety factors

$R_{P0.2}$ – yield strength, MPa;

A dedicated software code developed specially for the current work was used to determine the nozzle critical defect size.

3 Input data

The input data include drawings of the Dn300 nozzle (Dn – diameter nominal of the tube) and actual geometric size values; material data and thermo-hydraulic loads under different plant operating conditions. The results provided apply to the Dn300 nozzles on the hydraulic accumulators (HA) of the upper and lower plenum chambers of the reactor pressure vessel (RPV) with a thermal sleeve of the passive system for ECC. The function of the Emergency Core Cooling System (ECCS); is to quickly inject boric acid solution in the RPV to cool down the core, and flood it in case of a loss of coolant accident (LOCA) when the pressure (P_1) in the primary circuit drops below 5.9 MPa. The ECCS passive part comprises tanks (HA), Dn300 connecting pipelines between the HA and the RPV, and valves on the pipelines. The four Dn300 nozzles between the ECCS passive part and the RPV come in couples – two of which lie on the beltline diametrically opposite of the Dn 850 of the lower plenum chamber, and another two nozzles located on the beltline diametrically opposite of the Dn 850 of the upper mixing chamber. All the four Dn 300 nozzles have similar geometry. Each of them has a thermal mantle (jacket) from stainless Cr-Ni austenitic steel welded through bimetal Y-type welded joints.

Stress calculations are performed taking into account the internal pressure, dead weight and temperature gradients. Residual stresses in welding joints and in cladding are calculated for all operating conditions. The loads considerations in NOC are as follow [3]:1) Planned heating from cold condition to normal work condition at a rate of $20^0 C/h$; 2) Spurious actuation of the RPV emergency protection; 3) Planned cooling from normal work condition to cold condition at a rate of $30^0 C/h$; 4) Weight of the nozzle; 5) Total seismic load; 6) Heat transfer from pipeline system.

The loads considerations in DNOC are: 1) Shutdown of turbine valves; 2) Loss (total) of coolant; 3) Stopping of water feed to steam-generator; 4) Breaking of a steam-generator heat exchange tube; 5) Water injection (temperature $60 \div 70^0 C$) in pressuriser; 6) RPV cooling at a rate of $60^0 C/h$.

4 Results and discussion

The loads model and the strength calculations used 15 transients and conditions typical for ECCS for the $Dn300$ nozzle. Two emergency conditions regarded as representative for the defects growth were considered: a) in case $P_I < 5.9MPa$, and b) in case the passive component of ECCS should be actuated. The trend-curves of pressure $P_I(t)$ and temperature $T(t)$ are shown on Fig.2.

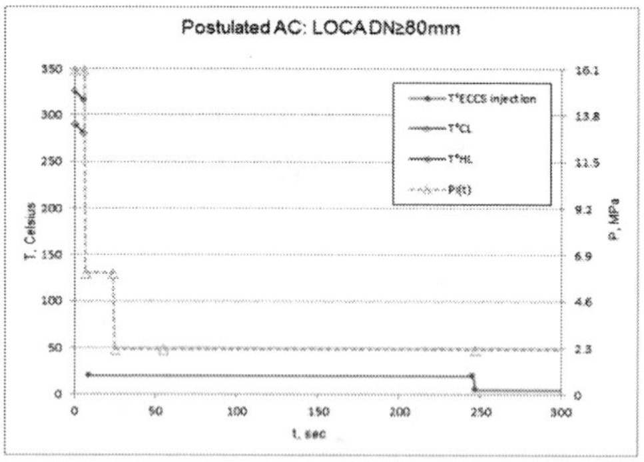

Fig. 2a. The trend-curves of pressure and temperature in LOCA: rupture of pipeline values of the coolant temperature, coming out from the RPV; values of the coolant temperature, coming in the RPV; injection: boric acid injected in the RPV (CL/HL – cold/hot legs)

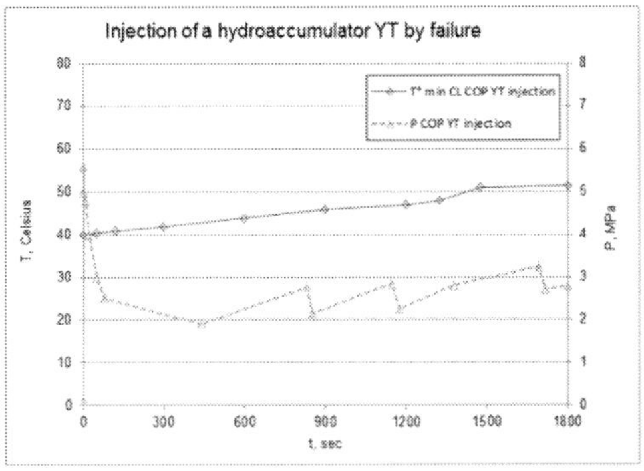

Fig. 2b. The trend-curves of temperature in EC: boric acid is wrongly injected into one of the accumulators YT. CL COP YT injection: values of the temperature of the coolant in the RPV (lower plenum chamber).

FEM of nozzle $Dn300$ is presented on Fig. 3.

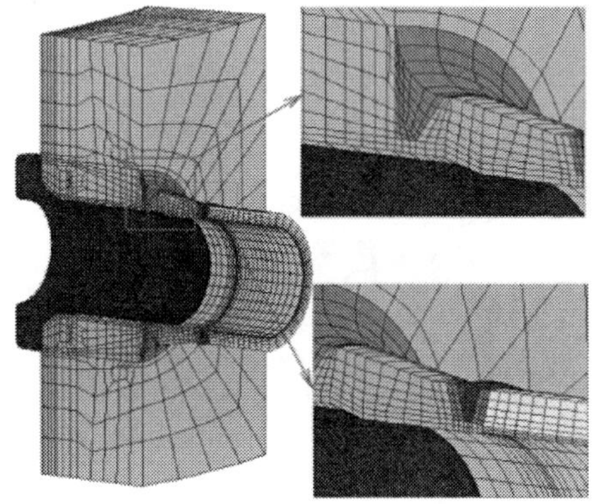

Fig. 3. FEM of nozzle $Dn300$. Ferritic welded joint (up) and bimetal welded joint (down)

The heat transfer transients of $Dn300$ nozzle FEM are shown on Fig. 4 . The initial and the boundary conditions for the thermal conductivity and structural calculations are illustrated. The process for determining the strength-stress condition of the $Dn300$ nozzle is demonstrated.

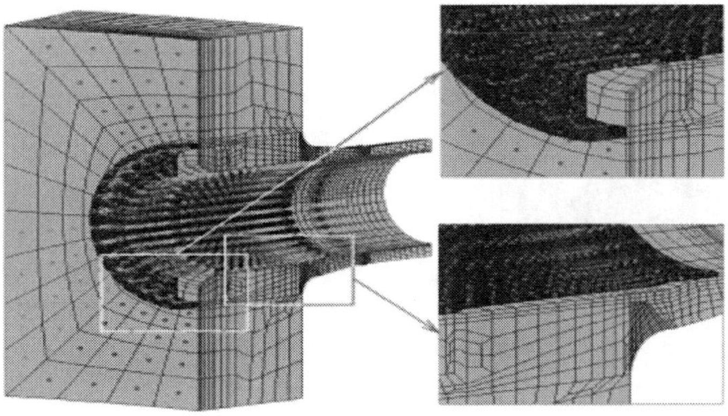

Fig. 4. $Dn300$ nozzle FEM – Boundary conditions applied for Heat transfer simulations (up) and for obtaining stress fields (down).

The stress zones with maximum stresses are received by large selection of typical transients. The assessment of stress zones was based on the analysis of the distribution of the axial (Szz) and circular (Syy) stresses due to thermal loads under LOCA with $Dn \geq 80mm$ rupture. Some typical transients were selected and the areas (cross-sections) with maximum stresses were identified, Fig.5.

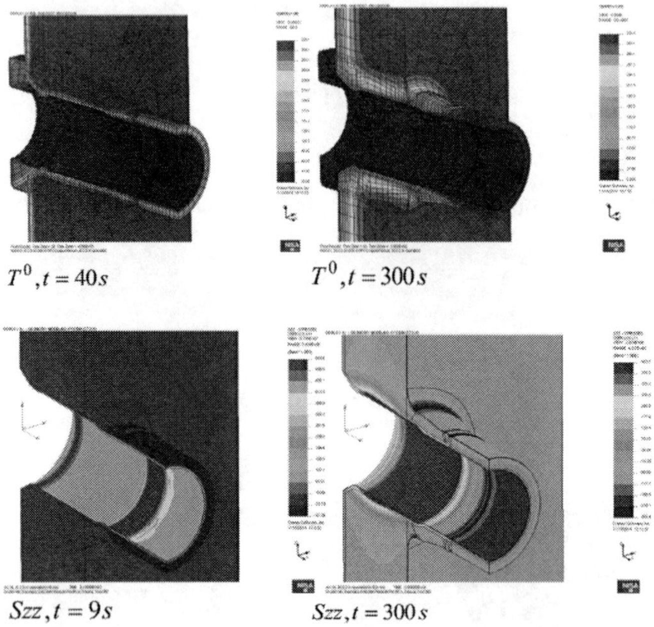

Fig. 5. Areas of ECCS Dn300 nozzle subjected to transient at LOCA $Dn \geq 80mm$. Distribution of the temperature T at $t = 40s$ and $t = 300s$ (up). Distributions of the axial (Szz) stresses due to thermal loads at $t = 9s$ and $t = 300s$ (down).

The cross-sections for which calculations provided according to [1] are shown on Fig.6. The defects position was postulated at cross-sections $1 \div 5$ and their size growth was calculated according to the Methodology [1]. The defects behaviour was assessed in terms of "worst-case scenario" for the defect's location and growth. The criterion for an admissible defect is $h/l = 0.15$,[3].

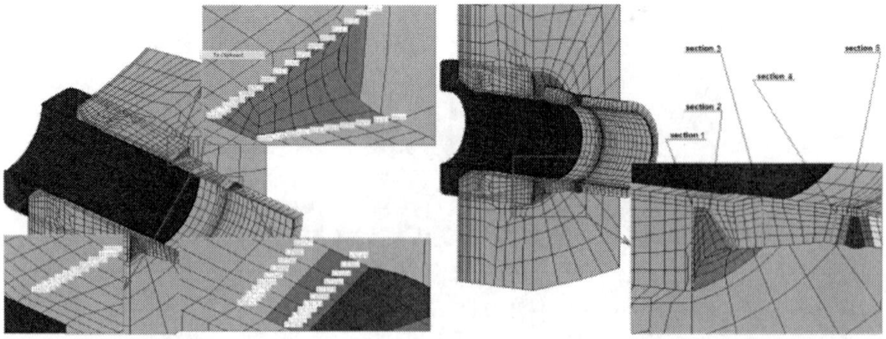

Fig. 6. Cross-sections for which calculations are provided according to [1]. Cross-section areas $1 \div 5$ with maximum axial (Szz) stress due to thermal loads.

The results demonstrate two types of "worst-case" defects: an axial defect, situated very closely to the ferritic welded joint (cross-section 2) and a circle defect, situated closely to the bimetal welded joint (cross-section 4).

Based on the calculations, the development for the postulated axial surface defect reaches values of, respectively 2.0 mm for the period of 4years and 9.4mm for the period of 30 years. The development for the postulated circle surface defect reaches respectively 0.8mm for the period of 4years and 4.4 mm for the period of 30 years.

Maximum growth for the surface defect is received for an axial defect, in ferritic cross-section 1 - defect reaches respectively 6.4mm for the period of 4 years and 26.1 mm for the period of 30 years. The results for defects' growth for a period of 4years are presented in Table 1.

Table 1. The sizes of critical defect growth after 4 years in nozzle $Dn300$, with ratio of semi-axis 0.3 and 0.7

For conditional surface defect, [mm]									
0.3					0.7				
a	$2c$	h/l	h	l	a	$2c$	h/l	h	l
6.0	40.4	0.15	6.0	40	7.7	22.0	0.35	7.7	22.0
For embedded defect, [mm]									
3.65	24.3	0.3	7.3	24.3	4.7	13.3	0.7	9.4	13.4

5 Conclusions

The research presents an approach for definition of critical zones in bimetallic welded joints on pipeline nozzles for which there is evidence to assume potential occurrence of defects after many years of operation. Presence of defects is postulated for these zones, and the critical defect sizes are calculated. The approach for the defects' critical size assessment is underpinned by regulatory requirements. Special assessments of strain-stress condition are made.

Based on the assessment of thermo-hydraulic loads, it can be possible to identify comparatively precisely the spots for non-destructive testing. The study is intended to assist in the development of the process of non-destructive testing of NPP equipment, as well as for technical diagnostics of the components.

6 References

1. VNiiAEC, *Methodology for definition of the allowable sizes of the defects in the metal of the equipment and pipelines in NPP*, **M-02-91** (Russia, 1991, rev. from 1997),
2. IAEA, *Unified Procedure for Lifetime Assessment of Components and Piping in WWER NPPs,* ***"VERLIFE Procedure"*** (Austria, 2012)
3. *Code of Strenght Calculation of Equipment and Pipelines of Nuclear Power Plant,* **PNAE G-7-002-86**, (Russia, Energoatomizdat, 1986)

Ageing management in NPP. Effectiveness of the methods for control, examination and monitoring in relation to mechanisms of degradation of mechanical properties

Galya Dimova [1,*]

[1]Kozloduy NPP EAD, ICC-DCC, 3321 Kozloduy, Bulgaria

Abstract: This paper describes ageing management activities for Long Term Operation (LTO) of components of Kozloduy NPP EAD. The stages of LTO Project are: Stage 1 - implementing a comprehensive assessment of the actual condition of the equipment; Stage 2 - providing of a Complex Analysis; Stage 3 – licensing of Plant Life Extension (PLEX) for long-term operation (LTO). The main activities are assessment of actual condition of the equipment and building; review of methods for control (non-destructive), examination and monitoring; assessment of effectiveness of methods; necessity of additional control/ examination/ service. The paper describes some mechanisms of degradation of mechanical properties, methods for control and criterions for their effectiveness.

1. Introduction

Many nuclear power plants (NPPs) around the world have been operating for over 30 years. As of October 2016, of all the 450 NPPs in operation, 79 have been operated for over 40 years, while the service life of 182 ones has exceeded 30 years, (https://www.iaea.org/pris), Fig.1. The operating design life of Nuclear Power Plants (NPPs) is normally 30-40 years, [1].

The possibilities for plant life extension (PLEX) of nuclear power plants are demonstrated through analyses, tests and lifetime management for the expected long-term operation (LTO). Thus, it is of great importance to perform assessment and upgrading of the safety measures for LTO.

The subject of this paper concerns lifetime management and plant life extension of the NPP power units. More specifically, it arises from the implementation of forthcoming PLEX activities on units 5 and 6 at Kozloduy NPP (KNPP). The design life of unit 5 expires in 2017, and of unit 6 – in 2019. Since 2010, the plant has undertaken a large-scale PLEX project for comprehensive assessment of equipment and lifetime extension of the power units 5 and 6, [2]. The PLEX project has been subdivided in the following stages:

- Stage 1 – performing of a comprehensive assessment of the current condition of equipment and buildings (ageing management review) (2010-2013), [3].

* Corresponding author: gdimova@npp.bg

➢ Stage 2 – performing of comprehensive analyses of equipment (structures, systems and components, SSCs) (2011-2015).

➢ Stage 3 – licensing by the Bulgarian Nuclear Regulatory Agency (BNRA) (licence application submitted in 2016 for unit 5).

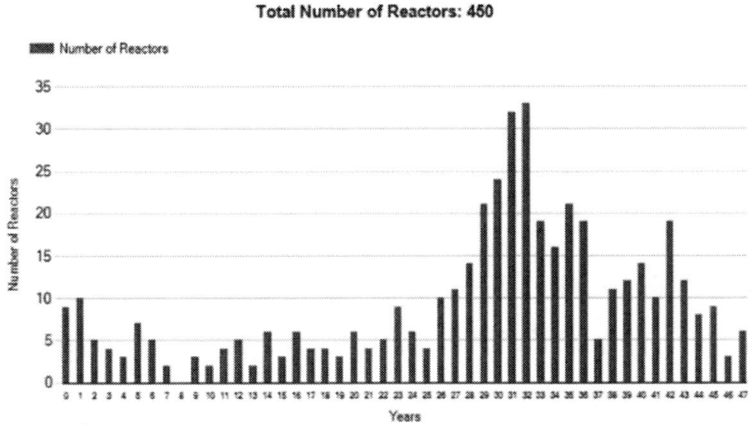

Fig. 1. Number of reactors / Years in operation.

For each power unit, additional service life of 30 years is envisaged. The standards and good practices of the International Atomic Energy Agency (IAEA) require that comprehensive activities be implemented for ageing management of SSCs, [4, 5]. The process of ageing management of components starts as early as a decision has been made for LTO, and lasts to the end of the service life of the nuclear power plant. New build NPPs are required to have ageing management in place as of the beginning of operation. The process of ageing management comprises the following stages:

➢ Selection (screening) is implemented of components important to safety and whose residual life needs to be assessed.

➢ The dominant ageing mechanisms are identified for the selected components, and evaluation follows of the existing efficient practical methods (or new ones are developed) for ageing process monitoring or mitigation.

➢ The selected components lifetime is managed through implementation of effective practices and initiatives associated with correct designing, manufacturing, storage, monitoring, surveillance, maintenance and operation.

Kozloduy NPP has in place a system of maintenance and repair used in carrying out the scheduled annual outages of equipment and pipelines, maintenance activities, condition-based maintenance and corrective (breakdown) maintenance, [6]. The maintenance and repair system serves to keep and enhance the reliability of equipment through limiting the degradation on the functioning systems, or the systems are restored to an acceptable level in order to proceed with the performance of their design functions. The methods utilised for maintenance and repair have to ensure support of the design functions assigned to the facility, and preserve its operability to the end of the service life envisaged. Evaluating the effectiveness of the maintenance and repair methods as regards the active degradation mechanisms focuses on whether they support/enhance the reliability of equipment. It is especially important for making a responsible decision for extending the operating life of the power units.

The purpose of this paper is to provide a general overview of the maintenance and repair activities, and on the manner of evaluating the effectiveness of the maintenance and repair methods. The body of this paper describes (some of the) degradation mechanisms (or the physical-chemical processes within materials) occurring under the impact of the operating conditions. Descriptions are also provided of the maintenance and repair approaches and methods used for periodic monitoring of the changes in materials' mechanical properties, and loss of resilience of the operating systems. There are no documented effectiveness criteria for the maintenance and repair methods on nuclear power plants, but a great deal of relevant operating experience has been accumulated. The facts reported in this paper regard the mechanical equipment of the main components and systems in the primary circuit.

2 Degradation of mechanical properties, maintenance and repair methods, and evaluation of their effectiveness

2.1 Corrosion-erosion processes

The corrosion-erosion processes that are typical for NPPs consist in corrosion destruction of materials followed by erosion wear caused by the flow of fluid, [7]. Factors affecting the process include the fluid composition, velocity and temperature, the component material and geometry, the active stresses, the periodicity of surface moisturising/drying. Local corrosion typically occurs in steam generators (SGs) and reactor sealing surfaces, pressurisers, and emergency core cooling systems (ECCS). Intergranular corrosion typically affects reactor pressure vessels (RPV) and steam generators; corrosion fatigue occurs in steam generators and pressurisers; stress corrosion - in reactors, SG heat exchanger tubes, and pipelines in the pressuriser system and ECCS. Stresses in the pipe bend areas will lead to a considerable change in the metal electrode potential. Tensile stresses (tensions) shift the electrode potential to the negative side, while the compressive stresses shift it to the positive side. The stretched sections act as anodes with regard to the rest of the metal and degrade most intensively. The equipment loss of function or change of its design functions is manifested through:

- ➢ Increased probability of a leak upon break of pipelines caused by reduced pipe-wall thickness.
- ➢ Elevated frequency of failures due to stress changes within the materials of the systems.
- ➢ Occurrence of abnormal operating conditions as a result of loss of tightness.
- ➢ Occurrence of abnormal operation of the heat exchangers resulting from heat exchanging tubes rupture.
- ➢ Deterioration of heat exchange in case of presence of sludge/deposits.
- ➢ Elevated radioactivity levels due to activation of the corrosion products.

The maintenance and repair activities comprise: 1) water chemistry measures implementation and maintenance; 2) performing a corrosion survey in search for the presence of sludge/deposits; 3) conducting of in-service inspection (ISI) through visual, dye penetrant, eddy current, and mechanical tests, ultrasonic thickness measurement of the pipe bends, bended sections, or at the outside formation of pipe bends. The effectiveness of the maintenance and repair methods regarding corrosion-erosion damage appears in terms of the following two focus areas: 1) water chemistry control (water chemistry metrics shall correspond to the normative values; all deviations from the norms shall be corrected within the timelines adopted; violations of water chemistry norms such that may lead to the power unit abnormal operation, shall be prevented; 2) conduct of non-destructive testing combined with timely finding and repairing of any discontinuities in the metal and welded joints, or wall thinning, to prevent failure of the element or avoid degradation of its functions.

2.2 Neutron embrittlement

Neutron embrittlement of the reactor pressure vessel metal occurs as a result of intensive neutron flux under high temperature and pressure conditions. Upon reacting with the substance, neutrons may either transfer their energy to atoms through elastic impacts, or serve as the source of charged particles formation, [8]. Such processes will impair the correct position of atoms within the metal crystal grid and this will cause defects formation. The density of radiation defects is subject to the type of radiation, its parameters and on the nuclear-physical characteristics of the material. The spot defects that occur - vacancies, internodal atoms, embedded atoms, etc., at sufficiently high temperatures can recombine, migrate to volumetric or surface leakages (dislocations, grain boundaries), formation of radiation stacking faults in the shape of pores and dislocation nodes. The irradiation of metal with fast neutrons results in microscopic areas of structural damages with high concentration of spot defects. Neutron embrittlement affects the reactor pressure vessel, barrel and core baffle, and reactor guard-tube bank. The influencing factors include the neutron flux values and direction, and the material chemical composition. The elevated levels of nickel (Ni) and manganese (Mn) in RPV steel grades enhance embrittlement due to the formation of Ni-Mn-Si clusters (dislocation nodes), while silicon (Si) reduces embrittlement. Fig. 2 shows microscope photos of samples with varying weight percentage of Ni (sample parts $0{,}22Cu - xNi - 1{,}6Mn$), subjected to neutron flux irradiation, with neutron energy exceeding $1{,}6MeV$ at temperature of $T = 290 \div 310^0C$, [9].

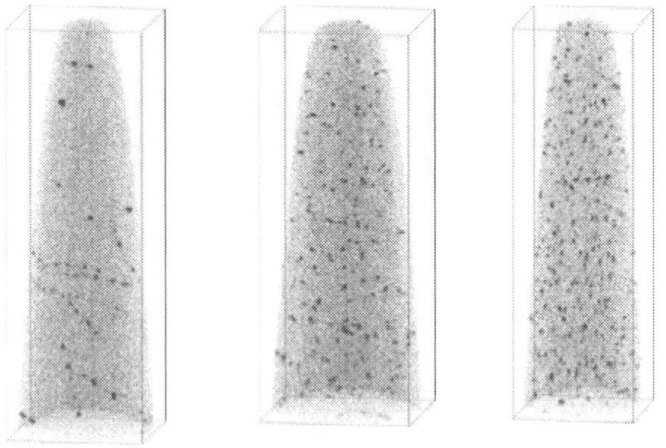

$0{,}02\%Ni$ $0{,}82\%Ni$ $1{,}59\%Ni$

Fig. 2. Dislocations forming in samples with varying weight percentage of Ni $(0{,}22Cu - xNi - 1{,}6Mn)$ at temperature of $T = 290 \div 310^0C$.

The neutron embrittlement of ferritic steels is manifested as increase of ΔT_{kF} of the critical temperature value of radiation embrittlement, T_κ, [8, 9]. Neutron embrittlement elevates the probability of brittle fracture of metals, especially of the RPV welded joints located opposite the core; it also increases the probability for changes occurring in the shape of components (core barrel and baffle). The maintenance and repair preventive measures include:

- Use of low-leak schemes of core refuelling.
- Calculating the neutron flux values following each fuel cycle.

For 15Kh2HMFAA steel types, the neutron flux limit is $5{,}7x10^{22} n / m^2$ for the RPV metal, and this is a limiting factor for the facilities' safe operation. Within certain time intervals, calculations and analyses are performed of the critical temperature of irradiation embrittlement $T_к$ and the mechanical characteristics are studied of units 5 and 6 RPV surveillance specimens. In-service inspection of metal is performed using visual, dye penetrant, and ultrasonic tests, [10]. Periodic thermohydraulic and strength analyses are conducted in order to assess the defects' propagation resistance. The maintenance and repair efficiency criteria are required to exclude the probability of RPV metal brittle fracture as a consequence of radiation embrittlement, or due to the presence of a defect. The radiation embrittlement critical temperature variation ΔT_{kF} needs to be within the design limits. Any discontinuities of metal or welded joints need to be detected and repaired in a timely fashion in a way such as not to lead to a failure of the element, or affect any of its functions.

2.3 Cyclic fatigue of metals

Degradation under the impact of cyclic (fatigue) loads occurs due to the action of variable stresses and deformations; gradually, changes of metal accrue in a way such that leads to change of properties, formation of cracks, their development resulting in the eventual degradation of the object. The fatigue degradations of metal of NPP facilities can be high-cycle fatigue and low-cycle fatigue, as well as thermal fatigue occurring due to the cyclic changes of the thermal field.

Cyclic fatigue affects reactors, steam generators, pressurisers, pipelines of the pressuriser system; main circulation pipelines and emergency core cooling systems. Corrosion fatigue can also be observed and it affects steam generators and pressurisers. The value and direction of the effective stresses, and the number of actually implemented cycles compared with the design ones are factors that influence fatigue. The growing number of cycles enhances the probability of metal degradation by fatigue. Cyclic fatigue preventive measures comprise:

- Regarding the RPV metal - analysis of surveillance specimens, [11].
- Performing of in-service inspection of metal using visual, dye penetrant, eddy current, and ultrasonic tests.
- Conduct of hydraulic testing.

The maintenance and repair effectiveness criteria require that:

- For each facility, the load cycles should be registered and controlled under different design modes, in a way such that no failure of the element or degradation of its functions occurs.
- Strength analyses should be conducted.
- Any discontinuities in metal or welded joints should be found and repaired in due time.

2.4 Thermal embrittlement

Materials feature different ageing mechanisms. In metal alloys, ageing depends on the degradation of saturated solid solutions. Thermal embrittlement affects reactors, steam generators, pressurisers, pipelines of the pressuriser system; main circulation pipelines and pipelines of the emergency core cooling systems. Thermal embrittlement is affected by factors such as the working temperature values and the time (duration) of the facility's operation. Fig. 3 illustrates the dependence of the critical temperature ΔTk(t), on the time of operation t for two number of units - a and b. Peak values of the critical temperature ΔTk(t) are observed with regard to the time period of the $3 \div 5$ years of the facility operation; after the elapse of this period, the impact on embrittlement of the thermal component decreases while the neutron one increases.

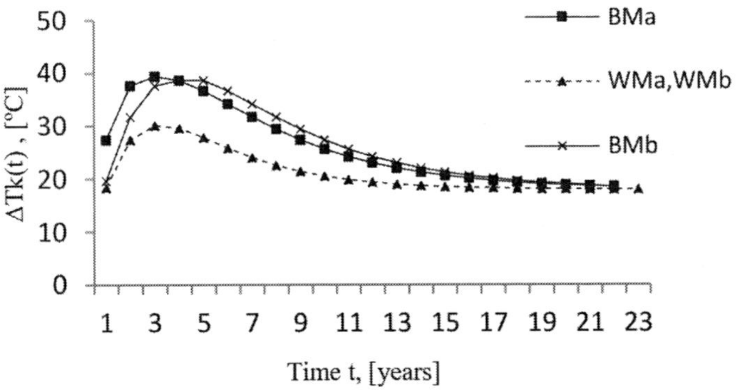

Fig. 3. Dependency of $\Delta T_K(t)$ on time t (or thermal inducted part of ΔT_K) for base metal, BMa and BMb, and weld metal, WMa and WMb.

The maintenance and repair preventive measures cover:

- Regarding the RPV metal, the mechanical properties of the surveillance specimens have to be studied.
- In-service inspection (ISI) of metal, using visual, radiographic and metallographic tests, needs to be undertaken.
- Hydraulic tests have to be performed on pressure vessels.

Thermal embrittlement increases the probability of the facilities brittle fracture. The maintenance and repair effectiveness criteria require that:

- The metals mechanical properties be retained.
- Any discontinuities of metal or welded joints need to be detected and repaired in a timely fashion, in a way such as not to lead to a failure of the element or affect any of its functions.

2.5 Wear

Multiple studies have demonstrated that the process of gradual loss of functionality of components in operation can be subdivided in three stages: 1) stage of adjustment, 2) normal operation stage, and 3) wear, caused by the facilities normal operation, Fig. 4, [7].

Throughout the stage of adjustment (1) mutual changes occur in the macro- and microgeometry of the working faces, and products of wear and oxidation are formed. The working faces wear rather intensively during this stage. Gradually, wear weakens and stabilises to a stage of normal operation wear (2). This stage (2) overlaps with the power unit design life. Once an energy limit has been exceeded the oversaturated solid solutions start to disintegrate, the wear value progressively increases (3), the components functioning deteriorates and the need of repair arises. Stage (3) usually occurs towards the end of the design life of the facilities.

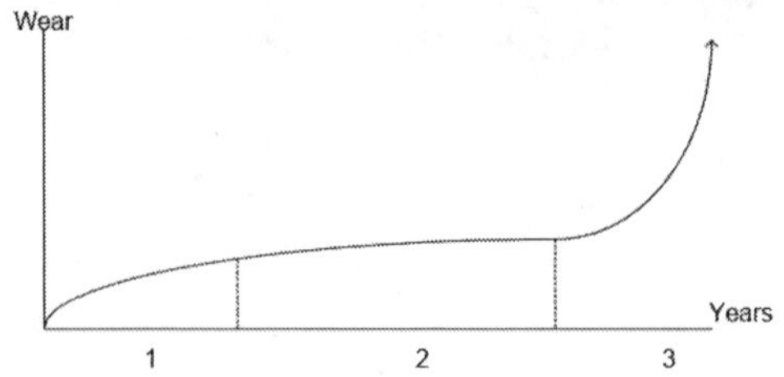

Fig. 4. Stages of wear in the course of the equipment operation, Stage (1) – wear during adjustment, Stage (2) - wear during normal operation, Stage (3) – wear in the end of the facility operating lifetime.

The process of wear affects hydraulic snubbers, sealing faces, fixing elements, internal parts of cylindrical vessels and pipelines. The rate of wear is affected by the following factors:

- Combination of materials for the working surfaces.
- Physical, chemical and mechanical properties of the surfaces subjected to friction.
- Interactions of the working surfaces with the environment.
- Clean processing of the friction surfaces.
- Type of friction (dry, border, semi liquid, liquid).
- Values of the normal pressure and the velocity of working surfaces one against the other.

Of the large number of wear types concerning the working surfaces of machine parts, major importance is attached to abrasive wear in the presence of grease, because the wear products that invariably arise from the machine components friction are oxidised and turn into a sort of abrasive materials. For instance, under normal operating conditions, the hydraulic snubbers components are subjected to negligible loads, and the hydraulic snubber may stop performing its functions. With degradation due to wear, a hydraulic snubber is incapable of performing its protective functions in case of strong vibrations, or an abrupt displacement of equipment caused by seismic loads. In the course of operation, wear affects pins (their cylindrical part) and pin sockets. The mechanical properties degradation deteriorates the fixing function and there exists a possibility that the facility be untight. Preventive measures in this case comprise visual inspection during maintenance. The criterion for effectiveness of the maintenance and repair methods in case of wear consists in preventing wear processes to a degree that may affect the equipment functionality.

3 Maintenance and repair programmes for NPP components and systems

The facilities and components of an NPP are covered in programmes for maintenance and repairs. They envisage activities for equipment control, testing and monitoring; the control and monitoring methods focus on prevention of the consequences from inherent degradation mechanisms on the metal mechanical properties. The programmes are usually developed in accordance with Russian rules and norms for safe operation of NPP facilities and pipelines, [12] and with type control programmes, [13]. The data obtained as a result of implementing the maintenance and repair programmes serve for assessment of the current state of components and equipment, and for operational purposes - to allow the facility to be returned to service; as well as for general purposes - implementing of design modification of the facility, modernisations, or for lifetime extension. Irreplaceable equipment and a great part of the replaceable equipment are subject to technical condition assessment and ageing management for further (long-term) operation.

The first stage of the PLEX project consists of a comprehensive assessment of the current condition (ageing management review) of the equipment, [11]. Table 1 shows monitored ageing mechanisms and respective parameters that determine the metal condition. On the basis of the key parameter a criterion is derived for assessment of the facility condition.

Table 1. Metal ageing mechanisms and parameters characterising its condition.

Monitored metal ageing mechanism	Key parameter characterising metal condition
Mechanical properties change	Strength limit, creep limit, relative elongation, relative shrinking, hardness number, KCV (impact toughness)
Structural changes	Content of phases, grains and micropores
Cracking	Geometrical dimensions of discontinuities
Erosion	Damaged area; wall thickness
Pitting	Number of damages per unit area and depth of damages

During stage 1 of the units 5 and 6 PLEX project an assessment was made of the effectiveness of maintenance and repair methods. The metal condition is evaluated by considering the resistance to the degradation mechanisms of mechanical properties typical for NPPs. During the second stage of the PLEX project, analyses are performed of the structural integrity of facilities. Fig. 5 demonstrates the complete process of comprehensive engineering assessment of components and equipment of KNPP units 5 and 6. The assessments input data are derived from equipment passport data, design requirements, discontinuities data, defects, failure data, accidents, operating conditions, ambient conditions and cycles of work. Analyses are generated for static and dynamic strength and seismic stability.

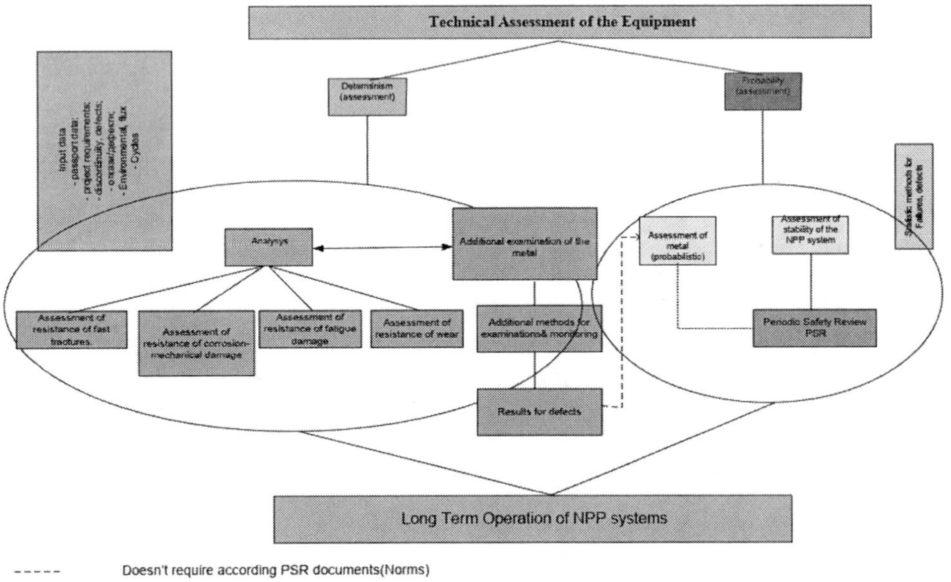

Fig. 5. Lifetime extension process for NPP components and facilities

The analyses should contain sufficient data to allow drawing of conclusions on the residual lifetime of the respective facility.

For all the main types of facilities, analyses need to be performed selecting the circulation loop or facility with the highest load. Based on the comprehensive condition assessments issued at PLEX project Stage 1, and the analyses completed during the project Stage 2, conclusions have been drawn on the facilities' residual lifetime, and final reports issued on the lifetime extension feasibility.

Following the analyses and before making the final decision for lifetime extension of the units, assessments should be conducted on the operational safety, and the probabilistic safety analysis (PSA) updated. The nuclear power plant performs a self assessment of safety, while the LTO licence is issued by the BNRA.

5 Conclusions

The PLEX project first and second stages are now complete for unit 5, while Stage 2 for unit 6 is currently under way. The maintenance and repair methods effectiveness has been assessed. The procedures and scope of control, testing and monitoring in place allow timely identification and recovery of defects that may lead to a failure of the element, or affect its operability. The maintenance and repair system ensures the required level of reliability and operability of elements. Before LTO, however, some components necessitate a further survey (the QF system buried pipelines, fire protection pipelines of the UJ system, supports, hydraulic snubbers, ventilation stack, reactor refuelling machine, etc.). For the equipment that is not part of the standard maintenance and repair programmes, additional programmes for a single-time metal condition survey are being prepared.

In documents such as conclusions on the condition of equipment, or final reports on PLEX, recommendations are provided for certain components to ensure future safe operation of the facility during the LTO period. Measures such as those listed below are being planned:

➢ Conduct of regular periodic control.

➢ Conduct of a single-time control of the mechanical characteristics after, for example, 10 years of operation. Such a measure is envisaged in case some of the operational load modes of the facility (hydraulic tests, or an emergency situation) cause the stresses to exceed the limits permitted.

6 References

1. R. Krivanek, J. Fiedler, Nucl Eng Des Journal, **316** (2017)

2. OAO OKB Gidropress, *General programme for comprehensive assessment of the actual condition and assessment of the residual lifetime of SSCs at Kozloduy NPP unit 5, within the design range of OAO OKB Gidropress* (Russia, 2012)

3. OAO OKB Gidropress, *Report on the results from the comprehensive assessment of the actual condition and assessment of the residual lifetime of equipment and pipelines of the reactor unit at Kozloduy NPP unit 5, within the design range of OAO OKB Gidropress*, (Russia, 2013)

4. IAEA, *IAEA Safety Standards Series* **NS-G-2.12** (Austria, 2009)

5. IAEA, *Ageing Management for Nuclear Power Plant: International Generic Ageing Lessons Learned, IGALL, Safety Reports Series* **82** (Austria, 2015)

6. KNPP, *Long-term schedule for preventive maintenance and repair of SSCs belonging to the safety systems and systems important to safety at units 5 and 6 and the common plant facilities*; **08.30.OCO.00.ГР.039/1** (Bulgaria, 2008)

7. V. Ostreykovskiy, *Materials ageing in nuclear energy* (Russia, 2000)

8. M. Georgiev, *Cracking resistance of metals under static loads* (Bulgaria, 2005)

9. A. Nomoto, *Understanding on the Mechanisms of Irradiation Embrittlement of RPV Steels and Development of Embrittlement Correlation Method*; Central Research Institute of Electric Power Industry (CRIEPI) (Japan, 2014)

10. KNPP, *ISI Procedure for base metal, welded-on surfaces and welded joints in equipment and components of WWER-1000 reactor types at Kozloduy NPP* (Bulgaria, 2016)

11. Kurchatovskij Institute, Close-out Report on Contract No 292000008, *Testing and study of surveillance specimens set 3L of unit 5 RPV, and set 2L of unit 6 RPV at Kozloduy NPP EAD, Stage I; 3L at unit 5* (Russia, 2015)

12. *НП–084-15 Federal norms and regulation for use of atomic energy. Regulations on the in-service inspection of base metal, welded joints and welded-on surfaces of equipment, pipelines and other elements of nuclear power plants* (Russia, 2015)

13. *Type programme for ISI of the condition of base metal and welded joints of equipment and pipelines at NPPs with WWER-1000* (Russia, 2009)

Improving of the water preparation systems in the industry thermal power plants

Daniela Choshnova [1,*]

[1] UCTM, Kliment Ohridski blvd.8, Sofia, 1756 Bulgaria

Abstract. Some modern possibilities for modernization of the water preparation in the heat power plants of metallurgical enterprises are presented. It is focused on the ecological and technological advantages of the methods for obtaining of ultra-pure water. This work is not intended to justify the need of usage of particular technology in an object. This is a separate task requiring in-depth consideration.

1 Introduction

Natural water can not be used in modern technological processes without pre-treatment. It needs to be softened and demineralized by ion exchange treatment in the industrial unit in heat power plants. In this way the necessary quality of the water circulating in the energy boilers is achieved, also the produced steam for technological needs is achieved. Some negative phenomena are preventing. They are causing:

- deposition of salts on the heating surfaces of the steam generators;
- deposition of salts on the blades, nozzles and guides of the steam turbines, which creates a risk of imbalance of the impeller, narrowing of the space between the blades, which leads to decreasing of the aggregates power;
- decreasing the flow rate of the steam from steam generators, respectively their productivity;
- creating conditions for overheating the heat exchange surfaces of the steam generators, etc .;

Despite the extensive researches in creating new generations resins and their usage in the existing ion-exchange installations, there are some significant drawbacks. For example, low hydrophilicity, which result in low diffusion rates of ions inside and low sorption rates and desorption. In practice, they are used in granular form whose bonding in the ion exchange equipment during the sorption process leads to the need for forced rupture and their gradual mechanical degradation during the operation. In addition, they require frequent regeneration to restore their ion-exchange capability.

Special procedures are required for the preparation of regeneration solutions, their storage and the protection of the personnel against possible leaks. The cost of reagents, as a rule, exceeds 2-3 times the stoichiometric quantities. Accordingly as many times, waste products (strongly acidic and strongly alkaline industrial waters) are increased, which must

* Corresponding author: daniela@uctm.edu

be neutralized before being discharged into the sewerage system. In the end, all this is reflected in large capital and operating costs.

For solving the above-mentioned technological problems, if possible, a series of apparatuses or their modified models are used, but recently they are created with original constructions for a particular process. A large number of variants for countercurrent regeneration with different ionites have been investigated, and according to the authors [1], with industrial applications are those in which a continuous layer is provided.

Article [2] presents a method for improving the control during regeneration of the filters in a chemical water treatment section in a heat power plant of a metallurgical plant. It reduces the consumption of sulfuric acid and sodium hydroxide and allows high-quality desalted water to be produced by minimizing the side unfavourable processes of steam generators.

The report in [3] is dedicated to the advantage of conjugated regeneration, which is an inexpensive option for decreasing reagent costs and reducing the waste water pollution. In [4] are presented results from comparative study of different schemes for combination of membrane and ion exchange technologies applied in Russian heat power plants. The impact of various factors (composition of water, type of equipment, etc.) on the sustainability of the end-product indicators is examined. Detailed analysis of reagent and power consumption for each installation is given.

In general, a number of complementary factors - reliability, environmental friendliness, capital and operating costs - have to be respected when choosing a water conditioning modernization scheme. The literature presents models that allow to evaluate other external factors that influence the efficient operation of ion-exchange treatment of water - the flow composition; a pre-cleaning method, etc. Separate publications on comparing the performance of different technologies, though fairly objective, carry the characteristics of a particular case and relate to the specific operating conditions.

This work is not intended to justify the need to use a particular technology in an object. This is a separate task requiring in-depth consideration. Here it is resented some modern opportunities for modernization of water treatment in the heat power plants of industrial enterprises.

2 Technological improvement of existing water treatment systems in industrial thermal power plants

The requirements for the quality of the treated water in the chemical treatment facilities for the heat power plants are shown in Table 1.

Table 1. Requirements for the quality of feed water for steam generators in industrial heat power plants

Parameter	Permissible norm
Specific conductivity	< 0,2 μSm/cm
Hardness	<5mg-eq/*l*
Sodium concentration	3-10mg/*l*
Silicone acid concentration	<0,2mg/*l*
Concentration of chlorides	<3mg/*l*
Concentration of organic carbon	100-300mg/*l*
Concentration of sulfates	<3mg/*l*

Conventional ion exchange systems shown in Fig.1 are not as complicated as construction and are reliable in operation. However, as mentioned, all their disadvantages are related to the need to use large quantities of ionic mass recovery reagents. Regenerative solutions are aggressive and contain components that go beyond environmental standards, and the process itself is not automated.

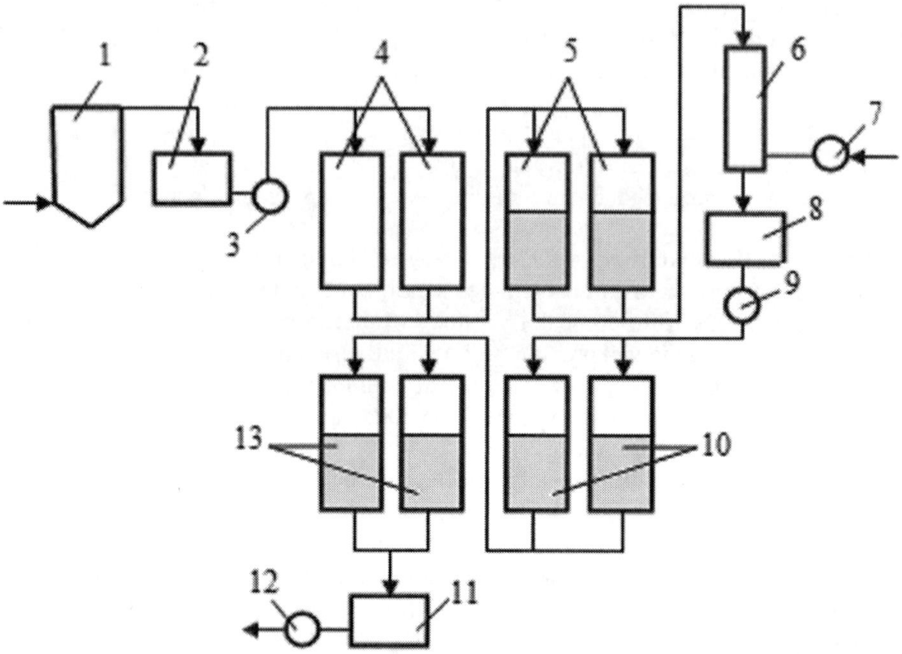

Fig. 1. Conventional ion-exchange installation: 1 - reactor-precipitator; 2 - clear water tank; 3 - clear water pump; 4 – sand filters; 5 - strong cationite filters; 6 - deaerator; 7 - air fan; 8 - cationic water tank; 9 - cationic water pump; 10 - weak anionic filters; 11 – demineralized water tank; - 12 - demineralized water pump; 13 - strong anionic filters

In some heat power plants these acidic and alkaline waters are mutually neutralized in the tanks in which they are mixed. Often, however, in the ion exchange plants the regenerations of the two types of filters do not take place simultaneously. In addition, sulfuric acid streams contain larger amounts of acid than the sodium hydroxide in the alkaline. For this reason, it is necessary to neutralize the waste water, which is collected in a common vessel and treated with a reagent [3]. A stock of chemicals is also needed, which is accompanied by additional costs.

In this connection, there is a need for increased control and monitoring of spent water, improvement of existing technologies or replacement with new more efficient ones.

Modern alternatives for ion exchange are ultrafiltration using mix-bad filters and compact layer membrane technologies including reverse osmosis and / or electro-deionization. They allow to improve the quality of treated water with minimal reagent consumption.

2.1 Installation of mix–bed filters after the main scheme

In Fig. 2 is a modernization scheme comprising installing mix-bed filters following the basic ion exchange treatment scheme. Deep downstream processing in them allows to achieve a specific conductivity below 0.3μS/cm and a silicate content of less than 0.02ppm.

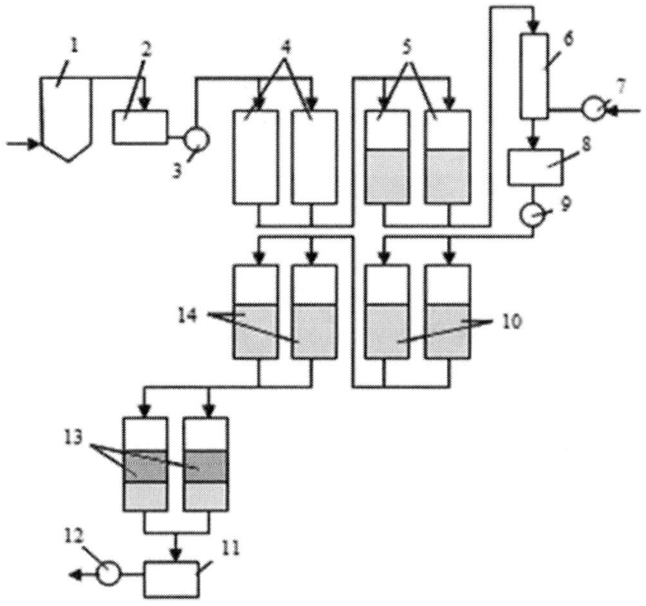

Fig. 2. Ion-exchange installation with purification of the technological flow in mix-bed filters: 1 - reactor-precipitator; 2 - clear water tank; 3 - clear water pump; 4 – sand filters; 5 - strong cationite filters; 6 - deaerator; 7 - air fan; 8 - cationic water tank; 9 - cationic water pump; 10 - weak anionic filter; 11 - ultra demineralized water tanks; 12 - ultra demineralized water pump; 13 - mix-bed filters; 14 - strong anionic filters

2.2 Final purification by electro-deionization of the stream

The electro-deionization is a highly efficient technology for ultrafiltration with 90-95% efficiency and features: constant quality of the filtrate, obtaining of ultrapure water with conductivity below 0.1µS/cm, silica content below 0.02ppm. The process is continuous and fully automated. The system is operated with a constant electric field and consists of ion-selective membranes located in a compact unit which is installed after the main scheme of the ion exchange system as shown in Fig.3.

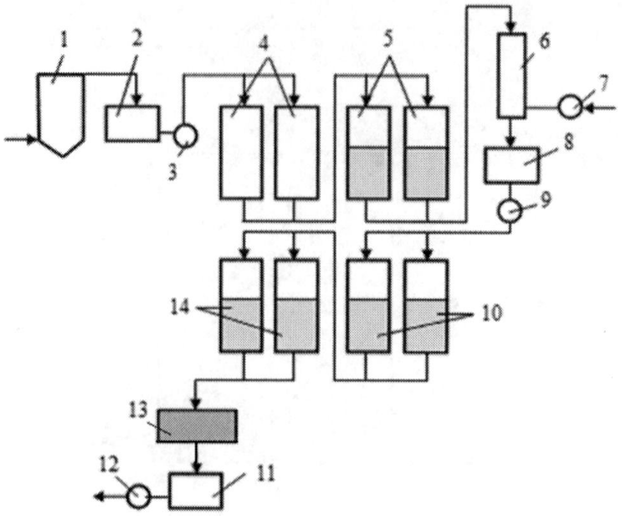

Fig. 3. Scheme of an ion-exchange installation with purification of the process stream by electro-deionization: 1 - reactor-precipitator; 2 - clear water tank; 3 - clear water pump; 4 – sand filters; 5 - strong cationite filters; 6 - deaerator; 7 - air fan; 8-cationic water tank; 9 - cationic water pump; 10 - weak anionic filter; 11 - ultra demineralized water tanks; 12 - ultra demineralized water pump; 13 – electro-deionization system; 14 - strong anionic filters

2.3 Combined installations with reverse osmosis membranes

Water purification by reverse osmosis systems involves the forced passage of the treated water through a semipermeable membrane. In most cases they are not universal. Typically, the scheme is developed taking into account a number of factors, including the business itself, the required volumes of purified water, and so on [6].

High purified water can be obtained by two-stage osmosis, in which the water stream is purified from salts to 97% and a specific electrical conductivity of 1-5 μSm/cm before final purification in mix-bed filters or electro-deionization module (Fig.4).

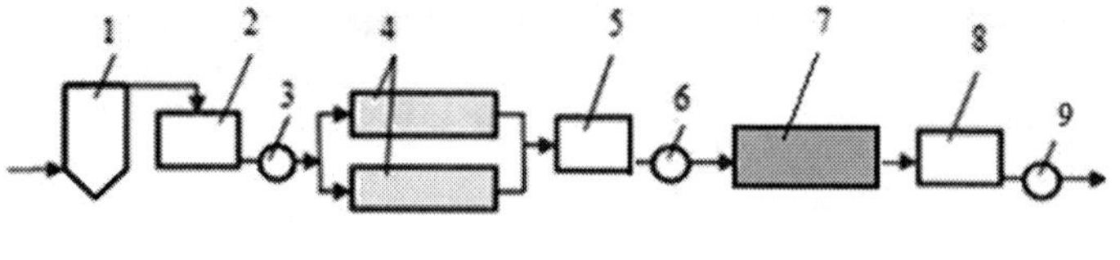

Fig.4. Scheme of an ion-exchange installation with reverse osmosis combined with electro-deionization system or mix-bed filters:1 - reactor-precipitator; 2 - clear water tank; 3 - clear water pump; 4 - reverse osmosis membranes; 5 - partial desalinated water tank; 6 - partial desalinated water pump; 7 – electro-deionization system or mix-bed filter; 8 – ultra demineralized water tank; 9 - ultra demineralized water pump

The advantages of membrane technologies based on reverse osmosis have been thoroughly studied. In [4,5] there are models for optimizing the consumption of chemicals and labor by adding a reverse osmosis module to existing installations. In [4] are presented 4 variants of a combined scheme, which are applied to a water flow with different mineralization. According to expert data, if raw water is highly mineralized and there is an increased consumption of sodium, a technological solution is the use of membrane technologies including reverse osmosis and purification by electro-deionization. But with increased electricity consumption and low-mineralized raw water the technological solution is reverse osmosis and mix bed filters. When the salt content in the raw water is variable, the second scheme is preferable.

3 Analysis of the different options

Conventional ion-exchange water treatment schemes for technological purposes are used in systems with high productivity and low salt content of raw water. When changing its composition and the requirements for steam production, the choice of a suitable method for its improvement is influenced by the following factors:

- expenditure of treated water in the power plant as a whole;
- way of removing and neutralizing waste streams;
- value of reagents for recovery of ion-exchange resins (mainly sodium base), incl. transport and storage costs;

Electro-deionization (EDI) is the newest method that can be successfully applied to heat power plants of industrial plants. Despite the high cost of equipment, it has some technological advantages that can not to be ignored:

- at low salt content of the water flow in the installation does not require preliminary treatment;
- uninterrupted work, with no need to use chemicals;
- it maintains constant water conductivity below 0.1 μS/cm;
- it requires low energy consumption for system performance and less labor input;
- equipment compactness;
- complete automation and full control of the process;
- safety for personnel and the environment.

The advantages of the EDI system to mix-bed filters are:

- there is no cost for regeneration chemicals and water for deionization and recovery;
- no waste treatment costs, incl. building special equipment for this;
- complete automation and process continuity.

According to the data from the practice, in a heat power plant at a metallurgical plant for the production of non-ferrous metals in our country, the costs for construction of the schemes presented in Figs. 2 and 3 are comparable (at the same flow rate of treated water). But in Fig. 2 contaminated waters are eliminated, neutralization of which requires the construction of neutralization facilities. Furthermore, as mentioned, costs for acid and base for regeneration of resins in mix bed filters, their servicing and maintenance are needed.

The problem of ensuring the fault-free operation of the installation comprising an electrodeionization module consists in its sensitivity to all sorts of disturbances in the operation of the precedent filters. This can leads to a deterioration in the quality of the water supplied to it. A basic requirement is that the residual iron content in inlet EDI modules water is below 0.01ppm, because it blocks the ion-exchange effect of the resins in it. Once deposited in them it isn't possible to be removed. When it is used in combination with reverse osmosis membranes and a highly mineralized primary stream, there is a possibility of accumulation of organic and mineral deposits polluting the intermembered spaces in the individual elements.

The combined schemes shown in Figs. 2-4 can be used for the reconstruction of existing installations where ion-exchange filters, collection and neutralization systems already exist. In their implementation, they will decrease, respectively eliminates (in variant reverse osmosis-electro deionization) the amount of waste streams and the reagent consumption.

The main components of industrial reverse osmosis systems include pre-mechanical water purification filters and a high pressure pump prior feeding the membranes. Due to the probability of its contamination, a dispenser for anti-caking agents and biocides should also be included to prevent organic deposition.

The economic efficiency of water treatment using membrane technologies is substantially higher than conventional ion-exchange installations. Significantly increased operational reliability, lower labor costs and reagents, and the role of human factor due to automation of the process.

For the successful realization of projects for modernization of ion exchange systems, it is necessary to choose the right equipment (pumps, valves and apparatus, control and measuring devices).

4 Conclusion

Obtaining of ultra-demineralized water in power plants is one of the main tasks in steam production for technological needs. Although the ion exchange installations have been in operation for a long time, their modernization is necessary. This will leads to decreasing of capital and operating costs increasing of the ecological efficiency, incl.

- decreasing the technological time for putting the equipment into operation;
- reduction of the environmental risk in servicing the waste streams neutralization system;
- decreasing the cost of resin regeneration reagents of the filters;
- improvement of working conditions;
- savings for servicing and repair due to automation of the process.

The presented article would be useful for designing of new or for reconstruction of existing schemes for water preparation.

Referenses

1. B. Riabchikov, A. Sibirev et al., J. En. and Wat. Prep. **5** , 43 (2006).
2. D. Choshnova, J. Sci. Proc. XXIV **1,** 187, 104 (2016).
3. D. Choshnova, J. Sci. Proc. XXIV **1,** 187, 107 (2016).
4. P. Surdilovski, N. Noev, P. Morin, J. En. and Wat. Prep. **2,** 58 (2009).
5. A. Nikitenko, J. En. and Wat. Prep. **5** , 61 (2009).
6. http://www.engineering-review.bg/

Bayesian reconstruction of images of objects with high-density inclusions with suppression of artifacts

Sergei Zolotarev[1], *Valery* Vengrinovich[1], *Mohsen* Mirzavand[2], *Mieteeg* Mukhtar [2] and *Ivan* Georgiev [3, 4,*]

[1]National Academy of Sciences of Belarus, Institute of Applied Physics, Minsk, Belarus
[2]Belarusian National Technical University, Minsk, Belarus
[3]Institute of Information and Communications Technologies, Bulgarian Academy of Sciences, Sofia, Bulgaria
[4]Instituteof Mathematics and Informatics, Bulgarian Academy of Sciences, Sofia, Bulgaria

Abstract. The technology of three-dimensional Bayesian tomographic reconstruction of homogeneous objects with high-density inclusions is developed. The approach is based on preliminary correction of projections by extracting the data corresponding to X-rays passing through a high-density region, and replacing it with synthesized data obtained by two-dimensional interpolation. An original method for selecting interpolation points is proposed and a mathematical algorithm is described that ensures the implementation of two-dimensional interpolation correction of projections.

1 Introduction

There is a wide class of problems of X-ray tomography, solutions of which are piecewise-constant functions and, accordingly, of industrial objects leading to the formulation of such tasks. Such control objects include, for example, so-called piecewise-homogeneous structures, objects or systems, which have different homogeneous parts with constant coefficients of linear attenuation. The boundaries between the heterogeneous parts of complex industrial products are usually sharp, without a smooth transition from one constructive part to another. Such heterogeneous media with a piecewise homogeneous structure can be adequately described by precisely defining the boundary surfaces that separate the homogeneous parts of such media.

We will consider control objects that have local phase inclusions inside, and we will assume that within each phase, the coefficient of linear attenuation of X-ray radiation is fairly homogeneous, and its numerical value differs significantly from the coefficient of linear attenuation of the main material, which, in general, can be heterophase structure. The number of local phase inclusions inside the main material of the object will be assumed to be arbitrary, but, of course, finite. We will also assume that the values of the attenuation

* Corresponding author: ivan.georgiev@parallel.bas.bg

coefficients of the desired phase inclusions may differ from each other, but they are all different from the coefficients of linear attenuation of the limiting material that limits them. At tomographic reconstruction of such industrial objects, for example, plastic products, inside which various small metal inserts are pressed, there are strong artifacts that create strong interference to the study of the internal structure and geometric configuration of the components of the product.

The main reasons for these artifacts are: 1) the effect of Beam Hardening of X-rays; 2) X-Ray scattering; and 3) a high level of noise component in the projection data and others. One of the known approaches is the method of preliminary correction of projections by extracting data corresponding to X-rays passing through a metal implant and replacing them with synthesized data obtained by linear or polynomial interpolation [1-4]. Next, we will briefly describe the technology of identifying homogeneous objects proposed by us, for example, the control of products made on 3D printers and considered a simple, but an effective enough way to overcome the above difficulties, which turned out to be useful for solving other problems.

Moreover, an original method of interpolation points selection will be offered and a mathematical algorithm providing implementation of interpolation projection correction will be described in steps.

2 Theory

2.1 The heuristic method of statistical iterative reconstruction

We develop statistical iterative algorithms of reconstruction using minimize total variations and OpenGL. Iterative methods for tomography based on solving a system of linear algebraic equations

$$Ax = p \tag{1}$$

We are experienced in this work a heuristic iterative algorithm on theoretical grounds close to the algorithm published in [5] but differs primarily by the fact that we use the correction of the current approach by minimizing the residuals for the radial integrals, not intensities. The proposed reconstruction algorithm (Heuristic Bayesian Inference Engine, HBIR) [6] can be written as

$$x_j^{(k+1)} = x_j^{(k)} \left[1 + \lambda^{(k)} \sum_{i=1}^{I} \frac{A_{ij}}{\sum_{j=1}^{J} A_{ij}} \frac{\left(p_i - \sum_{j=1}^{J} A_{ij} x_j^{(k)} \right)}{\sum_{j=1}^{J} A_{ij} x_j^{(k)}} \right], \tag{2}$$

where $x_j^{(k)}$ – the value of the *j*-th component of the vector of unknowns at the *k*-th iteration, A_{ij} – elements of the projection matrix, p_i – the value of the measured radial integral in the pixel number *i*, $\lambda^{(k)}$ – sequence relaxation parameter values ($0 < \lambda^{(k)} < 1$).

Correction algorithm represented by the formula (2) is nearly as convenient for parallelization using OpenGL graphics library, as well as the famous SART algorithm [7]. Direct projection of a three-dimensional texture, containing an image of the current approach is the same as for the SART algorithm, which is described in detail in [8]. The difference lies in the fact that its implementation must use two additional two-dimensional textures. In one of them contains a correction image. The second texture contains an image of layer of the corrected three-dimensional texture. Correcting the image is first projected perspective [9] to the place where is the corrected layer, and then multiplied with the corrected image of layer and recorded in its place. For multiplication of two-dimensional texture, use texture mapping to the object with GL_MODULATE regime.

2.1 The principle of extraction from projections of local regions with images of inhomogeneities

This method assumes that for each of the X-ray projections it is possible to segment local regions with images of inhomogeneities. Thus, the question arises: how, having projective views containing the projections of voxels with maximum density, it is correct to extract them from the original projection species containing the values of the ray integrals.

We describe the simple algorithm we proposed and tested in practice, which provides the solution to this problem. Consider the following Figure 1, which shows the projection, with the local area on it being the image of a high-density heterogeneity. Through a point inside the local region of inhomogeneity, a set of straight lines located through equal angular intervals is conducted.

The interpolation points necessary to obtain a new approximated value in the pixel of the digital image, are located at the ends of segments emerging from the point under consideration and terminating outside the boundary of the local inhomogeneity.

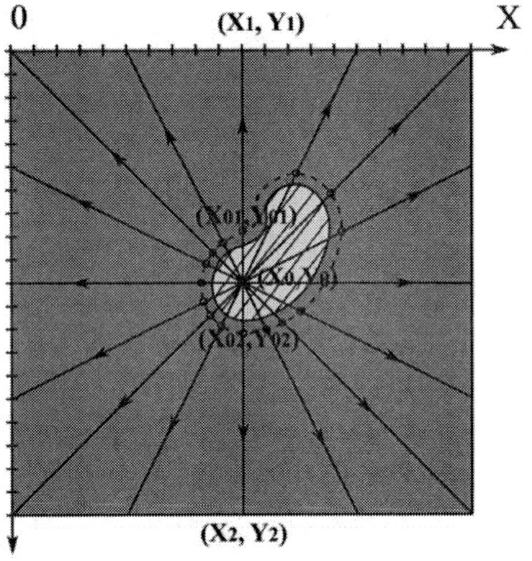

Fig 1. Projections with the local area of high-density heterogeneity

3 Interpolation algorithm for extractable heterogeneity values

Input: The number of two-dimensional projections – $\boldsymbol{n\_prj}$, Projections with Beam Integrals $\boldsymbol{P^{(l)}}$, $l=1,...,n\_prj$ with dimensions along the axes $\underline{OX}$ and $\underline{OY}$ – $\boldsymbol{size\_x}$, $\boldsymbol{size\_y}$, defined as one-dimensional vectors $\boldsymbol{P^{(l)}}=(P^{(l)}_1,..., P^{(l)}_{n\_pix})$, where $\boldsymbol{n\_pix}=\boldsymbol{size\_x}*\boldsymbol{size\_y}$ and binary projections with segmented heterogeneities $\boldsymbol{S^{(l)}}=(S^{(l)}_1,..., S^{(l)}_{n\_pix})$ $l=1,...,n\_prj$ and the same dimensions ***size_x*** и ***size_y, n_lin*** – the number of straight lines that will be used to estimate the interpolated value.

Output: Adjusted projections $P^{(l)}$, *l=1,...,n_prj* with extracted pixel values belonging to high-density heterogeneities.

1: We organize **cycle 1** in the number of projections **n_prj:** ***for(l=1;l<=n_prj;l++){Body of cycle 1}***

2: Inside the body of **cycle 1** simultaneously for a projection with ray integrals $P^{(l)}$, and the corresponding segmented (binary) projection $S^{(l)}$ organize **cycle 2** in terms of the number of pixels ***n_pix: for(n=1; n<= n_pix; n++){ Body of cycle 2}***

3: Inside the body of **cycle 2** for each pixel of the segmented projection $S^{(l)}{}_n$ we check that condition

4: if The pixel value of the segmented projection $S^{(l)}{}_n$ is nonzero **then**

5: From the value of the index n, we determine the discrete coordinates *(i0,j0)* this pixel *koord_pixel(n,size_x,&i0,&j0)*, and on them we further determine its Cartesian coordinates (X_0,Y_0).

6: We carry out the projections with ray integrals $P^{(l)}$ through the point (X_0,Y_0) *n_lin* direct lines up to the intersection with its boundaries at points $(X^{(k)}{}_1,Y^{(k)}{}_1)$ и $(X^{(k)}{}_2,Y^{(k)}{}_2)$, where $k\in\{1,\ldots, n\_lin\}$. In figure1 , the k-index for the first line is not shown. These straight lines are located at the same angular intervals relative to each other.

7: Inside the body of **cycle 2**, we organize **cycle 3** in the number of lines ***n_lin:*** ***for(k=1;k<=n_lin;k++){ Body of cycle 3}***

8: Inside the body of **cycle 3** for each straight line, determine the number of pixels *m_1*, lying between the points (X_0,Y_0) и $(X^{(k)}{}_1,Y^{(k)}{}_1)$, their discrete coordinates i_1_m , j_1_m , где $m=1,\ldots, m\_1$, indices $n\_1_1,\ldots, n\_1_{m\_1}$, where $n\_1_m\in\{1,\ldots, n\_pix\}$ and have them in the order from the point (X_0,Y_0) to the point $(X^{(k)}{}_1,Y^{(k)}{}_1)$. Similarly, for each straight line, determine the number of pixels *m_2*, lying between points (X_0,Y_0) и $(X^{(k)}{}_2,Y^{(k)}{}_2)$, their discrete coordinates and have them in the order from the point (X_0,Y_0) к точке $(X^{(k)}{}_2,Y^{(k)}{}_2)$ and so on.

9: Inside the body of **cycle 3,** arrange **loop 4** on the number of pixel points ***m_1***: ***for(m=1;m<=m_1;m++){ Body of cycle 4}***

10: Inside the body of **cycle 4,** we move along the pixels from the point (X_0,Y_0) to the point $(X^{(k)}{}_1,Y^{(k)}{}_1)$ and for each pixel of the segmented projection $S^{(l)}{}_n$ with the current index n_1_m further we verify that condition

11: if The pixel value of the segmented projection $S^{(l)}{}_n$ with the current index n_1_m is nonzero **then**

12: We keep the current values of the indices, discrete and Cartesian coordinates, which are determined by discrete coordinates, as well as the number of pixels passed

$$n\_1^{(k)}= n\_1_m;\ i\_1^{(k)} = i\_1_m;\ j\_1^{(k)} =j\_1_m;\ X\_1^{(k)}= X\_1_m;\ Y\_1^{(k)}= Y\_1_m;\ \ m\_1^{(k)}=m;$$

13: else We leave the body of **cycle 4**

14: end if

15: Inside the body of **cycle 3**, we organize **cycle 5** in terms of the number of pixel points ***m_2*** : ***for(m=1;m<=m_2;m++){*** body of **cycle 5}**

16: Inside the body of **cycle 5**, we move along the pixels from the point (X_0,Y_0) to the point $(X^{(k)}{}_2,Y^{(k)}{}_2)$ and for each pixel of the segmented projection $S^{(l)}{}_n$ with the current index n_2_m we check that condition

17: if The pixel value of the segmented projection $S^{(l)}{}_n$ with the current index n_2_m is nonzero **then**

18: Save the current values of the indexes, discrete and Cartesian coordinates, which are determined by discrete coordinates, as well as the number of traversed pixels

$$n\_2^{(k)}= n\_2_m;\ i\_2^{(k)} = i\_2_m;\ j\_2^{(k)} =j\_2_m;\ X\_2^{(k)}= X\_2_m;\ Y\_2^{(k)}= Y\_2_m;\ \ m\_2^{(k)}=m;$$

19: else We leave the body of **cycle 5**

20: end if
21: Shift by shifting pixels along the current line from the end point $(X\_1^{(k)}, Y\_1^{(k)})$ to the point $(X^{(k)}_1, Y^{(k)}_1)$ with the aim that the new point lies outside the region of local inhomogeneity and, thus, redefine the end point:

$X\_1^{(k)} = X_0 + (m\_1^{(k)} + shift) * parameter1 * (X^{(k)}_1 - X_0);$
$Y\_1^{(k)} = Y_0 + (m\_1^{(k)} + shift) * parameter1 * (Y^{(k)}_1 - Y_0);$

In accordance with the displacement of the end point, we also redefine the values of the indices and discrete coordinates for the end point $(X\_1^{(k)}, Y\_1^{(k)})$. Shift by shifting pixels along the current line from the end point $(X\_2^{(k)}, Y\_2^{(k)})$ to the point $(X^{(k)}_2, Y^{(k)}_2)$ with the aim that the new point lies outside the region of local inhomogeneity and, thus, redefine the end point:

$X\_2^{(k)} = X_0 + (m\_2^{(k)} + shift) * parameter2 * (X^{(k)}_2 - X_0);$
$Y\_2^{(k)} = Y_0 + (m\_2^{(k)} + shift) * parameter2 * (Y^{(k)}_2 - Y_0);$

In accordance with the displacement of the end point, we also redefine the values of the indices and discrete coordinates for the end point $(X\_2^{(k)}, Y\_2^{(k)})$.
22: Determine the distance between the end points and the distance from the end points to the point (X_0, Y_0)
$dl\_12^{(k)} = distance\,(X\_1^{(k)}, Y\_1^{(k)}, X\_2^{(k)}, Y\_2^{(k)});$
$dl\_1^{(k)} = distance\,(X\_1^{(k)}, Y\_1^{(k)}, X_0, Y_0);$
$dl\_2^{(k)} = distance\,(X\_2^{(k)}, Y\_2^{(k)}, X_0, Y_0);$
23: We finish **cycle 3** in the number of lines ***n_lin***.
24: For the current index *n*, we calculate the interpolated value in pixel

$$P^{(l)}_n = \frac{\sum_{k=1}^{n\_lin} \left[P^{(l)}_{n\_1^{(k)}} / dl\_1^{(k)} + P^{(l)}_{n\_2^{(k)}} / dl\_2^{(k)} \right]}{\sum_{k=1}^{n\_lin} \left[1/dl\_1^{(k)} + 1/dl\_2^{(k)} \right]}.$$

25: else (That is, inside the body of **cycle 2,** the pixel value of the segmented projection S (l) *n* is zero) We move on to the next pixel in **cycle 2.**
26: end if
27: Finish **cycle 2.**
28: Finish **cycle 1.**

4 Reconstruction image of a two-component object

The experimental verification of the feasibility of implementing the proposed technology, in particular, suppression of artifacts caused by high-density inclusions, was carried out on Real-Time X-Ray Inspection System of Seifert x|cube with 225kV. Figure 2 below shows the projections with ray integrals of the physical phantom of the knee joint, manufactured on a 3D printer for angles 450, 900 and 1350. Figure 3 shows the projections of segmented metal pins.

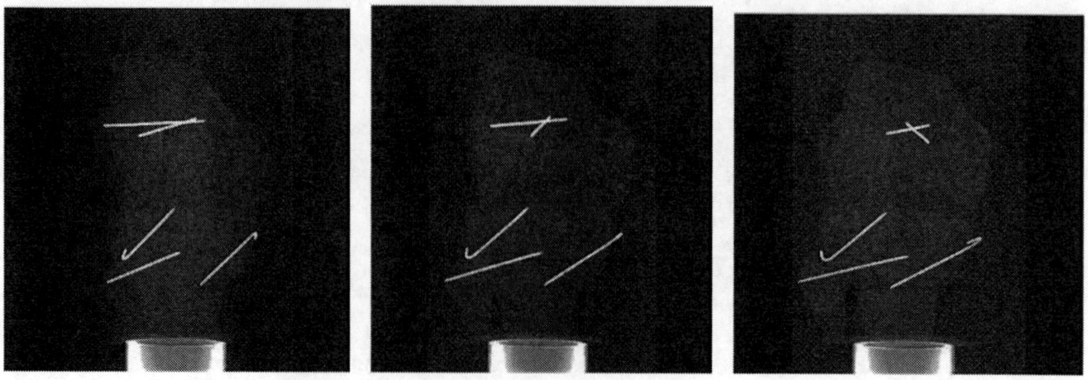

(a) (b) (c)

Fig 2. Projections with ray integrals for angles a) 45^0, b) 90^0 and c) 135^0

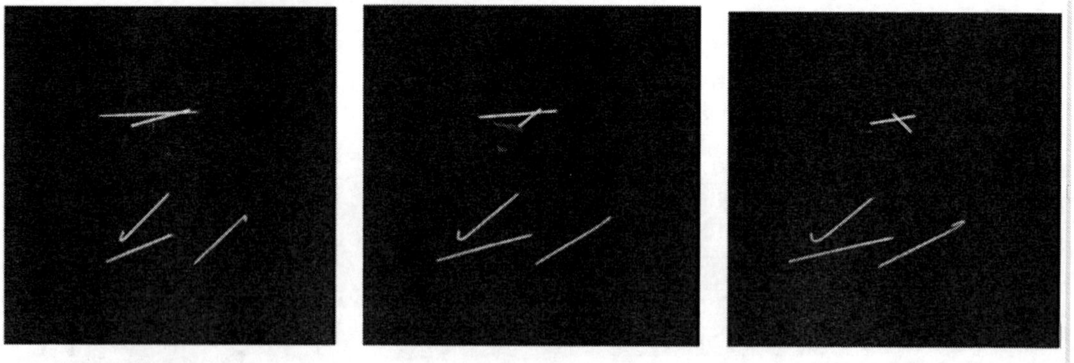

(a) (b) (c)

Fig. 3. Projections of segmented metal pins for angles a) 45^0, b) 90^0 and c) 135^0

When reconstructing the image of the knee joint, projections with ray integrals produced strong artifacts in the areas of the location of the metal pins that were caused, including, and the effect of "Beam hardening of X-rays." As a result of applying the algorithm described above, extraction of images of high-density materials from joint projections was carried out. Projections of the knee joint with extracted images of nuts for angles 0, 90 and 135 are shown in Figure 3.

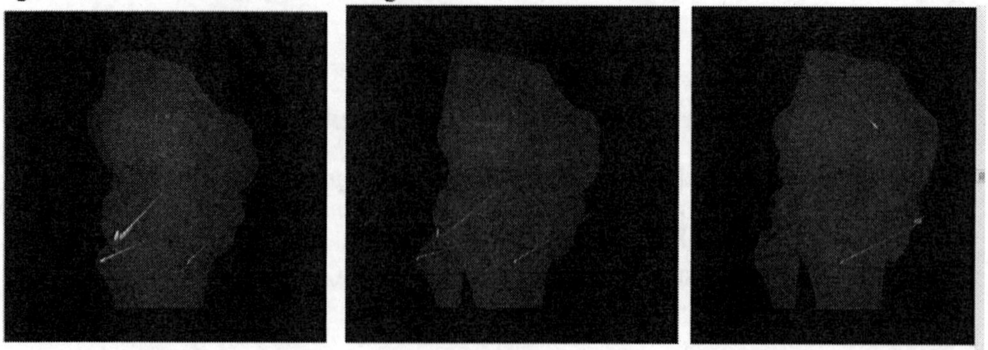

a) (b) (c)

Fig. 4. Filtered projections with ray integrals for angles a) 45^0, b) 90^0 and c) 135^0

As a result of the applied approach, it was possible to carry out a separate reconstruction of the plastic basis of the knee joint. Figure 5 shows a three-dimensional image of the plastic base of the knee joint. Figure 6 shows the cross sections of the 3D image of the knee joint by the planes XOY and YOZ and also XOZ.

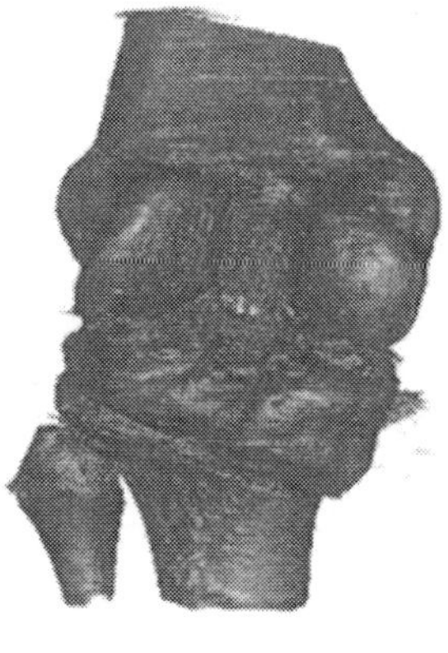

Fig. 5. Three-dimensional image of the plastic base of the knee joint

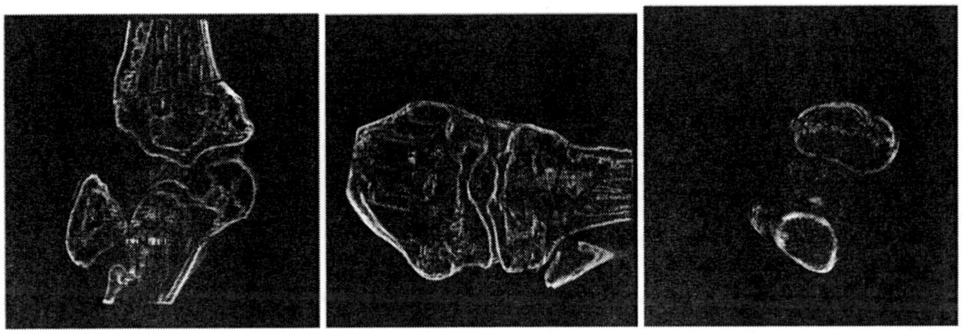

a) **(b)** **(c)**

Fig. 6. Cross sections of the three-dimensional image of the knee joint by the planes: a) XOY and b) YOZ and also c) XOZ

The proposed method of three-dimensional tomographic reconstruction of objects with high-density inclusions based on preliminary correction of projection images by extracting projection data corresponding to X-rays passing through high-density regions of the reconstruction object showed good efficiency both in tomographic reconstruction of the physical phantom of the knee joint and in the reconstruction of two-component industrial facilities. A new interpolation scheme is proposed for choosing the location of the interpolation points and a mathematical algorithm for its implementation is described in detail.

This work was partially supported by the Joint Research Project "Advanced Tomographic Reconstruction Methods" between IICT-BAS and IAP-NAS of Belarus and Bulgarian NSF Grant DFNI-I02/9.

References

1. G.H. Glover, Med. Phys. **8,** 6 , 799–807 (1981)
2. W.A. Kalender, Radiology **164**, 576–577 (1987)
3. E.Klotz, *Medical Imaging IV: PACS System Design and Evaluation, SPIE - The International Society for Optical Engineering* **1234,** 642–650 (1990)
4. R.M. Lewitt, Optik **50,** 3, 189–204 (1978)
5. K. Lange, Image Processing, IEEE Transactions **4**, 10, 430 – 1438 (1995)
6. S.A. Zolotarev, V.L. Vengrinovich, M.A. Mirzavand, Conf. Proc. NDT Days 2015, XXIII 2, 165 (2015).
7. A. H.Andersen, J.Ultras.Imaging **6**, 1, 81 – 94 (1984)
8. V. Vengrinovich, S Zolotarev, *Iterative methods for tomography* (Minsk ,Belarusian Science, 2009) (in Russian)
9. M. Segal, SIGGRAPH 92, **26,** 249–252, 1992

AUTHOR INDEX

AUTHOR INDEX